BEGINNING ALGEBRA

Donald Hutchison
Clackamas Community College

Barry Bergman
Clackamas Community College

Louis Hoelzle
Bucks County Community College

Stefan Baratto
Clackamas Community College

Mc Graw Hill **Higher Education**

Boston   Burr Ridge, IL   Dubuque, IA   Madison, WI   New York   San Francisco   St. Louis
Bangkok   Bogotá   Caracas   Kuala Lumpur   Lisbon   London   Madrid   Mexico City
Milan   Montreal   New Delhi   Santiago   Seoul   Singapore   Sydney   Taipei   Toronto

Higher Education

BEGINNING ALGEBRA, SIXTH EDITION

ISBN 0–07–254901–7
ISBN 0–07–286725–6 (Annotated Instructor's Edition)

Publisher, Mathematics and Statistics: *William K. Barter*
Publisher Developmental Mathematics: *Elizabeth J. Haefele*
Director of development: *David Dietz*
Developmental editor: *Peter Galuardi*
Senior marketing manager: *Mary K. Kittell*
Lead project manager: *Susan J. Brusch*
Senior production supervisor: *Laura Fuller*
Media project manager: *Sandra M. Schnee*
Senior media technology producer: *Jeff Huettman*
Senior designer: *David W. Hash*
Cover designer: *Rokusek Design*
Cover photo: *© Getty Images/Paul Chesly*
Lead photo research coordinator: *Carrie K. Burger*
Supplement producer: *Brenda A. Ernzen*
Compositor: *Interactive Composition Corporation*
Typeface: *10/12 Times Roman*
Printer: *Von Hoffmann Corporation*

Photo Credits
Page 1: © PhotoDisc Website; p. 75: © R. Lord/The Image Works; p. 159: © Vol. 80/PhotoDisc; p. 161: © Alan Levenson/Stone/Getty; p. 206: © Lifestyles Today/PhotoDisc; p. 218: © Lifestyles Today/PhotoDisc; p. 220: Business & Industry/PhotoDisc; p. 267: © Vol. 80/PhotoDisc; p. 271: © PhotoDisc website; p. 292: © Spacescapes/PhotoDisc; p. 294: © Earth in Focus/PhotoDisc; p. 347: © PhotoDisc Website; p. 409: © Banking & Finance/PhotoDisc; p. 419: © Brand X Pictures Vol. X128/Getty; p. 473: © Outdoor Celebrations and Lifestyles/PhotoDisc; p. 480: © American Vignette/Corbis CD; p. 501: © Memorable Moments/PhotoDisc; p. 521: People and Lifestyles/PhotoDisc; p. 564: © American Vignette/Corbis CD; p. 595: © Vol. 80/PhotoDisc; p. 603: Jose Luis Pelaez/Corbis; p. 613: © Far Eastern Business & Culture/PhotoDisc; p. 648: © Banking & Finance/PhotoDisc; p. 677: © Lester Lefkowitz/Corbis; p. 709: © Manufacturing & Industry/Corbis CD; p. 741: © Vol. 80/PhotoDisc; p. 747: © Small Business/Corbis CD; p. 782: © Nature, Wildlife, and Environment2/PhotoDisc; p. 796: © Homes and Gardens/PhotoDisc; p. 807: © Vol. 80/PhotoDisc; p. 811: © David Young-Wolff/Photo Edit; p. 835: © Homes and Gardens/PhotoDisc; p. 873: © Vol. 80/PhotoDisc.

www.mhhe.com

Table of Contents

To the Student   XI

Preface   XIII

Chapter 0

An Arithmetic Review   1

Pre-Test   2
0.1 Prime Factorization and Least
    Common Multiples   3
0.2 Fractions and Mixed Numbers   19
0.3 Decimals and Percents   35
0.4 Exponents and the Order
    of Operations   47
0.5 Positive and Negative Numbers   54

Summary   65
Summary Exercises   69
Self-Test   73

Chapter 1

The Language of Algebra   75

Pre-Test   76
1.1 From Arithmetic to Algebra   77
1.2 Properties of Real Numbers   87
1.3 Adding and Subtracting
    Real Numbers   95
1.4 Multiplying and Dividing
    Real Numbers   111
1.5 Evaluating Algebraic
    Expressions   126
1.6 Adding and Subtracting Terms   136
1.7 Multiplying and Dividing Terms   144

Summary   151
Summary Exercises   153
Self-Test   157
Activity 1: An Introduction to
Searching   159

Chapter 2

Equations and Inequalities   161

Pre-Test   162
2.1 Solving Equations by the
    Addition Property   163
2.2 Solving Equations by the
    Multiplication Property   177
2.3 Combining the Rules
    to Solve Equations   187
2.4 Formulas and Problem Solving   198

2.5 Applications of Linear
    Equations   215
2.6 Solving Percent Applications   229
2.7 Inequalities—An Introduction   243

Summary   259
Summary Exercises   261
Self-Test   265
Activity 2: Monetary
Conversions   267
Cumulative Review
Chapters 0–2   269

Chapter 3

Polynomials   271

Pre-Test   272
3.1 Exponents and Polynomials   273
3.2 Negative Exponents and
    Scientific Notation   287
3.3 Adding and Subtracting
    Polynomials   297
3.4 Multiplying Polynomials   307
3.5 Special Products   318
3.6 Dividing Polynomials   325

Summary   335
Summary Exercises   337
Self-Test   341
Activity 3: ISBNs and the
Check Digit   343
Cumulative Review
Chapters 0–3   345

Chapter 4

Factoring   347

Pre-Test   348
4.1 An Introduction to Factoring   349
4.2 Factoring Trinomials of the
    Form $x^2 + bx + c$ 357
4.3 Factoring Trinomials of the
    Form $ax^2 + bx + c$ 366
4.4 Difference of Squares and
    Perfect Square Trinomials   375
4.5 Factoring by Grouping   382
4.6 Using the ac Method to Factor   386
4.7 Strategies in Factoring   399
4.8 Solving Quadratic Equations
    by Factoring   404

Summary   411
Summary Exercises   413

Self-Test   415
Cumulative Review
Chapters 0–4   417

Chapter 5

RATIONAL EXPRESSIONS   **419**

Pre-Test   420
5.1 Simplifying Rational
     Expressions   421
5.2 Multiplying and Dividing
     Rational Expressions   431
5.3 Adding and Subtracting Like
     Rational Expressions   437
5.4 Adding and Subtracting
     Unlike Rational Expressions   444
5.5 Equations Involving Rational
     Expressions   457
5.6 Applications of Rational
     Expressions   470
5.7 Complex Rational
     Expressions   483

Summary   491
Summary Exercises   493
Self-Test   497
Cumulative Review
Chapters 0–5   499

Chapter 6

AN INTRODUCTION TO GRAPHING   **501**

Pre-Test   502
6.1 Solutions of Equations
     in Two Variables   504
6.2 The Rectangular Coordinate
     System   514
6.3 Graphing Linear Equations   526
6.4 The Slope of a Line   549
6.5 Reading Graphs   567

Summary   581
Summary Exercises   583
Self-Test   591
Activity 4: Graphing
with a Calculator   595
Cumulative Review
Chapters 0–6   599

Chapter 7

GRAPHING AND INEQUALITIES   **603**

Pre-Test   604
7.1 The Slope-Intercept Form   605
7.2 Parallel and Perpendicular
     Lines   618

7.3 The Point-Slope Form   629
7.4 Graphing Linear Inequalities   638
7.5 An Introduction to Functions   653

Summary   665
Summary Exercises   667
Self-Test   671
Activity 5: Graphing
with the Internet   673
Cumulative Review
Chapters 0–7   674

Chapter 8

SYSTEMS OF LINEAR EQUATIONS   **677**

Pre-Test   678
8.1 Systems of Linear Equations:
     Solving by Graphing   679
8.2 Systems of Linear Equations:
     Solving by Adding   690
8.3 Systems of Linear Equations:
     Solving by Substitution   711
8.4 Systems of Linear
     Inequalities   724

Summary   733
Summary Exercises   735
Self-Test   739
Activity 6: Growth of
Children—Fitting a Linear
Model to Data   741
Cumulative Review
Chapters 0–8   743

Chapter 9

EXPONENTS AND RADICALS   **747**

Pre-Test   748
9.1 Roots and Radicals   749
9.2 Simplifying Radical
     Expressions   760
9.3 Adding and Subtracting
     Radicals   769
9.4 Multiplying and Dividing
     Radicals   776
9.5 Solving Radical Equations   783
9.6 Applications of the
     Pythagorean Theorem   789

Summary   801
Summary Exercises   803
Self-Test   805
Activity 7: The Swing
of the Pendulum   807
Cumulative Review
Chapters 0–9   809

Chapter 10

QUADRATIC EQUATIONS **811**

Pre-Test 812
10.1 More on Quadratic
Equations 813
10.2 Completing the Square 821
10.3 The Quadratic Formula 827
10.4 Graphing Quadratic
Equations 837
10.5 Applications of Quadratic
Equations 855

Summary 865
Summary Exercises 867

Self-Test
Activity 8: Monetary
Exchange—Predicting the
Future 873
Cumulative Review
Chapters 0–10 875
Final Exam 879

**ANSWERS TO PRE-TESTS, SUMMARY
EXERCISES, SELF-TESTS, AND
CUMULATIVE REVIEWS** **A-1**
INDEX **I-1**

About the Authors

Don Hutchison   Don began teaching in a pre-school while he was an undergraduate. He subsequently taught children with disabilities, adults with disabilities, high school mathematics, and college mathematics. Although all of these positions were challenging and satisfying, it was breaking a challenging lesson into teachable components that he most enjoyed.

It was at Clackamas Community College that he found his professional niche. The community college allowed him to focus on teaching within a department that constantly challenged faculty and students to expect more. Under the guidance of Jim Streeter, Don learned to present his approach to teaching in the form of a textbook.

Don has also been an active member of many professional organizations. He has been President of ORMATYC, AMATYC committee chair, and ACM curriculum committee member. He has presented at AMATYC, ORMATYC, AACC, MAA, ICTCM, and numerous other conferences.

Barry Bergman   Barry has enjoyed teaching mathematics to a wide variety of students over the years. He began in the field of adult basic education, and moved into the teaching of high school mathematics in 1977. He taught at that level for 11 years, at which point he served as a K-12 mathematics specialist for his county. This work allowed him the opportunity to help promote the emerging NCTM Standards in his region.

In 1990 Barry began the present portion of his career, having been hired to teach at Clackamas Community College. He maintains a strong interest in the appropriate use of technology and visual models in the learning of mathematics.

Throughout the past 26 years, Barry has played an active role in professional organizations. As a member of OCTM, he contributed several articles and activities to the group's journal. He has made presentations at OCTM, NCTM, ORMATYC, and ICTCM conferences. Barry also served as an officer of ORMATYC for four years and participated on an AMATYC committee to provide feedback to revisions of NCTM's Standards.

Stefan Baratto   Stefan began teaching (math and science) in middle schools in New York. He also taught math at the University of Oregon, Southeast Missouri State University, and York County Technical College. Currently, Stefan is a member of the mathematics faculty at Clackamas Community College. Here, he has found a niche, delighting in the CCC faculty, staff, and students. Stefan's own education includes the University of Michigan (BGS, 1988), Brooklyn College (CUNY), and the University of Oregon (MS, 1996).

Stefan has been involved in numerous professional organizations including AMATYC, ORMATYC, and NCTM. He has applied his knowledge of math to various fields, using statistics, technology, and web-design.

More personally, Stefan and his wife, Peggy, try to spend time enjoying the wonders of Oregon and the Pacific Northwest. Their activities include scuba diving, self-defense training, and hiking.

Louis Hoelzle   Louis Hoelzle has been teaching at Bucks County Community College for 35 years. In 1989, Lou became chair of the Department of Mathematics, Computer and Information Science. He has taught the entire range of courses from arithmetic to calculus, giving him an excellent view of the current and future needs of developmental students.

Over the past 40 years, Lou has also taught physics courses at 4-year colleges, which has enabled him to have the perspective of the practical applications of mathematics.

Lou is also active in several professional organizations. He has served on various committees for AMATYC including the committee chair of the Placement and Assessment committee and has been President of PSYMATYC.

To the Student

You are about to begin a course in algebra. We made every attempt to provide a text that will help you understand what algebra is about and how to use it effectively. We made no assumptions about your previous experience with algebra. Your progress through the course will depend on the amount of time and effort you devote to the course and your previous background in math. There are some specific features in this book that will aid you in your studies. Here are some suggestions about how to use this book. (Keep in mind that a review of *all* the chapter and summary material will further enhance your ability to grasp later topics and to move more effectively through the text.)

1. If you are in a lecture class, make sure that you take the time to read the appropriate text section *before* your instructor's lecture on the subject. Then take careful notes on the examples that your instructor presents during class.

2. After class, work through similar examples in the text, making sure that you understand each of the steps shown. Examples are followed in the text by *Check Yourself* exercises. You can best learn algebra by being involved in the process, and that is the purpose of these exercises. Always have a pencil and paper at hand, and work out the problems presented and check your results immediately. If you have difficulty, go back and carefully review the previous exercises. Make sure you understand what you are doing and why. The best test of whether you do understand a concept lies in your ability to explain that concept to one of your classmates. Try working together.

3. At the end of each chapter section you will find a set of exercises. Work these carefully to check your progress on the section you have just finished. You will find the answers for the odd-numbered exercises following the problem set. If you have difficulties with any of the exercises, review the appropriate parts of the chapter section. If your questions are not completely cleared up, by all means do not become discouraged. Ask your instructor or an available tutor for further assistance. A word of caution: Work the exercises on a regular (preferably daily) basis. Again, learning algebra requires becoming involved. As is the case with learning any skill, the main ingredient is practice.

4. When you complete a chapter, review by using the *Summary.* You will find all the important terms and definitions in this section, along with examples illustrating all the techniques developed in the chapter. Following the *Summary* are *Summary Exercises* for further practice. The exercises are keyed to chapter sections, so you will know where to turn if you are still having problems.

5. When you finish with the *Summary Exercises,* try the *Self-Test* that appears at the end of each chapter. It is an actual practice test you can work on as you review for in-class testing. Again, answers with section references are provided.

6. Finally, an important element of success in studying algebra is the process of regular review. We provide a series of *Cumulative Reviews* throughout the textbook, beginning at the end of Chapter 2. These tests will help you review not only the concepts of the chapter that you have just completed but those of previous chapters. Use these tests in preparation for any midterm or final exams. If it appears that you have forgotten some concepts that are being tested, don't worry. Go back and review the sections where the idea was initially explained, or the appropriate chapter summary. That is the purpose of the *Cumulative Review.*

We hope that you will find our suggestions helpful as you work through this material, and we wish you the best of luck in the course.

Donald Hutchison
Barry Bergman
Stefan Baratto
Louis Hoelzle

New to the Sixth Edition

Global Changes

- Section objectives have been clearly identified within the sections and in the section exercises.
- Section references have been included with the Pre-Tests, Self-Tests, and Cumulative Reviews.
- Applications providing students with more real-world data and motivation have been added to sections and to exercise sets.
- Margin notes specifically asking students to recall previously learned content have been identified and clearly labeled.
- Revised instructions for the various text-features better facilitate their use by instructors and students.
- Application exercises have been titled to better reflect student interests.
- Chapter Activities have been added to this edition. These allow instructors to provide group work to be completed outside the classroom.
- More "thought-provoking" questions have been added throughout the text.

Chapter Specific Changes

Chapter 0

New examples and exercises related to Least Common Multiples, mixed numbers, decimals, percents, and advanced order of operations were added.

Chapter 1

Language was changed throughout to discuss "real" rather than "signed" numbers.

Chapter 3

Examples and exercises were added for raising a fraction to a negative power.

Chapter 4

A new section (4.7), "Strategies in Factoring," was added.

Chapter 5

The term "algebraic fraction" was changed to "rational expression."

The section "Multiplication and Division of Rational Expressions" was moved from Section 5.4 to become Section 5.2.

A new section (5.7), "Complex Rational Expressions," was added.

Chapter 6

Section 6.4 has been expanded to include direct variation and slope of a line.

Section 6.5, "Reading Graphs," is new to the sixth edition. This section meets the needs of instructors who include it as part of their curriculum.

Chapter 9

A new section (9.5), "Solving Radical Equations," was added.

Chapter 10

A new applications section, 10.5, was added.

Supplements for the Instructor

Annotated Instructor's Edition

This ancillary contains answers to problems and exercises in the text, including answers to all section exercises, all *Summary Exercises, Self-Tests,* and *Cumulative Reviews.* These answers are printed in a second color for ease of use by the instructor and are located on the appropriate pages throughout the text. Exercises, Self-Tests, Summary Exercises, and Cumulative Review Exercises are annotated with section references to aid the instructor who may have omitted certain sections from study.

Instructor's Testing and Resource CD

This cross-platform CD-ROM provides a wealth of resources for the instructor. Supplements featured on this CD-ROM include a computerized test bank utilizing Brownstone Diploma® testing software to quickly create customized exams. This user-friendly program allows instructors to search for questions by topic, format, or difficulty level; edit existing questions or add new ones; and scramble questions and answer keys for multiple versions of the same test.

Instructor's Solutions Manual

This supplement contains detailed solutions to all the exercises in the text. The methods used to solve the problems in the manual are the same as those used to solve the examples in the textbook.

 **www.mathzone.com***

* web-based product also available on CD ROM

Easy, Free, Has it all . . .

McGraw-Hill's MathZone is a complete, online tutorial and course management system for mathematics and statistics, designed for greater ease of use than any other system available. Free upon adoption of a McGraw-Hill title, instructors can create and share courses and assignments with colleagues and adjuncts in a matter of a few clicks of the mouse. All assignments, questions, e-Professors, online tutoring, and video lectures are directly tied to text-specific materials in *Beginning Algebra.* MathZone courses are customized to your textbook, but you can edit questions and algorithms, import your own content, create announcements, and due dates for assignments. MathZone has automatic grading and reporting of easy-to-assign algorithmically generated homework, quizzing, and testing. All student activity within MathZone is automatically recorded and available to you through a fully integrated grade book that can be downloaded to Excel.

PageOut

PageOut is McGraw-Hill's unique point-and-click course website tool, enabling you to create a full-featured, professional-quality course website without knowing HTML coding. With PageOut you can post your course syllabus, assign McGraw-Hill Online Learning Center content, add links to important off-site resources, and maintain student results in the online grade book. You can send class announcements, copy your course site to share with colleagues, and upload original files. PageOut is free for every McGraw-Hill user, and if you're short on time, we even have a team ready to help you create your site!

ALEKS v2.0

ALEKS® (**A**ssessment and **LE**arning in **K**nowledge **S**paces) is an artificial intelligence based system for individualized math learning, available over the World Wide Web. ALEKS delivers precise, qualitative diagnostic assessments of students' math knowledge, guides them in the selection of appropriate new study material, and records their progress toward mastery of curricular goals in a robust classroom management system. See page [xxvii] for more details regarding ALEKS.

Supplements for the Student

Student's Solutions Manual

The Student's Solutions Manual contains complete worked-out solutions to all the odd-numbered section exercises from the text. The procedures followed in the solutions in the manual match exactly those shown in worked examples in the text.

Streeter Video Series

The video series is composed of 11 videocassettes (one for each chapter of the text). An on-screen instructor introduces topics and works through examples using the methods presented in the text. The video series is also available on video CD-ROMs.

MathZone www.mathzone.com

McGraw-Hill's MathZone is a powerful new online tutorial for homework, quizzing, testing, and interactive applications. There are an unlimited number of exercises to allow for as much practice as needed. MathZone offers videos that feature classroom instructors giving a lecture and **e-Professor** takes you through animated, step-by-step instruction for solving problems in the book allowing you to digest each step at your own pace. NetTutor offers live, personalized tutoring via the internet. Every assignment, question, e-Professor, and video lecture is derived directly from *Beginning Algebra.*

NetTutor

NetTutor is a revolutionary system that enables students to interact with a live tutor over the World Wide Web. Students can receive instruction from live tutors using NetTutor's web-based, graphical chat capabilities. They can also submit questions and receive answers, browse previously answered questions, and view previous live chat sessions.

Math for the Anxious: Building Basic Skills, by Rosanne Proga

Math for the Anxious: Building Basic Skills is written to provide a practical approach to the problem of math anxiety. By combining strategies for success with a pain-free introduction to basic math content, students will overcome their anxiety and find greater success in their math courses. The first two chapters not only explain the sources of math anxiety, they more importantly outline pragmatic steps students can take to reduce it. In each of the following eight chapters, strategies are implemented for learning a particular topic such as fractions that may have frustrated students in the past but can now be digested and mastered through hints, patient explanations, and revelations of how students already encounter the topic on an everyday basis. The final chapter brings all the strategies together and prepares students to encounter future math topics with newfound confidence and finely tuned techniques at their disposal.

Acknowledgments

Throughout the writing process, the list of contributors to this text has grown steadily. Foremost are the reviewers. Every person on this list has helped by questioning, correcting, or complementing. Our thanks to each of them.

Beginning Algebra 6TH Edition Reviewers

Lisa G. Angelo, Bucks County Community College
A. Elena Bogardus, Camden County College
Dorothy Brown, Camden County College
Kirby Bunas, Santa Rosa Junior College
Kathleen Ciszewski , University of Akron
Emmett C. Dennis, Southern Connecticut State University
Matt Flacche, Camden County College
Ellen Freedman, Camden County College
Mary Ellen Gallegos, Santa Fe Community College
Dauhrice Gibson, Gulf Coast Community College
Renu A. Gupta, Louisiana State University-Alexandria
Joseph Lloyd Harris, Gulf Coast Community College
Virginia M. Licata, Camden County College
Gabrielle Michaelis, Cumberland County College
Kathryn Pletsch, Antelope Valley College
Brenda Reed, Lincoln University
Minnie Riley, Hinds Community College-Raymond
Adam Rubin, Harper College
Marvin L. Shubert, Hagerstown Community College
Barbara Jane Sparks, Camden County College
Karen M. Stein, University of Akron
Peter Stomieroski, Delaware Technical & Community College-Dover
Abolhassan S. Taghavy, Richard J. Daley College
Sandra Tannen, Camden County College
Dr. Sharon Testone, Onondaga Community College
Carol M. Walker, Hinds Community College-Raymond
Cheryll Wingard, Community College of Aurora
Kathryn Wright, Mineral Area College

Beginning Algebra 5TH Edition Reviewers

Sharon Abramson, Nassau Community College
Patricia Allaire, Queensborough Community College
Sharon Berrian, Northwest Shoals Community College
Matthews Chakkanakuzh, Palomar Community College
Alan Chutsky, Queensborough Community College
John Davidson, Southern State Community College
Katherine D'Orazio, Cumberland County Community College
Bill Dunn, Las Positas College
Ellen Freedman, Camden County Community College
Kelly Jackson, Camden County Community College
Karen Jensen, Southeastern Community College
Ginny Licata, Camden County Community College
S. Maheshwari, William Paterson University
Laurie McManus, St. Louis Community College-Meramec
Diane Metzger, Rend Lake College
Wayne Miller, Lee College

Jeff Mock, Diablo Valley College
Ellen Musen, Brookdale Community College
Larry Newman, Holyoke Community College
Lilia Orlova, Nassau Community College
Betty Pate, St. Petersburg Junior College
Kathryn Pletsch, Antelope Valley College
Larry Pontaski, Pueblo Community College
Donna Russo, Quincy College
Bruce Sisko, Belleville Area College
Barbara Jane Sparks, Camden County Community College
Peter Speier, Prince George's Community College
Sharon Testone, Onandaga Community College
Patricia Wake, San Jacinto College

Selecting the excellent list of reviewers was only one of the many contributions of our McGraw-Hill editors, David Dietz and Peter Galuardi. Thanks must go to everybody at McGraw-Hill with whom we worked. Their professional, caring attitude was always appreciated. We are especially appreciative for the keen eyes and good sense of Susan Brusch.

In this age of the high-tech mathematics lab, the supplements and the supplement authors have become as important as the text itself. We are proud that our names appear on the supplements together with these authors.

There are also people at Clackamas and Bucks County Community Colleges that deserve mention. We must thank all of the students that have been taught, talked to, questioned, and tested. This text was created for these students and shaped by them. We also want to recognize the contributions of the mathematics faculty at both colleges. Their interest in helping students has provided us with both the questions and the answers that are necessary to create a work like this text.

Don Hutchison
Barry Bergman
Stefan Baratto
Louis Hoelzle

ALEKS is an artificial intelligence-based system for individualized math learning, available for Higher Education from McGraw-Hill over the World Wide Web.

ALEKS delivers precise assessments of math knowledge, guides the student in the selection of appropriate new study material, and records student progress toward mastery of goals.

ALEKS interacts with a student much as a skilled human tutor would, moving between explanation and practice as needed, correcting and analyzing errors, defining terms and changing topics on request. By accurately assessing a student's knowledge, ALEKS can focus clearly on what the student is ready to learn next, helping to master the course content more quickly and easily.

ALEKS is:

- **A comprehensive course management system.** It tells the instructor exactly what students know and don't know.

- **Artificial intelligence.** It totally individualizes assessment and learning.

- **Customizable.** ALEKS can be set to cover the material in your course.

- **Web-based.** It uses a standard browser for easy Internet access.

- **Inexpensive.** There are no setup fees or site license fees.

ALEKS 2.0 adds the following new features:

- **Automatic Textbook Integration**

- **New Instructor Module**

- **Instructor-Created Quizzes**

- **New Message Center**

ALEKS maintains the features that have made it so popular including:

- **Web-Based Delivery** No complicated network or lab setup

- **Immediate Feedback** for students in learning mode

- **Integrated Tracking of Student Progress and Activity**

- **Individualized Instruction** which gives students problems they are *Ready to Learn*

For more information please contact your McGraw-Hill Sales Representative or visit ALEKS at http://www.highedmath.aleks.com.

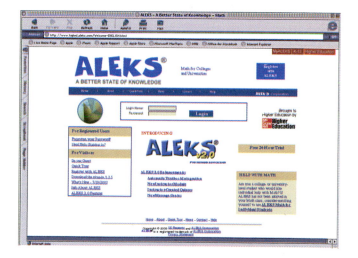

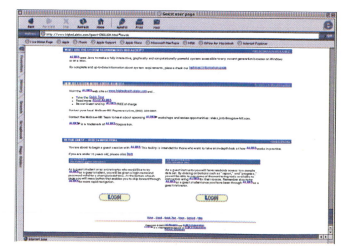

Applications Index

Business and Investments
advertising impact, 564
assets, relative to second person, 156
balance of trade, 46, 240
break-even point, 546, 658
car sales, by car size, 589
car sales decrease, 234
cash supply, by bill denomination, 226
checking account balance, 107
checking account service fees, 123
checking account withdrawals, 63
commission earned, 235, 236, 237, 481
commission rate, 266
copy machine costs, 257–258
cost of car rental, 614
cost of manufacturing, 284, 285, 508, 537, 545–546, 608–609, 636, 663
cost of office furniture, 721
cost of one item vs. another, 696–697
cost of operating restaurant, 609
cost of pens vs. pencils, 737
cost of postage stamps, predicting, 575–576
cost per item purchased, 46
cups of coffee per pound of coffee, 480
currency conversions, 267–268
currency value, trends in, 873–874
decrease in employees, 239
depreciation of assets, 239, 636
discount rate for stereo, 239
division of inheritance, 481
division of invested amount, 496
Dow Jones fluctuations, 63
earnings from paper drive, 613
electricity costs, 239, 677
enrollment decrease/increase at school, 108, 233, 234, 236, 239, 589
equilibrium price, 859, 863–864, 870, 878
evaluating business opportunities, 710
fixed cost, 537
fundraising success, 242
hiring of new employees, 234
hospital meals served, 641–642
hours worked, 205–206, 564
income vs. years of education, 590
interest problems
  assets after interest earned, 200
  borrowed amount, 108, 133, 232, 237
  compound interest, 240–241, 285
  formula for interest, 701
  interest earned, 85, 231, 236, 241, 481
  interest paid, 133, 231, 236
  interest rate, 133, 200, 208, 231, 232, 237
  invested amount, calculation of, 230, 701–702, 707, 709, 737, 746
International Standard Book Numbers (ISBN), 343–344
Internet encryption systems, 347

inventory management, 649
items sold, from total profit, 407
losses per investor, 123
markup on items for sale, 233, 238
mixing coffee beans to cost specifications, 707
mixing nuts to cost specifications, 707, 709
monthly loan payments, 46
payroll deductions, 236
percent of peanuts in nut mix, 235
percentage of faulty parts, 235, 238
personal computer sales, by year, 589–590
port shipping traffic monitoring, 503
Post Office package regulations, 271
product ratios per package, 481
production costs, 284, 285, 508, 537, 545–546, 608–609, 636, 663
production on fixed budget, 863, 870
production scheduling, 643, 721–722, 730
production to meet profit goal, 863
production volume at fixed cost, 508
profit
  net, calculation of, 142, 305
  vs. number of items sold, 407, 658, 663
  from sale of business, 118
  vs. sales per day, 409, 508
real estate value, 239
revenue per advertising dollar, 564
revenue per units sold, 285, 315, 316, 818, 826
sales growth over time, 637
sales needed for commission, 237, 257, 266
sales tax, 237
sales vs. advertising spending, 632–633
savings account deposits, 63
selling price, calculation of, 233, 409
shipping method analysis, 578
stamp types purchased, 219, 225, 706
stock value
  changes in, 103, 123
  loss calculation, 117
supply and demand curves, 859, 863–864, 870, 878
tickets sold, by month, 579
tickets sold, by type, 218–219, 225, 226, 699, 707, 708, 737, 746, 877
timber purchasing, 227
time required for printing, 675
tourism spending, by nation, 593
trees in orchard, 323
variable cost, 537
volume-based pricing, 511
wages
  calculating from deductions, 235, 238
  calculating from spending, 238
  calculation minus savings, 123
  change over time in, 522

compensation plan comparisons, 142, 512
increase in, 239, 240
increase, 239, 240
for partial day's work, 32
per-piece scale for, 563–564
vs. previous job, 213
productivity-based, 511
relative to others, 205, 210, 211, 212, 270
total, of hourly worker, 555–556, 562
weight of packages, 720
work hours scheduling, 648

Construction, Craft, and Home Improvement
building design, 799–800
checking architectural drawings, 623, 627
cutting cable, 420, 498, 737
cutting insulation, 796
cutting lumber, 28, 156, 212, 477, 697–698
cutting rope, 706, 740
cutting wire, 476
emergency exit design, 747, 799–800
garden design, 209, 224, 835, 857, 861, 870
guy wire length, 792, 795, 804, 855, 860
ladder height, estimation of, 792, 796, 855, 860
linoleum replacement, 33
mulch amount needed, 33
slope of roof, 613
solar panel bracket design, 796, 835
subdivision of land, 33
swimming pool tarp coverage, 857

Consumer Affairs
borrowed amount, calculation of, 108, 133, 232, 237
car age vs. price, 523
checking account balance, 107, 108
checking account service fees, 123
checking account withdrawals, 63
cost of apple types, 706
cost of appliances, 212, 258, 720, 721
cost of candy types, 706
cost of car rental, 614
cost of food per pound, 123
cost of gasoline for trip, 42
cost of meal before tip, 240
cost of one item vs. another, 696–697, 706, 708, 721, 737, 746
cost of postage stamps, predicting, 575–576
cost of travel, 212
cost of typing a paper, 284
credit card balance, 107
currency conversions, 267–268
currency value, trends in, 873–874
depreciation of car value, 239
discount rate for car, 270
discount rate for stereo, 239
downpayment amount, 237
electricity use, 213
fuel oil use, 213
gas used on trip, 475–476, 480
household budgets, 32, 238

increases in college costs, 633
loan interest rate calculations, 231, 232, 237
loan payments, 133, 231, 236
long distance costs, 257, 649
mean weight of candy bags, 186
median weight of candy bags, 186
monthly loan payments, 46
percent of peanuts in nut mix, 235
Post Office package regulations, 271
price
  calculated from discount price, 602
  calculated from tax paid, 235, 237
  before discount, 239, 240
  increases for vans, 239
product ratios per package, 481
property tax owed, 481
sales tax paid, 232, 237
savings account deposits, 63
scholarship spending by student, 570
stamp types purchased, 219, 225, 706
tax rates, 270

Ecology and Environment
carbon dioxide emissions, 241
erosion of topsoil, 62
forested land remaining, 257
formation of topsoil, 62
fungicide use, 436, 488
giant panda population, 257
herbicide use, 436, 488, 489
oil spill growth, 150
pesticide use, 436, 488, 489
recycling contest, 521, 544–545, 613
recycling profits, 545
reintroduction of endangered species, 124
species loss, 85
water per person on earth, 296
water use per person, 296

Economics
balance of trade, 46, 240
bankruptcy filings per year, 578–579
car production, global by year, 577
car sales, by car size, 589
cars sold annually, 571, 572
cost of electricity for city, 161
personal computer sales, by year, 589–590
social security recipients per year, 573
supply and demand curves, 859, 863–864, 870, 878
tourism spending, by nation, 593

Education
average student age, 572
average test scores, 108
correct answers needed on test, 238
cost of typing a paper, 284
dropout rate, 238
earnings from paper drive, 613
enrollment decrease/increase at school, 108, 233, 234, 236, 239, 589
exam grade vs. hours studied, 521

Education (continued)
  income vs. years of
      education, 590
  increases in college costs, 633
  language course enrollment, 238
  method of transportation to
      school, 569
  number of test items, from percent
      correct, 237
  percent of correct answers, 45
  percent receiving A grade, 235
  scholarship spending by
      student, 570
  school day activities analysis, 570
  student lunch practices, 570
  students per section, 213
  technology availability in U.S.
      public schools, 588
  test scores needed for grade
      average, 251–252, 257
Engineering
  electrical resistance in circuit,
      133, 489
  electricity generation, 161, 677
  fireworks design, 811
  guy wire length, 792, 795, 804,
      855, 860

Geography
  distance between two points,
      792–794, 798
  land area comparisons, 567–568
  map coordinates, 524
  map scale, 32
  population density per area, 569
  subdivision of land, 33
Geometry
  area of box bottom, 429
  area of circle of know radius, 134
  area of rectangle, 33, 315, 436,
      781–782
  area of square, 323, 324, 511, 767
  area of triangle, 133, 315
  circumference of circle of known
      diameter, 33
  diagonal of rectangle, length of,
      791, 794, 795, 804
  dimensions of box, 316
  dimensions of rectangle, 217–218,
      224, 225, 266, 270, 418, 500,
      708, 716–717, 721, 737, 740,
      798, 860, 872
  dimensions of square, 323, 324, 409
  dimensions of triangle, 225, 717,
      721, 737
  distance between two points,
      792–794, 798
  height of cylinder, 208
  height of rectangle, 208
  height of triangle, 183–184, 796
  length of rectangle, 156, 208, 429,
      795, 804, 806
  length of square, 757
  length of trapezoid base, 209
  length of triangle legs, 270,
      790–791, 797, 804, 856,
      860–861, 870
  perimeter of rectangle, 133, 141,
      142, 305, 442, 445, 774
  perimeter of square, 767
  perimeter of triangle, 142, 305,
      445, 774–775

radius of circle, 757
volume of cube, 52
volume of rectangular box, 33, 271
width of rectangle, 257, 436, 795

Medicine and Health
  body fat percentage vs. age, 614
  Body Mass Index (BMI), 419
  family doctors, decline in, 594
  growth rates of children, 740–741
  heart disease risk, 419
  hospital meals served, 641–642
  human growth hormone
      therapy, 501
  pharmaceutical manufacturing, 603
  programs for disabled, by year, 511
  weight gain in pets, 576
  weight loss from dieting, 123
Meteorology
  high temperatures over a period, 574
  mean temperature, 183, 184
  median temperature, 183, 184
  temperature change, 63, 107, 108,
      109, 123
    over time, 522
    rate of, 123, 613
  temperature conversions, Celsius
      to/from Fahrenheit, 134, 208,
      511, 636
  temperature range, 109
Mixture Problems
  amount of acid in solution,
      236, 237
  division of a liquid solution, 123
  mixing coffee beans to cost
      specifications, 707
  mixing nuts to cost specifications,
      707, 709
  number of coins by type, 156, 648,
      698–699, 706, 737, 740
  percent of peanuts in nut mix, 235
  percentage of faulty parts, 235, 238
  solution adjustment, 474–475, 480,
      496, 699–701, 707, 709,
      737, 877
  stamp types purchased, 219,
      225, 706
  tickets sold, by type, 218–219, 225,
      226, 699, 707, 708, 737,
      746, 877
Motion Problems
  descent of airplane, 613
  distance traveled
    at different rates, 226
    at given speed, 33
    in given time, 556
    as percent of total distance, 31
  falling, time required for, 757
  height
    of projectile dropped, 862, 872
    of projectile thrown down,
        861, 878
    of projectile thrown up,
        857–859, 861–863, 870
    of rocket, 811
  pendulum motion, 758–759,
      807–808
  slope of road, 613
  speed
    of airplane, subtracting wind
        speed from, 480, 496, 703,
        708, 737

of airplane in feet per
    second, 480
of airplane vs. train, 480
average, going vs. returning,
    219–220, 226, 420
of bike vs. car, 479
of boat, subtracting current
    speed from, 480, 702–703,
    709, 737, 740, 810
of express vs. local bus, 479
in feet per second, 500
going vs. returning, 498, 675
of passenger vs. freight
    train, 479
of train vs. bus, 473–474
of truck vs. car, 474
time
  of airplane flight, going vs.
      returning, 479
  driving time per day, 479
  driving time required for trip,
      472–473
  meeting time of two approach-
      ing vehicles, 221
  needed to overtake, 222,
      226–227
  of projectile dropped, 862, 872
  of projectile thrown down,
      861, 878
  of projectile thrown up,
      858–859, 861–863, 870

Numbers, Finding Unknown
  from formula, 216–217, 224, 266,
      270, 409, 418, 499–500,
      818, 826
  from formula and sum, 715–716,
      720, 737, 746
  from product, 409
  from ratio, 420, 470–471, 478–479,
      495–496, 498
  from sum, 201, 203–204, 210–211,
      266, 270
  from sum and difference, 706,
      737, 740
  from sum and product, 708

Physics and Astronomy
  astronomical distances, 290–291,
      292, 294
  kinetic energy of particle, 85
  light years, 290–291, 292, 296
  mass of sun, 294
Politics and Public Policy
  cat owners in city, 32
  dog owners in city, 33
  larceny-theft cases per year,
      574–576
  percent of eligible voters voting, 32
  programs for disabled, by year, 511
  registered vehicles, by type, 241
  social security recipients per
      year, 573
  votes cast, 204–205
  votes per candidate, 210, 211, 720

Science and Technology
  carbon dating, 75
  diameter of sand grain, 294
  distance above sea level, 108
  distance to horizon, 782
  erosion of topsoil, 62

formation of topsoil, 62
height of stacked items, 32
height of woman, vs. building, 481
Internet encryption systems, 347
molecules in a gas, 294
temperature conversions, Celsius
    to/from Fahrenheit, 134, 208,
    511, 636
water level in flood, 213
weight loss from drying, 603
Social Science
  dates of historical events, 75,
      109–110
  percent of left–handed
      persons, 238
  unemployment rates, 238
Sports and Leisure Activities
  baseball batting averages, 46
  bowling average, 257
  football yardage gained
    in series of plays, 107
    yearly changes in, 108
  gambling losses, 122
  garden dimensions, 209, 224, 835,
      857, 861, 870
  material required for making
      shirts, 31
  miles run, vs. others, 206
  patterns, enjoyment of, 134–135
  playing field dimensions, 225
  polls on favorite sport, 571
  runs scored per game in baseball,
      522–523
  team losing streak, 63
  total points scored, 122
  wins vs. total points in NHL, 523
Statistics and Demographics
  age of person, 156, 212, 266
  average student age, 572
  average test scores, 108
  baseball batting averages, 46
  body fat percentage by age, 614
  cat owners in city, 32
  dog owners in city, 33
  dropout rate, 238
  enrollment decrease/increase at
      school, 108, 233, 234, 236,
      239, 589
  exam grades vs. hours studied, 521
  fuel consumption, by fuel type, 241
  language course enrollment, 238
  percent of correct answers, 45
  percent of eligible voters voting, 32
  percent of left-handed persons, 238
  percent of registered vehicles, by
      type, 241
  percent of students receiving A
      grade, 235
  percent responding to survey, 238
  population doubling, 52, 85, 283
  population growth, 46, 233, 295,
      567–568, 579
  population growth factors, 283
  population growth predictions, 579
  programs for disabled, by year, 511
  social security recipients per
      year, 573
  students per section, 213
  team losing streak, 63
  unemployment rates, 238
  votes cast, 204–205
  votes per candidate, 210, 211, 720

AN ARITHMETIC REVIEW

Cultures from all over the world have developed number systems and ways to record patterns in their natural surroundings. The Mayans in Central America had one of the most sophisticated number systems in the world in the twelfth century A.D.* The Chinese numbering and recording system dates from around 1200 B.C.E. The oldest evidence of numerical record is in Africa. It is a bone notched in numerical patterns and dating from about 35,000 B.C.E. It was found in the Lebombo Mountains near modern-day Swaziland in southern Africa.

The roots of algebra developed among the Babylonians 4000 years ago in an area now part of the country of Iraq. The Babylonians developed ways to record useful numerical relationships so that they were easy to remember, easy to record, and helpful in solving problems. Archeologists have found many tables, such as one giving successive powers of a given number, 9^2, 9^3, 9^4, . . . , 9^n. The tables include instructions for solving problems in engineering, economics, city planning, and agriculture. The writing is on clay tablets. Some of the formulas developed by the Babylonians are still in use today.

You are about to embark on an exciting and useful endeavor: learning to use algebra to help you solve problems. It will take some time and effort, but do not be discouraged. Everyone can master this topic—people just like you have used it for many centuries! Today algebra is even more useful than in the past because it is used in nearly every field of human endeavor

---

* A.D. stands for the Latin *Anno Domini,* which means "in the year of the Lord."

Section _____ Date _____

ANSWERS

1. _____

2. _____

3. _____

4. _____

5. _____

6. _____ 7. _____

8. _____ 9. _____

10. _____ 11. _____

12. _____ 13. _____

14. _____ 15. _____

16. _____ 17. _____

18. _____

19. _____ _____

20. _____

21. _____

22. _____

23. _____

24. _____ 25. _____

26. _____ 27. _____

28. _____ 29. _____

30. _____

This pre-test will provide a preview of the type of exercises you will encounter in each section of this chapter. The answers for these exercises can be found in the back of the text. If you are working on your own, or ahead of the class, this pre-test can help you find the sections in which you should focus more of your time.

[0.1] 1. List all the factors of 42.

2. For the group of numbers 2, 3, 6, 7, 9, 17, 18, 21, and 23, list the prime numbers and the composite numbers.

3. Find the prime factorization for each of the following numbers.
   (a) 60   (b) 350

4. Find the greatest common factor (GCF) for each of the following groups of numbers.
   (a) 12 and 32   (b) 24, 36, and 42

5. Find the least common multiple (LCM) for each of the following groups of numbers.
   (a) 4, 5, and 10   (b) 36, 20, and 30

Perform the indicated operations.

[0.2] 6. $\dfrac{3}{5} \cdot \dfrac{25}{12}$ 7. $\dfrac{6}{7} \div \dfrac{12}{21}$ 8. $\dfrac{5}{6} + \dfrac{3}{4}$

9. $\dfrac{17}{18} - \dfrac{5}{9}$ **[0.3]** 10. $8.123 - 4.356$ 11. $7.16 \cdot 3.19$

[0.2] 12. $2\dfrac{2}{5} \cdot 5\dfrac{1}{4}$ 13. $1\dfrac{5}{6} \div 2\dfrac{4}{9}$ **[0.3]** 14. $3.896 \div 1.6$

Evaluate the following expressions.

[0.4] 15. $21 - 3 \cdot 5$ 16. $3 \cdot 4 - 2^2$ 17. $(18 \div 9) \cdot 2 + 3^2$

18. $(15 - 12 + 5) \div 2^2$

[0.3] 19. Write 23% as (a) a fraction and (b) a decimal.

20. Write 0.035 as a percent.

[0.5] 21. Represent the following integers on the number line shown: 6, -8, 4, -2, 10.

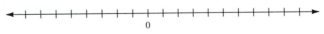

22. Place the following data set in ascending order: 5, -2, -4, 0, -1, 1.

23. Determine the maximum and minimum of the following data set: -4, 1, -5, 7, 3, 2.

Evaluate the following expressions.

[0.5] 24. $|-5|$ 25. $|6|$ 26. $|11 - 5|$

27. $|-11| - |5|$ 28. $|4 + 5| - |6 - 3|$

Find the opposite of each of the following.

29. -16 30. 23

0.1 Prime Factorization and Least Common Multiples

0.1 OBJECTIVES

1. Find the factors of a whole number
2. Determine whether a number is prime, composite, or neither
3. Find the prime factorization for a number
4. Find the GCF for two or more numbers
5. Find the LCM for two or more numbers

Overcoming Math Anxiety

Throughout this text, we will present you with a series of class-tested techniques that are designed to improve your performance in this math class.

Hint #1   Become familiar with your textbook.

Perform each of the following tasks.

1. Use the Table of Contents to find the title of Section 5.1.
2. Use the index to find the earliest reference to the term *mean.* (By the way, this term has nothing to do with the personality of either your instructor or the textbook author!)
3. Find the answer to the first Check Yourself exercise in Section 0.1.
4. Find the answers to the Self-Test for Chapter 1.
5. Find the answers to the odd-numbered exercises in Section 0.1.
6. In the margin notes for Section 0.1, find the definition for the term *relatively prime.*

Now you know where some of the most important features of the text are. When you have a moment of confusion, think about using one of these features to help you clear up that confusion.

How would you organize the following list of objects: cow, dog, daisy, fox, lily, sunflower, cat, tulip?

Although there are many ways to organize the objects, most people would break them into two groups, the animals and the flowers. In mathematics, we call a group of things that have something in common a **set.**

Definition: Set

A **set** is a collection of objects.

We generally use braces to enclose the elements of a set.

{cow, dog, fox, cat}   or   {daisy, lily, sunflower, tulip}

Of course, in mathematics many (but not all) of the sets we are interested in are sets of numbers.

The numbers used to count things—1, 2, 3, 4, 5, and so on—are called the **natural** (or **counting**) **numbers.** The **whole numbers** consist of the natural numbers and zero—0, 1, 2, 3, 4, 5, and so on. They can be represented on a number line like the one shown. Zero (0) is considered the origin.

The origin

0 1 2 3 4 5 6

The number line continues forever in both directions.

Any whole number can be written as a product of two whole numbers. For example, we say that $12 = 3 \cdot 4$. We call 3 and 4 **factors** of 12.

NOTE The centered dot represents multiplication.

Definition: Factor

A **factor** of a whole number is another whole number that will *divide exactly* into that number. This means that the division will have a remainder of 0.

OBJECTIVE 1

Example 1   Finding Factors

NOTE 3 and 6 can also be called *divisors* of 18. They divide 18 exactly.

NOTE This is a complete list of the factors. There are no other whole numbers that divide 18 exactly. Note that the factors of 18, except for 18 itself, are *smaller* than 18.

List all factors of 18.

$3 \cdot 6 = 18$ Because $3 \cdot 6 = 18$, 3 and 6 are factors (or divisors) of 18.

$2 \cdot 9 = 18$ 2 and 9 are also factors of 18.

$1 \cdot 18 = 18$ 1 and 18 are factors of 18.

1, 2, 3, 6, 9, and 18 are all the factors of 18.

 CHECK YOURSELF 1*

List all factors of 24.

NOTE A whole number greater than 1 will always have itself and 1 as factors. Sometimes these will be the *only* factors. For instance, 1 and 3 are the only factors of 3.

Listing factors leads us to an important classification of whole numbers. Any whole number larger than 1 is either a *prime* or a *composite* number. Let's look at the following definitions.

Definition: Prime Number

A **prime number** is any whole number greater than 1 that has only 1 and itself as factors.

NOTE How large can a prime number be? There is no largest prime number. To date, the largest *known* prime is $2^{13,466,917} - 1$. This is a number with 4,053,946 digits, if you are curious. Of course, a computer had to be used to verify that a number of this size is prime. By the time you read this, someone may very well have found an even larger prime number.

As examples, 2, 3, 5, and 7 are prime numbers. Their only factors are 1 and themselves.

To check whether a number is prime, one approach is simply to divide the smaller primes, 2, 3, 5, 7, and so on, into the given number. If no factors other than 1 and the given number are found, the number is prime.

* Check Yourself answers appear at the end of each section throughout the book.

Here is the method known as the sieve of Eratosthenes for identifying prime numbers.

1. Write down a series of counting numbers, starting with the number 2. In the example below, we stop at 50.
2. Start at the number 2. Delete every second number after the 2.
3. Move to the number 3. Delete every third number after 3 (some numbers will be deleted twice).
4. Continue this process, deleting every fourth number after 4, every fifth number after 5, and so on.
5. When you have finished, the undeleted numbers are the prime numbers.

	2	3	4	5	6	7	8	9	10
11	12	13	14	15	16	17	18	19	20
21	22	23	24	25	26	27	28	29	30
31	32	33	34	35	36	37	38	39	40
41	42	43	44	45	46	47	48	49	50

The prime numbers less than 50 are 2, 3, 5, 7, 11, 13, 17, 19, 23, 29, 31, 37, 41, 43, and 47.

OBJECTIVE 2

Example 2   Identifying Prime Numbers

Which of the numbers 17, 29, and 33 are prime?

17 is a prime number.   1 and 17 are the only factors.

29 is a prime number.   1 and 29 are the only factors.

33 is *not* prime.   1, 3, 11, and 33 are all factors of 33.

Note: For two-digit numbers, if the number is *not* a prime, it will have one or more of the numbers 2, 3, 5, or 7 as factors.

CHECK YOURSELF 2

Which of the following numbers are prime numbers?

2, 6, 9, 11, 15, 19, 23, 35, 41

We can now define a second class of whole numbers.

NOTE This definition tells us that a composite number *does* have factors other than 1 and itself.

Definition: Composite Number

A **composite number** is any whole number greater than 1 that is not prime.

> **Example 3   Identifying Composite Numbers**

Which of the following numbers are composite?

18 is a composite number.   1, 2, 3, 6, 9, and 18 are all factors of 18.

23 is not a composite number.   1 and 23 are the only factors. This means that 23 is a *prime number.*

25 is a composite number.   1, 5, and 25 are factors.

38 is a composite number.   1, 2, 19, and 38 are factors.

 CHECK YOURSELF 3 _____

Which of the following numbers are composite numbers?

2, 6, 10, 13, 16, 17, 22, 27, 31, 35

By the definitions of prime and composite numbers:

> **Rules and Properties: 0 and 1**
>
> The whole numbers 0 and 1 are neither prime nor composite.

To **factor a number** means to write the number as a product of whole-number factors.

OBJECTIVE 3   **Example 4   Factoring a Composite Number**

Factor the number 10.

$10 = 2 \cdot 5$ The order in which you write the factors does not matter, so $10 = 5 \cdot 2$ would also be correct.

Of course, $10 = 10 \cdot 1$ is also a correct statement. However, in this section we are interested in products that use factors other than 1 and the given number.

Factor the number 21.
$21 = 3 \cdot 7$

 CHECK YOURSELF 4 _____

Factor 35.

In writing composite numbers as a product of factors, there may be several different possible factorizations.

Example 5   Factoring a Composite Number

Find three ways to factor 72.

NOTE There have to be at least two different factorizations, because a composite number has factors other than 1 and itself.

$$72 = 8 \cdot 9 \quad (1)$$
$$= 6 \cdot 12 \quad (2)$$
$$= 3 \cdot 24 \quad (3)$$

 CHECK YOURSELF 5 _____

Find three ways to factor 42.

We now want to write composite numbers as a product of their **prime factors.** Look again at the first factored line of Example 5. The process of factoring can be continued until all the factors are prime numbers.

Example 6   Factoring a Composite Number into Prime Factors

NOTE This is often called a **factor tree.**

$$72 = \quad 8 \quad \cdot \quad 9$$
$$= \quad 2 \cdot 4 \quad \cdot 3 \cdot 3$$
$$= 2 \cdot 2 \cdot 2 \cdot 3 \cdot 3$$

4 is still not prime, and so we continue by factoring 4.

72 is now written as a product of prime factors.

NOTE Finding the prime factorization of a number is a skill that is used when adding fractions.

When we write 72 as $2 \cdot 2 \cdot 2 \cdot 3 \cdot 3$, no further factorization is possible. This is called the **prime factorization** of 72.

Now, what if we start with the second factored line from the same example, $72 = 6 \cdot 12$?

$$72 = \quad 6 \quad \cdot \quad 12$$
$$= 2 \cdot 3 \cdot \quad 3 \cdot 4$$
$$= 2 \cdot 3 \cdot 3 \cdot 2 \cdot 2$$

Continue to factor 6 and 12.

Continue again to factor 4. Other choices for the factors of 12 are possible. As we shall see, the end result will be the same.

No matter which pair of factors you start with, you will find the same prime factorization. In this case, there are three factors of 2 and two factors of 3. The order in which we write the factors does not matter.

 CHECK YOURSELF 6 _____

We could also write

$$72 = 2 \cdot 36$$

Continue the factorization.

> **Rules and Properties:**   The Fundamental Theorem of Arithmetic
>
> There is exactly one prime factorization for any composite number.

The method of Example 6 will always work. However, an easier method for factoring composite numbers exists. This method is particularly useful when numbers get large, in which case factoring with a number tree becomes unwieldy.

> **Rules and Properties:**   Factoring by Division
>
> To find the prime factorization of a number, divide the number by a series of primes until the final quotient is a prime number.

NOTE The prime factorization is the product of all the prime divisors and the final quotient.

> **Example 7**   **Finding Prime Factors**

To write 60 as a product of prime factors, divide 2 into 60 for a quotient of 30. Continue to divide by 2 again for the quotient 15. Because 2 won't divide exactly into 15, we try 3. Because the quotient 5 is prime, we are done.

$$\overset{30}{2\overline{)60}} \qquad \overset{15}{2\overline{)30}} \qquad \overset{5}{3\overline{)15}} \qquad \text{Prime}$$

Our factors are the prime divisors and the final quotient. We have

$$60 = 2 \cdot 2 \cdot 3 \cdot 5$$

> **CHECK YOURSELF 7**
>
> *Complete the process to find the prime factorization of* 90.
>
> $$\overset{45}{2\overline{)90}} \qquad \overset{?}{?\overline{)45}}$$
>
> *Remember to continue until the final quotient is prime.*

Writing composite numbers in their completely factored form can be simplified if we use a format called **continued division.**

> **Example 8**   **Finding Prime Factors Using Continued Division**

Use the continued-division method to divide 60 by a series of prime numbers.

NOTE In each short division, we write the quotient *below* rather than above the dividend. This is just a convenience for the next division.

$$\text{Primes} \begin{cases} 2\overline{)60} \\ 2\overline{)30} \\ 3\overline{)15} \\ 5 \end{cases}$$

Stop when the final quotient is prime.

To write the factorization of 60, we include each divisor used and the final prime quotient. In our example, we have

$$60 = 2 \cdot 2 \cdot 3 \cdot 5$$

CHECK YOURSELF 8 _____

Find the prime factorization of 234.

We know that a factor or a divisor of a whole number divides that number exactly.
The factors or divisors of 20 are

NOTE Again the factors of 20, other than 20 itself, are less than 20.

1, 2, 4, 5, 10, and 20

Each of these numbers divides 20 exactly, that is, with no remainder.
Our work in the rest of this section involves common factors or divisors. A **common factor** or **divisor** for two numbers is any factor that divides both the numbers exactly.

OBJECTIVE 4 **Example 9** **Finding Common Factors**

Look at the numbers 20 and 30. Is there a common factor for the two numbers?
First, we list the factors. Then we circle the ones that appear in both lists.

Factors

20: ①, ②, 4, ⑤, ⑩, 20

30: ①, ②, 3, ⑤, 6, ⑩, 15, 30

We see that 1, 2, 5, and 10 are common factors of 20 and 30. Each of these numbers divides both 20 and 30 exactly.
Our later work with fractions will require that we find the greatest common factor of a group of numbers.

Definition: Greatest Common Factor

The **greatest common factor** (GCF) of a group of numbers is the *largest* number that will divide each of the given numbers exactly.

In the first part of Example 9, the common factors of the numbers 20 and 30 were listed as

1, 2, 5, 10   Common factors of 20 and 30

The GCF of the two numbers is then 10, because 10 is the *largest* of the four common factors.

CHECK YOURSELF 9 _____

List the factors of 30 and 36, and then find the GCF.

The method of Example 9 will also work in finding the GCF of a group of more than two numbers.

Example 10 **Finding the GCF by Listing Factors**

Find the GCF of 24, 30, and 36. We list the factors of each of the three numbers.

NOTE Looking at the three lists, we see that 1, 2, 3, and 6 are common factors.

24: ①, ②, ③, 4, ⑥, 8, 12, 24

30: ①, ②, ③, 5, ⑥, 10, 15, 30

36: ①, ②, ③, 4, ⑥, 9, 12, 18, 36

6 is the GCF of 24, 30, and 36.

CHECK YOURSELF 10

Find the GCF of 16, 24, and 32.

The process shown in Example 10 is very time-consuming when larger numbers are involved. A better approach to the problem of finding the GCF of a group of numbers uses the prime factorization of each number. Let's outline the process.

Step by Step: Finding the GCF

NOTE If there are no common prime factors, the GCF is 1.

Step 1   Write the prime factorization for each of the numbers in the group.
Step 2   Locate the prime factors that appear in every prime factorization.
Step 3   The GCF will be the *product* of all the common prime factors.

Example 11   Finding the GCF

Find the GCF of 20 and 30.

Step 1   Write the prime factorizations of 20 and 30.

$20 = 2 \cdot 2 \cdot 5$

$30 = 2 \cdot 3 \cdot 5$

Step 2   Find the prime factors common to each number.

$20 = ②\cdot 2 \cdot ⑤$ 2 and 5 are the common prime factors.
$30 = ②\cdot 3 \cdot ⑤$

Step 3   Form the product of the common prime factors.

$2 \cdot 5 = 10$

10 is the GCF.

CHECK YOURSELF 11

Find the GCF of 30 and 36.

To find the GCF of a group of more than two numbers, we use the same process.

Example 12   Finding the GCF

Find the GCF of 24, 30, and 36.

$24 = ②\cdot 2 \cdot 2 \cdot ③$
$30 = ②\cdot ③\cdot 5$
$36 = ②\cdot 2 \cdot ③\cdot 3$

2 and 3 are the prime factors common to *all three numbers*.

$2 \cdot 3 = 6$ is the GCF.

CHECK YOURSELF 12 _____

Find the GCF of 15, 30, and 45.

Example 13   Finding the GCF

NOTE If two numbers, such as 15 and 28, have no common factor other than 1, they are called **relatively prime.**

Find the GCF of 15 and 28.

$15 = 3 \cdot 5$

$28 = 2 \cdot 2 \cdot 7$

There are no common prime factors listed. But remember that 1 is a factor of every whole number.

1 is the GCF.

CHECK YOURSELF 13 _____

Find the GCF of 30 and 49.

Another idea that will be important in our work with fractions is the concept of *multiples*. Every whole number has an associated group of multiples.

Definition: Multiples

The **multiples** of a number are the product of that number with the natural numbers 1, 2, 3, 4, 5, ...

OBJECTIVE 5   **Example 14   Listing Multiples**

List the multiples of 3.
    The multiples of 3 are

$3 \cdot 1, 3 \cdot 2, 3 \cdot 3, 3 \cdot 4, \ldots$

or

NOTE Notice that the multiples, except for 3 itself, are *larger* than 3.

3, 6, 9, 12, . . .   The three dots indicate that the list continues without stopping.

An easy way of listing the multiples of 3 is to think of *counting by threes*.

CHECK YOURSELF 14 _____

List the first seven multiples of 4.

Sometimes we need to find common multiples of two or more numbers.

Definition: Common Multiples

If a number is a multiple of each of a group of numbers, it is called a **common multiple** of the numbers; that is, it is a number that is evenly divisible by all the numbers in the group.

Example 15 Finding Common Multiples

NOTE 15, 30, 45, and 60 are multiples of *both* 3 and 5.

Find four common multiples of 3 and 5.
   Some common multiples of 3 and 5 are

15, 30, 45, 60

 CHECK YOURSELF 15 _____

List the first six multiples of 6. Then look at your list from Check Yourself 14 and list some common multiples of 4 and 6.

In our later work with fractions, we will use the *least common multiple* of a group of numbers.

Definition: Least Common Multiple

The **least common multiple** (LCM) of a group of numbers is the *smallest* number that is a multiple of each number in the group.

It is possible to simply list the multiples of each number and then find the LCM by inspection.

Example 16 Finding the LCM

Find the LCM of 6 and 8.

Multiples

NOTE 48 is also a common multiple of 6 and 8, but we are looking for the *smallest* such number.

6:   6, 12, 18, ⑳(24), 30, 36, 42, 48, . . .

8:   8, 16, (24), 32, 40, 48, . . .

We see that 24 is the smallest number common to both lists. So 24 is the LCM of 6 and 8.

 CHECK YOURSELF 16 _____

Find the LCM of 20 and 30 by listing the multiples of each number.

The technique of Example 16 will work for any group of numbers. However, it becomes tedious for larger numbers. Let's outline a different approach.

Step by Step: Finding the LCM

Step 1   Write the prime factorization for each of the numbers in the group.
Step 2   List all the prime factors that appear in any one of the prime factorizations.
Step 3   Form the product of those prime factors, using each factor the greatest number of times it occurs in any one factorization.

NOTE For instance, if a number appears three times in the factorization of a number, it will be included at least three times in forming the least common multiple.

Some students prefer a slightly different method of lining up the factors to help in remembering the process of finding the LCM of a group of numbers.

Example 17   Finding the LCM

To find the LCM of 10 and 18, we factor:

NOTE Line up the *like* factors vertically.

$$10 = 2 \qquad \cdot 5$$
$$18 = \underline{2 \cdot 3 \cdot 3}$$
$$ 2 \cdot 3 \cdot 3 \cdot 5 \qquad \text{Bring down the factors.}$$

2 and 5 appear, at most, one time in any one factorization. And 3 appears two times in one factorization.

$$2 \cdot 3 \cdot 3 \cdot 5 = 90$$

So 90 is the LCM of 10 and 18.

 CHECK YOURSELF 17

Use the method of Example 17 to find the LCM of 24 and 36.

The procedure is the same for a group of more than two numbers.

Example 18   Finding the LCM

To find the LCM of 12, 18, and 20, we factor:

$$12 = 2 \cdot 2 \cdot 3$$
$$18 = 2 \qquad \cdot 3 \cdot 3$$

NOTE The different factors that appear are 2, 3, and 5.

$$20 = \underline{2 \cdot 2 \qquad \cdot 5}$$
$$ 2 \cdot 2 \cdot 3 \cdot 3 \cdot 5$$

2 and 3 appear twice in one factorization, and 5 appears just once.

$$2 \cdot 2 \cdot 3 \cdot 3 \cdot 5 = 180$$

So 180 is the LCM of 12, 18, and 20.

 CHECK YOURSELF 18

Find the LCM of 3, 4, and 6.

CHECK YOURSELF ANSWERS

1. 1, 2, 3, 4, 6, 8, 12, and 24.   **2.** 2, 11, 19, 23, and 41 are prime numbers.

3. 6, 10, 16, 22, 27, and 35 are composite numbers.   **4.** $5 \cdot 7$

5. $2 \cdot 21, 3 \cdot 14, 6 \cdot 7$ **6.** $2 \cdot 2 \cdot 2 \cdot 3 \cdot 3$

7.

$$45 \searrow \qquad 15 \searrow \qquad 5$$
$$2\overline{)90} \qquad \longrightarrow 3\overline{)45} \qquad \longrightarrow 3\overline{)15}$$
$$90 = 2 \cdot 3 \cdot 3 \cdot 5$$

8. $2 \cdot 3 \cdot 3 \cdot 13$

9. 30: ①, ②, ③, 5, ⑥, 10, 15, 30

     36: ①, ②, ③, 4, ⑥, 9, 12, 18, 36

     The GCF is 6.

10. 16: ①, ②, ④, ⑧, 16

     24: ①, ②, 3, ④, 6, ⑧, 12, 24

     32: ①, ②, ④, ⑧, 16, 32

     The GCF is 8.

11. $30 = ② \cdot ③ \cdot 5$

     $36 = ② \cdot 2 \cdot ③ \cdot 3$

     The GCF is $2 \cdot 3 = 6$.

12. 15   **13.** The GCF is 1; 30 and 49 are relatively prime.

14. The first seven multiples of 4 are 4, 8, 12, 16, 20, 24, and 28.

15. 6, 12, 18, 24, 30, 36; some common multiples of 4 and 6 are 12, 24, and 36.

16. The multiples of 20 are 20, 40, 60, 80, 100, 120, . . . ; the multiples of 30 are 30, 60, 90, 120, 150, . . . ; the LCM of 20 and 30 is 60, the smallest number common to both lists.

17. $24 = 2 \cdot 2 \cdot 2 \cdot 3$

$$\frac{36 = 2 \cdot 2 \quad \cdot 3 \cdot 3}{2 \cdot 2 \cdot 2 \cdot 3 \cdot 3 = 72}$$

18. 12

0.1 Exercises

List the factors of each of the following numbers.

1. 8   **2.** 6

3. 10   **4.** 12

5. 15   **6.** 21

7. 24   **8.** 32

9. 64   **10.** 66

11. 13   **12.** 37

Use the following list of numbers for Exercises 13 and 14.

0, 1, 15, 19, 23, 31, 49, 55, 59, 87, 91, 97, 103, 105

13. Which of the given numbers are prime?

14. Which of the given numbers are composite?

15. List all the prime numbers between 30 and 50.

16. List all the prime numbers between 55 and 75.

Find the prime factorization of each number.

17. 20   **18.** 22

19. 30   **20.** 35

21. 51   **22.** 24

23. 63   **24.** 94

ANSWERS

1. _____
2. _____
3. _____
4. _____
5. _____
6. _____
7. _____
8. _____
9. _____
10. _____
11. _____
12. _____
13. _____
14. _____
15. _____
16. _____
17. _____
18. _____
19. _____
20. _____
21. _____
22. _____
23. _____
24. _____

25. _____

26. _____

27. _____

28. _____

29. _____

30. _____

31. _____

32. _____

33. _____

34. _____

35. _____

36. _____

37. _____

38. _____

39. _____

40. _____

41. _____

42. _____

43. _____

44. _____

45. _____

46. _____

47. _____

48. _____

49. _____

50. _____

51. _____

52. _____

25. 70   **26.** 90

27. 88   **28.** 100

29. 130   **30.** 66

31. 315   **32.** 400

33. 225   **34.** 132

35. 189   **36.** 330

In later mathematics courses, you often will want to find factors of a number with a given sum or difference. The following problems use this technique.

37. Find two factors of 24 with a sum of 10.

38. Find two factors of 15 with a difference of 2.

39. Find two factors of 30 with a difference of 1.

40. Find two factors of 28 with a sum of 11.

Find the GCF for each of the following groups of numbers.

41. 4 and 6   **42.** 6 and 9

43. 10 and 15   **44.** 12 and 14

45. 21 and 24   **46.** 22 and 33

47. 20 and 21   **48.** 28 and 42

49. 18 and 24   **50.** 35 and 36

51. 45, 60, and 75   **52.** 36, 54, and 180

53. 12, 36, and 60 **54.** 15, 45, and 90 **55.** 105, 140, and 175

56. 32, 80, and 112 **57.** 25, 75, and 150 **58.** 36, 72, and 144

Find the LCM for each of the following groups of numbers. Use whichever method you wish.

59. 12 and 15 **60.** 12 and 21 **61.** 18 and 36

62. 25 and 50 **63.** 25 and 40 **64.** 10 and 14

65. 3, 5, and 6 **66.** 2, 8, and 10 **67.** 18, 21, and 28

68. 8, 15, and 20 **69.** 20, 30, and 40 **70.** 12, 20, and 35

71. Prime numbers that differ by two are called *twin primes*. Examples are 3 and 5, 5 and 7, and so on. Find one pair of twin primes between 85 and 105.

72. The following questions refer to twin primes (see Exercise 71).

   (a) Search for, and make a list of several pairs of twin primes, in which the primes are greater than 3.
   (b) What do you notice about each number that lies *between* a pair of twin primes?
   (c) Write an explanation for your observation in part (b).

73. Obtain (or imagine that you have) a quantity of square tiles. Six tiles can be arranged in the shape of a rectangle in two different ways:

   (a) Record the dimensions of the rectangles shown.
   (b) If you use seven tiles, how many different rectangles can you form?
   (c) If you use ten tiles, how many different rectangles can you form?
   (d) What kind of number (of tiles) permits *only one* arrangement into a rectangle? *More than* one arrangement?

74. The number 10 has four factors: 1, 2, 5, and 10. We can say that 10 has an even number of factors. Investigate several numbers to determine which numbers have an *even number* of factors and which numbers have an *odd number* of factors.

53. _____
54. _____
55. _____
56. _____
57. _____
58. _____
59. _____
60. _____
61. _____
62. _____
63. _____
64. _____
65. _____
66. _____
67. _____
68. _____
69. _____
70. _____
71. _____
72. _____
73. _____
74. _____

75. _____

76. _____

77. _____

75. A natural number is said to be perfect if it is equal to the sum of its divisors, except itself.

    (a) Show that 28 is a perfect number.

    (b) Identify another perfect number less than 28.

76. Find the smallest natural number that is divisible by all of the following: 2, 3, 4, 6, 8, 9.

77. Suppose that a school has 1000 lockers and that they are all closed. A person passes through, opening every other locker, beginning with locker 2. Then another person passes through, changing every third locker (closing it if it is open, opening it if it is closed), starting with locker 3. Yet another person passes through, changing every fourth locker, beginning with locker 4. This process continues until 1000 people pass through.

    (a) At the end of this process, which locker numbers are closed?

    (b) Write an explanation for your answer to part (a).
    (Hint: It may help to attempt Exercise 74 first.)

Answers

We provide the answers for the odd-numbered problems at the end of each exercise set.

1. 1, 2, 4, 8   **3.** 1, 2, 5, 10   **5.** 1, 3, 5, 15

7. 1, 2, 3, 4, 6, 8, 12, 24   **9.** 1, 2, 4, 8, 16, 32, 64   **11.** 1, 13

13. 19, 23, 31, 59, 97, 103   **15.** 31, 37, 41, 43, 47   **17.** $2 \cdot 2 \cdot 5$

19. $2 \cdot 3 \cdot 5$ **21.** $3 \cdot 17$ **23.** $3 \cdot 3 \cdot 7$ **25.** $2 \cdot 5 \cdot 7$ **27.** $2 \cdot 2 \cdot 2 \cdot 11$

29. $2\overline{)130}$
     $5\overline{)65}$
       13
     $130 = 2 \cdot 5 \cdot 13$

31. $3 \cdot 3 \cdot 5 \cdot 7$ **33.** $3 \cdot 3 \cdot 5 \cdot 5$

35. $3\overline{)189}$
     $3\overline{)63}$
     $3\overline{)21}$
       7
     $189 = 3 \cdot 3 \cdot 3 \cdot 7$

37. 4, 6   **39.** 5, 6   **41.** 2   **43.** 5   **45.** 3

47. 1   **49.** 6   **51.** 15   **53.** 12   **55.** 35   **57.** 25   **59.** 60

61. 36   **63.** 200   **65.** 30   **67.** 252   **69.** 120

71.   **73.**   **75.**   **77.**

0.2 Fractions and Mixed Numbers

0.2 OBJECTIVES

1. Simplify a fraction
2. Multiply or divide two fractions
3. Add or subtract two fractions
4. Convert a fraction to a mixed number
5. Multiply or divide two mixed numbers
6. Add or subtract two mixed numbers

This section provides a review of the basic operations, addition, subtraction, division, and multiplication, on fractions and mixed numbers.

As mentioned in Section 0.1, the numbers used for counting are called the **natural numbers.** If we include zero in this group of numbers, we then call them the **whole numbers.** The **numbers of ordinary arithmetic** consist of all the whole numbers and all fractions. The fractions could be **proper fractions** (the numerator is less than the denominator) such as $\frac{1}{2}$ and $\frac{2}{3}$ or **improper fractions** (numerator is greater than or equal to the denominator) such as $\frac{7}{2}$ and $\frac{19}{5}$.

Every number of ordinary arithmetic can be written in fraction form $\frac{a}{b}$.

The number 1 has many different fractional forms. Any fraction in which the numerator and denominator are the same (and not zero) is another name for the number 1.

$$1 = \frac{2}{2} \qquad 1 = \frac{12}{12} \qquad 1 = \frac{257}{257}$$

Because these fractions are just different names for the same quantity, they are called **equivalent fractions.**

To write equivalent fractions, we use the **Fundamental Principle of Fractions (FPF).**

Rules and Properties: The Fundamental Principle of Fractions

$$\frac{a}{b} = \frac{a \cdot c}{b \cdot c} \qquad \text{or} \qquad \frac{a \cdot c}{b \cdot c} = \frac{a}{b}, \text{ in which neither } b \text{ nor } c \text{ can equal zero.}$$

OBJECTIVE 1

Example 1 Rewriting Fractions

Write three fractional representations for each number.

NOTE Each representation is a numeral, or name for the number. Each number has many names.

(a) $\frac{2}{3}$

We use the fundamental principle of fractions to multiply the numerator and denominator by the same number.

NOTE In each case, we have used the Fundamental Principle of Fractions with c equal to a different number.

$$\frac{2}{3} = \frac{2 \cdot 2}{3 \cdot 2} = \frac{4}{6}$$

$\frac{4}{6}$

$$\frac{2}{3} = \frac{2 \cdot 3}{3 \cdot 3} = \frac{6}{9}$$

$\frac{6}{9}$

$$\frac{2}{3} = \frac{2 \cdot 10}{3 \cdot 10} = \frac{20}{30}$$

$\frac{20}{30}$

(b) 5

$$5 = \frac{5 \cdot 2}{1 \cdot 2} = \frac{10}{2}$$

$$5 = \frac{5 \cdot 3}{1 \cdot 3} = \frac{15}{3}$$

$$5 = \frac{5 \cdot 100}{1 \cdot 100} = \frac{500}{100}$$

 CHECK YOURSELF 1

Write three fractional representations for each number.

(a) $\dfrac{5}{8}$ **(b)** $\dfrac{4}{3}$ **(c)** 3

The simplest fractional representation for a number has the smallest numerator and denominator. Fractions written in this form are said to be **simplified.**

Example 2   Simplifying Fractions

Simplify each fraction.

(a) $\dfrac{22}{55}$ **(b)** $\dfrac{35}{45}$ **(c)** $\dfrac{24}{36}$

In each case, we first find the prime factors for the numerator and for the denominator.

(a) $\dfrac{22}{55} = \dfrac{2 \cdot 11}{5 \cdot 11}$

We then use the fundamental principle of fractions.

$$\frac{22}{55} = \frac{2 \cdot 11}{5 \cdot 11} = \frac{2}{5}$$

(b) $\dfrac{35}{45} = \dfrac{7 \cdot 5}{3 \cdot 3 \cdot 5} = \dfrac{7 \cdot 5}{9 \cdot 5}$

Using the fundamental principle to remove the common factor of 5 yields

$$\frac{35}{45} = \frac{7}{9}$$

(c) $\dfrac{24}{36} = \dfrac{2 \cdot 2 \cdot 2 \cdot 3}{2 \cdot 2 \cdot 3 \cdot 3}$

Removing the common factor $2 \cdot 2 \cdot 3$ yields

$$\frac{24}{36} = \frac{2}{3}$$

CHECK YOURSELF 2

Simplify each fraction.

(a) $\dfrac{21}{33}$ (b) $\dfrac{15}{30}$ (c) $\dfrac{12}{54}$

Rules and Properties: Multiplication of Fractions

NOTE This is how two fractions, under the operation of multiplication, become one fraction.

$$\frac{a}{b} \cdot \frac{c}{d} = \frac{a \cdot c}{b \cdot d}$$

When multiplying two fractions, rewrite them in factored form, and then simplify before multiplying.

OBJECTIVE 2   **Example 3   Multiplying Fractions**

Find the product of the two fractions.

NOTE A product is the result from multiplication.

$$\frac{9}{2} \cdot \frac{4}{3}$$

$$\frac{9}{2} \cdot \frac{4}{3} = \frac{9 \cdot 4}{2 \cdot 3}$$

$$= \frac{3 \cdot 3 \cdot 2 \cdot 2}{2 \cdot 3}$$

$$= \frac{3 \cdot 2}{1}$$

$$= \frac{6}{1} \qquad \text{The denominator of 1 is not necessary.}$$

$$= 6$$

 CHECK YOURSELF 3

Multiply and simplify each pair of fractions.

(a) $\dfrac{3}{5} \cdot \dfrac{10}{7}$ (b) $\dfrac{12}{5} \cdot \dfrac{10}{6}$

Rules and Properties: Division of Fractions

NOTE This is how two fractions, under the operation of division, become one fraction.

$$\frac{a}{b} \div \frac{c}{d} = \frac{a}{b} \cdot \frac{d}{c} = \frac{a \cdot d}{b \cdot c}$$

To divide two fractions, the divisor is inverted, and then the fractions are multiplied.

Example 4   Dividing Fractions

Find the result from dividing the two fractions.

$$\frac{7}{3} \div \frac{5}{6}$$

$$\frac{7}{3} \div \frac{5}{6} = \frac{7}{3} \cdot \frac{6}{5} = \frac{7 \cdot 6}{3 \cdot 5}$$

$$= \frac{7 \cdot 2 \cdot 3}{3 \cdot 5} = \frac{7 \cdot 2}{5}$$

$$= \frac{14}{5}$$

CHECK YOURSELF 4

Find the result from dividing the two fractions.

$$\frac{9}{2} \div \frac{3}{5}$$

Rules and Properties: Addition of Fractions

NOTE This is how two fractions with the same denominator, become one fraction under the operation of addition.

$$\frac{a}{b} + \frac{c}{b} = \frac{a + c}{b}$$

NOTE Find the LCM of the set of denominators (as described in Section 0.1). The result is the LCD.

When adding two fractions with different denominators, find the **least common denominator (LCD)** first. The LCD is the smallest number that both denominators evenly divide. After rewriting the fractions with this denominator, add the numerators, and then simplify the result.

OBJECTIVE 3   ### Example 5   Adding Fractions

Find the sum of the two fractions.

NOTE A sum is the result from addition.

$$\frac{5}{8} + \frac{7}{12}$$

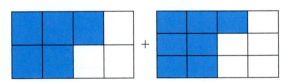

The LCD of 8 and 12 is 24. Each fraction should be rewritten as a fraction with that denominator.

$$\frac{5}{8} = \frac{15}{24}$$ Multiply the numerator and denominator by 3.

$$\frac{7}{12} = \frac{14}{24}$$ Multiply the numerator and denominator by 2.

$$\frac{5}{8} + \frac{7}{12} = \frac{15}{24} + \frac{14}{24} = \frac{29}{24}$$ This fraction cannot be simplified.

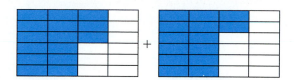

$+$

 CHECK YOURSELF 5

Find the sum for each pair of fractions.

(a) $\dfrac{4}{5} + \dfrac{7}{9}$

(b) $\dfrac{5}{6} + \dfrac{4}{15}$

Rules and Properties: Subtraction of Fractions

NOTE This is how two fractions with like denominators become one fraction under the operation of subtraction.

$$\frac{a}{b} - \frac{c}{b} = \frac{a-c}{b}$$

Subtracting fractions is treated exactly like adding them, except the numerator becomes the difference of the two numerators.

Example 6   Subtracting Fractions

Find the difference.

NOTE The difference is the result from subtraction.

$$\frac{7}{9} - \frac{1}{6}$$

The LCD is 18. We rewrite the fractions with that denominator.

$$\frac{7}{9} = \frac{14}{18}$$

$$\frac{1}{6} = \frac{3}{18}$$

$$\frac{7}{9} - \frac{1}{6} = \frac{14}{18} - \frac{3}{18} = \frac{11}{18}$$ This fraction cannot be simplified.

 CHECK YOURSELF 6

Find the difference $\dfrac{11}{12} - \dfrac{5}{8}$

Another way to write an improper fraction is as a *mixed number.*

Definition: Mixed Number

A **mixed number** is the sum of a whole number and a proper fraction.

$2\frac{3}{4}$ is a mixed number. It represents the sum of the whole number 2 and the fraction $\frac{3}{4}$.

For our later work it will be important to be able to change back and forth between improper fractions and mixed numbers. Because an improper fraction represents a number that is greater than or equal to 1, we have the following rule:

NOTE Both forms are correct, but in subsequent courses you will find that improper fractions are preferred to mixed numbers.

Rules and Properties: Improper Fractions to Mixed Numbers

An improper fraction can always be written as either a mixed number or as a whole number.

NOTE You can write the fraction $\frac{7}{5}$ as $7 \div 5$. We divide the numerator by the denominator.

To do this, remember that you can think of a fraction as indicating division. The numerator is divided by the denominator. This leads us to the following rule:

Step by Step: To Change an Improper Fraction to a Mixed Number

Step 1   Divide the numerator by the denominator.
Step 2   If there is a remainder, write the remainder over the original denominator.

NOTE In step 1, the quotient gives the whole-number portion of the mixed number. Step 2 gives the fractional portion of the mixed number.

OBJECTIVE 4   **Example 7   Converting a Fraction to a Mixed Number**

Convert $\frac{17}{5}$ to a mixed number.

Divide 17 by 5.

$$5\overline{)17} \quad \frac{17}{5} = 3\frac{2}{5}$$

Remainder / Original denominator / Quotient

   CHECK YOURSELF 7

Convert $\frac{32}{5}$ to a mixed number.

Example 8 illustrates how to convert a mixed number to an improper fraction.

Example 8   Converting Mixed Numbers to Improper Fractions

(a) Convert $3\frac{2}{5}$ to an improper fraction.

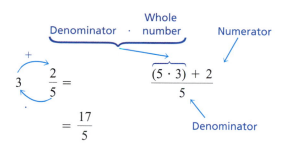

Multiply the denominator by the whole number ($5 \cdot 3 = 15$). Add the numerator. We now have 17.

$$= \frac{17}{5}$$

Write 17 over the original denominator.

NOTE Multiply the denominator, 7, by the whole number, 4, and add the numerator, 5.

(b) Convert $4\frac{5}{7}$ to an improper fraction.

$$4\frac{5}{7} = \frac{(7 \cdot 4) + 5}{7} = \frac{33}{7}$$

CHECK YOURSELF 8

Convert $5\frac{3}{8}$ to an improper fraction.

When multiplying two mixed numbers, it is usually easier to change the mixed numbers to improper fractions and then perform the multiplication. Example 9 illustrates this method.

OBJECTIVE 5   ### Example 9   Multiplying Two Mixed Numbers

Multiply.

$$3\frac{2}{3} \cdot 2\frac{1}{2} = \frac{11}{3} \cdot \frac{5}{2}$$ Change the mixed numbers to improper fractions.

$$= \frac{11 \cdot 5}{3 \cdot 2} = \frac{55}{6} = 9\frac{1}{6}$$

CAUTION

Be Careful! Students sometimes think of

$$3\frac{2}{3} \cdot 2\frac{1}{2} \quad \text{as} \quad (3 \cdot 2) + \left(\frac{2}{3} \cdot \frac{1}{2}\right)$$

This is *not* the correct multiplication pattern. You must first change the mixed numbers to improper fractions.

CHECK YOURSELF 9 _____

Multiply.

$$2\frac{1}{3} \cdot 3\frac{1}{2}$$

When dividing mixed numbers, simply change the mixed or whole numbers to improper fractions as the first step. Then proceed with the division. Example 10 illustrates this approach.

Example 10   Dividing Two Mixed Numbers

Divide.

$$2\frac{3}{8} \div 1\frac{3}{4} = \frac{19}{8} \div \frac{7}{4} \qquad \text{Write the mixed numbers as improper fractions.}$$

$$= \frac{19}{\overset{}{\underset{2}{8}}} \times \frac{\overset{1}{\cancel{4}}}{7} \qquad \text{Invert the divisor and multiply as before.}$$

$$= \frac{19}{14} = 1\frac{5}{14}$$

CHECK YOURSELF 10 _____

Divide $3\frac{1}{5} \div 2\frac{2}{5}$.

When adding or subtracting mixed numbers, first convert the mixed numbers to improper fractions and then proceed as you would when adding or subtracting fractions. Examples 11 and 12 illustrate this concept.

OBJECTIVE 6   ### Example 11   Adding Mixed Numbers

Add, and write the result as a mixed number.

$$3\frac{1}{6} + 2\frac{3}{8} = \frac{19}{6} + \frac{19}{8} \qquad \text{The LCD of the fractions is 24. Rename them with that denominator.}$$

$$= \frac{76}{24} + \frac{57}{24} \qquad \text{Then add as before.}$$

NOTE $\begin{array}{r} 5 \\ 24\overline{)133} \\ \underline{120} \\ 13 \end{array}$

$$= \frac{133}{24}$$

$$= 5\frac{13}{24}$$

CHECK YOURSELF 11 _____

Add $5\frac{7}{10} + 3\frac{5}{6}$. *Write the result as a mixed number.*

Example 12   Subtracting Mixed Numbers

Subtract.

$$8\frac{7}{10} - 3\frac{3}{8} = \frac{87}{10} - \frac{27}{8}$$ Write the fractions with denominator 40.

$$= \frac{348}{40} - \frac{135}{40}$$ Subtract as before.

$$= \frac{213}{40}$$

This can be written as $5\frac{13}{40}$.

 CHECK YOURSELF 12 _____

Subtract $7\frac{11}{12} - 3\frac{5}{8}$.

To subtract a mixed number from a whole number, we use the same techniques.

Example 13   Subtracting Mixed Numbers

Subtract.

$$6 - 2\frac{3}{4}$$

NOTE

$6 = \frac{6}{1} = \frac{24}{4}$

Multiply the numerator and denominator by 4 to form a common denominator.

$$6 - 2\frac{3}{4} = \frac{24}{4} - \frac{11}{4}$$ Write both the whole number and the mixed number as improper fractions with a common denominator.

$$= \frac{13}{4}$$

This can be written as $3\frac{1}{4}$.

 CHECK YOURSELF 13 _____

Subtract $7 - 3\frac{2}{5}$.

There are many applications in which fractions and mixed numbers are used. Examples 14 and 15 illustrate some of these.

OBJECTIVES 1–6   **Example 14   An Application of Fractions and Mixed Numbers**

Chair rail molding 136 inches (in.) long must be cut into pieces of $31\frac{1}{3}$ in. each. How many pieces can be cut from the molding?

$$136 \div 31\frac{1}{3} = 136 \div \frac{94}{3} = \frac{136}{1} \div \frac{94}{3} = \frac{136}{1} \cdot \frac{3}{94} = \frac{408}{94} = \frac{204}{47} = 4\frac{16}{47}$$

So four full-length pieces can be cut from the molding.

 CHECK YOURSELF 14 _____

After a family party, $10\frac{2}{3}$ cupcakes were left. If Amanda took $\frac{3}{8}$ of these, how many did she take?

Example 15   An Application of Fractions and Mixed Numbers

José must trim $2\frac{5}{16}$ feet (ft) from a board 8 ft long. How long will the board be after it is cut?

$$8 - 2\frac{5}{16} = 7\frac{16}{16} - 2\frac{5}{16} = 5\frac{11}{16}$$

The board will be $5\frac{11}{16}$ ft long after it is cut.

 CHECK YOURSELF 15 _____

Three pieces of lumber measure $5\frac{3}{8}$ ft, $7\frac{1}{2}$ ft, and $9\frac{3}{4}$ ft. What is the total length of the lumber?

CHECK YOURSELF ANSWERS _____

1. Answers will vary.   **2. (a)** $\frac{7}{11}$; **(b)** $\frac{1}{2}$; **(c)** $\frac{2}{9}$ **3. (a)** $\frac{6}{7}$; **(b)** 4

4. $\frac{15}{2}$ **5. (a)** $\frac{71}{45}$; **(b)** $\frac{11}{10}$ **6.** $\frac{7}{24}$ **7.** $6\frac{2}{5}$ **8.** $\frac{43}{8}$ **9.** $\frac{49}{6}$ or $8\frac{1}{6}$

10. $\frac{4}{3}$ or $1\frac{1}{3}$ **11.** $\frac{143}{15}$ or $9\frac{8}{15}$ **12.** $\frac{103}{24}$ or $4\frac{7}{24}$ **13.** $\frac{18}{5}$ or $3\frac{3}{5}$

14. 4   **15.** $22\frac{5}{8}$ ft

0.2   Exercises

In exercises 1 to 12, write three fractional representations for each number.

1. $\dfrac{3}{7}$

2. $\dfrac{2}{5}$

3. $\dfrac{4}{9}$

4. $\dfrac{7}{8}$

5. $\dfrac{5}{6}$

6. $\dfrac{11}{13}$

7. $\dfrac{10}{17}$

8. $\dfrac{3}{10}$

9. $\dfrac{9}{16}$

10. $\dfrac{6}{11}$

11. $\dfrac{7}{9}$

12. $\dfrac{15}{16}$

Write each fraction in simplest form.

13. $\dfrac{8}{12}$

14. $\dfrac{12}{15}$

15. $\dfrac{10}{14}$

16. $\dfrac{15}{50}$

17. $\dfrac{12}{18}$

18. $\dfrac{28}{35}$

19. $\dfrac{35}{40}$

20. $\dfrac{21}{24}$

21. $\dfrac{11}{44}$

22. $\dfrac{10}{25}$

23. $\dfrac{12}{36}$

24. $\dfrac{18}{48}$

25. $\dfrac{24}{27}$

26. $\dfrac{30}{50}$

27. $\dfrac{32}{40}$

28. $\dfrac{17}{51}$

29. $\dfrac{75}{105}$

30. $\dfrac{62}{93}$

31. $\dfrac{48}{60}$

32. $\dfrac{48}{66}$

33. $\dfrac{105}{135}$

34. $\dfrac{54}{126}$

35. $\dfrac{15}{44}$

36. $\dfrac{10}{63}$

Name _____

Section _____   Date _____

ANSWERS

1. _____
2. _____
3. _____
4. _____
5. _____
6. _____
7. _____
8. _____
9. _____
10. _____
11. _____
12. _____
13. ____ 14. ____ 15. ____
16. ____ 17. ____ 18. ____
19. ____ 20. ____ 21. ____
22. ____ 23. ____ 24. ____
25. ____ 26. ____ 27. ____
28. ____ 29. ____ 30. ____
31. ____ 32. ____ 33. ____
34. ____ 35. ____ 36. ____

SECTION 0.2   **29**

37. _____

38. _____

39. _____

40. _____

41. _____

42. _____

43. _____

44. _____

45. _____

46. _____

47. _____ 48. _____

49. _____ 50. _____

51. _____ 52. _____

53. _____ 54. _____

55. _____ 56. _____

57. _____ 58. _____

59. _____ 60. _____

61. _____ 62. _____

63. _____ 64. _____

65. _____ 66. _____

67. _____ 68. _____

69. _____ 70. _____

Multiply. Be sure to simplify each product.

37. $\dfrac{3}{4} \cdot \dfrac{7}{5}$

38. $\dfrac{2}{3} \cdot \dfrac{8}{5}$

39. $\dfrac{3}{5} \cdot \dfrac{5}{7}$

40. $\dfrac{6}{11} \cdot \dfrac{8}{6}$

41. $\dfrac{6}{13} \cdot \dfrac{4}{9}$

42. $\dfrac{5}{9} \cdot \dfrac{6}{11}$

43. $\dfrac{3}{11} \cdot \dfrac{7}{9}$

44. $\dfrac{7}{9} \cdot \dfrac{3}{5}$

45. $\dfrac{3}{10} \cdot \dfrac{5}{9}$

Divide. Write each result in simplest form.

46. $\dfrac{5}{21} \div \dfrac{25}{14}$

47. $\dfrac{1}{5} \div \dfrac{3}{4}$

48. $\dfrac{2}{5} \div \dfrac{1}{3}$

49. $\dfrac{2}{5} \div \dfrac{3}{4}$

50. $\dfrac{5}{8} \div \dfrac{3}{4}$

51. $\dfrac{8}{9} \div \dfrac{4}{3}$

52. $\dfrac{5}{9} \div \dfrac{8}{11}$

53. $\dfrac{7}{10} \div \dfrac{5}{9}$

54. $\dfrac{8}{9} \div \dfrac{11}{15}$

55. $\dfrac{8}{15} \div \dfrac{2}{5}$

56. $\dfrac{5}{27} \div \dfrac{15}{54}$

57. $\dfrac{5}{27} \div \dfrac{25}{36}$

58. $\dfrac{9}{28} \div \dfrac{27}{35}$

Add.

59. $\dfrac{2}{5} + \dfrac{1}{4}$

60. $\dfrac{2}{3} + \dfrac{3}{10}$

61. $\dfrac{2}{5} + \dfrac{7}{15}$

62. $\dfrac{3}{10} + \dfrac{7}{12}$

63. $\dfrac{3}{8} + \dfrac{5}{12}$

64. $\dfrac{5}{36} + \dfrac{7}{24}$

65. $\dfrac{2}{15} + \dfrac{9}{20}$

66. $\dfrac{9}{14} + \dfrac{10}{21}$

67. $\dfrac{7}{15} + \dfrac{13}{18}$

68. $\dfrac{12}{25} + \dfrac{19}{30}$

69. $\dfrac{1}{2} + \dfrac{1}{4} + \dfrac{1}{8}$

70. $\dfrac{1}{3} + \dfrac{1}{5} + \dfrac{1}{10}$

Subtract.

71. $\dfrac{8}{9} - \dfrac{3}{9}$

72. $\dfrac{9}{10} - \dfrac{6}{10}$

73. $\dfrac{5}{8} - \dfrac{1}{8}$

74. $\dfrac{11}{12} - \dfrac{7}{12}$

75. $\dfrac{7}{8} - \dfrac{2}{3}$

76. $\dfrac{5}{6} - \dfrac{3}{5}$

77. $\dfrac{11}{18} - \dfrac{2}{9}$

78. $\dfrac{5}{6} - \dfrac{1}{4}$

Convert the following fractions to mixed numbers.

79. $\dfrac{17}{4}$

80. $\dfrac{200}{11}$

Convert the following mixed numbers to fractions.

81. $3\dfrac{1}{4}$

82. $6\dfrac{3}{4}$

Perform the indicated operations.

83. $2\dfrac{2}{9} + 3\dfrac{5}{9}$

84. $5\dfrac{2}{9} + 6\dfrac{4}{9}$

85. $1\dfrac{1}{3} + 2\dfrac{1}{5}$

86. $2\dfrac{1}{4} + 1\dfrac{1}{6}$

87. $3\dfrac{2}{5} - 1\dfrac{4}{5}$

88. $5\dfrac{3}{7} - 2\dfrac{1}{7}$

89. $3\dfrac{2}{3} - 2\dfrac{1}{4}$

90. $5\dfrac{4}{5} - 1\dfrac{1}{6}$

91. $2\dfrac{2}{5} \cdot 3\dfrac{3}{4}$

92. $2\dfrac{2}{7} \cdot 2\dfrac{1}{3}$

93. $3\dfrac{1}{2} \div 2\dfrac{4}{5}$

94. $3\dfrac{3}{4} \div 1\dfrac{3}{8}$

Solve the following applications

95. Crafts. Roseann is making shirts for her three children. One shirt requires $\dfrac{1}{2}$ yard (yd) of material, a second shirt requires $\dfrac{1}{3}$ yd of material, and the third shirt requires $\dfrac{1}{4}$ yd of material. How much material is required for all three shirts?

96. Science. José rode his trail bike for 10 miles. Two-thirds of the distance was over a mountain trail. How long is the mountain trail?

97. _____

98. _____

99. _____

100. _____

101. _____

102. _____

103. _____

97. Business and finance. You make $240 a day on a job. What will you receive for working $\frac{2}{3}$ of a day?

98. Statistics. A survey has found that $\frac{3}{4}$ of the people in a city own pets. Of those who own pets, $\frac{2}{3}$ have cats. What fraction of those surveyed own cats?

99. Social science. The scale on a map is 1 in. = 200 miles (mi). What actual distance, in miles, does $\frac{3}{8}$ in. represent?

100. Business and finance. You make $90 a day on a job. What will you receive for working $\frac{3}{4}$ of a day?

101. Technology. A lumberyard has a stack of 80 sheets of plywood. If each sheet is $\frac{3}{4}$ in. thick, how high will the stack be?

102. Business and finance. A family uses $\frac{2}{5}$ of its monthly income for housing and utilities on average. If the family's monthly income is $1750, what is spent for housing and utilities? What amount remains?

103. Social science. Of the eligible voters in an election, $\frac{3}{4}$ were registered. Of those registered, $\frac{5}{9}$ actually voted. What fraction of those people who were eligible voted?

104. Statistics. A survey has found that $\frac{7}{10}$ of the people in a city own pets. Of those who own pets, $\frac{2}{3}$ have dogs. What fraction of those surveyed own dogs?

105. Geometry. A kitchen has dimensions $3\frac{1}{3}$ by $3\frac{3}{4}$ yards (yd). How many square yards (yd^2) of linoleum must be bought to cover the floor?

106. Science. If you drive at an average speed of 52 miles per hour (mi/h) for $1\frac{3}{4}$ h, how far will you travel?

107. Science. A jet flew at an average speed of 540 mi/h on a $4\frac{2}{3}$-h flight. What was the distance flown?

108. Geometry. A piece of land that has $11\frac{2}{3}$ acres is being subdivided for home lots. It is estimated that $\frac{2}{7}$ of the area will be used for roads. What amount remains to be used for lots?

109. Geometry. To find the approximate circumference or distance around a circle, we multiply its diameter by $\frac{22}{7}$. What is the circumference of a circle with a diameter of 21 in.?

110. Geometry. The length of a rectangle is $\frac{6}{7}$ yd, and its width is $\frac{21}{26}$ yd. What is its area in square yards?

111. Geometry. Find the volume of a box that measures $2\frac{1}{4}$ in. by $3\frac{7}{8}$ in. by $4\frac{5}{6}$ in.

112. Crafts. Nico wishes to purchase mulch to cover his garden. The garden measures $7\frac{7}{8}$ ft by $10\frac{1}{8}$ ft. He wants the mulch to be $\frac{1}{3}$ ft deep. How much mulch should Nico order if he must order a whole number of cubic feet?

113. Every fraction (rational number) has a corresponding decimal form that either terminates or repeats. For example, $\frac{5}{16} = 0.3125$ (the decimal form terminates), and $\frac{4}{11} = 0.363636\ldots$ (the decimal form repeats). Investigate a number of fractions to determine which ones terminate and which ones repeat. (Hint: You can focus on the denominator; study the prime factorizations of several denominators.)

104. _____

105. _____

106. _____

107. _____

108. _____

109. _____

110. _____

111. _____

112. _____

113. _____

114. Complete the following sums:

$$\frac{1}{2} + \frac{1}{4} =$$

$$\frac{1}{2} + \frac{1}{4} + \frac{1}{8} =$$

$$\frac{1}{2} + \frac{1}{4} + \frac{1}{8} + \frac{1}{16} =$$

Based on these, predict the sum:

$$\frac{1}{2} + \frac{1}{4} + \frac{1}{8} + \frac{1}{16} + \frac{1}{32} + \frac{1}{64} + \frac{1}{128}$$

Answers

For 1–11, answers will vary.

1. $\frac{6}{14}, \frac{9}{21}, \frac{12}{28}$ **3.** $\frac{8}{18}, \frac{16}{36}, \frac{40}{90}$ **5.** $\frac{10}{12}, \frac{15}{18}, \frac{50}{60}$ **7.** $\frac{20}{34}, \frac{30}{51}, \frac{100}{170}$

9. $\frac{18}{32}, \frac{27}{48}, \frac{90}{160}$ **11.** $\frac{14}{18}, \frac{35}{45}, \frac{140}{180}$ **13.** $\frac{2}{3}$ **15.** $\frac{5}{7}$ **17.** $\frac{2}{3}$ **19.** $\frac{7}{8}$

21. $\frac{1}{4}$ **23.** $\frac{1}{3}$ **25.** $\frac{8}{9}$ **27.** $\frac{4}{5}$ **29.** $\frac{5}{7}$ **31.** $\frac{4}{5}$ **33.** $\frac{7}{9}$

35. $\frac{15}{44}$ **37.** $\frac{21}{20}$ **39.** $\frac{3}{7}$ **41.** $\frac{8}{39}$ **43.** $\frac{7}{33}$ **45.** $\frac{1}{6}$ **47.** $\frac{4}{15}$

49. $\frac{8}{15}$ **51.** $\frac{2}{3}$ **53.** $\frac{63}{50}$ **55.** $\frac{4}{3}$ **57.** $\frac{4}{15}$ **59.** $\frac{13}{20}$ **61.** $\frac{13}{15}$

63. $\frac{19}{24}$ **65.** $\frac{7}{12}$ **67.** $\frac{107}{90}$ **69.** $\frac{7}{8}$ **71.** $\frac{5}{9}$ **73.** $\frac{1}{2}$ **75.** $\frac{5}{24}$

77. $\frac{7}{18}$ **79.** $4\frac{1}{4}$ **81.** $\frac{13}{4}$ **83.** $5\frac{7}{9}$ **85.** $3\frac{8}{15}$ **87.** $1\frac{3}{5}$

89. $1\frac{5}{12}$ **91.** 9 **93.** $1\frac{1}{4}$ **95.** $\frac{13}{12}$ yd **97.** $160

99. 75 mi **101.** 60 in. **103.** $\frac{5}{12}$ **105.** $12\frac{1}{2}$ yd^2 **107.** 2520 mi

109. 66 in. **111.** $42\frac{9}{64}$ in.^{3} **113.**

0.3 Decimals and Percents

Because a fraction can be interpreted as division, you can divide the numerator of the fraction by its denominator to convert a fraction to a decimal. The result is called a **decimal equivalent.**

OBJECTIVE 1

Example 1   Converting a Fraction to a Terminating Decimal

Write $\dfrac{5}{8}$ as a decimal.

RECALL 5 can be written as 5.0, 5.00, 5.000, and so on. In this case, we continue the division by adding zeros to the dividend until a 0 remainder is reached.

$$\begin{array}{r} 0.625 \\ 8\overline{)5.000} \\ \underline{4\,8} \\ 20 \\ \underline{16} \\ 40 \\ \underline{40} \\ 0 \end{array}$$

Because $\dfrac{5}{8}$ means $5 \div 8$, divide 8 into 5.

We see that $\dfrac{5}{8} = 0.625$; 0.625 is the decimal equivalent of $\dfrac{5}{8}$.

 CHECK YOURSELF 1

Find the decimal equivalent of $\dfrac{7}{8}$.

If a decimal equivalent does not terminate, you can round the result to approximate the fraction to some specified number of decimal places. Consider Example 2.

Example 2   Converting a Fraction to a Decimal

Write $\dfrac{3}{7}$ as a decimal. Round the answer to the nearest thousandth.

$$\begin{array}{r} 0.4285 \\ 7\overline{)3.0000} \\ \underline{2\,8} \\ 20 \\ \underline{14} \\ 60 \\ \underline{56} \\ 40 \\ \underline{35} \\ 5 \end{array}$$

In this example, we are choosing to round to three decimal places, so we must add enough zeros to carry the division to four decimal places.

So $\dfrac{3}{7} = 0.429$ (to the nearest thousandth).

CHECK YOURSELF 2

Find the decimal equivalent of $\dfrac{5}{11}$ to the nearest thousandth.

If a decimal equivalent does *not* terminate, it will *repeat* a sequence of digits. These decimals are called **repeating decimals.**

Example 3   Converting a Fraction to a Repeating Decimal

Write $\dfrac{5}{11}$ as a decimal.

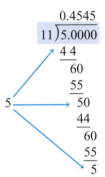

As soon as a remainder repeats itself, as 5 does here, the pattern of digits will repeat in the quotient.

$$\frac{5}{11} = 0.\overline{45}$$

$$= 0.4545\ldots$$

CHECK YOURSELF 3

Use the bar notation to write the decimal equivalent of $\dfrac{5}{7}$. (Be patient. You'll have to divide for a while to find the repeating pattern.)

To convert from a decimal to a fraction, write the decimal without the decimal point. This will be the **numerator** of the fraction. The **denominator** of the fraction is a 1 followed by as many zeros as there are places in the decimal. Examples 4 and 5 illustrate this process.

OBJECTIVE 2   Example 4   Converting a Decimal to a Fraction

$$0.7 = \frac{7}{10} \qquad\qquad 0.09 = \frac{9}{100} \qquad\qquad 0.257 = \frac{257}{1000}$$

One place   One zero   Two places   Two zeros   Three places   Three zeros

CHECK YOURSELF 4

Write as fractions.

(a) 0.3

(b) 0.311

When a decimal is converted to a fraction, the common fraction that results should be written in lowest terms.

Example 5   Converting a Decimal to a Fraction

Convert 0.395 to a fraction and write the result in lowest terms.

NOTE Divide the numerator and denominator by 5.

$$0.395 = \frac{395}{1000} = \frac{79}{200}$$

CHECK YOURSELF 5 _____

Write 0.275 as a fraction.

Example 6 illustrates the addition and subtraction of decimals.

OBJECTIVE 3   ### Example 6   Adding or Subtracting Two Decimals

Perform the indicated operation.

(a) Add 2.356 and 15.6.

Aligning the decimal points, we get

```
   2.356
+ 15.600
---------
  17.956
```
Although the zeros are not necessary, they ensure proper alignment.

(b) Subtract 3.84 from 8.1.

Again, we align the decimal points

```
   8.10
-  3.84
-------
   4.26
```
When subtracting, always add zeros so that the right columns line up.

CHECK YOURSELF 6 _____

Perform the indicated operation.

(a) 34.76 + 2.419   **(b)** 71.82 − 8.197

Example 7 illustrates the multiplication of two decimal fractions.

OBJECTIVE 4   ### Example 7   Multiplying Two Decimals

Multiply 4.6 and 3.27

```
      4.6
×    3.27
---------
      322
      920
    13800
---------
   15.042
```
It is not necessary to align decimals being multiplied. Note that the two factors have a total of three digits to the right of the decimal point.

The decimal point of the product is moved three digits to the left.

 **CHECK YOURSELF 7**

Multiply 5.8 and 9.62.

We want now to look at division *by* decimals. Here is an example using a fractional form.

> ### Example 8   Rewriting a Problem That Requires Dividing by a Decimal
>
> Rewrite the division so that the divisor is a whole number.

$$2.57 \div 3.4 = \frac{2.57}{3.4}$$ Write the division as a fraction.

$$= \frac{2.57 \times 10}{3.4 \times 10}$$ We multiply the numerator and denominator by 10 so the divisor is a whole number. This *does not change* the value of the fraction.

$$= \frac{25.7}{34}$$ Multiplying by 10, shift the decimal point in the numerator and denominator *one place to the right.*

$$= 25.7 \div 34$$ Our division problem is rewritten so that the divisor is a whole number.

NOTE It's always easier to rewrite a division problem so that you're dividing by a whole number. Dividing by a whole number makes it easy to place the decimal point in the quotient.

So

$$2.57 \div 3.4 = 25.7 \div 34$$ After we multiply the numerator and denominator by 10, we see that $2.57 \div 3.4$ is the same as $25.7 \div 34$.

 CHECK YOURSELF 8

Rewrite the division problem so that the divisor is a whole number.

$$3.42 \div 2.5$$

NOTE Of course, multiplying by any whole-number power of 10 greater than 1 is just a matter of shifting the decimal point to the right.

Do you see the rule suggested by Example 8? We multiplied the numerator and the denominator (the dividend and the divisor) by 10. We made the divisor a whole number without altering the actual digits involved. All we did was shift the decimal point in the divisor and dividend the same number of places. This leads us to the following procedure.

> ### Step by Step: To Divide by a Decimal
>
> **Step 1** Move the decimal point in the divisor *to the right,* making the divisor a whole number.
> **Step 2** Move the decimal point in the dividend to the right *the same number of places.* Add zeros if necessary.
> **Step 3** Place the decimal point in the quotient directly above the decimal point of the dividend.
> **Step 4** Divide as you would with whole numbers.

Let's look at an example of the use of our division rule.

Example 9   Rounding the Result of Dividing by a Decimal

Divide 1.573 by 0.48 and give the quotient to the nearest tenth.

Write

$$0.48\overline{\smash{)}1.57\,3}$$

Shift the decimal points two places to the right to make the divisor a whole number.

Now divide:

NOTE Once the division statement is rewritten, place the decimal point in the quotient above that in the dividend.

$$
\begin{array}{r}
3.27 \\
48\overline{\smash{)}157.30} \\
\underline{144} \\
13\,3 \\
\underline{9\,6} \\
3\,70 \\
\underline{3\,36} \\
34
\end{array}
$$

Add a 0 to carry the division to the hundredths place. In this case, we want to find the quotient to the nearest tenth.

Round 3.27 to 3.3. So

$1.573 \div 0.48 = 3.3$ (to the nearest tenth)

CHECK YOURSELF 9

Divide, rounding the quotient to the nearest tenth.

$3.4 \div 1.24$

We have used fractions and decimals to name parts of a whole. **Percents** can also be used to accomplish this. The word *percent* means "for each hundred." We can think of percents as fractions whose denominators are 100. So 25% would be written as $\dfrac{25}{100}$ or in simplified form $\dfrac{1}{4}$.

Because there are different ways of naming the parts of a whole, you need to know how to change from one of these ways to another. First let's look at changing a percent to a fraction. Because a percent is a fraction or a ratio with denominator 100, we can use the following rule.

Rules and Properties: Changing a Percent to a Fraction

To change a percent to a fraction, replace the percent symbol with $\dfrac{1}{100}$.

The use of this rule is shown in Example 10.

OBJECTIVE 5   **Example 10   Changing a Percent to a Fraction**

Change each percent to a fraction.

(a) $7\% = 7\left(\dfrac{1}{100}\right) = \dfrac{7}{100}$

NOTE You can choose to write $\dfrac{25}{100}$ in simplest form.

(b) $25\% = 25\left(\dfrac{1}{100}\right) = \dfrac{25}{100} = \dfrac{1}{4}$

 CHECK YOURSELF 10

Write 12% as a fraction.

In Example 10, we wrote percents as fractions by replacing the percent sign with $\dfrac{1}{100}$ and multiplying. How do we convert percents when we are working with decimals? Just move the decimal point two places to the left. This gives us a second rule for converting percents.

Rules and Properties: Changing a Percent to a Decimal

To change a percent to a decimal, replace the percent symbol with $\dfrac{1}{100}$. As a result of multiplying by $\dfrac{1}{100}$, the decimal point will move two places to the left.

Example 11   Changing a Percent to a Decimal

Change each percent to a decimal equivalent.

(a) $25\% = 25\left(\dfrac{1}{100}\right) = 0.25$ The decimal point in 25% is understood to be after the 5.

(b) $4.5\% = 4.5\left(\dfrac{1}{100}\right) = 0.045$ We must add a zero to move the decimal point.

NOTE A percent greater than 100 gives a decimal greater than 1.

(c) $130\% = 130\left(\dfrac{1}{100}\right) = 1.30$

 CHECK YOURSELF 11

Write as decimals.

(a) 5%   **(b)** 3.9%   **(c)** 115%

Changing a decimal to a percent is the opposite of changing from a percent to a decimal. We reverse the process. Here is the rule:

> **Rules and Properties:**   Changing a Decimal to a Percent
>
> To change a decimal to a percent, move the decimal point *two* places to the *right* and attach the percent symbol.

OBJECTIVE 6   **Example 12   Changing a Decimal to a Percent**

Change each decimal to a percent equivalent.

(a) $0.18 = 18\%$
(b) $0.03 = 3\%$
(c) $1.25 = 125\%$

CHECK YOURSELF 12

Change each decimal to a percent equivalent.

(a) 0.27   **(b)** 0.045   **(c)** 1.3

The following rule allows us to change fractions to percents.

> **Rules and Properties:**   Changing a Fraction to a Percent
>
> To change a fraction to a percent, write the decimal equivalent of the fraction. Move the decimal point two places to the right and attach the percent symbol.

Example 13   Changing a Fraction to a Percent

Change each fraction to a percent equivalent.

(a) $\dfrac{3}{5} = 0.60$ To find the decimal equivalent, just divide the denominator into the numerator.

Now write the percent.

$$\frac{3}{5} = 0.60 = 60\%$$

(b) $\dfrac{1}{8} = 0.125 = 12.5\%$ or $12\frac{1}{2}\%$

(c) $\dfrac{1}{3} = 0.\overline{3} = 0.333\overline{3} = 33.\overline{3}\%$ or $33\frac{1}{3}\%$

CHECK YOURSELF 13

Change each fraction to a percent equivalent.

(a) $\dfrac{3}{4}$ (b) $\dfrac{3}{8}$ (c) $\dfrac{2}{3}$

Example 14 illustrates one of the many applications using decimals.

Example 14   An Application of Decimals

Lucetia's car gets approximately 20 miles per gallon (mi/gal) of fuel. If 1 gal of fuel costs $1.93, how much does it cost her to drive 125 mi?

$125 \div 20 = 6.25$ gal

$6.25 \cdot 1.93 = \$12.06$ (rounded)

CHECK YOURSELF 14

The art department has a budget of $195.75 to purchase art supplies. After purchasing 35 paintbrushes for $1.92 each, six jars of paint remover for $0.93 each, and four cans of blue paint for $2.95 each, how much money was left in the budget?

CHECK YOURSELF ANSWERS

1. 0.875   **2.** 0.455   **3.** $0.\overline{714285}$ **4. (a)** $\dfrac{3}{10}$; **(b)** $\dfrac{311}{1000}$ **5.** $\dfrac{11}{40}$

6. (a) 37.179; **(b)** 63.623   **7.** 55.796   **8.** $34.2 \div 25$ **9.** 2.7

10. $\dfrac{12}{100}$ or $\dfrac{3}{25}$ **11. (a)** 0.05; **(b)** 0.039; **(c)** 1.15   **12. (a)** $\dfrac{27}{100} = 27\%$;

(b) 4.5%; **(c)** 130%   **13. (a)** 75%; **(b)** 37.5%; **(c)** $66.\overline{6}\%$ or $66\dfrac{2}{3}\%$

14. $111.17

0.3 Exercises

Find the decimal equivalents for each of the following fractions.

1. $\dfrac{3}{4}$

2. $\dfrac{4}{5}$

3. $\dfrac{9}{20}$

4. $\dfrac{3}{10}$

5. $\dfrac{1}{5}$

6. $\dfrac{1}{8}$

7. $\dfrac{5}{16}$

8. $\dfrac{11}{20}$

9. $\dfrac{7}{10}$

10. $\dfrac{7}{16}$

11. $\dfrac{27}{40}$

12. $\dfrac{17}{32}$

Find the decimal equivalents rounded to the indicated place.

13. $\dfrac{5}{6}$ thousandth

14. $\dfrac{7}{12}$ hundredth

15. $\dfrac{4}{15}$ thousandth

Write the decimal equivalents, using the bar notation.

16. $\dfrac{1}{18}$

17. $\dfrac{4}{9}$

18. $\dfrac{3}{11}$

Write each of the following as a fraction. Write your answer in lowest terms.

19. 0.9

20. 0.3

21. 0.8

22. 0.6

23. 0.37

24. 0.97

25. 0.587

26. 0.379

27. 0.48

28. 0.75

29. 0.58

30. 0.65

Perform the indicated operations.

31. 7.1562 + 14.78

32. 6.2358 + 3.14

33. 11.12 + 8.3792

34. 6.924 + 5.2

35. 9.20 − 2.85

36. 17.345 − 11.12

ANSWERS

1. _____
2. _____
3. _____
4. _____
5. _____
6. _____
7. _____
8. _____
9. _____
10. _____
11. _____
12. _____
13. _____ 14. _____
15. _____ 16. _____
17. _____ 18. _____
19. _____ 20. _____
21. _____ 22. _____
23. _____ 24. _____
25. _____ 26. _____
27. _____ 28. _____
29. _____ 30. _____
31. _____ 32. _____
33. _____ 34. _____
35. _____ 36. _____

37. _____

38. _____

39. _____

40. _____

41. _____

42. _____

43. _____

44. _____

45. _____ 46. _____

47. _____ 48. _____

49. _____ 50. _____

51. _____ 52. _____

53. _____ 54. _____

55. _____ 56. _____

57. _____ 58. _____

59. _____ 60. _____

61. _____ 62. _____

63. _____ 64. _____

65. _____ 66. _____

67. _____ 68. _____

69. _____ 70. _____

71. _____ 72. _____

73. _____ 74. _____

75. _____ 76. _____

77. _____ 78. _____

79. _____ 80. _____

37. $18.234 - 13.64$ **38.** $21.983 - 9.395$ **39.** $3.21 \cdot 2.1$

40. $15.6 \cdot 7.123$ **41.** $6.29 \cdot 9.13$ **42.** $8.245 \cdot 3.1$

Divide.

43. $16.68 \div 6$ **44.** $43.92 \div 8$ **45.** $1.92 \div 4$

46. $5.52 \div 6$ **47.** $5.48 \div 8$ **48.** $2.76 \div 8$

49. $13.89 \div 6$ **50.** $21.92 \div 5$ **51.** $185.6 \div 32$

52. $165.6 \div 36$ **53.** $79.9 \div 34$ **54.** $179.3 \div 55$

55. $52\overline{)13.78}$ **56.** $76\overline{)26.22}$ **57.** $0.6\overline{)11.07}$

58. $0.8\overline{)10.84}$ **59.** $3.8\overline{)7.22}$ **60.** $2.9\overline{)13.34}$

61. $5.2\overline{)11.622}$ **62.** $6.4\overline{)3.616}$

Write as fractions.

63. 6%   **64.** 17%   **65.** 75%

66. 20%   **67.** 65%   **68.** 48%

69. 50%   **70.** 52%   **71.** 46%

72. 35%   **73.** 66%   **74.** 4%

Write as decimals.

75. 20%   **76.** 70%   **77.** 35%

78. 75%   **79.** 39%   **80.** 27%

81. 5%   **82.** 7%   **83.** 135%

84. 250%   **85.** 240%   **86.** 160%

Write each decimal as a percent.

87. 4.40   **88.** 5.13   **89.** 0.065

90. 0.095   **91.** 0.025   **92.** 0.085

93. 0.002   **94.** 0.008

Write each fraction as a percent.

95. $\dfrac{1}{4}$ **96.** $\dfrac{4}{5}$ **97.** $\dfrac{2}{5}$

98. $\dfrac{1}{2}$ **99.** $\dfrac{1}{5}$ **100.** $\dfrac{3}{4}$

101. $\dfrac{5}{8}$ **102.** $\dfrac{7}{8}$

103. Statistics. On a math quiz, Adam answered 18 of 20 questions correctly, or $\dfrac{18}{20}$ of the quiz. Write the decimal equivalent of this fraction.

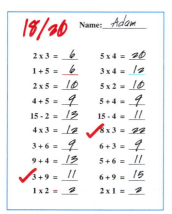

81. _____

82. _____

83. _____

84. _____

85. _____

86. _____

87. _____

88. _____

89. _____

90. _____

91. _____

92. _____

93. _____

94. _____

95. _____

96. _____

97. _____

98. _____

99. _____

100. _____

101. _____

102. _____

103. _____

104. _____

105. _____

106. _____

107. _____

108. _____

109. _____

104. Statistics. In a weekend baseball tournament, Joel had 4 hits in 13 times at bat. That is, he hit safely $\frac{4}{13}$ of the time. Write the decimal equivalent for Joel's hitting, rounding to three decimal places. (That number is Joel's batting average.)

105. Business and finance. A restaurant bought 50 glasses at a cost of $39.90. What was the cost per glass to the nearest cent?

106. Business and finance. The cost of a case of 48 items is $28.20. What is the cost of an individual item to the nearest cent?

107. Business and finance. An office bought 18 hand-held calculators for $284. What was the cost per calculator to the nearest cent?

108. Business and finance. Al purchased a new refrigerator that cost $736.12 with interest included. He paid $100 as a down payment and agreed to pay the remainder in 18 monthly payments. What amount will he be paying per month?

109. Business and finance. The cost of a television set with interest is $490.64. If you make a down payment of $50 and agree to pay the balance in 12 monthly payments, what will be the amount of each monthly payment?

Answers

1. 0.75   **3.** 0.45   **5.** 0.2   **7.** 0.3125   **9.** 0.7   **11.** 0.675

13. 0.833   **15.** 0.267   **17.** $0.\overline{4}$ **19.** $\frac{9}{10}$ **21.** $\frac{4}{5}$ **23.** $\frac{37}{100}$

25. $\frac{587}{1000}$ **27.** $\frac{12}{25}$ **29.** $\frac{29}{50}$ **31.** 21.9362   **33.** 19.4992

35. 6.35   **37.** 4.594   **39.** 6.741   **41.** 57.4277   **43.** 2.78

45. 0.48   **47.** 0.685   **49.** 2.315   **51.** 5.8   **53.** 2.35   **55.** 0.265

57. 18.45   **59.** 1.9   **61.** 2.235   **63.** $\frac{3}{50}$ **65.** $\frac{3}{4}$ **67.** $\frac{13}{20}$

69. $\frac{1}{2}$ **71.** $\frac{23}{50}$ **73.** $\frac{33}{50}$ **75.** 0.2   **77.** 0.35   **79.** 0.39

81. 0.05   **83.** 1.35   **85.** 2.4   **87.** 440%   **89.** 6.5%   **91.** 2.5%

93. 0.2%   **95.** 25%   **97.** 40%   **99.** 20%   **101.** 62.5%

103. 0.9   **105.** $0.80 or 80¢   **107.** $15.78   **109.** $36.72

0.4 Exponents and the Order of Operations

1. Write a product of factors in exponential form
2. Evaluate an expression involving several operations

Often in mathematics we define symbols that allow us to write a mathematical statement in a more compact or "shorthand" form. This is an idea that you have encountered before. For example, the repeated addition:

$$5 + 5 + 5$$

NOTE
$5 + 5 + 5 = 15$
and
$3 \cdot 5 = 15$

can be rewritten as

$$3 \cdot 5$$

Thus multiplication is shorthand for repeated addition.

In algebra, we frequently have a number or variable that is repeated as a factor in an expression several times. For instance, we might have

$$5 \cdot 5 \cdot 5$$

NOTE A factor is a number or a variable that is being multiplied by another number or variable.

To abbreviate this product, we write

$$5 \cdot 5 \cdot 5 = 5^3$$

This is called **exponential notation** or **exponential form.** The exponent or power, here 3, indicates the number of times that the factor or base, here 5, appears in a product.

C A U T I O N

Be careful: 5^3 is *not* the same as $5 \cdot 3$. Notice that
$5^3 = 5 \cdot 5 \cdot 5 = 125$ and
$5 \cdot 3 = 15$.

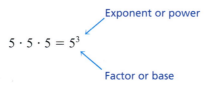

Exponent or power

$$5 \cdot 5 \cdot 5 = 5^3$$

Factor or base

This is read "5 to the third power" or "5 cubed."

OBJECTIVE 1

Example 1 Writing in Exponential Form

Write $3 \cdot 3 \cdot 3 \cdot 3$, using exponential form.
The number 3 appears four times in the product, so

Four factors of 3

$$3 \cdot 3 \cdot 3 \cdot 3 = 3^4$$

This is read "3 to the fourth power."

CHECK YOURSELF 1

Rewrite each expression using exponential form.

(a) $4 \cdot 4 \cdot 4 \cdot 4 \cdot 4 \cdot 4$ **(b)** $7 \cdot 7 \cdot 7 \cdot 7$

To evaluate an arithmetic expression, you need to know the order in which the operations are done. To see why, simplify the expression $5 + 2 \cdot 3$.

C A U T I O N

Only one of these results can be correct.

Method 1	or	Method 2

$\underbrace{5 + 2} \cdot 3$

Add first.

$= 7 \cdot 3$
$= 21$

$5 + \underbrace{2 \cdot 3}$

Multiply first.

$= 5 + 6$
$= 11$

Because we get different answers depending on how we do the problem, the language of mathematics would not be clear if there were no agreement on which method is correct. The following rules tell us the order in which operations should be done.

Step by Step: The Order of Operations

NOTE Parentheses and brackets are both grouping symbols. Later we will see that fraction bars and radicals are also grouping symbols.

Step 1	Evaluate all expressions inside grouping symbols first.
Step 2	Evaluate all expressions involving exponents.
Step 3	Do any multiplication or division in order, working from left to right.
Step 4	Do any addition or subtraction in order, working from left to right.

OBJECTIVE 2

Example 2 Evaluating Expressions

Evaluate $5 + 2 \cdot 3$.

There are no parentheses or exponents, so start with step 3: First multiply and then add.

$5 + 2 \cdot 3$

Multiply first.

$= 5 + 6$

Then add.

NOTE Method 2 shown above is the correct one.

$= 11$

CHECK YOURSELF 2

Evaluate the following expressions.

(a) $20 - 3 \cdot 4$ **(b)** $9 + 6 \div 3$

When there are no parentheses, evaluate the exponents first.

Example 3 Evaluating Expressions

Evaluate $5 \cdot 3^2$.

$5 \cdot 3^2 = 5 \cdot 9$

Evaluate the power first.

$= 45$

CHECK YOURSELF 3

Evaluate $4 \cdot 2^4$.

Both scientific and graphing calculators correctly interpret the order of operations. This is demonstrated in Example 4.

> ### Example 4   Using a Calculator to Evaluate Expressions

Use your scientific or graphing calculator to evaluate each expression. Round the answer to the nearest tenth.

(a) $24.3 + 6.2 \cdot 3.53$

When evaluating expressions by hand, you must consider the order of operations. In this case, the multiplication must be done first, and then the addition. With a calculator, you need only enter the expression correctly. The calculator is programmed to follow the order of operations.

Entering   24.3 $\boxed{+}$ 6.2 $\boxed{\times}$ 3.53 $\boxed{\text{ENTER}}$

NOTE With most graphing calculators, the final command is $\boxed{\text{ENTER}}$. With most scientific calculators, the key is marked $\boxed{=}$.

yields the evaluation 46.186. Rounding to the nearest tenth, we have 46.2.

(b) $2.45^3 - 49 \div 8000 + 12.2 \cdot 1.3$

Some calculators use the caret (^) to designate powers. Others use the symbol x^y (or y^x).

Entering 2.45 $\boxed{\wedge}$ 3 $\boxed{-}$ 49 $\boxed{\div}$ 8000 $\boxed{+}$ 12.2 $\boxed{\times}$ 1.3 $\boxed{\text{ENTER}}$

or   2.45 $\boxed{y^x}$ 3 $\boxed{-}$ 49 $\boxed{\div}$ 8000 $\boxed{+}$ 12.2 $\boxed{\times}$ 1.3 $\boxed{=}$

yields the evaluation 30.56. Rounding to the nearest tenth, we have 30.6.

CHECK YOURSELF 4

Use your scientific or graphing calculator to evaluate each expression.

(a) $67.89 - 4.7 \cdot 12.7$ **(b)** $4.3 \cdot 55.5 - 3.75^3 + 8007 \div 1600$

Operations inside grouping symbols are done first.

> ### Example 5   Evaluating Expressions

Evaluate $(5 + 2) \cdot 3$.
    Do the operation inside the parentheses as the first step.

$(5 + 2) \cdot 3 = 7 \cdot 3 = 21$

    Add.

CHECK YOURSELF 5

Evaluate $4 \cdot (9 - 3)$.

The principle is the same when more than two "levels" of operations are involved.

Example 6 Using Order of Operations

(a) Evaluate $4 \cdot (2 + 3)^3$.

Add inside the parentheses first.

$4 \cdot (2 + 3)^3 = 4 \cdot (5)^3$

Evaluate the power.

$= 4 \cdot 125$

Multiply.

$= 500$

(b) Evaluate $5 \cdot (7 - 3)^2 - 10$.

Evaluate the expression inside the parentheses.

$5 \cdot (7 - 3)^2 - 10 = 5(4)^2 - 10$

Evaluate the power.

$= 5 \cdot 16 - 10$

Multiply.

$= 80 - 10 = 70$

Subtract.

CHECK YOURSELF 6

Evaluate.

(a) $4 \cdot 3^3 - 8 \cdot 11$ **(b)** $12 + 4 \cdot (2 + 3)^2$

Example 7 Using Order of Operations

Evaluate $3 \cdot [(1 + 2)^2 - 5] + 8$

Do the operation inside the innermost grouping symbol first.

$3 \cdot [(1 + 2)^2 - 5] + 8 = 3 \cdot [(3)^2 - 5] + 8$
$= 3 \cdot [9 - 5] + 8$
$= 3 \cdot 4 + 8$
$= 12 + 8$
$= 20$

CHECK YOURSELF 7

Evaluate $8 - 2 \cdot [(5 - 3)^2 - 1]$.

CHECK YOURSELF ANSWERS

1. (a) 4^6; **(b)** 7^4 **2. (a)** 8; **(b)** 11 **3.** 64 **4. (a)** 8.2; **(b)** 190.92
5. 24 **6. (a)** 20; **(b)** 112 **7.** 2

0.4 Exercises

Write each expression in exponential form.

1. $7 \cdot 7 \cdot 7 \cdot 7$

2. $2 \cdot 2 \cdot 2 \cdot 2 \cdot 2 \cdot 2$

3. $6 \cdot 6 \cdot 6 \cdot 6 \cdot 6$

4. $4 \cdot 4 \cdot 4 \cdot 4 \cdot 4 \cdot 4 \cdot 4$

5. $8 \cdot 8 \cdot 8 \cdot 8 \cdot 8 \cdot 8 \cdot 8 \cdot 8 \cdot 8 \cdot 8$

6. $10 \cdot 10 \cdot 10$

7. $15 \cdot 15 \cdot 15 \cdot 15 \cdot 15 \cdot 15$

8. $31 \cdot 31 \cdot 31 \cdot 31 \cdot 31 \cdot 31 \cdot 31 \cdot 31 \cdot 31 \cdot 31$

Evaluate each of the following expressions.

9. $5 + 3 \cdot 4$

10. $10 - 4 \cdot 2$

11. $(7 + 2) \cdot 6$

12. $(10 - 4) \cdot 2$

13. $12 - 8 \div 4$

14. $20 + 10 \div 2$

15. $(24 - 12) \div 6$

16. $(10 + 20) \div 5$

17. $8 \cdot 7 + 2 \cdot 2$

18. $56 \div 7 - 8 \div 4$

19. $7 \cdot (8 + 3) \cdot 3$

20. $48 \div (8 - 4) \div 2$

21. $3 \cdot 5^2$

22. $5 \cdot 2^3$

23. $(2 \cdot 4)^2$

24. $(5 \cdot 2)^3$

25. $4 \cdot 3^2 - 2$

26. $3 \cdot 2^4 - 8$

27. _____

28. _____

29. _____

30. _____

31. _____

32. _____

33. _____

34. _____

35. _____

36. _____

37. _____

38. _____

39. _____

40. _____

41. _____

42. _____

43. _____

44. _____

45. _____

46. _____

47. _____

48. _____

49. _____

50. _____

27. $5 - [3 \cdot (4 - 2)^2] + (3 \cdot 5)$

28. $14 \div 7 \cdot [12 \div (4 - 2)^2 \cdot 5] - 3^3$

29. $3 \cdot 2^4 - 6 \cdot 2$

30. $4 \cdot 2^3 - 5 \cdot 6$

31. $4 \cdot (2 + 6)^2$

32. $3 \cdot (8 - 4)^2$

33. $(4 \cdot 2 + 6)^2$

34. $[25 \div (2^3 - 3)] \cdot 2$

35. $64 \div [(16 \div 2 \cdot 4) - 16]$

36. $5 \cdot (4 - 2)^3$

37. $3 \cdot 4 + 3^2$

38. $5 \cdot 4 - 2^3$

39. $4 \cdot (2 + 3)^2 - 25$

40. $8 + 2 \cdot (3 + 3)^2$

41. $(4 \cdot 2 + 3)^2 - 25$

42. $8 + (2 \cdot 3 + 3)^2$

43. $2 \cdot [16 - (1 + 3)^2]$

44. $[(2 + 3)^2 - 4 \cdot 5] + 7$

Evaluate using your calculator. Round your answer to the nearest tenth.

45. $(1.2)^3 \div 2.0736 \cdot 2.4 + 1.6935 - 2.4896$

46. $(5.21 \cdot 3.14 - 6.2154) \div 5.12 - 0.45625$

47. $1.23 \cdot 3.169 - 2.05194 + (5.128 \cdot 3.15 - 10.1742)$

48. $4.56 + (2.34)^4 \div 4.7896 \cdot 6.93 \div 27.5625 - 3.1269 + (1.56)^2$

49. Social science. Over the last 2000 years, the Earth's population has doubled approximately five times. Use exponential notation to write an expression that indicates doubling five times.

50. Geometry. The volume of a cube with each edge of length 9 in. is given by $9 \cdot 9 \cdot 9$. Write the volume using exponential notation.

Insert grouping symbols in the proper place so that the given value of the expression is obtained.

51. $36 \div 4 + 2 - 4; 2$

52. $48 \div 3 \cdot 2 - 2 \cdot 3; 2$

53. $6 + 9 \div 3 + 16 - 4 \cdot 2; 29$

54. $5 - 3 \cdot 2 + 8 \cdot 5 - 2; 28$

Answers

1. 7^4 **3.** 6^5 **5.** 8^{10} **7.** 15^6 **9.** 17 **11.** 54 **13.** 10
15. 2 **17.** 60 **19.** 231 **21.** 75 **23.** 64 **25.** 34
27. 8 **29.** 36 **31.** 256 **33.** 196 **35.** 4 **37.** 21
39. 75 **41.** 96 **43.** 0 **45.** 1.2 **47.** 7.8 **49.** 2^5
51. $36 \div (4 + 2) - 4$ **53.** $(6 + 9) \div 3 + (16 - 4) \cdot 2$

0.5 Positive and Negative Numbers

0.5 OBJECTIVES

1. Represent integers on a number line
2. Order real numbers
3. Find the opposite of a number
4. Evaluate numerical expressions involving absolute value

When numbers are used to represent physical quantities (altitudes, temperatures, and amounts of money are examples), it may be necessary to distinguish between *positive* and *negative* quantities. It is convenient to represent these quantities with plus (+) or minus (−) signs. For instance,

The altitude of Mount Whitney is 14,495 ft *above* sea level (+14,495).

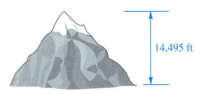

14,495 ft

Mount Whitney

The altitude of Death Valley is 282 ft *below* sea level (−282).

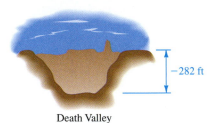

−282 ft

Death Valley

The temperature in Chicago is 10° *below* zero (−10°).

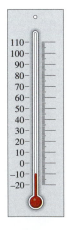

An account could show a *gain* of $100 (+100) or a *loss* of $100 (−100).

These numbers suggest the need to extend the number system to include both positive numbers (like +100) and negative numbers (like −282).

To represent the negative numbers, we extend the number line to the *left* of zero and name equally spaced points.

Numbers used to name points to the right of zero are positive numbers. They are written with a positive (+) sign or with no sign at all.

+6 and 9 are positive numbers

Numbers used to name points to the left of zero are negative numbers. They are always written with a negative (−) sign.

−3 and −20 are negative numbers

Read "negative 3."

Positive and negative numbers considered together are **signed numbers.**

Here is the number line extended to include both positive and negative numbers.

NOTE 0 is not considered a signed number.

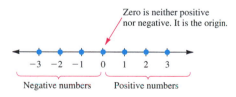

Zero is neither positive nor negative. It is the origin.

The numbers used to name the points shown on the number line above are called the **integers.** The integers consist of the natural numbers, their negatives, and the number 0. We can represent the set of integers by

NOTE The set of three dots is called *ellipses* and indicate that the pattern continues.

$$\{\ldots, -3, -2, -1, 0, 1, 2, 3, \ldots\}$$

OBJECTIVE 1   **Example 1   Representing Integers on the Number Line**

Represent the following integers on the number line shown.

$-3, -12, 8, 15, -7$

CHECK YOURSELF 1 _____

Represent the following integers on a number line.

$-1, -9, 4, -11, 8, 20$

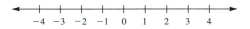

The set of numbers on the number line is *ordered*. The numbers get smaller moving to the left on the number line and larger moving to the right.

When a set of numbers is written from smallest to largest, the numbers are said to be in *ascending order.*

OBJECTIVE 2   **Example 2   Ordering Integers**

Place each set of numbers in ascending order.

(a) $9, -5, -8, 3, 7$

From smallest to largest, the numbers are

$-8, -5, 3, 7, 9$ Note that this is the order in which the numbers appear on a number line as we move from left to right.

(b) $3, -2, 18, -20, -13$

From smallest to largest, the numbers are

$-20, -13, -2, 3, 18$

CHECK YOURSELF 2 _____

Place each set of numbers in ascending order.

(a) $12, -13, 15, 2, -8, -3$ **(b)** $3, 6, -9, -3, 8$

The least and greatest numbers in a set are called the **extreme values.** The least element is called the **minimum,** and the greatest element is called the **maximum.**

Example 3   Labeling Extreme Values

For each set of numbers, determine the minimum and maximum values.

(a) $9, -5, -8, 3, 7$

From our previous ordering of these numbers, we see that -8, the least element, is the minimum, and 9, the greatest element, is the maximum.

(b) $3, -2, 18, -20, -13$

-20 is the minimum, and 18 is the maximum.

 CHECK YOURSELF 3 _____

For each set of numbers, determine the minimum and maximum values.

(a) $12, -13, 15, 2, -8, -3$ **(b)** $3, 6, -9, -3, 8$

Integers are not the only kind of signed numbers. Decimals and fractions can also be thought of as signed numbers.

Example 4   Identifying Numbers That are Integers

Which of the numbers $145, -28, 0.35,$ and $-\dfrac{2}{3}$ are integers?

(a) 145 is an integer.
(b) -28 is an integer.
(c) 0.35 is not an integer.
(d) $-\dfrac{2}{3}$ is not an integer.

 CHECK YOURSELF 4 _____

Which of the following numbers are integers?

$$-23 \qquad 1054 \qquad -0.23 \qquad 0 \qquad -500 \qquad -\dfrac{4}{5}$$

We refer to the negative of a number as its "opposite." But what is the opposite of the opposite of a number? It is the number itself. Example 5 illustrates this concept.

OBJECTIVE 3

Example 5 Finding Opposites

Find the opposite for each number.

(a) 5   The opposite of 5 is -5.

(b) -9 The opposite of -9 is 9.

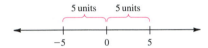

CHECK YOURSELF 5 _____

Find the opposite for each number.

(a) 17   **(b)** -12

An important idea for our work in this chapter is the **absolute value** of a number. This represents the distance of the point named by the number from the origin on the number line.

```
         5 units   5 units
       ⎧‾‾‾‾‾⎫ ⎧‾‾‾‾‾⎫
◄──────┼────┼────┼──────►
      −5    0    5
```

The absolute value of 5 is 5. The absolute value of -5 is also 5.

The absolute value of a positive number or zero is itself. The absolute value of a negative number is its opposite.

In symbols we write

$$|5| = 5 \quad\quad \text{and} \quad\quad |-5| = 5$$

Read "the absolute value of 5."   Read "the absolute value of negative 5."

The absolute value of a number does *not* depend on whether the number is to the right or to the left of the origin, but on its *distance* from the origin.

OBJECTIVE 4

Example 6 Simplifying Absolute Value Expressions

(a) $|7| = 7$

(b) $|-7| = 7$

(c) $-|-7| = -7$

This is the *negative,* or opposite, of the absolute value of negative 7.

(d) $|-10| + |10| = 10 + 10 = 20$

Absolute value bars serve as another set of grouping symbols, so do the operation *inside* first.

(e) $|8 - 3| = |5| = 5$

(f) $|8| - |3| = 8 - 3 = 5$

Here, evaluate the absolute values and then subtract.

CHECK YOURSELF 6

Evaluate.

(a) $|8|$

(b) $|-8|$

(c) $-|-8|$

(d) $|-9| + |4|$

(e) $|9 - 4|$

(f) $|9| - |4|$

CHECK YOURSELF ANSWERS

1.

2. (a) $-13, -8, -3, 2, 12, 15$

   (b) $-9, -3, 3, 6, 8$

3. (a) Minimum is -13, maximum is 15;   (b) Minimum is -9, maximum is 8

4. $-23, 1054, 0,$ and -500 5. (a) -17;   (b) 12

6. (a) 8;   (b) 8;   (c) -8;   (d) 13;   (e) 5;   (f) 5

Represent each quantity with an integer.

1. An altitude of 400 ft above sea level

2. An altitude of 80 ft below sea level

3. A loss of $200

4. A profit of $400

5. A decrease in population of 25,000

6. An increase in population of 12,500

Represent the integers on the number lines shown.

7. 5, −15, 18, −8, 3
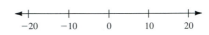

8. −18, 4, −5, 13, 9

Which numbers in the following sets are integers?

9. $\left\{5, -\dfrac{2}{9}, 175, -234, -0.64\right\}$

10. $\left\{-45, 0.35, \dfrac{3}{5}, 700, -26\right\}$

Place each of the following sets of numbers in ascending order.

11. 3, −5, 2, 0, −7, −1, 8

12. −2, 7, 1, −8, 6, −1, 0

13. 9, −2, −11, 4, −6, 1, 5

14. 23, −18, −5, −11, −15, 14, 20

15. −6, 7, −7, 6, −3, 3

16. 12, −13, 14, −14, 15, −15

For each set, determine the maximum and minimum values.

17. 5, −6, 0, 10, −3, 15, 1, 8

18. 9, −1, 3, 11, −4, 2, 5, −2

19. 21, −15, 0, 7, −9, 16, −3, 11

20. −22, 0, 22, −31, 18, −5, 3

21. 3, 0, 1, −2, 5, 4, −1

22. 2, 7, −3, 5, −10, −5

Find the opposite of each number.

23. 15

24. 18

25. 15

26. 34

27. −19

28. −6

29. −7

30. −54

Evaluate.

31. $|17|$

32. $|28|$

33. $|-19|$

34. $|-7|$

35. $-|21|$

36. $-|3|$

37. $-|-8|$

38. $-|-13|$

39. $|-2|+|3|$

40. $|4|+|-3|$

41. $|-9|+|9|$

42. $|11|+|-11|$

ANSWERS

17. _____
18. _____
19. _____
20. _____
21. _____
22. _____
23. _____
24. _____
25. _____
26. _____
27. _____
28. _____
29. _____
30. _____
31. _____
32. _____
33. _____
34. _____
35. _____
36. _____
37. _____
38. _____
39. _____
40. _____
41. _____
42. _____

43. _____
44. _____
45. _____
46. _____
47. _____
48. _____
49. _____
50. _____
51. _____
52. _____
53. _____
54. _____
55. _____
56. _____
57. _____
58. _____
59. _____
60. _____
61. _____
62. _____
63. _____
64. _____

43. $|-6|-|6|$

44. $|5|-|-5|$

45. $|15|-|8|$

46. $|11|-|3|$

47. $|15-8|$

48. $|11-3|$

49. $|-9|+|2|$

50. $|-7|+|4|$

51. $|7|-|-6|$

52. $|-9|-|-4|$

Label each statement as true or false.

53. All natural numbers are integers.

54. Zero is an integer.

55. All integers are whole numbers.

56. All real numbers are integers.

57. All negative integers are whole numbers.

58. Zero is neither positive nor negative.

Place absolute value bars in the proper location on the left side of the expression so that the equation is true.

59. $6+(-2)=4$

60. $8+(-3)=5$

61. $6+(-2)=8$

62. $8+(-3)=11$

Represent each quantity with a real number.

63. **Science and technology.** The erosion of 5 centimeters (cm) of topsoil from an Iowa corn field.

64. **Science and technology.** The formation of 2.5 cm of new topsoil on the African savanna.

65. Business and finance. The withdrawal of $50 from a checking account.

65. _____

66. _____

67. _____

68. _____

69. _____

70. _____

71. _____

72. _____

73. _____

66. Business and finance. The deposit of $200 in a savings account.

67. Science and technology. The temperature change pictured.

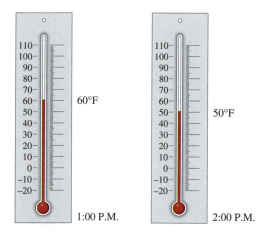

60°F   50°F

1:00 P.M.   2:00 P.M.

68. Business and finance. An increase of 75 points in the Dow-Jones average.

69. Statistics. An eight-game losing streak by the local baseball team.

70. Social science. An increase of 25,000 in the population of the city.

71. Business and finance. A country exported $90,000,000 more than it imported, creating a positive trade balance.

72. Business and finance. A country exported $60,000,000 less than it imported, creating a negative trade balance.

For each collection of numbers given in exercises 73 to 76, answer the following:

    (a) Which number is smallest?
    (b) Which number lies farthest from the origin?
    (c) Which number has the largest absolute value?
    (d) Which number has the smallest absolute value?

73. $-6, 3, 8, 7, -2$

74. _____

75. _____

76. _____

77. _____

74. $-8, 3, -5, 4, 9$

75. $-2, 6, -1, 0, 2, 5$

76. $-9, 0, -2, 3, 6$

77. Simplify each of the following:

$$-(-7) \qquad -(-(-7)) \qquad -(-(-(-7)))$$

Based on your answers, generalize your results.

Answers

1. 400 or $(+400)$ **3.** -200 **5.** $-25,000$

7.

```
        −15  −8        3 5         18
    ←———+————+————+————+——+————+————+————→
      −20    −10    0     10    20
```

9. $5, 175, -234$

11. $-7, -5, -1, 0, 2, 3, 8$ **13.** $-11, -6, -2, 1, 4, 5, 9$

15. $-7, -6, -3, 3, 6, 7$ **17.** Max: 15; min: -6 **19.** Max: 21, min: -15

21. Max: 5; min: -2 **23.** -15 **25.** -15 **27.** 19   **29.** 7   **31.** 17

33. 19   **35.** -21 **37.** -8 **39.** 5   **41.** 18   **43.** 0   **45.** 7

47. 7   **49.** 11   **51.** 1   **53.** True   **55.** False   **57.** False

59. $\left| 6 + (-2) \right| = 4$ **61.** $\left| 6 \right| + \left| -2 \right| = 8$ **63.** -5 **65.** -50

67. $-10°F$ **69.** -8 **71.** $+90,000,000$ **73.** $-6; 8; 8; -2$

75. $-2; 6; 6; 0$ **77.**

0 Summary

DEFINITION/PROCEDURE	EXAMPLE	REFERENCE
Prime Factorization and Least Common Multiples		**Section 0.1**
Factor A **factor** of a whole number is another whole number that will divide exactly into that number, leaving a remainder of zero.	The factors of 12 are 1, 2, 3, 4, 6, and 12.	*p. 4*
Prime Number Any whole number greater than 1 that has only 1 and itself as factors.	7, 13, 29, and 73 are prime numbers.	*p. 4*
Composite Number Any whole number greater than 1 that is not prime.	8, 15, 42, and 65 are composite numbers.	*p. 5*
Zero and 1 0 and 1 are classified as neither prime nor composite numbers.		*p. 6*
Greatest Common Factor (GCF) The GCF is the *largest* number that is a factor of each of a group of numbers. *To Find the GCF* **1.** Write the prime factorization for each of the numbers in the group. **2.** Locate the prime factors that appear in every prime factorization. **3.** The GCF will be the product of all the common prime factors. If there are no common prime factors, the GCF is 1.	To find the GCF of 24, 30, and 36: $24 = ②\cdot 2 \cdot 2 \cdot ③$ $30 = ②\cdot ③\cdot 5$ $36 = ②\cdot 2 \cdot ③\cdot 3$ The GCF is $2 \cdot 3 = 6$.	*p. 9*
Least Common Multiple (LCM) The LCM is the *smallest* number that is a multiple of each of a group of numbers. *To Find the LCM* **1.** Write the prime factorization for each of the numbers in the group. **2.** Find all the prime factors that appear in any one of the prime factorizations. **3.** Form the product of those prime factors, using each factor the greatest number of times it occurs in any one factorization.	To find the LCM of 12, 15, and 18: $12 = 2 \cdot 2 \cdot 3$ $15 = 3 \cdot 5$ $18 = 2 \cdot 3 \cdot 3$ $\overline{ 2 \cdot 2 \cdot 3 \cdot 3 \cdot 5}$ The LCM is $2 \cdot 2 \cdot 3 \cdot 3 \cdot 5$, or 180.	*p. 12*
Fractions and Mixed Numbers		**Section 0.2**
The Fundamental Principle of Fractions $\dfrac{a}{b} = \dfrac{a \cdot c}{b \cdot c}$ in which neither b nor c can equal zero.	$\dfrac{2}{3} = \dfrac{2 \cdot 3}{3 \cdot 3} = \dfrac{6}{9}$	*p. 19*
Multiplying Fractions **1.** Multiply numerator by numerator. This gives the numerator of the product. **2.** Multiply denominator by denominator. This gives the denominator of the product. **3.** Simplify the resulting fraction if possible. In multiplying fractions it is usually easiest to factor and simplify the numerator and denominator *before* multiplying.	$\dfrac{5}{8} \cdot \dfrac{3}{7} = \dfrac{5 \cdot 3}{8 \cdot 7} = \dfrac{15}{56}$ $\dfrac{5}{9} \cdot \dfrac{3}{10} = \dfrac{\overset{1}{\cancel{5}} \cdot \overset{1}{\cancel{3}}}{\underset{3}{\cancel{9}} \cdot \underset{2}{\cancel{10}}} = \dfrac{1}{6}$	*p. 21*

Continued

Fractions and Mixed Numbers (*continued*)		Section 0.2
Dividing Fractions Invert the divisor and multiply.	$\dfrac{3}{7} \div \dfrac{4}{5} = \dfrac{3}{7} \cdot \dfrac{5}{4} = \dfrac{15}{28}$	*p. 21*
To Add or Subtract Fractions with Different Denominators **1.** Find the LCD of the fractions. **2.** Change each fraction to an equivalent fraction with the LCD as a common denominator. **3.** Add or subtract the resulting like fractions as before.	$\dfrac{3}{4} + \dfrac{7}{10} = \dfrac{15}{20} + \dfrac{14}{20} = \dfrac{29}{20}$ $\dfrac{8}{9} - \dfrac{5}{6} = \dfrac{16}{18} - \dfrac{15}{18} = \dfrac{1}{18}$	*p. 22*
Mixed Number The sum of a whole number and a proper fraction.	$2\dfrac{1}{3}$ and $5\dfrac{7}{8}$ are mixed numbers. Note that $2\dfrac{1}{3}$ means $2 + \dfrac{1}{3}$.	*p. 24*
To Change an Improper Fraction to a Mixed Number **1.** Divide the numerator by the denominator. The quotient is the whole-number portion of the mixed number. **2.** If there is a remainder, write the remainder over the original denominator. This gives the fractional portion of the mixed number.	To change $\dfrac{22}{5}$ to a mixed number: $\begin{array}{r} 4 \\ 5\overline{)22} \\ \underline{20} \\ 2 \end{array}$ Quotient Remainder $\dfrac{22}{5} = 4\dfrac{2}{5}$	*p. 24*
To Convert a Mixed Number to an Improper Fraction **1.** Multiply the denominator of the fraction by the whole-number portion of the mixed number. **2.** Add the numerator of the fraction to that product. **3.** Write that sum over the original denominator to form the improper fraction.	Denominator Whole number Numerator $5\dfrac{3}{4} = \dfrac{(4 \cdot 5) + 3}{4} = \dfrac{23}{4}$ Denominator	*p. 25*
To Add or Subtract Mixed Numbers **1.** Rewrite as improper fractions. **2.** Add or subtract the fractions. **3.** Rewrite the results as a mixed number if required.	$5\dfrac{1}{2} - 3\dfrac{3}{4} = \dfrac{11}{2} - \dfrac{15}{4}$ $= \dfrac{22}{4} - \dfrac{15}{4}$ $= \dfrac{7}{4}$ $= 1\dfrac{3}{4}$	*p. 26*
Decimals and Percents		Section 0.3
To Convert a Fraction to a Decimal **1.** Divide the numerator of the fraction by its denominator. **2.** The quotient is the decimal equivalent of the common fraction.	To convert $\dfrac{1}{2}$ to a decimal: $\begin{array}{r} 0.5 \\ 2\overline{)1.0} \\ \underline{1\,0} \\ 0 \end{array}$	*p. 35*

Decimals and Percents (*continued*)		Section 0.3
To Convert a Terminating Decimal Less Than 1 to a Fraction 1. Write the digits of the decimal without the decimal point. This will be the numerator of the fraction. 2. The denominator of the fraction is a 1 followed by as many zeros as there are places in the decimal.	To convert 0.275 to a fraction: $0.275 = \dfrac{275}{1000} = \dfrac{11}{40}$	*p. 36*
To Add Decimals 1. Write the numbers being added in column form with their decimal points in a vertical line. 2. Add just as you would with whole numbers. 3. Place the decimal point of the sum in line with the decimal points of the addends.	To add 2.7, 3.15, and 0.48: $\begin{array}{r} 2.7 \\ 3.15 \\ +\ 0.48 \\ \hline 6.33 \end{array}$	*p. 37*
To Subtract Decimals 1. Write the numbers being subtracted in column form with their decimal points in a vertical line. You may have to place zeros to the right of the existing digits. 2. Subtract just as you would with whole numbers. 3. Place the decimal point of the difference in line with the decimal points of the numbers being subtracted.	To subtract 5.875 from 8.5: $\begin{array}{r} 8.500 \\ -\ 5.875 \\ \hline 2.625 \end{array}$	*p. 37*
To Multiply Decimals 1. Multiply the decimals as though they were whole numbers. 2. Add the number of decimal places in the factors. 3. Place the decimal point in the product so that the number of decimal places in the product is the sum of the number of decimal places in the factors.	To multiply 2.85×0.045: $\begin{array}{r} 2.85 \leftarrow \text{Two places} \\ \times\ 0.045 \leftarrow \text{Three places} \\ \hline 1425 \\ 11400 \\ \hline 0.12825 \leftarrow \text{Five places} \end{array}$	*p. 37*
To Divide by a Decimal 1. Move the decimal point to the right, making the divisor a whole number. 2. Move the decimal point in the dividend to the right the same number of places. Add zeros if necessary. 3. Place the decimal point in the quotient directly above the decimal point of the dividend. 4. Divide as you would with whole numbers.	To divide 16.5 by 5.5, move the decimal points: $\begin{array}{r} 3 \\ 5.5\overline{)16.5} \\ \underline{16\ 5} \\ 0 \end{array}$	*p. 38*
Percent Another way of naming parts of a whole. Percent means per hundred.	Fractions and decimals are other ways of naming parts of a whole. $21\% = 21\left(\dfrac{1}{100}\right) = \dfrac{21}{100} = 0.21$	*p. 39*
Changing a Percent to a Fraction or a Decimal 1. *To convert a percent to a fraction,* replace the percent symbol with $\dfrac{1}{100}$ and multiply. 2. *To convert a percent to a decimal,* remove the percent symbol, and move the decimal point two places to the left.	$37\% = 37\left(\dfrac{1}{100}\right) = \dfrac{37}{100}$ $37\% = 37\left(\dfrac{1}{100}\right) = 0.37$	*p. 40*

Continued

Decimals and Percents (*continued*)		**Section 0.3**
Changing a Decimal or Fraction to a Percent **1.** *To convert a decimal to a percent,* move the decimal point two places to the right, and attach the percent symbol. **2.** *To convert a fraction to a percent,* write the decimal equivalent of the fraction, and then change that decimal to a percent.	$$0.58 = 58\left(\frac{1}{100}\right) = 58\%$$ $$\frac{3}{5} = 0.60 = 60\left(\frac{1}{100}\right) = 60\%$$	*p. 41*
Exponents and the Order of Operations		**Section 0.4**
Using Exponents **Base** The number that is raised to a power. **Exponent** The exponent is written to the right and above the base. The exponent tells the number of times the base is to be used as a factor.	Exponent $$5^3 = 5 \cdot 5 \cdot 5 = 125$$ Base Three factors This is read "5 to the third power" or "5 cubed."	*p. 47*
The Order of Operations *Mixed operations* in an expression should be done in the following order: **1.** Do any operations inside grouping symbols. **2.** Evaluate any powers. **3.** Do all multiplication and division in order from left to right. **4.** Do all addition and subtraction in order from left to right.	$$4 \cdot (2 + 3)^2 - 7$$ $$= 4 \cdot 5^2 - 7$$ $$= 4 \cdot 25 - 7$$ $$= 100 - 7$$ $$= 93$$	*p. 48*
Positive and Negative Numbers		**Section 0.5**
Positive Numbers Numbers used to name points to the right of the origin on the number line. *Negative Numbers* Numbers used to name points to the left of the origin on the number line. *Integers* The natural (or counting) numbers, their negatives, and zero. The integers are $$\{\ldots, -3, -2, -1, 0, 1, 2, 3, \ldots\}$$	The origin $-3\ -2\ -1\quad 0\quad 1\quad 2\quad 3$ Negative numbers Positive numbers	*p. 55*
Absolute Value The distance (on the number line) between the point named by an integer and the origin. The absolute value of x is written $|x|$.	$$|7| = 7$$ $$|-10| = 10$$	*p. 58*
Opposite of a Number The opposite of a number is the negative of that number.	The opposite of 2 is -2. The opposite of -9 is 9.	*p. 58*

Summary Exercises

This summary exercise set is provided to give you practice with each of the objectives of this chapter. Each exercise is keyed to the appropriate chapter section. When you are finished, you can check your answers to the odd-numbered exercises against those presented in the back of the text. If you have difficulty with any of these questions, go back and reread the examples from that section. Your instructor will give you guidelines on how best to use these exercises in your instructional setting.

[0.1]   List all the factors of the given numbers.

1. 52

2. 41

3. 76

4. 315

Use the group of numbers 2, 5, 7, 11, 14, 17, 21, 23, 27, 39, and 43.

5. List the prime numbers; then list the composite numbers.

Find the prime factorization for the given numbers.

6. 48

7. 420

8. 60

9. 180

Find the greatest common factor (GCF).

10. 15 and 20

11. 30 and 31

12. 72 and 180

13. 240 and 900

Find the least common multiple (LCM).

14. 4 and 12

15. 8 and 16

16. 18 and 24

17. 12 and 18

[0.2] Write three fractional representations for each number.

18. $\dfrac{5}{7}$

19. $\dfrac{3}{11}$

20. $\dfrac{4}{9}$

21. Write the fraction $\dfrac{24}{64}$ in simplest form.

[0.3] Perform the indicated operations.

22. $\dfrac{7}{15} \cdot \dfrac{5}{21}$

23. $\dfrac{10}{27} \cdot \dfrac{9}{20}$

24. $\dfrac{5}{12} \div \dfrac{5}{8}$

25. $\dfrac{7}{15} \div \dfrac{14}{25}$

26. $\dfrac{5}{6} + \dfrac{11}{18}$

27. $\dfrac{5}{18} + \dfrac{7}{12}$

28. $\dfrac{11}{18} - \dfrac{2}{9}$

29. $\dfrac{11}{27} - \dfrac{5}{18}$

30. $5.123 + 6.4$

31. $10.127 - 5.49$

32. $5.26 \cdot 3.796$

33. $6\dfrac{5}{7} + 3\dfrac{4}{7}$

34. $5\dfrac{7}{10} + 3\dfrac{11}{12}$

35. $7\dfrac{7}{9} - 3\dfrac{4}{9}$

36. $6\dfrac{5}{12} - 3\dfrac{5}{8}$

37. $5\dfrac{1}{3} \cdot 1\dfrac{4}{5}$

38. $3\dfrac{2}{5} \cdot \dfrac{5}{8}$

39. $3\dfrac{3}{8} \div 2\dfrac{1}{4}$

Divide and round the quotient to the nearest thousandth.

40. $6\dfrac{1}{7} \div \dfrac{3}{14}$

41. $3.042 \div 0.37$

42. $0.2549 \div 2.87$

[0.3] Write the percent as a fraction or a mixed number.

43. 2%

44. 20%

45. 37.5%

46. 150%

47. $233\dfrac{1}{3}\%$

48. 300%

Write the percents as decimals.

49. 75%

50. 4%

51. 6.25%

52. 13.5%

53. 0.6%

54. 225%

Write as percents.

55. 0.06

56. 0.375

57. 2.4

58. 7

59. 0.035

60. 0.005

61. $\dfrac{43}{100}$

62. $\dfrac{7}{10}$

63. $\dfrac{2}{5}$

64. $1\dfrac{1}{4}$

65. $2\dfrac{2}{3}$

[0.4]   Evaluate each of the following expressions.

66. $18 - 3 \cdot 5$

67. $(18 - 3) \cdot 5$

68. $5 \cdot 4^2$

69. $(5 \cdot 4)^2$

70. $5 \cdot 3^2 - 4$

71. $5 \cdot (3^2 - 4)$

72. $5 \cdot (4 - 2)^2$

73. $5 \cdot 4 - 2^2$

74. $(5 \cdot 4 - 2)^2$

75. $3 \cdot (5 - 2)^2$

76. $3 \cdot 5 - 2^2$

77. $(3 \cdot 5 - 2)^2$

78. $8 \div 4 \cdot 2$

79. $19 - 14 + 2 \cdot 5$

80. $36 + 4 \cdot 2 - 7 \cdot 6$

[0.5]

81. Represent the following integers on the number line shown: 6, −18, −3, 2, 15, −9.

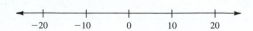

Place each of the following sets in ascending order.

82. 4, −3, 6, −7, 0, 1, −2

83. −7, 8, −8, 1, 2, −3, 3, 0, 7

For each data set, determine the maximum and minimum.

84. 4, −2, 5, 1, −6, 3, −4

85. −4, 2, 5, −9, 8, 1, −6

Find the opposite of each number.

86. 17

87. −63

Evaluate.

88. $|9|$

89. $|-9|$

90. $-|9|$

91. $-|-9|$

92. $|12-8|$

93. $|8| + |-12|$

94. $-|8+12|$

95. $|-18| - |-12|$

96. $|-7| - |-3|$

97. $|-9| + |-5|$

Self-Test for Chapter 0

Name _____

Section _____ Date _____

The purpose of this self-test is to help you check your progress and to review for the next in-class exam. Allow yourself about an hour to take this test. At the end of that hour check your answers against those given in the back of the text. Section references accompany the answers. If you missed any questions, go back to those sections and reread the examples until you master the concepts.

ANSWERS

1. Which of the numbers 5, 9, 13, 17, 22, 27, 31, and 45 are prime numbers? Which are composite numbers?

2. Find the prime factorization for 264.

Find the greatest common factor (GCF) for the given numbers.

3. 36 and 84   **4.** 16, 24, and 72

Find the least common multiple (LCM) for the given numbers.

5. 12 and 27   **6.** 3, 4, and 18

Perform the indicated operations.

7. $\dfrac{8}{21} \cdot \dfrac{3}{4}$ **8.** $\dfrac{7}{12} \div \dfrac{28}{36}$

9. $\dfrac{3}{4} + \dfrac{5}{6}$ **10.** $\dfrac{8}{21} - \dfrac{2}{7}$

11. $3.25 + 4.125$ **12.** $16.234 - 12.35$

13. $7.29 \cdot 3.15$ **14.** $6.10 \cdot 13.1$

15. $2\dfrac{2}{3} \cdot 1\dfrac{2}{7}$ **16.** $4\dfrac{1}{6} + 3\dfrac{3}{4}$

17. $3\dfrac{5}{6} - 2\dfrac{2}{9}$ **18.** $3.969 \div 0.54$

Write as fractions.

19. 7%   **20.** 72%

1. _____
2. _____
3. _____
4. _____
5. _____
6. _____
7. _____
8. _____
9. _____
10. _____
11. _____
12. _____
13. _____
14. _____
15. _____
16. _____
17. _____
18. _____
19. _____
20. _____

21. _____

22. _____

23. _____

24. _____

25. _____

26. _____

27. _____

28. _____

29. _____

30. _____

31. _____

32. _____

33. _____

34. _____

35. _____

36. _____

37. _____

38. _____

39. _____

40. _____

41. _____

42. _____

43. _____

44. _____

45. _____

Write as decimals.

21. 42%   **22.** 6%   **23.** 160%

Write as percents.

24. 0.03   **25.** 0.042

26. $\dfrac{2}{5}$ **27.** $\dfrac{5}{8}$

Write, using exponents.

28. $4 \cdot 4 \cdot 4 \cdot 4$ **29.** $9 \cdot 9 \cdot 9 \cdot 9 \cdot 9$

Evaluate the following expressions.

30. $23 - 4 \cdot 5$ **31.** $4 \cdot 5^2 - 35$

32. $4 \cdot (2 + 4)^2$ **33.** $16 \cdot 2 - 5^2$

34. $(3 \cdot 2 - 4)^3$ **35.** $8 - 3 \cdot 2 + 5$

36. Represent the following integers on the number line shown: 5, −12, 4, −7, 18, −17.

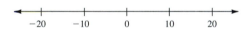

37. Place the following data set in ascending order: 4, −3, −6, 5, 0, 2, −2.

38. Determine the maximum and minimum of the following data set: 3, 2, −5, 6, 1, −2.

Evaluate.

39. $\left|7\right|$ **40.** $\left|-7\right|$

41. $\left|18 - 7\right|$ **42.** $\left|18\right| - \left|-7\right|$

43. $-\left|24 - 5\right|$

Find the opposite of each number.

44. 40   **45.** −19

1

THE LANGUAGE OF ALGEBRA

INTRODUCTION

Anthropologists and archeologists investigate modern human cultures and societies as well as cultures that existed so long ago that their characteristics must be inferred from objects found buried in lost cities or villages. When some interesting object is found, such as the Babylonian tablets mentioned in Chapter 0, often the first questions that arise are "How old is this? When did this culture flourish?" With methods such as carbon dating, it has been established that large, organized cultures existed around 3000 B.C.E. in Egypt, 2800 B.C.E. in India, no later than 1500 B.C.E. in China, and around 1000 B.C.E. in the Americas.

How long ago was 1500 B.C.E.? Which is older, an object from 3000 B.C.E. or an object from 500 A.D.? Using the Christian notation for dates, we have to count A.D. years and B.C.E. years differently. An object from 500 A.D. is about $2000 - 500$ years old, or about 1500 years old. But an object from 3000 B.C.E. is about $2000 + 3000$ years old, or about 5000 years old. Why subtract in the first case but add in the other? Because of the way years are counted before the Christian era (B.C.E.) and after the birth of Christ (A.D.), the B.C.E. dates must be considered as *negative* numbers.

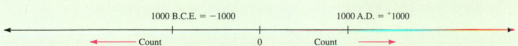

Very early on, the Chinese accepted the idea that a number could be negative; they used red calculating rods for positive numbers and black for negative numbers. Hindu mathematicians in India worked out the arithmetic of negative numbers as long ago as 400 A.D., but western mathematicians did not recognize this idea until the sixteenth century. It would be difficult today to think of measuring things such as temperature, altitude, and money without using negative numbers.

Pre-Test Chapter 1

1. _____

2. _____

3. _____

4. _____

5. _____

6. _____

7. _____

8. _____

9. _____

10. _____

11. _____

12. _____

13. _____

14. _____

15. _____

16. _____

17. _____

18. _____

19. _____

20. _____ 21. _____

22. _____

23. _____

24. _____

25. _____

This pre-test will provide a preview of the types of exercises you will encounter in each section of this chapter. The answers for these exercises can be found in the back of the text. If you are working on your own, or ahead of the class, this pre-test can help you identify the sections in which you should focus more of your time.

Write each of the following using symbols.

[1.1] **1.** 8 less than x

2. The quotient when w is divided by the product of x and 17

Identify which are expressions and which are not.

3. $7x - 5 = 11$

4. $3x - 2(x + 1)$

Identify the property that is illustrated by the following statements.

[1.2] **5.** $8 \cdot 9 = 9 \cdot 8$

6. $3(4 + 2) = 3 \cdot 4 + 3 \cdot 2$

7. $9 + (1 + 7) = (9 + 1) + 7$

Perform the indicated operations.

[1.3–1.4] **8.** $-7 + (-3)$ **9.** $8 + (-9)$ **10.** $(-3) + (-2)$

11. $-\dfrac{7}{4} + \dfrac{3}{4}$ **12.** $8 - 11$ **13.** $-8 - 11$

14. $9 - (-3)$ **15.** $6 + (-6)$ **16.** $(-7)(-3)$

17. $(3.5)(4)$ **18.** $(3)\left(\dfrac{1}{6}\right)$ **19.** $\dfrac{-27 + 6}{-3}$

Evaluate the following expressions.

[1.5] **20.** $5 - 4^2 \cdot 3 \div 6$ **21.** $(45 - 3 \cdot 5) + 5^2$

22. If $x = -2$, $y = 7$, and $w = -4$, evaluate the expression $\dfrac{x^2 y}{w}$.

Combine like terms.

[1.6] **23.** $5w^2 t + 3w^2 t$ **24.** $4a^2 - 3a + 5 + 7a - 2 - 5a^2$

Divide.

[1.7] **25.** $\dfrac{96x^3 y^5}{8x^2 y^3}$

1.1 From Arithmetic to Algebra

1.1 OBJECTIVES

1. Represent addition, subtraction, multiplication, and division using the symbols of algebra
2. Identify algebraic expressions

Overcoming Math Anxiety

Throughout this text, we will present you with a series of class-tested techniques that are designed to improve your performance in this math class.

Hint #2   Become familiar with your syllabus.

In the first class meeting, your instructor probably handed out a class syllabus. If you haven't already done so, you need to incorporate important information into your calendar and address book.

1. Write all important dates in your calendar. This includes homework due dates, quiz dates, test dates, and the date and time of the final exam. Never allow yourself to be surprised by any deadline!

2. Write your instructor's name, contact number, and office number in your address book. Also include the office hours. Make it a point to see your instructor early in the term. Although this is not the only person who can help clear up your confusion, your instructor is the most important person.

3. Make note of other resources that are made available to you. This includes CDs, video tapes, web pages, and tutoring.

Given all of these resources, it is important that you never let confusion or frustration mount. If you can't "get it" from the text, try another resource. All of the resources are there specifically for you, so take advantage of them!

In arithmetic, you learned how to do calculations with numbers using the basic operations of addition, subtraction, multiplication, and division.

In algebra, you will still use numbers and the same four operations. However, you will also use letters to represent numbers. Letters such as x, y, L, or W are called **variables** when they represent numerical values.

Here we see two rectangles whose lengths and widths are labeled with numbers.

If we need to represent the length and width of *any* rectangle, we can use the variables L and W.

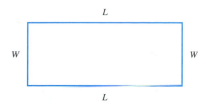

NOTE In arithmetic:
+ denotes addition
− denotes subtraction
× denotes multiplication
÷ denotes division.

You are familiar with the four symbols $(+, -, \times, \div)$ used to indicate the fundamental operations of arithmetic.

To see how these operations are indicated in algebra, we begin with addition.

Definition: Addition

$x + y$ means the *sum* of x and y or x *plus* y.

OBJECTIVE 1

Example 1 **Writing Expressions That Indicate Addition**

(a) *The sum of a and* 3 is written as $a + 3$.

(b) *L plus W* is written as $L + W$.

(c) 5 *more than m* is written as $m + 5$.

(d) *x increased by* 7 is written as $x + 7$.

(e) 15 added to x is written as $x + 15$.

 CHECK YOURSELF 1

Write, using symbols.

(a) The sum of y and 4   (b) a plus b

(c) 3 more than x (d) n increased by 6

Let's look at how subtraction is indicated in algebra.

Definition: Subtraction

$x - y$ means the *difference* of x and y or x *minus* y.

Example 2 **Writing Expressions That Indicate Subtraction**

(a) *r minus s* is written as $r - s$.

(b) *The difference of m and* 5 is written as $m - 5$.

(c) *x decreased by* 8 is written as $x - 8$.

(d) 4 *less than a* is written as $a - 4$.

 CHECK YOURSELF 2

Write, using symbols.

(a) w minus z (b) The difference of a and 7

(c) y decreased by 3   (d) 5 less than b

You have seen that the operations of addition and subtraction are written exactly the same way in algebra as in arithmetic. This is not true in multiplication because the sign $\times$ looks like the letter x. So in algebra we use other symbols to show multiplication to avoid any confusion. Here are some ways to write multiplication.

NOTE *x* and *y* are called the **factors** of the product *xy*.

> ### Definition: Multiplication
>
> | A centered dot | $x \cdot y$ | |
> |---|---|---|
> | Parentheses | $(x)(y)$ | These expressions all indicate the *product* of *x* and *y* or *x* times *y*. |
> | Writing the letters next to each other | xy | |

Example 3   Writing Expressions That Indicate Multiplication

NOTE You can place letters next to each other or numbers and letters next to each other to show multiplication. But you *cannot* place numbers side by side to show multiplication: 37 means the number "thirty-seven," not 3 times 7.

(a) The product of 5 and *a* is written as $5 \cdot a$, $(5)(a)$, or $5a$. The last expression, $5a$, is the shortest and the most common way of writing the product.

(b) 3 times 7 can be written as $3 \cdot 7$ or $(3)(7)$.

(c) Twice *z* is written as $2z$.

(d) The product of 2, *s*, and *t* is written as $2st$.

(e) 4 more than the product of 6 and *x* is written as $6x + 4$.

CHECK YOURSELF 3

Write, using symbols.

(a) *m* times *n*

(b) The product of *h* and *b*

(c) The product of 8 and 9

(d) The product of 5, *w*, and *y*

(e) 3 more than the product of 8 and *a*

Before we move on to division, we will look at how we can combine the symbols we have learned so far.

NOTE Not every collection of symbols is an expression.

> ### Definition: Expression
>
> An **expression** is a meaningful collection of numbers, variables, and symbols of operation.

OBJECTIVE 2   ### Example 4   Identifying Expressions

(a) $2m + 3$ is an expression. It means that we multiply 2 and *m*, and then add 3.

(b) $x + \cdot + 3$ is not an expression. The three operations in a row have no meaning.

(c) $y = 2x - 1$ is not an expression. The equal–sign is not an operation sign.

(d) $3a + 5b - 4c$ is an expression. Its meaning is clear.

CHECK YOURSELF 4

Identify which are expressions and which are not.

(a) $7 - \cdot x$

(b) $6 + y = 9$

(c) $a + b - c$

(d) $3x - 5yz$

To write more complicated products in algebra, we need some "punctuation marks." Parentheses ( ) mean that an expression is to be thought of as a single quantity. Brackets [ ] are used in exactly the same way as parentheses in algebra. Look at Example 5 which shows the use of these signs of grouping.

Example 5   Expressions with More Than One Operation

(a) 3 times the sum of a and b is written as

NOTE This can be read as "3 times the quantity a plus b."

$$3(a + b)$$

The sum of a and b is a single quantity, so it is enclosed in parentheses.

NOTE No parentheses are needed here because the 3 multiplies *only* the a.

(b) The sum of 3 times a and b is written as $3a + b$.

(c) 2 times the difference of m and n is written as $2(m - n)$.

(d) The product of s plus t and s minus t is written as $(s + t)(s - t)$.

(e) The product of b and 3 less than b is written as $b(b - 3)$.

CHECK YOURSELF 5

Write, using symbols.

(a) Twice the sum of p and q

(b) The sum of twice p and q

(c) The product of a and the quantity $b - c$

(d) The product of x plus 2 and x minus 2

(e) The product of x and 4 more than x

NOTE In algebra the fraction form is usually used.

Now we look at the operation of division. In arithmetic, we use the division sign $\div$, the long division symbol $\overline{)}$, and fraction notation. For example, to indicate the quotient when 9 is divided by 3, we may write

$$9 \div 3 \quad \text{or} \quad 3\overline{)9} \quad \text{or} \quad \frac{9}{3}$$

Definition: Division

$\dfrac{x}{y}$ means *x divided by y* or *the quotient of x and y.*

Example 6   Writing Expressions That Indicate Division

(a) m divided by 3 is written as $\dfrac{m}{3}$.

(b) The quotient when a plus b is divided by 5 is written as $\dfrac{a + b}{5}$.

(c) The sum p plus q divided by the difference p minus q is written as $\dfrac{p + q}{p - q}$.

CHECK YOURSELF 6

Write, using symbols.

(a) r divided by s

(b) The quotient when x minus y is divided by 7

(c) The difference a minus 2 divided by the sum a plus 2

Notice that we can use many different letters to represent variables. In Example 6 the letters $m, a, b, p,$ and q represented different variables. We often choose a letter that reminds us of what it represents, for example, L for *length* or W for *width*.

Example 7   Writing Geometric Expressions

(a) *Length* times *width* is written $L \cdot W$.

(b) One-half of *altitude* times *base* is written $\frac{1}{2} a \cdot b$. *area* △

(c) *Length* times *width* times *height* is written $L \cdot W \cdot H$. *Volume*

(d) Pi (π) times *diameter* is written πd.

CHECK YOURSELF 7

Write each geometric expression, using symbols.

(a) Two times *length* plus two times *width*   **(b)** Two times pi (π) times *radius*

CHECK YOURSELF ANSWERS

1. **(a)** $y + 4$; **(b)** $a + b$; **(c)** $x + 3$; **(d)** $n + 6$ **2.** **(a)** $w - z$; **(b)** $a - 7$; **(c)** $y - 3$;
(d) $b - 5$ **3.** **(a)** mn; **(b)** hb; **(c)** $8 \cdot 9$ or $(8)(9)$; **(d)** $5wy$; **(e)** $8a + 3$
4. **(a)** Not an expression; **(b)** not an expression; **(c)** an expression; **(d)** an expression
5. **(a)** $2(p + q)$; **(b)** $2p + q$; **(c)** $a(b - c)$; **(d)** $(x + 2)(x - 2)$; **(e)** $x(x + 4)$
6. **(a)** $\dfrac{r}{s}$; **(b)** $\dfrac{x - y}{7}$; **(c)** $\dfrac{a - 2}{a + 2}$ **7.** **(a)** $2L + 2W$; **(b)** $2\pi r$

1.1 Exercises

Write each of the following phrases, using symbols.

1. The sum of c and d

2. a plus 7

3. w plus z

4. The sum of m and n

5. x increased by 5

6. 4 more than c

7. 10 more than y

8. m increased by 4

9. b minus a

10. 5 less than w

11. b decreased by 4

12. r minus 3

13. 6 less than r

14. x decreased by 3

15. w times z

16. The product of 3 and c

17. The product of 5 and t

18. 8 times a

19. The product of 8, *m*, and *n*

20. The product of 7, *r*, and *s*

21. The product of 3 and the quantity *p* plus *q*

22. The product of 5 and the sum of *a* and *b*

23. Twice the sum of *x* and *y*

24. 7 times the sum of *m* and *n*

25. The sum of twice *x* and *y*

26. The sum of 3 times *m* and *n*

27. Twice the difference of *x* and *y*

28. 3 times the difference of *a* and *c*

29. The quantity *a* plus *b* times the quantity *a* minus *b*

30. The product of *x* plus *y* and *x* minus *y*

31. The product of *m* and 3 more than *m*

32. The product of *a* and 7 less than *a*

33. *x* divided by 5

34. The quotient when *b* is divided by 8

35. The quotient of *a* minus *b*, divided by 9

36. The difference *x* minus *y*, divided by 9

37. The sum of *p* and *q*, divided by 4

ANSWERS

19.

20.

21.

22.

23.

24.

25.

26.

27.

28.

29.

30.

31.

32.

33.

34.

35.

36.

37.

38. _____

39. _____

40. _____

41. _____

42. _____

43. _____

44. _____

45. _____

46. _____

47. _____

48. _____

49. _____

50. _____

51. _____

52. _____

53. _____

54. _____

55. _____

56. _____

38. The sum of a and 5, divided by 9

39. The sum of a and 3, divided by the difference of a and 3

40. The sum of m and n, divided by the difference of m and n

Write each of the following phrases, using symbols. Use the variable x to represent the number in each case.

41. 5 more than a number

42. A number increased by 8

43. 7 less than a number

44. A number decreased by 10

45. 9 times a number

46. Twice a number

47. 6 more than 3 times a number

48. 5 times a number, decreased by 10

49. Twice the sum of a number and 5

50. 3 times the difference of a number and 4

51. The product of 2 more than a number and 2 less than that same number

52. The product of 5 less than a number and 5 more than that same number

53. The quotient of a number and 7

54. A number divided by 3

55. The sum of a number and 5, divided by 8

56. The quotient when 7 less than a number is divided by 3

57. 6 more than a number divided by 6 less than that same number

58. The quotient when 3 more than a number is divided by 3 less than that same number

Write each of the following geometric expressions using the given symbols.

59. Four times the length of a side (s)

60. $\dfrac{4}{3}$ times π times the cube of the radius (r)

61. The radius (r) squared times the height (h) times π

62. Twice the length (L) plus twice the width (W)

63. One-half the product of the height (h) and the sum of two unequal sides (b_1 and b_2)

64. Six times the length of a side (s) squared

Identify which are expressions and which are not.

65. $2(x + 5)$

66. $4 - (x + 3)$

67. $m \div + 4$

68. $6 + a = 7$

69. $y(x + 3)$

70. $8 = 4b$

71. $2a + 5b$

72. $4x + \cdot 7$

73. Social science. The Earth's population has doubled in the last 40 years. If we let x represent the Earth's population 40 years ago, what is the population today?

74. Science and medicine. It is estimated that the Earth is losing 4000 species of plants and animals every year. If S represents the number of species living last year, how many species are on Earth this year?

75. Business and finance. The simple interest (I) earned when a principal (P) is invested at a rate (r) for a time (t) is calculated by multiplying the principal times the rate times the time. Write an expression for the interest earned.

ANSWERS

57. _____
58. _____
59. _____
60. _____
61. _____
62. _____
63. _____
64. _____
65. _____
66. _____
67. _____
68. _____
69. _____
70. _____
71. _____
72. _____
73. _____
74. _____
75. _____

76. _____

77. _____

a. _____

b. _____

c. _____

d. _____

e. _____

f. _____

76. **Science and medicine.** The kinetic energy of a particle of mass m is found by taking one-half of the product of the mass (m) and the square of the velocity (v). Write an expression for the kinetic energy of a particle.

77. Rewrite the following algebraic expressions in English phrases. Exchange papers with another student to edit your writing. Be sure the meaning in English is the same as in algebra. These expressions are not complete sentences, so your English does not have to be in complete sentences. An example follows.

Algebra: $2(x - 1)$

English: We could write "One less than a number is doubled." Or we might write "A number is diminished by one and then multiplied by two."

(a) $n + 3$ (b) $\dfrac{x + 2}{5}$ (c) $3(5 + a)$ (d) $3 - 4n$ (e) $\dfrac{x + 6}{x - 1}$

 *Getting Ready for Section 1.2 [Section 0.4]

Evaluate the following:

(a) $8 - (5 + 2)$ (b) $(8 - 5) + 2$ (c) $16 \div 2 \cdot 4$

(d) $16 \div (2 \cdot 4)$ (e) $6 \cdot 2$ (f) $2 \cdot 6$

Answers

1. $c + d$ **3.** $w + z$ **5.** $x + 5$ **7.** $y + 10$ **9.** $b - a$

11. $b - 24$ **13.** $r - 6$ **15.** wz **17.** $5t$ **19.** $8mn$

21. $3(p + q)$ **23.** $2(x + y)$ **25.** $2x + y$ **27.** $2(x - y)$

29. $(a + b)(a - b)$ **31.** $m(m + 3)$ **33.** $\dfrac{x}{5}$ **35.** $\dfrac{a - b}{9}$ **37.** $\dfrac{p + q}{4}$

39. $\dfrac{a + 3}{a - 3}$ **41.** $x + 5$ **43.** $x - 7$ **45.** $9x$ **47.** $3x + 6$

49. $2(x + 5)$ **51.** $(x + 2)(x - 2)$ **53.** $\dfrac{x}{7}$ **55.** $\dfrac{x + 5}{8}$ **57.** $\dfrac{x + 6}{x - 6}$

59. $4s$ **61.** $\pi r^2 h$ **63.** $\dfrac{1}{2}h(b_1 + b_2)$ **65.** Expression

67. Not an expression   **69.** Expression   **71.** Expression

73. $2x$ **75.** Prt **77.** **a.** 1   **b.** 5   **c.** 32

d. 2   **e.** 12   **f.** 12

* Exercises headed "Getting Ready for . . ." are designed to help you prepare for material in the next section of the text. If you have any difficulty with these exercises, please review the section referred to in brackets.

1.2 Properties of Real Numbers

1.2 OBJECTIVES

1. Recognize applications of the commutative properties
2. Recognize applications of the associative properties
3. Recognize applications of the distributive property

NOTE All integers, decimals, and fractions that we see in this course are real numbers.

Everything that we do in algebra is based on the rules for the operations introduced in Section 1.1. We call these rules **properties of the real numbers.** In this section we consider those properties that we will use in the remainder of this chapter.

The **commutative properties** tell us that we can add or multiply in any order.

Rules and Properties: The Commutative Properties

If a and b are any numbers,

1. $a + b = b + a$ Commutative property of addition
2. $a \cdot b = b \cdot a$ Commutative property of multiplication

---

OBJECTIVE 1

Example 1   Identifying the Commutative Properties

(a) $5 + 9 = 9 + 5$ and $x + 7 = 7 + x$

These are applications of the commutative property of addition.

(b) $5 \cdot 9 = 9 \cdot 5$

This is an application of the commutative property of multiplication.

CHECK YOURSELF 1 _____

Identify the property being applied.

(a) $7 + 3 = 3 + 7$ (b) $7 \cdot 3 = 3 \cdot 7$
(c) $a + 4 = 4 + a$ (d) $x \cdot 2 = 2 \cdot x$

We also want to be able to change the grouping when simplifying expressions. Regrouping is possible because of the **associative properties.** Numbers or variables can be grouped in any manner to find a sum or a product.

Rules and Properties: The Associative Properties

If a, b, and c are any numbers,

1. $a + (b + c) = (a + b) + c$ Associative property of addition
2. $a \cdot (b \cdot c) = (a \cdot b) \cdot c$ Associative property of multiplication

---

OBJECTIVE 2

| Example 2   **Demonstrating the Associative Properties** |

(a) Show that $2 + (3 + 8) = (2 + 3) + 8$.

$$2 + \underbrace{(3 + 8)} \qquad\qquad \underbrace{(2 + 3)} + 8$$
         Add first.   Add first.

RECALL As we saw in Section 0.4, we always do the operation in the parentheses first.

$$= 2 + 11 \qquad\qquad\qquad = 5 + 8$$
$$= 13 \qquad\qquad\qquad\qquad = 13$$

So

$$2 + (3 + 8) = (2 + 3) + 8$$

(b) Show that $\dfrac{1}{3} \cdot (6 \cdot 5) = \left(\dfrac{1}{3} \cdot 6\right) \cdot 5$.

$$\dfrac{1}{3} \cdot \underbrace{(6 \cdot 5)} \qquad\qquad \underbrace{\left(\dfrac{1}{3} \cdot 6\right)} \cdot 5$$
   Multiply first.   Multiply first.

$$= \dfrac{1}{3} \cdot (30) \qquad\qquad = (2) \cdot 5$$
$$\qquad\qquad\qquad\qquad\qquad = 10$$
$$= 10$$

So

$$\dfrac{1}{3} \cdot (6 \cdot 5) = \left(\dfrac{1}{3} \cdot 6\right) \cdot 5$$

 CHECK YOURSELF 2 _____

Show that the following statements are true.

(a) $3 + (4 + 7) = (3 + 4) + 7$

(b) $3 \cdot (4 \cdot 7) = (3 \cdot 4) \cdot 7$

(c) $\left(\dfrac{1}{5} \cdot 10\right) \cdot 4 = \dfrac{1}{5} \cdot (10 \cdot 4)$

The **distributive property** involves addition and multiplication together. We can illustrate this property with an application.

Suppose that we want to find the total of the two areas shown in the following figure.

RECALL The area of a rectangle is the product of its length and width:
$A = L \cdot W$

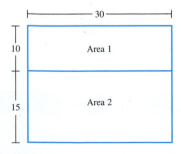

We can find the total area by multiplying the length by the overall width, which is found by adding the two widths.   [or]   We can find the total area as a sum of the two areas.

	(Area 1)	(Area 2)
Length Overall Width	Length · Width	Length · Width

$$30 \cdot (10 + 15)$$ $$30 \cdot 10 \quad + \quad 30 \cdot 15$$

$$= 30 \cdot 25$$ $$= 300 + 450$$

$$= 750$$ $$= 750$$

So

$$30 \cdot (10 + 15) = 30 \cdot 10 + 30 \cdot 15$$

This leads us to the following property.

Rules and Properties: The Distributive Property

NOTE Notice the pattern.

$a(b + c) = a \cdot b + a \cdot c$

We "distributed" the multiplication "over" the addition.

If *a*, *b*, and *c* are any numbers,

$$a(b + c) = a \cdot b + a \cdot c \quad \text{and} \quad (b + c)a = b \cdot a + c \cdot a$$

OBJECTIVE 3   **Example 3   Using the Distributive Property**

Use the distributive property to remove the parentheses in the following.

NOTE $5(3 + 4) = 5 \cdot 7 = 35$

or

$5 \cdot 3 + 5 \cdot 4 = 15 + 20 = 35$

(a) $5(3 + 4)$

$$5(3 + 4) = 5 \cdot 3 + 5 \cdot 4$$
$$= 15 + 20 = 35$$

NOTE Because the variables are different, $8x + 8y$ cannot be simplified further.

(b) $8(x + y)$

$$8(x + y) = 8x + 8y$$

(c) $2(3x + 5)$

$$2(3x + 5) = 2 \cdot 3x + 2 \cdot 5$$
$$= 6x + 10$$

NOTE It is also true that

$\frac{1}{3}(9 + 12) = \frac{1}{3}(21) = 7$

(d) $\frac{1}{3}(9 + 12) = \frac{1}{3} \cdot 9 + \frac{1}{3} \cdot 12$

$$= 3 + 4 = 7$$

CHECK YOURSELF 3

Use the distributive property to remove the parentheses.

(a) $4(6 + 7)$ **(b)** $9(m + n)$

(c) $3(5a + 7)$ **(d)** $\frac{1}{5}(10 + 15)$

Example 4 requires that you identify which property is being demonstrated. Look for patterns that will help you remember each of the properties.

Example 4 **Identifying Properties**

Name the property demonstrated.

(a) $3(x + 2) = 3x + 3 \cdot 2$ demonstrates the distributive property.

(b) $2 + (3 + 5) = (2 + 3) + 5$ demonstrates the associative property of addition.

(c) $3 \cdot 5 = 5 \cdot 3$ demonstrates the commutative property of multiplication.

 CHECK YOURSELF 4

Name the property demonstrated.

(a) $2 \cdot (3 \cdot 5) = (2 \cdot 3) \cdot 5$

(b) $4(a + b) = 4a + 4b$

(c) $x + 8 = 8 + x$

CHECK YOURSELF ANSWERS

1. (a) Commutative property of addition; (b) commutative property of multiplication; (c) commutative property of addition; (d) commutative property of multiplication

2. (a) $3 + (4 + 7) = 3 + 11 = 14$ (b) $3 \cdot (4 \cdot 7) = 3 \cdot 28 = 84$
$(3 + 4) + 7 = 7 + 7 = 14$ $(3 \cdot 4) \cdot 7 = 12 \cdot 7 = 84$

   (c) $\left(\dfrac{1}{5} \cdot 10\right) \cdot 4 = 2 \cdot 4 = 8$

   $\dfrac{1}{5} \cdot (10 \cdot 4) = \dfrac{1}{5} \cdot 40 = 8$

3. (a) $4 \cdot 6 + 4 \cdot 7 = 24 + 28 = 52$; (b) $9m + 9n$; (c) $15a + 21$;

   (d) $\dfrac{1}{5} \cdot 10 + \dfrac{1}{5} \cdot 15 = 2 + 3 = 5$

4. (a) Associative property of multiplication; (b) distributive property; (c) commutative property of addition

Identify the property that is illustrated by each of the following statements.

1. $5 + 9 = 9 + 5$

2. $6 + 3 = 3 + 6$

Ca

3. $2 \cdot (3 \cdot 5) = (2 \cdot 3) \cdot 5$

4. $3 \cdot (5 \cdot 6) = (3 \cdot 5) \cdot 6$

Am

5. $\dfrac{1}{4} \cdot \dfrac{1}{5} = \dfrac{1}{5} \cdot \dfrac{1}{4}$

6. $7 \cdot 9 = 9 \cdot 7$

Cm

7. $8 + 12 = 12 + 8$

8. $6 + 2 = 2 + 6$

Ca

9. $(5 \cdot 7) \cdot 2 = 5 \cdot (7 \cdot 2)$

10. $(8 \cdot 9) \cdot 2 = 8 \cdot (9 \cdot 2)$

Am

11. $7 \cdot (2 \cdot 5) = (7 \cdot 2) \cdot 5$

12. $\dfrac{1}{2} \cdot 6 = 6 \cdot \dfrac{1}{2}$

Cm

13. $2(3 + 5) = 2 \cdot 3 + 2 \cdot 5$

14. $5 \cdot (4 + 6) = 5 \cdot 4 + 5 \cdot 6$

15. $5 + (7 + 8) = (5 + 7) + 8$

16. $8 + (2 + 9) = (8 + 2) + 9$

17. $\left(\dfrac{1}{3} + 4\right) + \dfrac{1}{5} = \dfrac{1}{3} + \left(4 + \dfrac{1}{5}\right)$

18. $(5 + 5) + 3 = 5 + (5 + 3)$

Aa

19. $7 \cdot (3 + 8) = 7 \cdot 3 + 7 \cdot 8$

20. $5 \cdot (6 + 8) = 5 \cdot 6 + 5 \cdot 8$

Verify that each of the following statements is true by evaluating each side of the equation separately and comparing the results.

21. $7 \cdot (3 + 4) = 7 \cdot 3 + 7 \cdot 4$

22. $4 \cdot (5 + 1) = 4 \cdot 5 + 4 \cdot 1$

23. $2 + (9 + 8) = (2 + 9) + 8$

24. $6 + (15 + 3) = (6 + 15) + 3$

25. $\dfrac{1}{3} \cdot (6 \cdot 3) = \left(\dfrac{1}{3} \cdot 6\right) \cdot 3$

26. $2 \cdot (9 \cdot 10) = (2 \cdot 9) \cdot 10$

ANSWERS

1. _____
2. _____
3. _____
4. _____
5. _____
6. _____
7. _____
8. _____
9. _____
10. _____
11. _____
12. _____
13. _____
14. _____
15. _____
16. _____
17. _____
18. _____
19. _____
20. _____
21. _____
22. _____
23. _____
24. _____
25. _____
26. _____

27. _____

28. _____

29. _____

30. _____

31. _____

32. _____

33. _____

34. _____

35. _____

36. _____

37. _____

38. _____

39. _____

40. _____

41. _____

42. _____

43. _____

44. _____

45. _____

46. _____

47. _____

48. _____

49. _____

50. _____

51. _____

52. _____

53. _____

54. _____

27. $5 \cdot (2 + 8) = 5 \cdot 2 + 5 \cdot 8$

28. $\frac{1}{4} \cdot (10 + 2) = \frac{1}{4} \cdot 10 + \frac{1}{4} \cdot 2$

29. $(3 + 12) + 8 = 3 + (12 + 8)$

30. $(8 + 12) + 7 = 8 + (12 + 7)$

31. $(4 \cdot 7) \cdot 2 = 4 \cdot (7 \cdot 2)$

32. $(6 \cdot 5) \cdot 3 = 6 \cdot (5 \cdot 3)$

33. $\frac{1}{2} \cdot (2 + 6) = \frac{1}{2} \cdot 2 + \frac{1}{2} \cdot 6$

34. $\frac{1}{3} \cdot (6 + 9) = \frac{1}{3} \cdot 6 + \frac{1}{3} \cdot 9$

35. $\left(\frac{2}{3} + \frac{1}{6}\right) + \frac{1}{3} = \frac{2}{3} + \left(\frac{1}{6} + \frac{1}{3}\right)$

36 $\frac{3}{4} + \left(\frac{5}{8} + \frac{1}{2}\right) = \left(\frac{3}{4} + \frac{5}{8}\right) + \frac{1}{2}$

37. $(2.3 + 3.9) + 4.1 = 2.3 + (3.9 + 4.1)$

38. $(1.7 + 4.1) + 7.6 = 1.7 + (4.1 + 7.6)$

39. $\frac{1}{2} \cdot (2 \cdot 8) = \left(\frac{1}{2} \cdot 2\right) \cdot 8$

40. $\frac{1}{5} \cdot (5 \cdot 3) = \left(\frac{1}{5} \cdot 5\right) \cdot 3$

41. $\left(\frac{3}{5} \cdot \frac{5}{6}\right) \cdot \frac{4}{3} = \frac{3}{5} \cdot \left(\frac{5}{6} \cdot \frac{4}{3}\right)$

42. $\frac{4}{7} \cdot \left(\frac{21}{16} \cdot \frac{8}{3}\right) = \left(\frac{4}{7} \cdot \frac{21}{16}\right) \cdot \frac{8}{3}$

43. $2.5 \cdot (4 \cdot 5) = (2.5 \cdot 4) \cdot 5$

44. $4.2 \cdot (5 \cdot 2) = (4.2 \cdot 5) \cdot 2$

Use the distributive property to remove the parentheses in each of the following expressions. Then simplify your result where possible.

45. $3(2 + 6)$

46. $5(4 + 6)$

47. $3(x + 5)$

48. $5(y + 8)$

49. $4(w + v)$

50. $9(d + c)$

51. $3(2x + 7)$

52. $3(7a + 4)$

53. $\frac{1}{3} \cdot (15 + 9)$

54. $\frac{1}{6} \cdot (36 + 24)$

Use the properties of addition and multiplication to complete each of the following statements.

55. $5 + 7 = \quad + 5$

56. $(5 + 3) + 4 = 5 + (\quad + 4)$

57. $(8)(3) = (3)(\quad)$

58. $8(3 + 4) = 8 \cdot 3 + \quad \cdot 4$

59. $7(2 + 5) = 7 \cdot \quad + 7 \cdot 5$

60. $4 \cdot (2 \cdot 4) = (\quad \cdot 2) \cdot 4$

Use the indicated property to write an expression that is equivalent to each of the following expressions.

61. $3 + 7$ (commutative property of addition)

62. $2(3 + 4)$ (distributive property)

63. $5 \cdot (3 \cdot 2)$ (associative property of multiplication)

64. $(3 + 5) + 2$ (associative property of addition)

65. $2 \cdot 4 + 2 \cdot 5$ (distributive property)

66. $7 \cdot 9$ (commutative property of multiplication)

Evaluate each of the following pairs of expressions. Then answer the given question.

67. $8 - 5$ and $5 - 8$
Do you think subtraction is commutative?

68. $12 \div 3$ and $3 \div 12$
Do you think division is commutative?

69. $(12 - 8) - 4$ and $12 - (8 - 4)$
Do you think subtraction is associative?

70. $(48 \div 16) \div 4$ and $48 \div (16 \div 4)$
Do you think division is associative?

71. $3(6 - 2)$ and $3 \cdot 6 - 3 \cdot 2$
Do you think multiplication is distributive over subtraction?

ANSWERS

55. ____
56. ____
57. ____
58. ____
59. ____
60. ____
61. ____
62. ____
63. ____
64. ____
65. ____
66. ____
67. ____
68. ____
69. ____
70. ____
71. ____

72. _____

73. _____

74. _____

75. _____

76. _____

77. _____

78. _____

a. _____

b. _____

c. _____

d. _____

e. _____

f. _____

72. $\dfrac{1}{2}(16 - 10)$ and $\dfrac{1}{2} \cdot 16 - \dfrac{1}{2} \cdot 10$

Do you think multiplication is distributive over subtraction?

Complete the statement using the

(a) Distributive property

(b) Commutative property of addition

(c) Commutative property of multiplication

73. $5 \cdot (3 + 4) =$ **74.** $6 \cdot (5 + 4) =$

Identify the property that is used.

75. $5 + (6 + 7) = (5 + 6) + 7$ **76.** $5 + (6 + 7) = 5 + (7 + 6)$

77. $4 \cdot (3 + 2) = 4 \cdot (2 + 3)$ **78.** $4 \cdot (3 + 2) = (3 + 2) \cdot 4$

 Getting Ready for Section 1.3 [Section 0.2]

Find each sum or difference.

(a) $3 + (8 + 9)$ (b) $6 + (12 + 3)$

(c) $(3 + 8) + (9 + 4)$ (d) $15 - 11 - (2 + 1)$

(e) $\dfrac{3}{5} + \dfrac{4}{15}$ (f) $\dfrac{12}{27} - \dfrac{2}{9}$

Answers

1. Commutative property of addition   **3.** Associative property of multiplication
5. Commutative property of multiplication   **7.** Commutative property of
addition   **9.** Associative property of multiplication
11. Associative property of multiplication
13. Distributive property   **15.** Associative property of addition
17. Associative property of addition   **19.** Distributive property
21. $49 = 49$ **23.** $19 = 19$ **25.** $6 = 6$ **27.** $50 = 50$
29. $23 = 23$ **31.** $56 = 56$ **33.** $4 = 4$ **35.** $\dfrac{7}{6} = \dfrac{7}{6}$
37. $10.3 = 10.3$ **39.** $8 = 8$ **41.** $\dfrac{2}{3} = \dfrac{2}{3}$
43. $50 = 50$ **45.** 24 **47.** $3x + 15$ **49.** $4w + 4v$ **51.** $6x + 21$
53. 8 **55.** 7 **57.** 8 **59.** 2 **61.** $7 + 3$ **63.** $(5 \cdot 3) \cdot 2$
65. $2 \cdot (4 + 5)$ **67.** No   **69.** No   **71.** Yes   **73. (a)** $5 \cdot 3 + 5 \cdot 4$
(b) $5 \cdot (4 + 3)$ **(c)** $(3 + 4) \cdot 5$ **75.** Associative property of addition
77. Commutative property of addition   **a.** 20 **b.** 21 **c.** 24 **d.** 1
e. $\dfrac{13}{15}$ **f.** $\dfrac{2}{9}$

1.3 Adding and Subtracting Real Numbers

1.3 OBJECTIVES

1. Use a number line to find the sum of two real numbers
2. Find the median of a set of real numbers
3. Find the difference of two real numbers
4. Find the range of a set of real numbers

In Section 0.5 we introduced the idea of numbers with different signs. Now we will examine the four arithmetic operations (addition, subtraction, multiplication, and division) to see how those operations are performed when these numbers are involved. We start by considering addition.

An application may help. As before, we represent a gain of money as a positive number and a loss as a negative number.

If you gain $3 and then gain $4, the result is a gain of $7:

$$3 + 4 = 7$$

If you lose $3 and then lose $4, the result is a loss of $7:

$$-3 + (-4) = -7$$

If you gain $3 and then lose $4, the result is a loss of $1:

$$3 + (-4) = -1$$

If you lose $3 and then gain $4, the result is a gain of $1:

$$-3 + 4 = 1$$

The number line can be used to illustrate the addition of these numbers. Starting at the origin, we move to the *right* for positive numbers and to the *left* for negative numbers.

OBJECTIVE 1 **Example 1 Adding Two Negative Numbers**

(a) Add $-3 + (-4)$.

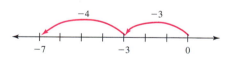

Start at the origin and move 3 units to the left. Then move 4 more units to the left to find the sum. From the number line we see that the sum is

$$-3 + (-4) = -7$$

(b) Add $-\dfrac{3}{2} + \left(-\dfrac{1}{2}\right)$.

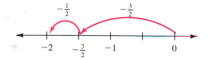

As before, we start at the origin. From that point move $\frac{3}{2}$ units left. Then move another $\frac{1}{2}$ unit left to find the sum. In this case

$$-\frac{3}{2} + \left(-\frac{1}{2}\right) = -2$$

CHECK YOURSELF 1

Add.

(a) $-4 + (-5)$

(b) $-3 + (-7)$

(c) $-5 + (-15)$

(d) $-\frac{5}{2} + \left(-\frac{3}{2}\right)$

You have probably noticed some helpful patterns in the previous examples. These patterns will allow you to do the work mentally without having to use the number line. Look at the following rule.

Rules and Properties: Adding Real Numbers Case 1: Same Sign

If two numbers have the same sign, add their absolute values. Give the sum the sign of the original numbers.

NOTE This means that the sum of two positive numbers is positive and the sum of two negative numbers is negative. We first encountered absolute values in Section 0.5.

Let's again use the number line to illustrate the addition of two numbers. This time the numbers will have *different* signs.

Example 2   Adding Two Numbers with Different Signs

(a) Add $3 + (-6)$.

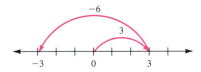

First move 3 units to the right of the origin. Then move 6 units to the left.

$3 + (-6) = -3$

(b) Add $-4 + 7$.

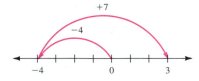

This time move 4 units to the left of the origin as the first step. Then move 7 units to the right.

$-4 + 7 = 3$

CHECK YOURSELF 2

Add.

(a) $7 + (-5)$ **(b)** $4 + (-8)$ **(c)** $-\dfrac{1}{3} + \dfrac{16}{3}$ **(d)** $-7 + 3$

You have no doubt noticed that, in adding a positive number and a negative number, sometimes the sum is positive and sometimes it is negative. This depends on which of the numbers has the larger absolute value. This leads us to the second part of our addition rule.

RECALL We first encountered absolute values in Section 0.5.

> **Rules and Properties:** Adding Real Numbers Case 2: Different Signs
>
> If two numbers have different signs, subtract their absolute values, the smaller from the larger. Give the result the sign of the number with the larger absolute value.

Example 3   Adding Positive and Negative Numbers

(a) $7 + (-19) = -12$

Because the two numbers have different signs, subtract the absolute values $(19 - 7 = 12)$. The sum has the sign $(-)$ of the number with the larger absolute value.

(b) $-\dfrac{13}{2} + \dfrac{7}{2} = -3$

Subtract the absolute values $\left(\dfrac{13}{2} - \dfrac{7}{2} = \dfrac{6}{2} = 3\right)$. The sum has the sign $(-)$ of the number with the larger absolute value, $-\dfrac{13}{2}$.

(c) $-8.2 + 4.5 = -3.7$

Subtract the absolute values $(8.2 - 4.5 = 3.7)$. The sum has the sign $(-)$ of the number with the larger absolute value, -8.2.

CHECK YOURSELF 3

Add mentally.

(a) $5 + (-14)$ **(b)** $-7 + (-8)$ **(c)** $-8 + 15$

(d) $7 + (-8)$ **(e)** $-\dfrac{2}{3} + \left(-\dfrac{7}{3}\right)$ **(f)** $5.3 + (-2.3)$

In Section 1.2 we discussed the commutative, associative, and distributive properties. There are two other properties of addition that we should mention. First, the sum of any number and 0 is always that number. In symbols,

NOTE No number loses its identity after addition with 0. Zero is called the **additive identity.**

> ## Rules and Properties: Additive Identity Property
>
> For any number a,
>
> $$a + 0 = 0 + a = a$$

Example 4   Adding the Identity

Add.

(a) $9 + 0 = 9$

(b) $0 + \left(-\dfrac{5}{4}\right) = -\dfrac{5}{4}$

(c) $(-25) + 0 = -25$

 CHECK YOURSELF 4

Add.

(a) $8 + 0$ **(b)** $0 + \left(-\dfrac{8}{3}\right)$ **(c)** $(-36) + 0$

NOTE The opposite of a number is also called the **additive inverse** of that number.

NOTE 3 and -3 are opposites.

Recall that every number has an *opposite*. It corresponds to a point the same distance from the origin as the given number, but in the opposite direction.

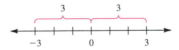

The opposite of 9 is -9.
The opposite of -15 is 15.

Our second property states that the sum of any number and its opposite is 0.

NOTE Here $-a$ represents the opposite of the number a. The sum of any number and its opposite, or additive inverse, is 0.

> ## Rules and Properties: Additive Inverse Property
>
> For any number a, there exists a number $-a$ such that
>
> $$a + (-a) = (-a) + a = 0$$

Example 5   Adding Inverses

(a) $9 + (-9) = 0$

(b) $-15 + 15 = 0$

(c) $(-2.3) + 2.3 = 0$

(d) $\dfrac{4}{5} + \left(-\dfrac{4}{5}\right) = 0$

CHECK YOURSELF 5 _____

Add.

(a) $(-17) + 17$ **(b)** $12 + (-12)$

(c) $\dfrac{1}{3} + \left(-\dfrac{1}{3}\right)$ **(d)** $-1.6 + 1.6$

In Section 0.5 we saw that the least and greatest elements of a set were called the minimum and maximum. The middle value of an ordered set is called the **median.** The median is sometimes used to represent an *average* of the set of numbers.

OBJECTIVE 2   **Example 6   Finding the Median**

Find the median for each set of numbers.

(a) $9, -5, -8, 3, 7$

First, rewrite the set in ascending order.

$-8, -5, 3, 7, 9$

The median is then the element that has just as many numbers to its right as it has to its left. In this set, 3 is the median, because there are two numbers that are larger (7 and 9) and two numbers that are smaller (-8 and -5).

(b) $3, -2, 18, -20, -13$

First, rewrite the set in ascending order.

$-20, -13, -2, 3, 18$

The median is then the element that is exactly in the middle. The median for this set is -2.

CHECK YOURSELF 6 _____

Find the median for each set of numbers.

(a) $-3, 2, 7, -6, -1$ **(b)** $5, 1, -10, 2, -20$

In Example 6, each set had an odd number of elements. If we had an even number of elements, there would be no single middle number.

To find the median from a set with an even number of elements, add the two middle numbers and divide their sum by 2.

Example 7   Finding the Median

Find the median for each set of numbers.

(a) $-3, 3, -8, 4, -1, -7, 5, 9$

First, rewrite the set in ascending order.

$-8, -7, -3, -1, 3, 4, 5, 9$

Add the middle two numbers (-1 and 3), and then divide their sum by 2.

$$\frac{-1 + (3)}{2} = \frac{2}{2} = 1$$

The median is 1.

(b) $8, 3, -2, 4, -5, -7$

Rewrite the set in ascending order.

$$-7, -5, -2, 3, 4, 8$$

The median is one-half the sum of the middle two numbers.

$$\frac{-2 + 3}{2} = \frac{1}{2} = 0.5$$

CHECK YOURSELF 7

Find the median for each set of numbers.

(a) $2, -5, 15, 8, 3, -4$ **(b)** $8, 3, 6, -8, 9, -7$

To begin our discussion of subtraction when signed numbers are involved, we can look back at a problem using natural numbers. Of course, we know that

$$8 - 5 = 3$$

From our work in adding real numbers, we know that it is also true that

$$8 + (-5) = 3$$

Comparing these equations, we see that the results are the same. This leads us to an important pattern. Any subtraction problem can be written as a problem in addition. Subtracting 5 is the same as adding the opposite of 5, or -5. We can write this fact as follows:

$$8 - 5 = 8 + (-5) = 3$$

This leads us to the following rule for subtracting real numbers.

Rules and Properties: Subtracting Real Numbers

1. Rewrite the subtraction problem as an addition problem.
   a. Change the minus sign to a plus sign.
   b. Replace the number being subtracted with its opposite.
2. Add the resulting signed numbers as before.
   In symbols,

$$a - b = a + (-b)$$

NOTE This is the *definition* of subtraction.

Example 8 illustrates the use of this definition while subtracting.

OBJECTIVE 3   **Example 8   Subtracting Real Numbers**

Simplify each expression.

Change the subtraction symbol ($-$)
to an addition symbol ($+$).

(a) $15 - 7 = 15 + (-7)$

Replace 7 with its opposite, -7.

$= 8$

(b) $9 - 12 = 9 + (-12) = -3$

(c) $-6 - 7 = -6 + (-7) = -13$

(d) $-\dfrac{3}{5} - \dfrac{7}{5} = -\dfrac{3}{5} + \left(-\dfrac{7}{5}\right) = -\dfrac{10}{5} = -2$

(e) $2.1 - 3.4 = 2.1 + (-3.4) = -1.3$

(f) Subtract 5 from -2. We write the statement as $-2 - 5$ and proceed as before:

$-2 - 5 = -2 + (-5) = -7$

CHECK YOURSELF 8

Subtract.

(a) $18 - 7$

(b) $5 - 13$

(c) $-7 - 9$

(d) $-\dfrac{5}{6} - \dfrac{7}{6}$

(e) $-2 - 7$

(f) $5.6 - 7.8$

The subtraction rule is used in the same way when the number being subtracted is negative. Change the subtraction to addition. Replace the negative number being subtracted with its opposite, which is positive. Example 9 illustrates this principle.

Example 9   Subtracting Real Numbers

Simplify each expression.

Change the subtraction
to an addition.

(a) $5 - (-2) = 5 + (+2) = 5 + 2 = 7$

Replace -2 with its opposite, $+2$ or 2.

(b) $7 - (-8) = 7 + (+8) = 7 + 8 = 15$

(c) $-9 - (-5) = -9 + 5 = -4$

(d) $-12.7 - (-3.7) = -12.7 + 3.7 = -9$

(e) $-\dfrac{3}{4} - \left(-\dfrac{7}{4}\right) = -\dfrac{3}{4} + \left(+\dfrac{7}{4}\right) = \dfrac{4}{4} = 1$

(f) Subtract -4 from -5. We write

$-5 - (-4) = -5 + 4 = -1$

CHECK YOURSELF 9 _____

Subtract.

(a) $8 - (-2)$ **(b)** $3 - (-10)$ **(c)** $-7 - (-2)$
(d) $-9.8 - (-5.8)$ **(e)** $7 - (-7)$

Given a set of numbers, the **range** is the difference between the maximum and the minimum.

OBJECTIVE 4

Example 10   Finding the Range

Find the range for each set of numbers.

(a) $5, -2, -7, 9, 3$

Rewrite the set in ascending order. The maximum is 9. The minimum is -7. The range is the difference.

$$9 - (-7) = 9 + 7 = 16$$

The range is 16.

(b) $3, 8, -17, 12, -2$

Rewrite the set in ascending order. The maximum is 12. The minimum is -17. The range is $12 - (-17) = 29$.

CHECK YOURSELF 10 _____

Find the range for each set of numbers.

(a) $2, -4, 7, -3, -1$ **(b)** $-3, 4, -7, 5, 9, -4$

Your scientific calculator can be used to do arithmetic with real numbers. Before we look at an example, there are some keys on your calculator with which you should become familiar.

There are two similar keys you must find on the calculator. The first is used for subtraction ($\boxed{-}$) and is usually found in the right column of calculator keys. The second will "change the sign" of a number. It is usually a $\boxed{+/-}$ and is found on the bottom row.

NOTE Some graphing calculators have a negative sign $\boxed{(-)}$ that acts to change the sign of a number.

We will use these keys in Example 11.

OBJECTIVE 3

Example 11   Subtracting Real Numbers

Using your calculator, find the difference.

(a) $-12.43 - 3.516$

Enter the 12.43 and press the $\boxed{+/-}$ to make it negative. Then press $\boxed{-}$ 3.516 $\boxed{=}$. The result should be -15.946.

NOTE If you have a graphing calculator, the key sequence will be
$\boxed{(-)}$ 12.43 $\boxed{-}$ 3.516 $\boxed{ENTER}$

(b) $23.56 - (-4.7)$

The key sequence is

23.56 ⊟ 4.7 +/− =

The answer should be 28.26.

CHECK YOURSELF 11 _____

Use your calculator to find the difference.

(a) −13.46 − 5.71   **(b)** −3.575 − (−6.825)

Example 12   An Application of Adding and Subtracting Real Numbers

Oscar owned four stocks. This year his holdings in Cisco went up $2250, in AT&T it went down $1345, in Texaco it went down $5215, and in IBM it went down $1525. How much less are his holdings worth at the end of the year compared to at the beginning of the year?
    To find the change in Oscar's holdings we add the amounts that went up and subtract the amounts that went down.

$2250 − $1345 − $5215 − $1525 = −$5835

Oscar's holdings are worth $5835 less at the end of the year.

CHECK YOURSELF 12 _____

A bus with fifteen people stopped at Avenue A. Nine people got off and five people got on. At Avenue B six people got off and eight people got on. At Avenue C four people got off the bus and six people got on. How many people were now on the bus?

CHECK YOURSELF ANSWERS _____

1. **(a)** −9; **(b)** −10; **(c)** −20; **(d)** −4   **2.** **(a)** 2; **(b)** −4; **(c)** 5; **(d)** −4

3. **(a)** −9; **(b)** −15; **(c)** 7; **(d)** −1; **(e)** −3; **(f)** 3   **4.** **(a)** 8; **(b)** $-\dfrac{8}{3}$; **(c)** −36

5. **(a)** 0; **(b)** 0; **(c)** 0; **(d)** 0   **6.** **(a)** −1; **(b)** 1   **7.** **(a)** 2.5; **(b)** 4.5

8. **(a)** 11; **(b)** −8; **(c)** −16; **(d)** −2; **(e)** −9; **(f)** −2.2   **9.** **(a)** 10; **(b)** 13; **(c)** −5;
(d) −4; **(e)** 14   **10.** **(a)** 7 − (−4) = 11; **(b)** 9 − (−7) = 16
11. **(a)** −19.17; **(b)** 3.25   **12.** 15 people

1.3   Exercises

Add.

1. $3 + 6$

2. $5 + 9$

3. $11 + 5$

4. $8 + 7$

5. $\dfrac{4}{5} + \dfrac{6}{5}$

6. $\dfrac{7}{3} + \dfrac{8}{3}$

7. $\dfrac{1}{2} + \dfrac{4}{5}$

8. $\dfrac{2}{3} + \dfrac{5}{9}$

9. $(-4) + (-1)$

10. $(-1) + (-9)$

11. $\left(-\dfrac{3}{5}\right) + \left(-\dfrac{7}{5}\right)$

12. $\left(-\dfrac{3}{5}\right) + \dfrac{12}{5}$

13. $\left(-\dfrac{1}{2}\right) + \left(-\dfrac{3}{8}\right)$

14. $\left(-\dfrac{4}{7}\right) + \left(-\dfrac{3}{14}\right)$

15. $(-1.6) + (-2.3)$

16. $(-3.5) + (-2.6)$

17. $3 + (-9)$

18. $11 + (-7)$

19. $\dfrac{3}{4} + \left(-\dfrac{1}{2}\right)$ $\frac{1}{4}$

20. $\dfrac{2}{3} + \left(-\dfrac{1}{6}\right)$ $\frac{1}{2}$

21. $\left(-\dfrac{4}{5}\right) + \dfrac{9}{20}$ $\frac{7}{20}$

22. $\left(-\dfrac{11}{6}\right) + \dfrac{5}{12}$ $-1\frac{5}{12}$

23. $-11.4 + 13.4$ 2

24. $-5.2 + 9.2$ 4

25. $-3.6 + 7.6$ 4

26. $-2.6 + 4.9$ 2.3

27. $0 + 8$ 8

28. $-15 + 0$ -15

29. $7 + (-7)$ 0

30. $12 + (-12)$ 0

31. $-4.5 + 4.5$ *0*

32. $\left(-\dfrac{2}{3}\right) + \dfrac{2}{3}$ *0*

33. $9 + (-7) + 6 + (-5)$ *3*

34. $(-4) + 6 + (-3) + 0$ *-1*

35. $6 + 9 + (-7) + (-5)$ *3*

36. $-\dfrac{6}{5} + \left(-\dfrac{13}{5}\right) + \dfrac{4}{5}$ *-3*

37. $-\dfrac{3}{2} + \left(-\dfrac{7}{4}\right) + \dfrac{1}{4}$ *-3*

38. $\left(-\dfrac{1}{2}\right) + \dfrac{1}{3} + \left(-\dfrac{5}{6}\right)$ *-1*

39. $2.3 + (-5.4) + (-2.9)$ *-6*

40. $-5.4 + (-2.1) + (-3.5)$ *-11*

Subtract.

41. $21 - 13$

42. $36 - 22$

43. $82 - 45$

44. $103 - 56$

45. $\dfrac{8}{7} - \dfrac{15}{7}$

46. $\dfrac{17}{8} - \dfrac{9}{8}$

47. $7.9 - 5.4$

48. $11.7 - 4.5$

49. $18 - 20$

50. $14 - 19$

51. $24 - 45$

52. $136 - 352$

53. $\dfrac{7}{6} - \dfrac{19}{6}$

54. $-\dfrac{4}{11} - \dfrac{29}{11}$

55. $7.8 - 11.6$

56. $14.3 - 25.5$

57. $-3 - 5$

58. $-15 - 8$

59. $-9 - 14$

60. $-8 - 12$

61. $-\dfrac{2}{5} - \dfrac{7}{10}$

62. $-\dfrac{5}{9} - \dfrac{7}{18}$

ANSWERS

31. _____
32. _____
33. _____
34. _____
35. _____
36. _____
37. _____
38. _____
39. _____
40. _____
41. _____
42. _____
43. _____
44. _____
45. _____
46. _____
47. _____
48. _____
49. _____
50. _____
51. _____
52. _____
53. _____
54. _____
55. _____ 56. _____
57. _____ 58. _____
59. _____ 60. _____
61. _____ 62. _____

63.

64.

65.

66.

67.

68.

69.

70.

71.

72.

73.

74.

75.

76.

77.

78. ____ 79.

80. ____ 81.

82.

83. ____ 84.

85. ____ 86.

87. ____ 88.

89.

90.

91.

92.

93. ____ 94.

95. ____ 96.

63. $-3.4 - 4.7$

64. $-8.1 - 7.6$

65. $5 - (-11)$

66. $8 - (-4)$

67. $12 - (-7)$

68. $3 - (-10)$

69. $\dfrac{3}{4} - \left(-\dfrac{3}{2}\right)$

70. $\dfrac{5}{6} - \left(-\dfrac{7}{6}\right)$

71. $\dfrac{6}{7} - \left(-\dfrac{5}{14}\right)$

72. $\dfrac{11}{16} - \left(-\dfrac{7}{8}\right)$

73. $8.3 - (-5.7)$

74. $6.5 - (-4.3)$

75. $8.9 - (-11.7)$

76. $14.5 - (-24.6)$

77. $-8 - 4 - 1 - (-2) - (-5)$

78. $-28 - (-11)$

79. $-27 - (-19)$

80. $-11 - (-16)$

81. $\left(-\dfrac{3}{4}\right) - \left(-\dfrac{11}{4}\right)$

82. $-\dfrac{1}{2} - \left(-\dfrac{5}{8}\right)$

83. $-12.7 - (-5.7)$

84. $-5.6 - (-2.6)$

85. $-6.9 - (-10.1)$

86. $-3.4 - (-7.6)$

Use your calculator to evaluate each expression.

87. $-4.1967 - 5.2943$

88. $5.3297 - (-4.1897)$

89. $-4.1623 - (-3.1468)$

90. $(-3.6829) - 4.5687$

91. $-6.3267 + 8.6789$

92. $-6.6712 + 5.3245$

Find the median for each of the following sets.

93. 2, 4, 6, 7, 9

94. 1, 3, 5, 6, 8

95. 8, 7, 2, 25, 5, 13, 3

96. 53, 23, 34, 21, 32, 30, 32

Determine the range for each of the following sets.

97. 2, 7, 9, 15, 24

98. 4, 8, 11, 15, 27

99. −4, −3, 2, 7, 9

100. −7, −2, 1, 8, 11

101. $\frac{7}{8}$, 2, $-\frac{1}{2}$, −8, $\frac{3}{4}$

$-8\frac{1}{2}, 3/4, 7/8, 2, \quad 2$

102. 3, $\frac{5}{6}$, −7, $-\frac{1}{3}$, $\frac{2}{3}$

103. 3, 2, −5, 6, −3

104. 1, −9, 7, −2, 3

Solve the following problems.

105. Business and finance. Amir has $100 in his checking account. He writes a check for $23 and makes a deposit of $51. What is his new balance?

106. Business and finance. Olga has $250 in her checking account. She deposits $52 and then writes a check for $77. What is her new balance?

Bal: _____ 250
Dep: _____ 52
CK # 1111: _____ 77

107. Statistics. On four consecutive running plays, Duce Staley of the Philadelphia Eagles gained 23 yards, lost 5 yards, gained 15 yards, and lost 10 yards. What was his net yardage change for the series of plays?

108. Business and finance. Ramon owes $780 on his VISA account. He returns three items costing $43.10, $36.80, and $125.00 and receives credit on his account. Next, he makes a payment of $400. He then makes a purchase of $82.75. How much does Ramon still owe?

109. Science and medicine. The temperature at noon on a June day was 82°. It fell by 12° over the next 4 h. What was the temperature at 4:00 P.M.?

ANSWERS

97. _____

98. _____

99. _____

100. _____

101. _____

102. _____

103. _____

104. _____

105. _____

106. _____

107. _____

108. _____

109. _____

110.

111.

112.

113.

114.

115.

116.

110. Statistics. Chia is standing at a point 6000 ft above sea level. She descends to a point 725 ft lower. What is her distance above sea level?

111. Business and finance. Omar's checking account was overdrawn by $72. He wrote another check for $23.50. How much was his checking account overdrawn after writing the check?

112. Business and finance. Angelo owed his sister $15. He later borrowed another $10. What integer represents his current financial condition?

113. Statistics. A local community college had a decrease in enrollment of 750 students in the fall of 1999. In the spring of 2000, there was another decrease of 425 students. What was the total decrease in enrollment for both semesters?

114. Science and medicine. At 7 A.M., the temperature was −15°F. By 1 P.M., the temperature had increased by 18°F. What was the temperature at 1 P.M.?

115. Statistics. Ezra's scores on five tests taken in a mathematics class were 87, 71, 95, 81, and 90. What was the range of his scores?

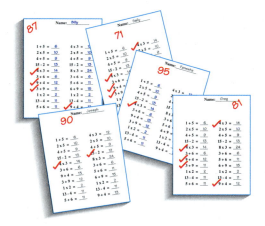

116. Statistics. The bar chart shown represents the league-leading regular season passing yardage in the NFL from 1998 to 2002. Use a real number to represent the change in total passing yardage from one year to the next.

(a) from 1998 to 1999   (b) from 1999 to 2000
(c) from 2000 to 2001   (d) from 2001 to 2002

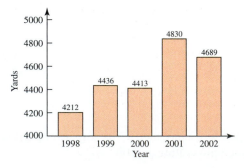

117. In this chapter, it is stated that "Every number has an opposite." The opposite of 9 is −9. This corresponds to the idea of an opposite in English. In English an opposite is often expressed by a prefix, for example, *un-* or *ir-*.

(a) Write the opposite of these words: unmentionable, uninteresting, irredeemable, irregular, uncomfortable.

(b) What is the meaning of these expressions: not uninteresting, not irredeemable, not irregular, not unmentionable?

(c) Think of other prefixes that *negate* or change the meaning of a word to its *opposite*. Make a list of words formed with these prefixes, and write a sentence with three of the words you found. Make a sentence with two words and phrases from each of the lists above. Look up the meaning of the word *irregardless*.

   What is the value of $-[-(-5)]$? What is the value of $-(-6)$? How does this relate to the above examples? Write a short description about this relationship.

118. The temperature on the plains of North Dakota can change rapidly, falling or rising many degrees in the course of an hour. Here are some temperature changes during each day over a week.

Day	Mon.	Tues.	Wed.	Thurs.	Fri.	Sat.	Sun.
Temp. Change from 10 A.M. to 3 P.M.	+13°	+20°	−18°	+10°	−25°	−5°	+15°

Write a short speech for the TV weather reporter that summarizes the daily temperature change. Use the median as you characterize the average daily midday change.

119. Science. The daily average temperatures in degrees Fahrenheit for a week in February were −1, 3, 5, −2, 4, 12, and 10. What was the range of temperatures for that week?

120. How long ago was the year 1250 B.C.E.? What year was 3300 years ago? Make a number line and locate the following events, cultures, and objects on it. How long ago was each item in the list? Which two events are the closest to each other? You may want to learn more about some of the cultures in the list and the mathematics and science developed by that culture.

Inca culture in Peru—1400 A.D.
The *Ahmes Papyrus,* a mathematical text from Egypt—1650 B.C.E.
Babylonian arithmetic develops the use of a zero symbol—300 B.C.E.
First Olympic Games—776 B.C.E.
Pythagoras of Greece dies—580 B.C.E.
Mayans in Central America independently develop use of zero—500 A.D.
The *Chou Pei,* a mathematics classic from China—1000 B.C.E.

121. _____

122. _____

a. _____

b. _____

c. _____

d. _____

e. _____

f. _____

The *Aryabhatiya,* a mathematics work from India—499 A.D.
Trigonometry arrives in Europe via the Arabs and India—1464 A.D.
Arabs receive algebra from Greek, Hindu, and Babylonian sources and
    develop it into a new systematic form—850 A.D.
Development of calculus in Europe—1670 A.D.
Rise of abstract algebra—1860 A.D.
Growing importance of probability and development of statistics—1902 A.D.

121. Complete the following statement: "$3 - (-7)$ is the same as _____ because . . ."
Write a problem that might be answered by doing this subtraction.

122. Explain the difference between the two phrases: "a number subtracted from 5"
and "a number less than 5." Use algebra and English to explain the meaning of
these phrases. Write other ways to express subtraction in English. Which ones are
confusing?

Getting Ready for Section 1.4 [Section 0.4]

Add.

(a) $(-1) + (-1) + (-1) + (-1)$

(b) $3 + 3 + 3 + 3 + 3$

(c) $9 + 9 + 9$

(d) $(-10) + (-10) + (-10)$

(e) $(-5) + (-5) + (-5) + (-5) + (-5)$

(f) $(-8) + (-8) + (-8) + (-8)$

Answers

1. 9   **3.** 16   **5.** 2   **7.** $\dfrac{13}{10}$ **9.** -5 **11.** -2 **13.** $-\dfrac{7}{8}$

15. -3.9 **17.** -6 **19.** $\dfrac{1}{4}$ **21.** $-\dfrac{7}{20}$ **23.** 2   **25.** 4   **27.** 8

29. 0   **31.** 0   **33.** 3   **35.** 3   **37.** -3 **39.** -6 **41.** 8
43. 37   **45.** -1 **47.** 2.5   **49.** -2 **51.** -21 **53.** -2

55. -3.8 **57.** -8 **59.** -23 **61.** $-\dfrac{11}{10}$ **63.** -8.1 **65.** 16

67. 19   **69.** $\dfrac{9}{4}$ **71.** $\dfrac{17}{14}$ **73.** 14   **75.** 20.6   **77.** -6 **79.** -8

81. 2   **83.** -7 **85.** 3.2   **87.** -9.491 **89.** -1.0155 **91.** 2.3522
93. 6   **95.** 7   **97.** 22   **99.** 13   **101.** 10   **103.** 11   **105.** \$128
107. 23 yards   **109.** 70°   **111.** \$95.50   **113.** 1175   **115.** 24
117.   **119.** 14°F   **121.**   **a.** -4 **b.** 15   **c.** 27

d. -30 **e.** -25 **f.** -32

1.4 Multiplying and Dividing Real Numbers

1.4 OBJECTIVES

1. Find the product of two real numbers
2. Find the quotient of two real numbers
3. Use order of operations to evaluate expressions with real numbers

When you first considered multiplication in arithmetic, it was thought of as repeated addition. Let's see what our work with the addition of numbers with different signs can tell us about multiplication when real numbers are involved.

$$3 \cdot 4 = \underline{4 + 4 + 4} = 12$$

We interpret multiplication as repeated addition to find the product, 12.

Now, consider the product $(3)(-4)$:

$$(3)(-4) = (-4) + (-4) + (-4) = -12$$

Looking at this product suggests the first portion of our rule for multiplying numbers with different signs. The product of a positive number and a negative number is negative.

> **Rules and Properties:** Multiplying Real Numbers Case 1: Different Signs
>
> The product of two numbers with different signs is negative.

To use this rule in multiplying two numbers with different signs, multiply their absolute values and attach a negative sign.

OBJECTIVE 1

Example 1 Multiplying Numbers with Different Signs

Multiply.

(a) $(5)(-6) = -30$

The product is negative.

(b) $(-10)(10) = -100$

(c) $(8)(-12) = -96$

NOTE Multiply numerators together and then denominators and simplify.

(d) $\left(-\dfrac{3}{4}\right)\left(\dfrac{2}{5}\right) = -\dfrac{3}{10}$

CHECK YOURSELF 1

Multiply.

(a) $(-7)(5)$ **(b)** $(-12)(9)$ **(c)** $(-15)(8)$ **(d)** $\left(-\dfrac{4}{7}\right)\left(\dfrac{14}{5}\right)$

The product of two negative numbers is harder to visualize. The following pattern may help you see how we can determine the sign of the product.

$$(3)(-2) = -6$$
$$(2)(-2) = -4$$
$$(1)(-2) = -2$$
$$(0)(-2) = 0$$
$$(-1)(-2) = 2$$

Do you see that the product is *increasing* by 2 each time?

NOTE This number is decreasing by 1.

NOTE $(-1)(-2)$ is the opposite of -2.

What should the product $(-2)(-2)$ be? Continuing the pattern shown, we see that

$$(-2)(-2) = 4$$

This suggests that the product of two negative numbers is positive. We can extend our multiplication rule.

NOTE If you would like a more detailed explanation, see the discussion at the end of this section.

Rules and Properties: Multiplying Real Numbers Case 2: Same Sign

The product of two numbers with the same sign is positive.

Example 2 **Multiplying Real Numbers with the Same Sign**

Multiply.

(a) $9 \cdot 7 = 63$ The product of two positive numbers (same sign, $+$) is positive.

(b) $(-8)(-5) = 40$ The product of two negative numbers (same sign, $-$) is positive.

(c) $\left(-\dfrac{1}{2}\right)\left(-\dfrac{1}{3}\right) = \dfrac{1}{6}$

 CHECK YOURSELF 2

Multiply.

(a) $10 \cdot 12$ **(b)** $(-8)(-9)$ **(c)** $\left(-\dfrac{2}{3}\right)\left(-\dfrac{6}{7}\right)$

Two numbers, 0 and 1, have special properties in multiplication.

Rules and Properties: Multiplicative Identity Property

The product of 1 and any number is that number. In symbols,

$$a \cdot 1 = 1 \cdot a = a$$

NOTE The number 1 is called the **multiplicative identity** for this reason.

Rules and Properties: Multiplicative Property of Zero

The product of 0 and any number is 0. In symbols,

$$a \cdot 0 = 0 \cdot a = 0$$

Example 3   Multiplying Real Numbers Involving 0 and 1

Find each product.

(a) $(1)(-7) = -7$

(b) $(15)(1) = 15$

(c) $(-7)(0) = 0$

(d) $0 \cdot 12 = 0$

(e) $\left(-\dfrac{4}{5}\right)(0) = 0$

CHECK YOURSELF 3

Multiply.

(a) $(-10)(1)$ (b) $(0)(-17)$ (c) $\left(\dfrac{5}{7}\right)(1)$ (d) $(0)\left(\dfrac{3}{4}\right)$

Before we continue, consider the following equivalent fractions:

$$-\frac{1}{a} = \frac{-1}{a} = \frac{1}{-a}$$

Any of these forms can occur in the course of simplifying an expression. The first form is generally preferred.

To complete our discussion of the properties of multiplication, we state the following.

Rules and Properties: Multiplicative Inverse Property

NOTE $\dfrac{1}{a}$ Is called the **multiplicative inverse**, or the **reciprocal**, of a. The product of any nonzero number and its reciprocal is 1.

For any number a, where $a \neq 0$, there is a number $\dfrac{1}{a}$ such that

$$a \cdot \frac{1}{a} = 1$$

Example 4 illustrates this property.

Example 4   Multiplying Reciprocals

(a) $3 \cdot \dfrac{1}{3} = 1$ The reciprocal of 3 is $\dfrac{1}{3}$.

(b) $-5\left(-\dfrac{1}{5}\right) = 1$ The reciprocal of -5 is $\dfrac{1}{-5}$ or $-\dfrac{1}{5}$.

(c) $\dfrac{2}{3} \cdot \dfrac{3}{2} = 1$ The reciprocal of $\dfrac{2}{3}$ is $\dfrac{1}{\frac{2}{3}}$, or $\dfrac{3}{2}$.

CHECK YOURSELF 4

Find the multiplicative inverse (or the reciprocal) of each of the following numbers.

(a) 6   **(b)** -4 **(c)** $\dfrac{1}{4}$ **(d)** $-\dfrac{3}{5}$

You know from your work in arithmetic that multiplication and division are related operations. We can use that fact, and our work in the earlier part of this section, to determine rules for the division of numbers with different signs. Every equation involving division can be stated as an equivalent equation involving multiplication. For instance,

$\dfrac{15}{5} = 3$ can be restated $15 = 5 \cdot 3$

$\dfrac{-24}{6} = -4$ can be restated $-24 = (6)(-4)$

$\dfrac{-30}{-5} = 6$ can be restated $-30 = (-5)(6)$

These examples illustrate that because the two operations are related, the rule of signs that we stated in the earlier part of this section for multiplication is also true for division.

Rules and Properties: Dividing Real Numbers

1. The quotient of two numbers with different signs is negative.
2. The quotient of two numbers with the same sign is positive.

Again, the rule is easy to use. To divide two numbers with different signs, divide their absolute values. Then attach the proper sign according to the rule above.

OBJECTIVE 2

Example 5   Dividing Real Numbers

Divide.

(a) Positive → $\dfrac{28}{7} = 4$ ← Positive
Positive →

(b) Negative → $\dfrac{-36}{-4} = 9$ ← Positive
Negative →

(c) Negative $\longrightarrow \dfrac{-42}{7} = -6 \longleftarrow$ Negative
Positive $\longrightarrow$

(d) Positive $\longrightarrow \dfrac{75}{-3} = -25 \longleftarrow$ Negative
Negative $\longrightarrow$

(e) Positive $\longrightarrow \dfrac{15.2}{-3.8} = -4 \longleftarrow$ Negative
Negative $\longrightarrow$

CHECK YOURSELF 5

Divide.

(a) $\dfrac{-55}{11}$ (b) $\dfrac{80}{20}$ (c) $\dfrac{-48}{-8}$ (d) $\dfrac{144}{-12}$ (e) $\dfrac{-13.5}{-2.7}$

You should be very careful when 0 is involved in a division problem. Remember that 0 divided by any nonzero number is just 0. Recall that

$$\dfrac{0}{-7} = 0 \qquad \text{because} \qquad 0 = (-7)(0)$$

However, if zero is the *divisor,* we have a special problem. Consider

$$\dfrac{9}{0} = \,?$$

This means that $9 = 0 \cdot \,?$.

Can 0 times a number ever be 9? No, so there is no solution.

Because $\dfrac{9}{0}$ cannot be replaced by any number, we agree that *division by 0 is not allowed.* We say that

Rules and Properties: Division by Zero

Division by 0 is undefined.

Example 6 **Dividing Numbers Involving Zero**

Divide, if possible.

(a) $\dfrac{7}{0}$ is undefined.

(b) $\dfrac{-9}{0}$ is undefined.

(c) $\dfrac{0}{5} = 0$

(d) $\dfrac{0}{-8} = 0$

Note: The expression $\dfrac{0}{0}$ is called an **indeterminate form.** You will learn more about this in later mathematics classes.

CHECK YOURSELF 6 _____

Divide if possible.

(a) $\dfrac{0}{3}$ (b) $\dfrac{5}{0}$ (c) $\dfrac{-7}{0}$ (d) $\dfrac{0}{-9}$

The fraction bar serves as a *grouping symbol.* This means that all operations in the numerator and denominator should be performed separately. Then the division is done as the last step. Example 7 illustrates this property.

OBJECTIVE 3   **Example 7   Operations with Grouping Symbols**

Evaluate each expression.

(a) $\dfrac{(-6)(-7)}{3} = \dfrac{42}{3} = 14$ Multiply in the numerator, and then divide.

(b) $\dfrac{3 + (-12)}{3} = \dfrac{-9}{3} = -3$ Add in the numerator, and then divide.

(c) $\dfrac{-4 + (2)(-6)}{-6 - 2} = \dfrac{-4 + (-12)}{-6 - 2}$ Multiply in the numerator. Then add in the numerator and subtract in the denominator.

$= \dfrac{-16}{-8} = 2$ Divide as the last step.

CHECK YOURSELF 7 _____

Evaluate each expression.

(a) $\dfrac{-4 + (-8)}{6}$ (b) $\dfrac{3 - (2)(-6)}{-5}$ (c) $\dfrac{(-2)(-4) - (-6)(-5)}{(-4)(11)}$

Evaluating fractions with a calculator poses a special problem. Example 8 illustrates this problem.

Example 8   Using a Calculator to Divide

Use your scientific calculator to evaluate each fraction.

(a) $\dfrac{4}{2 - 3}$

As you can see, the correct answer should be -4. To get this answer with your calculator, you must place the denominator in parentheses. The key stroke sequence will be

$4 \div (\ 2\ -\ 3\ )\ =$

(b) $\dfrac{-7 - 7}{3 - 10}$

In this problem, the correct answer is 2. This result can be found on your calculator by placing the numerator in parentheses and then placing the denominator in parentheses. The key stroke sequence will be

$\boxed{(}\ \boxed{7}\ \boxed{+/-}\ \boxed{-}\ \boxed{7}\ \boxed{)}\ \boxed{\div}\ \boxed{(}\ \boxed{3}\ \boxed{-}\ \boxed{10}\ \boxed{)}\ \boxed{=}$

When evaluating a fraction with a calculator, it is safest to use parentheses in both the numerator and the denominator.

 CHECK YOURSELF 8

Evaluate using your calculator.

(a) $\dfrac{-8}{5 - 7}$

(b) $\dfrac{-3 - 2}{-13 + 23}$

Example 9   Order of Operations

Evaluate each expression.

(a) $7(-9 + 12)$ Evaluate inside the parentheses first.

   $= 7(3) = 21$

(b) $(-8)(-7) - 40$ Multiply first, then subtract.

   $= 56 - 40 = 16$

NOTE Note that
$(-5)^2 = (-5)(-5) = 25$ but that
$-5^2 = -25$. The power applies
only to the 5.

(c) $(-5)^2 - 3$ Evaluate the power first.

   $= (-5)(-5) - 3$

   $= 25 - 3 = 22$

(d) $-5^2 - 3$

   $= -25 - 3$

   $= -28$

 CHECK YOURSELF 9

Evaluate each expression.

(a) $8(-9 + 7)$ **(b)** $(-3)(-5) + 7$

(c) $(-4)^2 - (-4)$ **(d)** $-4^2 - (-4)$

Example 10   An Application of Multiplying and Dividing Real Numbers

Three partners own stock worth $4680. One partner sells it for $3678. How much did each partner lose?

   First find the total loss: $4680 − $3678 = $1002

Then divide the total loss by 3: $\dfrac{\$1002}{3} = \$334$

Each person lost $334.

CHECK YOURSELF 10 _____

Sal and Vinnie invested $8500 in a business. Ten years later they sold the business for $22,000. How much profit did each make? 6,750

NOTE Here is a more detailed explanation of why the product of two negative numbers is positive.

> **Rules and Properties:**   The Product of Two Negative Numbers
>
> From our earlier work, we know that the sum of a number and its opposite is 0:
>
> $5 + (-5) = 0$
>
> Multiply both sides of the equation by -3:
>
> $(-3)[5 + (-5)] = (-3)(0)$
>
> Because the product of 0 and any number is 0, on the right we have 0.
>
> $(-3)[5 + (-5)] = 0$
>
> We use the distributive property on the left.
>
> $(-3)(5) + (-3)(-5) = 0$
>
> We know that $(-3)(5) = -15$, so the equation becomes
>
> $-15 + (-3)(-5) = 0$
>
> We now have a statement of the form
>
> $-15 + \square = 0$
>
> in which $\square$ is the value of $(-3)(-5)$. We also know that $\square$ is the number that must be added to -15 to get 0, so $\square$ is the opposite of -15, or 15. This means that
>
> $(-3)(-5) = 15$ The product is positive!
>
> It doesn't matter what numbers we use in this argument. The resulting product of two negative numbers will always be positive.

CHECK YOURSELF ANSWERS _____

1. **(a)** -35; **(b)** -108; **(c)** -120; **(d)** $-\dfrac{8}{5}$ **2.** **(a)** 120; **(b)** 72; **(c)** $\dfrac{4}{7}$

3. **(a)** -10; **(b)** 0; **(c)** $\dfrac{5}{7}$; **(d)** 0 **4.** **(a)** $\dfrac{1}{6}$; **(b)** $-\dfrac{1}{4}$; **(c)** 4; **(d)** $-\dfrac{5}{3}$

5. **(a)** -5; **(b)** 4; **(c)** 6; **(d)** -12; **(e)** 5 **6.** **(a)** 0; **(b)** undefined; **(c)** undefined;

(d) 0 **7.** **(a)** -2; **(b)** -3; **(c)** $\dfrac{1}{2}$ **8.** **(a)** 4; **(b)** -0.5 **9.** **(a)** -16; **(b)** 22;

(c) 20; **(d)** -12 **10.** $\$6750$

1.4 Exercises

Multiply.

1. $4 \cdot 10$

2. $3 \cdot 14$ *42*

3. $(-6)(10)$

4. $(10)(-2)$

5. $(-8)(9)$

6. $(-12)(3)$

7. $(4)\left(-\dfrac{3}{2}\right)$

8. $(9)\left(-\dfrac{2}{3}\right)$

9. $\left(-\dfrac{1}{4}\right)(8)$

10. $\left(\dfrac{-5}{3} \cdot 6\right)$

11. $(3.25)(-4)$

12. $(5.4)(-5)$

13. $(-9)(-11)$

14. $(-9)(-8)$

15. $(-5)(-12)$

16. $(-7)(-3)$

17. $(-9)\left(-\dfrac{2}{3}\right)$

18. $\left(-\dfrac{2}{7}\right)(-21)$ $\dfrac{6}{1} = 6$

19. $(-1.25)(-12)$ *30*

20. $(-1.5)(-20)$

21. $(0)(-18)$ *0*

22. $(-17)(0)$ *0*

23. $(15)(0)$ *0*

24. $(0)(25)$ *0*

25. $\left(-\dfrac{11}{12}\right)(0)$ *0*

26. $\left(-\dfrac{8}{9}\right)(0)$ *0*

27. $(-3.57)(0)$ *0*

28. $(-2.37)(0)$ *0*

29. $\left(-\dfrac{3}{2}\right)\left(-\dfrac{2}{3}\right)$ *1*

30. $\left(-\dfrac{4}{5}\right)\left(-\dfrac{5}{4}\right)$ *1*

ANSWERS

1. _____
2. _____
3. _____
4. _____
5. _____
6. _____
7. _____
8. _____
9. _____
10. _____
11. _____
12. _____
13. _____
14. _____
15. _____
16. _____
17. _____
18. _____
19. _____
20. _____
21. _____
22. _____

23.	24.
25.	26.
27.	28.
29.	30.

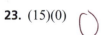

31. _____

32. _____

33. _____

34. _____

35. _____

36. _____

37. _____

38. _____

39. _____

40. _____

41. _____

42. _____

43. _____

44. _____

45. _____

46. _____

47. _____

48. _____

49. _____

50. _____

51. _____

52. _____

53. _____

54. _____

55. _____

56. _____

31. $\left(\dfrac{4}{7}\right)\left(-\dfrac{7}{4}\right)$

32. $\left(\dfrac{8}{9}\right)\left(-\dfrac{9}{8}\right)$ -1

Divide.

33. $\dfrac{-35}{-7}$

34. $\dfrac{70}{14}$ 5

35. $\dfrac{48}{6}$

36. $\dfrac{-24}{8}$ -3

37. $\dfrac{50}{-5}$

38. $\dfrac{-48}{-12}$ 4

39. $\dfrac{-52}{4}$

40. $\dfrac{56}{-7}$ -8

41. $\dfrac{-125}{-5}$

42. $\dfrac{-60}{15}$ -4

43. $\dfrac{0}{-8}$

44. $\dfrac{-125}{-25}$ 5

45. $\dfrac{-11}{-1}$

46. $\dfrac{-10}{0}$ undefined

47. $\dfrac{-96}{-8}$

48. $\dfrac{-20}{2}$ -10

49. $\dfrac{-18}{0}$

50. $\dfrac{0}{-8}$ 0

51. $\dfrac{-17}{1}$

52. $\dfrac{-27}{-1}$ 27

53. $\dfrac{-144}{-16}$

54. $\dfrac{-150}{6}$ -25

55. $\dfrac{-29.4}{4.9}$

56. $\dfrac{-25.9}{-3.7}$ 7

57. $\dfrac{-8}{32}$

58. $\dfrac{-6}{-30}$ $\dfrac{1}{5}$

59. $\dfrac{24}{-16}$

60. $\dfrac{25}{-10}$ -2.5

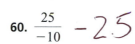

61. $\dfrac{-28}{-42}$

62. $\dfrac{-125}{-75}$ $1.\overline{6}$

Perform the indicated operations.

63. $\dfrac{(-6)(-3)}{2}$

64. $\dfrac{(-9)(5)}{-3}$ 15

65. $\dfrac{(-8)(2)}{-4}$

66. $\dfrac{(7)(-8)}{-14}$ 4

67. $\dfrac{24}{-4-8}$

68. $\dfrac{36}{-7+3}$ $\dfrac{36}{-4}=-9$

69. $\dfrac{-15-15}{-5}$

70. $\dfrac{-14-4}{-6}$ 3

71. $\dfrac{55-19}{-12-6}$

72. $\dfrac{-11-7}{-14+8}$ $\dfrac{-18}{-6}$ 3

73. $\dfrac{5-7}{4-4}$

74. $\dfrac{10-6}{4-4}$ $\dfrac{4}{0}$ undefined

Perform the indicated operations. Remember the rules for the order of operations.

75. $5(7-2)$

76. $7(5-8)$ $7(-3)$ -21

77. $2(5-8)$

78. $6(14-16)$ -12

79. $-3(9-7)$

80. $-6(12+9)$ 18

81. $-3(-2-5)$

82. $-2(-7-3)$ 20

83. $(-2)(3)-5$

84. $(-8)(6)-27$ -75

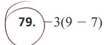

ANSWERS

57. _____
58. _____
59. _____
60. _____
61. _____
62. _____
63. _____
64. _____
65. _____
66. _____
67. _____
68. _____
69. _____
70. _____
71. _____
72. _____
73. _____
74. _____
75. _____
76. _____
77. _____
78. _____
79. _____
80. _____
81. _____
82. _____
83. _____
84. _____

85.	
86.	
87.	
88.	
89.	
90.	
91.	
92.	
93.	
94.	
95.	
96.	
97.	
98.	
99.	
100.	
101.	
102.	
103.	
104.	
105.	
106.	
107.	
108.	
109.	
110.	
111.	
112.	

85. $(-5)(-2)(0)(-3) - 10$

86. $(-3)(-9) - 11$

87. $(-5)(-2) - 12$

88. $(-7)(-3) - 25$

89. $(3)(-7) + 20$

90. $(2)(-6) + 8$

91. $4 \cdot 8 \div 2 - 5^2$

92. $-5 + (-2)(3)$

93. $36 \div 4 \cdot 3 - (-25)$

94. $9 - (-2)(-7)$

95. $(-7)^2 - 17$

96. $(-6)^2 - 20$

97. $(-5)^2 + 18$

98. $(-2)^2 + 10$

99. $-6^2 - 4$

100. $-5^2 - 3$

101. $14 + 3 \cdot 9 - 28 \div 7 \cdot 2$

102. $(-3)^3 - (-8)(-2)$

103. $(-8)^2 - 5^2$

104. $-8 + 14 - 2 \cdot 4 - 3$

105. $8 + (2 \cdot 3 + 3)^2$

106. $(-8)^2 - (-4)^2$

107. $-8^2 - 5^2 + 8 \div 4 \cdot 2$

108. $-6^2 - 3^2$

109. $-8^2 - (-5)^2$

110. $-9^2 - (-6)^2$

111. Statistics. You score 23 points a game for 11 straight games. What is the total number of points that you scored?

112. Business and finance. In Atlantic City, Nick played the slot machines for 12 h. He lost $45 an hour. Use real numbers to represent the change in Nick's financial status at the end of the 12 h.

113. Business and finance. Suppose you own 35 shares of stock. If the price increases $0.75 per share, how much money have you made?

114. Business and finance. Your bank charges a flat service charge of $7.25 per month on your checking account. You have had the account for 3 years. How much have you paid in service charges? (7.25) ⊗ (36)

115. Science and medicine. The temperature is −6°F at 5:00 in the evening. If the temperature drops 2°F every hour, what is the temperature at 1:00 A.M.?

116. Science and medicine. A woman lost 42 pounds (lbs) while dieting. If she lost 3 lbs each week, how long has she been dieting? $\frac{42}{3}$ = 14 weeks

117. Business and finance. Patrick worked all day mowing lawns and was paid $9 per hour. If he had $125 at the end of a 9-h day, how much did he have before he started working?

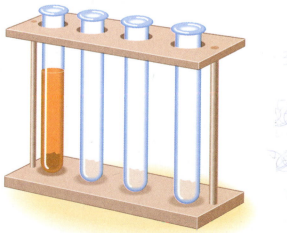

118. Business and finance. A 4.5-lb can of food costs $11.25. What is the cost per pound? $2.50 per pound

119. Business and finance. Suppose that you and your two brothers bought equal shares of an investment for a total of $20,000 and sold it later for $16,232. How much did each person lose?

120. Science and medicine. Suppose that the temperature outside is dropping at a constant rate. At noon, the temperature is 70°F and it drops to 58°F at 5:00 P.M. How much did the temperature change each hour? 70−58=12÷5=2.4   12   2.4 per hr

121. Science and medicine. A chemist has 84 ounces (oz) of a solution. He pours the solution into test tubes. Each test tube holds $\frac{2}{3}$ oz. How many test tubes can he fill?

ANSWERS

113. _____

114. _____

115. _____

116. _____

117. _____

118. _____

119. _____

120. _____

121. _____

122. _____

123. _____

124. _____

125. _____

126. _____

127. _____

128. _____

129. _____

130. _____

Use your calculator to evaluate each expression.

122. $\dfrac{7}{4 - 5}$

123. $\dfrac{-8}{-4 + 2}$

124. $\dfrac{-6 - 9}{-4 + 1}$

125. $\dfrac{-10 + 4}{-7 + 10}$

126. Some animal ecologists in Minnesota are planning to reintroduce a group of animals into a wilderness area. The animals, a mammal on the endangered species list, will be released into an area where they once prospered and where there is an abundant food supply. But, the animals will face predators. The ecologists expect the number of mammals to grow about 25 percent each year but that 30 of the animals will die from attacks by predators and hunters.

The ecologists need to decide how many animals they should release to establish a stable population. Work with other students to try several beginning populations and follow the numbers through 8 years. Is there a number of animals that will lead to a stable population? Write a letter to the editor of your local newspaper explaining how to decide what number of animals to release. Include a formula for the number of animals next year based on the number this year. Begin by filling out this table to track the number of animals living each year after the release:

No. Initially Released	Year							
	1	2	3	4	5	6	7	8
20	+___ -___ =_____							
100	+___ -___ =_____							
200	+___ -___ =_____							

Fill in the blank with the most appropriate word from the choices: *never, sometimes,* or *always*.

127. A product made up of an odd number of negative factors is _____ negative.

128. A product of an even number of negative factors is _____ negative.

129. The quotient $\dfrac{x}{y}$ is _____ positive.

130. The quotient $\dfrac{x}{y}$ is _____ negative.

Getting Ready for Section 1.5 [Sections 1.3 and 1.4]

Simplify.

(a) $\dfrac{6 \cdot 2 + 8}{5 - 3}$

(b) $\dfrac{4 \cdot 5 - 8}{8 - 4 \div 2}$

(c) $\dfrac{-8 + 3 \cdot 2}{-12 \div 6}$

(d) $\dfrac{-3^2 - (-4 - 1)}{-2 \cdot 2}$

(e) $8 \div 4 - 3 \cdot 2$

(f) $6^2 - 18 \div 2 \cdot 3$

Answers

1. 40   **3.** −60   **5.** −72   **7.** −6   **9.** −2   **11.** −13   **13.** 99
15. 60   **17.** 6   **19.** 15   **21.** 0   **23.** 0   **25.** 0   **27.** 0
29. 1   **31.** −1   **33.** 5   **35.** 8   **37.** −10   **39.** −13   **41.** 25
43. 0   **45.** 11   **47.** 12   **49.** Undefined   **51.** −17   **53.** 9
55. −6   **57.** $-\dfrac{1}{4}$ **59.** $-\dfrac{3}{2}$ **61.** $\dfrac{2}{3}$ **63.** 9   **65.** 4   **67.** −2
69. 6   **71.** −2   **73.** Undefined   **75.** 25   **77.** −6   **79.** −6
81. 21   **83.** −11   **85.** −10   **87.** −2   **89.** −1   **91.** −9
93. 52   **95.** 32   **97.** 43   **99.** −40   **101.** 33   **103.** 39
105. 89   **107.** −85   **109.** −89   **111.** 253 points   **113.** $26.25
115. −22°F   **117.** $44   **119.** $1256   **121.** 126   **123.** 4
125. −2   **127.** Always   **129.** Sometimes   **a.** 10   **b.** 2   **c.** 1
d. 1   **e.** −4   **f.** 9

1.5 Evaluating Algebraic Expressions

In applying algebra to problem solving, you will often want to find the value of an algebraic expression when you know certain values for the letters (or variables) in the expression. Finding the value of an expression is called *evaluating the expression* and uses the following steps.

Step by Step: To Evaluate an Algebraic Expression

Step 1   Replace each variable by the given number value.
Step 2   Do the necessary arithmetic operations, following the rules for order of operations.

OBJECTIVE 1

Example 1   Evaluating Algebraic Expressions

Suppose that $a = 5$ and $b = 7$.

(a) To evaluate $a + b$, we replace a with 5 and b with 7.

$$a + b = (5) + (7) = 12$$

(b) To evaluate $3ab$, we again replace a with 5 and b with 7.

$$3ab = 3 \cdot (5) \cdot (7) = 105$$

CHECK YOURSELF 1

If $x = 6$ and $y = 7$, evaluate.

(a) $y - x$ **(b)** $5xy$

We are now ready to evaluate algebraic expressions that require following the rules for the order of operations.

Example 2   Evaluating Algebraic Expressions

Evaluate the following expressions if $a = 2$, $b = 3$, $c = 4$, and $d = 5$.

CAUTION

This is different from
$(3c)^2 = (3 \cdot 4)^2$
$= 12^2 = 144$

(a) $5a + 7b = 5(2) + 7(3)$ Multiply first.
$= 10 + 21 = 31$ Then add.

(b) $3c^2 = 3(4)^2$ Evaluate the power.
$= 3 \cdot 16 = 48$ Then multiply.

(c) $7(c + d) = 7[(4) + (5)]$ Add inside the brackets.

$= 7 \cdot 9 = 63$

(d) $5a^4 - 2d^2 = 5(2)^4 - 2(5)^2$ Evaluate the powers.

$= 5 \cdot 16 - 2 \cdot 25$ Multiply.

$= 80 - 50 = 30$ Subtract.

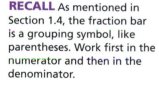

CHECK YOURSELF 2

If x = 3, y = 2, z = 4, and w = 5, evaluate the following expressions.

(a) $4x^2 + 2$ (b) $5(z + w)$ (c) $7(z^2 - y^2)$

To evaluate algebraic expressions when a fraction bar is used, do the following: Start by doing all the work in the numerator, then do all the work in the denominator. Divide the numerator by the denominator as the last step.

Example 3   Evaluating Algebraic Expressions

If $p = -2$, $q = 3$, and $r = 4$, evaluate:

(a) $\dfrac{8p}{r}$

RECALL As mentioned in Section 1.4, the fraction bar is a grouping symbol, like parentheses. Work first in the numerator and then in the denominator.

Replace p with -2 and r with 4.

$\dfrac{8p}{r} = \dfrac{8(-2)}{(4)} = \dfrac{-16}{4} = -4$ Divide as the last step.

(b) $\dfrac{7q + r}{p + q} = \dfrac{7(3) + (4)}{(-2) + (3)}$ Now evaluate the top and bottom separately.

$= \dfrac{21 + 4}{(-2) + (3)}$

$= \dfrac{25}{1} = 25$

CHECK YOURSELF 3

Evaluate the following if c = -5, d = 8, and e = 3.

(a) $\dfrac{6c}{e}$ (b) $\dfrac{4d + e}{c}$ (c) $\dfrac{10d - e}{d + e}$

Example 4 shows how a scientific calculator can be used to evaluate algebraic expressions.

OBJECTIVE 2

Example 4   Using a Calculator to Evaluate Expressions

Use a scientific calculator to evaluate the following expressions.

(a) $\dfrac{4x + y}{z}$ if $x = 2$, $y = 1$, and $z = 3$

Replace x with 2, y with 1, and z with 3:

$$\dfrac{4x + y}{z} = \dfrac{4(2) + (1)}{3}$$

Now, use the following keystrokes:

$$(\;4\; \times\; 2\; +\; 1\;)\; \div\; 3\; =$$

The display will read 3.

(b) $\dfrac{7x - y}{3z - x}$ if $x = 2$, $y = 6$, and $z = 2$

$$\dfrac{7x - y}{3z - x} = \dfrac{7(2) - (6)}{3(2) - (2)}$$

Use the following keystrokes:

$$(\;7\; \times\; 2\; -\; 6\;)\; \div\; (\; 3\; \times\; 2\; -\; 2\;)\; =$$

The display will read 2.

CHECK YOURSELF 4

Use a scientific calculator to evaluate the following if $x = 2$, $y = 6$, and $z = 5$.

(a) $\dfrac{2x + y}{z}$ (b) $\dfrac{4y - 2z}{x}$

OBJECTIVE 1   **Example 5   Evaluating Expressions**

Evaluate $5a + 4b$ if $a = -2$ and $b = 3$.

Replace a with -2 and b with 3.

RECALL The rules for the order of operations call for us to multiply first, and then add.

$$5a + 4b = 5(-2) + 4(3)$$
$$= -10 + 12$$
$$= 2$$

CHECK YOURSELF 5

Evaluate $3x + 5y$ if $x = -2$ and $y = -5$.

We follow the same rules no matter how many variables are in the expression.

Example 6   Evaluating Expressions

Evaluate the following expressions if $a = -4$, $b = 2$, $c = -5$, and $d = 6$.

This becomes $-(-20)$, or $+20$.

(a) $7a - 4c = 7(-4) - 4(-5)$

$= -28 + 20$

$= -8$

Evaluate the power first, and then multiply by 7.

CAUTION

When a squared variable is replaced by a negative number, square the negative.

$(-5)^2 = (-5)(-5) = 25$

The exponent applies to -5!

$-5^2 = -(5 \cdot 5) = -25$

The exponent applies only to 5!

(b) $7c^2 = 7(-5)^2 = 7 \cdot 25$

$= 175$

(c) $b^2 - 4ac = (2)^2 - 4(-4)(-5)$

$= 4 - 4(-4)(-5)$

$= 4 - 80$

$= -76$

Add inside the brackets first.

(d) $b(a + d) = (2)[(-4) + (6)]$

$= 2(2)$

$= 4$

 CHECK YOURSELF 6

Evaluate if $p = -4$, $q = 3$, and $r = -2$.

(a) $5p - 3r$ (b) $2p^2 + q$ (c) $p(q + r)$

(d) $-q^2$ (e) $(-q)^2$

If an expression involves a fraction, remember that the fraction bar is a grouping symbol. This means that you should do the required operations first in the numerator and then in the denominator. Divide as the last step.

Example 7   Evaluating Expressions

Evaluate the following expressions if $x = 4$, $y = -5$, $z = 2$, and $w = -3$.

(a) $\dfrac{z - 2y}{x} = \dfrac{(2) - 2(-5)}{(4)} = \dfrac{2 + 10}{4}$

$= \dfrac{12}{4} = 3$

(b) $\dfrac{3x - w}{2x + w} = \dfrac{3(4) - (-3)}{2(4) + (-3)} = \dfrac{12 + 3}{8 + (-3)}$

$= \dfrac{15}{5} = 3$

CHECK YOURSELF 7

Evaluate if m = −6, n = 4, and p = −3.

(a) $\dfrac{m + 3n}{p}$

(b) $\dfrac{4m + n}{m + 4n}$

When an expression is evaluated by a calculator, the same order of operations that we introduced in Section 0.4 is followed.

	Algebraic Notation	Calculator Notation
Addition	$6 + 2$	6 $\boxed{+}$ 2
Subtraction	$4 - 8$	4 $\boxed{-}$ 8
Multiplication	$(3)(-5)$	3 $\boxed{\times}$ $\boxed{(-)}$ 5 or 3 $\boxed{\times}$ 5 $\boxed{+/-}$
Division	$\dfrac{8}{6}$	8 $\boxed{\div}$ 6
Exponential	3^4	3^4 or 3 $\boxed{y^x}$ 4

In many applications, you need to find the sum of a set of numbers that you are working with. In mathematics, the shorthand symbol for "sum of" is the Greek letter Σ (capital sigma, the "S" of the Greek alphabet). The expression Σx, in which x refers to all the numbers in a given set, means the sum of all the numbers in that set.

OBJECTIVE 3   **Example 8   Summing a Set**

Find Σx for the following set of numbers:

$-2, -6, 3, 5, -4$

$$\Sigma x = -2 + (-6) + 3 + 5 + (-4)$$
$$= (-8) + 3 + 5 + (-4)$$
$$= (-8) + 8 + (-4)$$
$$= -4$$

CHECK YOURSELF 8

Find Σx for each set of numbers.

(a) $-3, 4, -7, -9, 8$

(b) $-2, 6, -5, -3, 4, 7$

CHECK YOURSELF ANSWERS

1. (a) 1; (b) 210   **2.** (a) 38; (b) 45; (c) 84   **3.** (a) −10; (b) −7; (c) 7
4. (a) 2; (b) 7   **5.** −31   **6.** (a) −14; (b) 35; (c) −4; (d) −9; (e) 9
7. (a) −2; (b) −2   **8.** (a) −7; (b) 7

1.5   Exercises

Evaluate each of the expressions if $a = -2$, $b = 5$, $c = -4$, and $d = 6$.

1. $3c - 2b$

2. $4c - 2b$

$4(-4) - 2(5) \quad -16 \quad -10$
$\quad\quad\quad\quad\quad\quad -20$

3. $8b + 2c$

4. $7a - 2c$

$-14 - (2)(-4)$

5. $-b^2 + b$

6. $(-b)^2 + b$

7. $3a^2$

8. $6c^2$

9. $c^2 - 2d$

10. $3b^2 + 4c$

11. $2a^2 + 3b^2$

12. $4b^2 - 2c^2$

13. $2(a + b)$

14. $5(b - c)$

15. $-4(2c - a)$

16. $6(3c - d)$

17. $a(b + 3c)$

18. $c(3a - d)$

19. $\dfrac{6d}{c}$

20. $\dfrac{8c}{2a}$

21. $\dfrac{3d + 2c}{b}$

22. $\dfrac{2b + 3d}{2a}$

23. $\dfrac{2b - 3a}{c + 2d}$

24. $\dfrac{3d - 2b}{5a + d}$

25. $d^2 - b^2$

26. $c^2 - a^2$

27. $(d - b)^2$

28. $(c - a)^2$

29. $(d - b)(d + b)$

30. $(c - a)(c + a)$

31. $d^3 - b^3$

32. $c^3 + a^3$

ANSWERS

1. _____
2. _____
3. _____
4. _____
5. _____
6. _____
7. _____
8. _____
9. _____
10. _____
11. _____
12. _____
13. _____
14. _____
15. _____
16. _____
17. _____
18. _____

19.	20.
21.	22.
23.	24.
25.	26.
27.	28.
29.	30.
31.	32.

33. _____

34. _____

35. _____

36. _____

37. _____

38. _____

39. _____

40. _____

41. _____

42. _____

43. _____

44. _____

45. _____

46. _____

47. _____

48. _____

49. _____

50. _____

51. _____

52. _____

53. _____

54. _____

55. _____

56. _____

57. _____

58. _____

59. _____

60. _____

33. $(d - b)^3$

34. $(c + a)^3$

35. $(d - b)(d^2 + db + b^2)$

36. $(c + a)(c^2 - ac + a^2)$

37. $-(b + a)^2$

38. $(d - a)^2$

39. $3a - 2b + \dfrac{2d}{c}$

40. $4b + 5d - \dfrac{c}{a}$

41. $a^2 + 2ad + d^2$

42. $b^2 - 2bc + c^2$

Use your calculator to evaluate each expression if $x = -2.34$, $y = -3.14$, and $z = 4.12$. Round your answer to the nearest tenth.

43. $x + yz$

44. $y - 2z$

45. $x^2 - z^2$

46. $x^2 + y^2$

47. $\dfrac{xy}{z - x}$

48. $\dfrac{y^2}{zy}$

49. $\dfrac{2x + y}{2x + z}$

50. $\dfrac{x^2y^2}{xz}$

For the following data sets, evaluate Σx.

51. 1, 2, 3, 7, 8, 9, 11

52. 2, 4, 5, 6, 10, 11, 12

53. −5, −3, −1, 2, 3, 4, 8

54. −4, −2, −1, 5, 7, 8, 10

55. 3, 2, −1, −4, −3, 8, 6

56. 3, −4, 2, −1, 2, −7, 9

57. $-\dfrac{1}{2}, -\dfrac{3}{4}, 2, 3, \dfrac{1}{4}, \dfrac{3}{2}, -1$

58. $-\dfrac{1}{3}, -\dfrac{5}{3}, -1, 1, 3, \dfrac{2}{3}, \dfrac{5}{3}$

59. −2.5, −3.2, 2.6, −1, 2, 4, −3

60. −2.4, −3.1, −1.7, 3, 1, 2, 5

In each of the following problems, decide if the given values make the statement true or false.

61. $x - 7 = 2y + 5$; $x = 22$, $y = 5$

62. $3(x - y) = 6$; $x = 5$, $y = -3$

63. $2(x + y) = 2x + y$; $x = -4$, $y = -2$

64. $x^2 - y^2 = x - y$; $x = 4$, $y = -3$

65. Science and medicine. The formula for the total resistance in a parallel circuit is given by the formula $R_T = R_1 R_2/(R_1 + R_2)$. Find the total resistance if $R_1 = 6$ ohms (Ω) and $R_2 = 10\ \Omega$.

66. Geometry. The formula for the area of a triangle is given by $A = \dfrac{1}{2}bh$. Find the area of a triangle if $b = 4$ cm and $h = 8$ cm.

67. Geometry. The perimeter of a rectangle of length L and width W is given by the formula $P = 2L + 2W$. Find the perimeter when $L = 10$ in. and $W = 5$ in.

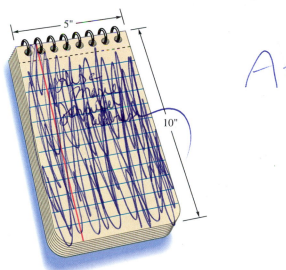

68. Business and finance. The simple interest I on a principal of P dollars at interest rate r for time t, in years, is given by $I = Prt$. Find the simple interest on a principal of $6000 at 3% for 2 years. (**Hint:** 3% = 0.03)

69. Business and finance. Use the simple interest formula to find the principal if the total interest earned was $150 and the rate of interest was 4% for 2 years.

70. Business and finance. Use the simple interest formula to find the rate of interest if $5000 earns $300 interest in 3 years.

ANSWERS

61. _____

62. _____

63. _____

64. _____

65. _____

66. _____

67. _____

68. _____

69. _____

70. _____

71. _____

72. _____

73. _____

74. _____

75. _____

71. Science and medicine. A formula that relates Celsius and Fahrenheit temperature is $F = \frac{9}{5}C + 32$. If the temperature of the day is $-10°C$, what is the Fahrenheit temperature?

72. Geometry. If the area of a circle whose radius is r is given by $A = \pi r^2$, in which $\pi \approx 3.14$, find the area when $r = 3$ meters (m).

$A = \pi(3)$ $9 \times 3.14 =$

73. Write an English interpretation of each of the following algebraic expressions.

(a) $(2x^2 - y)^3$ (b) $3n - \dfrac{n - 1}{2}$ (c) $(2n + 3)(n - 4)$

74. Is $a^n + b^n = (a + b)^n$? Try a few numbers and decide if you think this is true for all numbers, for some numbers, or never true. Write an explanation of your findings and give examples.

75. Enjoyment of patterns in art, music, and language is common to all cultures, and many cultures also delight in and draw spiritual significance from patterns in numbers. One such set of patterns is that of the "magic" square. One of these squares appears in a famous etching by Albrecht Dürer, who lived from 1471 to 1528 in Europe. He was one of the first artists in Europe to use geometry to give perspective, a feeling of three dimensions, in his work.

The magic square in his work is this one:

16	3	2	13
5	10	11	8
9	6	7	12
4	15	15	1

Why is this square "magic"? It is magic because every row, every column, and both diagonals add to the same number. In this square there are sixteen spaces for the numbers 1 through 16.

Part 1: What number does each row and column add to?

Write the square that you obtain by adding −17 to each number. Is this still a magic square? If so, what number does each column and row add to? If you add 5 to each number in the original magic square, do you still have a magic square? You have been studying the operations of addition, multiplication, subtraction, and division with integers and with rational numbers. What operations can you perform on this magic square and still have a magic square? Try to find something that will not work. Use algebra to help you decide what will work and what won't. Write a description of your work and explain your conclusions.

Part 2: Here is the oldest published magic square. It is from China, about 250 B.C.E. Legend has it that it was brought from the River Lo by a turtle to the Emperor Yii, who was a hydraulic engineer.

4	9	2
3	5	7
8	1	6

Check to make sure that this is a magic square. Work together to decide what operation might be done to every number in the magic square to make the sum of each row, column, and diagonal the *opposite* of what it is now. What would you do to every number to cause the sum of each row, column, and diagonal to equal zero?

Getting Ready for Section 1.6 [Sections 1.3 and 1.4]

Simplify.

(a) $(8 + 9) - 5$

(b) $15 - 4 - 11$

(c) $5(4 + 3) - 9$

(d) $-3(5 - 7) + 11$

(e) $-6(-9 + 7) - 4$

(f) $8 - 7(-2 - 6)$

Answers

1. −22   **3.** 32   **5.** −20   **7.** 12   **9.** 4   **11.** 83   **13.** 6

15. 24   **17.** 14   **19.** −9   **21.** 2   **23.** 2   **25.** 11   **27.** 1

29. 11   **31.** 91   **33.** 1   **35.** 91   **37.** −9   **39.** −19   **41.** 16

43. −15.3   **45.** −11.5   **47.** 1.1   **49.** 14.0   **51.** 41   **53.** 8

55. 11   **57.** $\dfrac{9}{2}$ **59.** −1.1   **61.** True   **63.** False   **65.** 3.75 Ω

67. 30 in.   **69.** $1875   **71.** 14°F   **73.**   **75.**

a. 12   **b.** 0   **c.** 26   **d.** 17   **e.** 8   **f.** 64

1.6 Adding and Subtracting Terms

1.6 OBJECTIVES

1. Identify terms and like terms
2. Combine like terms

To find the perimeter of (or the distance around) a rectangle, we add 2 times the length and 2 times the width. In the language of algebra, this can be written as

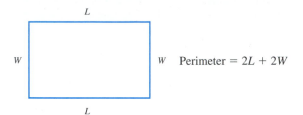

Perimeter $= 2L + 2W$

We call $2L + 2W$ an **algebraic expression,** or more simply an **expression.** Recall from Section 1.1 that an expression allows us to write a mathematical idea in symbols. It can be thought of as a meaningful collection of letters, numbers, and operation signs.

Some expressions are

$$5x^2 \qquad\qquad 3a + 2b \qquad\qquad 4x^3 + (-2y) + 1$$

In algebraic expressions, the addition and subtraction signs break the expressions into smaller parts called *terms.*

> **Definition:** Term
>
> A **term** is a number, or the product of a number and one or more variables, raised to a power.

In an expression, each sign ($+$ or $-$) is a part of the term that follows the sign.

OBJECTIVE 1

Example 1 Identifying Terms

(a) $5x^2$ has one term.

(b) $\underbrace{3a}_{\text{Term}} + \underbrace{2b}_{\text{Term}}$ has two terms: $3a$ and $2b$.

NOTE This could also be written as $4x^3 - 2y + 1$

(c) $\underbrace{4x^3}_{\text{Term}} + \underbrace{(-2y)}_{\text{Term}} + \underbrace{1}_{\text{Term}}$ has three terms: $4x^3$, $-2y$, and 1.

CHECK YOURSELF 1

List the terms of each expression.

(a) $2b^4$ **(b)** $5m + 3n$ **(c)** $2s^2 - 3t - 6$

Note that a term in an expression may have any number of factors. For instance, $5xy$ is a term. It has factors of 5, x, and y. The number factor of a term is called the **numerical coefficient.** So for the term $5xy$, the numerical coefficient is 5.

Example 2   Identifying the Numerical Coefficient

(a) $4a$ has the numerical coefficient 4.
(b) $6a^3b^4c^2$ has the numerical coefficient 6.
(c) $-7m^2n^3$ has the numerical coefficient -7.
(d) Because $1 \cdot x = x$, the numerical coefficient of x is understood to be 1.

 CHECK YOURSELF 2

Give the numerical coefficient for each of the following terms.

(a) $8a^2b$ **(b)** $-5m^3n^4$ **(c)** y

If terms contain exactly the *same letters* (or variables) raised to the *same powers,* they are called **like terms.**

Example 3   Identifying Like Terms

(a) The following are like terms.

$6a$ and $7a$
$5b^2$ and b^2
$10x^2y^3z$ and $-6x^2y^3z$
$-3m^2$ and m^2

Each pair of terms has the same letters, with each letter raised to the same power—the numerical coefficients can be any number.

(b) The following are *not* like terms.

Different letters
$6a$ and $7b$

Different exponents
$5b^2$ and $5b^3$

Different exponents
$3x^2y$ and $4xy^2$

 CHECK YOURSELF 3

Circle the like terms.

$5a^2b$ ab^2 a^2b $-3a^2$ $4ab$ $3b^2$ $-7a^2b$

Like terms of an expression can always be combined into a single term. Look at the following:

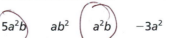

$2x + 5x = 7x$
$\underbrace{x+x+x+x+x+x+x}\ \underbrace{x+x+x+x+x+x+x}$

Rather than having to write out all those x's, try

$$2x + 5x = (2 + 5)x = 7x$$

In the same way,

$$9b + 6b = (9 + 6)b = 15b$$

and $10a + (-4a) = (10 + (-4))a = 6a$

This leads us to the following rule.

Step by Step: To Combine Like Terms

To combine like terms, use the following steps.

Step 1   Add or subtract the numerical coefficients.

Step 2   Attach the common variables.

OBJECTIVE 2

Example 4   Combining Like Terms

Combine like terms.*

(a) $8m + 5m = (8 + 5)m = 13m$

(b) $5pq^3 - 4pq^3 = 5pq^3 + (-4pq^3) = 1pq^3 = pq^3$

(c) $7a^3b^2 - 7a^3b^2 = 7a^3b^2 + (-7a^3b^2) = 0a^3b^2 = 0$

 CHECK YOURSELF 4

Combine like terms.

(a) $6b + 8b$

(b) $12x^2 - 3x^2$

(c) $8xy^3 - 7xy^3$

(d) $9a^2b^4 - 9a^2b^4$

Let's look at some expressions involving more than two terms. The idea is the same.

Example 5   Combining Like Terms

Combine like terms.

(a) $5ab - 2ab + 3ab$

$$= 5ab + (-2ab) + 3ab$$

$$= (5 + (-2) + 3)ab = 6ab$$

Only like terms can be combined.

(b) $\overbrace{8x - 2x} + 5y$

$$= (8 + (-2))x + 5y$$

$$= 6x \qquad + 5y$$

* When an example requires simplification of an expression, that expression will be screened. The simplification will then follow the equal–sign.

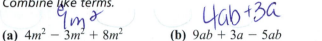

Like terms   Like terms

NOTE With practice you won't be writing out these steps, but will be doing it mentally.

(c) $5m + 8n \quad + 4m - 3n$

$= (5m + 4m) + (8n + (-3n))$

$= \quad 9m \quad + \quad 5n$

Here we have used both the associative and commutative properties.

(d) $4x^2 + 2x - 3x^2 + x$

$= (4x^2 + (-3x^2)) + (2x + x)$

$= x^2 + 3x$

As these examples illustrate, combining like terms often means changing the grouping and the order in which the terms are written. Again all this is possible because of the properties of addition that we introduced in Section 1.2.

 CHECK YOURSELF 5

Combine like terms.

(a) $4m^2 - 3m^2 + 8m^2$ **(b)** $9ab + 3a - 5ab$ **(c)** $4p + 7q + 5p - 3q$

As you have seen in arithmetic, subtraction can be performed directly. As this is the form used for most of mathematics, we will use that form throughout this text. Just remember, by using negative numbers, you can always rewrite a subtraction problem as an addition problem.

Example 6   Combining Like Terms

Combine the like terms.

(a) $2xy - 3xy + 5xy$

$= (2 - 3 + 5)xy$

$= 4xy$

(b) $5a - 2b + 7b - 8a$

$= (5a - 8a) + (-2b + 7b)$

$= -3a + 5b$

 CHECK YOURSELF 6

Combine like terms.

(a) $4ab + 5ab - 3ab - 7ab$ **(b)** $2x - 7y - 8x - y$

CHECK YOURSELF ANSWERS

1. (a) $2b^4$; **(b)** $5m, 3n$; **(c)** $2s^2, -3t, -6$ **2. (a)** 8; **(b)** -5; **(c)** 1

3. The like terms are $5a^2b, a^2b$, and $-7a^2b$ **4. (a)** $14b$; **(b)** $9x^2$; **(c)** xy^3; **(d)** 0

5. (a) $9m^2$; **(b)** $4ab + 3a$; **(c)** $9p + 4q$ **6. (a)** $-ab$; **(b)** $-6x - 8y$

1.6 **Exercises**

List the terms of the following expressions.

1. $5a + 2$

2. $7a - 4b$

3. $4x^3$

4. $3x^2$

5. $3x^2 + 3x - 7$

6. $2a^3 - a^2 + a$

Circle the like terms in the following groups of terms.

7. $(5ab,$ $3b,$ $3a,$ $(4ab)$

8. $(9m^2),$ $8mn,$ $(5m^2),$ $7m$

9. $4xy^2,$ $(2x^2y,)$ $5x^2,$ $(-3x^2y,)$ $5y,$ $(6x^2y)$

10. $(8a^2b,)$ $4a^2,$ $3ab^2,$ $(-5a^2b,)$ $3ab,$ $(5a^2b)$

Combine the like terms.

11. $4m + 6m$

12. $6a^2 + 8a^2$

13. $7b^3 + 10b^3$

14. $7rs + 13rs$

15. $21xyz + 7xyz$

16. $-3mn^2 + 9mn^2$

17. $9z^2 - 3z^2$

18. $7m - 6m$

19. $9a^5 - 9a^5$

20. $13xy - 9xy$

21. $19n^2 - 18n^2$

22. $7cd - 7cd$

23. $21p^2q - 6p^2q$

24. $17r^3s^2 - 8r^3s^2$

25. $5x^2 - 3x^2 + 9x^2$

26. $13uv + uv - 12uv$

27. $11b - 9a - 6b$

28. $5m^2 - 3m + 6m^2$

29. $7x + 5y - 4x - 4y$

30. $6a^2 + 11a + 7a^2 - 9a$

31. $4a + 7b + 3 - 2a + 3b - 2$

32. $5p^2 + 2p + 8 + 4p^2 + 5p - 6$

ANSWERS

1. $5a, 2$
2. $7a, -4b$
3. $4, x^3$
4. $3, x^2$
5. $3x^2, 3x, -7$
6. $2a^3, -1a^2, a$
7.
8.
9.
10.
11. $10m$ 12. $14a^2$
13. $17b^3$ 14. $20rs$
15. $28xyz$ 16. $6mn^2$
17. $6z^3$ 18. $-m$
19. a^5 20. $4xy$
21. n^2 22. cd
23. $15p^2q$
24. $9r^3s^2$
25. $11x^2$
26. $2uv$
27. $5b - 9a$
28. $11m^2 - 3m$
29. $3x + y$
30. $13a^2 + 2a$
31. $2a + 10b - 1$
32. $9p^2 + 7p + 2$

33. $\dfrac{2}{3}m + 3 + \dfrac{4}{3}m$

34. $\dfrac{1}{5}a - 2 + \dfrac{4}{5}a$

35. $\dfrac{13}{5}x + 2 - \dfrac{3}{5}x + 5$

36. $\dfrac{17}{12}y + 7 + \dfrac{7}{12}y - 3$

37. $2.3a + 7 + 4.7a + 3$

38. $5.8m + 4 - 2.8m + 11$

8.6

Rewrite as an algebraic expression.

39. Find the sum of $5a^4$ and $8a^4$.

40. Find the sum of $9p^2$ and $12p^2$.

41. Subtract $12a^3$ from $15a^3$.

42. Subtract $5m^3$ from $18m^3$.

43. Subtract $6x$ from the sum of $11x$ and $2x$.

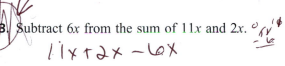

$11x + 2x - 6x$

44. Subtract $8ab$ from the sum of $7ab$ and $5ab$.

$12ab - 8ab$

45. Subtract $3mn^2$ from the sum of $9mn^2$ and $5mn^2$.

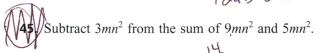

46. Subtract $4x^2y$ from the sum of $6x^2y$ and $12x^2y$.

Use the distributive property to remove the parentheses in each expression. Then simplify by combining like terms.

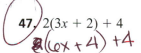

47. $2(3x + 2) + 4$

$2(6x + 4) + 4$

48. $3(4z + 5) - 9$

$12z + 15 - 9$

$12z + 6$

49. $5(6a - 2) + 12a$

50. $7(4w - 3) - 25w$

51. $4s + 2(s + 4) + 4$

52. $5p + 4(p + 3) - 8$

53. Geometry. Find an expression for the perimeter of the given rectangle.

$(2x^2 - x + 1 \text{ cm})$

$(3x - 2 \text{ cm})$

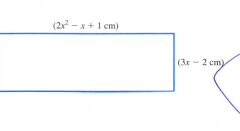

ANSWERS

33. $8\dfrac{6}{3}m + 3$

34. $a - 2$

35. $2x + 7$

36. $2y + 4$

37. $7a + 10$

38. $3.8m + 15$

39. $13a^4$

40. $21p^2$

41. $3a^3$

42. $13m^3$

43. $5x$

44. $4ab$

45. $11mn^2$

46. $14x^2y$

47. $6x + 8$

48. $12z + 6$

49. _____

50. _____

51. _____

52. _____

53. _____

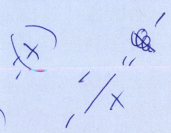

ANSWERS

54. $2x^2 - x + 4$

55. $7x - 1$

56. ____

57. $-x^2 + 65x - 150$

58. ____

59.

60.

61.

62.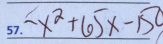

63. ____

---

54. Geometry. Find an expression for the perimeter of the given triangle.

x ft   3x + 3 ft

$2x^2 - 5x + 1$ ft

55. Geometry. A rectangle has sides of $8x + 9$ and $6x - 7$. Find the expression that represents its perimeter.

56. Geometry. A triangle has sides $3x + 7$, $4x - 9$, and $5x + 6$. Find the expression that represents its perimeter.

57. Business and finance. The cost of producing x units of an item is $C = 150 + 25x$. The revenue for selling x units is $R = 90x - x^2$. The profit is given by the revenue minus the cost. Find the expression that represents profit.

58. Business and finance. The revenue for selling y units is $R = 3y^2 - 2y + 5$ and the cost of producing y units is $C = y^2 + y - 3$. Find the expression that represents profit.

59. Write a paragraph explaining the difference between n^2 and $2n$.

60. Complete the explanation: "x^3 and $3x$ are not the same because . . ."

61. Complete the statement: "$x + 2$ and $2x$ are different because . . ."

62. Write an English phrase for each given algebraic expression:

    (a) $2x^3 + 5x$ (b) $(2x + 5)^3$ (c) $6(n + 4)^2$

63. Work with another student to complete this exercise. Place $>$, $<$, or $=$ in the blank in these statements.

1^2 ____ 2^1

2^3 ____ 3^2 What happens as the table of numbers is extended? Try more examples.

3^4 ____ 4^3 What sign seems to occur the most in your table? $>$, $<$, or $=$?

4^5 ____ 5^4

Write an algebraic statement for the pattern of numbers in this table. Do you think this is a pattern that continues? Add more lines to the table and extend the pattern to the general case by writing the pattern in algebraic notation. Write a short paragraph stating your conjecture.

64. Work with other students on this exercise.

Part 1: Evaluate the three expressions $\dfrac{n^2 - 1}{2}, n, \dfrac{n^2 + 1}{2}$ using odd values of n: 1, 3, 5, 7, etc. Make a chart like the one below and complete it.

n	$a = \dfrac{n^2 - 1}{2}$	$b = n$	$c = \dfrac{n^2 + 1}{2}$	a^2	b^2	c^2
1						
3						
5						
7						
9						
11						
13						

Part 2: The numbers a, b, and c that you get in each row have a surprising relationship to each other. Complete the last three columns and work together to discover this relationship. You may want to find out more about the history of this famous number pattern.

Getting Ready for Section 1.7 [Section 0.4]

Write the following using exponential notation.

(a) $4 \cdot 4 \cdot 4$ (b) $6 \cdot 6 \cdot 6 \cdot 6 \cdot 6 \cdot 6$

(c) $3 \cdot 3 \cdot 3 \cdot 3 \cdot 3$ (d) $(-2) \cdot (-2) \cdot (-2)$

(e) $(-8) \cdot (-8) \cdot (-8) \cdot (-8)$ (f) $9 \cdot 9 \cdot 9 \cdot 9 \cdot 9 \cdot 9 \cdot 9 \cdot 9$

Answers

1. $5a, 2$ **3.** $4x^3$ **5.** $3x^2, 3x, -7$ **7.** $5ab, 4ab$ **9.** $2x^2y, -3x^2y, 6x^2y$

11. $10m$ **13.** $17b^3$ **15.** $28xyz$ **17.** $6z^2$ **19.** 0 **21.** n^2

23. $15p^2q$ **25.** $11x^2$ **27.** $-9a + 5b$ **29.** $3x + y$

31. $2a + 10b + 1$ **33.** $2m + 3$ **35.** $2x + 7$ **37.** $7a + 10$

39. $13a^4$ **41.** $3a^3$ **43.** $7x$ **45.** $11mn^2$ **47.** $6x + 8$

49. $42a - 10$ **51.** $6s + 12$ **53.** $4x^2 + 4x - 2$

55. $28x + 4$ **57.** $-x^2 + 65x - 150$

59.   **61.**   **63.**   **a.** 4^3 **b.** 6^6 **c.** 3^5

d. $(-2)^3$ **e.** $(-8)^4$ **f.** 9^8

1.7 Multiplying and Dividing Terms

1.7 OBJECTIVES

1. Find the product of two algebraic terms using the first property of exponents
2. Find the quotient of two algebraic terms using the second property of exponents

In Section 0.4, we introduced exponential notation. Remember that the exponent tells us how many times the base is to be used as a factor.

Exponent

$$2^5 = 2 \cdot 2 \cdot 2 \cdot 2 \cdot 2 = 32$$

Base   The fifth power of 2

NOTE In general,

$$x^m = \underbrace{x \cdot x \cdots \cdot x}_{m \text{ factors}}$$

where m is a natural number. **Natural numbers** are the numbers we use for counting: 1, 2, 3, and so on.

The notation can also be used when working with letters or variables.

$$x^4 = \underbrace{x \cdot x \cdot x \cdot x}_{4 \text{ factors}}$$

Now look at the product $x^2 \cdot x^3$.

$$x^2 \cdot x^3 = \underbrace{(x \cdot x)}\underbrace{(x \cdot x \cdot x)} = \underbrace{x \cdot x \cdot x \cdot x \cdot x} = x^5$$

2 factors + 3 factors = 5 factors

So

NOTE Note that the exponent of x^5 is the *sum* of the exponents in x^2 and x^3.

$$x^2 \cdot x^3 = x^{2+3} = x^5$$

This leads us to the following property of exponents.

Rules and Properties: Property 1 of Exponents

For any positive integers m and n and any real number a,

$$a^m \cdot a^n = a^{m+n}$$

In words, to multiply expressions with the same base, keep the base and add the exponents.

OBJECTIVE 1

Example 1   Using the First Property of Exponents

C A U T I O N

The product is *not* 9^6. The base does not change.

(a) $a^5 \cdot a^7 = a^{5+7} = a^{12}$

(b) $x \cdot x^8 = x^1 \cdot x^8 = x^{1+8} = x^9$ $x = x^1$

(c) $3^2 \cdot 3^4 = 3^{2+4} = 3^6$

(d) $y^2 \cdot y^3 \cdot y^5 = y^{2+3+5} = y^{10}$

(e) $x^3 \cdot y^4$ *cannot* be simplified. The bases are not the same.

144

CHECK YOURSELF 1

Multiply.

(a) $b^6 \cdot b^8$ **(b)** $y^7 \cdot y$ **(c)** $2^3 \cdot 2^4$ **(d)** $a^2 \cdot a^4 \cdot a^3$

Suppose that numerical coefficients (other than 1) are involved in a product. To find the product, multiply the coefficients and then use the first property of exponents to combine the variables.

NOTE Note that although we have several factors, this is still a single term.

$$2x^3 \cdot 3x^5 = (2 \cdot 3)(x^3 \cdot x^5) \qquad \text{Multiply the coefficients.}$$
$$= 6x^{3+5} \qquad\qquad\qquad \text{Add the exponents.}$$
$$= 6x^8$$

You may have noticed that we have again changed the order and grouping. This method uses the commutative and associative properties of Section 1.2.

Example 2   Using the First Property of Exponents

Multiply.

NOTE Again we have written out all the steps. You can do the multiplication mentally with practice.

(a) $5a^4 \cdot 7a^6 = (5 \cdot 7)(a^4 \cdot a^6) = 35a^{10}$

(b) $y^2 \cdot 3y^3 \cdot 6y^4 = (1 \cdot 3 \cdot 6)(y^2 \cdot y^3 \cdot y^4) = 18y^9$

(c) $2x^2y^3 \cdot 3x^5y^2 = (2 \cdot 3)(x^2 \cdot x^5)(y^3 \cdot y^2) = 6x^7y^5$

CHECK YOURSELF 2

Multiply.

(a) $4x^3 \cdot 7x^5$ **(b)** $3a^2 \cdot 2a^4 \cdot 2a^5$ **(c)** $3m^2n^4 \cdot 5m^3n$

What about dividing expressions when exponents are involved? For instance, what if we want to divide x^5 by x^2? We can use the following approach to division:

$$\frac{x^5}{x^2} = \frac{\overbrace{x \cdot x \cdot x \cdot x \cdot x}^{5 \text{ factors}}}{\underbrace{x \cdot x}_{2 \text{ factors}}} = \frac{x \cdot x \cdot x \cdot x \cdot x}{x \cdot x}$$

We can divide by 2 factors of *x*.

$$= \overbrace{x \cdot x \cdot x}^{3 \text{ factors}} = x^3$$

So

NOTE Note that the exponent of x^3 is the *difference* of the exponents in x^5 and x^2.

$$\frac{x^5}{x^2} = x^{5-2} = x^3$$

This leads us to a second property of exponents.

> **Rules and Properties:**   Property 2 of Exponents
>
> For any positive integers m and n, in which m is greater than n, and any real number a, in which a is not equal to zero,
>
> $$\frac{a^m}{a^n} = a^{m-n}$$
>
> In words, to divide expressions with the same base, keep the base and subtract the exponents.

OBJECTIVE 2   **Example 3   Using the Second Property of Exponents**

Divide the following.

(a) $\dfrac{y^7}{y^3} = y^{7-3} = y^4$

(b) $\dfrac{m^6}{m} = \dfrac{m^6}{m^1} = m^{6-1} = m^5$ Apply the second property to each variable separately.

(c) $\dfrac{a^3 b^5}{a^2 b^2} = a^{3-2} \cdot b^{5-2} = ab^3$

CHECK YOURSELF 3

Divide.

(a) $\dfrac{m^9}{m^6}$ **(b)** $\dfrac{a^8}{a}$ **(c)** $\dfrac{a^3 b^5}{a^2}$ **(d)** $\dfrac{r^5 s^6}{r^3 s^2}$

If numerical coefficients are involved, just divide the coefficients and then use the second law of exponents to divide the variables. Look at Example 4.

Example 4   Using the Second Property of Exponents

Divide the following.

Subtract the exponents.

(a) $\dfrac{6x^5}{3x^2} = 2x^{5-2} = 2x^3$

6 divided by 3

20 divided by 5

(b) $\dfrac{20a^7b^5}{5a^3b^4} = 4a^{7-3} \cdot b^{5-4}$

Again apply the second property to each variable separately.

$= 4a^4b$

 CHECK YOURSELF 4

Divide.

(a) $\dfrac{4x^3}{2x}$　　　　　　　　**(b)** $\dfrac{20a^6}{5a^2}$　　　　　　　　**(c)** $\dfrac{24x^5y^3}{4x^2y^2}$

CHECK YOURSELF ANSWERS

1. (a) b^{14}; **(b)** y^8; **(c)** 2^7; **(d)** a^9　　**2. (a)** $28x^8$; **(b)** $12a^{11}$; **(c)** $15m^5n^5$
3. (a) m^3; **(b)** a^7; **(c)** ab^5; **(d)** r^2s^4　　　**4. (a)** $2x^2$; **(b)** $4a^4$; **(c)** $6x^3y$

1.7 Exercises

Multiply.

1. $x^5 \cdot x^7$ **2.** $b^2 \cdot b^4$

3. $3^2 \cdot 3^6$ **4.** $y^6 \cdot y^4$

5. $a^9 \cdot a$ **6.** $3^4 \cdot 3^5$

7. $z^{10} \cdot z^3$ **8.** $x^6 \cdot x^3$

9. $p^5 \cdot p^7$ **10.** $s^6 \cdot s^9$

11. $x^3y \cdot x^2y^4$ **12.** $m^2n^3 \cdot mn^4$

13. $w^3 \cdot w^4 \cdot w^2$ **14.** $x^5 \cdot x^4 \cdot x^6$

15. $m^3 \cdot m^2 \cdot m^4$ **16.** $r^3 \cdot r \cdot r^5$

17. $a^3b \cdot a^2b^2 \cdot ab^3$ **18.** $w^2z^3 \cdot wz \cdot w^3z^4$

19. $p^2q \cdot p^3q^5 \cdot pq^4$ **20.** $c^3d \cdot c^4d^2 \cdot cd^5$

21. $2a^5 \cdot 3a^2$ **22.** $5x^3 \cdot 3x^2$

23. $x^2 \cdot 3x^5$ **24.** $2m^4 \cdot 6m^7$

25. $5m^3n^2 \cdot 4mn^3$ **26.** $7x^2y^5 \cdot 6xy^4$

27. $4x^5y \cdot 3xy^2$ **28.** $5a^3b \cdot 10ab^4$

29. $2a^2 \cdot a^3 \cdot 3a^7$ **30.** $2x^3 \cdot 3x^4 \cdot x^5$

31. $3c^2d \cdot 4cd^3 \cdot 2c^5d$ **32.** $5p^2q \cdot p^3q^2 \cdot 3pq^3$

33. $5m^2 \cdot m^3 \cdot 2m \cdot 3m^4$ **34.** $3a^3 \cdot 2a \cdot a^4 \cdot 2a^5$

35. $2r^3s \cdot rs^2 \cdot 3r^2s \cdot 5rs$ **36.** $6a^2b \cdot ab \cdot 3ab^3 \cdot 2a^2b$

Divide.

Zero power = 1 (handwritten)

37. $\dfrac{a^{10}}{a^7}$

38. $\dfrac{m^8}{m^2}$

39. $\dfrac{y^{10}}{y^4}$

40. $\dfrac{b^9}{b^4}$

41. $\dfrac{p^{15}}{p^{10}}$

42. $\dfrac{s^{15}}{s^9}$

43. $\dfrac{x^5 y^3}{x^2 y^2}$

44. $\dfrac{s^5 t^4}{s^3 t^2}$

45. $\dfrac{10m^6}{5m^4}$

46. $\dfrac{8x^5}{4x}$

47. $\dfrac{24a^7}{6a^4}$

48. $\dfrac{25x^9}{5x^8}$

49. $\dfrac{26m^8 n}{13m^6}$

50. $\dfrac{30a^4 b^5}{6b^4}$

51. $\dfrac{35w^4 z^6}{5w^2 z}$

52. $\dfrac{48p^6 q^7}{8p^4 q}$

53. $\dfrac{48x^4 y^5 z^9}{24x^2 y^3 z^6}$

54. $\dfrac{25a^5 b^4 c^3}{5a^4 b c^2}$

Simplify each of the following expressions where possible.

55. $3a^4 b^3 \cdot 2a^2 b^4$

56. $2xy^3 \cdot 3xy^2$

57. $2a^3 b + 3a^2 b$

58. $2xy^3 + 3xy^2$

59. $2x^2 y^3 \cdot 3x^2 y^3$

60. $5a^3 b^2 \cdot 10a^3 b^2$

61. $2x^3 y^2 + 3x^3 y^2$

62. $5a^3 b^2 + 10a^3 b^2$

63. $\dfrac{8a^2 b \cdot 6a^3 b}{2ab}$ (handwritten: $\dfrac{^4 8a^2 b \cdot 6a^3 b}{2ab}$ $12a^4 b^2$)

64. $\dfrac{6x^2 y^3 \cdot 9x^2 y^3}{3x^2 y^2}$ (handwritten: $\dfrac{\cancel{6}x^{2}y^{3}\cdot \cancel{9}x^{2}y^{3}}{\cancel{3x^{2}y^{2}}}$ $x^0 y^1$ $24a^3 b^1$ $6\ \dfrac{54x^4 y^6}{3x^2 y^2}$)

ANSWERS (handwritten)

37. a^3

38. m^6

39. y^6

40. b^5

41. p^5

42. s^6

43. $x^3 y^1$

44. $s^2 t^2$

45. $2m^2$

46. $2x^4$

47. $4a^3$

48. $5x^1$

49. $2m^2 n$

50. $5a^4 b^1$

51. $7w^2 z^5$

52. $6p^2 q^6$

53. $2x^2 y^2 z^3$

54. $5a b^3 c^1$

55. $6a^6 b^7$

56. $6x^2 y^5$

57. $5a$

58. $5x^2 y^5$

59. $6x^4 y^6$

60. $50a^6 b^4$

61. $5x^6 y^4$

62. $15a^6 b^4$

63. (scribbled out)

64. $18x^2 y^8 4$

65. $\dfrac{8a^2b + 6a^2b}{2ab}$

66. $\dfrac{6x^2y^3 + 9x^2y^3}{3x^2y^2}$

67. Complete the following statements:

   (a) a^n is negative when _____ because _____.

   (b) a^n is positive when _____ because _____.

   (give all possibilities)

68. "Earn Big Bucks!" reads an ad for a job. "You will be paid 1 cent for the first day and 2 cents for the second day, 4 cents for the third day, 8 cents for the fourth day, and so on, doubling each day. Apply now!" What kind of deal is this—where is the big money offered in the headline? The fine print at the bottom of the ad says: "Highly qualified people may be paid \$1,000,000 for the first 28 working days if they choose." Well, *that* does sound like big bucks! Work with other students to decide which method of payment is better and how much better. You may want to make a table and try to find a formula for the first offer.

69. An oil spill from a tanker in pristine Prince Williams Sound in Alaska begins in a circular shape only 2 ft across. The area of the circle is $A = \pi r^2$. Make a table to decide what happens to the area if the diameter is doubling each hour. How large will the spill be in 24 h? (**Hint:** The radius is one-half the diameter.)

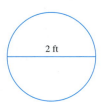

2 ft

Answers

1. x^{12} **3.** 3^8 **5.** a^{10} **7.** z^{13} **9.** p^{12} **11.** x^5y^5 **13.** w^9
15. m^9 **17.** a^6b^6 **19.** p^6q^{10} **21.** $6a^7$ **23.** $3x^7$ **25.** $20m^4n^5$
27. $12x^6y^3$ **29.** $6a^{12}$ **31.** $24c^8d^5$ **33.** $30m^{10}$ **35.** $30r^7s^5$ **37.** a^3
39. y^6 **41.** p^5 **43.** x^3y **45.** $2m^2$ **47.** $4a^3$ **49.** $2m^2n$
51. $7w^2z^5$ **53.** $2x^2y^2z^3$ **55.** $6a^6b^7$ **57.** Cannot simplify **59.** $6x^4y^6$
61. $5x^3y^2$ **63.** $24a^3b$ **65.** $7a$ **67.** **69.**

1 Summary

DEFINITION/PROCEDURE	EXAMPLE	REFERENCE
From Arithmetic to Algebra		**Section 1.1**
Addition $x + y$ means the **sum** of x **and** y or x **plus** y. Some other words indicating addition are "more than" and "increased by."	The sum of x and 5 is $x + 5$. 7 more than a is $a + 7$. b increased by 3 is $b + 3$.	p. 78
Subtraction $x - y$ means the **difference** of x **and** y or x **minus** y. Some other words indicating subtraction are "less than" and "decreased by."	The difference of x and 3 is $x - 3$. 5 less than p is $p - 5$. a decreased by 4 is $a - 4$.	p. 78
Multiplication $\left.\begin{array}{l} x \cdot y \\ (x)(y) \\ xy \end{array}\right\}$ These all mean the *product* of x and y or x *times* y.	The product of m and n is mn. The product of 2 and the sum of a and b is $2(a + b)$.	p. 79
Expressions An expression is a meaningful collection of numbers, variables, and signs of operation.	$3x + y$ is an expression $3x = y$ is not an expression	p. 79
Division $\dfrac{x}{y}$ means x *divided by* y or the *quotient* when x is divided by y.	n divided by 5 is $\dfrac{n}{5}$. The sum of a and b, divided by 3, is $\dfrac{a + b}{3}$.	p. 80
Properties of Real Numbers		**Section 1.2**
The Commutative Properties If a and b are any numbers, **1.** $a + b = b + a$ **2.** $a \cdot b = b \cdot a$	$3 + 8 = 8 + 3$ $2 \cdot 5 = 5 \cdot 2$	p. 87
The Associative Properties If a, b, and c are any numbers, **1.** $a + (b + c) = (a + b) + c$ **2.** $a \cdot (b \cdot c) = (a \cdot b) \cdot c$	$3 + (7 + 12) = (3 + 7) + 12$ $2 \cdot (5 \cdot 12) = (2 \cdot 5) \cdot 12$	p. 87
The Distributive Property If a, b, and c are any numbers, $a(b + c) = a \cdot b + a \cdot c$	$6 \cdot (8 + 15) = 6 \cdot 8 + 6 \cdot 15$	p. 89
Adding and Subtracting Real Numbers		**Section 1.3**
Addition **1.** If two numbers have the same sign, add their absolute values. Give the sum the sign of the original numbers. **2.** If two numbers have different signs, subtract their absolute values, the smaller from the larger. Give the result the sign of the number with the larger absolute value.	$9 + 7 = 16$ $(-9) + (-7) = -16$ $15 + (-10) = 5$ $(-12) + 9 = -3$	p. 96 p. 97

Continued

Adding and Subtracting Numbers with Different Signs		Section 1.3

Subtraction
1. Rewrite the subtraction problem as an addition problem by
  a. Changing the subtraction symbol to an addition symbol
  b. Replacing the number being subtracted with its opposite
2. Add the resulting signed numbers as before.
	$16 - 8 = 16 + (-8)$
	$= 8$
	$8 - 15 = 8 + (-15)$
	$= -7$
	$-9 - (-7) = -9 + 7$
	$= -2$
		p. 100

Multiplying and Dividing Real Numbers		Section 1.4
Multiplication Multiply the absolute values of the two numbers. **1.** If the numbers have different signs, the product is negative. **2.** If the numbers have the same sign, the product is positive.	$5(-7) = -35$ $(-10)(9) = -90$ $8 \cdot 7 = 56$ $(-9)(-8) = 72$	*p. 111* *p. 112*
Division Divide the absolute values of the two numbers. **1.** If the numbers have different signs, the quotient is negative. **2.** If the numbers have the same sign, the quotient is positive.	$\dfrac{-32}{4} = -8$ $\dfrac{75}{-5} = -15$ $\dfrac{20}{5} = 4$ $\dfrac{-18}{-9} = 2$	*p. 114*

Evaluating Algebraic Expressions		Section 1.5
Evaluating Algebraic Expressions To evaluate an algebraic expression: **1.** Replace each variable or letter with its number value. **2.** Do the necessary arithmetic, following the rules for the order of operations.	Evaluate $2x + 3y$ if $x = 5$ and $y = -2$. $2x + 3y$ $= 2(5) + (3)(-2)$ $= 10 - 6 = 4$	*p. 126*

Adding and Subtracting Terms		Section 1.6
Term A number or the product of a number and one or more variables.		*p. 136*
Combining Like Terms To combine like terms: **1.** Add or subtract the coefficients (the numbers multiplying the variables). **2.** Attach the common variable.	$5x + 2x = 7x$ $5 + 2$ $8a - 5a = 3a$ $8 - 5$	*p. 138*

Multiplying and Dividing Terms		Section 1.7
Property 1 of Exponents $$a^m \cdot a^n = a^{m+n}$$	$x^7 \cdot x^3 = x^{7+3} = x^{10}$	*p. 144*
Property 2 of Exponents $$\dfrac{a^m}{a^n} = a^{m-n}$$	$\dfrac{y^7}{y^3} = y^{7-3} = y^4$	*p. 146*

Summary Exercises

This summary exercise set is provided to give you practice with each of the objectives of this chapter. Each exercise is keyed to the appropriate chapter section. When you are finished, you can check your answers to the odd-numbered exercises against those presented in the back of the text. If you have difficulty with any of these questions, go back and reread the examples from that section. Your instructor will give you guidelines on how to best use these exercises in your instructional setting.

[1.1]   Write, using symbols.

1. 5 more than y

2. c decreased by 10

3. The product of 8 and a

4. The quotient when y is divided by 3

5. 5 times the product of m and n

6. The product of a and 5 less than a

7. 3 more than the product of 17 and x

8. The quotient when a plus 2 is divided by a minus 2

Identify which are expressions and which are not.

9. $4(x + 3)$

10. $7 \div \cdot 8$

11. $y + 5 = 9$

12. $11 + 2(3x - 9)$

[1.2]   Identify the property that is illustrated by each of the following statements.

13. $5 + (7 + 12) = (5 + 7) + 12$

14. $2(8 + 3) = 2 \cdot 8 + 2 \cdot 3$

15. $4 \cdot (5 \cdot 3) = (4 \cdot 5) \cdot 3$

16. $4 \cdot 7 = 7 \cdot 4$

Verify that each of the following statements is true by evaluating each side of the equation separately and comparing the results.

17. $8(5 + 4) = 8 \cdot 5 + 8 \cdot 4$

18. $2(3 + 7) = 2 \cdot 3 + 2 \cdot 7$

19. $(7 + 9) + 4 = 7 + (9 + 4)$

20. $(2 + 3) + 6 = 2 + (3 + 6)$

21. $(8 \cdot 2) \cdot 5 = 8(2 \cdot 5)$

22. $(3 \cdot 7) \cdot 2 = 3 \cdot (7 \cdot 2)$

Use the distributive law to remove parentheses.

23. $3(7 + 4)$

24. $4(2 + 6)$

25. $4(w + v)$

26. $6(x + y)$

27. $3(5a + 2)$

28. $2(4x^2 + 3x)$

[1.3] Add.

29. $-3 + (-8)$

30. $10 + (-4)$

31. $6 + (-6)$

32. $-16 + (-16)$

33. $-18 + 0$

34. $\dfrac{3}{8} + \left(-\dfrac{11}{8}\right)$

35. $5.7 + (-9.7)$

36. $-18 + 7 + (-3)$

Subtract.

37. $8 - 13$

38. $-7 - 10$

39. $10 - (-7)$

40. $-5 - (-1)$

41. $-9 - (-9)$

42. $0 - (-2)$

43. $-\dfrac{5}{4} - \left(-\dfrac{17}{4}\right)$

44. $7.9 - (-8.1)$

Find the median for each of the following sets.

45. $2, 4, 9, 10, 15$

46. $-7, -3, 2, 4, 5$

47. $-3, -8, 4, 1, 6$

48. $6, -3, 2, -5, 1$

49. $2, 4, 1, 8, 6, 7$

50. $-3, -1, -5, 3, 4, 1$

Determine the range for each of the following sets.

51. $3, 5, 1, 8, 9$

52. $-4, -5, 6, 4, 2, 1$

53. $-5, 2, -1, 3, 8$

54. $7, 3, 5, 3, -4$

[1.4] Multiply.

55. $(10)(-7)$

56. $(-8)(-5)$

57. $(-3)(-15)$

58. $(1)(-15)$

59. $(0)(-8)$

60. $\left(\dfrac{2}{3}\right)\left(-\dfrac{3}{2}\right)$

61. $(-4)\left(\dfrac{3}{8}\right)$

62. $\left(-\dfrac{5}{4}\right)(-1)$

Divide.

63. $\dfrac{80}{16}$

64. $\dfrac{-63}{7}$

65. $\dfrac{-81}{-9}$

66. $\dfrac{0}{-5}$

67. $\dfrac{32}{-8}$

68. $\dfrac{-7}{0}$

Perform the indicated operations.

69. $\dfrac{-8 + 6}{-8 - (-10)}$

70. $\dfrac{-6 - 1}{5 - (-2)}$

71. $\dfrac{25 - 4}{-5 - (-2)}$

[1.5] Evaluate each of the following expressions.

72. $18 - 3 \cdot 5$

73. $(18 - 3) \cdot 5$

74. $5 \cdot 4^2$

75. $(5 \cdot 4)^2$

76. $5 \cdot 3^2 - 4$

77. $5(3^2 - 4)$

78. $5(4 - 2)^2$

79. $5 \cdot 4 - 2^2$

80. $(5 \cdot 4 - 2)^2$

81. $3(5 - 2)^2$

82. $3 \cdot 5 - 2^2$

83. $(3 \cdot 5 - 2)^2$

Evaluate the expressions if $x = -3$, $y = 6$, $z = -4$, and $w = 2$.

84. $3x + w$

85. $5y - 4z$

86. $x + y - 3z$

87. $5z^2$

88. $3x^2 - 2w^2$

89. $3x^3$

90. $5(x^2 - w^2)$

91. $\dfrac{6z}{2w}$

92. $\dfrac{2x - 4z}{y - z}$

93. $\dfrac{3x - y}{w - x}$

94. $\dfrac{x(y^2 - z^2)}{(y + z)(y - z)}$

95. $\dfrac{y(x - w)^2}{x^2 - 2xw + w^2}$

[1.6] List the terms of the expressions.

96. $4a^3 - 3a^2$

97. $5x^2 - 7x + 3$

Circle like terms.

98. $5m^2, -3m, -4m^2, 5m^3, m^2$

99. $4ab^2, 3b^2, -5a, ab^2, 7a^2, -3ab^2, 4a^2b$

Combine like terms.

100. $5c + 7c$

101. $2x + 5x$

102. $4a - 2a$

103. $6c - 3c$

104. $9xy - 6xy$

105. $5ab^2 + 2ab^2$

106. $7a + 3b + 12a - 2b$

107. $6x - 2x + 5y - 3x$

108. $5x^3 + 17x^2 - 2x^3 - 8x^2$

109. $3a^3 + 5a^2 + 4a - 2a^3 - 3a^2 - a$

110. Subtract $4a^3$ from the sum of $2a^3$ and $12a^3$.

111. Subtract the sum of $3x^2$ and $5x^2$ from $15x^2$.

[1.7]   Simplify.

112. $\dfrac{x^{10}}{x^3}$

113. $\dfrac{a^5}{a^4}$

114. $\dfrac{x^2 \cdot x^3}{x^4}$

115. $\dfrac{m^2 \cdot m^3 \cdot m^4}{m^5}$

116. $\dfrac{18p^7}{9p^5}$

117. $\dfrac{24x^{17}}{8x^{13}}$

118. $\dfrac{30m^7n^5}{6m^2n^3}$

119. $\dfrac{108x^9y^4}{9xy^4}$

120. $\dfrac{48p^5q^3}{6p^3q}$

121. $\dfrac{52a^5b^3c^5}{13a^4c}$

122. $(4x^3)(5x^4)$

123. $(3x)^2(4xy)$

124. $(8x^2y^3)(3x^3y^2)$

125. $(-2x^3y^3)(-5xy)$

126. $(6x^4)(2x^2y)$

Write the algebraic expression that answers the question.

127. Construction. If x ft are cut off the end of a board that is 23 ft long, how much is left?

128. Business and finance. Joan has 25 nickels and dimes in her pocket. If x of these are dimes, how many of the coins are nickels?

129. Social science. Sam is 5 years older than Angela. If Angela is x years old now, how old is Sam?

130. Business and finance. Margaret has $5 more than twice as much money as Gerry. Write an expression for the amount of money that Margaret has.

131. Geometry. The length of a rectangle is 4 m more than the width. Write an expression for the length of the rectangle.

132. Number problem. A number is 7 less than 6 times the number n. Write an expression for the number.

133. Construction. A 25-ft plank is cut into two pieces. Write expressions for the length of each piece.

134. Business and finance. Bernie has x dimes and q quarters in his pocket. Write an expression for the amount of money that Bernie has in his pocket.

Self-Test for Chapter 1

The purpose of this self-test is to help you check your progress and to review for the next in-class exam. Allow yourself about an hour to take this test. At the end of that hour check your answers against those given in the back of the text. Section references accompany the answers. If you missed any questions, go back to those sections and reread the examples until you master the concepts.

Write, using symbols.

1. 5 less than *a*

2. The product of 6 and *m*

3. 4 times the sum of *m* and *n*

4. The quotient when the sum of *a* and *b* is divided by 3

Identify the property that is illustrated by each of the following statements.

5. $6 \cdot 7 = 7 \cdot 6$

6. $2(6 + 7) = 2 \cdot 6 + 2 \cdot 7$

7. $4 + (3 + 7) = (4 + 3) + 7$

Use the distributive property to remove parentheses. Then simplify your result.

8. $3(5 + 2)$

9. $4(5x + 3)$

Identify which are expressions and which are not.

10. $5x + 6 = 4$

11. $4 + (6 + x)$

Add.

12. $-8 + (-5)$

13. $6 + (-9)$

14. $(-9) + (-12)$

15. $-\dfrac{5}{3} + \dfrac{8}{3}$

Subtract.

16. $9 - 15$

17. $-10 - 11$

18. $5 - (-4)$

19. $-7 - (-7)$

Find the median of each of the following sets.

20. 2, 4, 5, 7, 8, 3, 10

21. −4, 6, −1, −9, 3, 7, −6, 11

ANSWERS

1. _____
2. _____
3. _____
4. _____
5. _____
6. _____
7. _____
8. _____
9. _____
10. _____
11. _____
12. _____
13. _____
14. _____
15. _____
16. _____
17. _____
18. _____
19. _____
20. _____
21. _____

22. _____

23. _____

24. _____

25. _____

26. _____

27. _____

28. _____

29. _____

30. _____

31. _____

32. _____

33. _____

34. _____

35. _____

36. _____

37. _____

38. _____

39. _____

40. _____

41. _____

42. _____

43. _____

44. _____

45. _____

Multiply.

22. $(8)(-5)$ **23.** $(-9)(-7)$

24. $(4.5)(-6)$ **25.** $(6)(-4)$

26. Determine the range for the following set: $4, -1, 6, 3, -6, 2, 8, 5$

Evaluate each expression.

27. $\dfrac{-100}{4}$ **28.** $\dfrac{-36 + 9}{-9}$

29. $\dfrac{(-15)(-3)}{-9}$ **30.** $\dfrac{9}{0}$

Evaluate the following expressions.

31. $29 - 3 \cdot 4$ **32.** $4 \cdot 5^2 - 35$

33. $4(2 + 4)^2$ **34.** $16 \div (-4) + (-5)$

35. If $x = 2$, $y = -1$, and $z = 3$, evaluate the expression $\dfrac{9x^2y}{3z}$.

Combine like terms.

36. $9a + 4a$ **37.** $10x + 8y + 9x - 3y$

38. Subtract $9a^2$ from the sum of $12a^2$ and $5a^2$.

Multiply.

39. $a^5 \cdot a^9$ **40.** $2x^3y^2 \cdot 4x^4y$

Divide.

41. $\dfrac{9x^9}{3x^3}$ **42.** $\dfrac{20a^3b^5}{5a^2b^2}$ **43.** $\dfrac{x^{10} \cdot x^5}{x^6}$

44. Social science. Tom is 8 years younger than twice Moira's age. Write an expression for Tom's age. Let x represent Moira's age.

45. Geometry. The length of a rectangle is 4 more than twice the width. Write an expression for the length of the rectangle.

ACTIVITY 1: AN INTRODUCTION TO SEARCHING

Each activity in this text is designed to either enhance your understanding of the topics of the preceding chapter, provide you with a mathematical extension of those topics, or both. The activities can be undertaken by one student, but they are better suited for a small group project. Occasionally it is only through discussion that different facets of the activity become apparent. For material related to this activity, visit the text website at www.mhhe.com/streeter.

As you work through this text, you will encounter several activities. Many (though not all) of these activities will require you to use the Internet. This activity is designed to familiarize you with searching the Internet for information.

If you are new to computers or the Internet, your instructor (or perhaps a friend) can help you get started. You will need to access the Internet through one of the many web browsers, such as Microsoft's Internet Explorer, Netscape Navigator, Opera, or AOL's browser.

We will ask that you "bookmark" various websites—some browsers refer to bookmarks as "favorites," or "preferences." Bookmarking a website allows you to return to the site at a later time by choosing it from a list of bookmarked sites. This way, you do not have to search for it every time you wish to visit that site.

Most web browsers provide you with a way to bookmark a website. When you are looking at a website that you wish to bookmark, click on the menu that says Bookmarks (or Favorites, Preferences, or whatever name your browser uses), and click on the menu item Add Bookmark (or Add to Favorites, Create Preference, etc.). The next time you want to look at the website that you bookmarked, you need only click on the Bookmark menu and click on the title of the site saved.

Activity: First, you will need to connect to the Internet. Next, you will need to access a page containing a *search engine* that searches the World Wide Web at large (rather than just doing an internal search). Many *default* home pages (such as www.msn.com and www.aol.com) contain a *search* field. If yours does not, several of the more popular sites that do are listed here:

> http://www.altavista.com
> http://www.google.com
> http://www.lycos.com
> http://www.yahoo.com

Access one of these websites or use some other search engine, as you see fit. Now we are ready to continue.

I.

1. Locate your state government home page. You may accomplish this by typing your state name into the search field and clicking on Go or Find. A (long) list of "matches" will appear. Read the blurbs to determine which belongs to the state government. **Note:** You should not need to go through more than the first page or two of listings. Bookmark this site (as previously described).
2. Find an official e-mail address for the governor of your state. Find the names of the senators for your state. See if you can locate e-mail addresses for the senators of your state. Note: If their e-mail addresses are not provided, then we will find them later.

3. Locate the home page of your college or university, or the flagship university in your state.

4. Repeat steps 1 to 3 for a different state.

II.

1. Search for the White House in Washington, D.C. Find an official e-mail address for the President of the United States.

2. Find the home page for the U.S. Congress. If you could not find e-mail addresses for senators on the state pages (step I.2), then find one through the U.S. Congress website.

3. Locate the home page of the U.S. Census Bureau and bookmark this site. Find the census data for your state. How many people live in your state, according to the most recent census? How many people live in your city or hometown?

4. Repeat step 3 for the second state that you used in part I (choose the state capital as the city).

III.

1. Search for weather data for your city or hometown (the National Weather Service is a good starting point). What is the predicted weather for tomorrow? What was the weather yesterday? What was the weather 1 month ago? What was the weather 1 year ago, today?

2. Repeat step 1 for the capital of the second state that you used in part I.

IV.

1. Choose a country on some other continent. Can you locate the name and e-mail address of that nation's leader? Census data? Weather?

2. Locate a home page for a university in the country found in step 1.

3. Locate the home page for the United Nations (UN) and bookmark this site. Does the UN site provide some of the information asked for in step 1? Does the UN data differ from the data provided by the nation itself?

V.

1. Locate two different images of the so-called Mandelbrot fractal. These can be the same image with different colors, or they can be two different views of the fractal.

2. Locate an image of a different fractal.

3. Locate information on fractals in general. (What are they?) Write a paragraph describing fractals in general and express your thoughts on the fractals you found in steps 1 and 2.

VI.

1. Search for a business or an organization that hires people in your field; if you are not already on a career path, determine a potential career for yourself. Bookmark this site.

2. Write a paragraph describing the organization as it has presented itself on its website.

3. Find the name and an e-mail address of a person whom you would contact if you were applying for a job.

4. Find information describing the application process for someone seeking employment with this business or organization.

5. Are there forms that the site provides for job seekers? Can you download them (using Save As or a link)? If forms exist, what format are these forms in (Adobe pdf files, Microsoft Word documents, or some other format)?

You have now searched through the World Wide Web for information. Later, you will be asked to use the skills gained here to complete some of the other activities in this text.

EQUATIONS AND INEQUALITIES

INTRODUCTION

Many engineers, economists, and environmental scientists are working on the problem of meeting the increasing energy demands of a growing global population. One promising solution to this problem is power generated by wind-driven turbines.

The cost of wind-generated power has fallen from $0.30 per kilowatt hour (kWh) in the early 1980s to about $0.05 per kWh in 2003; thus, using this form of power production is becoming economically feasible. And compared to the cost of pollution from burning coal and oil, wind-generated power may be less expensive.

An economist for a city might use this equation to try to compute the cost for electricity for his city:

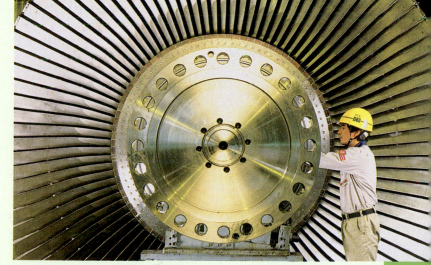

$$C = P(0.05)(1000)$$

in which C = cost in dollars

P = power in megawatts

The city engineer who is investigating the possibility of building turbines to supply the city with electricity knows that each turbine can produce 1.1 million kWh of power, so she uses the equation

$$P = n(1.1)$$

in which P = power in megawatts

n = number of turbines required to produce power

The equation is an ancient tool for solving problems and writing numerical relationships clearly and accurately. In this chapter you will learn methods to solve linear equations and practice writing equations that accurately describe problem situations.

Pre-Test Chapter 2

This pre-test will provide a preview of the types of exercises you will encounter in each section of this chapter. The answers for these exercises can be found in the back of the text. If you are working on your own, or ahead of the class, this pre-test can help you identify the sections in which you should focus more of your time.

Tell whether the number shown in parentheses is a solution for the given equation.

[2.1] **1.** $4x - 9 = 15$ (7)

2. $7x - 5 = 3x + 31$ (9)

Solve the following equations and check your results.

[2.2–2.3] **3.** $8x - 23 = 7x - 15$

4. $6x + 15 - 2x = 18 + 8x - 5x - 15$

5. $7x - 15 = 4x + 6$

6. $\dfrac{5}{7}x = 25$

7. $9x - (x - 5) = -(6 - x) - 3$

8. $2(3 - x) + 15 = 8(4x - 5) - 7$

Solve for the indicated variable.

[2.4] **9.** $P = 2L + 2W$ for W

10. $5x - 3y = 14$ for y

Translate each statement into an algebraic equation.

[2.5] **11.** 5 more than 4 times a number is 17.

12. 4 times the sum of a number and 6 is 6 more than 10 times the number.

Solve the following inequalities.

[2.7] **13.** $x - 7 \le 8$

14. $6 - 2x \ge 9 + x$

[2.5–2.7] Solve the following word problems.

15. Number problem. 7 times a number decreased by 5 is 37. Find the number.

16. Number problem. The sum of 2 consecutive odd integers is 32. Find the integers.

17. Geometry. The perimeter of a rectangle is 34 cm. If the length is 1 cm more than 3 times the width, what are the dimensions of the rectangle?

18. Business and finance. A state sales tax rate is 2.5%. If the tax on a purchase is $13.50, what was the amount of the purchase?

19. Business and finance. A house sells for $250,000 and the rate of commission is 5.25%. How much will the salesperson make for the sale?

20. Business and finance. A stereo system is marked down from $790 to $671.50. What was the rate of discount?

2.1 Solving Equations by the Addition Property

2.1 OBJECTIVES

1. Determine whether a given number is a solution for an equation
2. Identify expressions and equations
3. Use the addition property to solve an equation
4. Use the distributive property in solving equations

Overcoming Math Anxiety

Throughout this text, we will present you with a series of class-tested techniques that are designed to improve your performance in this math class.

Hint #3   Don't procrastinate!

1. Do your math homework while you're still fresh. If you wait until too late at night, your tired mind will have much more difficulty understanding the concepts.

2. Do your homework the day it is assigned. The more recent the explanation, the easier it is to recall.

3. When you've finished your homework, try reading the next section through one time. This will give you a sense of direction when you next hear the material. This works in a lecture or lab setting.

Remember that, in a typical math class, you are expected to do two or three hours of homework for each weekly class hour. This means two or three hours per night. Schedule the time and stick to your schedule.

In this chapter you will begin working with one of the most important tools of mathematics, the equation. The ability to recognize and solve various types of equations is probably the most useful algebraic skill you will learn. We will continue to build upon the methods of this chapter throughout the text. To start, let's describe what we mean by an *equation*.

Definition: Equation

An **equation** is a mathematical statement that two expressions are equal.

Some examples are $3 + 4 = 7$, $x + 3 = 5$, $P = 2L + 2W$.

As you can see, an equal–sign ($=$) separates the two equal expressions. These expressions are usually called the *left side* and the *right side* of the equation.

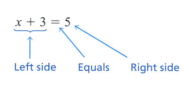

$$x + 3 = 5$$

Left side   Equals   Right side

NOTE An equation such as

$x + 3 = 5$

is called a **conditional equation** because it can be either true or false depending on the value given to the variable.

Just as the balance scale may be in balance or out of balance, an equation may be either true or false. For instance, $3 + 4 = 7$ is true because both sides name the same number. What about an equation such as $x + 3 = 5$ that has a letter or variable on one side? Any number can replace x in the equation. However, only one number will make this equation a true statement.

$$\text{If } x = 2 \begin{cases} 1 & 1 + 3 = 5 \text{ is false} \\ 2 & 2 + 3 = 5 \text{ is true} \\ 3 & 3 + 3 = 5 \text{ is false} \end{cases}$$

The number 2 is called the **solution** (or *root*) of the equation $x + 3 = 5$ because substituting 2 for x gives a true statement.

Definition: Solution

A **solution** for an equation is any value for the variable that makes the equation a true statement.

OBJECTIVE 1

Example 1   Verifying a Solution

(a) Is 3 a solution for the equation $2x + 4 = 10$?

To find out, replace x with 3 and evaluate $2x + 4$ on the left.

Left side		*Right side*
$2(3) + 4$	$\overset{?}{=}$	10
$6 + 4$	$\overset{?}{=}$	10
10	$=$	10

NOTE We could represent the solution using set notation as {3}.

Because $10 = 10$ is a true statement, 3 is a solution of the equation.

(b) Is 5 a solution of the equation $3x - 2 = 2x + 1$?

To find out, replace x with 5 and evaluate each side separately.

RECALL Remember the rules for the order of operations. Multiply first; then add or subtract.

Left side		*Right side*
$3(5) - 2$	$\overset{?}{=}$	$2 \cdot 5 + 1$
$15 - 2$	$\overset{?}{=}$	$10 + 1$
13	$\neq$	11

Because the two sides do not name the same number, we do not have a true statement, and 5 is not a solution.

CHECK YOURSELF 1

For the equation

$2x - 1 = x + 5$

(a) Is 4 a solution?   **(b)** Is 6 a solution?

You may be wondering whether an equation can have more than one solution. It certainly can. For instance,

NOTE This is an example of a **quadratic equation.** We will consider methods of solution in Chapter 4 and then again in Chapter 10.

$$x^2 = 9$$

has two solutions. They are 3 and -3 because

$$3^2 = 9 \quad \text{and} \quad (-3)^2 = 9$$

In this chapter, however, we will always work with *linear equations in one variable.* These are equations that can be put into the form

$$ax + b = 0$$

in which the variable is x, a and b are any numbers, and a is not equal to 0. In a linear equation, the variable can appear only to the first power. No other power (x^2, x^3, etc.) can appear. Linear equations are also called **first-degree equations.** The degree of an equation in one variable is the highest power to which the variable appears.

Rules and Properties: Linear Equations

Linear equations in one variable are equations that can be written in the form

$$ax + b = 0 \qquad a \neq 0$$

Every such equation will have exactly one solution.

OBJECTIVE 2   **Example 2   Identifying Expressions and Equations**

Label each of the following as an expression, a linear equation, or an equation that is not linear.

(a) $4x + 5$ is an expression.

(b) $2x + 8 = 0$ is a linear equation.

(c) $3x^2 - 9 = 0$ is not a linear equation.

(d) $5x = 15$ is a linear equation.

(e) $5 - \dfrac{7}{x} = 4x$ is not a linear equation.

 CHECK YOURSELF 2

Label each as an expression, a linear equation, or an equation that is not linear.

(a) $2x^2 = 8$ **(b)** $2x - 3 = 0$ **(c)** $5x - 10$ **(d)** $2x + 1 = 7$ **(e)** $\dfrac{3}{x} - 4 = x$

It is not difficult to find the solution for an equation such as $x + 3 = 8$ by guessing the answer to the question "What plus 3 is 8?" Here the answer to the question is 5, and that is also the solution for the equation. But for more complicated equations you will need something more than guesswork. A better method is to transform the given equation to an *equivalent equation* whose solution can be found by inspection.

> **Definition:** Equivalent Equations
>
> Equations that have the same solution are called **equivalent equations.**

The following are all equivalent equations:

$$2x + 3 = 5 \qquad 2x = 2 \qquad \text{and} \qquad x = 1$$

They all have the same solution, 1. We say that a linear equation is *solved* when it is transformed to an equivalent equation of the form

NOTE In some cases we'll write the equation in the form

$\square = x$

The number will be our solution when the equation has the variable isolated on the left or on the right.

$x = \square$

The variable is alone on the left side.　The right side is some number, the solution.

The addition property of equality is the first property you will need to transform an equation to an equivalent form.

> **Rules and Properties:** The Addition Property of Equality
>
> If $\qquad a = b$
>
> then $\qquad a + c = b + c$
>
> In words, adding the same quantity to both sides of an equation gives an equivalent equation.

RECALL An equation is a statement that the two sides are equal. Adding the same quantity to both sides does not change the equality or "balance."

Recall that we said that a true equation was like a scale in balance.

The addition property is equivalent to adding the same weight to both sides of the scale. It will remain in balance.

NOTE This scale represents

$a + b = b + c$

OBJECTIVE 3   **Example 3   Using the Addition Property to Solve an Equation**

Solve

$$x - 3 = 9$$

Remember that our goal is to isolate x on one side of the equation. Because 3 is being subtracted from x, we can add 3 to remove it. We must use the addition property to add 3 to both sides of the equation.

NOTE To check, replace x with 12 in the original equation:

$$x - 3 \overset{?}{=} 9$$
$$(12) - 3 \overset{?}{=} 9$$
$$9 = 9$$

Because we have a true statement, 12 is the solution.

$$
\begin{array}{rl}
x - 3 = & 9 \\
+\ 3 & +3 \\
\hline
x\ \ \ = & 12
\end{array}
$$

⎧ Adding 3 "undoes" the
⎨ subtraction and leaves
⎩ x alone on the left.

Because 12 is the solution for the equivalent equation $x = 12$, it is the solution for our original equation.

 CHECK YOURSELF 3

Solve and check.

$$x - 5 = 4$$

The addition property also allows us to add a negative number to both sides of an equation. This is really the same as subtracting the same quantity from both sides.

Example 4   Using the Addition Property to Solve an Equation

Solve

$$x + 5 = 9$$

In this case, 5 is *added* to x on the left. We can use the addition property to add a -5 to both sides. Because $5 + (-5) = 0$, this will "undo" the addition and leave the variable x alone on one side of the equation.

RECALL Remember our comment that we could write an equation in the equivalent forms $x = \square$ or $\square = x$, in which $\square$ represents some number. Suppose we have an equation like

$$12 = x + 7$$

Adding -7 will isolate x on the right:

$$
\begin{array}{rl}
12 = & x + 7 \\
-7 & -\ 7 \\
\hline
5 = & x
\end{array}
$$

and the solution is 5.

$$
\begin{array}{rl}
x + 5 = & 9 \\
-\ 5 & -5 \\
\hline
x\ \ \ = & 4
\end{array}
$$

The solution is 4. To check, replace x with 4:

$$(4 + 5) = 9 \quad \text{(True)}$$

 CHECK YOURSELF 4

Solve and check.

$$x + 6 = 13$$

What if the equation has a variable term on both sides? We have to use the addition property to add or subtract a term involving the variable to get the desired result.

Example 5 Using the Addition Property to Solve an Equation

Solve

$$5x = 4x + 7$$

We will start by adding $-4x$ to both sides of the equation. Do you see why? Remember that an equation is solved when we have an equivalent equation of the form $x = \square$.

RECALL Adding $-4x$ is identical to subtracting $4x$.

$$
\begin{array}{r}
5x = 4x + 7 \\
\underline{-4x -4x } \\
x = 7
\end{array}
$$

$\left\{ \begin{array}{l} \text{Adding } -4x \text{ to both} \\ \text{sides } \textit{removes } 4x \\ \text{from the right.} \end{array} \right.$

To check: Because 7 is a solution for the equivalent equation $x = 7$, it should be a solution for the original equation. To find out, replace x with 7.

$$5(7) \stackrel{?}{=} 4(7) + 7$$
$$35 \stackrel{?}{=} 28 + 7$$
$$35 = 35 \qquad \text{(True)}$$

CHECK YOURSELF 5

Solve and check.

$$7x = 6x + 3$$

You may have to apply the addition property more than once to solve an equation. Look at Example 6.

Example 6 Using the Addition Property to Solve an Equation

Solve

$$7x - 8 = 6x$$

We want all variables on *one* side of the equation. If we choose the left, we add $-6x$ to both sides of the equation. This will remove $6x$ from the right:

$$
\begin{array}{r}
7x - 8 = 6x \\
\underline{-6x -6x} \\
x - 8 = 0
\end{array}
$$

We want the variable alone, so we add 8 to both sides. This isolates x on the left.

$$
\begin{array}{r}
x - 8 = 0 \\
\underline{+\,8 +8} \\
x = 8
\end{array}
$$

The solution is 8. We'll leave it to you to check this result.

CHECK YOURSELF 6

Solve and check.

$9x + 3 = 8x$

Often an equation has more than one variable term *and* more than one number. You have to apply the addition property twice to solve these equations.

Example 7   Using the Addition Property to Solve an Equation

Solve

$$5x - 7 = 4x + 3$$

We would like the variable terms on the left, so we start by adding $-4x$ to remove the $4x$ term from the right side of the equation:

$$
\begin{array}{rcl}
5x - 7 = & 4x + 3 \\
-4x & -4x \\
\hline
x - 7 = & 3
\end{array}
$$

Now, to isolate the variable, we add 7 to both sides.

$$
\begin{array}{rcl}
x - 7 = & 3 \\
+ 7 & +7 \\
\hline
x \quad = & 10
\end{array}
$$

NOTE You could just as easily have added 7 to both sides and *then* added $-4x$. The result would be the same. In fact, some students prefer to combine the two steps.

The solution is 10. To check, replace x with 10 in the original equation:

$$5(10) - 7 \stackrel{?}{=} 4(10) + 3$$
$$43 = 43 \quad \text{(True)}$$

CHECK YOURSELF 7

Solve and check.

(a) $4x - 5 = 3x + 2$ **(b)** $6x + 2 = 5x - 4$

RECALL By *simplify* we mean to combine all like terms.

In solving an equation, you should always simplify each side as much as possible before using the addition property.

Example 8   Combining Like Terms and Solving the Equation

Solve

Like terms   Like terms

$$5 + 8x - 2 = 2x - 3 + 5x$$

Because like terms appear on each side of the equation, we start by combining the numbers on the left (5 and -2). Then we combine the like terms ($2x$ and $5x$) on the right. We have

$$3 + 8x = 7x - 3$$

Now we can apply the addition property, as before.

$$
\begin{array}{rcl}
3 + 8x = & 7x - 3 & \\
\underline{-7x = -7x} & & \text{Add } -7x. \\
3 + x = & -3 & \\
\underline{-3 \qquad -3} & & \text{Add } -3 \text{ to isolate } x. \\
x = & -6 &
\end{array}
$$

The solution is -6. To check, always return to the original equation. That will catch any possible errors in simplifying. Replacing x with -6 gives

$$5 + 8(-6) - 2 \overset{?}{=} 2(-6) - 3 + 5(-6)$$
$$5 - 48 - 2 \overset{?}{=} -12 - 3 - 30$$
$$-45 = -45 \quad \text{(True)}$$

CHECK YOURSELF 8 _____

Solve and check.

(a) $3 + 6x + 4 = 8x - 3 - 3x$ **(b)** $5x + 21 + 3x = 20 + 7x - 2$

We may have to apply some of the properties discussed in Section 1.2 in solving equations. Example 9 illustrates the use of the distributive property to clear an equation of parentheses.

OBJECTIVE 4   **Example 9**   **Using the Distributive Property and Solving Equations**

Solve

NOTE $2(3x + 4)$
$= 2(3x) + 2(4)$
$= 6x + 8$

$$2(3x + 4) = 5x - 6$$

Applying the distributive property on the left, we have

$$6x + 8 = 5x - 6$$

We can then proceed as before:

$$
\begin{array}{rcl}
6x + 8 = & 5x - 6 & \\
\underline{-5x \qquad -5x} & & \text{Add } -5x. \\
x + 8 = & -6 & \\
\underline{-8 \qquad -8} & & \text{Add } -8. \\
x = & -14 &
\end{array}
$$

The solution is -14. We will leave the checking of this result to the reader.

Remember: Always return to the original equation to check.

 CHECK YOURSELF 9

Solve and check each of the following equations.

(a) $4(5x - 2) = 19x + 4$

$$20x - 8 = 19x + 4$$
$$20x = 19x + 12$$
$$-19x \qquad 19x$$
$$\boxed{x - 2} $$

(b) $3(5x + 1) = 2(7x - 3) - 4$

$$15x + 3 = (14x - 6) - 4$$
$$15x + 3 = 14x - 6 - 4$$
$$-14x \qquad -14x$$
$$\overline{}$$
$$x + 3 = -10$$
$$-3 \quad -3$$
$$\boxed{x = -13}$$

$$x + 3 = -6 - 4$$
$$x + 3 = -10$$
$$-3$$
$$x = -13$$

Given an expression such as

$$-2(x - 5)$$

the distributive property can be used to create the equivalent expression

$$-2x + 10$$

The distribution of a negative number is used in Example 10.

Example 10   **Distributing a Negative Number**

Solve each of the following equations.

(a) $-2(x - 5) = -3x + 2$

$-2x + 10 = -3x + 2$	Distribute the -2.
$+3x \qquad\qquad +3x$	Add $3x$.

$$\overline{}$$

$x + 10 = \qquad 2$	
$-10 = \quad -10$	Add -10.

$$\overline{}$$

$$x \qquad = \qquad -8$$

(b) $-3(3x + 5) = -5(2x - 2)$

$-9x - 15 = -5(2x - 2)$	Distribute the -3.
$-9x - 15 = -10x + 10$	Distribute the -5.
$+10x \qquad\qquad +10x$	Add $10x$.

$$\overline{}$$

$x - 15 = \qquad 10$	
$+15 \qquad\qquad +15$	Add 15.

$$\overline{}$$

$$x \qquad = \qquad 25$$

 CHECK YOURSELF 10

Solve each of the following.

(a) $-2(x - 3) = -x + 5$

(b) $-4(2x - 1) = -3(3x + 2)$

When parentheses are preceded only by a negative, or by the minus sign, we say that we have a silent negative one. Example 11 illustrates this case.

Example 11 **Distributing the Silent Negative One**

Solve

$$-(2x + 3) = -3x + 7$$

$$-1(2x + 3) = -3x + 7$$

$$(-1)(2x) + (-1)(3) = -3x + 7$$

$$
\begin{array}{rcl}
-2x - 3 &=& -3x + 7 \\
+3x & & +3x \qquad \text{Add } 3x. \\
\hline
x - 3 &=& 7 \\
+3 & & +3 \qquad \text{Add } 3. \\
\hline
x &=& 10
\end{array}
$$

CHECK YOURSELF 11

Solve $-(3x + 2) = -2x - 6.$

CHECK YOURSELF ANSWERS

1. **(a)** 4 is not a solution; **(b)** 6 is a solution

2. **(a)** Not a linear equation; **(b)** linear equation; **(c)** expression; **(d)** linear equation;
(e) Not a linear equation   **3.** 9   **4.** 7   **5.** 3   **6.** −3   **7.** **(a)** 7; **(b)** −6
8. **(a)** −10; **(b)** −3   **9.** **(a)** 12; **(b)** −13   **10.** **(a)** 1; **(b)** −10   **11.** 4

2.1   Exercises

Is the number shown in parentheses a solution for the given equation?

1. $x + 7 = 12$ (5)

2. $x + 2 = 11$ (8)

3. $x - 15 = 6$ (−21)

4. $x - 11 = 5$ (16)

5. $5 - x = 2$ (4)

6. $10 - x = 7$ (3)

7. $8 - x = 5$ (−3)

8. $5 - x = 6$ (−3)

9. $3x + 4 = 13$ (8)

10. $5x + 6 = 31$ (5)

11. $4x - 5 = 7$ (2)

12. $4x - 3 = 9$ (3)

13. $7 - 3x = 10$ (−1)

14. $4 - 5x = 9$ (−2)

15. $4x - 5 = 2x + 3$ (4)

16. $5x + 4 = 2x + 10$ (4)

17. $x + 3 + 2x = 5 + x + 8$ (5)

18. $5x - 3 + 2x = 3 + x - 12$ (−2)

19. $\dfrac{2}{3}x = 9$ (15)

20. $\dfrac{3}{5}x = 24$ (40)

21. $\dfrac{3}{5}x + 5 = 11$ (10)

22. $\dfrac{2}{3}x + 8 = -12$ (−6)

Label each of the following as an expression, a linear equation, or an equation that is not linear.

23. $2x + 1 = 9$
$-1\ -1$
$2x = 8/2$

24. $7x + 14$

25. $\dfrac{2x}{2} - \dfrac{8}{2}$

26. $5x - 3 = 12$
$+3\ +3$
$5x = 15$

27. $2x^2 - 8 = 0$
$+8\ +8$
$\dfrac{2x^2}{2} = \dfrac{8}{2}\ x^2 = 4$

28. $x + 5 = 13$
$-5\ -5$

29. $2x - 8 = 3$
$+8\ +8$
$\dfrac{2x}{2} = \dfrac{11}{2}$

30. $\dfrac{2}{x} - 4 = 3x + 4$
$+4$
$\dfrac{2}{x} = 3x + 4$

ANSWERS

1. _____
2. _____
3. _____
4. _____
5. _____
6. _____
7. _____
8. _____
9. _____
10. _____
11. _____
12. _____
13. _____ 14. _____
15. _____ 16. _____
17. _____ 18. _____
19. _____ 20. _____
21. _____ 22. _____
23. $X = 4$
24. $X = 2$
25. $X = -4$
26. $X = 3$
27. $X^2 = 4$
28. $X = 8$
29. $X^{11}/_2$
30. _____

SECTION 2.1   **173**

31. $x = 2$ 32. $x = 10$

33. $x = -4$ 34. $x = 4$

35. $x = 2$ 36. $x = 3$

37. $x = -7$ 38. $x = 1$

39. $x = 6$ 40. $x = -7$

41. $x = 4$ 42. $x = -8$

43. $x = 12$ 44. $x = 5$

45. $x = -3$

46. _____

47. _____

48. _____

49. _____

50. _____

51. _____

52. _____

53. _____

54. _____

55. _____

56. _____

57. _____

58. _____

59. _____

60. _____

61. _____

62. _____

63. _____

64. _____

Solve and check the following equations.

31. $x + 9 = 11$ **32.** $x - 4 = 6$

33. $x - 5 = -9$ **34.** $x + 11 = 15$

35. $x - 8 = -10$ **36.** $x + 5 = 2$

37. $x + 4 = -3$ **38.** $x - 6 = -5$

39. $17 = x + 11$ **40.** $x + 7 = 0$

41. $4x = 3x + 4$ **42.** $7x = 6x - 8$

43. $9x = 8x - 12$ **44.** $9x = 8x + 5$

45. $6x + 3 = 5x$ **46.** $12x - 6 = 11x$

47. $7x - 5 = 6x$ **48.** $9x - 7 = 8x$

49. $2x + 3 = x + 5$ **50.** $5x - 6 = 4x + 2$

51. $4x - \dfrac{3}{5} = 3x + \dfrac{1}{10}$ **52.** $5\left(x - \dfrac{3}{4}\right) = 4x + \dfrac{3}{8}$

53. $\dfrac{7}{8}(x - 2) = \dfrac{3}{4} - \dfrac{1}{8}x$ **54.** $\dfrac{5}{6}(3x - 2) = \dfrac{3}{2}(x + 1)$

55. $3x - 0.54 = 2(x - 0.15)$ **56.** $7x + 0.125 = 6x - 0.289$

57. $6x + 3(x - 0.2789) = 4(2x + 0.3912)$ **58.** $9x - 2(3x - 0.124) = 2x + 0.965$

59. $5x - 7 + 6x - 9 - x = 2x - 8 + 7x$ **60.** $5x + 8 + 3x - x + 5 = 6x - 3$

61. $5x - (0.345 - x) = 5x + 0.8713$ **62.** $-3(0.234 - x) = 2(x + 0.974)$

63. $3(7x + 2) = 5(4x + 1) + 17$ **64.** $5(5x + 3) = 3(8x - 2) + 4$

65. $\dfrac{5}{4}x - 1 = \dfrac{1}{4}x + 7$

66. $\dfrac{7}{5}x + 3 = \dfrac{2}{5}x - 8$

67. $\dfrac{9}{2}x - \dfrac{3}{4} = \dfrac{7}{2}x + \dfrac{5}{4}$

68. $\dfrac{11}{3}x + \dfrac{1}{6} = \dfrac{8}{3}x + \dfrac{19}{6}$

69. Which of the following is equivalent to the equation $5x - 7 = 4x - 12$?

    a. $9x = 19$ **b.** $x - 7 = -12$

    c. $x = -18$ **d.** $4x - 5 = 8$

70. Which of the following is equivalent to the equation $12x - 6 = 8x + 14$?

    a. $4x - 6 = 14$ **b.** $x = 20$

    c. $20x = 20$ **d.** $4x = 8$

71. Which of the following is equivalent to the equation $7x + 5 = 12x - 10$?

    a. $5x = -15$ **b.** $7x - 5 = 12x$

    c. $-5 = 5x$ **d.** $7x + 15 = 12x$

True or false?

72. Every linear equation with one variable has exactly one solution.

73. Isolating the variable on the right side of an equation will result in a negative solution.

74. An algebraic equation is a complete sentence. It has a subject, a verb, and a predicate. For example, $x + 2 = 5$ can be written in English as "Two more than a number is five." Or, "A number added to two is five." Write an English version of the following equations. Be sure you write complete sentences and that the sentences express the same idea as the equations. Exchange sentences with another student, and see if your interpretations of each other's sentences result in the same equation.

    (a) $2x - 5 = x + 1$ **(b)** $2(x + 2) = 14$

    (c) $n + 5 = \dfrac{n}{2} - 6$ **(d)** $7 - 3a = 5 + a$

75. Complete the following explanation in your own words: "The difference between $3(x - 1) + 4 - 2x$ and $3(x - 1) + 4 = 2x$ is. . . ."

ANSWERS

65. _____

66. _____

67. _____

68. _____

69. _____

70. _____

71. _____

72. _____

73. _____

74. _____

75. _____

76.

a.

b.

c.

d.

e.

f.

g.

h.

76. "Surprising Results!" Work with other students to try this experiment. Each person should do the following six steps mentally, not telling anyone else what their calculations are:

(a) Think of a number.

(b) Add 7.

(c) Multiply by 3.

(d) Add 3 more than the original number.

(e) Divide by 4.

(f) Subtract the original number.

What number do you end up with? Compare your answer with everyone else's. Does everyone have the same answer? Make sure that everyone followed the directions accurately. How do you explain the results? Algebra makes the explanation clear. Work together to do the problem again, using a variable for the number. Make up another series of computations that yields "surprising results."

Getting Ready for Section 2.2 [Section 1.4]

Multiply.

(a) $\left(\dfrac{1}{3}\right)(3)$

(b) $(-6)\left(-\dfrac{1}{6}\right)$

(c) $(7)\left(\dfrac{1}{7}\right)$

(d) $\left(-\dfrac{1}{4}\right)(-4)$

(e) $\left(\dfrac{3}{5}\right)\left(\dfrac{5}{3}\right)$

(f) $\left(\dfrac{7}{8}\right)\left(\dfrac{8}{7}\right)$

(g) $\left(-\dfrac{4}{7}\right)\left(-\dfrac{7}{4}\right)$

(h) $\left(-\dfrac{6}{11}\right)\left(-\dfrac{11}{6}\right)$

Answers

1. Yes   **3.** No   **5.** No   **7.** No   **9.** No   **11.** No   **13.** Yes
15. Yes   **17.** Yes   **19.** No   **21.** Yes   **23.** Linear equation
25. Expression   **27.** Not a linear equation   **29.** Linear equation   **31.** 2
33. -4 **35.** -2 **37.** -7 **39.** 6   **41.** 4   **43.** -12 **45.** -3
47. 5   **49.** 2   **51.** $\dfrac{7}{10}$ **53.** $\dfrac{5}{2}$ **55.** 0.24   **57.** 2.4015   **59.** 8
61. 1.2163   **63.** 16   **65.** 8   **67.** 2   **69.** b   **71.** d   **73.** False
75.   **a.** 1   **b.** 1   **c.** 1   **d.** 1   **e.** 1   **f.** 1   **g.** 1

h. 1

2.2 Solving Equations by the Multiplication Property

2.2 OBJECTIVES

1. Use the multiplication property to solve equations
2. Find the mean for a given set

Let's look at a different type of equation. For instance, what if we want to solve an equation like the following?

$$6x = 18$$

Using the addition property of Section 2.1 won't help. We will need a second property for solving equations.

Rules and Properties: The Multiplication Property of Equality

NOTE Do you see why the number cannot be 0? Multiplying by 0 gives 0 = 0. We have lost the variable!

If $a = b$ then $ac = bc$ where $c \neq 0$

In words, multiplying both sides of an equation by the same nonzero number gives an equivalent equation.

RECALL Again, as long as you do the *same* thing to *both* sides of the equation, the "balance" is maintained.

Again, we return to the image of the balance scale. We start with the assumption that a and b have the same weight.

The multiplication property tells us that the scale will be in balance as long as we have the same number of "a weights" as we have of "b weights."

NOTE The scale represents the equation $5a = 5b$.

Let's work through some examples, using this second rule.

OBJECTIVE 1

Example 1 **Solving Equations Using the Multiplication Property**

Solve

$$6x = 18$$

Here the variable x is multiplied by 6. So we apply the multiplication property and multiply both sides by $\frac{1}{6}$. Keep in mind that we want an equation of the form

$$x = \square$$

NOTE

$$\frac{1}{6}(6x) = \left(\frac{1}{6} \cdot 6\right)x$$

$$= 1 \cdot x, \text{ or } x$$

We then have x alone on the left, which is what we want.

$$\frac{1}{6}(6x) = \left(\frac{1}{6}\right)18$$

We can now simplify.

$$1 \cdot x = 3 \qquad \text{or} \qquad x = 3$$

The solution is 3. To check, replace x with 3:

$$6 \cdot 3 \stackrel{?}{=} 18$$
$$18 = 18 \qquad \text{(True)}$$

 CHECK YOURSELF 1 _____

Solve and check.

8x = 32

In Example 1 we solved the equation by multiplying both sides by the reciprocal of the coefficient of the variable.

Example 2 illustrates a slightly different approach to solving an equation by using the multiplication property.

Example 2 **Solving Equations Using the Multiplication Property**

Solve

$$5x = -35$$

NOTE Because division is defined in terms of multiplication, we can also divide both sides of an equation by the same nonzero number.

The variable x is multiplied by 5. We *divide* both sides by 5 to "undo" that multiplication:

$$\frac{5x}{5} = \frac{-35}{5}$$

$$x = -7 \quad \left\{\begin{array}{l}\text{Note that the right side} \\ \text{reduces to } -7. \text{ Be careful} \\ \text{with the rules for signs.}\end{array}\right.$$

We leave it to you to check the solution.

CHECK YOURSELF 2

Solve and check.

$7x = -42$

Example 3   Solving Equations by Using the Multiplication Property

Solve

$$-9x = 54$$

In this case, x is multiplied by -9, so we divide both sides by -9 to isolate x on the left:

$$\frac{-9x}{-9} = \frac{54}{-9}$$

$$x = -6$$

The solution is -6. To check:

$$(-9)(-6) \stackrel{?}{=} 54$$

$$54 = 54 \quad \text{(True)}$$

CHECK YOURSELF 3

Solve and check.

$-10x = -60$

Example 4 illustrates the use of the multiplication property when fractions appear in an equation.

Example 4   Solving Equations by Using the Multiplication Property

(a) Solve

$$\frac{x}{3} = 6$$

Here x is *divided* by 3. We will use multiplication to isolate x.

$$3\left(\frac{x}{3}\right) = 3 \cdot 6$$

This leaves x alone on the left because

$$3\left(\frac{x}{3}\right) = \frac{3}{1} \cdot \frac{x}{3} = \frac{x}{1} = x$$

$$x = 18$$

To check:

$$\frac{18}{3} \stackrel{?}{=} 6$$

$$6 = 6 \quad \text{(True)}$$

(b) Solve

$$\frac{x}{5} = -9$$

$$5\left(\frac{x}{5}\right) = 5(-9)$$ Because x is divided by 5, multiply both sides by 5

$$x = -45$$

The solution is -45. To check, we replace x with -45:

$$\frac{-45}{5} \overset{?}{=} -9$$

$$-9 = -9 \quad \text{(True)}$$

The solution is verified.

 CHECK YOURSELF 4 _____

Solve and check.

(a) $\dfrac{x}{7} = 3$ **(b)** $\dfrac{x}{4} = -8$

When the variable is multiplied by a fraction that has a numerator other than 1, there are two approaches to finding the solution.

Example 5   Solving Equations Using Reciprocals

Solve

$$\frac{3}{5}x = 9$$

One approach is to multiply by 5 as the first step.

$$5\left(\frac{3}{5}x\right) = 5 \cdot 9$$

$$3x = 45$$

Now we divide by 3.

$$\frac{3x}{3} = \frac{45}{3}$$

$$x = 15$$

To check:

$$\frac{3}{5}(15) \overset{?}{=} 9$$

$$9 = 9 \quad \text{(True)}$$

A second approach combines the multiplication and division steps and is generally more efficient. We multiply by $\frac{5}{3}$.

RECALL $\frac{5}{3}$ is the *reciprocal* of $\frac{3}{5}$, and the product of a number and its reciprocal is just 1! So $\left(\frac{5}{3}\right)\left(\frac{3}{5}\right) = 1$

$$\frac{5}{3}\left(\frac{3}{5}x\right) = \frac{5}{3} \cdot 9$$

$$x = \frac{5}{\overset{1}{\cancel{3}}} \cdot \frac{\overset{3}{\cancel{9}}}{1} = 15$$

So $x = 15$, as before.

 CHECK YOURSELF 5 _____

Solve and check.

$$\frac{2}{3}x = 18$$

You may have to simplify an equation before applying the methods of this section. Example 6 illustrates this property.

> ### Example 6   Combining Like Terms and Solving Equations

Solve and check:

$$3x + 5x = 40$$

Using the distributive property, we can combine the like terms on the left to write

$$8x = 40$$

We can now proceed as before.

$$\frac{8x}{8} = \frac{40}{8} \qquad \text{Divide by 8.}$$

$$x = 5$$

The solution is 5. To check, we return to the original equation. Substituting 5 for x yields

$$3 \cdot (5) + 5 \cdot (5) \overset{?}{=} 40$$
$$15 + 25 \overset{?}{=} 40$$
$$40 = 40 \qquad \text{(True)}$$

The solution is verified.

 CHECK YOURSELF 6 _____

Solve and check.

$$7x + 4x = -66$$

An **average** is a value that is representative of a set of numbers. One kind of average is the *mean*.

> **Definition: Mean**
>
> The **mean** of a set is the sum of the set divided by the number of elements in the set. The mean is written as $\bar{x}$ (sometimes called "*x*-bar"). In mathematical symbols, we say
>
> $$\bar{x} = \frac{\Sigma x}{n}$$ ←The sum of the set
>   ←The number of elements in the set

OBJECTIVE 2

Example 7   Finding the Mean

Find the mean for each set of numbers.

(a) $2, -3, 5, 4, 7$

We begin by finding Σx.

$$\Sigma x = 2 + (-3) + 5 + 4 + 7 = 15$$

Next we determine n.

$n = 5$ Remember that n is the number of elements in the set.

Finally, we substitute our numbers into the equation.

$$\bar{x} = \frac{\Sigma x}{n} = \frac{15}{5} = 3$$

The mean of the set is 3.

(b) $-4, 7, 9, -3, 6, -2, -3, 8$

First find Σx.

$$\Sigma x = (-4) + 7 + 9 + (-3) + 6 + (-2) + (-3) + 8 = 18$$

Next determine n.

$n = 8$

Substitute these numbers into the equation

$$\bar{x} = \frac{\Sigma x}{n} = \frac{18}{8} = \frac{9}{4} \ (\text{or } 2.25)$$

The mean of this set is $\frac{9}{4}$ or 2.25.

CHECK YOURSELF 7

Find the mean for each set of numbers.

(a) $5, -2, 6, 3, -2$ **(b)** $6, -2, 3, 8, 5, -6, 1, -3$

Example 8   Finding the Mean

During a week in February the low temperature in Fargo, North Dakota, was recorded each day. The results are presented in the following table. Find both the median and the mean for the set of numbers.

M	T	W	Th	F	Sa	Su
-11	-17	-15	-18	-20	-2	20

RECALL We discussed the median in Section 1.3.

To find the median we place the numbers in ascending order:

$$-20 \quad -18 \quad -17 \quad -15 \quad -11 \quad -2 \quad 20$$

The median is the middle value, so the median is $-15°$.

To find the mean, we first find Σx.

$$\Sigma x = (-11) + (-17) + (-15) + (-18) + (-20) + (-2) + 20 = -63$$

Then, given that $n = 7$, we use the equation for the mean.

$$\bar{x} = \frac{\Sigma x}{n} = \frac{-63}{7} = -9$$

The mean is -9.

Which average was more appropriate? There is really no "right" answer to that question. In this case, the median would probably be preferred by most statisticians. It yields a temperature that was actually the low temperature on Wednesday of that week, so it is more representative of the set of low temperatures.

CHECK YOURSELF 8 _____

The low temperatures in Anchorage, Alaska, for one week in January are given in the following table. Compute both the median and the mean low temperature for that week.

M	T	W	Th	F	Sa	Su
6	−10	−12	−22	−28	−26	−27

CHECK YOURSELF ANSWERS _____

1. 4   **2.** −6   **3.** 6   **4. (a)** 21; **(b)** −32   **5.** 27   **6.** −6

7. (a) 2; **(b)** 1.5   **8.** Median = −22, mean = −17

Name _____

Section _____ Date _____

Solve for x and check your result.

1. $5x = 20$

2. $6x = 30$

3. $8x = 48$

4. $6x = -42$

5. $77 = 11x$

6. $66 = 6x$

7. $4x = -16$

8. $-3x = 27$

9. $-9x = 72$

10. $10x = -100$

11. $6x = -54$

12. $-7x = 49$

13. $-5x = -15$

14. $52 = -4x$

15. $-42 = 6x$

16. $-7x = -35$

17. $-6x = -54$

18. $-7x = -42$

19. $\dfrac{x}{2} = 4$

20. $\dfrac{x}{3} = 2$

21. $\dfrac{x}{5} = 3$

22. $\dfrac{x}{8} = 5$

23. $5 = \dfrac{x}{8}$

24. $6 = \dfrac{x}{3}$

25. $\dfrac{x}{5} = -4$

26. $\dfrac{x}{7} = -5$

27. $-\dfrac{x}{3} = 8$

28. $-\dfrac{x}{6} = -2$

29. $\dfrac{2}{3}x = 0.9$

30. $\dfrac{3}{7}x = 15$

31. $\dfrac{3}{4}x = -15$

32. $\dfrac{3}{5}x = 10 - \dfrac{6}{5}$

33. $-\dfrac{6}{5}x = -18$

34. $5x + 4x = 36$

35. $16x - 9x = -16.1$

36. $4x - 2x + 7x = 36$

Once again, certain equations involving decimal fractions can be solved by the methods of this section. For instance, to solve $2.3x = 6.9$ we simply use our multiplication property to divide both sides of the equation by 2.3. This will isolate x on the left as desired. Use this idea to solve each of the following equations for x.

37. $5.6x = 22.4$

38. $5.1x = -15.3$

ANSWERS

1. _____ 2. _____

3. _____ 4. _____

5. _____ 6. _____

7. _____ 8. _____

9. _____ 10. _____

11. _____ 12. _____

13. _____ 14. _____

15. _____ 16. _____

17. _____ 18. _____

19. _____ 20. _____

21. _____ 22. _____

23. _____ 24. _____

25. _____ 26. _____

27. _____ 28. _____

29. _____ 30. _____

31. _____ 32. _____

33. _____ 34. _____

35. _____ 36. _____

37. _____ 38. _____

ANSWERS

39. _____

40. _____

41. _____

42. _____

43. _____

44. _____

45. _____

46. _____

47. _____

48. _____

49. _____

50. _____

a. _____

b. _____

c. _____

d. _____

e. _____

f. _____

g. _____

h. _____

39. $-4.5x = 3.51$ **40.** $-8.2x = -31.078$

41. $1.3x + 2.8x = 12.3$ **42.** $2.7x + 5.4x = -16.2$

Find the mean and the median of each data set.

43. 1, 4, 7, 10, 13   **44.** 1, 3, 8, 10, 18

45. $-3, -1, 2, 4, 6, 10$ **46.** $-5, -2, 1, 4, 6, 8$

47. $-\dfrac{3}{2}, -1, 2, \dfrac{5}{2}, 3, 7$ **48.** $-\dfrac{4}{3}, -\dfrac{1}{3}, \dfrac{2}{3}, 5, 6$

49. Statistics. Kareem bought four bags of candy. The weights of the bags were 16, 21, 18, and 15 oz. Find the median and the mean weight of the bags of candy.

50. Business and finance. Jose has savings accounts for each of his five children. They contain $215, $156, $318, $75, and $25. Find the median and the mean amount of money per account. *mean 157.80   median 156.00*

 Getting Ready for Section 2.3 [Section 1.2]

Use the distributive property to remove the parentheses in the following expressions.

(a) $2(x - 3)$ (b) $3(a + 4)$ (c) $5(2b + 1)$ (d) $3(3p - 4)$
(e) $7(3x - 4)$ (f) $-4(5x + 4)$ (g) $-3(4x - 3)$ (h) $-5(3y - 2)$

Answers

1. 4   **3.** 6   **5.** 7   **7.** -4 **9.** -8 **11.** -9 **13.** 3
15. -7 **17.** 9   **19.** 8   **21.** 15   **23.** 40   **25.** -20 **27.** -24
29. 1.35   **31.** -20 **33.** 15   **35.** -2.3 **37.** 4   **39.** -0.78
41. 3   **43.** Mean: 7; median: 7   **45.** Mean: 3; median: 3
47. Mean: 2; median: $\dfrac{9}{4}$ **49.** Mean: 17.5; median: 17 oz.

a. $2x - 6$ **b.** $3a + 12$ **c.** $10b + 5$ **d.** $9p - 12$ **e.** $21x - 28$
f. $-20x - 16$ **g.** $-12x + 9$ **h.** $-15y + 10$

2.3 Combining the Rules to Solve Equations

2.3 OBJECTIVES

1. Combine the addition and multiplication properties to solve an equation
2. Solve equations that contain parentheses
3. Solve equations that contain fractions
4. Recognize identities
5. Recognize equations with no solutions

In each example thus far, either the addition property or the multiplication property was used in solving an equation. Often, finding a solution will require the use of both properties.

OBJECTIVE 1

Example 1 Solving Equations

(a) Solve

$$4x - 5 = 7$$

Here x is *multiplied* by 4. The result, $4x$, then has 5 subtracted from it (or -5 added to it) on the left side of the equation. These two operations mean that both properties must be applied in solving the equation.

Because the variable term is already on the left, we start by adding 5 to both sides:

$$
\begin{array}{rcl}
4x - 5 & = & 7 \\
+\,5 & & +5 \\
\hline
4x & = & 12
\end{array}
$$

We now divide both sides by 4:

$$\frac{4x}{4} = \frac{12}{4}$$

$$x = 3$$

The solution is 3. To check, replace x with 3 in the original equation. Be careful to follow the rules for the order of operations.

$$4 \cdot 3 - 5 \overset{?}{=} 7$$
$$12 - 5 \overset{?}{=} 7$$
$$7 = 7 \quad \text{(True)}$$

(b) Solve

$$
\begin{array}{rcl}
3x + 8 & = & -4 \\
-\,8 & & -8 \qquad \text{Add } -8 \text{ to both sides.} \\
\hline
3x & = & -12
\end{array}
$$

Now divide both sides by 3 to isolate x on the left.

$$\frac{3x}{3} = \frac{-12}{3}$$

$$x = -4$$

The solution is -4. We leave the check of this result to you.

✔ CHECK YOURSELF 1

Solve and check.

(a) $6x + 9 = -15$ **(b)** $5x - 8 = 7$

The variable may appear in any position in an equation. Just apply the rules carefully as you try to write an equivalent equation, and you will find the solution.

Example 2   Solving Equations

Solve

$$
\begin{array}{rl}
3 - 2x = & 9 \\
-3 \qquad & -3 \qquad \text{\textcolor{blue}{First add } -3 \text{ \textcolor{blue}{to both sides.}}} \\
\hline
-2x = & 6
\end{array}
$$

NOTE $\dfrac{-2}{-2} = 1$, so we divide by -2 to isolate x on the left.

Now divide both sides by -2. This will leave x alone on the left.

$$
\frac{-2x}{-2} = \frac{6}{-2}
$$

$$
x = -3
$$

The solution is -3. We leave it to you to check this result.

✔ CHECK YOURSELF 2

Solve and check.

$10 - 3x = 1$

You may also have to combine multiplication with addition or subtraction to solve an equation. Consider Example 3.

Example 3   Solving Equations

(a) Solve

$$
\frac{x}{5} - 3 = 4
$$

To get the x term alone, we first add 3 to both sides.

$$
\begin{array}{rl}
\dfrac{x}{5} - 3 = & 4 \\
+ 3 \quad +3 & \\
\hline
\dfrac{x}{5} \quad = & 7
\end{array}
$$

Now, to undo the division multiply both sides of the equation by 5.

$$
5\left(\frac{x}{5}\right) = 5 \cdot 7
$$

$$
x = 35
$$

The solution is 35. Return to the original equation to check the result.

$$\frac{35}{5} - 3 \stackrel{?}{=} 4$$

$$7 - 3 \stackrel{?}{=} 4$$

$$4 = 4 \qquad \text{(True)}$$

(b) Solve

$$\frac{2}{3}x + 5 = 13$$

$$\underline{\quad -5 \quad -5 \quad} \qquad \text{First add } -5 \text{ to both sides.}$$

$$\frac{2}{3}x \quad = \quad 8$$

Now multiply both sides by $\frac{3}{2}$, the reciprocal of $\frac{2}{3}$.

$$\left(\frac{3}{2}\right)\left(\frac{2}{3}x\right) = \left(\frac{3}{2}\right)8$$

or

$$x = 12$$

The solution is 12. We leave it to you to check this result.

 CHECK YOURSELF 3 _____

Solve and check.

(a) $\frac{x}{6} + 5 = 3$ ~~-5 -5~~ $X = -2$ $\boxed{6 \cdot -12}$

(b) $\frac{3}{4}x - 8 = 10$ $\boxed{X = 24}$ $\frac{3}{4}x = 18 / \frac{3}{4}$

   In Section 2.1, you learned how to solve certain equations when the variable appeared on both sides. Example 4 will show you how to extend that work when using the multiplication and addition properties of equality.

| **Example 4**   **Solving an Equation** |

Solve

$$6x - 4 = 3x - 2$$

   First add 4 to both sides. This will undo the subtraction on the left.

$$6x - 4 = 3x - 2$$
$$\underline{\quad +4 \qquad +4 \quad}$$
$$6x \quad = 3x + 2$$

Now add $-3x$ so that the terms in x will only be on the left side.

$$6x = \quad 3x + 2$$
$$\underline{-3x \quad -3x \quad}$$
$$3x = \qquad 2$$

Finally divide by 3.

$$\frac{3x}{3} = \frac{2}{3}$$

$$x = \frac{2}{3}$$

Check:

$$6\left(\frac{2}{3}\right) - 4 \overset{?}{=} 3\left(\frac{2}{3}\right) - 2$$

$$4 - 4 \overset{?}{=} 2 - 2$$

$$0 = 0 \quad \text{(True)}$$

The basic idea is to use our two properties to form an equivalent equation with the x isolated. Here we added 4 and then subtracted $3x$. You can do these steps in either order. Try it for yourself the other way. In either case, the multiplication property is then used as the *last step* in finding the solution.

 CHECK YOURSELF 4

Solve and check.

$7x - 5 = 3x + 5$

Next, we look at two approaches to solving equations in which the coefficient on the right side is greater than the coefficient on the left side.

Example 5   Solving an Equation (Two Methods)

Solve $4x - 8 = 7x + 7$.

Method 1

$$
\begin{array}{rl}
4x - 8 = & 7x + 7 \\
+ 8 & + 8 \\
\hline
4x \;\;\; = & 7x + 15 \\
-7x & -7x \\
\hline
-3x \;\;\; = & \;\;\;\;\; 15
\end{array}
$$

Adding 8 will leave the x term alone on the left.

Adding $-7x$ will get the variable terms on the left.

$$\frac{-3x}{-3} = \frac{15}{-3}$$

$$x = -5$$

Dividing by -3 will isolate x on the left.

We'll let you check this result.

To avoid a negative coefficient (in this example, -3), some students prefer a different approach.

This time we work toward having the number on the *left* and the x term on the *right,* or $\square = x$.

NOTE It is usually easier to isolate the variable term on the side that will result in a positive coefficient.

Method 2

$$4x - 8 = 7x + 7$$
$$\underline{ - 7 \qquad\quad -7} \qquad \text{Add } -7.$$
$$4x - 15 = 7x$$
$$\underline{-4x \qquad\qquad -4x} \qquad \text{Add } -4x \text{ to get the variables}$$
$$-15 = 3x \qquad\qquad \text{on the right.}$$

$$\frac{-15}{3} = \frac{3x}{3} \qquad \text{Divide by 3 to isolate } x \text{ on the right.}$$

$$-5 = x$$

Because $-5 = x$ and $x = -5$ are equivalent equations, it really makes no difference; the solution is still -5! You can use whichever approach you prefer.

CHECK YOURSELF 5

Solve $5x + 3 = 9x - 21$ by finding equivalent equations of the form $x = \square$ and $\square = x$ to compare the two methods of finding the solution.

It may also be necessary to remove grouping symbols in solving an equation.

OBJECTIVE 2 | **Example 6   Solving Equations That Contain Parentheses**

Solve.

NOTE

$5(x - 3)$

$= 5(x + (-3))$

$= 5x + 5(-3)$

$= 5x + (-15)$

$= 5x - 15$

$$5(x - 3) - 2x = x + 7 \qquad \text{First, apply the distributive property.}$$
$$5x - 15 - 2x = x + 7 \qquad \text{Combine like terms.}$$
$$3x - 15 = x + 7$$
$$\underline{+ 15 \qquad\qquad + 15} \qquad \text{Add 15.}$$
$$3x = x + 22$$
$$\underline{-x \quad -x} \qquad\qquad \text{Add } -x.$$
$$2x = 22 \qquad\qquad \text{Divide by 2.}$$
$$x = 11$$

The solution is 11. To check, substitute 11 for x in the original equation. Again note the use of our rules for the order of operations.

$$5((11) - 3) - 2 \cdot (11) \stackrel{?}{=} (11) + 7 \qquad \text{Simplify terms in parentheses.}$$
$$5 \cdot 8 - 2 \cdot 11 \stackrel{?}{=} 11 + 7 \qquad \text{Multiply.}$$
$$40 - 22 \stackrel{?}{=} 11 + 7 \qquad \text{Add and subtract.}$$
$$18 = 18 \qquad \text{A true statement.}$$

CHECK YOURSELF 6

Solve and check.

$$7(x + 5) - 3x = x - 7$$

We will now look at equations that contain fractions with different denominators. To solve an equation involving fractions, the first step is to multiply both sides of the equation by the **least common multiple (LCM)** of all denominators in the equation. Recall that the **LCM** of a set of numbers is the *smallest* number into which all the numbers will divide evenly.

OBJECTIVE 3

Example 7   Solving an Equation That Contains Fractions

Solve

$$\frac{x}{2} - \frac{2}{3} = \frac{5}{6}$$

First, multiply each side by 6, the LCM of 2, 3, and 6.

$$6\left(\frac{x}{2} - \frac{2}{3}\right) = 6\left(\frac{5}{6}\right) \qquad \text{Apply the distributive property.}$$

$$6\left(\frac{x}{2}\right) - 6\left(\frac{2}{3}\right) = 6\left(\frac{5}{6}\right) \qquad \text{Simplify.}$$

$$3x - 4 = 5$$

Next, isolate the variable x on the left side.

$$3x = 9$$
$$x = 3$$

The solution can be checked by returning to the original equation.

CHECK YOURSELF 7

Solve and check.

$$\frac{x}{4} - \frac{4}{5} = \frac{19}{20} \qquad +4/5 \qquad \frac{x}{4} = 1\frac{3}{4} \qquad x=7$$

Example 8   Solving an Equation That Contains Fractions

Solve

$$\frac{2x - 1}{5} + 1 = \frac{x}{2}$$

First multiply each side by 10, the LCM of 5 and 2.

$$10\left(\frac{2x - 1}{5} + 1\right) = 10\left(\frac{x}{2}\right) \qquad \text{Apply the distributive property on the left and reduce.}$$

$$10\left(\frac{2x - 1}{5}\right) + 10(1) = 10\left(\frac{x}{2}\right)$$

$$2(2x - 1) + 10 = 5x$$

$$4x - 2 + 10 = 5x \qquad \text{Next, isolate } x \text{ on the right side.}$$

$$4x + 8 = 5x$$

$$8 = x \qquad \text{The solution to the original equation is 8.}$$

CHECK YOURSELF 8

Solve and check.

$$\frac{3x + 1}{4} - 2 = \frac{x + 1}{3}$$

An equation that is true for any value of x is called an **identity.**

OBJECTIVE 4 | **Example 9   Solving an Equation** |

Solve the equation $2(x - 3) = 2x - 6$.

$$2(x - 3) = \quad 2x - 6$$
$$2x - 6 = \quad 2x - 6$$
$$\underline{-2x \qquad\quad -2x}$$
$$-6 = \qquad -6$$

NOTE We could ask the question "For what values of *x* does −6 = −6?"

The statement $-6 = -6$ is true for any value of *x*. The original equation is an identity.

CHECK YOURSELF 9

Solve the equation $3(x - 4) - 2x = x - 12$.

There are also equations for which there are no solutions.

OBJECTIVE 5 | **Example 10   Solving an Equation** |

Solve the equation $3(2x - 5) - 4x = 2x + 1$.

$$3(2x - 5) - 4x = \quad 2x + 1$$
$$6x - 15 \; - 4x = \quad 2x + 1$$
$$2x \; - 15 = \quad 2x + 1$$
$$\underline{-2x \qquad\qquad -2x}$$
$$-15 = 1$$

NOTE We could ask the question "For what values of *x* does −15 = 1?"

These two numbers are never equal. The original equation has no solutions.

CHECK YOURSELF 10

Solve the equation $2(x - 5) + x = 3x - 3$.

NOTE Such an outline of steps is sometimes called an **algorithm**.

Step by Step: Solving Linear Equations

Step 1 Use the distributive property to remove any grouping symbols. Then simplify by combining like terms on each side of the equation.

Step 2 Add or subtract the same term on each side of the equation until the variable term is on one side and a number is on the other.

Step 3 Multiply or divide both sides of the equation by the same nonzero number so that the variable is alone on one side of the equation. If no variable remains, determine whether the original equation is an identity or whether it has no solutions.

Step 4 Check the solution in the original equation.

CHECK YOURSELF ANSWERS

1. (a) -4; **(b)** 3 **2.** 3 **3. (a)** -12; **(b)** 24 **4.** $\dfrac{5}{2}$ **5.** 6 **6.** -14

7. 7 **8.** 5 **9.** The equation is an identity, x can be any real number.

10. There are no solutions.

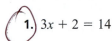

2.3   Exercises

Solve for x and check your result.

1. $3x + 2 = 14$

2. $3x - 1 = 17$
$$+ 1 \quad +18$$
$$3x = \frac{18}{3} \quad x = 6$$

3. $3x - 2 = 7$

4. $7x + 9 = 37$
$$-9 \quad -9$$
$$7x = 28/7 = x = 4$$

5. $4x + 7 = 35$

6. $7x - 8 = 13$
$$-8 \quad +8$$
$$7x = 21/7 \quad x = 3$$

7. $2x + 9 = 5$

8. $6x + 25 = -5$
$$-25 \quad -25$$
$$6x = -30 \quad x = -5$$

9. $4 - 7x = 18$

10. $8 - 5x = -7$
$$-8 \quad -8$$
$$-5x = -15/-5 \quad x = 3$$

11. $5 - 3x = 11$

12. $5 - 4x = 25$
$$-5 \quad -5$$
$$-4x = 20/-4 \quad x = -5$$

13. $\dfrac{x}{2} + 1 = 5$

14. $\dfrac{x}{5} - 3 = 2$
$$+3 \quad +3$$
$$\frac{x}{5} = 5 \quad x = 25$$

15. $\dfrac{x}{5} - 3 = 4$

16. $\dfrac{x}{5} + 3 = 8$
$$-3 \quad -3$$
$$\frac{x}{5} = 5 \quad x = 25$$

17. $\dfrac{2}{3}x + 5 = 17$

18. $\dfrac{3}{4}x - 5 = 4$
$$+5 \quad +5$$
$$3/4x = 9/3/4 \quad x = 12$$

19. $\dfrac{3}{4}x - 2 = 16$

20. $\dfrac{5}{7}x + 4 = 14$
$$-4 \quad -4$$
$$5/7x = 10/5/7 \quad x = 14$$

21. $5x = 2x + 9$

22. $18 - 2x = 7x$
$$+2x \quad +2x$$
$$18 = 9x \quad x = 2$$

23. $3x = 10 - 2x$

24. $11x = 7x + 20$
$$-7x \quad -7x$$
$$4x = 20/4 \quad x = 5$$

25. $9x + 2 = 3x + 38$

26. $8x - 3 = 4x + 17$
$$+3 \quad +3$$
$$4x = 20 \quad x = 5$$

27. $4x - 8 = x - 14$

28. $6x - 5 = 3x - 29$
$$3x = -24/3 \quad x = 8 \quad \boxed{x = -8}$$

29. $5x + 7 = 2x - 3$

30. $9x + 7 = 5x - 3$
$$+3 \quad +3 \quad 9x + 10 = 5x$$
$$9x + 10 = 5x \quad \boxed{x = -25}$$
$$10 = \cancel{8} \quad \frac{10}{-4} = \frac{-4000}{-4}$$

31. $7x - 3 = 9x + 5$

32. $5x - 2 = 8x - 11$
$$+2 \quad +2$$
$$5x = 8x - 9$$
$$-8x \quad -8x$$
$$-3x = -9/-3 \quad x = 3$$

ANSWERS

1. _____
2. _____
3. _____
4. _____
5. _____
6. _____
7. _____
8. _____
9. _____
10. _____
11. _____
12. _____
13. _____
14. _____
15. _____
16. _____

17.	18.
19.	20.
21.	22.
23.	24.
25.	26.
27.	28.
29.	30.
31.	32.

33. _____

34. _____

35. _____

36. _____

37. _____

38. _____

39. _____

40. _____

41. _____

42. _____

43. _____

44. _____

45. _____

46. _____

47. _____ 48. _____

49. _____ 50. _____

51. _____ 52. _____

53. _____ 54. _____

55. _____

56. _____

57. _____

58. _____

59. _____

60. _____

61. _____

62. _____

63. _____

33. $5x + 4 = 7x - 8$

34. $2x + 23 = 6x - 5$ $\quad$ x=7
$\qquad$ +5 $\quad$ +5 $\qquad$ 28 = 4x
$\qquad$ 2x + 28 = 6x

35. $2x - 3 + 5x = 7 + 4x + 2$

36. $8x - 7 - 2x = 2 + 4x - 5$

37. $6x + 7 - 4x = 8 + 7x - 26$

38. $7x - 2 - 3x = 5 + 8x + 13$

39. $9x - 2 + 7x + 13 = 10x - 13$

40. $5x + 3 + 6x - 11 = 8x + 25$

41. $2(x + 3) = 8$

42. $-3(x - 1) = 4(x + 2) + 2$

43. $7(2x - 1) - 5x = x + 25$

44. $9(3x + 2) - 10x = 12x - 7$

45. $3x + 2(4x - 3) = 6x - 9$

46. $7x + 3(2x + 5) = 10x + 17$

47. $\dfrac{8}{3}x - 3 = \dfrac{2}{3}x + 15$

48. $\dfrac{12}{5}x + 7 = 31 - \dfrac{3}{5}x$

49. $\dfrac{2x}{5} - \dfrac{x}{3} = \dfrac{7}{15}$

50. $\dfrac{2x}{7} - \dfrac{3x}{5} = \dfrac{6}{35}$ $\quad$ 10x -21x = 6
$\qquad \dfrac{10x}{35} - \dfrac{21x}{35} = \dfrac{6}{35}$ $\quad$ -11x = 6
$\qquad$ $\dfrac{}{-11}$
$\qquad$ x = -0.54

51. $5.3x - 7 = 2.3x + 5$

52. $9.8x + 2 = 3.8x + 20$

53. $\dfrac{5x - 3}{4} - 2 = \dfrac{x}{3}$

54. $\dfrac{6x - 1}{5} - \dfrac{2x}{3} = 3$ $\quad$ 5
$\qquad \dfrac{18x-3}{15} - \dfrac{10x}{15} = \dfrac{5}{15}$ $\quad$ 18x-3-10x=5
$\qquad$ 8x-3=5
$\qquad$ r 3+3
$\qquad$ $\dfrac{8x=8}{}$ x=1

55. $5(x + 1) - 4x = x - 5$

56. $-4(2x - 3) = -8x + 5$

57. $6x - 4x + 1 = 12 + 2x - 11$

58. $-2x + 5x - 9 = 3(x - 4) - 5$

59. $-4(x + 2) - 11 = 2(-2x - 3) - 13$

60. $4(-x - 2) + 5 = -2(2x + 7)$

61. Create an equation of the form $ax + b = c$ that has 2 as a solution.

62. Create an equation of the form $ax + b = c$ that has 7 as a solution.

63. The equation $3x = 3x + 5$ has no solution, whereas the equation $7x + 8 = 8$ has zero as a solution. Explain the difference between a solution of zero and no solution.

64. Construct an equation for which every real number is a solution.

In exercises 65 to 68, find the length of each side of the figure for the given perimeter.

65.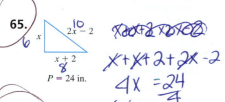

$P = 24$ in.

handwritten: $x + x + 2 + 2x - 2$
$4x = \dfrac{24}{4}$
$x = 6$

66.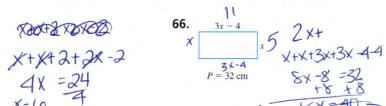

$P = 32$ cm

handwritten: $2x +$
$x + x + 3x + 3x - 4 - 4$
$8x - 8 = 32$
$+8 \quad +8$
$8x = 40$
$x = 5$

67.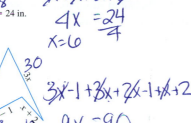

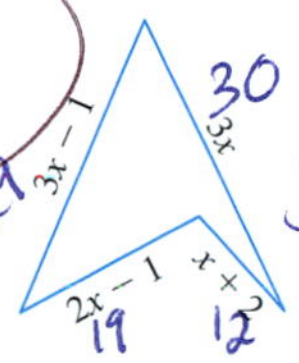

$P = 90$ in.

handwritten: $3x - 1 + 3x + 2x - 1 + x + 2 = 90$
$9x = \dfrac{90}{9}$
$x = 10$

68.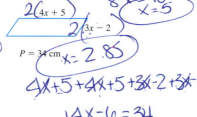

$2(4x + 5)$
$2(3x - 2)$
$P = 34$ cm $x = 2.85$

handwritten: $4x + 5 + 4x + 5 + 3x - 2 + 3x - 2$
$14x - 6 = 34$
$+6$
$14 = 40/14$

 Getting Ready for Section 2.4 [Section 1.7]

Divide.

(a) $\dfrac{3b}{3}$ b

(b) $\dfrac{5x}{5}$ x

(c) $\dfrac{4xy}{4x}$ y

(d) $\dfrac{6a^2b}{6a^2}$ b

(e) $\dfrac{7mn^2}{7n^2}$ m

(f) $\dfrac{\pi ab}{\pi a}$ b

(g) $\dfrac{srt}{sr}$ t

(h) $\dfrac{x^2yz}{x^2z}$ y

handwritten (right):
$8x + 10 + 6x - 4$
$14x + 6 = 32$
-6
$14x = \dfrac{26}{14}$
$x = 1.85$

$8x + 10 + 6x - 4 = 34$
$14x + 6 = 34$
-6
$14x = \dfrac{28}{14} \quad x = 2$

Answers

1. 4 **3.** 3 **5.** 7 **7.** -2 **9.** -2 **11.** -2 **13.** 8

15. 35 **17.** 18 **19.** 24 **21.** 3 **23.** 2 **25.** 6 **27.** -2

29. $-\dfrac{10}{3}$ **31.** -4 **33.** 6 **35.** 4 **37.** 5 **39.** -4 **41.** 1

43. 4 **45.** $-\dfrac{3}{5}$ **47.** 9 **49.** 7 **51.** 4 **53.** 3

55. No solution **57.** Identity **59.** Identity **61.** $6x + 5 = 17$

63. ✎ **65.** 6 in., 8 in., 10 in. **67.** 12 in., 19 in., 29 in., 30 in.

a. b **b.** x **c.** y **d.** b **e.** m **f.** b **g.** t **h.** y

handwritten (bottom center):
$36x + 24 = 12x + 12$
$36x - 12 = 12x + 12$
$12x$
$24x - 12 = 12$
2^0

ANSWERS

64. _____

65. _____

66. $11, 5$

67. _____

68. $13, 4$

a. _____

b. _____

c. _____

d. _____

e. _____

f. _____

g. _____

h. _____

handwritten (right margin):
$12\left(\dfrac{3x+1}{4}\right) - (2) = \left(\dfrac{x+1}{3}\right)$

$\dfrac{9x+3}{12} - \dfrac{6}{12} = \dfrac{4x+4}{12}$

$9x + 3 - 6 = 4x + 4$
$-4x \qquad -4x$

$5x + 3 - 6 = 16 \qquad 5x = 7$
$+6$
$5x + 3 = 10$

2.4 Formulas and Problem Solving

1. Solve a literal equation for one of its variables
2. Solve an application involving a literal equation
3. Translate a word phrase to an expression or an equation
4. Use an equation to solve an application

Formulas are extremely useful tools in any field in which mathematics is applied. Formulas are simply equations that express a relationship between more than one letter or variable. You are no doubt familiar with all kinds of formulas, such as

$$A = \frac{1}{2} bh \qquad \text{The area of a triangle}$$

$$I = Prt \qquad \text{Interest}$$

$$V = \pi r^2 h \qquad \text{The volume of a cylinder}$$

Actually a formula is also called a **literal equation** because it involves several letters or variables. For instance, our first formula or literal equation, $A = \frac{1}{2} bh$, involves the three letters A (for area), b (for base), and h (for height).

Unfortunately, formulas are not always given in the form needed to solve a particular problem. Then algebra is needed to change the formula to a more useful equivalent equation, which is solved for a particular letter or variable. The steps used in the process are very similar to those you used in solving linear equations. Consider an example.

OBJECTIVE 1

Example 1 Solving a Literal Equation Involving a Triangle

Suppose that we know the area A and the base b of a triangle and want to find its height h.

We are given

$$A = \frac{1}{2} bh$$

Our job is to find an equivalent equation with h, the unknown, by itself on one side. We call $\frac{1}{2} b$ the **coefficient** of h. We can remove the two *factors* of that coefficient, $\frac{1}{2}$ and b, separately.

$$2A = 2\left(\frac{1}{2} bh\right) \qquad \text{\color{blue}Multiply both sides by 2 to clear the equation of fractions.}$$

NOTE

$$2\left(\frac{1}{2} bh\right) = \left(2 \cdot \frac{1}{2}\right)(bh)$$
$$= 1(bh)$$
$$= bh$$

or

$$2A = bh$$

$$\frac{2A}{b} = \frac{bh}{b} \qquad \text{\color{blue}Divide by b to isolate h.}$$

$$\frac{2A}{b} = h$$

or

$$h = \frac{2A}{b} \qquad \text{\color{blue}Reverse the sides to write h on the left.}$$

We now have the height h in terms of the area A and the base b. This is called **solving the equation for h** and means that we are rewriting the formula as an equivalent equation of the form

NOTE Here $\square$ means an expression containing all the numbers or letters *other* than h.

$$h = \square.$$

CHECK YOURSELF 1

Solve $V = \dfrac{1}{3}Bh$ for h.

You have already learned the methods needed to solve most literal equations or formulas for some specified variable. As Example 1 illustrates, the rules of Sections 2.1 and 2.2 are applied in exactly the same way as they were applied to equations with one variable.

You may have to apply both the addition and the multiplication properties when solving a formula for a specified variable. Example 2 illustrates this property.

Example 2   Solving a Literal Equation

Solve $y = mx + b$ for x.

Remember that we want to end up with x alone on one side of the equation. Start by subtracting b from both sides to undo the addition on the right.

$$\begin{array}{r} y = mx + b \\ \underline{-b \qquad\quad -b} \\ y - b = mx \end{array}$$

If we now divide both sides by m, then x will be alone on the right-hand side.

$$\frac{y - b}{m} = \frac{mx}{m}$$

$$\frac{y - b}{m} = x$$

or

$$x = \frac{y - b}{m}$$

CHECK YOURSELF 2

Solve $v = v_0 + gt$ for t.

Let's summarize the steps illustrated by our examples.

Step by Step: Solving a Formula or Literal Equation

Step 1 If necessary, multiply both sides of the equation by the same term to clear it of fractions.

Step 2 Add or subtract the same term on each side of the equation so that all terms involving the variable that you are solving for are on one side of the equation and all other terms are on the other side.

Step 3 Divide both sides of the equation by the coefficient of the variable that you are solving for.

Look at one more example using these steps.

Example 3 Solving a Literal Equation Involving Money

NOTE This is a formula for the *amount* of money in an account after interest has been earned.

Solve $A = P + Prt$ for r.

$$\begin{array}{rcl} A & = & P + Prt \\ -P & & -P \\ \hline A - P & = & Prt \end{array}$$

Adding $-P$ to both sides will leave the term involving r alone on the right.

$$\frac{A - P}{Pt} = \frac{Prt}{Pt}$$

Dividing both sides by Pt will isolate r on the right.

$$\frac{A - P}{Pt} = r$$

or

$$r = \frac{A - P}{Pt}$$

CHECK YOURSELF 3 _____

Solve $2x + 3y = 6$ for y.

Now look at an application of solving a literal equation.

OBJECTIVE 2

Example 4 Solving a Literal Equation Involving Money

Suppose that the amount in an account, 3 years after a principal of $5000 was invested, is $6050. What was the interest rate?

From Example 3,

$$A = P + Prt$$

in which A is the amount in the account, P is the principal, r is the interest rate, and t is the time that the money has been invested. By the result of Example 3 we have

$$r = \frac{A - P}{Pt}$$

and we can substitute the known values into this equation:

NOTE Do you see the advantage of having our equation solved for the desired variable?

$$r = \frac{(6050) - (5000)}{(5000)(3)}$$

$$= \frac{1050}{15,000} = 0.07 = 7\%$$

The interest rate is 7 percent.

CHECK YOURSELF 4 _____

Suppose that the amount in an account, 4 years after a principal of $3000 was invested, is $3480. What was the interest rate?

The main reason for learning how to set up and solve algebraic equations is so that we can use them to solve word problems. In fact, algebraic equations were *invented* to make solving word problems much easier. The first word problems that we know about are over 4000 years old. They were literally "written in stone," on Babylonian tablets, about 500 years before the first algebraic equation made its appearance.

Before algebra, people solved word problems primarily by "guess-and-check," which is a method of finding unknown numbers by using trial and error in a logical way. Example 5 shows how to solve a word problem using substitution.

Example 5   Solving a Word Problem by Substitution

The sum of two consecutive integers is 37. Find the two integers.

If the two integers were 20 and 21, their sum would be 41. Because that's more than 37, the integers must be smaller. If the integers were 15 and 16, the sum would be 31. More trials yield that the sum of 18 and 19 is 37.

CHECK YOURSELF 5 _____

The sum of two consecutive integers is 91. *Find the two integers.*

Most word problems are not so easily solved by the guess-and-check method. For more complicated word problems, a five-step procedure is used. Using this step-by-step approach will, with practice, allow you to organize your work. Organization is the key to solving word problems. Here are the five steps.

Step by Step: To Solve Word Problems

Step 1   Read the problem carefully. Then reread it to decide what you are asked to find.

Step 2   Choose a letter to represent one of the unknowns in the problem. Then represent all other unknowns of the problem with expressions that use the same letter.

Step 3   Translate the problem to the language of algebra to form an equation.

Step 4   Solve the equation and answer the question of the original problem.

Step 5   Check your solution by returning to the original problem.

RECALL We discussed these translations in Section 1.1. You might find it helpful to review that section before going on.

The third step is usually the hardest part. We must translate words to the language of algebra. Before we look at a complete example, the following table may help you review that translation step.

Translating Words to Algebra

Words	Algebra
The sum of x and y	$x + y$
3 plus a	$3 + a$ or $a + 3$
5 more than m	$m + 5$
b increased by 7	$b + 7$
The difference of x and y	$x - y$
4 less than a	$a - 4$
s decreased by 8	$s - 8$
The product of x and y	$x \cdot y$ or xy
5 times a	$5 \cdot a$ or $5a$
Twice m	$2m$
The quotient of x and y	$\dfrac{x}{y}$
a divided by 6	$\dfrac{a}{6}$
One-half of b	$\dfrac{b}{2}$ or $\dfrac{1}{2}b$

Now let's look at some typical examples of translating phrases to algebra.

OBJECTIVE 3 **Example 6 Translating Statements**

Translate each statement to an algebraic expression.

(a) The sum of a and twice b $a + 2b$

       Sum   Twice b

(b) 5 times m increased by 1 $5m + 1$

       5 times m Increased by 1

(c) 5 less than 3 times x $3x - 5$

       3 times x 5 less than

(d) The product of x and y, divided by 3 $\dfrac{xy}{3}$

       The product of x and y

       Divided by 3

CHECK YOURSELF 6

Translate to algebra.

(a) 2 more than twice x

(b) 4 less than 5 times n

(c) The product of twice a and b

(d) The sum of s and t, divided by 5

Now let's work through a complete example. Although this problem could be solved by substitution, it is presented here to help you practice the five-step approach.

OBJECTIVE 4

Example 7   Solving an Application

The sum of a number and 5 is 17. What is the number?

Step 1   *Read carefully.*   You must find the unknown number.

Step 2   *Choose letters or variables.*   Let x represent the unknown number. There are no other unknowns.

Step 3   *Translate.*

The sum of

$$x + 5 = 17$$

is

Step 4   *Solve.*

$$
\begin{array}{r}
x + 5 = 17 \\
\underline{-5 \quad -5} \quad \text{Add } -5. \\
x = 12
\end{array}
$$

So the number is 12.

NOTE Always return to the *original problem* to check your result and *not* to the equation of step 3. This will prevent possible errors!

Step 5   *Check.*   Is the sum of 12 and 5 equal to 17? Yes ($12 + 5 = 17$). We have checked our solution.

CHECK YOURSELF 7

The sum of a number and 8 is 35. What is the number?

$n + 8 = 35$
$\quad\quad -8$
$n = 27$

Definition: Consecutive Integers

Consecutive integers are integers that follow one another, like 10, 11, and 12. To represent them in algebra:

If x is an integer, then $x + 1$ is the next consecutive integer, $x + 2$ is the next, and so on.

We need this idea in Example 8.

Example 8   Solving an Application

REMEMBER THE STEPS!
Read the problem carefully. What do you need to find?

Assign letters to the unknown or unknowns.

Write an equation.

The sum of two consecutive integers is 41. What are the two integers?

Step 1   We want to find the two consecutive integers.

Step 2   Let x be the first integer. Then $x + 1$ must be the next.

Step 3

The first integer   The second integer

$$x + (x + 1) = 41$$

The sum   Is

NOTE Solve the equation.

Step 4

$$x + x + 1 = 41$$
$$2x + 1 = 41$$
$$2x = 40$$
$$x = 20$$

The first integer (x) is 20, and the next integer ($x + 1$) is 21.

NOTE Check.

Step 5   The sum of the two integers 20 and 21 is 41.

 CHECK YOURSELF 8 _____

The sum of three consecutive integers is 51. What are the three integers?

Sometimes algebra is used to reconstruct missing information. Example 9 does just that with some election information.

Example 9   Solving an Application

There were 55 more yes votes than no votes on an election measure. If 735 votes were cast in all, how many yes votes were there? How many no votes?

NOTE What do you need to find?

NOTE Assign letters to the unknowns.

Step 1   We want to find the number of yes votes and the number of no votes.

Step 2   Let x be the number of no votes. Then

$$\underline{x + 55}$$

55 more than x

is the number of yes votes.

NOTE Write an equation.

Step 3

$$x + \underline{x + 55} = 735$$

No votes   Yes votes

NOTE Solve the equation.

Step 4

$$x + x + 55 = 735$$
$$2x + 55 = 735$$
$$2x = 680$$
$$x = 340$$
$$\text{No votes } (x) = 340$$
$$\text{Yes votes } (x + 55) = 395$$

NOTE Check.

Step 5   Thus 340 no votes plus 395 yes votes equals 735 total votes. The solution checks.

 CHECK YOURSELF 9 ⎯⎯⎯⎯⎯⎯⎯

Francine earns $120 per month more than Rob. If they earn a total of $2680 per month, what are their monthly salaries? Rob=1280   Fran = 1400

Similar methods will allow you to solve a variety of word problems. Example 10 includes three unknown quantities but uses the same basic solution steps.

Example 10   Solving an Application

NOTE There are other choices for *x*, but choosing the smallest quantity will usually give the easiest equation to write and solve.

Juan worked twice as many hours as Jerry. Marcia worked 3 more hours than Jerry. If they worked a total of 31 hours, find out how many hours each worked.

Step 1   We want to find the hours each worked, so there are three unknowns.

Step 2   Let *x* be the hours that Jerry worked.

Twice Jerry's hours

Then $2x$ is Juan's hours worked

3 more hours than Jerry worked

and $x + 3$ is Marcia's hours.

Step 3

Jerry   Juan   Marcia

$$x + 2x + (x + 3) = 31$$

Sum of their hours

Fran + Rob = 2680

Rob + 120 = Fran.

$x + 120 = y$

Rob = x
Fran x+120

1280
+ 120
1400

$2x + 120 = 2680$
$-120 \quad -120$

$$2x = \frac{2560}{2}$$

Step 4

$$x + 2x + x + 3 = 31$$
$$4x + 3 = 31$$
$$4x = 28$$
$$x = 7$$

Jerry's hours $(x) = 7$

Juan's hours $(2x) = 14$

Marcia's hours $(x + 3) = 10$

Step 5   The sum of their hours $(7 + 14 + 10)$ is 31, and the solution is verified.

CHECK YOURSELF 10 ⎯⎯⎯⎯⎯⎯⎯⎯⎯⎯

Paul jogged half as many miles (mi) as Lucy and 7 less than Isaac. If the three ran a total of 23 mi, how far did each person run?

CHECK YOURSELF ANSWERS ⎯⎯⎯⎯⎯⎯⎯⎯⎯

1. $h = \dfrac{3V}{B}$ **2.** $t = \dfrac{v - v_0}{g}$ **3.** $y = \dfrac{6 - 2x}{3}$ or $y = -\dfrac{2}{3}x + 2$

4. 4 percent   **5.** 45 and 46   **6.** **(a)** $2x + 2$; **(b)** $5n - 4$; **(c)** $2ab$; **(d)** $\dfrac{s + t}{5}$

7. The equation is $x + 8 = 35$. The number is 27.

8. The equation is $x + x + 1 + x + 2 = 51$. The integers are 16, 17, and 18.

9. The equation is $x + x + 120 = 2680$. Rob's salary is $1280, and Francine's is $1400.   **10.** Paul: 4 mi; Lucy: 8 mi; Isaac: 11 mi

Name _____

Section _____ Date _____

Solve each literal equation for the indicated variable.

1. $p = 4s$ (for s)   Perimeter of a square

$\dfrac{}{4}$ $\dfrac{}{4}$

2. $V = Bh$ (for B)   Volume of a prism

3. $E = IR$ (for R)   Voltage in an electric circuit

$\dfrac{}{I}$ $\dfrac{}{I}$

4. $I = Prt$ (for r)   Simple interest

$\dfrac{}{Pt}$ $\dfrac{}{Pt}$

5. $V = LWH$ (for H)   Volume of a rectangular solid

$\dfrac{}{LW}$ $\dfrac{}{LW}$

6. $V = \pi r^2 h$ (for h)   Volume of a cylinder

$\dfrac{}{\pi r^2}$ $\dfrac{}{\pi r^2}$

7. $A + B + C = 180$ (for B)   Measure of angles in a triangle

$A + B + C = 180 - B$

8. $P = I^2R$ (for R)   Power in an electric circuit

9. $ax + b = 0$ (for x)   Linear equation in one variable

10. $y = mx + b$ (for m)   Slope-intercept form for a line

11. $s = \dfrac{1}{2}gt^2$ (for g)   Distance

12. $K = \dfrac{1}{2}mv^2$ (for m)   Energy

13. $x + 5y = 15$ (for y)   Linear equation

14. $2x + 3y = 6$ (for x)   Linear equation

15. $P = 2L + 2W$ (for L)   Perimeter of a rectangle

16. $ax + by = c$ (for y)   Linear equation in two variables

ANSWERS

1. $\dfrac{P}{4} = S$

2. $\dfrac{V}{h} = B$

3. $\dfrac{E}{I} = R$

4. $\dfrac{1}{Pt} = R$

5. $\dfrac{V}{LW} = H$

6. $\dfrac{V}{\pi r^2} = h$

7. _____

8. _____

9. _____

10. _____

11. _____

12. _____

13. _____

14. _____

15. _____

16. _____

17. _____

18. _____

19. _____

20. _____

21. _____

22. _____

23. _____

24. _____

25. _____

26. _____

27. _____

28. _____

29. _____

17. $V = \dfrac{KT}{P}$ (for T)   Volume of a gas

18. $V = \dfrac{1}{3}\pi r^2 h$ (for h)   Volume of a cone

19. $x = \dfrac{a + b}{2}$ (for b)   Mean of two numbers

20. $D = \dfrac{C - s}{n}$ (for s)   Depreciation

21. $F = \dfrac{9}{5}C + 32$ (for C)   Celsius/Fahrenheit

22. $A = P + Prt$ (for t)   Amount at simple interest

23. $S = 2\pi r^2 + 2\pi rh$ (for h)   Total surface area of a cylinder

24. $A = \dfrac{1}{2}h(B + b)$ (for b)   Area of a trapezoid

25. Geometry. A rectangular solid has a base with length 8 cm and width 5 cm. If the volume of the solid is 120 cm³, find the height of the solid. (See exercise 5.)

26. Geometry. A cylinder has a radius of 4 in. If the volume of the cylinder is 48π in.³, what is the height of the cylinder? (See exercise 6.)

27. Business and finance. A principal of $3000 was invested in a savings account for 3 years. If the interest earned for the period was $450, what was the interest rate? (See exercise 4.)

28. Geometry. If the perimeter of a rectangle is 60 ft and the width is 12 ft, find its length. (See exercise 15.)

29. Science and medicine. The high temperature in New York for a particular day was reported at 77°F. How would the same temperature have been given in degrees Celsius? (See exercise 21.)

30. Crafts. Rose's garden is in the shape of a trapezoid. If the height of the trapezoid is 16 m, one base is 20 m, and the area is 224 m², find the length of the other base. (See exercise 24.)

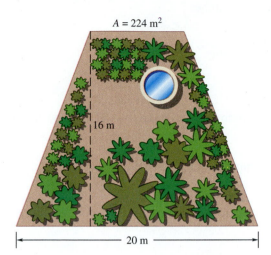

$A = 224$ m²

16 m

20 m

30. _____

31. _____

32. _____

33. _____

34. _____

35. _____

36. _____

37. _____

38. _____

39. _____

40. _____

41. _____

Translate each statement to an algebraic equation. Let x represent the number in each case.

31. 3 more than a number is 7.

$$3+n=7$$

32. 5 less than a number is 12.

$$n-5=12$$

33. 7 less than 3 times a number is twice that same number.

$$3n-7=2n$$

34. 4 more than 5 times a number is 6 times that same number.

$$5n+4=6n$$

35. 2 times the sum of a number and 5 is 18 more than that same number.

$$2(n+5)=18+n$$

36. 3 times the sum of a number and 7 is 4 times that same number.

$$3(7+n)=4n$$

37. 3 more than twice a number is 7.

$$3+2n=7$$

38. 5 less than 3 times a number is 25.

$$3n-5=25$$

$$3x-5=25$$
$$+5\quad\quad$$
$$3x=30 \quad \boxed{x=10}$$

39. 7 less than 4 times a number is 41.

$$4n-7=41$$

40. 10 more than twice a number is 44.

$$2n+10=44$$

41. 5 more than two-thirds of a number is 21.

$$\tfrac{2}{3}n+5=21$$

© 2005 McGraw-Hill Companies

42. 3 less than three-fourths of a number is 24.

43. 3 times a number is 12 more than that number.

44. 5 times a number is 8 less than that number.

Solve the following word problems. Be sure to label the unknowns and to show the equation you use for the solution.

45. **Number problem.** The sum of a number and 7 is 33. What is the number?

46. **Number problem.** The sum of a number and 15 is 22. What is the number?

47. **Number problem.** The sum of a number and -15 is 7. What is the number?

48. **Number problem.** The sum of a number and -8 is 17. What is the number?

49. **Social science.** In an election, the winning candidate has 1840 votes. If the total number of votes cast was 3260, how many votes did the losing candidate receive?

55

50. **Business and finance.** Mike and Stefanie work at the same company and make a total of $2760 per month. If Stefanie makes $1400 per month, how much does Mike earn every month?

51. **Number problem.** The sum of twice a number and 5 is 35. What is the number?

52. **Number problem.** 3 times a number, increased by 8, is 50. Find the number.

53. **Number problem.** 5 times a number, minus 12, is 78. Find the number.

54. **Number problem.** 4 times a number, decreased by 20, is 44. What is the number?

55. **Number problem.** The sum of two consecutive integers is 47. Find the two integers.

56. **Number problem.** The sum of two consecutive integers is 145. Find the two integers.

$$n + n + 1 = 145 \qquad 2n + 144$$
$$2n + 1 = 145 \quad n = 72, 73 \quad 2$$

57. **Number problem.** The sum of three consecutive integers is 63. What are the three integers?

58. Number problem. If the sum of three consecutive integers is 93, find the three integers.

$$n + n + 1 + n + 2 = 93$$
$$3n + 3 = 93 \qquad 3n = 90$$
$$n = 30, 31, 32$$

59. Number problem. The sum of two consecutive even integers is 66. What are the two integers? (*Hint:* Consecutive even integers such as 10, 12, and 14 can be represented by x, $x + 2$, $x + 4$, and so on.)

60. Number problem. If the sum of two consecutive even integers is 114, find the two integers.

61. Number problem. If the sum of two consecutive odd integers is 52, what are the two integers? (*Hint:* Consecutive odd integers such as 21, 23, and 25 can be represented by x, $x + 2$, $x + 4$, and so on.)

62. Number problem. The sum of two consecutive odd integers is 88. Find the two integers.

$$n + n + 2 = 88$$
$$2n + 2 = 88 \qquad \boxed{43, 45} \quad \frac{2n = 86}{2}$$

63. Number problem. The sum of three consecutive odd integers is 63. What are the three integers?

64. Number problem. The sum of three consecutive even integers is 126. What are the three integers?

65. Number problem. The sum of four consecutive integers is 86. What are the four integers?

$$n + n + 1 + n + 2 + n + 3 = 86 \qquad 20, 21, 22, 23$$
$$4n + 6 = 86 \qquad 4n = \frac{80}{4}$$

66. Number problem. The sum of four consecutive integers is 62. What are the four integers?

$$n + n + 1 + n + 2 + n + 3 \qquad 4n = \frac{56}{4} \qquad 14, 15, 16, 17$$
$$2n + 6 = 62$$

67. Number problem. 4 times an integer is 9 more than 3 times the next consecutive integer. What are the two integers?

68. Number problem. 4 times an integer is 30 less than 5 times the next consecutive even integer. Find the two integers.

69. Social science. In an election, the winning candidate had 160 more votes than the loser. If the total number of votes cast was 3260, how many votes did each candidate receive?

70. Business and finance. Jody earns $140 more per month than Frank. If their monthly salaries total $2760, what amount does each earn?

58. 30, 31, 32
59.
60.
61.
62. 43, 45
63.
64.
65. 20, 21, 22, 23
66. 14, 15, 16, 17
67.
68.
69.
70.

71. _____

72. _____

73. _____

74. _____

75. _____

76. _____

77. _____

78. _____

71. Business and finance. A washer-dryer combination costs $650. If the washer costs $70 more than the dryer, what does each appliance cost?

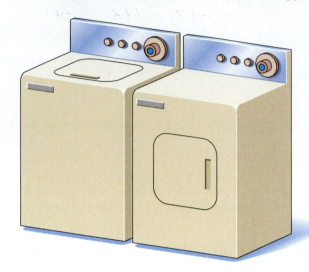

72. Crafts. Yuri has a board that is 98 in. long. He wishes to cut the board into two pieces so that one piece will be 10 in. longer than the other. What should the length of each piece be?

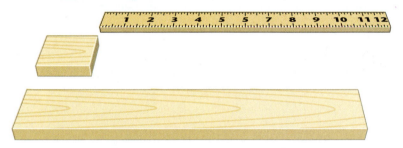

73. Social science. Yan Ling is 1 year less than twice as old as his sister. If the sum of their ages is 14 years, how old is Yan Ling?

74. Social science. Diane is twice as old as her brother Dan. If the sum of their ages is 27 years, how old are Diane and her brother?

$$X + 2X = 27 \qquad 3x = 27/3$$

75. Social science. Maritza is 3 years less than 4 times as old as her daughter. If the sum of their ages is 37, how old is Maritza?

76. Social science. Mrs. Jackson is 2 years more than 3 times as old as her son. If the difference between their ages is 22 years, how old is Mrs. Jackson?

77. Business and finance. On her vacation in Europe, Jovita's expenses for food and lodging were $60 less than twice as much as her airfare. If she spent $2400 in all, what was her airfare?

78. Business and finance. Rachel earns $6000 less than twice as much as Tom. If their two incomes total $48,000, how much does each earn?

79. Statistics. There are 99 students registered in three sections of algebra. There are twice as many students in the 10 A.M. section as the 8 A.M. section and 7 more students at 12 P.M. than at 8 A.M. How many students are in each section?

80. Business and finance. The Randolphs used 12 more gal of fuel oil in October than in September and twice as much oil in November as in September. If they used 132 gal for the 3 months, how much was used during each month?

81. "I make $2.50 an hour more in my new job." If x = the amount I used to make per hour and y = the amount I now make, which equation(s) below say the same thing as the statement above? Explain your choice(s) by translating the equation into English and comparing with the original statement.

(a) $x + y = 2.50$ **(b)** $x - y = 2.50$

(c) $x + 2.50 = y$ **(d)** $2.50 + y = x$

(e) $y - x = 2.50$ **(f)** $2.50 - x = y$

82. "The river rose 4 feet above flood stage last night." If a = the river's height at flood stage, b = the river's height now (the morning after), which equations below say the same thing as the statement? Explain your choices by translating the equations into English and comparing the meaning with the original statement.

(a) $a + b = 4$ **(b)** $b - 4 = a$

(c) $a - 4 = b$ **(d)** $a + 4 = b$

(e) $b + 4 = b$ **(f)** $b - a = 4$

83. Maxine lives in Pittsburgh, Pennsylvania, and pays 8.33 cents per kilowatt hour (kWh) for electricity. During the 6 months of cold winter weather, her household uses about 1500 kWh of electric power per month. During the two hottest summer months, the usage is also high because the family uses electricity to run an air conditioner. During these summer months, the usage is 1200 kWh per month; the rest of the year, usage averages 900 kWh per month.

(a) Write an expression for the total yearly electric bill.

(b) Maxine is considering spending $2000 for more insulation for her home so that it is less expensive to heat and to cool. The insulation company claims that "with proper installation the insulation will reduce your heating and cooling bills by 25 percent." If Maxine invests the money in insulation, how long will it take her to get her money back by saving on her electric bill? Write to her about what information she needs to answer this question. Give her your opinion about how long it will take to save $2000 on heating and cooling bills, and explain your reasoning. What is your advice to Maxine?

ANSWERS

79. _____

80. _____

81. _____

82. _____

83. _____

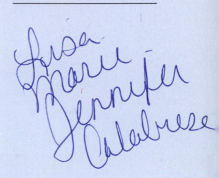

Getting Ready for Section 2.5 [Section 1.4]

Perform the indicated operations.

(a) $4 \cdot (8 - 6) + 7$ (b) $3 \cdot (8 - 5) + 10 \div 2$

(c) $4(6 - 8) \div 4 + 5$ (d) $8(7 - 3 \cdot 4) + 12(5 - 3)$

(e) $-2(16 \div 4 \cdot 2) - 5(3 - 5)$ (f) $8 \cdot (4 \cdot 5 - 2)(10 \div 2 \cdot 5)$

(g) $-7(13 - 4 \cdot 2) \div (25 - 10 \cdot 2)$ (h) $2(-25 + 5 \cdot 3) \cdot (4 - 3 \cdot 2)$

Answers

1. $S = \dfrac{p}{4}$ **3.** $R = \dfrac{E}{I}$ **5.** $H = \dfrac{V}{LW}$ **7.** $B = 180 - A - C$

9. $x = -\dfrac{b}{a}$ **11.** $g = \dfrac{2s}{t^2}$ **13.** $y = \dfrac{15 - x}{5}$ or $-\dfrac{1}{5}x + 3$

15. $L = \dfrac{P - 2W}{2}$ or $\dfrac{P}{2} - W$ **17.** $T = \dfrac{PV}{K}$ **19.** $b = 2x - a$

21. $C = \dfrac{5}{9}(F - 32)$ or $C = \dfrac{5(F - 32)}{9}$

23. $h = \dfrac{S - 2\pi r^2}{2\pi r}$ or $h = \dfrac{S}{2\pi r} - r$ **25.** 3 cm   **27.** 5%   **29.** 25°C

31. $x + 3 = 7$ **33.** $3x - 7 = 2x$ **35.** $2(x + 5) = x + 18$

37. $2x + 3 = 7$ **39.** $4x - 7 = 41$ **41.** $\dfrac{2}{3}x + 5 = 21$ **43.** $3x = x + 12$

45. $x + 7 = 33$; 26   **47.** $x - 15 = 7$; 22   **49.** $x + 1840 = 3260$; 1420
51. $2x + 5 = 35$; 15   **53.** $5x - 12 = 78$; 18   **55.** $x + x + 1 = 47$; 23, 24
57. $x + x + 1 + x + 2 = 63$; 20, 21, 22   **59.** $x + x + 2 = 66$; 32, 34
61. $x + x + 2 = 52$; 25, 27   **63.** $x + x + 2 + x + 4 = 63$; 19, 21, 23
65. $x + x + 1 + x + 2 + x + 3 = 86$; 20, 21, 22, 23
67. $4x = 3(x + 1) + 9$; 12, 13   **69.** $x + x + 160 = 3260$; 1550, 1710
71. $x + x + 70 = 650$; Dryer, \$290; washer, \$360
73. $x + 2x - 1 = 9$; 9 years old   **75.** $x + 4x - 3 = 37$; 29 years old
77. $x + 2x - 60 = 2400$; \$820
79. $x + 2x + x + 7 = 99$; 8 A.M.: 23; 10 A.M.: 46, 12 P.M.: 30
81.   **83.**   **a.** 15   **b.** 14   **c.** 3   **d.** -16

e. -6 **f.** 3600   **g.** -7 **h.** 40

Applications of Linear Equations

2.5 OBJECTIVES

1. Solve linear equations when signs of grouping are present
2. Solve applications involving numbers
3. Solve geometry problems
4. Solve mixture problems
5. Solve motion problems

In Section 1.2, we looked at the distributive property. That property is used when solving equations in which parentheses are involved.

Let's start by reviewing an example similar to those we considered earlier. We will then solve other equations involving grouping symbols.

OBJECTIVE 1

Example 1   Solving Equations with Parentheses

Solve for x:

$$5(2x - 1) = 25$$

First, multiply on the left to remove the parentheses, and then solve as before.

RECALL Again, returning to the *original equation* will catch any possible errors in the removal of the parentheses.

Left side	Right side

$5(2 \cdot (3) - 1) \stackrel{?}{=} 25$
$5(6 - 1) \stackrel{?}{=} 25$
$5 \cdot 5 \stackrel{?}{=} 25$
$25 = 25$ (True)

$$
\begin{aligned}
10x - 5 &= 25 \\
+5 \quad &+5 \qquad \text{Add 5.} \\
\hline
10x &= 30 \\
\frac{10x}{10} &= \frac{30}{10} \qquad \text{Divide by 10.} \\
x &= 3
\end{aligned}
$$

The answer is 3. To check, return to the *original equation*. Substitute 3 for x. Then evaluate the left and right sides separately.

✔ **CHECK YOURSELF 1** _____

Solve for x.

$$8(3x + 5) = 16$$

NOTE Given an expression such as $a - (b + c)$ you could rewrite it as
$a + (-(b + c))$.

Be especially careful if a negative sign precedes a grouping symbol. The sign of each term inside the grouping symbol must be changed.

Example 2   Solving Equations with Parentheses

Solve $8 - (3x + 1) = -8$.

First, remove the parentheses. The original equation then becomes

NOTE Remember,
$-(3x + 1) = -3x - 1$
↑ ↑
Change *both* signs.

$$
\begin{aligned}
8 - 3x - 1 &= -8 \\
-3x + 7 &= -8 \qquad \text{Combine like terms.} \\
-7 \quad &-7 \qquad \text{Add } -7 \text{ to each side.} \\
\hline
-3x &= -15 \\
x &= 5 \qquad \text{Divide by } -3.
\end{aligned}
$$

The solution is 5. You should verify this result.

CHECK YOURSELF 2 _____

Solve for x.

$7 - (4x - 3) = 22$

Example 3 illustrates the solution process when more than one grouping symbol is involved in an equation.

Example 3   Solving Equations with Parentheses

Solve $2(3x - 1) - 3(x + 5) = 4$.

$2(3x - 1) - 3(x + 5) = 4$	Use the distributive property to remove the parentheses.
$6x - 2 - 3x - 15 = 4$	Combine like terms on the left.
$3x - 17 = 4$	Add 17.
$3x = 21$	Divide by 3.
$x = 7$	

The solution is 7.
To check, return to the original equation to replace x with 7.

NOTE Notice how the rules for the order of operations are applied.

$$2(3 \cdot (7) - 1) - 3((7) + 5) \overset{?}{=} 4$$
$$2(21 - 1) - 3(7 + 5) \overset{?}{=} 4$$
$$2 \cdot 20 - 3 \cdot 12 \overset{?}{=} 4$$
$$40 - 36 \overset{?}{=} 4$$
$$4 = 4 \quad \text{(True)}$$

The solution is verified.

CHECK YOURSELF 3 _____

Solve for x.

$5(2x + 4) = 7 - 3(1 - 2x)$

Many applications lead to equations involving parentheses. That means the methods of Examples 2 and 3 will have to be applied during the solution process. Before we look at examples, you should review the five-step process for solving word problems found in Section 2.4.

These steps are illustrated in Example 4.

OBJECTIVE 2   ### Example 4   Solving a Number Problem

One number is 5 more than a second number. If 3 times the smaller number plus 4 times the larger is 104, find the two numbers.

Step 1   What are you asked to find? You must find the two numbers.

Step 2   Represent the unknowns. Let x be the smaller number. Then

NOTE "5 more than" x

$x + 5$

is the larger number.

Step 3   Write an equation.

$3x + 4(x + 5) = 104$

NOTE Note that the parentheses are *essential* in writing the correct equation.

3 times   Plus   4 times
the smaller   the larger

Step 4   Solve the equation.

$$3x + 4(x + 5) = 104$$
$$3x + 4x + 20 = 104$$
$$7x + 20 = 104$$
$$7x = 84$$
$$x = 12$$

The smaller number (x) is 12, and the larger number ($x + 5$) is 17.

Step 5   Check the solution: 12 is the smaller number, and 17 is the larger number.

$3 \cdot (12) + 4 \cdot ((12) + 5) = 104$ (True)

CHECK YOURSELF 4 ⎯⎯⎯⎯⎯⎯⎯⎯⎯⎯⎯⎯⎯⎯⎯⎯

One number is 4 more than another. If 6 times the smaller minus 4 times the larger is 4, what are the two numbers?

The solutions for many problems from geometry will also yield equations involving parentheses. Consider Example 5.

OBJECTIVE 3

Example 5   Solving a Geometry Application

NOTE Whenever you are working on an application involving geometric figures, you should draw a sketch of the problem, including the labels assigned in step 2.

The length of a rectangle is 1 cm less than 3 times the width. If the perimeter is 54 cm, find the dimensions of the rectangle.

Step 1   You want to find the dimensions (the width and length).

Step 2   Let x be the width.

Then $3x - 1$ is the length.

3 times   1 less than
the width

Length $3x - 1$

Width
x

Step 3   To write an equation, we use this formula for the perimeter of a rectangle:

$P = 2W + 2L$

So

$$2x + 2(3x - 1) = 54$$

Twice the width   Twice the length   Perimeter

Step 4 Solve the equation.

$$2x + 2(3x - 1) = 54$$
$$2x + 6x - 2 = 54$$
$$8x = 56$$
$$x = 7$$

NOTE Be sure to return to the original statement of the problem when checking your result.

The width x is 7 cm, and the length, $3x - 1$, is 20 cm. We leave step 5, the check, to you.

CHECK YOURSELF 5

The length of a rectangle is 5 in. more than twice the width. If the perimeter of the rectangle is 76 in., what are the dimensions of the rectangle?

You will also often use parentheses in solving *mixture problems*. Mixture problems involve combining things that have a different value, rate, or strength. Look at Example 6.

OBJECTIVE 4

Example 6 Solving a Mixture Problem

Four hundred tickets were sold for a school play. General admission tickets were $4, and student tickets were $3. If the total ticket sales were $1350, how many of each type of ticket were sold?

Step 1 You want to find the number of each type of ticket sold.

Step 2 Let x be the number of general admission tickets.

Then $\underline{400 - x}$ student tickets were sold.

NOTE We subtract x, the number of general admission tickets, from 400, the total number of tickets, to find the number of student tickets.

400 tickets were sold in all.

Step 3 The revenue from each kind of ticket is found by multiplying the price of the ticket by the number sold.

General admission tickets: $4x$ $4 for each of the x tickets

Student tickets: $3(400 - x)$ $3 for each of the $400 - x$ tickets

So to form an equation, we have

$$4x + 3(400 - x) = 1350$$

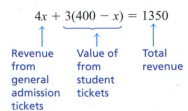

Revenue from general admission tickets   Value of from student tickets   Total revenue

Step 4   Solve the equation.

$$4x + 3(400 - x) = 1350$$
$$4x + 1200 - 3x = 1350$$
$$x + 1200 = 1350$$
$$x = 150$$

So 150 general admission and $400 - 150$ or 250 student tickets were sold. We leave the check to you.

 CHECK YOURSELF 6 _____

Beth bought 40¢ stamps and 3¢ stamps at the post office. If she purchased 92 stamps at a cost of $22, how many of each kind did she buy?

The next group of applications we look at in this section are *motion problems*. They involve a distance traveled, a rate or speed, and time. To solve motion problems, we need a relationship among these three quantities.

Suppose you travel at a rate of 50 mi/h on a highway for 6 h. How far (what distance) will you have gone? To find the distance, you multiply:

NOTE Be careful to make your units consistent. If a rate is given in *miles per hour*, then the time must be given in *hours* and the distance in *miles*.

$$(50 \text{ mi/h})(6 \text{ h}) = 300 \text{ mi}$$

Speed   Time   Distance
or rate

> **Definition:**   Relationship for Motion Problems
>
> In general, if *r* is a rate, *t* is the time, and *d* is the distance traveled,
>
> $$d = r \cdot t$$

This is the key relationship, and it will be used in all motion problems. Let's see how it is applied in Example 7.

OBJECTIVE 5   **Example 7   Solving a Motion Problem**

On Friday morning Ricardo drove from his house to the beach in 4 h. In coming back on Sunday afternoon, heavy traffic slowed his speed by 10 mi/h, and the trip took 5 h. What was his average speed (rate) in each direction?

Step 1   We want the speed or rate in each direction.

Step 2   Let *x* be Ricardo's speed to the beach. Then $x - 10$ is his return speed.
It is always a good idea to sketch the given information in a motion problem. Here we would have

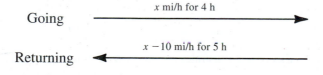

Going   *x* mi/h for 4 h

Returning   *x* −10 mi/h for 5 h

NOTE Distance (going) = distance (returning)

or

Time · rate (going) = time · rate (returning)

Step 3 Because we know that the distance is the same each way, we can write an equation, using the fact that the product of the rate and the time each way must be the same.
So

$$4x = 5(x - 10)$$

↑ Time · rate (going)   ↑ Time · rate (returning)

An alternate method is to use a chart, which can help summarize the given information. We begin by filling in the information given in the problem.

	Distance	Rate	Time
Going		x	4
Returning		$x - 10$	5

Now we fill in the missing information. Here we use the fact that $d = rt$ to complete the chart.

	Distance	Rate	Time
Going	$4x$	x	4
Returning	$5(x - 10)$	$x - 10$	5

From here we set the two distances equal to each other and solve as before.

Step 4   Solve.

$$4x = 5(x - 10)$$
$$4x = 5x - 50$$
$$-x = -50$$
$$x = 50 \text{ mi/h}$$

NOTE x was his rate going, $x - 10$ his rate returning.

So Ricardo's rate going to the beach was 50 mi/h, and his rate returning was 40 mi/h.

Step 5   To check, you should verify that the product of the time and the rate is the same in each direction.

✔ **CHECK YOURSELF 7** _____

A plane made a flight (with the wind) between two towns in 2 h. Returning against the wind, the plane's speed was 60 mi/h slower, and the flight took 3 h. What was the plane's speed in each direction?

Example 8 illustrates another way of using the distance relationship.

Example 8   Solving a Motion Problem

Katy leaves Las Vegas for Los Angeles at 10 A.M., driving at 50 mi/h. At 11 A.M. Jensen leaves Los Angeles for Las Vegas, driving at 55 mi/h along the same route. If the cities are 260 mi apart, at what time will Katy and Jensen meet?

Step 1   Find the time that Katy travels until they meet.

Step 2   Let x be Katy's time.

Then $x - 1$ is Jensen's time.

Jensen left 1 h later!

Again, you should draw a sketch of the given information.

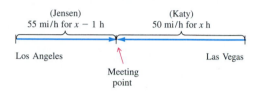

Step 3   To write an equation, we again need the relationship $d = rt$. From this equation, we can write

Katy's distance $= 50x$

Jensen's distance $= 55(x - 1)$

As before, we can use a chart to solve.

	Distance	Rate	Time
Katy	$50x$	50	x
Jensen	$55(x - 1)$	55	$x - 1$

From the original problem, the sum of those distances is 260 mi, so

$$50x + 55(x - 1) = 260$$

Step 4

$$50x + 55(x - 1) = 260$$
$$50x + 55x - 55 = 260$$
$$105x - 55 = 260$$
$$105x = 315$$
$$x = 3 \text{ h}$$

NOTE Be sure to answer the question asked in the problem.

Finally, because Katy left at 10 A.M., the two will meet at 1 P.M. We leave the check of this result to you.

CHECK YOURSELF 8 _____

At noon a jogger leaves one point, running at 8 mi/h. One hour later a bicyclist leaves the same point, traveling at 20 mi/h in the opposite direction. At what time will they be 36 mi apart?

CHECK YOURSELF ANSWERS _____

1. -1 **2.** -3 **3.** -4 **4.** The numbers are 10 and 14.   **5.** The width is 11 in.; the length is 27 in.   **6.** 52 @ 40¢, and 40 @ 3¢   **7.** 180 mi/h with the wind and 120 mi/h against the wind   **8.** At 2 P.M.

2.5   Exercises

Solve each of the following equations for *x*, and check your results.

1. $3(x - 5) = 6$

2. $2(x + 3) = -6$

3. $3(2x - 4) = 18$

4. $4(3x - 5) = 88$

5. $7(5x + 8) = -84$

6. $3(6x - 1) = 69$

7. $10 - (x - 2) = 15$

8. $12 - (x + 3) = 3$

9. $6 - (3x - 1) = 19$

10. $9 - (3x - 2) = 2$

11. $7 - (3x - 5) = 13$

12. $5 - (4x + 3) = 4$

13. $5x = 3(x - 6)$

14. $5x = 2(x + 12)$

15. $7(2x - 3) = 20x$

16. $3(4x - 2) = 15x$

17. $6(6 - x) = 3x$

18. $5(8 - x) = 3x$

19. $2(2x - 1) = 3(x + 1)$

20. $3(3x - 1) = 4(2x + 1)$

21. $4(5x - 3) = 3(2x - 1)$

22. $4(6x - 1) = 7(3x + 2)$

23. $9(8x - 1) = 5(4x + 6)$

24. $7(3x + 11) = 9(3 - 6x)$

25. $-3(2x + 5) + 2(3x - 1) = x - 5$

26. $7(3x + 4) = 8(2x + 5) + 13$

27. $5(2x - 1) - 3(x - 4) = 4(x + 4)$

28. $2(x - 3) - 3(x + 5) = 3(x - 2) - 7$

29. $3(3 - 4x) + 30 = 5x - 2(6x - 7)$

30. $3x - 5(3x - 7) = 2(x + 9) + 45$

ANSWERS

1. _____   2. _____

3. _____   4. _____

5. _____   6. _____

7. _____   8. _____

9. _____   10. _____

11. _____   12. _____

13. _____   14. _____

15. _____

16. _____

17. _____

18. _____

19. _____

20. _____

21. _____

22. _____

23. _____

24. _____

25. _____

26. _____

27. _____

28. _____

29. _____

30. _____

31. _____

32. _____

33. _____

34. _____

35. _____

36. _____

37. _____

38. _____

39. _____

40. _____

41. _____

42. _____

43. _____

31. $-2x + [3x - (-2x + 5)] = -(15 + 2x)$

32. $-3x + [5x - (-x + 4)] = -2(x - 3)$

33. $3x^2 - 2(x^2 + 2) = x^2 - 4$

34. $5x^2 - [2(2x^2 + 3)] - 3 = x^2 - 9$

Solve the following word problems. Be sure to show the equation you use for the solution.

35. Number problem. One number is 8 more than another. If the sum of the smaller number and twice the larger number is 46, find the two numbers.

36. Number problem. One number is 3 less than another. If 4 times the smaller number minus 3 times the larger number is 4, find the two numbers.

37. Number problem. One number is 7 less than another. If 4 times the smaller number plus 2 times the larger number is 62, find the two numbers.

38. Number problem. One number is 10 more than another. If the sum of twice the smaller number and 3 times the larger number is 55, find the two numbers.

39. Number problem. Find two consecutive integers such that the sum of twice the first integer and 3 times the second integer is 28. (*Hint:* If x represents the first integer, $x + 1$ represents the next consecutive integer.)

40. Number problem. Find two consecutive odd integers such that 3 times the first integer is 5 more than twice the second. (*Hint:* If x represents the first integer, $x + 2$ represents the next consecutive odd integer.)

41. Geometry. The length of a rectangle is 1 in. more than twice its width. If the perimeter of the rectangle is 74 in., find the dimensions of the rectangle.

42. Geometry. The length of a rectangle is 5 cm less than 3 times its width. If the perimeter of the rectangle is 46 cm, find the dimensions of the rectangle.

43. Geometry. The length of a rectangular garden is 4 m more than 3 times its width. The perimeter of the garden is 56 m. What are the dimensions of the garden?

44. Geometry. The length of a rectangular playing field is 5 ft less than twice its width. If the perimeter of the playing field is 230 ft, find the length and width of the field.

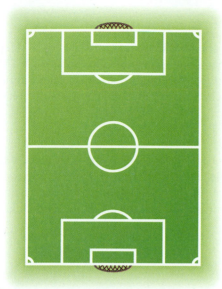

45. Geometry. The base of an isosceles triangle is 3 cm less than the length of the equal sides. If the perimeter of the triangle is 36 cm, find the length of each of the sides.

46. Geometry. The length of one of the equal legs of an isosceles triangle is 3 in. less than twice the length of the base. If the perimeter is 29 in., find the length of each of the sides.

47. Business and finance. Tickets for a play cost $8 for the main floor and $6 in the balcony. If the total receipts from 500 tickets were $3600, how many of each type of ticket were sold?

48. Business and finance. Tickets for a basketball tournament were $6 for students and $9 for nonstudents. Total sales were $10,500, and 250 more student tickets were sold than nonstudent tickets. How many of each type of ticket were sold?

49. Business and finance. Maria bought 56 stamps at the post office in 37¢ and 20¢ denominations. If she paid $18 for the stamps, how many of each denomination did she buy?

ANSWERS

44. _____

45. _____

46. _____

47. _____

48. _____

49. _____

$$n \qquad n-7$$

$$4n^2 + 2(n-7) = 62$$

$$6n - 14 = 62$$

$$6n = 78/6 \qquad \boxed{n = 13}$$

50. _____

51. _____

52. _____

53. _____

54. _____

55. _____

56. _____

57. _____

50. Business and finance. A bank teller had a total of 125 $10 bills and $20 bills to start the day. If the value of the bills was $1650, how many of each denomination did he have?

51. Business and finance. Tickets for a train excursion were $120 for a sleeping room, $80 for a berth, and $50 for a coach seat. The total ticket sales were $8600. If there were 20 more berth tickets sold than sleeping room tickets and 3 times as many coach tickets as sleeping room tickets, how many of each type of ticket were sold?

52. Business and finance. Admission for a college baseball game is $6 for box seats, $5 for the grandstand, and $3 for the bleachers. The total receipts for one evening were $9000. There were 100 more grandstand tickets sold than box seat tickets. Twice as many bleacher tickets were sold as box seat tickets. How many tickets of each type were sold?

53. Science and medicine. Patrick drove 3 h to attend a meeting. On the return trip, his speed was 10 mi/h less and the trip took 4 h. What was his speed each way?

54. Science and medicine. A bicyclist rode into the country for 5 h. In returning, her speed was 5 mi/h faster and the trip took 4 h. What was her speed each way?

55. Science and medicine. A car leaves a city and goes north at a rate of 50 mi/h at 2 P.M. One hour later a second car leaves, traveling south at a rate of 40 mi/h. At what time will the two cars be 320 mi apart?

56. Science and medicine. A bus leaves a station at 1 P.M., traveling west at an average rate of 44 mi/h. One hour later a second bus leaves the same station, traveling east at a rate of 48 mi/h. At what time will the two buses be 274 mi apart?

57. Science and medicine. At 8:00 A.M., Catherine leaves on a trip at 45 mi/h. One hour later, Max decides to join her and leaves along the same route, traveling at 54 mi/h. When will Max catch up with Catherine?

ANSWERS

58. _____

59. _____

60. _____

61. _____

62. _____

63. _____

64. _____

65. _____

58. Science and medicine. Martina leaves home at 9 A.M., bicycling at a rate of 24 mi/h. Two hours later, John leaves, driving at the rate of 48 mi/h. At what time will John catch up with Martina?

59. Science and medicine. Mika leaves Boston for Baltimore at 10:00 A.M., traveling at 45 mi/h. One hour later, Hiroko leaves Baltimore for Boston on the same route, traveling at 50 mi/h. If the two cities are 425 mi apart, when will Mika and Hiroko meet?

60. Science and medicine. A train leaves town A for town B, traveling at 35 mi/h. At the same time, a second train leaves town B for town A at 45 mi/h. If the two towns are 320 mi apart, how long will it take for the two trains to meet?

61. Business and finance. There are a total of 500 Douglas fir and hemlock trees in a section of forest bought by Hoodoo Logging Co. The company paid an average of $250 for each Douglas fir and $300 for each hemlock. If the company paid $132,000 for the trees, how many of each kind did the company buy?

62. Business and finance. There are 850 Douglas fir and ponderosa pine trees in a section of forest bought by Sawz Logging Co. The company paid an average of $300 for each Douglas fir and $225 for each ponderosa pine. If the company paid $217,500 for the trees, how many of each kind did the company buy?

63. There is a universally agreed on "order of operations" used to simplify expressions. Explain how the order of operations is used in solving equations. Be sure to use complete sentences.

64. A common mistake when solving equations is the following:

The equation: $2(x - 2) = x + 3$
First step in solving: $2x - 2 = x + 3$

Write a clear explanation of what error has been made. What could be done to avoid this error?

65. Another very common mistake is in the equation below:

The equation: $6x - (x + 3) = 5 + 2x$
First step in solving: $6x - x + 3 = 5 + 2x$

Write a clear explanation of what error has been made and what could be done to avoid the mistake.

66.

a. _____

b. _____

c. _____

d. _____

e. _____

f. _____

66. Write an algebraic equation for the English statement "Subtract 5 from the sum of *x* and 7 times 3 and the result is 20." Compare your equation with other students. Did you all write the same equation? Are all the equations correct even though they don't look alike? Do all the equations have the same solution? What is wrong? The English statement is *ambiguous*. Write another English statement that leads correctly to more than one algebraic equation. Exchange with another student and see if the other student thinks the statement is ambiguous. Notice that the algebra is *not* ambiguous!

Getting Ready for Section 2.6 [Section 1.4]

Evaluate the following.

(a) $\dfrac{270 \cdot 10}{90}$

(b) $\dfrac{660 \cdot 100}{11}$

(c) $\dfrac{120 \cdot 100}{4000}$

(d) $\dfrac{320 \cdot 100}{2.5}$

(e) $\dfrac{23 \cdot 4.5}{10}$

(f) $\dfrac{46 \cdot 15}{100}$

Answers

1. 7   **3.** 5   **5.** −4   **7.** −3   **9.** −4   **11.** $-\dfrac{1}{3}$ **13.** −9

15. $-\dfrac{7}{2}$ **17.** 4   **19.** 5   **21.** $\dfrac{9}{14}$ **23.** $\dfrac{3}{4}$ **25.** −12   **27.** 3

29. 5   **31.** −2   **33.** All real numbers   **35.** 10, 18   **37.** 8, 15

39. 5, 6   **41.** 12 in., 25 in.   **43.** 6 m, 22 m

45. Legs, 13 cm; base, 10 cm   **47.** 200 $6 tickets, 300 $8 tickets

49. 40 37¢ stamps 16 20¢ stamps

51. 60 coach, 40 berth, and 20 sleeping room   **53.** 40 mi/h, 30 mi/h

55. 6 P.M.   **57.** 2 P.M.   **59.** 3 P.M.   **61.** 360 Douglas fir, 140 hemlock

63.   **65.**   **a.** 30   **b.** 6000   **c.** 3   **d.** 12,800

e. 10.35   **f.** 6.9

2.6 Solving Percent Applications

2.6 OBJECTIVES

1. Convert rational numbers to percents
2. Identify base, rate, and percent
3. Solve a percent application for the amount
4. Solve a percent application for the rate
5. Solve a percent application for the base

The word **percent** is Latin for "for each hundred." Look at the following drawing.

NOTE The rectangle is divided into 100 parts, 25 of them are shaded.

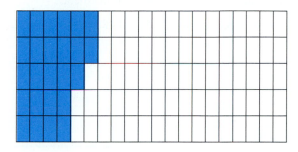

How can we describe how much of the rectangle is shaded? We can say that

1. $\frac{1}{4}$ of the rectangle is shaded,

2. 0.25 of the rectangle is shaded, or

3. 25% of the rectangle is shaded.

OBJECTIVE 1

Example 1   Converting Rational Numbers to Percents

Convert each rational number to a percent.

(a) $0.36 = 36\%$ To convert a decimal number to a percent, we move the decimal point two places to the right.

(b) $1.575 = 157.5\%$ Again, we move the decimal point two places to the right.

(c) $\frac{2}{5} = 0.4 = 40\%$ When converting a fraction to a percent, rewrite the fraction as a decimal and move the decimal point two places to the right.

CHECK YOURSELF 1

Convert each rational number to a percent.

(a) 0.45   **(b)** 2.1   **(c)** $\frac{1}{8}$

There are many practical applications that use percents. Every complete percent statement has three parts that need to be identified. Let's look at some definitions that will help with that process.

> **Definition:** Base, Amount, and Rate
>
> The **base** is the whole in a statement. It is the standard used for comparison.
> The **amount** is the part of the whole that is being compared to the base.
> The **rate** is the ratio of the amount to the base written as a percent.

OBJECTIVE 2   **Example 2   Identifying Base, Amount, and Rate**

Identify the indicated quantity in each statement.

(a) The rate in the statement "50% of 480 is 240." The rate is 50%.

(b) The amount in the statement "20% of 400 is 80." The amount is 80.

(c) The base in the statement "125% of 200 is 250." The base is 200. Note that the base is almost always preceded by the word *of.*

 CHECK YOURSELF 2 _____

Identify the indicated quantity in each statement.

(a) The rate in the statement "40% of 80 is 32."

(b) The amount in the statement "150% of 300 is 450."

(c) The base in the statement "5% of 800 is 40."

Given an application of percents, it is frequently easiest to rewrite the statement in the form

R% of B is A

to identify the three parts.

Example 3   Identifying Rate, Base, and Amount

Identify the rate, base, and amount in the following statement.

Delia borrows $10,000 for 1 year at 11% interest. How much interest will she pay?

We can rewrite the statement as "11% of $10,000 is what amount?"

11% is the rate, $10,000 is the base, and the amount is unknown (at least at the moment).

 CHECK YOURSELF 3 _____

Identify the rate, base, and amount in the following statement.

Melina earned $140 last year from a certificate of deposit that paid 6%. How much did she invest?

Definition: Percent Proportion

To solve such problems as those in Example 3, we use the **percent proportion.**

$$\frac{Amount}{Base} = \frac{Rate}{100}$$

This proportion could also be written as the equation

$100 \cdot \text{amount} = \text{rate} \cdot \text{base}$ or $100A = R \cdot B$

OBJECTIVE 3   **Example 4   Solving a Problem Involving an Unknown Amount**

Find the interest you must pay if you borrow \$2000 for 1 year with an interest rate of $9\frac{1}{2}\%$.

The base (the principal) is \$2000, the rate is $9\frac{1}{2}\%$, and we want to find the interest (the amount). Using the given percent proportion

RECALL

$9\frac{1}{2}\% = 9.5\%$

$$\frac{A}{2000} = \frac{9.5}{100}$$

so

$100A = 9.5 \cdot 2000$

or

$$A = \frac{19{,}000}{100} = 190$$

The interest (amount) is \$190.

 CHECK YOURSELF 4

You invest \$5000 for 1 year at $8\frac{1}{2}\%$. How much interest will you earn?

Let's look at an application that requires finding the rate.

OBJECTIVE 4   **Example 5   Solving a Problem Involving an Unknown Rate**

You borrow \$2000 from a bank for 1 year and are charged \$150 interest. What is the interest rate?

The base is the amount of the loan (the principal). The amount is the interest paid. To find the interest rate, we again use the percent proportion.

$$\frac{150}{2000} = \frac{R}{100}$$

Then

$100 \cdot 150 = R \cdot 2000$

$$R = \frac{15{,}000}{2000} = 7.5$$

The interest rate is 7.5%.

CHECK YOURSELF 5

Xian borrowed $3200 and was charged $352 in interest for 1 year. What was the interest rate?

Now look at an application that requires finding the base.

OBJECTIVE 5   **Example 6   Solving a Problem Involving an Unknown Base**

Ms. Hobson agrees to pay 11% interest on a loan for her new automobile. She is charged $550 interest on a loan for 1 year. How much did she borrow?

The rate is 11%. The amount, or interest, is $550. We want to find the base, which is the principal, or the size of the loan. To solve the problem, we have

$$\frac{550}{B} = \frac{11}{100}$$

$$100 \cdot 550 = 11B$$

$$B = \frac{55{,}000}{11} = 5000$$

She borrowed $5000.

CHECK YOURSELF 6

Sue pays $210 interest for a 1-year loan at 10.5%. What was the size of her loan?

Percents are used in too many ways for us to list. Look at the variety in the following examples, which illustrate some additional situations in which you will find percents.

Another common application of percents involves tax rates.

Example 7   Solving a Percent Problem

A state taxes sales at 5.5%. How much sales tax will you pay on a purchase of $48?

The tax you pay is the amount (the part of the whole). Here the base is the purchase price, $48, and the rate is the tax rate, 5.5%.

NOTE In an application involving taxes, the tax paid is always the amount.

$$\frac{A}{48} = \frac{5.5}{100} \qquad \text{or} \qquad 100A = 5.5 \cdot 48$$

Now

NOTE $48 \cdot 5.5 = 264$

$$A = \frac{264}{100} = 2.64$$

The sales tax paid is $2.64.

CHECK YOURSELF 7

Suppose that a state has a sales tax rate of $6\frac{1}{2}$%. If you buy a used car for $1200, how much sales tax must you pay?

Percents are also used to deal with store markups or discounts. Consider Example 8.

Example 8 Solving a Percent Problem

A store marks up items to make a 30% profit. If an item cost $2.50 from the supplier, what will the selling price be?

The base is the cost of the item, $2.50, and the rate is 30%. In the percent proportion, the markup is the amount in this application.

$$\frac{A}{2.50} = \frac{30}{100} \qquad \text{or} \qquad 100A = 30 \cdot 2.50$$

Then

$$A = \frac{75}{100} = 0.75$$

The markup is $0.75. Finally we have

NOTE
Selling price = original cost + markup

Selling price = $2.50 + $0.75 = $3.25   Add the cost and the markup to find the selling price.

CHECK YOURSELF 8 _____

A store wants to discount (or mark down) an item by 25% for a sale. If the original price of the item was $45, find the sale price. [Hint: Find the discount (the amount the item will be marked down) and subtract that from the original price.]

Our final examples illustrate increases and decreases stated in terms of percents.

Example 9 Solving a Percent Problem

The population of a town increased 15% in a 3-year period. If the original population was 12,000, what was the population at the end of the period?

First we find the increase in the population. That increase is the amount in the problem.

$$\frac{A}{12,000} = \frac{15}{100} \qquad \text{so} \qquad 100A = 15 \cdot 12,000$$

$$A = \frac{180,000}{100}$$

$$= 1800$$

To find the population at the end of the period, we add

12,000 + 1800 = 13,800

Original population   Increase   New population

CHECK YOURSELF 9 _____

A school's enrollment decreased by 8% from one year to the next. If the enrollment was 550 students the first year, how many students were enrolled the second year?

Example 10   Solving a Percent Problem

Enrollment at a school increased from 800 to 888 students from one year to the next. What was the rate of increase?

First we must subtract to find the actual increase.

Increase: $888 - 800 = 88$ students

Now to find the rate, we have

NOTE We use the *original* enrollment, 800, as our base.

$$\frac{88}{800} = \frac{R}{100} \quad \text{so} \quad 100 \cdot 88 = R \cdot 800$$

$$R = \frac{8800}{800} = 11$$

The enrollment increased at a rate of 11%.

CHECK YOURSELF 10

Car sales at a dealership decreased from 350 units one year to 322 units the next. What was the rate of decrease? [Hint: Use the original sales as the base.]

Example 11   Solving a Percent Problem

A company hired 18 new employees in 1 year. If this was a 15% increase, how many employees did the company have before the increase?

The rate (R) is 15%. The amount (A) is 18, the number of new employees. The base in this problem is the number of employees *before the increase.* So

$$\frac{18}{B} = \frac{15}{100}$$

$$100 \cdot 18 = 15B \quad \text{or} \quad B = \frac{1800}{15} = 120$$

The company had 120 employees before the increase.

CHECK YOURSELF 11

A school had 54 new students in one term. If this was a 12% increase over the previous term, how many students were there before the increase?

CHECK YOURSELF ANSWERS

1. (a) 45%; **(b)** 210%; **(c)** 12.5%   **2. (a)** 40%; **(b)** 450; **(c)** 800
3. Rate = 6%; base unknown; amount = $140   **4.** $425   **5.** 11%   **6.** $2000
7. $78   **8.** $33.75   **9.** 506   **10.** 8%   **11.** 450

Convert each rational number to a percent.

1. 0.23   **2.** 0.31   **3.** 2.5   **4.** 1.8   **5.** $\dfrac{3}{8}$ **6.** $\dfrac{7}{16}$

ANSWERS

Identify the indicated quantity in each statement.

7. The rate in the statement "23% of 400 is 92."

8. The base in the statement "40% of 600 is 240."

9. The amount in the statement "200 is 40% of 500."

10. The rate in the statement "480 is 60% of 800."

11. The base in the statement "16% of 350 is 56."

12. The amount in the statement "150 is 75% of 200."

Identify the rate, base, and amount in the following applications. *Do not solve* the applications at this point.

13. Business and finance. Jan has a 5% commission rate on all her sales. If she sells $40,000 worth of merchandise in 1 month, what commission will she earn?

14. Business and finance. 22% of Shirley's monthly salary is deducted for withholding. If those deductions total $209, what is her salary?

15. Science and medicine. In a chemistry class of 30 students, 5 received a grade of A. What percent of the students received A's?

16. Business and finance. A can of mixed nuts contains 80% peanuts. If the can holds 16 oz, how many ounces of peanuts does it contain?

17. Business and finance. The sales tax rate in a state is 5.5%. If you pay a tax of $3.30 on an item that you purchase, what is its selling price?

18. Business and finance. In a shipment of 750 parts, 75 were found to be defective. What percent of the parts were faulty?

ANSWERS
1. _____
2. _____
3. _____
4. _____
5. _____
6. _____
7. _____
8. _____
9. _____
10. _____
11. _____
12. _____
13. _____
14. _____
15. _____
16. _____
17. _____
18. _____

19. _____

20. _____

21. _____

22. _____

23. _____

24. _____

25. _____

19. Statistics. A college had 9000 students at the start of a school year. If there is an enrollment increase of 6% by the beginning of the next year, how many additional students were there?

20. Business and finance. Paul invested $5000 in a time deposit. What interest will he earn for 1 year if the interest rate is 6.5%?

Solve each of the following applications.

21. Business and finance. What interest will you pay on a $3400 loan for 1 year if the interest rate is 12%?

22. Science and medicine. A chemist has 300 milliliters (mL) of solution that is 18% acid. How many milliliters of acid are in the solution?

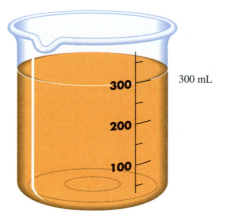

300 mL

23. Business and finance. Roberto has 26% of his pay withheld for deductions. If he earns $550 per week, what amount is withheld?

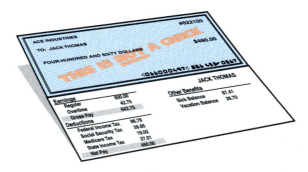

24. Business and finance. A real estate agent's commission rate is 6%. What will the amount of the commission on the sale be for a $185,000 home?

25. Business and finance. If a salesman is paid a $140 commission on the sale of a $2800 sailboat, what is his commission rate?

26. Business and finance. Ms. Jordan has been given a loan of $2500 for 1 year. If the interest charged is $275, what is the interest rate on the loan?

27. Business and finance. Joan was charged $18 interest for 1 month on a $1200 credit card balance. What was the monthly interest rate?

28. Science and medicine. There is 117 grams (g) of acid in 900 g of a solution of acid and water. What percent of the solution is acid?

29. Statistics. On a test, Alice had 80% of the problems right. If she had 20 problems correct, how many questions were on the test?

30. Business and finance. A state sales tax rate is 3.5%. If the tax on a purchase is $7, what was the amount of the purchase?

31. Business and finance. Patty pays $525 interest for a 1-year loan at 10.5%. What was the amount of her loan?

32. Business and finance. A saleswoman is working on a 5% commission basis. If she wants to make $1800 in 1 month, how much must she sell?

33. Business and finance. A state sales tax is levied at a rate of 6.4%. How much tax would one pay on a purchase of $260?

34. Business and finance. Betty must make a $9\frac{1}{2}$% down payment on the purchase of a $2000 motorcycle. How much must she pay down?

35. Business and finance. If a house sells for $125,000 and the commission rate is $6\frac{1}{2}$%, how much will the salesperson make for the sale?

ANSWERS

26. _____

27. _____

28. _____

29. _____

30. _____

31. _____

32. _____

33. _____

34. _____

35. _____

36. _____

37. _____

38. _____

39. _____

40. _____

41. _____

42. _____

43. _____

44. _____

45. _____

36. Statistics. Marla needs 70% on a final test to receive a C for a course. If the exam has 120 questions, how many questions must she answer correctly?

37. Social science. A study has shown that 102 of the 1200 people in the workforce of a small town are unemployed. What is the town's unemployment rate?

38. Statistics. A survey of 400 people found that 66 were left-handed. What percent of those surveyed were left-handed?

39. Statistics. Of 60 people who start a training program, 45 complete the course. What is the dropout rate?

40. Business and finance. In a shipment of 250 parts, 40 are found to be defective. What percent of the parts are faulty?

41. Statistics. In a recent survey, 65% of those responding were in favor of a freeway improvement project. If 780 people were in favor of the project, how many people responded to the survey?

42. Statistics. A college finds that 42% of the students taking a foreign language are enrolled in Spanish. If 1512 students are taking Spanish, how many foreign language students are there?

43. Business and finance. 22% of Samuel's monthly salary is deducted for withholding. If those deductions total $209, what is his salary?

44. Business and finance. The Townsend's budget 36% of their monthly income for food. If they spend $864 on food, what is their monthly income?

45. Business and finance. An appliance dealer marks up refrigerators 22% (based on cost). If the cost of one model was $600, what will its selling price be?

46. Statistics. A school had 900 students at the start of a school year. If there is an enrollment increase of 7% by the beginning of the next year, what is the new enrollment?

47. Business and finance. A home lot purchased for $125,000 increased in value by 25% over 3 years. What was the lot's value at the end of the period?

48. Business and finance. New cars depreciate an average of 28% in their first year of use. What would an $18,000 car be worth after 1 year?

49. Statistics. A school's enrollment was up from 950 students in 1 year to 1064 students in the next. What was the rate of increase?

50. Business and finance. Under a new contract, the salary for a position increases from $31,000 to $33,635. What rate of increase does this represent?

51. Business and finance. A stereo system is marked down from $450 to $382.50. What is the discount rate?

52. Business and finance. The electricity costs of a business decreased from $12,000 one year to $10,920 the next. What is the rate of decrease?

53. Business and finance. The price of a new van has increased $4830, which amounts to a 14% increase. What was the price of the van before the increase?

54. Business and finance. A television set is marked down $75, for a sale. If this is a 12.5% decrease from the original price, what was the selling price before the sale?

55. Statistics. A company had 66 fewer employees in July 2005 than in July 2004. If this represents a 5.5% decrease, how many employees did the company have in July 2004?

ANSWERS

46. _____

47. _____

48. _____

49. _____

50. _____

51. _____

52. _____

53. _____

54. _____

55. _____

56. _____

57. _____

58. _____

59. _____

60. _____

61. _____

56. Business and finance. Carlotta received a monthly raise of $162.50. If this represented a 6.5% increase, what was her monthly salary before the raise?

57. Business and finance. A pair of shorts, advertised for $48.75, is being sold at 25% off the original price. What was the original price?

58. Business and finance. If the total bill at a restaurant, including a 15% tip, is $65.32, what was the cost of the meal alone?

The following chart shows U.S. trade with Mexico from 1992–1997. Use this information for exercises 59 to 62.

U.S. Trade with Mexico, 1992–97
Source: Office of Trade and Economic Analysis,
U.S. Dept. of Commerce
(millions of dollars)

MEXICO

Year	Exports	Imports	Trade Balance[1]
1992	$40,592	$35,211	$5,381
1993	41,581	39,917	1,664
1994[2]	50,844	49,494	1,350
1995	46,292	61,685	−15,393
1996	56,792	74,297	−17,506
1997	71,388	85,938	−14,549

(1) Totals may not add due to rounding
(2) NAFTA provisions began to take effect: Jan. 1, 1994

59. What is the rate of increase (to the nearest whole percent) of exports from 1992 to 1997?

60. What is the rate of increase (to the nearest whole percent) of imports from 1992 to 1997?

61. By what percent did exports exceed imports in 1992?

62. By what percent did imports exceed exports in 1997?

Many percent problems involve calculating what is known as **compound interest.**

Suppose that you invest $1000 at 5% in a savings account for 1 year. For year 1, the interest is 5% of $1000, or 0.05 × $1000 = $50. At the end of year 1, you will have $1050 in the account.

$1000 $\xrightarrow{\text{At 5\%}}$ $1050
Start   Year 1

Now, if you leave that amount in the account for a second year, the interest will be calculated on the original principal, $1000, plus the first year's interest, $50. This is called *compound interest*.

For year 2, the interest is 5% of $1050, or 0.05 × $1050 = $52.50. At the end of year 2, you will have $1102.50 in the account.

$1000 $\xrightarrow{\text{At 5\%}}$ $1050 $\xrightarrow{\text{At 5\%}}$ $1102.50
Start   Year 1   Year 2

Assume the interest is compounded annually (at the end of each year), and find the amount in an account with the given interest rate and principal.

63. $4000, 6%, 2 years

64. $3000, 7%, 2 years

65. $4000, 5%, 3 years

66. $5000, 6%, 3 years

67. Statistics. In 1990, there were an estimated 145.0 million passenger cars registered in the United States. The total number of vehicles registered in the United States for 1990 was estimated at 194.5 million. What percent of the vehicles registered were passenger cars?

68. Statistics. Gasoline accounts for 85% of the motor fuel consumed in the United States every day. If 8882 thousand barrels (bbl) of motor fuel are consumed each day, how much gasoline is consumed each day in the United States?

69. Statistics. In 1999, transportation accounted for 63% of U.S. petroleum consumption. Assuming that same rate applies now, and 10.85 million bbl of petroleum are used each day for transportation in the United States, what is the total daily petroleum consumption by all sources in the United States?

70. Statistics. Each year, 540 million metric tons (t) of carbon dioxide are added to the atmosphere by the United States. Burning gasoline and other transportation fuels is responsible for 35% of the carbon dioxide emissions in the United States. How much carbon dioxide is emitted each year by the burning of transportation fuels in the United States?

62. _____
63. _____
64. _____
65. _____
66. _____
67. _____
68. _____
69. _____
70. _____

71. _____

72. _____

73. _____

74. _____

75. _____

76. _____

a. _____

b. _____

c. _____

d. _____

e. _____

f. _____

g. _____

h. _____

71. Statistics. The progress of the local Lion's club is shown below. What percent of the goal has been achieved so far?

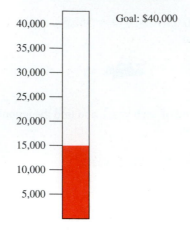

In exercises 72 to 76, use the following number line.

72. Length *AC* is what percent of length *AB*?   **73.** Length *AD* is what percent of *AB*?

74. Length *AE* is what percent of *AB*?   **75.** Length *AE* is what percent of *AD*?

76. Length *AC* is what percent of *AE*?

Getting Ready for Section 2.7 [Section 1.2]

Locate each of the following numbers on the number line.

(a) 4   (b) −5   (c) −3   (d) 2

(e) $-\dfrac{7}{2}$ (f) $\dfrac{2}{3}$ (g) 2.5   (h) −1.1

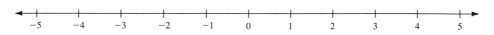

Answers

1. 23%   **3.** 250%   **5.** 37.5%   **7.** 23%   **9.** 200   **11.** 350
13. R: 5%; A: commission; B: $40,000   **15.** A: 5; R: percent; B: 30
17. A: $3.30; R: 5.5%; B: selling price   **19.** A: students; R: 6%; B: 9000
21. $408   **23.** $143   **25.** 5%   **27.** 1.5%   **29.** 25 questions
31. $5000   **33.** $16.64   **35.** $8125   **37.** 8.5%   **39.** 25%
41. 1200 people   **43.** $950   **45.** $732   **47.** $156,250   **49.** 12%
51. 15%   **53.** $34,500   **55.** 1200 employees   **57.** $65   **59.** 76%
61. ≈15%   **63.** $4494.40   **65.** $4630.50   **67.** ≈74.6%
69. ≈17.22 million bbl   **71.** 37.5%   **73.** 75%   **75.** 50%
a.–h.

mult/Div by negative Switch sign

2.7 Inequalities—An Introduction

2.7 OBJECTIVES

1. Use the notation of inequalities
2. Graph the solution set of an inequality
3. Solve an inequality and graph the solution set
4. Solve an application using inequalities

As pointed out in the introduction to this chapter, an equation is just a statement that two expressions are equal. In algebra, an **inequality** is a statement that one expression is less than or greater than another. Four new symbols are used in writing inequalities. The use of two of them is illustrated in Example 1.

OBJECTIVE 1

Example 1   Reading the Inequality Symbol

NOTE To help you remember, the "arrowhead" always points toward the smaller quantity.

$5 < 8$ is an inequality read "5 is less than 8."
$9 > 6$ is an inequality read "9 is greater than 6."

 CHECK YOURSELF 1 _____

Fill in the blanks, using the symbols $<$ and $>$.

(a) 12 _____ 8   **(b)** 20 _____ 25

Like an equation, an inequality can be represented by a balance scale. Note that, in each case, the inequality arrow points to the side that is "lighter."

$2x < 4x - 3$

NOTE The $2x$ side is less than the $4x - 3$ side, so it is "lighter."

$5x - 6 > 9$

243

Just as was the case with equations, inequalities that involve variables may be either true or false depending on the value that we give to the variable. For instance, consider the inequality

$x < 6$

$$\text{If } x = \begin{cases} 3 & 3 < 6 \text{ is true} \\ 5 & 5 < 6 \text{ is true} \\ -10 & -10 < 6 \text{ is true} \\ 8 & 8 < 6 \text{ is false} \end{cases}$$

Therefore 3, 5, and -10 are some *solutions* for the inequality $x < 6$; they make the inequality a true statement. You should see that 8 is *not* a solution. We call the set of all solutions the **solution set** for the inequality. Of course, there are many possible solutions.

Because there are so many solutions (an infinite number, in fact), we certainly do not want to try to list them all! A convenient way to show the solution set of an inequality is with the use of a number line.

OBJECTIVE 2

Example 2   Solving Inequalities

To graph the solution set for the inequality $x < 6$, we want to include all real numbers that are "less than" 6. This means all numbers *to the left* of 6 on the number line. We then start at 6 and draw an arrow extending left, as shown:

NOTE The colored arrow indicates the direction of the *solution set.*

Note: The **open circle** at 6 means that we do not include 6 in the solution set (6 is not less than itself). The colored arrow shows all the numbers in the solution set, with the arrowhead indicating that the solution set continues indefinitely to the left.

CHECK YOURSELF 2 _____

Graph the solution set of $x < -2$.

Two other symbols are used in writing inequalities. They are used with inequalities such as

$x \geq 5$ and $x \leq 2$

Here $x \geq 5$ is really a combination of the two statements $x > 5$ and $x = 5$. It is read "x is greater than or equal to 5." The solution set includes 5 in this case.

The inequality $x \leq 2$ combines the statements $x < 2$ and $x = 2$. It is read "x is less than or equal to 2."

Example 3   Graphing Inequalities

NOTE Here the filled-in circle means that we want to include 5 in the solution set. This is often called a **closed** circle.

The solution set for $x \geq 5$ is graphed as follows.

✔ **CHECK YOURSELF 3**

Graph the solution sets.

(a) $x \leq -4$ **(b)** $x \geq 3$

You have learned how to graph the solution sets of some simple inequalities, such as $x < 8$ or $x \geq 10$. Now we will look at more complicated inequalities, such as

$$2x - 3 < x + 4$$

This is called a **linear inequality in one variable.** Only one variable is involved in the inequality, and it appears only to the first power. Fortunately, the methods used to solve this type of inequality are very similar to those we used earlier in this chapter to solve linear equations in one variable. Here is our first property for inequalities.

NOTE Equivalent inequalities have exactly the same solution sets.

> **Rules and Properties: The Addition Property of Inequality**
>
> If $a < b$ then $a + c < b + c$
>
> In words, adding the same quantity to both sides of an inequality gives an **equivalent inequality.**

NOTE Because $a < b$, the scale shows b to be heavier.

Again, we can use the idea of a balance scale to see the significance of this property. If we add the same weight to both sides of an unbalanced scale, it stays unbalanced.

NOTE This scale represents $a + c < b + c$

OBJECTIVE 3 | **Example 4   Solving Inequalities**

NOTE The inequality is solved when an equivalent inequality has the form

$x < \square$ or $x > \square$

Solve and graph the solution set for $x - 8 < 7$.
     To solve $x - 8 < 7$, add 8 to both sides of the inequality by the addition property.

$$\begin{array}{rl} x - 8 < & 7 \\ + 8 & +8 \\ \hline x & < 15 \end{array}$$ (The inequality is solved.)

The graph of the solution set is

 CHECK YOURSELF 4 _____

Solve and graph the solution set.

$x - 9 > -3$

As with equations, the addition property allows us to subtract the same quantity from both sides of an inequality.

Example 5 Solving Inequalities

Solve and graph the solution set for $4x - 2 \geq 3x + 5$.
First, we add $-3x$ to both sides of the inequality.

NOTE We added $-3x$ and then added 2 to both sides. If these steps are done in the other order, the resulting inequality will be the same.

$$
\begin{array}{rl}
4x - 2 \geq & 3x + 5 \\
-3x & -3x \\
\hline
x - 2 \geq & 5 \\
+ 2 & + 2 \qquad \text{Now we add 2 to both sides.} \\
\hline
x \quad \geq & 7
\end{array}
$$

The graph of the solution set is

 CHECK YOURSELF 5 _____

Solve and graph the solution set.

$7x - 8 \leq 6x + 2$

You will also need a rule for multiplying on both sides of an inequality. Here you have to be a bit careful. There is a difference between the multiplication property for inequalities and that for equations. Look at the following:

$2 < 7$ **(A true inequality)**

Multiply both sides by 3.

$2 < 7$
$3 \cdot 2 < 3 \cdot 7$
$6 < 21$ **(A true inequality)**

Now we multiply both sides by -3.

$2 < 7$
$(-3)(2) < (-3)(7)$
$-6 < -21$ **(*Not* a true inequality)**

Let's try something different.

$$
\begin{array}{l}
2 \boxed{<} 7 \\
(-3)(2) \boxed{>} (-3)(7) \\
-6 \boxed{>} -21
\end{array}
$$

Change the direction of the inequality: $<$ becomes $>$. **(This is now a true inequality.)**

This suggests that multiplying both sides of an inequality by a negative number changes the direction of the inequality.

We can state the following general property.

Rules and Properties: The Multiplication Property of Inequality

If $a < b$ then $ac < bc$ when $c > 0$
          and $ac > bc$ when $c < 0$

In words, multiplying both sides of an inequality by the same *positive* number gives an equivalent inequality.

When both sides of an inequality are multiplied by the same *negative* number, it is necessary to *reverse the direction* of the inequality to give an equivalent inequality.

Example 6   Solving and Graphing Inequalities

(a) Solve and graph the solution set for $5x < 30$.

Multiplying both sides of the inequality by $\frac{1}{5}$ gives

$$\frac{1}{5}(5x) < \frac{1}{5}(30)$$

Simplifying, we have

$$x < 6$$

The graph of the solution set is

(b) Solve and graph the solution set for $-4x \geq 28$.

In this case we want to multiply both sides of the inequality by $-\frac{1}{4}$ to leave x alone on the left.

$$\left(-\frac{1}{4}\right)(-4x) \leq \left(-\frac{1}{4}\right)(28)$$ Reverse the direction of the inequality because you are multiplying by a negative number!

or $x \leq -7$

The graph of the solution set is

CHECK YOURSELF 6

Solve and graph the solution sets.

(a) $7x > 35$

(b) $-8x \leq 48$

Example 7 illustrates the use of the multiplication property when fractions are involved in an inequality.

Example 7   Solving and Graphing Inequalities

(a) Solve and graph the solution set for

$$\frac{x}{4} > 3$$

Here we multiply both sides of the inequality by 4. This will isolate x on the left.

$$4\left(\frac{x}{4}\right) > 4(3)$$

$$x > 12$$

The graph of the solution set is

(b) Solve and graph the solution set for

$$-\frac{x}{6} \geq -3$$

In this case, we multiply both sides of the inequality by -6:

NOTE Note that we reverse the direction of the inequality because we are multiplying by a negative number.

$$(-6)\left(-\frac{x}{6}\right) \leq (-6)(-3)$$

$$x \leq 18$$

The graph of the solution set is

0   18

CHECK YOURSELF 7

Solve and graph the solution sets.

(a) $\frac{x}{5} \leq 4$

(b) $-\frac{x}{3} < -7$

Example 8   Solving and Graphing Inequalities

(a) Solve and graph the solution set for $5x - 3 < 2x$.
First, add 3 to both sides to undo the subtraction on the left.

$$\begin{array}{rl} 5x - 3 & < 2x \\ + 3 & \quad + 3 \end{array}$$ Add 3 to both sides to undo the subtraction.

$$5x \quad < 2x + 3$$

Now add $-2x$, so that only the number remains on the right.

$$\begin{array}{rl} 5x & < \quad 2x + 3 \\ +(-2x) & +(-2x) \end{array}$$ Add $-2x$ to isolate the number on the right.

$$3x \quad < \quad 3$$

NOTE Note that the multiplication property also allows us to divide both sides by a nonzero number.

Next *divide* both sides by 3.

$$\frac{3x}{3} < \frac{3}{3}$$

$$x < 1$$

The graph of the solution set is

(b) Solve and graph the solution set for $2 - 5x < 7$.

$$\begin{array}{r} 2 - 5x < \quad 7 \\ -2 \qquad\quad -2 \end{array} \qquad \text{Add } -2.$$

$$-5x < \quad 5$$

$$\frac{-5x}{-5} > \frac{5}{-5} \qquad \begin{array}{l}\text{Divide by } -5.\text{ Be sure to reverse the} \\ \text{direction of the inequality.}\end{array}$$

or $\qquad x > -1$

The graph is

 CHECK YOURSELF 8

Solve and graph the solution sets.

(a) $4x + 9 \geq x$ 　　　　　　　　　　**(b)** $5 - 6x < 41$

As with equations, we collect all variable terms on one side and all constant terms on the other.

Example 9   Solving and Graphing Inequalities

Solve and graph the solution set for $5x - 5 \geq 3x + 4$.

$$\begin{array}{r} 5x - 5 \geq \quad 3x + 4 \\ + 5 \qquad\quad + 5 \end{array} \qquad \text{Add } 5.$$

$$\begin{array}{r} 5x \quad\geq\quad 3x + 9 \\ -3x \qquad\quad -3x \end{array} \qquad \text{Add } -3x.$$

$$2x \quad\geq\quad 9$$

$$\frac{2x}{2} \geq \frac{9}{2} \qquad \text{Divide by 2.}$$

$$x \geq \frac{9}{2}$$

The graph of the solution set is

CHECK YOURSELF 9

Solve and graph the solution set.

$8x + 3 < 4x - 13$

[handwritten work:]
$-3 \quad -3$
$-8x < 4x - 16$
$-4x \quad -4x$
$4x < -16$ $\quad x < -4$

[handwritten graph at top right]
$-4 \ -3 \ -2 \ \ 0$

Be especially careful when negative coefficients occur in the process of solving.

Example 10   Solving and Graphing Inequalities

Solve and graph the solution set for $2x + 4 < 5x - 2$.

$$
\begin{array}{rcl}
2x + 4 & < & 5x - 2 \\
-\ 4 & & -\ 4 \\
\hline
2x & < & 5x - 6 \\
-5x & & -5x \\
\hline
-3x & < & -\ 6 \\
\end{array}
$$
Add -4.

Add $-5x$.

$$
\dfrac{-3x}{-3} \quad > \quad \dfrac{-6}{-3}
$$
Divide by -3, and reverse the direction of the inequality.

$$x > 2$$

The graph of the solution set is

CHECK YOURSELF 10

Solve and graph the solution set.

$5x + 12 \geq 10x - 8$

Solving inequalities may also require the use of the distributive property.

Example 11   Solving and Graphing Inequalities

Solve and graph the solution set for

$5(x - 2) \geq -8$

Applying the distributive property on the left yields

$5x - 10 \geq -8$

Solving as before yields

$$
\begin{array}{rcl}
5x - 10 & \geq & -\ 8 \\
+\ 10 & & +10 \\
\hline
5x & \geq & 2 \\
\end{array}
$$
Add 10.

$$x \geq \dfrac{2}{5}$$
Divide by 5.

The graph of the solution set is

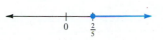

 CHECK YOURSELF 11

Solve and graph the solution set.

$4(x + 3) < 9$

Some applications are solved by using an inequality instead of an equation. Example 12 illustrates such an application.

OBJECTIVE 4   **Example 12   Solving an Application with Inequalities**

Mohammed needs a mean score of 92 or higher on four tests to get an A. So far his scores are 94, 89, and 88. What score on the fourth test will get him an A?

Name: Mohammed

88

2 x 3 = 6	5 x 4 = 20
1 + 5 = 6	3 x 4 = 12
2 x 5 = 10	5 x 2 = 10
4 + 5 = 9	5 + 4 = 9
15 - 2 = 13	15 - 4 = 11
4 x 3 = 12	✓ 8 x 3 = 22
3 + 6 = 9	6 + 3 = 9
9 + 4 = 13	5 + 6 = 11
✓ 3 + 9 = 11	6 + 9 = 15
1 x 2 = 2	2 x 1 = 2
13 - 4 = 9	13 - 3 = 10
5 + 6 = 11	✓ 9 + 4 = 12
8 x 4 = 32	

NOTE What do you need to find?

NOTE Assign a letter to the unknown.

NOTE Write an inequality.

Step 1   We are looking for the score that will, when combined with the other scores, give Mohammed an A.

Step 2   Let x represent a fourth-test score that will get him an A.

Step 3   The inequality will have the mean on the left side, which must be greater than or equal to the 92 on the right.

$$\frac{94 + 89 + 88 + x}{4} \geq 92$$

NOTE Solve the inequality.

Step 4   First, multiply both sides by 4:

$94 + 89 + 88 + x \geq 368$

Then add the test scores:

$183 + 88 + x \geq 368$

$271 + x \geq 368$

Subtracting 271 from both sides,

$x \geq 97$

Step 5   To check the solution, we find the mean of the four test scores, 94, 89, 88, and 97.

$$\frac{94 + 89 + 88 + (97)}{4} = \frac{368}{4} = 92$$

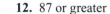

CHECK YOURSELF 12

Felicia needs a mean score of at least 75 on five tests to get a passing grade in her health class. On her first four tests she has scores of 68, 79, 71, and 70. What score on the fifth test will give her a passing grade?

The following outline (or algorithm) summarizes our work in this section.

Step by Step: Solving Linear Inequalities

Step 1   Remove any grouping symbols and combine any like terms appearing on either side of the inequality.

Step 2   Apply the addition property to write an equivalent inequality with the variable term on one side of the inequality and the number on the other.

Step 3   Apply the multiplication property to write an equivalent inequality with the variable isolated on one side of the inequality. Be sure to reverse the direction of the inequality if you multiply or divide by a negative number. The set of solutions derived in step 3 can then be graphed on a number line.

CHECK YOURSELF ANSWERS

1. (a) $>$; **(b)** $<$ **2.**

3. (a) ———————→ ; **(b)** ←——————————
                     -4 0 0 3

4. $x > 6$ (0 6)   **5.** $x \leq 10$ (0 10)

6. (a) $x > 5$ (0 5) ;   **(b)** $x \geq -6$ (-6 0)

7. (a) $x \leq 20$ (0 20) ;   **(b)** $x > 21$ (0 21)

8. (a) $x \geq -3$ (-3 0) ;   **(b)** $x > -6$ (-6 0)

9. $x < -4$ (-4 0)   **10.** $x \leq 4$ (0 4)

11. $x < -\dfrac{3}{4}$ ($-\frac{3}{4}$ 0)   **12.** 87 or greater

2.7 Exercises

Complete the statements, using the symbol < or >.

1. 9 _____ 6

2. 9 _____ 8

3. 7 _____ −2

4. 0 _____ −5

5. 0 _____ 4

6. −12 _____ −7

7. −2 _____ −5

8. −4 _____ −11

Write each inequality in words.

9. $x < 3$

10. $x \leq -5$

11. $x \geq -4$

12. $x < -2$

13. $-5 \leq x$

14. $2 < x$

Graph the solution set of each of the following inequalities.

15. $x > 2$

16. $x < -3$

17. $x < 10$

18. $x > 4$

19. $x > 1$

20. $x < -2$

21. $x < 8$

22. $x > 5$

23. $x > -7$

24. $x < -4$

ANSWERS

1. _____

2. _____

3. _____

4. _____

5. _____

6. _____

7. _____

8. _____

9. _____

10. _____

11. _____

12. _____

13. _____

14. _____

15. _____

16. _____

17. _____

18. _____

19. _____

20. _____

21. _____

22. _____

23. _____

24. _____

25. _____

26. _____

27. _____

28. _____

29. _____

30. _____

31. _____

32. _____

33. _____

34. _____

35. _____

36. _____

37. _____

38. _____

39. _____

40. _____

41. _____

42. _____

43. _____

44. _____

45. _____

46. _____

25. $x \geq 11$

26. $x \geq 0$

27. $x < 0$

28. $x \leq -3$

Solve and graph the solution set of each of the following inequalities.

29. $x + 9 < 22$

30. $x + 5 \leq 4$

31. $x + 8 \geq 10$

32. $x - 14 > -17$

33. $5x < 4x + 7$

34. $3x \geq 2x - 4$

35. $6x - 8 \leq 5x$

36. $3x + 2 > 2x$

37. $6x + 5 \geq 5x + 19$

38. $5x + 2 \leq 4x - 6$

39. $7x + 5 < 6x - 4$

40. $8x - 7 > 7x + 3$

41. $4x \leq 12$

42. $5x > 20$

43. $5x > -35$

44. $8x \leq -24$

45. $-6x \geq 18$

46. $-9x < 45$

47. $-12x < -72$

48. $-12x \geq -48$

49. $\dfrac{x}{4} > 5$

50. $\dfrac{x}{3} \leq -3$

51. $-\dfrac{x}{2} \geq -3$

52. $-\dfrac{x}{4} < 5$

53. $\dfrac{2x}{3} < 6$

54. $\dfrac{3x}{4} \geq -9$

55. $6x > 3x + 12$

56. $4x \leq x - 9$

57. $5x - 2 > 3x$

58. $7x + 3 \geq 2x$

59. $3 - 2x > 5$

60. $-7 - 5x \leq 18$

61. $2x \geq 5x + 18$

62. $3x < 7x - 28$

63. $5x - 3 \leq 3x + 15$

64. $8x + 7 > 5x + 34$

65. $11x + 8 > 4x - 6$

66. $10x - 5 \leq 8x - 25$

67. $7x - 5 < 3x + 2$

68. $5x - 2 \geq 2x - 7$

ANSWERS

47. _____

48. _____

49. _____

50. _____

51. _____

52. _____

53. _____

54. _____

55. _____

56. _____

57. _____

58. _____

59. _____

60. _____

61. _____

62. _____

63. _____

64. _____

65. _____

66. _____

67. _____

68. _____

69. _____

70. _____

71. _____

72. _____

73. _____

74. _____

75. _____

76. _____

77. _____

78. _____

79. _____

80. _____

81. _____

82. _____

83. _____

84. _____

85. _____

86. _____

87. _____

88. _____

69. $5x + 7 > 8x - 17$

$\longleftarrow\!\!\!\!\longrightarrow$

70. $4x - 3 \leq 9x + 27$

$\longleftarrow\!\!\!\!\longrightarrow$

71. $3x - 2 \leq 5x + 3$

$\longleftarrow\!\!\!\!\longrightarrow$

72. $2x + 3 > 8x - 2$

$\longleftarrow\!\!\!\!\longrightarrow$

73. $4(x + 7) \leq 2x + 31$

$\longleftarrow\!\!\!\!\longrightarrow$

74. $7(x - 3) > 5x - 14$

$\longleftarrow\!\!\!\!\longrightarrow$

75. $2(x - 7) > 5x - 12$

$\longleftarrow\!\!\!\!\longrightarrow$

76. $3(x + 4) \leq 7x + 7$

$\longleftarrow\!\!\!\!\longrightarrow$

Translate the following statements into inequalities. Let x represent the number in each case.

77. 6 more than a number is greater than 5.

78. 3 less than a number is less than or equal to 5.

79. 4 less than twice a number is less than or equal to 7.

80. 10 more than a number is greater than negative 2.

81. 4 times a number, decreased by 15, is greater than that number.

82. 2 times a number, increased by 28, is less than or equal to 6 times that number.

Match each inequality on the right with a statement on the left.

83. x is nonnegative

a. $x \geq 0$

84. x is negative

b. $x \geq 5$

85. x is no more than 5

c. $x \leq 5$

86. x is positive

d. $x > 0$

87. x is at least 5

e. $x < 5$

88. x is less than 5

f. $x < 0$

89. _____

90. _____

91. _____

92. _____

93. _____

94. _____

95. _____

96. _____

89. Social science. There are fewer than 1000 wild giant pandas left in the bamboo forests of China. Write an inequality expressing this relationship.

90. Science and medicine. Let C represent the amount of Canadian forest and M represent the amount of Mexican forest. Write an inequality showing the relationship of the forests of Mexico and Canada if Canada contains at least 9 times as much forest as Mexico.

91. Statistics. To pass a course with a grade of B or better, Liza must have an average of 80 or more. Her grades on three tests are 72, 81, and 79. Write an inequality representing the score that Liza must get on the fourth test to obtain a B average or better for the course.

92. Statistics. Sam must have an average of 70 or more in his summer course to obtain a grade of C. His first three test grades were 75, 63, and 68. Write an inequality representing the score that Sam must get on the last test to get a C grade.

93. Business and finance. Juanita is a salesperson for a manufacturing company. She may choose to receive $500 or 5% commission on her sales as payment for her work. How much does she need to sell to make the 5% offer a better deal?

94. Business and finance. The cost for a long-distance telephone call is $0.36 for the first minute and $0.21 for each additional minute or portion thereof. Write an inequality representing the number of minutes a person could talk without exceeding $3.

95. Geometry. The perimeter of a rectangle is to be no greater than 250 cm and the length must be 105 cm. Find the maximum width of the rectangle.

$$\overset{\longleftarrow \quad 105 \text{ cm} \quad \longrightarrow}{\boxed{}} \updownarrow x \text{ cm}$$

96. Statistics. Sarah bowled 136 and 189 in her first two games. What must she bowl in her third game to have an average of at least 170?

97. You are the office manager for a small company. You need to acquire a new copier for the office. You find a suitable one that leases for $250 a month from the copy machine company. It costs 2.5¢ per copy to run the machine. You purchase paper for $3.50 a ream (500 sheets). If your copying budget is no more than $950 per month, is

ANSWERS

97. _____

98. _____

this machine a good choice? Write a brief recommendation to the Purchasing Department. Use equations and inequalities to explain your recommendation.

98. Your aunt calls to ask for your help in making a decision about buying a new refrigerator. She says that she found two that seem to fit her needs, and both are supposed to last at least 14 years, according to *Consumer Reports.* The initial cost for one refrigerator is $712, but it only uses 88 kilowatt-hours (kWh) per month. The other refrigerator costs $519 and uses an estimated 100 kWh/per month. You do not know the price of electricity per kilowatt-hour where your aunt lives, so you will have to decide what in cents per kilowatt-hour will make the first refrigerator cheaper to run during its 14 years of expected usefulness. Write your aunt a letter explaining what you did to calculate this cost, and tell her to make her decision based on how the kilowatt-hour rate she has to pay in her area compares with your estimation.

Answers

1. $9 > 6$ **3.** $7 > -2$ **5.** $0 < 4$ **7.** $-2 > -5$ **9.** x is less than 3
11. x is greater than or equal to -4 **13.** -5 is less than or equal to x

15. [number line: open circle at 2, shaded right; marks 0, 2]

17. [number line: open circle at 10, shaded left; marks 0, 10]

19. [number line: open circle at 1, shaded right; marks 0, 1]

21. [number line: open circle at 8, shaded left; marks 0, 8]

23. [number line: open circle at -7, shaded left; marks -7, 0]

25. [number line: closed dot at 11, shaded left; marks 0, 11]

27. [number line: open circle at 0, shaded right]

29. $x < 13$ [number line: open circle at 13, shaded left; marks 0, 13]

31. $x \geq 2$ [number line: closed dot at 2, shaded right; marks 0, 2]

33. $x < 7$ [number line: open circle at 7, shaded left; marks 0, 7]

35. $x \leq 8$ [number line: closed dot at 8, shaded left; marks 0, 8]

37. $x \geq 14$ [number line: closed dot at 14, shaded right; marks 0, 14]

39. $x < -9$ [number line: open circle at -9, shaded left; marks -9, 0]

41. $x \leq 3$ [number line: closed dot at 3, shaded left; marks 0, 3]

43. $x > -7$ [number line: open circle at -7, shaded right; marks -7, 0]

45. $x \leq -3$ [number line: closed dot at -3, shaded left; marks -3, 0]

47. $x > 6$ [number line: open circle at 6, shaded right; marks 0, 6]

49. $x > 20$ [number line: open circle at 20, shaded right; marks 0, 20]

51. $x \leq 6$ [number line: closed dot at 6, shaded left; marks 0, 6]

53. $x < 9$ [number line: open circle at 9, shaded left; marks 0, 9]

55. $x > 4$ [number line: open circle at 4, shaded right; marks 0, 4]

57. $x > 1$ [number line: open circle at 1, shaded right; marks 0, 1]

59. $x < -1$ [number line: open circle at -1, shaded left; marks -1, 0]

61. $x \leq -6$ [number line: closed dot at -6, shaded left; marks -6, 0]

63. $x \leq 9$ [number line: closed dot at 9, shaded left; marks 0, 9]

65. $x > -2$ [number line: open circle at -2, shaded right; marks -2, 0]

67. $x < \dfrac{7}{4}$ [number line: open circle at $\frac{7}{4}$, shaded left; marks 0, $\frac{7}{4}$]

69. $x < 8$ [number line: open circle at 8, shaded left; marks 0, 8]

71. $x \geq -\dfrac{5}{2}$ [number line: closed dot at $-\frac{5}{2}$, shaded right; marks $-\frac{5}{2}$, 0]

73. $x \leq \dfrac{3}{2}$ [number line: closed dot at $\frac{3}{2}$, shaded left; marks 0, $\frac{3}{2}$]

75. $x < -\dfrac{2}{3}$ [number line: open circle at $-\frac{2}{3}$, shaded left; marks $-\frac{2}{3}$, 0]

77. $x + 6 > 5$ **79.** $2x - 4 \leq 7$

81. $4x - 15 > x$ **83.** a   **85.** c   **87.** b   **89.** $P < 1000$
91. $x \geq 88$ **93.** $> \$10,000$ **95.** 20 cm   **97.**

2　Summary

DEFINITION/PROCEDURE	EXAMPLE	REFERENCE
Solving Equations by the Addition Property		**Section 2.1**
Equation A statement that two expressions are equal	$2x - 3 = 5$ is an equation.	*p. 163*
Solution A value for a variable that makes an equation a true statement	4 is a solution for the above equation because $2(4) - 3 = 5$.	*p. 164*
Equivalent Equations Equations that have exactly the same solutions	$2x - 3 = 5$ and $x = 4$ are equivalent equations.	*p. 166*
The Addition Property of Equality If $a = b$, then $a + c = b + c$.	If $2x - 3 = 7$, then $2x - 3 + 3 = 7 + 3$.	*p. 166*
Solving Equations by the Multiplication Property		**Section 2.2**
The Multiplication Property of Equality If $a = b$, then $a \cdot c = b \cdot c$.	If $\frac{1}{2}x = 7$, then $2\left(\frac{1}{2}x\right) = 2(7)$.	*p. 177*
The mean of a set is the sum of that set divided by the number of elements in the set.	Given the set $-2, -1, 6, 9$, the mean is $\frac{12}{4} = 3$.	*p. 182*
Combining the Rules to Solve Equations		**Section 2.3**
Solving Linear Equations The steps of solving a linear equation are as follows: 1. Use the distributive property to remove any grouping symbols. Then simplify by combining like terms. 2. Add or subtract the same term on each side of the equation until the variable term is on one side and a number is on the other. 3. Multiply or divide both sides of the equation by the same nonzero number so that the variable is alone on one side of the equation. 4. Check the solution in the original equation.	Solve: $3(x - 2) + 4x = 3x + 14$ $3x - 6 + 4x = 3x + 14$ $7x - 6 = 3x + 14$ $\underline{+6 \qquad +6}$ $7x = 3x + 20$ $\underline{-3x \qquad -3x}$ $4x = 20$ $\frac{4x}{4} = \frac{20}{4}$ $x = 5$	*p. 194*
Formulas and Problem Solving		**Section 2.4**
Literal Equation An equation that involves more than one letter or variable.	$a = \frac{2b + c}{3}$	*p. 198*
Solving Literal Equations 1. Multiply both sides of the equation by the same term to clear it of fractions. 2. Add or subtract the same term on both sides of the equation so that all terms containing the variable you are solving for are on one side. 3. Divide both sides by the coefficient of the variable that you are solving for.	Solve for b: $a = \frac{2b + c}{3}$ $3a = \left(\frac{2b + c}{3}\right)3$ $3a = 2b + c$ $3a - c = 2b$ $\frac{3a - c}{2} = b$	*p. 199*

Continued

259

Solving Percent Applications		Section 2.6
The base is the whole in a statement.	14 is 25% of 56. 56 is the base.	p. 230
The amount is the part being compared to the base.	14 is the amount.	p. 230
The rate is the ratio of the amount to the base.	25% is the rate.	p. 230
The proportion used for solving most percent applications is $\dfrac{A}{B} = \dfrac{R}{100}$ $\dfrac{\text{Amount}}{\text{Base}} = \dfrac{\text{Rate}}{100}$	$\dfrac{14}{56} = \dfrac{25}{100}$	p. 231

Inequalities—An Introduction		Section 2.7
Inequality A statement that one quantity is less than (or greater than) another. Four symbols are used: $a < b$ $a > b$ $a \le b$ $a \ge b$ ↑ ↑ ↑ ↑ *a* is less than *b* *a* is greater than *b* *a* is less than or equal to *b* *a* is greater than or equal to *b*	$-4 < -1$ $x^2 + 1 \ge x + 1$	p. 243
Graphing Inequalities To graph $x < a$, we use an open circle and an arrow pointing left. The heavy arrow indicates all numbers less than (or to the left of) *a*. The open circle means *a* is not included.	Graph $x < 3$.	p. 244
To graph $x \ge b$, we use a closed circle and an arrow pointing right. The closed circle means that in this case *b* is included.	Graph $x \ge -1$.	p. 244

Solving Inequalities
 An inequality is "solved" when it is in the form $x < \square$ or $x > \square$.
 	
     Proceed as in solving equations by using the following properties.
 	
 1. If $a < b$, then $a + c < b + c$.
 	
     Adding (or subtracting) the same quantity to each side of an inequality gives an equivalent inequality.
 	
 2. If $a < b$, then $ac < bc$ when $c > 0$ and $ac > bc$ when $c < 0$.
 	
     Multiplying both sides of an inequality by the same *positive number* gives an equivalent inequality. When both sides of an inequality are multiplied by the same *negative number, you must reverse the direction* of the inequality to give an equivalent inequality.
	$2x - 3 > 5x + 6$
	 $\underline{+3 \qquad\quad +3}$
	 $2x > 5x + 9$
	 $\underline{-5x \qquad -5x}$
	 $-3x > 9$
	
	 $\dfrac{-3x}{-3} < \dfrac{9}{-3}$
	 $x < -3$
		p. 245

Summary Exercises

This summary exercise set is provided to give you practice with each of the objectives of this chapter. Each exercise is keyed to the appropriate chapter section. When you are finished, you can check your answers to the odd-numbered exercises against those presented in the back of the text. If you have difficulty with any of these questions, go back and reread the examples from that section. Your instructor will give you guidelines on how to best use these exercises in your instructional setting.

[2.1] Tell whether the number shown in parentheses is a solution for the given equation.

1. $7x + 2 = 16$ (2)

2. $5x - 8 = 3x + 2$ (4)

3. $7x - 2 = 2x + 8$ (2)

4. $4x + 3 = 2x - 11$ (−7)

5. $x + 5 + 3x = 2 + x + 23$ (6)

6. $\frac{2}{3}x - 2 = 10$ (21)

[2.1–2.3] Solve the following equations and check your results.

7. $x + 5 = 7$

8. $x - 9 = 3$

9. $7 + 6x = 5x$

10. $3x - 9 = 2x$

11. $5x - 3 = 4x + 2$

12. $9x + 2 = 8x - 7$

13. $7x - 5 = 6x - 4$

14. $3 + 4x - 1 = x - 7 + 2x$

15. $4(2x + 3) = 7x + 5$

16. $5(5x - 3) = 6(4x + 1)$

17. $6x = 42$

18. $7x = -28$

19. $-6x = 24$

20. $-9x = -63$

21. $\frac{x}{8} = 4$

22. $-\frac{x}{3} = -5$

23. $\frac{2}{3}x = 18$

24. $\frac{3}{4}x = 24$

25. $5x - 3 = 12$

26. $4x + 3 = -13$

27. $7x + 8 = 3x$

28. $3 - 5x = -17$

29. $3x - 7 = x$

30. $2 - 4x = 5$

31. $\frac{x}{3} - 5 = 1$

32. $\frac{3}{4}x - 2 = 7$

33. $6x - 5 = 3x + 13$

34. $3x + 7 = x - 9$

35. $7x + 4 = 2x + 6$

36. $9x - 8 = 7x - 3$

37. $2x + 7 = 4x - 5$

38. $3x - 15 = 7x - 10$

39. $\dfrac{10}{3}x - 5 = \dfrac{4}{3}x + 7$

40. $\dfrac{11}{4}x - 15 = 5 - \dfrac{5}{4}x$

41. $3.7x + 8 = 1.7x + 16$

42. $5.4x - 3 = 8.4x + 9$

43. $3x - 2 + 5x = 7 + 2x + 21$

44. $8x + 3 - 2x + 5 = 3 - 4x$

45. $5(3x - 1) - 6x = 3x - 2$

[2.4]   Solve for the indicated variable.

46. $V = LWH$ (for L)

47. $P = 2L + 2W$ (for L)

48. $ax + by = c$ (for y)

49. $A = \dfrac{1}{2}bh$ (for h)

50. $A = P + Prt$ (for t)

51. $m = \dfrac{n - p}{q}$ (for n)

[2.4–2.6]   Solve the following word problems. Be sure to label the unknowns and to show the equation you used.

52. Number problem. The sum of 3 times a number and 7 is 25. What is the number?

53. Number problem. 5 times a number, decreased by 8, is 32. Find the number.

54. Number problem. If the sum of two consecutive integers is 85, find the two integers.

55. Number problem. The sum of three consecutive odd integers is 57. What are the three integers?

56. Business and finance. Rafael earns $35 more per week than Andrew. If their weekly salaries total $715, what amount does each earn?

57. Number problem. Larry is 2 years older than Susan, and Nathan is twice as old as Susan. If the sum of their ages is 30 years, find each of their ages.

58. Business and finance. Joan works on a 4% commission basis. She sold $45,000 in merchandise during 1 month. What was the amount of her commission?

59. Business and finance. David buys a dishwasher that is marked down $77 from its original price of $350. What is the discount rate?

60. Science and medicine. A chemist prepares a 400-milliliter (400-mL) acid-water solution. If the solution contains 30 mL of acid, what percent of the solution is acid?

61. Business and finance. The price of a new compact car has increased $819 over the previous year. If this amounts to a 4.5% increase, what was the price of the car before the increase?

62. Business and finance. A store advertises, "Buy the red-tagged items at 25% off their listed price." If you buy a coat marked $136, what will you pay for the coat during the sale?

63. Business and finance. Tom has 6% of his salary deducted for a retirement plan. If that deduction is $168, what is his monthly salary?

64. Statistics. A college finds that 35% of its science students take biology. If there are 252 biology students, how many science students are there altogether?

65. Business and finance. A company finds that its advertising costs increased from $72,000 to $76,680 in 1 year. What was the rate of increase?

66. Business and finance. A savings bank offers 3.25% on 1-year time deposits. If you place $900 in an account, how much will you have at the end of the year?

67. Business and finance. Maria's company offers her a 4% pay raise. This will amount to a $126 per month increase in her salary. What is her monthly salary before and after the raise?

68. Statistics. A computer has 8 gigabytes (GB) of storage space. Arlene is going to add 16 GB of storage space. By what percent will the available storage space be increased?

69. Statistics. A virus scanning program is checking every file for viruses. It has completed 30% of the files in 150 s. How long should it take to check all the files?

70. Business and finance. If the total bill at a restaurant for 10 people is $572.89, including an 18% tip, what was the cost of the food?

71. Business and finance. A pair of running shoes is advertised at 30% off the original price for $80.15. What was the original price?

[2.7]   Solve and graph the solution sets for the following inequalities.

72. $x - 4 \leq 7$

73. $x + 3 > -2$

74. $5x > 4x - 3$

75. $4x \geq -12$

76. $-12x < 36$

77. $-\dfrac{x}{5} \geq 3$

78. $2x \leq 8x - 3$

79. $2x + 3 \geq 9$

80. $4 - 3x > 8$

81. $5x - 2 \leq 4x + 5$

82. $7x + 13 \geq 3x + 19$

83. $4x - 2 < 7x + 16$

Self-Test for Chapter 2

Name _____

Section _____ Date _____

The purpose of this self-test is to help you check your progress and to review for the next in-class exam. Allow yourself about an hour to take this test. At the end of that hour check your answers against those given in the back of the text. Section references accompany the answers. If you missed any questions, go back to those sections and reread the examples until you master the concepts.

ANSWERS

Tell whether the number shown in parentheses is a solution for the given equation.

1. $7x - 3 = 25$ (5)

2. $8x - 3 = 5x + 9$ (4)

Solve the following equations and check your results.

3. $x - 7 = 4$

4. $7x - 12 = 6x$

5. $9x - 2 = 8x + 5$

Solve the following equations and check your results.

6. $7x = 49$

7. $\frac{1}{4}x = -3$

8. $\frac{4}{5}x = 20$

Solve the following equations and check your results.

9. $7x - 5 = 16$

10. $10 - 3x = -2$

11. $7x - 3 = 4x - 5$

12. $\frac{3x}{2} - 5 = 4x + \frac{5}{8}$

Solve for the indicated variable.

13. $C = 2\pi r$ (for r)

14. $V = \frac{1}{3}Bh$ (for h)

15. $3x + 2y = 6$ (for y)

Solve and graph the solution sets for the following inequalities.

16. $x - 5 \le 9$

17. $5 - 3x > 17$

1. _____

2. _____

3. _____

4. _____

5. _____

6. _____

7. _____

8. _____

9. _____

10. _____

11. _____

12. _____

13. _____

14. _____

15. _____

16. _____

17. _____

18. _____

19. _____

20. _____

21. _____

22. _____

23. _____

24. _____

25. _____

18. $5x + 13 \geq 2x + 17$

$\longleftarrow\!\!\!\!\longrightarrow$

19. $2x - 3 < 7x + 2$

$\longleftarrow\!\!\!\!\longrightarrow$

Solve the following word problems. Be sure to show the equation you used for the solution.

20. Number problem. 5 times a number, decreased by 7, is 28. What is the number?

21. Number problem. The sum of three consecutive integers is 66. Find the three integers.

22. Number problem. Jan is twice as old as Juwan, and Rick is 5 years older than Jan. If the sum of their ages is 35 years, find each of their ages.

23. Geometry. The perimeter of a rectangle is 62 in. If the length of the rectangle is 1 in. more than twice its width, what are the dimensions of the rectangle?

24. Business and finance. Mrs. Moore made a $450 commission on the sale of a $9000 pickup truck. What was her commission rate?

25. Business and finance. Cynthia makes a 5% commission on all her sales. She earned $1750 in commissions during 1 month. What were her gross sales for the month?

ACTIVITY 2: MONETARY CONVERSIONS

Each activity in this text is designed to either enhance your understanding of the topics of the preceding chapter, provide you with a mathematical extension of those topics, or both. The activities can be undertaken by one student, but they are better suited for a small group project. Occasionally it is only through discussion that different facets of the activity become apparent. For material related to this activity, visit the text website at www.mhhe.com/streeter.

Every year, millions of people travel to other countries for business and pleasure. When traveling to another country, there are many important considerations to take into account: passports and visas, immunizations, local sights, restaurants and hotels, and language, to name just a few.

Another consideration when traveling internationally is currency. Nearly every country has its own money, which you need to use when doing things like eating out, purchasing gifts, or riding in taxis. For example, the Japanese currency is the yen (¥), the British use the pound (£), and Canadians use Canadian dollars (CAN$), whereas in the United States of America we use US$.

When you visit another country, you need to acquire the local currency. We do this by giving our money to a bank, exchange booth, or even an ATM machine. In exchange, we receive the equivalent amount in local currency that we can spend in the country in whatever manner we wish.

This activity uses exchange rates to explore the idea of variables. As stated in this chapter, a **variable** is a symbol (such as a letter) that we use to represent an unknown quantity.

The *Wall Street Journal* publishes exchange rates for currency on a daily basis (as do numerous other sources). For example, on September 11, 2003, the exchange rate for CAN$ was 1.327. This means that US$1 is equivalent to CAN$1.561. That is, if we exchange US$100, we would receive CAN$156.10. We compute this as follows:

$$US\$ \times \text{Exchange Rate} = CAN\$$$

Activity

I.

1. Choose a country that you would like to visit. Use a search engine (as we learned to do in Activity 1) to find the exchange rate between US$ and the currency of your chosen country. Bookmark this page (as described in Activity 1).
2. If you are only visiting for a short time, you may not need too much money. Determine how much of the local currency you will receive in exchange for US$250.
3. If you stay for an extended period, you will need more money. How much would you receive in exchange for US$900?

In part I, we treated the amount (US$) as a *variable*. This quantity varied depending upon our needs. If we visit Canada and let x = the amount exchanged in US$ and y = the amount received in CAN$, then, using the exchange rate previously given, we have the equation

$$y = 1.327x$$

You may ask, "Isn't y (CAN\$ received) a variable too?" The answer to this question is yes. In fact, all three quantities are variables. The *Wall Street Journal* reported that the exchange rate for CAN\$ on November 18, 2001, was 1.4988. This quantity varies on a daily basis. If we let r = the exchange rate, then the equation can be written as

$$y = rx$$

II.

1. Consider the country you chose to visit in part I. Find the exchange rate for another date and repeat steps I.2 and I.3 for this other exchange rate.
2. Choose another nation that you would like to visit. Repeat the steps in part I for this country.

Data Set

Currency Last Trade	U.S.\$ N/A	¥en 2:39 P.M.	Euro 2:39 P.M.	CAN\$ 2:39 P.M.	U.K.£ 2:39 P.M.	Aust\$ 2:39 P.M.
U.S.\$	1	0.008538	1.12	0.7287	1.594	0.6594
¥en	117.1	1	131.2	85.35	186.8	77.24
Euro	0.8928	0.007622	1	0.6505	1.423	0.5887
CAN\$	1.327	0.01172	1.537	1	2.188	0.905
U.K.£	0.6272	0.005355	0.7025	0.457	1	0.4136
Aust\$	1.517	0.01295	1.699	1.105	2.418	1

Source: Quote data provided by Reuters 12/14/01.

I.1 We have chosen to visit Canada and will use the 12/14/01 exchange rate of 1.561 for our sample data set.

I.2
$$(US\$) \cdot (\text{Exchange Rate}) = CAN\$$$
$$US\$250 \cdot 1.327 = CAN\$331.75$$

So, if we exchange US\$250 for CAN\$, we would receive CAN\$331.75.

I.3
$$US\$900 \cdot 1.327 = CAN\$1194.30$$

II.1 We choose the 11/18/01 exchange rate of 1.4988.

$$US\$250 \cdot 1.4988 = CAN\$374.70$$
$$US\$900 \cdot 1.4988 = CAN\$1348.92$$

II.2 Choosing to visit Japan, we would receive 127.3 Yen (¥) for each US\$ (12/14/01).

$$US\$250 \cdot 127.3 = ¥31,825$$
$$US\$900 \cdot 127.3 = ¥114,570$$

Cumulative Review
Chapters 0–2

Name _____

Section _____ Date _____

The following exercises are presented to help you review concepts from earlier chapters. This is meant as a review and not as a comprehensive exam. The answers are presented in the back of the text. Section references accompany the exercises. If you have difficulty with any of these questions, be certain to at least read through the Summary related to those sections.

Perform the indicated operations.

1. $8 + (-4)$

2. $-7 + (-5)$

3. $6 - (-2)$

4. $-4 - (-7)$

5. $(-6)(3)$

6. $(-11)(-4)$

7. $20 \div (-4)$

8. $(-50) \div (-5)$

9. $0 \div (-26)$

10. $15 \div 0$

Evaluate the expressions if $x = 5$, $y = 2$, $z = -3$, and $w = -4$.

11. $2xy$

12. $2x + 7z$

13. $3z^2$

14. $4(x + 3w)$

15. $\dfrac{2w}{y}$

16. $\dfrac{2x - w}{2y - z}$

Simplify each of the following expressions.

17. $14x^2y - 11x^2y$

18. $2x^3(3x - 5y)$

19. $\dfrac{x^2y - 2xy^2 + 3xy}{xy}$

20. $10x^2 + 5x + 2x^2 - 2x$

Solve the following equations and check your results.

21. $9x - 5 = 8x$

22. $-\dfrac{3}{4}x = 18$

23. $6x - 8 = 2x - 3$

24. $2x + 3 = 7x + 5$

25. $\dfrac{4}{3}x - 6 = 4 - \dfrac{2}{3}x$

ANSWERS

1. _____
2. _____
3. _____
4. _____
5. _____
6. _____
7. _____
8. _____
9. _____
10. _____
11. _____
12. _____
13. _____
14. _____
15. _____
16. _____
17. _____
18. _____
19. _____
20. _____
21. _____
22. _____
23. _____
24. _____
25. _____

© 2003 McGraw-Hill Companies

26.

27.

28.

29.

30.

31.

32.

33.

34.

35.

36.

37.

38.

39.

40.

Solve the following equations for the indicated variable.

26. $I = Prt$ (for r)   **27.** $A = \frac{1}{2}bh$ (for h)   **28.** $ax + by = c$ (for y)

Solve and graph the solution sets for the following inequalities.

29. $3x - 5 < 4$ **30.** $7 - 2x \geq 10$

31. $7x - 2 > 4x + 10$ **32.** $2x + 5 \leq 8x - 3$

Solve the following word problems. Be sure to show the equation used for the solution.

33. Number problem. If 4 times a number decreased by 7 is 45, find that number.

34. Number problem. The sum of two consecutive integers is 85. What are those two integers?

35. Number problem. If 3 times an integer is 12 more than the next consecutive odd integer, what is that integer?

36. Business and finance. Michelle earns $120 more per week than Dmitri. If their weekly salaries total $720, how much does Michelle earn?

37. Geometry. The length of a rectangle is 2 cm more than 3 times its width. If the perimeter of the rectangle is 44 cm, what are the dimensions of the rectangle?

38. Geometry. One side of a triangle is 5 in. longer than the shortest side. The third side is twice the length of the shortest side. If the triangle perimeter is 37 in., find the length of each leg.

39. Business and finance. Jesse paid $1562.50 in state income tax last year. If his salary was $62,500, what was the rate of tax?

40. Business and finance. A car is marked down from $31,500 to $29,137.50. What was the discount rate?

POLYNOMIALS

INTRODUCTION

The U.S. Post Office limits the size of rectangular boxes it accepts for mailing. The regulations state that "length plus girth cannot exceed 108 inches." *Girth* means the distance around a cross section; in this case, this measurement is $2h + 2w$. Using the polynomial $l + 2w + 2h$ to describe the measurement required by the Post Office, the regulations say that $l + 2w + 2h \leq 108$ in.

The volume of a rectangular box is expressed by another polynomial: $V = lwh$.

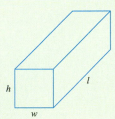

A company that wishes to produce boxes for use by postal patrons must use these formulas and do a statistical survey about the shapes that are most useful to their customers. The surface area, expressed by another polynomial expression, $2lw + 2wh + 2lh$, is also used so each box can be manufactured with the least amount of material, to help lower costs.

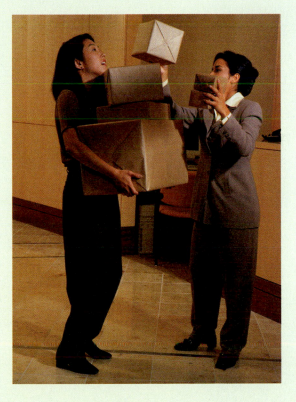

Pre-Test Chapter 3

This pre-test will provide a preview of the types of exercises you will encounter in each section of this chapter. The answers for these exercises can be found in the back of the text. If you are working on your own, or ahead of the class, this pre-test can help you identify the sections in which you should focus more of your time.

Simplify each of the following expressions. Write your answers with positive exponents only.

[3.1]   **1.** $x^5 x^7$

2. $(2x^3 y^2)(4x^2 y^5)$

3. $\dfrac{9x^5 y^2}{3x^2 y}$

[3.2]   **4.** $(2x^3 y^4)^2 (x^{-4} y^{-6})^0$

5. $\dfrac{(x^{-2})^{-4}}{x^{-8}}$

6. $2x^{-5} y^3$

Classify each polynomial as a monomial, binomial, or trinomial.

[3.3]   **7.** $6x^2 - 7x$

8. $-4x^3 + 5x - 9$

Add.

9. $4x^2 - 7x + 5$ and $-2x^2 + 5x - 7$

Subtract.

10. $-2x^2 + 3x - 1$ from $7x^2 - 8x + 5$

Multiply.

[3.4]   **11.** $3xy(4x^2 y^2 - 2xy + 7xy^3)$

12. $(3x + 2)(2x - 5)$

[3.5]   **13.** $(x + 2y)(x - 2y)$

14. $(4m + 5)^2$

15. $(3x - 2y)(x^2 - 4xy + 3y^2)$

16. $x(3x - 5y)^2$

Divide.

[3.6]   **17.** $\dfrac{28x^2 y^3 - 35x^4 y^5}{7x^2 y^2}$

18. $\dfrac{x^2 - x - 6}{x - 3}$

19. $\dfrac{x^3 - 2x - 3x^2 + 6}{x^2 - 2}$

20. $\dfrac{3x^2 + 7x - 25}{x + 4}$

3.1 Exponents and Polynomials

3.1 OBJECTIVES

1. Use the properties of exponents to simplify expressions
2. Identify types of polynomials
3. Find the degree of a polynomial
4. Write a polynomial in descending–exponent form
5. Evaluate a polynomial

Overcoming Math Anxiety

Hint #4   Preparing for a Test

Preparation for a test really begins on the first day of class. Everything you have done in class and at home has been part of that preparation. However, there are a few things that you should focus on in the last few days before a scheduled test.

1. Plan your test preparation to end at least 24 h before the test. The last 24 h is too late, and besides, you will need some rest before the test.

2. Go over your homework and class notes with pencil and paper in hand. Write down all of the problem types, formulas, and definitions that you think might give you trouble on the test.

3. The day before the test, take the page(s) of notes from step 2, and transfer the most important ideas to a 3 × 5 card.

4. Just before the test, review the information on the card. You will be surprised at how much you remember about each concept.

5. Understand that, if you have been successful at completing your homework assignments, you can be successful on the test. This is an obstacle for many students, but it is an obstacle that can be overcome. Truly anxious students are often surprised that they scored as well as they did on a test. They tend to attribute this to blind luck. It is not. It is the first sign that you really do "get it." Enjoy the success.

In Chapter 0, we reviewed the idea of exponents. Recall that exponent notation indicates repeated multiplication and the exponent tells us how many times the base is to be used as a factor.

Exponent

$$3^5 = \underbrace{3 \cdot 3 \cdot 3 \cdot 3 \cdot 3}_{5 \text{ factors}} = 243$$

Base

Now, we will look at the properties of exponents.

The first property is used when multiplying two values with the same base.

Rules and Properties: Property 1 of Exponents

For any real number a and positive integers m and n,

$a^m \cdot a^n = a^{m+n}$

For example,

$2^5 \cdot 2^7 = 2^{5+7} = 2^{12}$

The second property is used when dividing two values with the same base.

> **Rules and Properties:** Property 2 of Exponents
>
> For any real number a and positive integers m and n, with $m > n$,
>
> $$\frac{a^m}{a^n} = a^{m-n}$$
>
> For example,
>
> $$\frac{2^{12}}{2^7} = 2^{12-7} = 2^5$$

NOTE This means that the base, x^2, is used as a factor 4 times.

Consider the following:

$$(x^2)^4 = x^2 \cdot x^2 \cdot x^2 \cdot x^2 = x^8$$

This leads us to our third property for exponents.

> **Rules and Properties:** Property 3 of Exponents
>
> For any real number a and positive integers m and n,
>
> $$(a^m)^n = a^{m \cdot n}$$
>
> For example,
>
> $$(2^3)^2 = 2^{3 \cdot 2} = 2^6$$

The use of this new property is illustrated in Example 1.

OBJECTIVE 1

C A U T I O N

Be careful! Be sure to distinguish between the correct use of Property 1 and Property 3.
$(x^4)^5 = x^{4 \cdot 5} = x^{20}$
but
$x^4 \cdot x^5 = x^{4+5} = x^9$

Example 1 Using the Third Property of Exponents

Simplify each expression.

(a) $(x^4)^5 = x^{4 \cdot 5} = x^{20}$

(b) $(2^3)^4 = 2^{3 \cdot 4} = 2^{12}$ Multiply the exponents.

CHECK YOURSELF 1

Simplify each expression.

(a) $(m^5)^6$ **(b)** $(m^5)(m^6)$ **(c)** $(3^2)^4$ **(d)** $(3^2)(3^4)$

NOTE Here the base is $3x$.

Suppose we now have a product raised to a power, such as $(3x)^4$.

We know that

NOTE Here we have applied the commutative and associative properties.

$$(3x)^4 = (3x)(3x)(3x)(3x)$$
$$= (3 \cdot 3 \cdot 3 \cdot 3)(x \cdot x \cdot x \cdot x)$$
$$= 3^4 \cdot x^4 = 81x^4$$

Note that the power, here 4, has been applied to each factor, 3 and x. In general, we have

---

Rules and Properties: Property 4 of Exponents

For any real numbers a and b and positive integer m,

$$(ab)^m = a^m b^m$$

For example,

$$(3x)^3 = 3^3 \cdot x^3 = 27x^3$$

The use of this property is shown in Example 2.

Example 2   Using the Fourth Property of Exponents

NOTE $(2x)^5$ and $2x^5$ are entirely different expressions. For $(2x)^5$, the base is $2x$, so we raise each factor to the fifth power. For $2x^5$, the base is x, and so the exponent applies only to x.

Simplify each expression.

(a) $(2x)^5 = 2^5 \cdot x^5 = 32x^5$

(b) $(3ab)^4 = 3^4 \cdot a^4 \cdot b^4 = 81a^4b^4$

(c) $5(2r)^3 = 5 \cdot 2^3 \cdot r^3 = 40r^3$

 CHECK YOURSELF 2

Simplify each expression.

(a) $(3y)^4$ (b) $(2mn)^6$ (c) $3(4x)^2$ (d) $5x^3$

We may have to use more than one property when simplifying an expression involving exponents. Consider Example 3.

Example 3   Using the Properties of Exponents

NOTE To help you understand each step of the simplification, we refer to the property being applied. Make a list of the properties now to help you as you work through the remainder of this section and Section 3.2.

Simplify each expression.

(a) $(r^4s^3)^3 = (r^4)^3 \cdot (s^3)^3$ Property 4
$$= r^{12}s^9$$ Property 3

(b) $(3x^2)^2 \cdot (2x^3)^3$
$$= 3^2(x^2)^2 \cdot 2^3 \cdot (x^3)^3$$ Property 4
$$= 9x^4 \cdot 8x^9$$ Property 3
$$= 72x^{13}$$ Multiply the coefficients and apply Property 1.

(c) $\dfrac{(a^3)^5}{a^4} = \dfrac{a^{15}}{a^4}$ Property 3

$\qquad\qquad = a^{11}$ Property 2

CHECK YOURSELF 3

Simplify each expression.

(a) $(m^5 n^2)^3$ **(b)** $(2p)^4(4p^2)^2$ **(c)** $\dfrac{(s^4)^3}{s^5}$

We have one final exponent property to develop. Suppose we have a quotient raised to a power. Consider the following:

$$\left(\frac{x}{3}\right)^3 = \frac{x}{3} \cdot \frac{x}{3} \cdot \frac{x}{3} = \frac{x \cdot x \cdot x}{3 \cdot 3 \cdot 3} = \frac{x^3}{3^3}$$

Note that the power, here 3, has been applied to the numerator x and to the denominator 3. This gives us our fifth property of exponents.

Rules and Properties: Property 5 of Exponents

For any real numbers a and b, when b is not equal to 0, and positive integer m,

$$\left(\frac{a}{b}\right)^m = \frac{a^m}{b^m}$$

For example,

$$\left(\frac{2}{5}\right)^3 = \frac{2^3}{5^3} = \frac{8}{125}$$

Example 4 illustrates the use of this property. Again note that the other properties may also be applied when simplifying an expression.

Example 4   Using the Fifth Property of Exponents

Simplify each expression.

(a) $\left(\dfrac{3}{4}\right)^3 = \dfrac{3^3}{4^3} = \dfrac{27}{64}$ Property 5

(b) $\left(\dfrac{x^3}{y^2}\right)^4 = \dfrac{(x^3)^4}{(y^2)^4}$ Property 5

$\qquad\qquad = \dfrac{x^{12}}{y^8}$ Property 3

(c) $\left(\dfrac{r^2 s^3}{t^4}\right)^2 = \dfrac{(r^2 s^3)^2}{(t^4)^2}$ Property 5

$\qquad\qquad = \dfrac{(r^2)^2(s^3)^2}{(t^4)^2}$ Property 4

$\qquad\qquad = \dfrac{r^4 s^6}{t^8}$ Property 3

✔ **CHECK YOURSELF 4**

Simplify each expression.

(a) $\left(\dfrac{2}{3}\right)^4$ **(b)** $\left(\dfrac{m^3}{n^4}\right)^5$ **(c)** $\left(\dfrac{a^2b^3}{c^5}\right)^2$

The following table summarizes the five properties of exponents that were discussed in this section:

General Form	Example
1. $a^m a^n = a^{m+n}$	$x^2 \cdot x^3 = x^5$
2. $\dfrac{a^m}{a^n} = a^{m-n} \; (m > n)$	$\dfrac{5^7}{5^3} = 5^4$
3. $(a^m)^n = a^{mn}$	$(z^5)^4 = z^{20}$
4. $(ab)^m = a^m b^m$	$(4x)^3 = 4^3 x^3 = 64x^3$
5. $\left(\dfrac{a}{b}\right)^m = \dfrac{a^m}{b^m}$	$\left(\dfrac{2}{3}\right)^3 = \dfrac{2^3}{3^3} = \dfrac{8}{27}$

Our work in this chapter deals with the most common kind of algebraic expression, a *polynomial*. To define a polynomial, let's recall our earlier definition of the word *term*.

Definition: Term

A **term** is a number or the product of a number and one or more variables.

For example, x^5, $3x$, $-4xy^2$, and 8 are terms. A **polynomial** consists of one or more terms in which the only allowable exponents are the whole numbers, 0, 1, 2, 3, . . . and so on. These terms are connected by addition or subtraction signs.

Definition: Numerical Coefficient

NOTE In a polynomial, terms are separated by $+$ and $-$ signs.

In each term of a polynomial, the number is called the **numerical coefficient,** or more simply the **coefficient,** of that term.

OBJECTIVE 2   **Example 5   Identifying Polynomials**

State whether each expression is a polynomial. List the terms and the coefficient of each term.

(a) $x + 3$ is a polynomial. The terms are x and 3. The coefficients are 1 and 3.

(b) $3x^2 - 2x + 5$, or $3x^2 + (-2x) + 5$, is also a polynomial. Its terms are $3x^2$, $-2x$, and 5. The coefficients are 3, -2, and 5.

(c) $5x^3 + 2 - \dfrac{3}{x}$ is *not* a polynomial because of the division by x in the third term.

CHECK YOURSELF 5 _____

Which of the following are polynomials?

(a) $5x^2$ (b) $3y^3 - 2y + \dfrac{5}{y}$ (c) $4x^2 - 2x + 3$

Certain polynomials are given special names because of the number of terms that they have.

NOTE The prefix *mono-* means 1. The prefix *bi-* means 2. The prefix *tri-* means 3. There are no special names for polynomials with four or more terms.

> ### Definition: Monomial, Binomial, and Trinomial
>
> A polynomial with one term is called a **monomial.**
> A polynomial with two terms is called a **binomial.**
> A polynomial with three terms is called a **trinomial.**

Example 6 Identifying Types of Polynomials

(a) $3x^2y$ is a monomial. It has one term.
(b) $2x^3 + 5x$ is a binomial. It has two terms, $2x^3$ and $5x$.
(c) $5x^2 - 4x + 3$, or $5x^2 + (-4x) + 3$, is a trinomial. Its three terms are $5x^2$, $-4x$, and 3.

CHECK YOURSELF 6 _____

Classify each of these as a monomial, binomial, or trinomial.

(a) $5x^4 - 2x^3$ (b) $4x^7$ (c) $2x^2 + 5x - 3$

RECALL In a polynomial the allowable exponents are the whole numbers 0, 1, 2, 3, and so on. The degree will be a whole number.

We also classify polynomials by their *degree*. The **degree** of a polynomial that has only one variable is the highest power appearing in any one term.

OBJECTIVE 3 ### Example 7 Classifying Polynomials by Their Degree

The highest power

(a) $5x^3 - 3x^2 + 4x$ has degree 3.

The highest power

(b) $4x - 5x^4 + 3x^3 + 2$ has degree 4.

NOTE We will see in the next section that $x^0 = 1$.

(c) $8x$ has degree 1.   (Because $8x = 8x^1$)
(d) 7 has degree 0.   The degree of any nonzero constant expression is zero.

Note: Polynomials can have more than one variable, such as $4x^2y^3 + 5xy^2$. The degree is then the highest sum of the powers in any single term (here $2 + 3$, or 5). In general, we will be working with polynomials in a single variable, such as x.

CHECK YOURSELF 7

Find the degree of each polynomial.

(a) $6x^5 - 3x^3 - 2$ **(b)** $5x$ **(c)** $3x^3 + 2x^6 - 1$ **(d)** 9

Working with polynomials is much easier if you get used to writing them in **descending-exponent form** (sometimes called *descending-power form*). This simply means that the term with the highest exponent is written first, then the term with the next highest exponent, and so on.

OBJECTIVE 4   **Example 8   Writing Polynomials in Descending-Exponent Form**

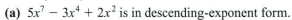

The exponents get smaller from left to right.

(a) $5x^7 - 3x^4 + 2x^2$ is in descending-exponent form.

(b) $4x^4 + 5x^6 - 3x^5$ is *not* in descending-exponent form. The polynomial should be written as

$5x^6 - 3x^5 + 4x^4$

Notice that the degree of the polynomial is the power of the *first*, or *leading*, term once the polynomial is arranged in descending-exponent form.

CHECK YOURSELF 8

Write the following polynomials in descending-exponent form.

(a) $5x^4 - 4x^5 + 7$ **(b)** $4x^3 + 9x^4 + 6x^8$

A polynomial can represent any number. Its value depends on the value given to the variable.

OBJECTIVE 5   **Example 9   Evaluating Polynomials**

Given the polynomial

$3x^3 - 2x^2 - 4x + 1$

(a) Find the value of the polynomial when $x = 2$.

Substituting 2 for x, we have

RECALL Note how the rules for the order of operations are applied. See Section 0.3 for a review.

$3(2)^3 - 2(2)^2 - 4(2) + 1$
$= 3(8) - 2(4) - 4(2) + 1$
$= 24 - 8 - 8 + 1$
$= 9$

(b) Find the value of the polynomial when $x = -2$.

Now we substitute -2 for x.

$3(-2)^3 - 2(-2)^2 - 4(-2) + 1$

$= 3(-8) - 2(4) - 4(-2) + 1$

$= -24 - 8 + 8 + 1$

$= -23$

CHECK YOURSELF 9

Find the value of the polynomial

$4x^3 - 3x^2 + 2x - 1$

when

(a) $x = 3$ **(b)** $x = -3$

CHECK YOURSELF ANSWERS

1. (a) m^{30}; **(b)** m^{11}; **(c)** 3^8; **(d)** 3^6 **2. (a)** $81y^4$; **(b)** $64m^6n^6$; **(c)** $48x^2$; **(d)** $5x^3$

3. (a) $m^{15}n^6$; **(b)** $256p^8$; **(c)** s^7 **4. (a)** $\dfrac{16}{81}$; **(b)** $\dfrac{m^{15}}{n^{20}}$; **(c)** $\dfrac{a^4b^6}{c^{10}}$

5. (a) polynomial; **(b)** not a polynomial; **(c)** polynomial

6. (a) binomial; **(b)** monomial; **(c)** trinomial   **7. (a)** 5; **(b)** 1; **(c)** 6; **(d)** 0

8. (a) $-4x^5 + 5x^4 + 7$; **(b)** $6x^8 + 9x^4 + 4x^3$ **9. (a)** 86; **(b)** -142

3.1 Exercises

Use Property 3 of exponents to simplify each of the following expressions.

1. $(x^2)^3$ x^6

2. $(a^5)^3$ a^{15}

3. $(m^4)^4$ m^{16}

4. $(p^7)^2$ p^{14}

5. $(2^4)^2$ 2^8

6. $(3^3)^2$ 3^6

7. $(5^3)^5$ 5^{15}

8. $(7^2)^4$ 7^8

Use the five properties of exponents to simplify each of the following expressions.

9. $(3x)^3$ $3^3 x^3$ $27x^3$

10. $(4m)^2$

11. $(2xy)^4$ $2^4 x^4 y^4$

12. $(5pq)^3$

13. $5(3ab)^3$ $5(3^3 a^3 b^3)$ $15^3 a^5 b^{15}$

14. $4(2rs)^4$ $4(2^4 r^4 s^4)$

15. $\left(\dfrac{3}{4}\right)^2$ $\dfrac{3^2}{4^2}$

16. $\left(\dfrac{2}{3}\right)^3$

17. $\left(\dfrac{x}{5}\right)^3$ $\dfrac{x^3}{5^3}$ x^3 16

18. $\left(\dfrac{a}{2}\right)^5$

19. $(2x^2)^4$ $2^4 x^8$

20. $(3y^2)^5$

21. $(a^8b^6)^2$ $a^{14}b^{12}$

22. $(p^3q^4)^2$

23. $(4x^2y)^3$ $4^3 x^6 y^3$

24. $(4m^4n^4)^2$ $4^2 m^8 n^8$

25. $(3m^2)^4(m^3)^2$ $3^4 m^8 \cdot m^6$

26. $(y^4)^3(4y^3)^2$ $y^{12} \cdot 4^2 y^6$ $16y^{18}$

27. $\dfrac{(x^4)^3}{x^2}$ $\dfrac{x^{12}}{x^2}$ x^{10}

28. $\dfrac{(m^5)^3}{m^6}$ $\dfrac{m^{15}}{m^6}$ M^9

29. $\dfrac{(s^3)^2(s^2)^3}{(s^5)^2}$ $\dfrac{s^6 \cdot s^6}{s^{10}}$ $16y^6 y^{12}$

30. $\dfrac{(y^5)^3(y^3)^2}{(y^4)^4}$ $\dfrac{y^{15} y^6 y^{21}}{y^{16}}$ y^{16}

31. $\left(\dfrac{m^3}{n^2}\right)^3$ $\dfrac{m^9}{n^2}$

32. $\left(\dfrac{a^4}{b^3}\right)^4$ $\dfrac{a^{16}}{b^3}$

33. $\left(\dfrac{a^3b^2}{c^4}\right)^2$ $\dfrac{a^6 b^4}{c^8}$

34. $\left(\dfrac{x^5y^2}{z^4}\right)^3$ $\dfrac{x^{15} y^6}{z^{12}}$

Which of the following expressions are polynomials?

35. $7x^3$ ✓

36. $5x^3 - \dfrac{3}{x}$ ✓

37. $4x^4y^2 - 3x^3y$ ✓

38. 7 ✗

39. -7 ✓

40. $4x^3 + x$ ✓

41. $\dfrac{3 + x}{x^2}$

42. $5a^2 - 2a + 7$

ANSWERS

1. _____   2. _____
3. _____   4. _____
5. _____   6. _____
7. _____   8. _____
9. _____   10. _____
11. _____   12. _____
13. _____   14. _____
15. _____   16. _____
17. _____   18. _____
19. _____   20. _____
21. _____   22. _____
23. _____   24. _____
25. _____   26. $16y^{18}$
27. _____   28. _____
29. _____   30. _____
31. _____   32. _____
33. _____   34. _____
35. _____
36. _____
37. _____
38. _____
39. _____
40. _____
41. _____
42. _____

43. _____

44. _____

45. _____

46. _____

47. _____

48. _____

49. _____

50. _____

51. _____

52. _____

53. _____

54. _____

55. _____

56. _____

57. _____

58. _____

59. _____

60. _____

61. _____

62. _____

63. _____

64. _____

65. _____ 66. _____

67. _____ 68. _____

69. _____ 70. _____

71. _____ 72. _____

For each of the following polynomials, list the terms and the coefficients.

43. $2x^2 - 3x$

44. $5x^3 + x$

45. $4x^3 - 3x + 2$

46. $7x^2$

Classify each of the following as a monomial, binomial, or trinomial where possible.

47. $7x^3 - 3x^2$

48. $4x^7$

49. $7y^2 + 4y + 5$

50. $2x^2 + 3xy + y^2$

51. $2x^4 - 3x^2 + 5x - 2$

52. $x^4 + \dfrac{5}{x} + 7$

53. $6y^8$

54. $4x^4 - 2x^2 + 5x - 7$

55. $x^5 - \dfrac{3}{x^2}$

56. $4x^2 - 9$

Arrange in descending-exponent form if necessary, and give the degree of each polynomial.

57. $4x^5 - 3x^2$

58. $5x^2 + 3x^3 + 4$

59. $7x^7 - 5x^9 + 4x^3$

60. $2 + x$

61. $4x$

62. $x^{17} - 3x^4$

63. $5x^2 - 3x^5 + x^6 - 7$

64. 5

Find the values of each of the following polynomials for the given values of the variable.

65. $6x + 1$, $x = 1$ and $x = -1$

66. $5x - 5$, $x = 2$ and $x = -2$

67. $x^3 - 2x$, $x = 2$ and $x = -2$

68. $3x^2 + 7$, $x = 3$ and $x = -3$

69. $3x^2 + 4x - 2$, $x = 4$ and $x = -4$

70. $2x^2 - 5x + 1$, $x = 2$ and $x = -2$

$2(2) - 5(2) + 1 \quad 8 - 10 + 1$

71. $-x^2 - 2x + 3$, $x = 1$ and $x = -3$

72. $-x^2 - 5x - 6$, $x = -3$ and $x = -2$

$2(-2^2) - 5(-2) + 1$

$2(4)$ $8 + 10$ 19

Indicate whether each of the following statements is always true, sometimes true, or never true.

73. A monomial is a polynomial.

74. A binomial is a trinomial.

75. The degree of a trinomial is 3.

76. A trinomial has three terms.

77. A polynomial has four or more terms.

78. A binomial must have two coefficients.

Solve the following problems.

79. Write x^{12} as a power of x^2.

80. Write y^{15} as a power of y^3.

81. Write a^{16} as a power of a^2.

82. Write m^{20} as a power of m^5.

83. Write each of the following as a power of 8. (Remember that $8 = 2^3$.)

$2^{12}, 2^{18}, (2^5)^3, (2^7)^6$

84. Write each of the following as a power of 9.

$3^8, 3^{14}, (3^5)^8, (3^4)^7$

85. What expression raised to the third power is $-8x^6y^9z^{15}$?

86. What expression raised to the fourth power is $81x^{12}y^8z^{16}$?

The formula $(1 + R)^y = G$ gives us useful information about the growth of a population. Here R is the rate of growth expressed as a decimal, y is the time in years, and G is the growth factor. If a country has a 2% growth rate for 35 years, then its population will double:

$(1.02)^{35} \approx 2$

87. Social science.

a. With a 2% growth rate, how many doublings will occur in 105 years? How much larger will the country's population be?

b. The less-developed countries of the world had an average growth rate of 2% in 1986. If their total population was 3.8 billion, what will their population be in 105 years if this rate remains unchanged?

88. Social science. The United States has a growth rate of 0.7%. What will be its growth factor after 35 years?

89. Write an explanation of why $(x^3)(x^4)$ is *not* x^{12}.

73.
74.
75.
76.
77.
78.
79.
80.
81.
82.
83.
84.
85.
86.
87.
88.
89.

90. Your algebra study partners are confused. "Why isn't $x^2 \cdot x^3 = 2x^5$?" they ask you. Write an explanation that will convince them.

Capital italic letters such as P or Q are often used to name polynomials. For example, we might write $P(x) = 3x^3 - 5x^2 + 2$ in which $P(x)$ is read "P of x." The notation permits a convenient shorthand. We write $P(2)$, read "P of 2," to indicate the value of the polynomial when $x = 2$. Here

$$P(2) = 3(2)^3 - 5(2)^2 + 2$$
$$= 3 \cdot 8 - 5 \cdot 4 + 2$$
$$= 6$$

Use the information above in the following problems.

If $P(x) = x^3 - 2x^2 + 5$ and $Q(x) = 2x^2 + 3$, find:

91. $P(1)$ **92.** $P(-1)$ **93.** $Q(2)$ **94.** $Q(-2)$

95. $P(3)$ **96.** $Q(-3)$ **97.** $P(0)$

98. $Q(0)$ **99.** $P(2) + Q(-1)$ **100.** $P(-2) + Q(3)$

101. $P(3) - Q(-3) \div Q(0)$ **102.** $Q(-2) \div Q(2) \cdot P(0)$

103. $\left| Q(4) \right| - \left| P(4) \right|$ **104.** $\dfrac{P(-1) + Q(0)}{P(0)}$

105. Business and finance. The cost, in dollars, of typing a term paper is given as 3 times the number of pages plus 20. Use y as the number of pages to be typed and write a polynomial to describe this cost. Find the cost of typing a 50-page paper.

106. Business and finance. The cost, in dollars, of making suits is described as 20 times the number of suits plus 150. Use s as the number of suits and write a polynomial to describe this cost. Find the cost of making seven suits.

107. Business and finance. The revenue, in dollars, when x pairs of shoes are sold is given by $3x^2 - 95$. Find the revenue when 12 pairs of shoes are sold. What is the average revenue per pair of shoes?

108. Business and finance. The cost in dollars of manufacturing w wing nuts is given by the expression $0.07w + 13.3$. Find the cost when 375 wing nuts are made. What is the average cost to manufacture one wing nut?

109. Business and finance. Suppose that when you were born, a rich uncle put $500 in the bank for you. He never deposited money again, but the bank paid 5% interest on the money every year on your birthday. How much money was in the bank after 1 year? After 2 years? After 1 year (as you know), the amount is $500 + 500(0.05)$, which can be written as $500(1 + 0.05)$ because of the distributive property. $1 + 0.05 = 1.05$, so after 1 year the amount in the bank was $500(1.05)$. After 2 years, this amount was again multiplied by 1.05. How much is in the bank today? Complete the following chart.

Birthday	Computation	Amount
0 (Day of Birth)		$500
1	$500(1.05)	
2	$500(1.05)(1.05)	
3	$500(1.05)(1.05)(1.05)	
4	$500(1.05)^4$	
5	$500(1.05)^5$	
6		
7		
8		

Write a formula for the amount in the bank on your nth birthday. About how many years does it take for the money to double? How many years for it to double again? Can you see any connection between this and the rules for exponents? Explain why you think there may or may not be a connection.

110. Work with another student to correctly complete the statements:

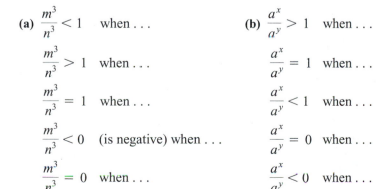

(a) $\dfrac{m^3}{n^3} < 1$ when . . .

$\dfrac{m^3}{n^3} > 1$ when . . .

$\dfrac{m^3}{n^3} = 1$ when . . .

$\dfrac{m^3}{n^3} < 0$ (is negative) when . . .

$\dfrac{m^3}{n^3} = 0$ when . . .

(b) $\dfrac{a^x}{a^y} > 1$ when . . .

$\dfrac{a^x}{a^y} = 1$ when . . .

$\dfrac{a^x}{a^y} < 1$ when . . .

$\dfrac{a^x}{a^y} = 0$ when . . .

$\dfrac{a^x}{a^y} < 0$ when . . .

Getting Ready for Section 3.2 [Section 1.7]

Reduce each of the following fractions to simplest form.

(a) $\dfrac{2^3}{2^5}$ (b) $\dfrac{3^7}{3^{10}}$ (c) $\dfrac{4^3}{4^9}$ (d) $\dfrac{5^4}{5^8}$

(e) $\dfrac{2^3}{2^3}$ (f) $\dfrac{3^5}{3^5}$ (g) $\dfrac{4^7}{4^7}$ (h) $\dfrac{5^{10}}{5^{10}}$

Answers

1. x^6 **3.** m^{16} **5.** 2^8 **7.** 5^{15} **9.** $27x^3$ **11.** $16x^4y^4$

13. $135a^3b^3$ **15.** $\dfrac{9}{16}$ **17.** $\dfrac{x^3}{125}$ **19.** $16x^8$ **21.** $a^{16}b^{12}$ **23.** $64x^6y^3$

25. $81m^{14}$ **27.** x^{10} **29.** s^2 **31.** $\dfrac{m^9}{n^6}$ **33.** $\dfrac{a^6b^4}{c^8}$ **35.** Polynomial

37. Polynomial **39.** Polynomial **41.** Not a polynomial
43. $2x^2, -3x$; 2, -3 **45.** $4x^3, -3x, 2$; 4, $-3, 2$ **47.** Binomial
49. Trinomial **51.** Not classified **53.** Monomial **55.** Not a polynomial
57. $4x^5 - 3x^2$; 5 **59.** $-5x^9 + 7x^7 + 4x^3$; 9 **61.** $4x$; 1
63. $x^6 - 3x^5 + 5x^2 - 7$; 6 **65.** 7, -5 **67.** 4, -4 **69.** 62, 30
71. 0, 0 **73.** Always **75.** Sometimes **77.** Sometimes **79.** $(x^2)^6$
81. $(a^2)^8$ **83.** $8^4, 8^6, 8^5, 8^{14}$ **85.** $-2x^2y^3z^5$
87. (a) Three doublings, 8 times as large; **(b)** 30.4 billion **89.**

91. 4 **93.** 11 **95.** 14 **97.** 5 **99.** 10 **101.** 7 **103.** -2

105. $3y + 20$, \$170 **107.** \$337, \$28.08 **109.** **a.** $\dfrac{1}{2^2}$

b. $\dfrac{1}{3^3}$ **c.** $\dfrac{1}{4^6}$ **d.** $\dfrac{1}{5^4}$ **e.** 1 **f.** 1 **g.** 1 **h.** 1

3.2 Negative Exponents and Scientific Notation

3.2 OBJECTIVES

1. Evaluate expressions involving a zero or negative exponent
2. Simplify expressions involving a zero or negative exponent
3. Write a decimal number in scientific notation
4. Solve an application of scientific notation

In Section 3.1, we discussed exponents.

We now want to extend our exponent notation to include 0 and negative integers as exponents.

First, what do we do with x^0? It will help to look at a problem that gives us x^0 as a result. What if the numerator and denominator of a fraction have the same base raised to the same power and we extend our division rule? For example,

$$\frac{a^5}{a^5} = a^{5-5} = a^0$$

RECALL By Property 2,

$$\frac{a^m}{a^n} = a^{m-n}$$

when $m > n$. Here m and n are *both* 5, so $m = n$.

But from our experience with fractions we know that

$$\frac{a^5}{a^5} = 1$$

By comparing these equations, it seems reasonable to make the following definition:

NOTE As was the case with $\frac{0}{0}$, 0^0 will be discussed in a later course.

Definition: Zero Power

For any number a, $a \neq 0$,

$$a^0 = 1$$

In words, any expression, except 0, raised to the 0 power is 1.

Example 1 illustrates the use of this definition.

OBJECTIVE 1

Example 1 Raising Expressions to the Zero Power

Evaluate each expression. Assume all variables are nonzero.

(a) $5^0 = 1$

(b) $27^0 = 1$

(c) $(x^2 y)^0 = 1$ if $x \neq 0$ and $y \neq 0$

(d) $6x^0 = 6 \cdot 1 = 6$ if $x \neq 0$

CAUTION

In part (d) the 0 exponent applies only to the x and *not* to the factor 6, because the base is x.

CHECK YOURSELF 1

Evaluate each expression. Assume all variables are nonzero.

(a) 7^0

(b) $(-8)^0$

(c) $(xy^3)^0$

(d) $3x^0$

The second property of exponents allows us to define a negative exponent. Suppose that the exponent in the denominator is *greater than* the exponent in the numerator. Consider the expression $\dfrac{x^2}{x^5}$.

Our previous work with fractions tells us that

$$\frac{x^2}{x^5} = \frac{x \cdot x}{x \cdot x \cdot x \cdot x \cdot x} = \frac{1}{x^3}$$

However, if we extend the second property to let *n* be greater than *m*, we have

RECALL $\dfrac{a^m}{a^n} = a^{m-n}$

$$\frac{x^2}{x^5} = x^{2-5} = x^{-3}$$

Now, by comparing these equations, it seems reasonable to define x^{-3} as $\dfrac{1}{x^3}$.

In general, we have this result:

Definition: Negative Powers

NOTE John Wallis (1616–1703), an English mathematician, was the first to fully discuss the meaning of 0 and negative exponents.

For any number *a*, $a \neq 0$, and any positive integer *n*,

$$a^{-n} = \frac{1}{a^n}$$

OBJECTIVE 2

Example 2 Rewriting Expressions That Contain Negative Exponents

Rewrite each expression, using only positive exponents. Simplify when possible.

Negative exponent in numerator

(a) $x^{-4} = \dfrac{1}{x^4}$

Positive exponent in denominator

(b) $m^{-7} = \dfrac{1}{m^7}$

(c) $3^{-2} = \dfrac{1}{3^2}$ or $\dfrac{1}{9}$

(d) $10^{-3} = \dfrac{1}{10^3}$ or $\dfrac{1}{1000}$

CAUTION

$2x^{-3}$ is not the same as $(2x)^{-3}$.

(e) $2x^{-3} = 2 \cdot \dfrac{1}{x^3} = \dfrac{2}{x^3}$

The −3 exponent applies only to *x*, because *x* is the base.

(f) $\left(\dfrac{2}{5}\right)^{-1} = \dfrac{1}{\frac{2}{5}} = \dfrac{5}{2}$

(g) $-4x^{-5} = -4 \cdot \dfrac{1}{x^5} = -\dfrac{4}{x^5}$

CHECK YOURSELF 2

Write each expression using only positive exponents.

(a) a^{-10} **(b)** 4^{-3} **(c)** $3x^{-2}$ **(d)** $\left(\dfrac{3}{2}\right)^{-2}$

We will now allow negative integers as exponents in our first property for exponents. Consider Example 3.

Example 3   Simplifying Expressions Containing Exponents

NOTE $a^m \cdot a^n = a^{m+n}$ for *any* integers *m* and *n*. So add the exponents.

Simplify (write an equivalent expression that uses only positive exponents).

(a) $x^5 x^{-2} = x^{5+(-2)} = x^3$

Note: An alternative approach would be

NOTE By definition

$x^{-2} = \dfrac{1}{x^2}$

$$x^5 x^{-2} = x^5 \cdot \frac{1}{x^2} = \frac{x^5}{x^2} = x^3$$

(b) $a^7 a^{-5} = a^{7+(-5)} = a^2$

(c) $y^5 y^{-9} = y^{5+(-9)} = y^{-4} = \dfrac{1}{y^4}$

 CHECK YOURSELF 3

Simplify (write an equivalent expression that uses only positive exponents).

(a) $x^7 x^{-2}$ **(b)** $b^3 b^{-8}$

Example 4 shows that all the properties of exponents introduced in the last section can be extended to expressions with negative exponents.

Example 4   Simplifying Expressions Containing Exponents

Simplify each expression.

(a) $\dfrac{m^{-3}}{m^4} = m^{-3-4}$ Property 2

$$= m^{-7} = \frac{1}{m^7}$$

(b) $\dfrac{a^{-2}b^6}{a^5 b^{-4}} = a^{-2-5}b^{6-(-4)}$ Apply Property 2 to each variable.

$$= a^{-7}b^{10} - \frac{b^{10}}{a^7}$$

NOTE This could also be done by using Property 4 first, so
$$(2x^4)^{-3} = 2^{-3} \cdot (x^4)^{-3} = 2^{-3}x^{-12}$$
$$= \frac{1}{2^3 x^{12}}$$
$$= \frac{1}{8x^{12}}$$

(c) $(2x^4)^{-3} = \dfrac{1}{(2x^4)^3}$ Definition of the negative exponent

$$= \frac{1}{2^3(x^4)^3}$$ Property 4

$$= \frac{1}{8x^{12}}$$ Property 3

(d) $\dfrac{(y^{-2})^4}{(y^3)^{-2}} = \dfrac{y^{-8}}{y^{-6}}$ Property 3

$$= y^{-8-(-6)}$$ Property 2

$$= y^{-2} = \frac{1}{y^2}$$

CHECK YOURSELF 4

Simplify each expression.

(a) $\dfrac{x^5}{x^{-3}}$ **(b)** $\dfrac{m^3 n^{-5}}{m^{-2} n^3}$ **(c)** $(3a^3)^{-4}$ **(d)** $\dfrac{(r^3)^{-2}}{(r^{-4})^2}$

Let us now take a look at an important use of exponents, scientific notation.

We begin the discussion with a calculator exercise. On most calculators, if you multiply 2.3 times 1000, the display will read

2300

Multiply by 1000 a second time. Now you will see

2300000.

Multiplying by 1000 a third time will result in the display

NOTE This must equal 2,300,000,000.

2.3   09   or   2.3   E09

And multiplying by 1000 again yields

NOTE Consider the following table:

2.3 = 2.3×10^0
23 = 2.3×10^1
230 = 2.3×10^2
2300 = 2.3×10^3
23,000 = 2.3×10^4
230,000 = 2.3×10^5

2.3   12   or   2.3   E12

Can you see what is happening? This is the way calculators display very large numbers. The number on the left is always between 1 and 10, and the number on the right indicates the number of places the decimal point must be moved to the right to put the answer in standard (or decimal) form.

This notation is used frequently in science. It is not uncommon in scientific applications of algebra to find yourself working with very large or very small numbers. Even in the time of Archimedes (287–212 B.C.E.), the study of such numbers was not unusual. Archimedes estimated that the universe was 23,000,000,000,000,000 m in diameter, which is the approximate distance light travels in $2\frac{1}{2}$ years. By comparison, Polaris (the North Star) is actually 680 light-years from the Earth. Example 6 will discuss the idea of light-years.

In scientific notation, Archimedes's estimate for the diameter of the universe would be

2.3×10^{16} m

In general, we can define scientific notation as follows.

> **Definition:** Scientific Notation
>
> Any number written in the form
>
> $a \times 10^n$
>
> in which $1 \leq a < 10$ and n is an integer, is written in scientific notation.

OBJECTIVE 3

Example 5   Using Scientific Notation

Write each of the following numbers in scientific notation.

(a) $120{,}000. = 1.2 \times 10^5$

5 places   The power is 5.

(b) $88{,}000{,}000. = 8.8 \times 10^7$

7 places   The power is 7.

NOTE The exponent on 10 shows the *number of places* we must move the decimal point. A positive exponent tells us to move right, and a negative exponent indicates to move left.

(c) $520{,}000{,}000. = 5.2 \times 10^8$

8 places

(d) $4{,}000{,}000{,}000. = 4 \times 10^9$

9 places

(e) $0.0005 = 5 \times 10^{-4}$ If the decimal point is to be moved to the left, the exponent will be negative.

4 places

NOTE To convert back to standard or decimal form, the process is simply reversed.

(f) $0.0000000081 = 8.1 \times 10^{-9}$

9 places

✔ **CHECK YOURSELF 5**

Write in scientific notation.

(a) $212{,}000{,}000{,}000{,}000{,}000$ 2.12×10^{17}

(b) 0.00079 7.9×10^{-4}

(c) $5{,}600{,}000$ 5.6×10^6

(d) 0.0000007 7.0×10^{-7}

OBJECTIVE 4

> **Example 6** **An Application of Scientific Notation**
>
> **(a)** Light travels at a speed of 3.0×10^8 meters per second (m/s). There are approximately 3.15×10^7 s in a year. How far does light travel in a year?
>
> We multiply the distance traveled in 1 s by the number of seconds in a year. This yields
>
> $$(3.0 \times 10^8)(3.15 \times 10^7) = (3.0 \cdot 3.15)(10^8 \cdot 10^7)$$ Multiply the coefficients,
> $$= 9.450 \times 10^{15}$$ and add the exponents.
>
> For our purposes we round the distance light travels in 1 year to 10^{16} m. This unit is called a **light-year,** and it is used to measure astronomical distances.
>
> **(b)** The distance from the Earth to the star Spica (in Virgo) is 2.2×10^{18} m. How many light-years is Spica from the Earth?

NOTE Notice that
$9.6075 \times 10^{15} \approx 10 \times 10^{15} = 10^{16}$

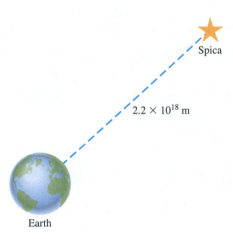

Spica

2.2×10^{18} m

Earth

NOTE We divide the distance (in meters) by the number of meters in 1 light-year.

$$\frac{2.2 \times 10^{18}}{10^{16}} = 2.2 \times 10^{18-16}$$

$$= 2.2 \times 10^2 = 220 \text{ light-years}$$

 CHECK YOURSELF 6 _____

The farthest object that can be seen with the unaided eye is the Andromeda galaxy. This galaxy is 2.3×10^{22} m from the Earth. What is this distance in light-years?

CHECK YOURSELF ANSWERS _____

1. (a) 1; **(b)** 1; **(c)** 1; **(d)** 3   **2. (a)** $\dfrac{1}{a^{10}}$; **(b)** $\dfrac{1}{4^3}$ or $\dfrac{1}{64}$; **(c)** $\dfrac{3}{x^2}$; **(d)** $\dfrac{4}{9}$

3. (a) x^5; **(b)** $\dfrac{1}{b^5}$ **4. (a)** x^8; **(b)** $\dfrac{m^5}{n^8}$; **(c)** $\dfrac{1}{81a^{12}}$; **(d)** r^2

5. (a) 2.12×10^{17}; **(b)** 7.9×10^{-4}; **(c)** 5.6×10^6; **(d)** 7×10^{-7}

6. 2,300,000 light-years

3.2 Exercises

Evaluate (assume the variables are nonzero).

1. 4^0

2. $(-7)^0$

3. $(-29)^0$

4. 75^0

5. $(x^3 y^2)^0$

6. $7m^0$

7. $11x^0$

8. $(2a^3 b^7)^0$

9. $(-3p^6 q^8)^0$

10. $-7x^0$

Write each of the following expressions using positive exponents; simplify when possible.

11. b^{-8}

12. p^{-12}

13. 3^{-4}

14. 2^{-5}

15. $\left(\dfrac{1}{5}\right)^{-2}$

16. $\left(\dfrac{1}{4}\right)^{-3}$

17. 10^{-4}

18. 10^{-5}

19. $5x^{-1}$

20. $3a^{-2}$

21. $(5x)^{-1}$

22. $(3a)^{-2}$

23. $-2x^{-5}$

24. $3x^{-4}$

25. $(-2x)^{-5}$

26. $(3x)^{-4}$

Use Properties 1 and 2 to simplify each of the following expressions. Write your answers with only positive exponents.

27. $a^5 a^3$

28. $m^5 m^7$

29. $x^8 x^{-2}$

30. $a^{12} a^{-8}$

31. $b^7 b^{-11}$

32. $y^5 y^{-12}$

33. $x^0 x^5$

34. $r^{-3} r^0$

35. $\dfrac{a^8}{a^5}$

36. $\dfrac{m^9}{m^4}$

37. $\dfrac{x^7}{x^9}$

38. $\dfrac{a^3}{a^{10}}$

39. $\dfrac{r^{-3}}{r^5}$

ANSWERS

1. _____   2. _____
3. _____   4. _____
5. _____   6. _____
7. _____   8. _____
9. _____   10. _____
11. _____   12. _____
13. _____   14. _____
15. _____   16. _____
17. _____   18. _____
19. _____   20. _____
21. _____   22. _____
23. _____   24. _____
25. _____   26. _____
27. _____   28. _____
29. _____   30. _____
31. _____   32. _____
33. _____   34. _____
35. _____   36. _____
37. _____   38. _____
39. _____

40. _____

41. _____

42. _____

43. _____

44. _____

45. _____

46. _____

47. _____

48. _____

49. _____

50. _____

51. _____

52. _____

53. _____

54. _____

55. _____

56. _____

57. _____

58. _____

59. _____

60. _____

Simplify each of the following expressions. Write your answers with positive exponents only.

40. $\dfrac{x^3y^2}{x^{-5}y^2}$

41. $\dfrac{x^{-4}yz}{x^{-5}yz}$

42. $\dfrac{p^{-6}q^{-3}}{p^{-3}q^{-6}}$

43. $\dfrac{m^5n^{-3}}{m^{-4}n^5}$

44. $\dfrac{p^{-3}q^{-2}}{p^4q^{-3}}$

45. $(2a^{-3})^4$

46. $(3x^2)^{-3}$

47. $(x^{-2}y^3)^{-2}$

48. $(a^5b^{-3})^{-3}$

49. $\dfrac{(r^{-2})^3}{r^{-4}}$

50. $\dfrac{(y^3)^{-4}}{y^{-6}}$

51. $\dfrac{(x^{-3})^3}{(x^4)^{-2}}$

52. $\dfrac{(m^4)^{-3}}{(m^{-2})^4}$

53. $\dfrac{(a^{-3})^2(a^4)}{(a^{-3})^{-3}}$

54. $\dfrac{(x^2)^{-3}(x^{-2})}{(x^2)^{-4}}$

In exercises 55 to 58, express each number in scientific notation.

55. Science and medicine. The distance from the Earth to the sun: 93,000,000 mi.

56. Science and medicine. The diameter of a grain of sand: 0.000021 m.

57. Science and medicine. The diameter of the sun: 130,000,000,000 cm.

58. Science and medicine. The number of molecules in 22.4 L of a gas: 602,000,000,000,000,000,000,000 (Avogadro's number).

59. Science and medicine. The mass of the sun is approximately 1.99×10^{30} kg. If this were written in standard or decimal form, how many 0s would follow the digit 8?

60. Science and medicine. Archimedes estimated the universe to be 2.3×10^{19} millimeters (mm) in diameter. If this number were written in standard or decimal form, how many 0s would follow the digit 3?

In exercises 61 to 64, write each expression in standard notation.

61. 8×10^{-3} **62.** 7.5×10^{-6} **63.** 2.8×10^{-5} **64.** 5.21×10^{-4}

In exercises 65 to 68, write each of the following in scientific notation.

65. 0.0005   **66.** 0.000003   **67.** 0.00037   **68.** 0.000051

In exercises 69 to 72, compute the expressions using scientific notation, and write your answer in that form.

69. $(4 \times 10^{-3})(2 \times 10^{-5})$

70. $(1.5 \times 10^{-6})(4 \times 10^2)$ $(.0000015)(400)$ 6×10^{-4} $.0006$

71. $\dfrac{9 \times 10^3}{3 \times 10^{-2}}$

72. $\dfrac{7.5 \times 10^{-4}}{1.5 \times 10^2}$ $\dfrac{7.5 \times 10^{-4}}{150}$ 5×10^{-6}

In exercises 73 to 78, perform the indicated calculations. Write your result in scientific notation.

73. $(2 \times 10^5)(4 \times 10^4)$

74. $(2.5 \times 10^7)(3 \times 10^5)$

75. $\dfrac{6 \times 10^9}{3 \times 10^7}$

76. $\dfrac{4.5 \times 10^{12}}{1.5 \times 10^7}$

77. $\dfrac{(3.3 \times 10^{15})(6 \times 10^{15})}{(1.1 \times 10^8)(3 \times 10^6)}$

78. $\dfrac{(6 \times 10^{12})(3.2 \times 10^8)}{(1.6 \times 10^7)(3 \times 10^2)}$

In 1975 the population of the Earth was approximately 4 billion and doubling every 35 years. The formula for the population P in year Y for this doubling rate is

P (in billions) $= 4 \times 2^{(y-1975)/35}$

79. Social science. What was the approximate population of the Earth in 1960?

80. Social science. What will the Earth's population be in 2025?

The U.S. population in 1990 was approximately 250 million, and the average growth rate for the past 30 years gives a doubling time of 66 years. The above formula for the United States then becomes

P (in millions) $= 250 \times 2^{(y-1990)/66}$

81. Social science. What was the approximate population of the United States in 1960?

82. Social science. What will the population of the United States be in 2025 if this growth rate continues?

ANSWERS

61. _____
62. _____
63. _____
64. _____
65. _____
66. _____
67. _____
68. _____
69. _____
70. _____
71. _____
72. _____
73. _____
74. _____
75. _____
76. _____
77. _____
78. _____
79. _____
80. _____
81. _____
82. _____

83. _____

84. _____

85. _____

86. _____

87. _____

a. _____

b. _____

c. _____

d. _____

e. _____

f. _____

g. _____

h. _____

83. **Science and medicine.** Megrez, the nearest of the Big Dipper stars, is 6.6×10^{17} m from the Earth. Approximately how long does it take light, traveling at 10^{16} m/year, to travel from Megrez to the Earth?

84. **Science and medicine.** Alkaid, the most distant star in the Big Dipper, is 2.1×10^{18} m from the Earth. Approximately how long does it take light to travel from Alkaid to the Earth?

85. **Social science.** The number of liters of water on the Earth is 15,500 followed by 19 zeros. Write this number in scientific notation. Then use the number of liters of water on the Earth to find out how much water is available for each person on the Earth. The population of the Earth is 6 billion.

86. **Social science.** If there are 6×10^9 people on the Earth and there is enough fresh-water to provide each person with 8.79×10^5 L, how much freshwater is on the Earth?

87. **Social science.** The United States uses an average of 2.6×10^6 L of water per person each year. The United States has 3.2×10^8 people. How many liters of water does the United States use each year?

 Getting Ready for Section 3.3 [Section 1.6]

Combine like terms where possible.

(a) $8m + 7m$ (b) $9x - 5x$

(c) $9m^2 - 8m$ (d) $8x^2 - 7x^2$

(e) $5c^3 + 15c^3$ (f) $9s^3 + 8s^3$

(g) $8c^2 - 6c + 2c^2$ (h) $8r^3 - 7r^2 + 5r^3$

Answers

1. 1   **3.** 1   **5.** 1   **7.** 11   **9.** 1   **11.** $\dfrac{1}{b^8}$ **13.** $\dfrac{1}{81}$

15. 25   **17.** $\dfrac{1}{10,000}$ **19.** $\dfrac{5}{x}$ **21.** $\dfrac{1}{5x}$ **23.** $-\dfrac{2}{x^5}$

25. $-\dfrac{1}{32x^5}$ **27.** a^8 **29.** x^6 **31.** $\dfrac{1}{b^4}$ **33.** x^5 **35.** a^3 **37.** $\dfrac{1}{x^2}$

39. $\dfrac{1}{r^8}$ **41.** x **43.** $\dfrac{m^9}{n^8}$ **45.** $\dfrac{16}{a^{12}}$ **47.** $\dfrac{x^4}{y^6}$ **49.** $\dfrac{1}{r^2}$ **51.** $\dfrac{1}{x}$

53. $\dfrac{1}{a^{11}}$ **55.** 9.3×10^7 mi   **57.** 1.3×10^{11} cm   **59.** 28   **61.** 0.008

63. 0.000028   **65.** 5×10^{-4} **67.** 3.7×10^{-4} **69.** 8×10^{-8}

71. 3×10^5 **73.** 8×10^9 **75.** 2×10^2 **77.** 6×10^{16}

79. 2.97 billion   **81.** 182 million   **83.** 66 years

85. 1.55×10^{23} L; 2.58×10^{13} L   **87.** 8.32×10^{14} L   **a.** $15m$ **b.** $4x$

c. $9m^2 - 8m$ **d.** x^2 **e.** $20c^3$ **f.** $17s^3$ **g.** $10c^2 - 6c$

h. $13r^3 - 7r^2$

3.3 Adding and Subtracting Polynomials

3.3 OBJECTIVES

1. Add two polynomials
2. Distribute a negative over a binomial
3. Subtract two polynomials

Addition is always a matter of combining like quantities (two apples plus three apples, four books plus five books, and so on). If you keep that basic idea in mind, adding polynomials will be easy. It is just a matter of combining like terms. Suppose that you want to add

$$5x^2 + 3x + 4 \quad \text{and} \quad 4x^2 + 5x - 6$$

Parentheses are sometimes used in adding, so for the sum of these polynomials, we can write

$$(5x^2 + 3x + 4) + (4x^2 + 5x - 6)$$

NOTE The plus sign between the parentheses indicates the addition.

Now what about the parentheses? You can use the following rule.

> **Rules and Properties:** Removing Signs of Grouping Case 1
>
> When summing two polynomials, if a plus sign (+) or nothing at all appears in front of parentheses, just remove the parentheses. No other changes are necessary.

Now let's return to the addition.

NOTE Just remove the parentheses. No other changes are necessary.

$$(5x^2 + 3x + 4) + (4x^2 + 5x - 6)$$
$$= 5x^2 + 3x + 4 + 4x^2 + 5x - 6$$

Like terms   Like terms   Like terms

NOTE Note the use of the associative and commutative properties in reordering and regrouping.

Collect like terms. (*Remember:* Like terms have the same variables raised to the same power).

$$= (5x^2 + 4x^2) + (3x + 5x) + (4 - 6)$$

Combine like terms for the result:

NOTE Here we use the distributive property. For example,
$$5x^2 + 4x^2 = (5 + 4)x^2 = 9x^2$$

$$= 9x^2 + 8x - 2$$

As should be clear, much of this work can be done mentally. You can then write the sum directly by locating like terms and combining. Example 1 illustrates this property.

OBJECTIVE 1

NOTE We call this the "horizontal method" because the entire problem is written on one line.
$3 + 4 = 7$ is the horizontal method.

$$\begin{array}{r} 3 \\ + 4 \\ \hline 7 \end{array}$$

is the vertical method.

Example 1   Combining Like Terms

Add $3x - 5$ and $2x + 3$.

Write the sum.

$$(3x - 5) + (2x + 3)$$
$$= 3x - 5 + 2x + 3 = 5x - 2$$

Like terms   Like terms

CHECK YOURSELF 1

Add $6x^2 + 2x$ and $4x^2 - 7x$.

The same technique is used to find the sum of two trinomials.

Example 2   Adding Polynomials Using the Horizontal Method

Add $4a^2 - 7a + 5$ and $3a^2 + 3a - 4$.

Write the sum.

RECALL Only the like terms are combined in the sum.

$(4a^2 - 7a + 5) + (3a^2 + 3a - 4)$

$= 4a^2 - 7a + 5 + 3a^2 + 3a - 4 = 7a^2 - 4a + 1$

Like terms

Like terms

Like terms

 CHECK YOURSELF 2

Add $5y^2 - 3y + 7$ and $3y^2 - 5y - 7$.

Example 3   Adding Polynomials Using the Horizontal Method

Add $2x^2 + 7x$ and $4x - 6$.

Write the sum.

$(2x^2 + 7x) + (4x - 6)$

$= 2x^2 + \underline{7x + 4x} - 6$

These are the only like terms; $2x^2$ and -6 cannot be combined.

$= 2x^2 + 11x - 6$

 CHECK YOURSELF 3

Add $5m^2 + 8$ and $8m^2 - 3m$.

As we mentioned in Section 3.1 writing polynomials in descending-exponent form usually makes the work easier. Look at Example 4.

Example 4   Adding Polynomials Using the Horizontal Method

Add $3x - 2x^2 + 7$ and $5 + 4x^2 - 3x$.

Write the polynomials in descending-exponent form and then add.

$(-2x^2 + 3x + 7) + (4x^2 - 3x + 5)$

$= 2x^2 + 12$

CHECK YOURSELF 4

Add $8 - 5x^2 + 4x$ and $7x - 8 + 8x^2$.

Subtracting polynomials requires another rule for removing signs of grouping.

> **Rules and Properties:** Removing Signs of Grouping Case 2
>
> When finding the difference of two polynomials, if a minus sign ($-$) appears in front of a set of parentheses, the parentheses can be removed by changing the sign of each term inside the parentheses.

The use of this rule is illustrated in Example 5.

OBJECTIVE 2   **Example 5   Removing Parentheses**

In each of the following, remove the parentheses.

NOTE This uses the distributive property, because

$-(2x + 3y) = (-1)(2x + 3y)$
$= -2x - 3y$

(a) $-(2x + 3y) = -2x - 3y$ Change each sign to remove the parentheses.

(b) $m - (5n - 3p) = m - 5n + 3p$
                          Sign changes.

(c) $2x - (-3y + z) = 2x + 3y - z$
                          Sign changes.

CHECK YOURSELF 5

(a) $-(3m + 5n)$ **(b)** $-(5w - 7z)$

(c) $3r - (2s - 5t)$ **(d)** $5a - (-3b - 2c)$

Subtracting polynomials is now a matter of using the previous rule to remove the parentheses and then combining the like terms. Consider Example 6.

OBJECTIVE 3   **Example 6   Subtracting Polynomials Using the Horizontal Method**

(a) Subtract $5x - 3$ from $8x + 2$.

Write

NOTE The expression following "from" is written first in the problem.

$(8x + 2) - (5x - 3)$

$= 8x + 2 - 5x + 3$ Recall that subtracting $5x$ is the same as adding $-5x$.
                Sign changes.

$= 3x + 5$

(b) Subtract $4x^2 - 8x + 3$ from $8x^2 + 5x - 3$.

Write

$(8x^2 + 5x - 3) - (4x^2 - 8x + 3)$

$= 8x^2 + 5x - 3 \underbrace{- 4x^2 + 8x - 3}$

Sign changes.

$= 4x^2 + 13x - 6$

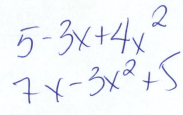

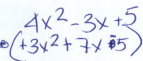

 CHECK YOURSELF 6

(a) Subtract $7x + 3$ from $10x - 7$.
(b) Subtract $5x^2 - 3x + 2$ from $8x^2 - 3x - 6$.

Again, writing all polynomials in descending-exponent form will make locating and combining like terms much easier. Look at Example 7.

Example 7   Subtracting Polynomials Using the Horizontal Method

(a) Subtract $4x^2 - 3x^3 + 5x$ from $8x^3 - 7x + 2x^2$.

Write

$(8x^3 + 2x^2 - 7x) - (-3x^3 + 4x^2 + 5x)$

$= 8x^3 + 2x^2 - 7x \underbrace{+ 3x^3 - 4x^2 - 5x}$

Sign changes.

$= 11x^3 - 2x^2 - 12x$

(b) Subtract $8x - 5$ from $-5x + 3x^2$.

Write

$(3x^2 - 5x) - (8x - 5)$

$= 3x^2 \underbrace{- 5x - 8x} + 5$

Only the like terms can be combined.

$= 3x^2 - 13x + 5$

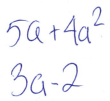

 CHECK YOURSELF 7

(a) Subtract $7x - 3x^2 + 5$ from $5 - 3x + 4x^2$.
(b) Subtract $3a - 2$ from $5a + 4a^2$.

If you think back to addition and subtraction in arithmetic, you'll remember that the work was arranged vertically. That is, the numbers being added or subtracted were placed under one another so that each column represented the same place value. This meant that in adding or subtracting columns you were always dealing with "like quantities."

It is also possible to use a vertical method for adding or subtracting polynomials. First rewrite the polynomials in descending-exponent form, and then arrange them one under another, so that each column contains like terms. Then add or subtract in each column.

Example 8   Adding Using the Vertical Method

Add $2x^2 - 5x$, $3x^2 + 2$, and $6x - 3$.

Like terms

$$
\begin{array}{r}
2x^2 - 5x \\
3x^2 \qquad + 2 \\
6x - 3 \\
\hline
5x^2 + \ x - 1
\end{array}
$$

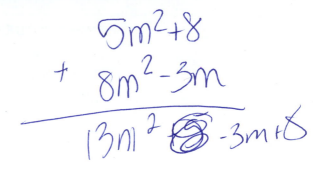

CHECK YOURSELF 8

Add $3x^2 + 5$, $x^2 - 4x$, and $6x + 7$.

Example 9 illustrates subtraction by the vertical method.

Example 9   Subtracting Using the Vertical Method

(a) Subtract $5x - 3$ from $8x - 7$.

Write

$$
\begin{array}{r}
8x - 7 \\
(-)\ (5x - 3) \\
\hline
3x - 4
\end{array}
$$
To subtract, change each sign of $5x - 3$ to get $-5x + 3$ and then add.

$$
\begin{array}{r}
8x - 7 \\
-5x + 3 \\
\hline
3x - 4
\end{array}
$$

(b) Subtract $5x^2 - 3x + 4$ from $8x^2 + 5x - 3$.

Write

$$
\begin{array}{r}
8x^2 + 5x - 3 \\
(-)\ (5x^2 - 3x + 4) \\
\hline
3x^2 + 8x - 7
\end{array}
$$
To subtract, change each sign of $5x^2 - 3x + 4$ to get $-5x^2 + 3x - 4$ and then add.

$$
\begin{array}{r}
8x^2 + 5x - 3 \\
-5x^2 + 3x - 4 \\
\hline
3x^2 + 8x - 7
\end{array}
$$

Subtracting using the vertical method takes some practice. Take time to study the method carefully. You'll be using it in long division in Section 3.6.

CHECK YOURSELF 9 _____

Subtract, using the vertical method.

(a) $4x^2 - 3x$ from $8x^2 + 2x$ **(b)** $8x^2 + 4x - 3$ from $9x^2 - 5x + 7$

CHECK YOURSELF ANSWERS _____

1. $10x^2 - 5x$ **2.** $8y^2 - 8y$ **3.** $13m^2 - 3m + 8$ **4.** $3x^2 + 11x$

5. **(a)** $-3m - 5n$; **(b)** $-5w + 7z$; **(c)** $3r - 2s + 5t$; **(d)** $5a + 3b + 2c$

6. **(a)** $3x - 10$; **(b)** $3x^2 - 8$ **7.** **(a)** $7x^2 - 10x$; **(b)** $4a^2 + 2a + 2$

8. $4x^2 + 2x + 12$ **9.** **(a)** $4x^2 + 5x$; **(b)** $x^2 - 9x + 10$

$$4x^2 - 3x$$
$$-(-8x^2 - 2x)$$
$$-4x^2 + 5x$$

3.3   Exercises

Add.

1. $6a - 5$ and $3a + 9$

2. $9x + 3$ and $3x - 4$

3. $8b^2 - 11b$ and $5b^2 - 7b$

4. $2m^2 + 3m$ and $6m^2 - 8m$

5. $3x^2 - 2x$ and $-5x^2 + 2x$

6. $3p^2 + 5p$ and $-7p^2 - 5p$

7. $2x^2 + 5x - 3$ and $3x^2 - 7x + 4$

8. $4d^2 - 8d + 7$ and $5d^2 - 6d - 9$

9. $2b^2 + 8$ and $5b + 8$

10. $4x - 3$ and $3x^2 - 9x$

11. $8y^3 - 5y^2$ and $5y^2 - 2y$

12. $9x^4 - 2x^2$ and $2x^2 + 3$

13. $2a^2 - 4a^3$ and $3a^3 + 2a^2$

14. $9m^3 - 2m$ and $-6m - 4m^3$

15. $4x^2 - 2 + 7x$ and $5 - 8x - 6x^2$

16. $5b^3 - 8b + 2b^2$ and $3b^2 - 7b^3 + 5b$

Remove the parentheses in each of the following expressions and simplify when possible.

17. $-(2a + 3b)$

$-2a - 3b$

18. $-(7x - 4y)$

19. $5a - (2b - 3c)$

20. $7x - (4y + 3z)$

21. $9r - (3r + 5s)$

22. $10m - (3m - 2n)$

23. $5p - (-3p + 2q)$

24. $8d - (-7c - 2d)$

ANSWERS

1. _____
2. _____
3. _____
4. _____
5. _____
6. _____
7. _____
8. _____
9. _____
10. _____
11. _____
12. _____
13. _____
14. _____
15. _____
16. _____
17. _____
18. _____
19. _____
20. _____
21. _____
22. _____
23. _____
24. _____

25.

26.

27.

28.

29.

30.

31.

32.

33.

34.

35.

36.

37.

38.

39.

40.

41.

42.

43.

44.

45.

46.

47.

48.

Subtract.

25. $x + 4$ from $2x - 3$

26. $x - 2$ from $3x + 5$

27. $3m^2 - 2m$ from $4m^2 - 5m$

28. $9a^2 - 5a$ from $11a^2 - 10a$

29. $6y^2 + 5y$ from $4y^2 + 5y$

30. $9n^2 - 4n$ from $7n^2 - 4n$

31. $x^2 - 4x - 3$ from $3x^2 - 5x - 2$

32. $3x^2 - 2x + 4$ from $5x^2 - 8x - 3$

33. $3a + 7$ from $8a^2 - 9a$

34. $3x^3 + x^2$ from $4x^3 - 5x$

35. $4b^2 - 3b$ from $5b - 2b^2$

36. $7y - 3y^2$ from $3y^2 - 2y$

37. $x^2 - 5 - 8x$ from $3x^2 - 8x + 7$

38. $4x - 2x^2 + 4x^3$ from $4x^3 + x - 3x^2$

Perform the indicated operations.

39. Subtract $3b + 2$ from the sum of $4b - 2$ and $5b + 3$.

40. Subtract $5m - 7$ from the sum of $2m - 8$ and $9m - 2$.

41. Subtract $3x^2 + 2x - 1$ from the sum of $x^2 + 5x - 2$ and $2x^2 + 7x - 8$.

42. Subtract $4x^2 - 5x - 3$ from the sum of $x^2 - 3x - 7$ and $2x^2 - 2x + 9$.

43. Subtract $2x^2 - 3x$ from the sum of $4x^2 - 5$ and $2x - 7$.

44. Subtract $5a^2 - 3a$ from the sum of $3a - 3$ and $5a^2 + 5$.

45. Subtract the sum of $3y^2 - 3y$ and $5y^2 + 3y$ from $2y^2 - 8y$.

46. Subtract the sum of $7r^3 - 4r^2$ and $-3r^3 + 4r^2$ from $2r^3 + 3r^2$.

Add, using the vertical method.

47. $2w^2 + 7$, $3w - 5$, and $4w^2 - 5w$

48. $3x^2 - 4x - 2$, $6x - 3$, and $2x^2 + 8$

49. $3x^2 + 3x - 4$, $4x^2 - 3x - 3$, and $2x^2 - x + 7$

50. $5x^2 + 2x - 4$, $x^2 - 2x - 3$, and $2x^2 - 4x - 3$

Subtract, using the vertical method.

51. $3a^2 - 2a$ from $5a^2 + 3a$

52. $6r^3 + 4r^2$ from $4r^3 - 2r^2$

53. $5x^2 - 6x + 7$ from $8x^2 - 5x + 7$

54. $8x^2 - 4x + 2$ from $9x^2 - 8x + 6$

55. $5x^2 - 3x$ from $8x^2 - 9$

56. $7x^2 + 6x$ from $9x^2 - 3$

Perform the indicated operations.

57. $[(9x^2 - 3x + 5) - (3x^2 + 2x - 1)] - (x^2 - 2x - 3)$

58. $[(5x^2 + 2x - 3) - (-2x^2 + x - 2)] - (2x^2 + 3x - 5)$

Find values for a, b, c, and d so that the following equations are true.

59. $3ax^4 - 5x^3 + x^2 - cx + 2 = 9x^4 - bx^3 + x^2 - 2d$

60. $(4ax^3 - 3bx^2 - 10) - 3(x^3 + 4x^2 - cx - d) = x^2 - 6x + 8$

61. Geometry. A rectangle has sides of $8x + 9$ and $6x - 7$. Find the polynomial that represents its perimeter.

6x − 7

8x + 9

62. Geometry. A triangle has sides $3x + 7$, $4x - 9$, and $5x + 6$. Find the polynomial that represents its perimeter.

5x + 6   3x + 7

4r − 9

$5x + 6 + 3x + 7 + 4x - 9$

$12x + 4$

63. Business and finance. The cost of producing x units of an item is $C = 150 + 25x$. The revenue for selling x units is $R = 90x - x^2$. The profit is given by the revenue minus the cost. Find the polynomial that represents profit.

64. Business and finance. The revenue for selling y units is $R = 3y^2 - 2y + 5$ and the cost of producing y units is $C = y^2 + y - 3$. Find the polynomial that represents profit.

$3y^2 - 2y + 3$

$- \quad y^2 + y - 3$

$y^2 - 3y + 8$

ANSWERS

49. _____

50. _____

51. _____

52. _____

53. _____

54. _____

55. _____

56. _____

57. _____

58. _____

59. _____

60. _____

61. _____

62. _____

63. _____

64. _____

$P = R - C$

Profit   Revenue   Cost

a. _____

b. _____

c. _____

d. _____

e. _____

f. _____

g. _____

h. _____

 Getting Ready for Section 3.4 [Section 1.7]

Multiply.

(a) $2^5 \cdot 2^7$ (b) $3^8 \cdot 3^{12}$ (c) $2 \cdot 4^3 \cdot 4^4$

(d) $3 \cdot 5^5 \cdot 5^2$ (e) $4 \cdot 2^5 \cdot 3 \cdot 2$ (f) $6 \cdot 3^2 \cdot 5 \cdot 3^3$

(g) $(-2 \cdot 4^2)(8 \cdot 4^7)$ (h) $(-10 \cdot 5)(-3 \cdot 5^5)$

Answers

1. $9a + 4$ **3.** $13b^2 - 18b$ **5.** $-2x^2$ **7.** $5x^2 - 2x + 1$

9. $2b^2 + 5b + 16$ **11.** $8y^3 - 2y$ **13.** $-a^3 + 4a^2$ **15.** $-2x^2 - x + 3$

17. $-2a - 3b$ **19.** $5a - 2b + 3c$ **21.** $6r - 5s$ **23.** $8p - 2q$

25. $x - 7$ **27.** $m^2 - 3m$ **29.** $-2y^2$ **31.** $2x^2 - x + 1$

33. $8a^2 - 12a - 7$ **35.** $-6b^2 + 8b$ **37.** $2x^2 + 12$ **39.** $6b - 1$

41. $10x - 9$ **43.** $2x^2 + 5x - 12$ **45.** $-6y^2 - 8y$ **47.** $6w^2 - 2w + 2$

49. $9x^2 - x$ **51.** $2a^2 + 5a$ **53.** $3x^2 + x$ **55.** $3x^2 + 3x - 9$

57. $5x^2 - 3x + 9$ **59.** $a = 3, b = 5, c = 0, d = -1$ **61.** $28x + 4$

63. $-x^2 + 65x - 150$ **a.** 2^{12} **b.** 3^{20} **c.** $2(4)^7$ **d.** $3(5)^7$

e. $12(2)^6$ **f.** $30(3)^5$ **g.** $-16(4)^9$ **h.** $30(5)^6$

3.4 Multiplying Polynomials

3.4 OBJECTIVES

1. Find the product of a monomial and a polynomial
2. Find the product of two binomials
3. Find the product of two polynomials

You have already had some experience in multiplying polynomials. In Section 1.7 we stated the first property of exponents and used that property to find the product of two monomial terms.

Step by Step: To Find the Product of Monomials

RECALL The first property of exponents:

$x^m \cdot x^n = x^{m+n}$

Step 1	Multiply the coefficients.
Step 2	Use the first property of exponents to combine the variables.

OBJECTIVE 1   **Example 1   Multiplying Monomials**

Multiply $3x^2y$ and $2x^3y^5$.

Write

RECALL We used the commutative and associative properties to rewrite the problem.

$$(3x^2y)(2x^3y^5)$$
$$= (3 \cdot 2)(x^2 \cdot x^3)(y \cdot y^5)$$

Multiply the coefficients.   Add the exponents.

$$= 6x^5y^6$$

 CHECK YOURSELF 1

Multiply.

(a) $(5a^2b)(3a^2b^4)$ **(b)** $(-3xy)(4x^3y^5)$

RECALL You might want to review Section 1.2 before going on.

Our next task is to find the product of a monomial and a polynomial. Here we use the distributive property, which we introduced in Section 1.2. That property leads us to the following rule for multiplication.

Rules and Properties: To Multiply a Polynomial by a Monomial

NOTE Distributive property:

$a(b + c) = ab + ac$

Use the distributive property to multiply each term of the polynomial by the monomial.

Example 2 Multiplying a Monomial and a Binomial

(a) Multiply $2x + 3$ by x.

Write

NOTE With practice you will do this step mentally.

$x(2x + 3)$

$= x \cdot 2x + x \cdot 3$

$= 2x^2 + 3x$

Multiply x by $2x$ and then by 3 (the terms of the polynomial). That is, "distribute" the multiplication over the sum.

(b) Multiply $2a^3 + 4a$ by $3a^2$.

Write

$3a^2(2a^3 + 4a)$

$= 3a^2 \cdot 2a^3 + 3a^2 \cdot 4a = 6a^5 + 12a^3$

 CHECK YOURSELF 2 _____

Multiply.

(a) $2y(y^2 + 3y)$

(b) $3w^2(2w^3 + 5w)$

$2y^3 + 6y^2$

$6w^5 + 15w^3$

The patterns of Example 2 extend to *any* number of terms.

Example 3 Multiplying a Monomial and a Polynomial

Multiply the following.

(a) $3x(4x^3 + 5x^2 + 2)$

$= 3x \cdot 4x^3 + 3x \cdot 5x^2 + 3x \cdot 2 = 12x^4 + 15x^3 + 6x$

NOTE We show all the steps of the process. With practice, you will be able to write the product directly, and you should try to do so.

(b) $5y^2(2y^3 - 4)$

$= 5y^2 \cdot 2y^3 - 5y^2 \cdot 4 = 10y^5 - 20y^2$

(c) $-5c(4c^2 - 8c)$

$= (-5c)(4c^2) - (-5c)(8c) = -20c^3 + 40c^2$

(d) $3c^2d^2(7cd^2 - 5c^2d^3)$

$= 3c^2d^2 \cdot 7cd^2 - 3c^2d^2 \cdot 5c^2d^3 = 21c^3d^4 - 15c^4d^5$

 CHECK YOURSELF 3 _____

Multiply.

(a) $3(5a^2 + 2a + 7)$ $15a^2 + 6a + 21$

(b) $4x^2(8x^3 - 6)$ $32x^5 - 24x^2$

(c) $-5m(8m^2 - 5m)$ $-40m^3 + 25m^2$

(d) $9a^2b(3a^3b - 6a^2b^4)$ $27a^5b^2 - 54a^4b^5$

OBJECTIVE 2 | Example 4   Multiplying Binomials |

(a) Multiply $x + 2$ by $x + 3$.

NOTE Note that this ensures that each term, x and 2, of the first binomial is multiplied by each term, x and 3, of the second binomial.

We can think of $x + 2$ as a single quantity and apply the distributive property.

$\overbrace{(x + 2)}(x + 3)$ Multiply $x + 2$ by x and then by 3.

$= (x + 2)x + (x + 2)3$

$= x \cdot x + 2 \cdot x + x \cdot 3 + 2 \cdot 3$

$= x^2 + 2x + 3x + 6$

$= x^2 + 5x + 6$

(b) Multiply $a - 3$ by $a - 4$. (Think of $a - 3$ as a single quantity and distribute.)

$(a - 3)(a - 4)$

$= (a - 3)a - (a - 3)(4)$

$= a \cdot a - 3 \cdot a - [(a \cdot 4) - (3 \cdot 4)]$

$= a^2 - 3a - (4a - 12)$ Note that the parentheses are needed

$= a^2 - 3a - 4a + 12$ here because a *minus sign* precedes

$= a^2 - 7a + 12$ the binomial.

 CHECK YOURSELF 4

Multiply.

(a) $(x + 2)(x + 5)$ 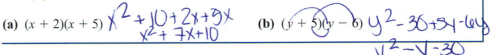 $x^2 + 10 + 2x + 5x$
$x^2 + 7x + 10$

(b) $(y + 5)(y - 6)$ $y^2 - 30 + 5y - 6y$
$y^2 - y - 30$

Fortunately, there is a pattern to this kind of multiplication that allows you to write the product of the two binomials without going through all these steps. We call it the **FOIL method** of multiplying. The reason for this name will be clear as we look at the process in more detail.

To multiply $(x + 2)(x + 3)$:

1. $(x + 2)(x + 3)$ Find the product of the *first* terms of the factors.

NOTE Remember this by F! $x \cdot x$

2. $(x + 2)(x + 3)$ Find the product of the *outer* terms.

NOTE Remember this by O! $x \cdot 3$

3. $(x + 2)(x + 3)$ Find the product of the *inner* terms.

NOTE Remember this by I! $2 \cdot x$

4. $(x + 2)(x + 3)$ Find the product of the *last* terms.

NOTE Remember this by L! $2 \cdot 3$

NOTE Of course these are the same four terms found in Example 4a.

Combining the four steps, we have

$(x + 2)(x + 3)$

$= x^2 + 3x + 2x + 6$

$= x^2 + 5x + 6$

NOTE It's called FOIL to give you an easy way of remembering the steps: *First, Outer, Inner,* and *Last.*

With practice, you can use the FOIL method to write products quickly and easily. Consider Example 5, which illustrates this approach.

Example 5 **Using the FOIL Method**

Find each product using the FOIL method.

(a) $(x + 4)(x + 5)$

$$= x^2 + 5x + 4x + 20$$

$$= x^2 + 9x + 20$$

NOTE When possible, you should combine the outer and inner products mentally and write just the final product.

(b) $(x - 7)(x + 3)$

Combine the outer and inner products as $-4x$.

$$= x^2 - 4x - 21$$

CHECK YOURSELF 5

Multiply.

(a) $(x + 6)(x + 7)$ (b) $(x + 3)(x - 5)$ (c) $(x - 2)(x - 8)$

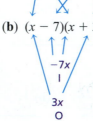

$$x^2 + 42 + 6x + 7x$$

Using the FOIL method, you can also find the product of binomials with coefficients other than 1 or with more than one variable.

$$x^2 + 13x + 42$$

Example 6 **Using the FOIL Method**

Find each product using the FOIL method.

(a) $(4x - 3)(3x + 2)$

Combine:
$-9x + 8x = -x$

$$= 12x^2 - x - 6$$

(b) $(3x - 5y)(2x - 7y)$

$6x^2$ $35y^2$

$-10xy$

$-21xy$

Combine:
$-10xy - 21xy = -31xy$

$= 6x^2 - 31xy + 35y^2$

The following rule summarizes our work in multiplying binomials.

> **Step by Step: To Multiply Two Binomials**
>
> **Step 1** Find the first term of the product of the binomials by multiplying the first terms of the binomials (F).
>
> **Step 2** Find the middle term of the product as the sum of the outer and inner products (O + I).
>
> **Step 3** Find the last term of the product by multiplying the last terms of the binomials (L).

CHECK YOURSELF 6

Multiply.

(a) $(5x + 2)(3x - 7)$ **(b)** $(4a - 3b)(5a - 4b)$ **(c)** $(3m + 5n)(2m + 3n)$

[handwritten: $20a^2 + 12b^2 - 15ab - 16ab$ $20a^2 \, ab + 12b^2$]

Sometimes, especially with larger polynomials, it is easier to use the vertical method to find their product. This is the same method you originally learned when multiplying two large integers.

[handwritten: $20a^2 - 31ab + 12b^2$]

> **Example 7   Multiplying Using the Vertical Method**

Use the vertical method to find the product of $(3x + 2)(4x - 1)$.
   First, we rewrite the multiplication in vertical form.

$3x + 2$
$4x + (-1)$

[handwritten: $-3x + (-2)$ $+12x^2$]

Multiplying the quantity $3x + 2$ by -1 yields

$3x + 2$
$\underline{4x + (-1)}$
$-3x + (-2)$

[handwritten: $(3x+2)(4x-1)$ $3x+2$ $4x(-1)$ $-3x+(-2)$]

Note that we maintained the columns of the original binomial when we found the product. We will continue with those columns as we multiply by the $4x$ term.

$3x + 2$
$\underline{4x + (-1)}$
$ -3x + (-2)$
$\underline{12x^2 + 8x}$
$12x^2 + 5x + (-2)$

[handwritten: $6m^2 + 15n^2 + 16nm$ $+9nm$ $6m + 19nm + 15n^2$]

We could write the product as $(3x + 2)(4x - 1) = 12x^2 + 5x - 2$.

CHECK YOURSELF 7

Use the vertical method to find the product of $(5x - 3)(2x + 1)$.

We use the vertical method again in Example 8. This time, we will multiply a binomial and a trinomial. Note that the FOIL method can never work for anything but the product of two binomials.

OBJECTIVE 3   **Example 8   Using the Vertical Method to Multiply Polynomials**

Multiply $x^2 - 5x + 8$ by $x + 3$.

Step 1
$$
\begin{array}{r}
x^2 - 5x + 8 \\
x + 3 \\
\hline
3x^2 - 15x + 24
\end{array}
$$
Multiply each term of $x^2 - 5x + 8$ by 3.

Step 2
$$
\begin{array}{r}
x^2 - 5x + 8 \\
x + 3 \\
\hline
3x^2 - 15x + 24 \\
x^3 - 5x^2 + 8x
\end{array}
$$
Now multiply each term by x.

Note that this line is shifted over so that like terms are in the same columns.

NOTE Using the vertical method ensures that each term of one factor multiplies each term of the other. That's why it works!

Step 3
$$
\begin{array}{r}
x^2 - 5x + 8 \\
x + 3 \\
\hline
3x^2 - 15x + 24 \\
x^3 - 5x^2 + 8x \\
\hline
x^3 - 2x^2 - 7x + 24
\end{array}
$$
Now combine like terms to write the product.

CHECK YOURSELF 8

Multiply $2x^2 - 5x + 3$ by $3x + 4$.

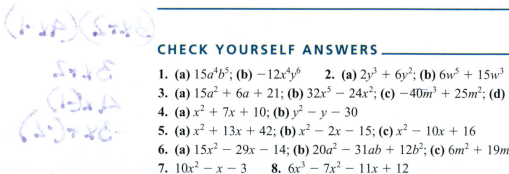

CHECK YOURSELF ANSWERS

1. (a) $15a^4b^5$; (b) $-12x^4y^6$ **2.** (a) $2y^3 + 6y^2$; (b) $6w^5 + 15w^3$
3. (a) $15a^2 + 6a + 21$; (b) $32x^5 - 24x^2$; (c) $-40m^3 + 25m^2$; (d) $27a^5b^2 - 54a^4b^5$
4. (a) $x^2 + 7x + 10$; (b) $y^2 - y - 30$
5. (a) $x^2 + 13x + 42$; (b) $x^2 - 2x - 15$; (c) $x^2 - 10x + 16$
6. (a) $15x^2 - 29x - 14$; (b) $20a^2 - 31ab + 12b^2$; (c) $6m^2 + 19mn + 15n^2$
7. $10x^2 - x - 3$ **8.** $6x^3 - 7x^2 - 11x + 12$

Multiply.

1. $(5x^2)(3x^3)$

2. $(7a^5)(4a^6)$

3. $(-2b^2)(14b^8)$

4. $(14y^4)(-4y^6)$

5. $(-10p^6)(-4p^7)$

6. $(-6m^8)(9m^7)$

7. $(4m^5)(-3m)$

8. $(-5r^7)(-3r)$

9. $(4x^3y^2)(8x^2y)$

10. $(-3r^4s^2)(-7r^2s^5)$

11. $(-3m^5n^2)(2m^4n)$

12. $(7a^3b^5)(-6a^4b)$

13. $5(2x + 6)$

14. $4(7b - 5)$

15. $3a(4a + 5)$

16. $5x(2x - 7)$

17. $3s^2(4s^2 - 7s)$

18. $9a^2(3a^3 + 5a)$

19. $2x(4x^2 - 2x + 1)$

20. $5m(4m^3 - 3m^2 + 2)$

21. $3xy(2x^2y + xy^2 + 5xy)$

22. $5ab^2(ab - 3a + 5b)$

23. $6m^2n(3m^2n - 2mn + mn^2)$

24. $8pq^2(2pq - 3p + 5q)$

Multiply.

25. $(x + 3)(x + 2)$

26. $(a - 3)(a - 7)$

27. $(m - 5)(m - 9)$

28. $(b + 7)(b + 5)$

29. $(p - 8)(p + 7)$

30. $(x - 10)(x + 9)$

31. $(w + 10)(w + 20)$

32. $(s - 12)(s - 8)$

ANSWERS

1. _____ 2. _____
3. _____ 4. _____
5. _____ 6. _____
7. _____ 8. _____
9. _____ 10. _____
11. _____ 12. _____
13. _____
14. _____
15. _____
16. _____
17. _____
18. _____
19. _____
20. _____
21. _____
22. _____
23. _____
24. _____
25. _____
26. _____
27. _____
28. _____
29. _____
30. _____
31. _____
32. _____

33. _____

34. _____

35. _____

36. _____

37. _____

38. _____

39. _____

40. _____

41. _____

42. _____

43. _____

44. _____

45. _____

46. _____

47. _____

48. _____

49. _____

50. _____

51. _____

52. _____

53. _____

54. _____

55. _____

56. _____

57. _____

58. _____

59. _____

60. _____

61. _____

62. _____

33. $(3x - 5)(x - 8)$ **34.** $(w + 5)(4w - 7)$

35. $(2x - 3)(3x + 4)$ **36.** $(5a + 1)(3a + 7)$

37. $(3a - b)(4a - 9b)$ **38.** $(7s - 3t)(3s + 8t)$

39. $(3p - 4q)(7p + 5q)$ **40.** $(5x - 4y)(2x - y)$

41. $(2x + 5y)(3x + 4y)$ **42.** $(4x - 5y)(4x + 3y)$

43. $(x + 5)(x + 5)$ **44.** $(y + 8)(y + 8)$

45. $(y - 9)(y - 9)$ **46.** $(2a + 3)(2a + 3)$

47. $(6m + n)(6m + n)$ **48.** $(7b - c)(7b - c)$

49. $(a - 5)(a + 5)$ **50.** $(x - 7)(x + 7)$

51. $(x - 2y)(x + 2y)$ **52.** $(7x + y)(7x - y)$

53. $(5s + 3t)(5s - 3t)$ **54.** $(9c - 4d)(9c + 4d)$

Multiply, using the vertical method.

55. $(x + 2)(3x + 5)$ **56.** $(a - 3)(2a + 7)$

57. $(2m - 5)(3m + 7)$ **58.** $(5p + 3)(4p + 1)$

59. $(3x + 4y)(5x - 2y)$ **60.** $(7a - 2b)(2a + 4b)$

61. $(a^2 + 3ab - b^2)(a^2 - 5ab + b^2)$ **62.** $(m^2 - 5mn + 3n^2)(m^2 + 4mn - 2n^2)$

63. $(x - 2y)(x^2 + 2xy + 4y^2)$

64. $(m + 3n)(m^2 - 3mn + 9n^2)$

65. $(3a + 4b)(9a^2 - 12ab + 16b^2)$

66. $(2r - 3s)(4r^2 + 6rs + 9s^2)$

Multiply.

67. $2x(3x - 2)(4x + 1)$

68. $3x(2x + 1)(2x - 1)$

69. $5a(4a - 3)(4a + 3)$

70. $6m(3m - 2)(3m - 7)$

71. $3s(5s - 2)(4s - 1)$

72. $7w(2w - 3)(2w + 3)$

73. $(x - 2)(x + 1)(x - 3)$

74. $(y + 3)(y - 2)(y - 4)$

75. $(a - 1)^3$

76. $(x + 1)^3$

Multiply the following.

77. $\left(\dfrac{x}{2} + \dfrac{2}{3}\right)\left(\dfrac{2x}{3} - \dfrac{2}{5}\right)$

78. $\left(\dfrac{x}{3} + \dfrac{3}{4}\right)\left(\dfrac{3x}{4} - \dfrac{3}{5}\right)$

79. $[x + (y - 2)][x - (y - 2)]$

80. $[x + (3 - y)][x - (3 - y)]$

Label the following as true or false.

81. $(x + y)^2 = x^2 + y^2$ T

82. $(x - y)^2 = x^2 - y^2$ T

83. $(x + y)^2 = x^2 + 2xy + y^2$ F

84. $(x - y)^2 = x^2 - 2xy + y^2$ F

85. Geometry. The length of a rectangle is given by $3x + 5$ cm and the width is given by $2x - 7$ cm. Express the area of the rectangle in terms of x.

86. Geometry. The base of a triangle measures $3y + 7$ in. and the height is $2y - 3$ in. Express the area of the triangle in terms of y.

87. Business and finance. The price of an item is given by $p = 2x - 10$. If the revenue generated is found by multiplying the number of items (x) sold by the price of an item, find the polynomial which represents the revenue.

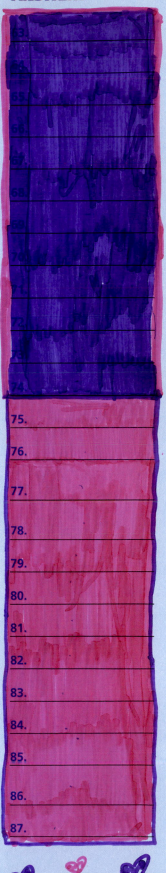

ANSWERS

75.

76.

77.

78.

79.

80.

81.

82.

83.

84.

85.

86.

87.

SECTION 3.4   **315**

88. Business and finance. The price of an item is given by $p = 2x^2 - 100$. Find the polynomial that represents the revenue generated from the sale of x items.

89. Geometry. Work with another student to complete this table and write the polynomial. A paper box is to be made from a piece of cardboard 20 in. wide and 30 in. long. The box will be formed by cutting squares out of each of the four corners and folding up the sides to make a box.

If x is the dimension of the side of the square cut out of the corner, when the sides are folded up, the box will be x in. tall. You should use a piece of paper to try this to see how the box will be made. Complete the following chart.

Length of Side of Corner Square	Length of Box	Width of Box	Depth of Box	Volume of Box
1 in.				
2 in.				
3 in.				
n in.				

Write a general formula for the width, length, and height of the box and a general formula for the *volume* of the box, and simplify it by multiplying. The variable will be the height, the side of the square cut out of the corners. What is the highest power of the variable in the polynomial you have written for the volume?

90. (a) Multiply $(x - 1)(x + 1)$.

(b) Multiply $(x - 1)(x^2 + x + 1)$.

(c) Multiply $(x - 1)(x^3 + x^2 + x + 1)$.

(d) Based on your results to (a), (b), and (c), find the product $(x - 1)(x^{29} + x^{28} + \cdots + x + 1)$.

Getting Ready for Section 3.5 [Section 1.4]

Simplify.

(a) $(3a)(3a)$ (b) $(3a)^2$

(c) $(5x)(5x)$ (d) $(5x)^2$

(e) $(-2w)(-2w)$ (f) $(-2w)^2$

(g) $(-4r)(-4r)$ (h) $(-4r)^2$

Answers

1. $15x^5$ **3.** $-28b^{10}$ **5.** $40p^{13}$ **7.** $-12m^6$ **9.** $32x^5y^3$

11. $-6m^9n^3$ **13.** $10x + 30$ **15.** $12a^2 + 15a$

17. $12s^4 - 21s^3$ **19.** $8x^3 - 4x^2 + 2x$

21. $6x^3y^2 + 3x^2y^3 + 15x^2y^2$ **23.** $18m^4n^2 - 12m^3n^2 + 6m^3n^3$

25. $x^2 + 5x + 6$ **27.** $m^2 - 14m + 45$ **29.** $p^2 - p - 56$

31. $w^2 + 30w + 200$ **33.** $3x^2 - 29x + 40$ **35.** $6x^2 - x - 12$

37. $12a^2 - 31ab + 9b^2$ **39.** $21p^2 - 13pq - 20q^2$

41. $6x^2 + 23xy + 20y^2$ **43.** $x^2 + 10x + 25$

45. $y^2 - 18y + 81$ **47.** $36m^2 + 12mn + n^2$ **49.** $a^2 - 25$ **51.** $x^2 - 4y^2$

53. $25s^2 - 9t^2$ **55.** $3x^2 + 11x + 10$ **57.** $6m^2 - m - 35$

59. $15x^2 + 14xy - 8y^2$ **61.** $a^4 - 2a^3b - 15a^2b^2 + 8ab^3 - b^4$

63. $x^3 - 8y^3$ **65.** $27a^3 + 64b^3$ **67.** $24x^3 - 10x^2 - 4x$

69. $80a^3 - 45a$ **71.** $60s^3 - 39s^2 + 6s$ **73.** $x^3 - 4x^2 + x + 6$

75. $a^3 - 3a^2 + 3a - 1$ **77.** $\dfrac{x^2}{3} + \dfrac{11x}{45} - \dfrac{4}{15}$ **79.** $x^2 - y^2 + 4y - 4$

81. False   **83.** True   **85.** $6x^2 - 11x - 35 \text{ cm}^2$ **87.** $2x^2 - 10x$

89.   **a.** $9a^2$ **b.** $9a^2$ **c.** $25x^2$ **d.** $25x^2$ **e.** $4w^2$ **f.** $4w^2$

g. $16r^2$ **h.** $16r^2$

3.5 Special Products

3.5 OBJECTIVES

1. Square a binomial
2. Find the product of two binomials that differ only in the sign between terms

Certain products occur frequently enough in algebra that it is worth learning special formulas for dealing with them. First, let's look at the **square of a binomial,** which is the product of two equal binomial factors.

$$(x + y)^2 = (x + y)(x + y)$$
$$= x^2 + 2xy + y^2$$

$$(x - y)^2 = (x - y)(x - y)$$
$$= x^2 - 2xy + y^2$$

The patterns above lead us to the following rule.

Step by Step: To Square a Binomial

Step 1 Find the first term of the square by squaring the first term of the binomial.

Step 2 Find the middle term of the square as twice the product of the two terms of the binomial.

Step 3 Find the last term of the square by squaring the last term of the binomial.

OBJECTIVE 1

Example 1 Squaring a Binomial

CAUTION

A very common mistake in squaring binomials is to forget the middle term.

(a) $(x + 3)^2 = x^2 + 2 \cdot x \cdot 3 + 3^2$

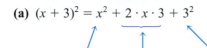

| Square of first term | Twice the product of the two terms | Square of the last term |

$$= x^2 + 6x + 9$$

(b) $(3a + 4b)^2 = (3a)^2 + 2(3a)(4b) + (4b)^2$
$$= 9a^2 + 24ab + 16b^2$$

(c) $(y - 5)^2 = y^2 + 2 \cdot y \cdot (-5) + (-5)^2$
$$= y^2 - 10y + 25$$

(d) $(5c - 3d)^2 = (5c)^2 + 2(5c)(-3d) + (-3d)^2$
$$= 25c^2 - 30cd + 9d^2$$

Again we have shown all the steps. With practice you can write just the square.

CHECK YOURSELF 1

Multiply.

(a) $(2x + 1)^2$ **(b)** $(4x - 3y)^2$

Example 2   Squaring a Binomial

Find $(y + 4)^2$.

NOTE You should see that $(2 + 3)^2 \neq 2^2 + 3^2$ because $5^2 \neq 4 + 9$.

$(y + 4)^2$ is *not* equal to $y^2 + 4^2$ or $y^2 + 16$

The correct square is

$(y + 4)^2 = y^2 + \boxed{8y} + 16$

The middle term is twice the product of y and 4.

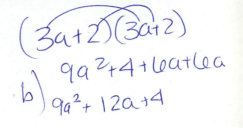

$(3a + 2)(3a + 2)$

$9a^2 + 4 + 6a + 6a$

b) $9a^2 + 12a + 4$

CHECK YOURSELF 2

Multiply.

(a) $(x + 5)^2$ **(b)** $(3a + 2)^2$ **(c)** $(y - 7)^2$ **(d)** $(5x - 2y)^2$

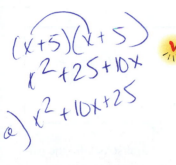

$(x+5)(x+5)$

$x^2 + 25 + 10x$

a) $x^2 + 10x + 25$

A second special product will be very important in Chapter 4, which deals with factoring. Suppose the form of a product is

$(x + y)(x - y)$

The two terms differ only in sign.

$(y-7)(y-7)$

$y^2 + 49 - 7y - 7y$

c) $y^2 - 14y + 49$

Let's see what happens when we multiply these two terms.

$(x + y)(x - y)$
$= x^2 \underbrace{- xy + xy}_{= 0} - y^2$

$= x^2 - y^2$

$(5x - 2y)(5x - 2y)$

$25x^2 + 4y^2 - 10xy$
$\qquad\qquad - 10xy$

Because the middle term becomes 0, we have the following rule.

d) $25x^2 - 20xy + 4y^2$

Rules and Properties: Special Product

The product of two binomials that differ only in the sign between the terms is the square of the first term minus the square of the second term.

Let's look at examples of this rule.

OBJECTIVE 2   ## Example 3   Finding a Special Product

Multiply each pair of binomials.

(a) $(x + 5)(x - 5) = x^2 - 5^2$

Square of the first term   Square of the second term

$= x^2 - 25$

RECALL

$(2y)^2 = (2y)(2y)$
$= 4y^2$

(b) $(x + 2y)(x - 2y) = x^2 - (2y)^2$

Square of
the first term

Square of
the second term

$= x^2 - 4y^2$

(c) $(3m + n)(3m - n) = 9m^2 - n^2$

(d) $(4a - 3b)(4a + 3b) = 16a^2 - 9b^2$

✔ **CHECK YOURSELF 3**

Find the products.

(a) $(a - 6)(a + 6)$ $a^2 - 36$

(b) $(x - 3y)(x + 3y)$ $x^2 - 9y^2$

(c) $(5n + 2p)(5n - 2p)$ $25n^2 - 4p^2$

(d) $(7b - 3c)(7b + 3c)$ $49b^2 - 9c^2$

When finding the product of three or more factors, it is useful to first look for the pattern in which two binomials differ only in their sign. Finding this product first will make it easier to find the product of all the factors.

Example 4   Multiplying Polynomials

(a) $x(x - 3)(x + 3)$ These binomials differ only in the sign.

$= x(x^2 - 9)$

$= x^3 - 9x$

(b) $(x + 1)(x - 5)(x + 5)$ These binomials differ only in the sign.

$= (x + 1)(x^2 - 25)$ With two binomials, use the FOIL method.

$= x^3 + x^2 - 25x - 25$

(c) $(2x - 1)(x + 3)(2x + 1)$ These two binomials differ only in the sign of
                        the second term. We can use the commutative
$= (x + 3)(2x - 1)(2x + 1)$ property to rearrange the terms.

$= (x + 3)(4x^2 - 1)$

$= 4x^3 + 12x^2 - x - 3$

✔ **CHECK YOURSELF 4**

Multiply.

(a) $3x(x - 5)(x + 5)$ **(b)** $(x - 4)(2x + 3)(2x - 3)$

(c) $(x - 7)(3x - 1)(x + 7)$ $3x^3 - x^2 - 147x + 49$

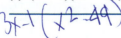

CHECK YOURSELF ANSWERS

1. (a) $4x^2 + 4x + 1$; **(b)** $16x^2 - 24xy + 9y^2$

2. (a) $x^2 + 10x + 25$; **(b)** $9a^2 + 12a + 4$; **(c)** $y^2 - 14y + 49$; **(d)** $25x^2 - 20xy + 4y^2$

3. (a) $a^2 - 36$; **(b)** $x^2 - 9y^2$; **(c)** $25n^2 - 4p^2$; **(d)** $49b^2 - 9c^2$

4. (a) $3x^3 - 75x$; **(b)** $4x^3 - 16x^2 - 9x + 36$; **(c)** $3x^3 - x^2 - 147x + 49$

3.5 Exercises

Find each of the following squares.

1. $(x + 5)^2$

2. $(y + 9)^2$

3. $(w - 6)^2$

4. $(a - 8)^2$

5. $(z + 12)^2$

6. $(p - 20)^2$

7. $(2a - 1)^2$

8. $(3x - 2)^2$

9. $(6m + 1)^2$

10. $(7b - 2)^2$

11. $(3x - y)^2$

12. $(5m + n)^2$

13. $(2r + 5s)^2$

14. $(3a - 4b)^2$

15. $(8a - 9b)^2$

16. $(7p + 6q)^2$

17. $\left(x + \dfrac{1}{2}\right)^2$

18. $\left(w - \dfrac{1}{4}\right)^2$

Find each of the following products.

19. $(x - 6)(x + 6)$

20. $(y + 8)(y - 8)$

21. $(m + 12)(m - 12)$

22. $(w - 10)(w + 10)$

23. $\left(x - \dfrac{1}{2}\right)\left(x + \dfrac{1}{2}\right)$

24. $\left(x + \dfrac{2}{3}\right)\left(x - \dfrac{2}{3}\right)$

ANSWERS

1. _____
2. _____
3. _____
4. _____
5. _____
6. _____
7. _____
8. _____
9. _____
10. _____
11. _____
12. _____
13. _____
14. _____
15. _____
16. _____
17. _____
18. _____
19. _____
20. _____
21. _____
22. _____
23. _____
24. _____

25. $(p - 0.4)(p + 0.4)$ **26.** $(m - 0.6)(m + 0.6)$

27. $(a - 3b)(a + 3b)$ **28.** $(p + 4q)(p - 4q)$

29. $(4r - s)(4r + s)$ **30.** $(7x - y)(7x + y)$

31. $(8w + 5z)(8w - 5z)$ **32.** $(7c + 2d)(7c - 2d)$

33. $(5x - 9y)(5x + 9y)$ **34.** $(6s - 5t)(6s + 5t)$

35. $x(x - 2)(x + 2)$ **36.** $a(a + 5)(a - 5)$

37. $2s(s - 3r)(s + 3r)$ **38.** $5w(2w - z)(2w + z)$

$5w(2w^2 - 2z) \quad 10w^3 - 5wz^2$

39. $5r(r + 3)^2$ **40.** $3x(x - 2)^2$

For each of the following problems, let x represent the number and then write an expression for the product.

41. The product of 6 more than a number and 6 less than that number

42. The square of 5 more than a number

43. The square of 4 less than a number

44. The product of 5 less than a number and 5 more than that number

Note that $(28)(32) = (30 - 2)(30 + 2) = 900 - 4 = 896$. Use this pattern to find each of the following products.

45. $(49)(51)$ **46.** $(27)(33)$

47. $(34)(26)$ **48.** $(98)(102)$

49. $(55)(65)$ **50.** $(64)(56)$

51. **Science and medicine.** Suppose an orchard is planted with trees in straight rows. If there are $5x - 4$ rows with $5x - 4$ trees in each row, how many trees are there in the orchard?

52. **Geometry.** A square has sides of length $3x - 2$ cm. Express the area of the square as a polynomial.

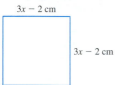

$3x - 2$ cm

$3x - 2$ cm

53. Complete the following statement: $(a + b)^2$ is not equal to $a^2 + b^2$ because.... But, wait! Isn't $(a + b)^2$ *sometimes* equal to $a^2 + b^2$? What do you think?

54. Is $(a + b)^3$ ever equal to $a^3 + b^3$? Explain.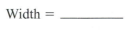

55. **Geometry.** In the following figures, identify the length, width, and area of the square:

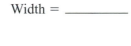

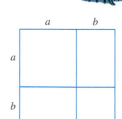

Length = _____

Width = _____

Area = _____

Length = _____

Width = _____

Area = _____

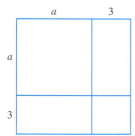

Length = _____

Width = _____

Area = _____

$(2x+3)(2x-3)$

$4x^2$

56. _____

a. _____

b. _____

c. _____

d. _____

e. _____

f. _____

g. _____

h. _____

56. Geometry. The square below is x units on a side. The area is _____.

Draw a picture of what happens when the sides are doubled. The area is _____.

Continue the picture to show what happens when the sides are tripled. The area is _____.

If the sides are quadrupled, the area is _____.

In general, if the sides are multiplied by n, the area is _____.

If each side is increased by 3, the area is increased by _____.

If each side is decreased by 2, the area is decreased by _____.

In general, if each side is increased by n, the area is increased by _____, and if each side is decreased by n, the area is decreased by _____.

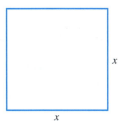

 Getting Ready for Section 3.6 [Section 1.7]

Divide.

(a) $\dfrac{2x^2}{2x}$ (b) $\dfrac{3a^3}{3a}$ (c) $\dfrac{6p^3}{2p^2}$

(d) $\dfrac{10m^4}{5m^2}$ (e) $\dfrac{20a^3}{5a^3}$ (f) $\dfrac{6x^2y}{3xy}$

(g) $\dfrac{12r^3s^2}{4rs}$ (h) $\dfrac{49c^4d^6}{7cd^3}$

Answers

1. $x^2 + 10x + 25$ **3.** $w^2 - 12w + 36$ **5.** $z^2 + 24z + 144$
7. $4a^2 - 4a + 1$ **9.** $36m^2 + 12m + 1$ **11.** $9x^2 - 6xy + y^2$

13. $4r^2 + 20rs + 25s^2$ **15.** $64a^2 - 144ab + 81b^2$ **17.** $x^2 + x + \dfrac{1}{4}$

19. $x^2 - 36$ **21.** $m^2 - 144$ **23.** $x^2 - \dfrac{1}{4}$ **25.** $p^2 - 0.16$

27. $a^2 - 9b^2$ **29.** $16r^2 - s^2$ **31.** $64w^2 - 25z^2$ **33.** $25x^2 - 81y^2$
35. $x^3 - 4x$ **37.** $2s^3 - 18r^2s$ **39.** $5r^3 + 30r^2 + 45r$ **41.** $x^2 - 36$
43. $x^2 - 8x + 16$ **45.** 2499 **47.** 884 **49.** 3575
51. $25x^2 - 40x + 16$ **53.**   **55.** **a.** x **b.** a^2
c. $3p$ **d.** $2m^2$ **e.** 4
f. $2x$ **g.** $3r^2s$ **h.** $7c^3d^3$

3.6 Dividing Polynomials

3.6 OBJECTIVES

1. Find the quotient when a polynomial is divided by a monomial.
2. Find the quotient when a polynomial is divided by a binomial.

In Section 1.7, we introduced the second property of exponents, which was used to divide one monomial by another monomial.

Step by Step: To Divide a Monomial by a Monomial

Step 1 Divide the coefficients.

Step 2 Use the second property of exponents to combine the variables.

OBJECTIVE 1

Example 1 Dividing by a Monomial

Divide: $\dfrac{8}{2} = 4$

RECALL The second property says: If x is not zero then

$$\frac{x^m}{x^n} = x^{m-n}$$

(a) $\dfrac{8x^4}{2x^2} = 4x^{4-2}$

Subtract the exponents.

$$= 4x^2$$

(b) $\dfrac{45a^5b^3}{9a^2b} = 5a^3b^2$

✔ **CHECK YOURSELF 1**

Divide.

(a) $\dfrac{16a^5}{8a^3}$ $2a^2$

(b) $\dfrac{28m^4n^3}{7m^3n}$ $4mn^2$

Now look at how this can be extended to divide any polynomial by a monomial. For example, to divide $12a^3 + 8a^2$ by $4a$, proceed as follows:

NOTE This step depends on the distributive property and the definition of division.

$$\frac{12a^3 + 8a^2}{4a} = \frac{12a^3}{4a} + \frac{8a^2}{4a}$$

Divide each term in the numerator by the denominator, $4a$.

Now do each division.

$$= 3a^2 + 2a$$

The work above leads us to the following rule.

Step by Step: To Divide a Polynomial by a Monomial

1. Divide each term of the polynomial by the monomial.
2. Simplify the results.

Example 2 **Dividing by Monomials**

Divide each term by 2.

(a) $\dfrac{4a^2 + 8}{2} = \dfrac{4a^2}{2} + \dfrac{8}{2}$

$= 2a^2 + 4$

Divide each term by $6y$.

(b) $\dfrac{24y^3 + (-18y^2)}{6y} = \dfrac{24y^3}{6y} + \dfrac{-18y^2}{6y}$

$= 4y^2 - 3y$

Remember the rules for signs in division.

(c) $\dfrac{15x^2 + 10x}{-5x} = \dfrac{15x^2}{-5x} + \dfrac{10x}{-5x}$

$= -3x - 2$

NOTE With practice you can just write the quotient.

(d) $\dfrac{14x^4 + 28x^3 - 21x^2}{7x^2} = \dfrac{14x^4}{7x^2} + \dfrac{28x^3}{7x^2} - \dfrac{21x^2}{7x^2}$

$= 2x^2 + 4x - 3$

(e) $\dfrac{9a^3b^4 - 6a^2b^3 + 12ab^4}{3ab} = \dfrac{9a^3b^4}{3ab} - \dfrac{6a^2b^3}{3ab} + \dfrac{12ab^4}{3ab}$

$= 3a^2b^3 - 2ab^2 + 4b^3$

CHECK YOURSELF 2

Divide.

(a) $\dfrac{20y^3 - 15y^2}{5y}$

(b) $\dfrac{8a^3 - 12a^2 + 4a}{-4a}$

(c) $\dfrac{16m^4n^3 - 12m^3n^2 + 8mn}{4mn}$

We are now ready to look at dividing one polynomial by another polynomial (with more than one term). The process is very much like long division in arithmetic, as Example 3 illustrates.

OBJECTIVE 2   **Example 3   Dividing by a Binomial**

Compare the steps in these two divisions.

Divide $x^2 + 7x + 10$ by $x + 2$.   *Divide 2176 by 32.*

NOTE The first term in the dividend, x^2, is divided by the first term in the divisor, x.

Step 1 $x + 2 \overline{)x^2 + 7x + 10}$ with x above   Divide x^2 by x to get x.

$$\dfrac{6}{32\overline{)2176}}$$

Step 2 $x + 2 \overline{)x^2 + 7x + 10}$
                    $\underline{x^2 + 2x}$

Multiply the divisor, $x + 2$, by x.

$$\begin{array}{r} 6 \\ 32\overline{)2176} \\ \underline{192} \end{array}$$

RECALL To subtract $x^2 + 2x$, mentally change each sign to $-x^2 - 2x$ and add. Take your time and be careful here. It's where most errors are made.

Step 3 $x + 2 \overline{)x^2 + 7x + 10}$
                    $\underline{x^2 + 2x}$
                    $\qquad\quad 5x + 10$

Subtract and bring down 10.

$$\begin{array}{r} 6 \\ 32\overline{)2176} \\ \underline{192} \\ 256 \end{array}$$

Step 4 $\qquad\qquad x + 5$
             $x + 2 \overline{)x^2 + 7x + 10}$
                    $\underline{x^2 + 2x}$
                    $\qquad\quad 5x + 10$ Divide $5x$ by x to get 5.

$$\begin{array}{r} 68 \\ 32\overline{)2176} \\ \underline{192} \\ \overline{256} \end{array}$$

NOTE We repeat the process until the degree of the remainder is less than that of the divisor or until there is no remainder.

Step 5 $\qquad\qquad x + 5$
             $x + 2 \overline{)x^2 + 7x + 10}$
                    $\underline{x^2 + 2x}$
                    $\qquad\quad 5x + 10$
                    $\qquad\quad \underline{5x + 10}$
                    $\qquad\qquad\qquad 0$

Multiply $x + 2$ by 5 and then subtract.

$$\begin{array}{r} 68 \\ 32\overline{)2176} \\ \underline{192} \\ 256 \\ \underline{256} \\ 0 \end{array}$$

The quotient is $x + 5$.

✔ CHECK YOURSELF 3 _____

Divide $x^2 + 9x + 20$ by $x + 4$.

In Example 3, we showed all the steps separately to help you see the process. In practice, the work can be shortened.

Example 4   Dividing by Binomials

Divide $x^2 + x - 12$ by $x - 3$.

NOTE You might want to write out a problem like $408 \div 17$ to compare the steps.

$$
\begin{array}{r}
x + 4 \\
x - 3{\overline{\smash{\big)}\,x^2 + x - 12}} \\
\underline{x^2 - 3x} \\
4x - 12 \\
\underline{4x - 12} \\
0
\end{array}
$$

Step 1 Divide x^2 by x to get x, the first term of the quotient.
Step 2 Multiply $x - 3$ by x.
Step 3 Subtract and bring down -12. Remember to mentally change the signs to $-x^2 + 3x$ and add.
Step 4 Divide $4x$ by x to get 4, the second term of the quotient.
Step 5 Multiply $x - 3$ by 4 and subtract.

The quotient is $x + 4$.

CHECK YOURSELF 4

Divide.

$(x^2 + 2x - 24) \div (x - 4)$

You may have a remainder in algebraic long division just as in arithmetic. Consider Example 5.

Example 5   Dividing by Binomials

Divide $4x^2 - 8x + 11$ by $2x - 3$.

$$
\begin{array}{r}
2x - 1 \quad \text{Quotient}\\
2x - 3{\overline{\smash{\big)}\,4x^2 - 8x + 11}} \\
\underline{4x^2 - 6x} \\
-\,2x + 11 \\
\underline{-\,2x + 3} \\
8
\end{array}
$$

Divisor

Remainder

This result can be written as

$$
\frac{4x^2 - 8x + 11}{2x - 3}
$$

$$
= 2x - 1 + \frac{8}{2x - 3} \quad \begin{array}{l}\leftarrow \text{Remainder}\\ \leftarrow \text{Divisor}\end{array}
$$

Quotient

CHECK YOURSELF 5

Divide.

$(6x^2 - 7x + 15) \div (3x - 5)$

The division process shown in our previous examples can be extended to dividends of a higher degree. The steps involved in the division process are exactly the same, as Example 6 illustrates.

Example 6   Dividing by Binomials

Divide $6x^3 + x^2 - 4x - 5$ by $3x - 1$.

$$
\begin{array}{r}
2x^2 + x - 1 \\
3x - 1\overline{)6x^3 + x^2 - 4x - 5} \\
\underline{6x^3 - 2x^2} \\
3x^2 - 4x \\
\underline{3x^2 - x} \\
-3x - 5 \\
\underline{-3x + 1} \\
-6
\end{array}
$$

The result can be written as

$$
\frac{6x^3 + x^2 - 4x - 5}{3x - 1} = 2x^2 + x - 1 + \frac{-6}{3x - 1}
$$

CHECK YOURSELF 6

Divide $4x^3 - 2x^2 + 2x + 15$ by $2x + 3$.

Suppose that the dividend is "missing" a term in some power of the variable. You can use 0 as the coefficient for the missing term. Consider Example 7.

Example 7   Dividing by Binomials

Divide $x^3 - 2x^2 + 5$ by $x + 3$.

$$
\begin{array}{r}
x^2 - 5x + 15 \\
x + 3\overline{)x^3 - 2x^2 + 0x + 5} \\
\underline{x^3 + 3x^2} \\
-5x^2 + 0x \\
\underline{-5x^2 - 15x} \\
15x + 5 \\
\underline{15x + 45} \\
-40
\end{array}
$$

Write 0x for the "missing" term in x.

This result can be written as

$$
\frac{x^3 - 2x^2 + 5}{x + 3} = x^2 - 5x + 15 + \frac{-40}{x + 3}
$$

✔ **CHECK YOURSELF 7**

Divide.

$(4x^3 + x + 10) \div (2x - 1)$

You should always arrange the terms of the divisor and dividend in descending-exponent form before starting the long-division process, as illustrated in Example 8.

Example 8   Dividing by Binomials

Divide $5x^2 - x + x^3 - 5$ by $-1 + x^2$.

Write the divisor as $x^2 - 1$ and the dividend as $x^3 + 5x^2 - x - 5$.

$$
\begin{array}{r}
x + 5 \\
x^2 - 1 \overline{)\, x^3 + 5x^2 - x - 5} \\
\underline{x^3 - x} \\
5x^2 - 5 \\
\underline{5x^2 - 5} \\
0
\end{array}
$$

Write $x^3 - x$, the product of x and $x^2 - 1$, so that like terms fall in the same columns.

The quotient is $x + 5$.

✔ **CHECK YOURSELF 8**

Divide:

$(5x^2 + 10 + 2x^3 + 4x) \div (2 + x^2)$

CHECK YOURSELF ANSWERS

1. (a) $2a^2$; **(b)** $4mn^2$ **2. (a)** $4y^2 - 3y$; **(b)** $-2a^2 + 3a - 1$; **(c)** $4m^3n^2 - 3m^2n + 2$

3. $x + 5$ **4.** $x + 6$ **5.** $2x + 1 + \dfrac{20}{3x - 5}$ **6.** $2x^2 - 4x + 7 + \dfrac{-6}{2x + 3}$

7. $2x^2 + x + 1 + \dfrac{11}{2x - 1}$ **8.** $2x + 5$

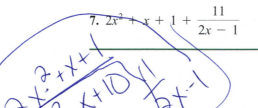

Name _____

Section _____ Date _____

Divide.

1. $\dfrac{18x^6}{9x^2}$ $2x^4$

2. $\dfrac{20a^7}{5a^5}$

3. $\dfrac{35m^3n^2}{7mn^2}$

4. $\dfrac{42x^5y^2}{6x^3y}$

5. $\dfrac{3a + 6}{3}$

6. $\dfrac{4x - 8}{4}$

7. $\dfrac{9b^2 - 12}{3}$

8. $\dfrac{10m^2 + 5m}{5}$

9. $\dfrac{16a^3 - 24a^2}{4a}$

10. $\dfrac{9x^3 + 12x^2}{3x}$

11. $\dfrac{12m^2 + 6m}{-3m}$

12. $\dfrac{20b^3 - 25b^2}{-5b}$

13. $\dfrac{18a^4 + 12a^3 - 6a^2}{6a}$

14. $\dfrac{21x^5 - 28x^4 + 14x^3}{7x}$

15. $\dfrac{20x^4y^2 - 15x^2y^3 + 10x^3y}{5x^2y}$

16. $\dfrac{16m^3n^3 + 24m^2n^2 - 40mn^3}{8mn^2}$

17. $\dfrac{27a^5b^5 + 9a^4b^4 - 3a^2b^3}{3a^2b^3}$

18. $\dfrac{7x^5y^5 - 21x^4y^4 + 14x^3y^3}{7x^3y^3}$

19. $\dfrac{3a^6b^4c^2 - 2a^4b^2c + 6a^3b^2c}{a^3b^2c}$

20. $\dfrac{2x^4y^4z^4 + 3x^3y^3z^3 - xy^2z^3}{xy^2z^3}$

Perform the indicated divisions.

21. $\dfrac{x^2 + 5x + 6}{x + 2}$

22. $\dfrac{x^2 + 8x + 15}{x + 3}$

23. $\dfrac{x^2 - x - 20}{x + 4}$

24. $\dfrac{x^2 - 2x - 35}{x + 5}$

ANSWERS

1. _____
2. _____
3. _____
4. _____
5. _____
6. _____
7. _____
8. _____
9. _____
10. _____
11. _____
12. _____
13. _____
14. _____
15. _____
16. _____
17. _____
18. _____
19. _____
20. _____
21. _____
22. _____
23. _____
24. _____

25. _____

26. _____

27. _____

28. _____

29. _____

30. _____

31. _____

32. _____

33. _____

34. _____

35. _____

36. _____

37. _____

38. _____

39. _____

40. _____

41. _____

42. _____

43. _____

44. _____

45. _____

46. _____

47. _____

48. _____

25. $\dfrac{2x^2 - 3x - 5}{x - 3}$

26. $\dfrac{3x^2 + 17x - 12}{x + 6}$

27. $\dfrac{6x^2 - x - 10}{3x - 5}$

28. $\dfrac{4x^2 + 6x - 25}{2x + 7}$

29. $\dfrac{x^3 + x^2 - 4x - 4}{x + 2}$

30. $\dfrac{x^3 - 2x^2 + 4x - 21}{x - 3}$

31. $\dfrac{4x^3 + 7x^2 + 10x + 5}{4x - 1}$

32. $\dfrac{2x^3 - 3x^2 + 4x + 4}{2x + 1}$

33. $\dfrac{x^3 - x^2 + 5}{x - 2}$

34. $\dfrac{x^3 + 4x - 3}{x + 3}$

35. $\dfrac{25x^3 + x}{5x - 2}$

36. $\dfrac{8x^3 - 6x^2 + 2x}{4x + 1}$

37. $\dfrac{2x^2 - 8 - 3x + x^3}{x - 2}$

38. $\dfrac{x^2 - 18x + 2x^3 + 32}{x + 4}$

39. $\dfrac{x^4 - 1}{x - 1}$

40. $\dfrac{x^4 + x^2 - 16}{x + 2}$

41. $\dfrac{x^3 - 3x^2 - x + 3}{x^2 - 1}$

42. $\dfrac{x^3 + 2x^2 + 3x + 6}{x^2 + 3}$

43. $\dfrac{x^4 + 2x^2 - 2}{x^2 + 3}$

44. $\dfrac{x^4 + x^2 - 5}{x^2 - 2}$

45. $\dfrac{y^3 + 1}{y + 1}$

46. $\dfrac{y^3 - 8}{y - 2}$

47. $\dfrac{x^4 - 1}{x^2 - 1}$

48. $\dfrac{x^6 - 1}{x^3 - 1}$

49. Find the value of c so that $\dfrac{y^2 - y + c}{y + 1} = y - 2$.

50. Find the value of c so that $\dfrac{x^3 + x^2 + x + c}{x^2 + 1} = x + 1$.

51. Write a summary of your work with polynomials. Explain how a polynomial is recognized and explain the rules for the arithmetic of polynomials—how to add, subtract, multiply, and divide. What parts of this chapter do you feel you understand very well, and what parts do you still have questions about or feel unsure of? Exchange papers with another student and compare your questions.

52. A funny (and useful) thing about division of polynomials: To find out about this funny thing, do this division. Compare your answer with another student's.

$(x - 2)\overline{)2x^2 + 3x - 5}$ Is there a remainder?

Now, evaluate the polynomial $2x^2 + 3x - 5$ when $x = 2$. Is this value the same as the remainder?

Try $(x + 3)\overline{)5x^2 - 2x + 1}$ Is there a remainder?

Evaluate the polynomial $5x^2 - 2x + 1$ when $x = -3$. Is this value the same as the remainder?
    What happens when there is no remainder?

Try $(x - 6)\overline{)3x^3 + 14x^2 - 23x + 6}$ Is the remainder zero?

Evaluate the polynomial $3x^3 + 14x^2 - 23x + 6$ when $x = 6$. Is this value zero? Write a description of the patterns you see. When does the pattern hold? Make up several more examples and test your conjecture.

53. (a) Divide $\dfrac{x^2 - 1}{x - 1}$.   (b) Divide $\dfrac{x^3 - 1}{x - 1}$.   (c) Divide $\dfrac{x^4 - 1}{x - 1}$.

(d) Based on your results on parts (a), (b), and (c), predict $\dfrac{x^{50} - 1}{x - 1}$.

54. (a) Divide $\dfrac{x^2 + x + 1}{x - 1}$.   (b) Divide $\dfrac{x^3 + x^2 + x + 1}{x - 1}$.

(c) Divide $\dfrac{x^4 + x^3 + x^2 + x + 1}{x - 1}$.

(d) Based on your results to (a), (b), and (c), predict $\dfrac{x^{10} + x^9 + x^8 + \cdots + x + 1}{x - 1}$.

Answers

1. $2x^4$ **3.** $5m^2$ **5.** $a + 2$ **7.** $3b^2 - 4$ **9.** $4a^2 - 6a$

11. $-4m - 2$ **13.** $3a^3 + 2a^2 - a$ **15.** $4x^2y - 3y^2 + 2x$

17. $9a^3b^2 + 3a^2b - 1$ **19.** $3a^3b^2c - 2a + 6$ **21.** $x + 3$

23. $x - 5$ **25.** $2x + 3 + \dfrac{4}{x - 3}$ **27.** $2x + 3 + \dfrac{5}{3x - 5}$

29. $x^2 - x - 2$ **31.** $x^2 + 2x + 3 + \dfrac{8}{4x - 1}$

33. $x^2 + x + 2 + \dfrac{9}{x - 2}$ **35.** $5x^2 + 2x + 1 + \dfrac{2}{5x - 2}$

37. $x^2 + 4x + 5 + \dfrac{2}{x - 2}$ **39.** $x^3 + x^2 + x + 1$ **41.** $x - 3$

43. $x^2 - 1 + \dfrac{1}{x^2 + 3}$ **45.** $y^2 - y + 1$ **47.** $x^2 + 1$ **49.** $c = -2$

51. ✎   **53. (a)** $x + 1$; **(b)** $x^2 + x + 1$; **(c)** $x^3 + x^2 + x + 1$;

(d) $x^{49} + x^{48} + \cdots + x + 1$

$x^2 + 8x + 15$

$x + 3$

$\boxed{x + 5}$

$x + 3 \overline{)\, x^2 + 8x + 15}$

$(x^2 + 3x)$

$5x + 15$

$5x + 15$

0

$(x + 5)(x + 3)$

$x^2 + 15 + 8x$

$2x - 3$

$3x - 5 \overline{)\, 6x^2 - 7x + 15}$

$6x^2 + 5x$

$-2x + 15$

$x + 5$

$x + 4 \overline{)\, x^2 + 9x + 20}$

$-x^2 - 4x$

$5x + 20$

$- \ 5x + 20$

0

$\dfrac{x^2}{x} \quad x(x + 4)$

$x^2 + 4x$

$\dfrac{5x}{x}$

$x + 6$

$x - 4 \overline{)\, x^2 - 2x - 24}$

$x^2 + 4x$

$2x - 24$

$\dfrac{x^2}{x} \quad x(x - 4)$

$x^2 - 4x$

$\dfrac{2x}{x}$

3 Summary

DEFINITION/PROCEDURE	EXAMPLE	REFERENCE
Exponents and Polynomials		**Section 3.1**
Properties of Exponents 1. $a^m \cdot a^n = a^{m+n}$	$3^3 \cdot 3^4 = 3^7$	*p. 273*
2. $\dfrac{a^m}{a^n} = a^{m-n}$	$\dfrac{x^6}{x^2} = x^4$	*p. 274*
3. $(a^m)^n = a^{mn}$	$(x^3)^5 = x^{15}$	*p. 274*
4. $(ab)^m = a^m b^m$	$(3x)^2 = 9x^2$	*p. 275*
5. $\left(\dfrac{a}{b}\right)^m = \dfrac{a^m}{b^m}$	$\left(\dfrac{2}{3}\right)^3 = \dfrac{8}{27}$	*p. 276*
Term A number, or the product of a number and variables. *Polynomial* An algebraic expression made up of terms in which the exponents are whole numbers. These terms are connected by plus or minus signs. Each sign (+ or −) is attached to the term following that sign.	$4x^3 - 3x^2 + 5x$ is a polynomial. The terms of $4x^3 - 3x^2 + 5x$ are $4x^3$, $-3x^2$, and $5x$.	*p. 277*
Coefficient In each term of a polynomial, the number is called the *numerical coefficient* or, more simply, the *coefficient* of that term.	The coefficients of $4x^3 - 3x^2$ are 4 and −3.	*p. 277*
Types of Polynomials A polynomial can be classified according to the number of terms it has. A *mono*mial has one term. A *bi*nomial has two terms. A *tri*nomial has three terms.	$2x^3$ is a monomial. $3x^2 - 7x$ is a binomial. $5x^5 - 5x^3 + 2$ is a trinomial.	*p. 278*
Degree The highest power of the variable appearing in any one term.	The degree of $4x^5 - 5x^3 + 3x$ is 5.	*p. 278*
Descending-Exponent Form The form of a polynomial when it is written with the highest-degree term first, the next highest-degree term second, and so on.	$4x^5 - 5x^3 + 3x$ is written in descending-exponent form.	*p. 279*
Negative Exponents and Scientific Notation		**Section 3.2**
The Zero Power Any expression raised to the zero power equals one.	$3^0 = 1$ $(5x)^0 = 1$	*p. 287*
Negative Powers An expression raised to a negative power equals its reciprocal taken to the absolute value of its power.	$\left(\dfrac{x}{3}\right)^{-4} = \left(\dfrac{3}{x}\right)^4 = \dfrac{3^4}{x^4}$	*p. 288*
Scientific Notation Any number written in the form $a \times 10^n$ in which $1 \le a < 10$ and n is an integer, is written in scientific notation.	6.2×10^{23}	*p. 291*

Continued

Adding and Subtracting Polynomials		Section 3.3
Removing Signs of Grouping 1. If a plus sign ($+$) or no sign at all appears in front of parentheses, just remove the parentheses. No other changes are necessary. 2. If a minus sign ($-$) appears in front of parentheses, the parentheses can be removed by changing the sign of each term inside the parentheses.	$3x + (2x - 3)$ $= 3x + 2x - 3$ $= 5x - 3$ $2x - (x - 4)$ $= 2x - x + 4$ $= x + 4$	p. 297 p. 299
Adding Polynomials Remove the signs of grouping. Then collect and combine any like terms.	$(2x + 3) + (3x - 5)$ $= 2x + 3 + 3x - 5 = 5x - 2$	p. 297
Subtracting Polynomials Remove the signs of grouping by changing the sign of each term in the polynomial being subtracted. Then combine any like terms.	$(3x^2 + 2x) - (2x^2 + 3x - 1)$ $= 3x^2 + 2x - 2x^2 - 3x + 1$ Sign changes $= 3x^2 - 2x^2 + 2x - 3x + 1$ $= x^2 - x + 1$	p. 299
Multiplying Polynomials		Section 3.4
To Multiply a Polynomial by a Monomial Multiply each term of the polynomial by the monomial and simplify the results.	$3x(2x + 3)$ $= 3x \cdot 2x + 3x \cdot 3$ $= 6x^2 + 9x$	p. 307
To Multiply a Binomial by a Binomial Use the FOIL method: $\quad\quad$ F $\quad$ O $\quad$ I $\quad$ L $(a + b)(c + d) = a \cdot c + a \cdot d + b \cdot c + b \cdot d$	$(2x - 3)(3x + 5)$ $= 6x^2 + 10x - 9x - 15$ $\quad$ F $\quad$ O $\quad$ I $\quad$ L $= 6x^2 + x - 15$	p. 309
To Multiply a Polynomial by a Polynomial Arrange the polynomials vertically. Multiply each term of the upper polynomial by each term of the lower polynomial and add the results.	$\begin{array}{r} x^2 - 3x + 5 \\ 2x - 3 \\ \hline -3x^2 + 9x - 15 \\ 2x^3 - 6x^2 + 10x \\ \hline 2x^3 - 9x^2 + 19x - 15 \end{array}$	p. 312
Special Products		Section 3.5
The Square of a Binomial $(a + b)^2 = a^2 + 2ab + b^2$	$(2x - 5)^2$ $= 4x^2 + 2 \cdot 2x \cdot (-5) + 25$ $= 4x^2 - 20x + 25$	p. 318
The Product of Binomials That Differ Only in Sign Subtract the square of the second term from the square of the first term. $(a + b)(a - b) = a^2 - b^2$	$(2x - 5y)(2x + 5y)$ $= (2x)^2 - (5y)^2$ $= 4x^2 - 25y^2$	p. 319
Dividing Polynomials		Section 3.6
To Divide a Polynomial by a Monomial 1. Divide each term of the polynomial by the monomial. 2. Simplify the result.	$\dfrac{27x^2y^2 + 9x^3y^4}{3xy^2}$ $= \dfrac{27x^2y^2}{3xy^2} + \dfrac{9x^3y^4}{3xy^2}$ $= 9x + 3x^2y^2$	p. 326

Summary Exercises

This summary exercise set is provided to give you practice with each of the objectives of this chapter. Each exercise is keyed to the appropriate chapter section. When you are finished, you can check your answer to the odd-numbered exercises against those presented in the back of the text. If you have difficulty with any of these questions, go back and reread the examples from that section. Your instructor will give you guidelines on how to best use these exercises in your instructional setting.

[3.1] Simplify each of the following expressions.

1. $\dfrac{x^{10}}{x^3}$

2. $\dfrac{a^5}{a^4}$

3. $\dfrac{x^2 \cdot x^3}{x^4}$

4. $\dfrac{m^2 \cdot m^3 \cdot m^4}{m^5}$

5. $\dfrac{18p^7}{9p^5}$

6. $\dfrac{24x^{17}}{8x^{13}}$

7. $\dfrac{30m^7n^5}{6m^2n^3}$

8. $\dfrac{108x^9y^4}{9xy^4}$

9. $\dfrac{48p^5q^3}{6p^3q}$

10. $\dfrac{52a^5b^3c^5}{13a^4c}$

11. $(2ab)^2$

12. $(p^2q^3)^3$

13. $(2x^2y^2)^3(3x^3y)^2$

14. $\left(\dfrac{p^2q^3}{t^4}\right)^2$

15. $\dfrac{(x^5)^2}{(x^3)^3}$

16. $(4w^2t)^2\,(3wt^2)^3$

17. $(y^3)^2(3y^2)^3$

18. $\left(\dfrac{4x^4}{3y}\right)^2$

Find the value of each of the following polynomials for the given value of the variable.

19. $5x + 1;\ x = -1$

20. $2x^2 + 7x - 5;\ x = 2$

21. $-x^2 + 3x - 1;\ x = 6$

22. $4x^2 + 5x + 7;\ x = -4$

Classify each polynomial as a monomial, binomial, or trinomial, where possible.

23. $5x^3 - 2x^2$

24. $7x^5$

25. $4x^5 - 8x^3 + 5$

26. $x^3 + 2x^2 - 5x + 3$

27. $9a^3 - 18a^2$

Arrange in descending-exponent form, if necessary, and give the degree of each polynomial.

28. $5x^5 + 3x^2$

29. $9x$

30. $6x^2 + 4x^4 + 6$

31. $5 + x$

32. -8

33. $9x^4 - 3x + 7x^6$

[3.2] Evaluate each of the following expressions.

34. 4^0

35. $(3a)^0$

36. $6x^0$

37. $(3a^4b)^0$

Write, using positive exponents. Simplify when possible.

38. x^{-5}

39. 3^{-3}

40. 10^{-4}

41. $4x^{-4}$

42. $\dfrac{x^6}{x^8}$

43. m^7m^{-9}

44. $\dfrac{a^{-4}}{a^{-9}}$

45. $\dfrac{x^2y^{-3}}{x^{-3}y^2}$

46. $(3m^{-3})^2$

47. $\dfrac{(a^4)^{-3}}{(a^{-2})^{-3}}$

In Exercises 48 to 50, express each number in scientific notation.

48. The average distance from the Earth to the sun is 150,000,000,000 m.

49. A bat emits a sound with a frequency of 51,000 cycles per second.

50. The diameter of a grain of salt is 0.000062 m.

In Exercises 51 to 54, compute the expression using scientific notation and express your answers in that form.

51. $(2.3 \times 10^{-3})(1.4 \times 10^{12})$

52. $(4.8 \times 10^{-10})(6.5 \times 10^{34})$

53. $\dfrac{(8 \times 10^{23})}{(4 \times 10^6)}$

54. $\dfrac{(5.4 \times 10^{-12})}{(4.5 \times 10^{16})}$

[3.3] Add.

55. $9a^2 - 5a$ and $12a^2 + 3a$

56. $5x^2 + 3x - 5$ and $4x^2 - 6x - 2$

57. $5y^3 - 3y^2$ and $4y + 3y^2$

Subtract.

58. $4x^2 - 3x$ from $8x^2 + 5x$

59. $2x^2 - 5x - 7$ from $7x^2 - 2x + 3$

60. $5x^2 + 3$ from $9x^2 - 4x$

Perform the indicated operations.

61. Subtract $5x - 3$ from the sum of $9x + 2$ and $-3x - 7$.

62. Subtract $5a^2 - 3a$ from the sum of $5a^2 + 2$ and $7a - 7$.

63. Subtract the sum of $16w^2 - 3w$ and $8w + 2$ from $7w^2 - 5w + 2$.

Add, using the vertical method.

64. $x^2 + 5x - 3$ and $2x^2 + 4x - 3$

65. $9b^2 - 7$ and $8b + 5$

66. $x^2 + 7$, $3x - 2$, and $4x^2 - 8x$

Subtract, using the vertical method.

67. $5x^2 - 3x + 2$ from $7x^2 - 5x - 7$

68. $8m - 7$ from $9m^2 - 7$

[3.4]   Multiply.

69. $(5a^3)(a^2)$

70. $(2x^2)(3x^5)$

71. $(-9p^3)(-6p^2)$

72. $(3a^2b^3)(-7a^3b^4)$

73. $5(3x - 8)$

74. $4a(3a + 7)$

75. $(-5rs)(2r^2s - 5rs)$

76. $7mn(3m^2n - 2mn^2 + 5mn)$

77. $(x + 5)(x + 4)$

78. $(w - 9)(w - 10)$

79. $(a - 7b)(a + 7b)$

80. $(p - 3q)^2$

81. $(a + 4b)(a + 3b)$

82. $(b - 8)(2b + 3)$

83. $(3x - 5y)(2x - 3y)$

84. $(5r + 7s)(3r - 9s)$

85. $(y + 2)(y^2 - 2y + 3)$

86. $(b + 3)(b^2 - 5b - 7)$

87. $(x - 2)(x^2 + 2x + 4)$

88. $(m^2 - 3)(m^2 + 7)$

89. $2x(x + 5)(x - 6)$

90. $a(2a - 5b)(2a - 7b)$

[3.5]   Find the following products.

91. $(x + 7)^2$

92. $(a - 8)^2$

93. $(2w - 5)^2$

94. $(3p + 4)^2$

95. $(a + 7b)^2$

96. $(8x - 3y)^2$

97. $(x - 5)(x + 5)$

98. $(y + 9)(y - 9)$

99. $(2m + 3)(2m - 3)$

100. $(3r - 7)(3r + 7)$

101. $(5r - 2s)(5r + 2s)$

102. $(7a + 3b)(7a - 3b)$

103. $2x(x - 5)^2$

104. $3c(c + 5d)(c - 5d)$

[3.6] Divide.

105. $\dfrac{9a^5}{3a^2}$

106. $\dfrac{24m^4n^2}{6m^2n}$

107. $\dfrac{15a - 10}{5}$

108. $\dfrac{32a^3 + 24a}{8a}$

109. $\dfrac{9r^2s^3 - 18r^3s^2}{-3rs^2}$

110. $\dfrac{35x^3y^2 - 21x^2y^3 + 14x^3y}{7x^2y}$

Perform the indicated long division.

111. $\dfrac{x^2 - 2x - 15}{x + 3}$

112. $\dfrac{2x^2 + 9x - 35}{2x - 5}$

113. $\dfrac{x^2 - 8x + 17}{x - 5}$

114. $\dfrac{6x^2 - x - 10}{3x + 4}$

115. $\dfrac{6x^3 + 14x^2 - 2x - 6}{6x + 2}$

116. $\dfrac{4x^3 + x + 3}{2x - 1}$

117. $\dfrac{3x^2 + x^3 + 5 + 4x}{x + 2}$

118. $\dfrac{2x^4 - 2x^2 - 10}{x^2 - 3}$

Self-Test for Chapter 3

Name _____

Section _____ Date _____

The purpose of this self-test is to help you check your progress and to review for the next in-class exam. Allow yourself about an hour to take this test. At the end of that hour check your answers against those given in the back of the text. Section references accompany the answers. If you missed any questions, go back to those sections and reread the examples until you master the concepts.

Simplify each of the following expressions.

1. $a^5 \cdot a^9$

2. $3x^2y^3 \cdot 5xy^4$

3. $\dfrac{4x^5}{2x^2}$

4. $\dfrac{20a^3b^5}{5a^2b^2}$

5. $(3x^2y)^3$

6. $\left(\dfrac{2w^2}{3t^3}\right)^2$

7. $(2x^3y^2)^4(x^2y^3)^3$

8. Find the value of the polynomial $y = -3x^2 - 5x + 8$ if $x = -2$.

Classify each of the following polynomials as a monomial, binomial, or trinomial.

9. $6x^2 + 7x$

10. $5x^2 + 8x - 8$

Arrange in descending-exponent form then give the coefficients and degree of the polynomial.

11. $-3x^2 + 8x^4 - 7$

Evaluate (assume the variables are nonzero).

12. 8^0

13. $6x^0$

Rewrite, using positive exponents. Simplify when possible.

14. y^{-5}

15. $3b^{-7}$

16. y^4y^{-8}

17. $\dfrac{p^{-5}}{p^5}$

Add.

18. $3x^2 - 7x + 2$ and $7x^2 - 5x - 9$

19. $7a^2 - 3a$ and $7a^3 + 4a^2$

ANSWERS

1. _____
2. _____
3. _____
4. _____
5. _____
6. _____
7. _____
8. _____
9. _____
10. _____
11. _____
12. _____
13. _____
14. _____
15. _____
16. _____
17. _____
18. _____
19. _____

20. _____

21. _____

22. _____

23. _____

24. _____

25. _____

26. _____

27. _____

28. _____

29. _____

30. _____

31. _____

32. _____

33. _____

34. _____

35. _____

36. _____

37. _____

38. _____

39. _____

40. _____

Subtract.

20. $5x^2 - 2x + 5$ from $8x^2 + 9x - 7$ **21.** $2b^2 + 5$ from $3b^2 - 7b$

22. $5a^2 + a$ from the sum of $3a^2 - 5a$ and $9a^2 - 4a$

Add, using the vertical method.

23. $x^2 + 3$, $5x - 7$, and $3x^2 - 2$

Subtract, using the vertical method.

24. $3x^2 - 5$ from $5x^2 - 7x$

Multiply.

25. $5ab(3a^2b - 2ab + 4ab^2)$ **26.** $(x - 2)(3x + 7)$

27. $(2x + y)(x^2 + 3xy - 2y^2)$ **28.** $(4x + 3y)(2x - 5y)$

29. $x(3x - y)(4x + 5y)$

30. $(3m + 2n)^2$ **31.** $(a - 7b)(a + 7b)$

Divide.

32. $\dfrac{14x^3y - 21xy^2}{7xy}$ **33.** $\dfrac{20c^3d - 30cd + 45c^2d^2}{5cd}$

34. $(x^2 - 2x - 24) \div (x + 4)$ **35.** $(2x^2 + x + 4) \div (2x - 3)$

36. $(6x^3 - 7x^2 + 3x + 9) \div (3x + 1)$ **37.** $(x^3 - 5x^2 + 9x - 9) \div (x - 1)$

Compute and answer using scientific notation.

38. $(2.1 \times 10^7)(8 \times 10^{12})$

39. $(6 \times 10^{-23})(5.2 \times 10^{12})$

40. $\dfrac{7.28 \times 10^3}{1.4 \times 10^{-16}}$

ACTIVITY 3: ISBNS AND THE CHECK DIGIT

Each activity in this text is designed to either enhance your understanding of the topics of the preceding chapter, provide you with a mathematical extension of those topics, or both. The activities can be undertaken by one student, but they are better suited for a small group project. Occasionally it is only through discussion that different facets of the activity become apparent. For material related to this activity, visit the text website at www.mhhe.com/streeter.

Have you ever noticed the long number, usually accompanied by a bar code, that can be found on the back of a book? This is called the International Standard Book Number (ISBN). Each book has a unique ISBN, which is 10 digits in length. The most common form for an ISBN is X-XX-XXXXXX-X.

Each ISBN has four blocks of digits.

• The first block of digits on the left represents the language of the book (0 is used to represent English). This block is usually one digit in length.
• The second block of digits represents the publisher. This block is usually two or three digits in length.
• The third block of digits represents the number assigned to the book by the publishing company. This is usually five or six digits in length.
• The fourth block consists of the check digit.

780697 069795

For the purposes of this document, we will consider the ISBN 0-07-229654-2 which is the ISBN for *Elementary and Intermediate Algebra* by Hutchison, Bergman, and Hoelzle.

The digit of most interest is the final digit, the **check digit.** When an ISBN is entered into a computer, the computer looks at the check digit to make certain that the numbers were properly entered. We will look at the algorithm used to generate the check digit.

Activity

I. To determine the check digit for the ISBN of a text, follow these steps.

1. Multiply each of the nine assigned digits by a weighted value. The weighted values are 1 for the first digit from the left, 2 for the second digit, 3 for third digit, etc. In our case, we have

$$1 \cdot 0 + 2 \cdot 0 + 3 \cdot 7 + 4 \cdot 2 + 5 \cdot 2 + 6 \cdot 9 + 7 \cdot 6 + 8 \cdot 5 + 9 \cdot 4 = 211$$

2. In mathematics, we occasionally are interested in only the remainder after we do division. We refer to the remainder as the **modular** (or mod) of the divisor. The ISBN uses (mod 11) to determine the check digit.

$$211 \div 11 = 19 \text{ with a remainder of 2. We say } 211 = 2 \pmod{11}$$

That is how we get the 2 in the ISBN 0-07-229654-2.

II. If we are given an ISBN and we want to check its validity, we can follow a similar algorithm.

1. Multiply each of the 10 digits by a weighted value. The weighted values are still 1 for the first digit from the left, 2 for the second digit, etc., but we also multiply 10 times the tenth (check) digit.

In our case, we now have

$$1 \cdot 0 + 2 \cdot 0 + 3 \cdot 7 + 4 \cdot 2 + 5 \cdot 2 + 6 \cdot 9 + 7 \cdot 6 + 8 \cdot 5 + 9 \cdot 4 + 10 \cdot 2 = 231$$

2. For any valid ISBN, the result will have a remainder of zero when divided by 11. In other words, we will have 0 (mod 11). Note that $231 \div 11 = 21$.

Determine whether each of the following is a valid ISBN.

0-07-038023-6
0-15-249584-2
0-553-34948-1
0-07-000317-3

For those numbers above that are valid, go on-line and find the books to which they refer.

Cumulative Review
Chapters 0–3

The following questions are presented to help you review concepts from earlier chapters. This is meant as a review and not as a comprehensive exam. The answers are presented in the back of the text. Section references accompany the answers. If you have difficulty with any of these questions, be certain to at least read through the summary related to those sections.

ANSWERS

Perform the indicated operations.

1. $8 - (-9)$ **2.** $-26 + 32$ **3.** $(-25)(-6)$ **4.** $(-48) \div (-12)$

Evaluate the expressions if $x = -2$, $y = 5$, and $z = -2$.

5. $-5(-3y - 2z)$

6. $\dfrac{3x - 4y}{2z + 5y}$

Use the properties of exponents to simplify each of the following expressions.

7. $(3x^2)^2 (x^3)^4$ **8.** $\left(\dfrac{x^5}{y^3}\right)^2$ **9.** $(2x^3y)^3$

10. $7y^0$ **11.** $(3x^4y^5)^0$

Simplify each expression using positive exponents only.

12. x^{-4} **13.** $3x^{-2}$ **14.** x^5x^{-9} **15.** $\dfrac{x^{-3}}{y^3}$

Simplify each of the following expressions.

16. $21x^5y - 17x^5y$ **17.** $(3x^2 + 4x - 5) - (2x^2 - 3x - 5)$

18. $3x + 2y - x - 4y$ **19.** $(x + 3)(x - 5)$

20. $(x + y)^2$ **21.** $(3x - 4y)^2$

22. $\dfrac{x^2 + 2x - 8}{x - 2}$ **23.** $x(x + y)(x - y)$

ANSWERS	
1.	
2.	
3.	
4.	
5.	
6.	
7.	
8.	
9.	
10.	
11.	
12.	
13.	
14.	
15.	
16.	
17.	
18.	
19.	
20.	
21.	
22.	
23.	

ANSWERS

24. _____

25. _____

26. _____

27. _____

28. _____

29. _____

30. _____

31. _____

32. _____

33. _____

34. _____

Solve the following equations.

24. $7x - 4 = 3x - 12$

25. $3x + 2 = 4x + 4$

26. $\dfrac{3}{4}x - 2 = 5 + \dfrac{2}{3}x$

27. $6(x - 1) - 3(1 - x) = 0$

28. Solve the equation $A = \dfrac{1}{2}(b + B)$ for B.

Solve the following inequalities.

29. $-5x - 7 \le 3x + 9$

30. $-3(x + 5) > -2x + 7$

Solve the following problems.

31. Sam made \$10 more than twice what Larry earned in one month. If together they earned \$760, how much did each earn that month?

32. The sum of two consecutive odd integers is 76. Find the two integers.

33. Two-fifths of a woman's income each month goes to taxes. If she pays \$848 in taxes each month, what is her monthly income?

34. The retail selling price of a sofa is \$806.25. What is the cost to the dealer if she sells at 25% markup on the cost?

FACTORING

INTRODUCTION

Developing secret codes is big business because of the widespread use of computers and the Internet. Corporations all over the world sell encryption systems that are supposed to keep data secure and safe.

In 1977, three professors from the Massachusetts Institute of Technology developed an encryption system they call RSA, a name derived from the first letters of their last names. They offered a $100 reward to anyone who could break their security code, which was based on a number that has *129 digits*. They called the code RSA-129. For the code to be broken, the 129-digit number must be factored into two prime numbers; that is, two prime numbers must be found that when multiplied together give the 129-digit number. The three professors predicted that it would take *40 quadrillion* years to find the two numbers.

In April 1994, a research scientist, three computer hobbyists, and more than 600 volunteers from the Internet, using 1600 computers, found the two numbers after 8 months of work and won the $100.

A data security company says that people who are using their system are safe because as yet no truly efficient algorithm for finding prime factors of massive numbers has been found, although one may someday exist. This company, hoping to test its encrypting system, now sponsors contests challenging people to factor more very large numbers into two prime numbers. RSA-576 up to RSA-2048 are being worked on now.

Software companies are waging a legal battle against the U.S. government because the government does not allow any codes to be used for which it does not have the key. The software firms claim that this prohibition is costing them about $60 billion in lost sales because many companies will not buy an encryption system knowing they can be monitored by the U.S. government.

Pre-Test Chapter 4

This pre-test will provide a preview of the types of exercises you will encounter in each section of this chapter. The answers for these exercises can be found in the back of the text. If you are working on your own, or ahead of the class, this pre-test can help you identify the sections in which you should focus more of your time.

1. _____

2. _____

Factor each of the following polynomials.

3. _____

4. _____

[4.1]　**1.** $15c + 35$　　　　　　　　**2.** $8q^4 - 20q^3$

5. _____

3. $6x^2 - 12x + 24$　　　　　　**4.** $7c^3d^2 - 21cd + 14cd^3$

6. _____

Factor each of the following trinomials.

7. _____

[4.2]　**5.** $b^2 + 2b - 15$　　　　　　**6.** $x^2 + 10x + 24$

8. _____

7. $x^2 - 14x + 45$　　　　　　**8.** $a^2 + 7ab + 12b^2$

9. _____

Factor each of the following trinomials completely.

10. _____

[4.3]　**9.** $3y^2 + 5y - 12$　　　　　**10.** $5w^2 + 23w + 12$

11. _____

11. $6x^2 + 5xy - 21y^2$　　　　**12.** $2x^3 - 7x^2 - 15x$

12. _____

Factor each of the following polynomials completely.

13. _____

[4.4]　**13.** $b^2 - 49$　　　　　　　　**14.** $36p^2 - q^2$

14. _____

15. $9x^2 - 12xy + 4y^2$　　　　**16.** $27xy^2 - 48x^3$

15. _____

16. _____

Solve each of the following equations.

17. _____

[4.8]　**17.** $x^2 - 11x + 28 = 0$　　　　**18.** $x^2 - 5x = 14$

18. _____

19. $5x^2 + 7x - 6 = 0$　　　　**20.** $9p^2 - 18p = 0$

19. _____

20. _____

4.1 An Introduction to Factoring

Overcoming Math Anxiety

Hint #5

Working Together

How many of your classmates do you know? Whether you are by nature gregarious or shy, you have much to gain by getting to know your classmates.

1. It is important to have someone to call when you miss class or are unclear on an assignment.

2. Working with another person is almost always beneficial to both people. If you don't understand something, it helps to have someone to ask about it. If you do understand something, nothing will cement that understanding more than explaining the idea to another person.

3. Sometimes we need to commiserate. If an assignment is particularly frustrating, it is reassuring to find that it is also frustrating for other students.

4. Have you ever thought you had the right answer, but it doesn't match the answer in the text? Frequently the answers are equivalent, but that's not always easy to see. A different perspective can help you see that. Occasionally there is an error in a textbook (here we are talking about *other* textbooks). In such cases it is wonderfully reassuring to find that someone else has the same answer you do.

In Chapter 3 you were given factors and asked to find a product. We are now going to reverse the process. You will be given a polynomial and asked to find its factors. This is called **factoring.**

Let's start with an example from arithmetic. To *multiply* $5 \cdot 7$, you write

$$5 \cdot 7 = 35$$

To *factor* 35, you would write

$$35 = 5 \cdot 7$$

Factoring is the *reverse* of multiplication.

Now let's look at factoring in algebra. You have used the distributive property as

$$a(b + c) = ab + ac$$

For instance,

NOTE 3 and $x + 5$ are the factors of $3x + 15$.

$$3(x + 5) = 3x + 15$$

To use the distributive property in factoring, we reverse that property as

$$ab + ac = a(b + c)$$

The property lets us remove the common factor a from the terms of $ab + ac$. To use this in factoring, the first step is to see whether each term of the polynomial has a common monomial factor. In our earlier example,

$$3x + 15 = 3 \cdot x + 3 \cdot 5$$

Common factor

So, by the distributive property,

$$3x + 15 = 3(x + 5)$$ The original terms are each divided by the greatest common factor to determine the terms in parentheses.

NOTE Again, factoring is the reverse of multiplication.

To check this, multiply $3(x + 5)$.

Multiplying

$$3(x + 5) = 3x + 15$$

Factoring

The first step in factoring is to identify the *greatest common factor* (GCF) of a set of terms. This factor is the product of the largest common numerical coefficient and the largest common factor of each variable.

NOTE Factoring out the GCF is the *first* method to try in any of the factoring problems we will discuss.

Definition: Greatest Common Factor

NOTE If a variable is not in a term, its exponent is zero.

The **greatest common factor (GCF)** of a polynomial is the factor that is a product of the largest numerical coefficient factor of the polynomial and each variable with the smallest exponent in any term.

OBJECTIVE 1

Example 1 Finding the GCF

Find the GCF for each set of terms.

(a) 9 and 12   The largest number that is a factor of both is 3.
(b) 10, 25, 150   The GCF is 5.
(c) x^4 and x^7

$$x^4 = \boxed{x} \cdot \boxed{x} \cdot \boxed{x} \cdot \boxed{x}$$
$$x^7 = \boxed{x} \cdot \boxed{x} \cdot \boxed{x} \cdot \boxed{x} \cdot x \cdot x \cdot x$$

The largest power that divides both terms is x^4.

(d) $12a^3$ and $18a^2$

$$12a^3 = 2 \cdot \boxed{2} \cdot \boxed{3} \cdot \boxed{a} \cdot \boxed{a} \cdot a$$
$$18a^2 = \boxed{2} \cdot \boxed{3} \cdot 3 \cdot \boxed{a} \cdot \boxed{a}$$

The GCF is $6a^2$.

CHECK YOURSELF 1

Find the GCF for each set of terms.

(a) 14, 24   **(b)** 9, 27, 81   **(c)** a^9, a^5 **(d)** $10x^5, 35x^4$

Step by Step: To Factor a Monomial from a Polynomial

NOTE Checking your answer is always important and perhaps is never easier than after you have factored.

Step 1 Find the GCF for all the terms.
Step 2 Use the GCF to factor each term and then apply the distributive property.
Step 3 Mentally check your factoring by multiplication.

Example 2   Finding the GCF of a Binomial

(a) Factor $8x^2 + 12x$.

The largest common numerical factor of 8 and 12 is 4, and x is the common variable factor with the largest power. So $4x$ is the GCF. Write

$$8x^2 + 12x = 4x \cdot 2x + 4x \cdot 3$$

⎣_____⎦ GCF

NOTE It is always a good idea to check your answer by multiplying to make sure that you get the **original** polynomial. Try it here. Multiply $4x$ by $2x + 3$.

Now, by the distributive property, we have

$$8x^2 + 12x = 4x(2x + 3)$$

(b) Factor $6a^4 - 18a^2$.

The GCF in this case is $6a^2$. Write

$$6a^4 + (-18a^2) = 6a^2 \cdot a^2 + 6a^2 \cdot (-3)$$

⎣_____⎦ GCF

NOTE It is also true that $6a^4 + (-18a^2) = 3a(2a^3 + (-6a))$. However, this is *not completely factored.* Do you see why? You want to find the common monomial factor with the *largest possible* coefficient and the *largest* exponent, in this case $6a^2$.

Again, using the distributive property yields

$$6a^4 - 18a^2 = 6a^2(a^2 - 3)$$

You should check this by multiplying.

CHECK YOURSELF 2

Factor each of the following polynomials.

(a) $5x + 20$ **(b)** $6x^2 - 24x$ **(c)** $10a^3 - 15a^2$

The process is exactly the same for polynomials with more than two terms. Consider Example 3.

Example 3   Finding the GCF of a Polynomial

(a) Factor $5x^2 - 10x + 15$.

NOTE The GCF is 5.

$$5x^2 - 10x + 15 = 5 \cdot x^2 - 5 \cdot 2x + 5 \cdot 3$$

⎣_____⎦ GCF

$$= 5(x^2 - 2x + 3)$$

(b) Factor $6ab + 9ab^2 - 15a^2$.

NOTE The GCF is $3a$.

$$6ab + 9ab^2 - 15a^2 = 3a \cdot 2b + 3a \cdot 3b^2 - 3a \cdot 5a$$

⎣_____⎦ GCF

$$= 3a(2b + 3b^2 - 5a)$$

(c) Factor $4a^4 + 12a^3 - 20a^2$.

NOTE The GCF is $4a^2$.

$$4a^4 + 12a^3 - 20a^2 = 4a^2 \cdot a^2 + 4a^2 \cdot 3a - 4a^2 \cdot 5$$

⎣_____⎦ GCF

$$= 4a^2(a^2 + 3a - 5)$$

NOTE In each of these examples, you should check the result by multiplying the factors.

(d) Factor $6a^2b + 9ab^2 + 3ab$.

Mentally note that 3, a, and b are factors of each term, so

$6a^2b + 9ab^2 + 3ab = 3ab(2a + 3b + 1)$

CHECK YOURSELF 3 _____

Factor each of the following polynomials.

(a) $8b^2 + 16b - 32$ **(b)** $4xy - 8x^2y + 12x^3$
(c) $7x^4 - 14x^3 + 21x^2$ **(d)** $5x^2y^2 - 10xy^2 + 15x^2y$

We can have two or more terms that have a binomial factor in common, as is the case in Example 4.

OBJECTIVE 2   **Example 4**   **Finding a Common Factor**

(a) Factor $3x(x + y) + 2(x + y)$.

We see that *the binomial $x + y$ is a common factor* and can be removed.

NOTE Because of the commutative property, the factors can be written in either order.

$3x(x + y) + 2(x + y)$
$= (x + y) \cdot 3x + (x + y) \cdot 2$
$= (x + y)(3x + 2)$

(b) Factor $3x^2(x - y) + 6x(x - y) + 9(x - y)$.

We note that here the GCF is $3(x - y)$. Factoring as before, we have

$3(x - y)(x^2 + 2x + 3)$

CHECK YOURSELF 4 _____

Completely factor each of the polynomials.

(a) $7a(a - 2b) + 3(a - 2b)$ **(b)** $4x^2(x + y) - 8x(x + y) - 16(x + y)$

CHECK YOURSELF ANSWERS _____

1. (a) 2; **(b)** 9; **(c)** a^5; **(d)** $5x^4$ **2. (a)** $5(x + 4)$; **(b)** $6x(x - 4)$; **(c)** $5a^2(2a - 3)$
3. (a) $8(b^2 + 2b - 4)$; **(b)** $4x(y - 2xy + 3x^2)$; **(c)** $7x^2(x^2 - 2x + 3)$;
(d) $5xy(xy - 2y + 3x)$ **4. (a)** $(a - 2b)(7a + 3)$; **(b)** $4(x + y)(x^2 - 2x - 4)$

4.1 Exercises

Name _____

Section _____ Date _____

Find the greatest common factor for each of the following sets of terms.

1. 10, 12

2. 15, 35

3. 16, 32, 88

4. 55, 33, 132

5. x^2, x^5

6. y^7, y^9

7. a^3, a^6, a^9

8. b^4, b^6, b^8

9. $5x^4, 10x^5$

10. $8y^9, 24y^3$

11. $8a^4, 6a^6, 10a^{10}$

12. $9b^3, 6b^5, 12b^4$

13. $9x^2y, 12xy^2, 15x^2y^2$

14. $12a^3b^2, 18a^2b^3, 6a^4b^4$

15. $15ab^3, 10a^2bc, 25b^2c^3$

16. $9x^2, 3xy^3, 6y^3$

17. $15a^2bc^2, 9ab^2c^2, 6a^2b^2c^2$

18. $18x^3y^2z^3, 27x^4y^2z^3, 81xy^2z$

19. $(x + y)^2, (x + y)^3$

20. $12(a + b)^4, 4(a + b)^3$

Factor each of the following polynomials.

21. $8a + 4$

22. $5x - 15$

23. $24m - 32n$

24. $7p - 21q$

25. $12m + 8$

26. $24n - 32$

27. $10s^2 + 5s$

28. $12y^2 - 6y$

29. $12x^2 + 12x$

30. $14b^2 + 14b$

31. $15a^3 - 25a^2$

32. $36b^4 + 24b^2$

33. $6pq + 18p^2q$

34. $8ab - 24ab^2$

ANSWERS

1. _____   2. _____
3. _____   4. _____
5. _____   6. _____
7. _____   8. _____
9. _____   10. _____
11. _____ 12. _____
13. _____ 14. _____
15. _____ 16. _____
17. _____
18. _____
19. _____
20. _____
21. _____
22. _____
23. _____
24. _____
25. _____
26. _____
27. _____
28. _____
29. _____
30. _____
31. _____
32. _____
33. _____
34. _____

35. _____

36. _____

37. _____

38. _____

39. _____

40. _____

41. _____

42. _____

43. _____

44. _____

45. _____

46. _____

47. _____

48. _____

49. _____

50. _____

51. _____

52. _____

53. _____

54. _____

55. _____

56. _____

57. _____

58. _____

59. _____

60. _____

35. $7m^3n - 21mn^3$

36. $36p^2q^2 - 9pq$

37. $6x^2 - 18x + 30$

38. $7a^2 + 21a - 42$

39. $3a^3 + 6a^2 - 12a$

40. $5x^3 - 15x^2 + 25x$

41. $6m + 9mn - 12mn^2$

42. $4s + 6st - 14st^2$

43. $10x^2y + 15xy - 5xy^2$

44. $3ab^2 + 6ab - 15a^2b$

45. $10r^3s^2 + 25r^2s^2 - 15r^2s^3$

46. $28x^2y^3 - 35x^2y^2 + 42x^3y$

47. $9a^5 - 15a^4 + 21a^3 - 27a$

48. $8p^6 - 40p^4 + 24p^3 - 16p^2$

49. $a(a + 2) - 3(a + 2)$

50. $b(b - 5) + 2(b - 5)$

51. $x(x - 2) + 3(x - 2)$

52. $y(y + 5) - 3(y + 5)$

53. The GCF of $2x - 6$ is 2. The GCF of $5x + 10$ is 5. Find the GCF of the product $(2x - 6)(5x + 10)$.

54. The GCF of $3z + 12$ is 3. The GCF of $4z + 8$ is 4. Find the GCF of the product $(3z + 12)(4z + 8)$.

55. The GCF of $2x^3 - 4x$ is $2x$. The GCF of $3x + 6$ is 3. Find the GCF of the product $(2x^3 - 4x)(3x + 6)$.

56. State, in a sentence, the rule that exercises 53 to 55 illustrated.

Find the GCF of each product.

57. $(2a + 8)(3a - 6)$

58. $(5b - 10)(2b + 4)$

59. $(2x^2 + 5x)(7x - 14)$

60. $(6y^2 - 3y)(y + 7)$

61. Geometry. The area of a rectangle with width t is given by $33t - t^2$. Factor the expression and determine the length of the rectangle in terms of t.

62. Geometry. The area of a rectangle of length x is given by $3x^2 + 5x$. Find the width of the rectangle.

63. Number problem. For centuries, mathematicians have found factoring numbers into prime factors a fascinating subject. A prime number is a number that cannot be written as a product of any numbers but 1 and itself. The list of primes begins with 2 because 1 is not considered a prime number and then goes on: 3, 5, 7, 11, . . . What are the first 10 primes? What are the primes less than 100? If you list the numbers from 1 to 100 and then cross out all numbers that are multiples of 2, 3, 5, and 7, what is left? Are all the numbers not crossed out prime? Write a paragraph to explain why this might be so. You might want to investigate the sieve of Eratosthenes, a system from 230 B.C.E. for finding prime numbers.

64. Number problem. If we could make a list of all the prime numbers, what number would be at the end of the list? Because there are an infinite number of prime numbers, there is no "largest prime number." But is there some formula that will give us all the primes? Here are some formulas proposed over the centuries:

$$n^2 + n + 17 \qquad 2n^2 + 29 \qquad n^2 - n + 11$$

In all these expressions, $n = +1, 2, 3, 4, \ldots$, that is, a positive integer beginning with 1. Investigate these expressions with a partner. Do the expressions give prime numbers when they are evaluated for these values of n? Do the expressions give *every* prime in the range of resulting numbers? Can you put in *any* positive number for n?

65. Number problem. How are primes used in coding messages and for security? Work together to decode the messages. The messages are coded using this code: After the numbers are factored into prime factors, the power of 2 gives the number of the letter in the alphabet. This code would be easy for a code breaker to figure out. Can you make up code that would be more difficult to break?

a. 1310720, 229376, 1572864, 1760, 460, 2097152, 336

b. 786432, 286, 4608, 278528, 1344, 98304, 1835008, 352, 4718592, 5242880

c. Code a message using this rule. Exchange your message with a partner to decode it.

Getting Ready for Section 4.2 [Section 3.4]

Multiply.

(a) $(a - 1)(a + 4)$

(b) $(x - 1)(x + 3)$

(c) $(x - 3)(x - 3)$

(d) $(y - 11)(y + 3)$

(e) $(x + 5)(x + 7)$

(f) $(y + 1)(y - 13)$

61. _____

62. _____

63. _____

64. _____

65. _____

a. _____

b. _____

c. _____

d. _____

e. _____

f. _____

Answers

1. 2   **3.** 8   **5.** x^2 **7.** a^3 **9.** $5x^4$ **11.** $2a^4$ **13.** $3xy$

15. $5b$ **17.** $3abc^2$ **19.** $(x + y)^2$ **21.** $4(2a + 1)$ **23.** $8(3m - 4n)$

25. $4(3m + 2)$ **27.** $5s(2s + 1)$ **29.** $12x(x + 1)$ **31.** $5a^2(3a - 5)$

33. $6pq(1 + 3p)$ **35.** $7mn(m^2 - 3n^2)$ **37.** $6(x^2 - 3x + 5)$

39. $3a(a^2 + 2a - 4)$ **41.** $3m(2 + 3n - 4n^2)$ **43.** $5xy(2x + 3 - y)$

45. $5r^2s^2(2r + 5 - 3s)$ **47.** $3a(3a^4 - 5a^3 + 7a^2 - 9)$

49. $(a - 3)(a + 2)$ **51.** $(x + 3)(x - 2)$ **53.** 10   **55.** $6x$

57. 6   **59.** $7x$ **61.** $t(33 - t); 33 - t$ **63.**   **65.**

a. $a^2 + 3a - 4$ **b.** $x^2 + 2x - 3$ **c.** $x^2 - 6x + 9$ **d.** $y^2 - 8y - 33$

e. $x^2 + 12x + 35$ **f.** $y^2 - 12y - 13$

Factoring Trinomials of the Form $x^2 + bx + c$

4.2 OBJECTIVES

1. Factor a trinomial of the form $x^2 + bx + c$
2. Factor a trinomial containing a common factor

NOTE The process used to factor here is frequently called the *trial-and-error method.* You'll see the reason for the name as you work through this section.

You learned how to find the product of any two binomials by using the FOIL method in Section 3.4. Because factoring is the reverse of multiplication, we now want to use that pattern to find the factors of certain trinomials.

Recall that when we multiply the binomials $x + 2$ and $x + 3$, our result is

$$(x + 2)(x + 3) = x^2 + 5x + 6$$

The product of the first terms $(x \cdot x)$.

The sum of the products of the outer and inner terms ($3x$ and $2x$).

The product of the last terms ($2 \cdot 3$).

CAUTION

Not every trinomial can be written as the product of two binomials.

Suppose now that you are given $x^2 + 5x + 6$ and want to find its factors. First, you know that the factors of a trinomial may be two binomials. So write

$$x^2 + 5x + 6 = (\quad\quad)(\quad\quad)$$

Because the first term of the trinomial is x^2, the first terms of the binomial factors must be x and x. We now have

$$x^2 + 5x + 6 = (x \quad\quad)(x \quad\quad)$$

The product of the last terms must be 6. Because 6 is positive, the factors must have *like* signs. Here are the possibilities:

$$6 = 1 \cdot 6$$
$$= 2 \cdot 3$$
$$= (-1)(-6)$$
$$= (-2)(-3)$$

This means that the possible factors of the trinomial are

$$(x + 1)(x + 6)$$
$$(x + 2)(x + 3)$$
$$(x - 1)(x - 6)$$
$$(x - 2)(x - 3)$$

How do we tell which is the correct pair? From the FOIL pattern we know that the sum of the outer and inner products must equal the middle term of the trinomial, in this case $5x$. This is the crucial step!

Possible Factorizations	Middle Terms
$(x + 1)(x + 6)$	$7x$
$(x + 2)(x + 3)$	$5x$
$(x - 1)(x - 6)$	$-7x$
$(x - 2)(x - 3)$	$-5x$

The correct middle term!

357

So we know that the correct factorization is

$$x^2 + 5x + 6 = (x + 2)(x + 3)$$

Are there any clues so far that will make this process quicker? Yes, there is an important one that you may have spotted. We started with a trinomial that had a positive middle term and a positive last term. The negative pairs of factors for 6 led to negative middle terms. So you don't need to bother with the negative factors if the middle term and the last term of the trinomial are both positive.

OBJECTIVE 1

Example 1   Factoring a Trinomial

(a) Factor $x^2 + 9x + 8$.

Because the middle term and the last term of the trinomial are both positive, consider only the positive factors of 8, that is, $8 = 1 \cdot 8$ or $8 = 2 \cdot 4$.

Possible Factorizations	Middle Terms
$(x + 1)(x + 8)$	$9x$
$(x + 2)(x + 4)$	$6x$

NOTE If you are wondering why we didn't list $(x + 8)(x + 1)$ as a possibility, remember that multiplication is commutative. The order doesn't matter!

Because the first pair gives the correct middle term,

$$x^2 + 9x + 8 = (x + 1)(x + 8)$$

(b) Factor $x^2 + 12x + 20$.

Possible Factorizations	Middle Terms
$(x + 1)(x + 20)$	$21x$
$(x + 2)(x + 10)$	$12x$
$(x + 4)(x + 5)$	$9x$

NOTE The factor–pairs of 20 are
$20 = 1 \cdot 20$
$ = 2 \cdot 10$
$ = 4 \cdot 5$

So

$$x^2 + 12x + 20 = (x + 2)(x + 10)$$

CHECK YOURSELF 1

Factor.

(a) $x^2 + 6x + 5$ **(b)** $x^2 + 10x + 16$

What if the middle term of the trinomial is negative but the first and last terms are still positive? Consider

Positive   Positive

$$x^2 - 11x + 18$$

Negative

Because we want a negative middle term $(-11x)$ and a positive last term, we use *two negative factors* for 18. Recall that the product of two negative numbers is positive.

Example 2   Factoring a Trinomial

(a) Factor $x^2 - 11x + 18$.

NOTE The negative factors of 18 are

$18 = (-1)(-18)$

$\quad = (-2)(-9)$

$\quad = (-3)(-6)$

Possible Factorizations	Middle Terms
$(x - 1)(x - 18)$	$-19x$
$(x - 2)(x - 9)$	$-11x$
$(x - 3)(x - 6)$	$-9x$

So

$$x^2 - 11x + 18 = (x - 2)(x - 9)$$

(b) Factor $x^2 - 13x + 12$.

NOTE The negative factors of 12 are

$12 = (-1)(-12)$

$\quad = (-2)(-6)$

$\quad = (-3)(-4)$

Possible Factorizations	Middle Terms
$(x - 1)(x - 12)$	$-13x$
$(x - 2)(x - 6)$	$-8x$
$(x - 3)(x - 4)$	$-7x$

So

$$x^2 - 13x + 12 = (x - 1)(x - 12)$$

A few more clues: We have listed all the possible factors in the above examples. It really isn't necessary. Just work until you find the right pair. Also, with practice much of this work can be done mentally.

CHECK YOURSELF 2

Factor.

(a) $x^2 - 10x + 9$ **(b)** $x^2 - 10x + 21$

Now let's look at the process of factoring a trinomial whose last term is negative. For instance, to factor $x^2 + 2x - 15$, we can start as before:

$$x^2 + 2x - 15 = (x \quad ?)(x \quad ?)$$

Note that the product of the last terms must be negative (-15 here). So we must choose factors that have different signs.

What are our choices for the factors of -15?

$$
\begin{aligned}
-15 &= (1)(-15) \\
&= (-1)(15) \\
&= (3)(-5) \\
&= (-3)(5)
\end{aligned}
$$

This means that the possible factors and the resulting middle terms are

NOTE Another clue: Some students prefer to look at the list of numerical factors rather than looking at the actual algebraic factors. Here you want the pair whose sum is 2, the coefficient of the middle term of the trinomial. That pair is -3 and 5, which leads us to the correct factors.

Possible Factorizations	Middle Terms
$(x + 1)(x - 15)$	$-14x$
$(x - 1)(x + 15)$	$14x$
$(x + 3)(x - 5)$	$-2x$
$(x - 3)(x + 5)$	$2x$

So $x^2 + 2x - 15 = (x - 3)(x + 5)$.

Let's work through some examples in which the constant term is negative.

Example 3   Factoring a Trinomial

(a) Factor $x^2 - 5x - 6$.

First, list the factors of -6. Of course, one factor will be positive, and one will be negative.

NOTE You may be able to pick the factors directly from this list. You want the pair whose sum is -5 (the coefficient of the middle term).

$$
\begin{aligned}
-6 &= (1)(-6) \\
&= (-1)(6) \\
&= (2)(-3) \\
&= (-2)(3)
\end{aligned}
$$

For the trinomial, then, we have

Possible Factorizations	Middle Terms
$(x + 1)(x - 6)$	$-5x$
$(x - 1)(x + 6)$	$5x$
$(x + 2)(x - 3)$	$-x$
$(x - 2)(x + 3)$	x

So $x^2 - 5x - 6 = (x + 1)(x - 6)$.

(b) Factor $x^2 + 8xy - 9y^2$.

The process is similar if two variables are involved in the trinomial. Start with

$$x^2 + 8xy - 9y^2 = (x \qquad ?)(x \qquad ?).$$

The product of the last terms
must be $-9y^2$.

$$-9y^2 = (-y)(9y)$$
$$= (y)(-9y)$$
$$= (3y)(-3y)$$

Possible Factorizations	Middle Terms
$(x - y)(x + 9y)$	$8xy$
$(x + y)(x - 9y)$	$-8xy$
$(x + 3y)(x - 3y)$	0

So $x^2 + 8xy - 9y^2 = (x - y)(x + 9y)$.

CHECK YOURSELF 3

Factor.

(a) $x^2 + 7x - 30$ **(b)** $x^2 - 3xy - 10y^2$

As was pointed out in Section 4.1, any time that we have a common factor, that factor should be factored out *before* we try any other factoring technique. Consider Example 4.

OBJECTIVE 2   **Example 4   Factoring a Trinomial**

(a) Factor $3x^2 - 21x + 18$.

$$3x^2 - 21x + 18 = 3(x^2 - 7x + 6) \qquad \text{Factor out the common factor of 3.}$$

We now factor the remaining trinomial. For $x^2 - 7x + 6$:

Possible Factorizations	Middle Terms	
$(x - 1)(x - 6)$	$-7x$	The correct middle term
$(x - 2)(x - 3)$	$-5x$	

CAUTION

A common mistake is to forget
to write the 3 that was factored
out as the first step.

So $3x^2 - 21x + 18 = 3(x - 1)(x - 6)$.

(b) Factor $2x^3 + 16x^2 - 40x$.

$$2x^3 + 16x^2 - 40x = 2x(x^2 + 8x - 20) \qquad \text{Factor out the common factor of 2x.}$$

To factor the remaining trinomial, which is $x^2 + 8x - 20$, we have

Possible Factorizations	Middle Terms	
$(x + 2)(x - 10)$	$-8x$	
$(x - 2)(x + 10)$	$8x$	The correct middle term

NOTE Once we have found the desired middle term, there is no need to continue.

So $2x^3 + 16x^2 - 40x = 2x(x - 2)(x + 10)$.

 CHECK YOURSELF 4 _____

Factor.

(a) $3x^2 - 3x - 36$ **(b)** $4x^3 + 24x^2 + 32x$

One further comment: Have you wondered if all trinomials are factorable? Look at the trinomial

$$x^2 + 2x + 6$$

The only possible factors are $(x + 1)(x + 6)$ and $(x + 2)(x + 3)$. Neither pair is correct (you should check the middle terms), and so this trinomial does not have factors with integer coefficients. Of course, there are many other trinomials which can not be factored. Can you find one?

CHECK YOURSELF ANSWERS _____

1. **(a)** $(x + 1)(x + 5)$; **(b)** $(x + 2)(x + 8)$ **2.** **(a)** $(x - 9)(x - 1)$; **(b)** $(x - 3)(x - 7)$
3. **(a)** $(x + 10)(x - 3)$; **(b)** $(x + 2y)(x - 5y)$
4. **(a)** $3(x - 4)(x + 3)$; **(b)** $4x(x + 2)(x + 4)$

4.2 Exercises

$x^2 + 6x + 5$
$\frac{5}{3}$

Complete each of the following statements.

1. $x^2 - 8x + 15 = (x - 3)(\quad)$ **2.** $y^2 - 3y - 18 = (y - 6)(\quad)$

3. $m^2 + 8m + 12 = (m + 2)(\quad)$ **4.** $x^2 - 10x + 24 = (x - 6)(\quad)$

5. $p^2 - 8p - 20 = (p + 2)(\quad)$ **6.** $a^2 + 9a - 36 = (a + 12)(\quad)$

7. $x^2 - 16x + 64 = (x - 8)(\quad)$ **8.** $w^2 - 12w - 45 = (w + 3)(\quad)$

9. $x^2 - 7xy + 10y^2 = (x - 2y)(\quad)$ **10.** $a^2 + 18ab + 81b^2 = (a + 9b)(\quad)$

Factor each of the following trinomials completely.

11. $x^2 + 8x + 15$ **12.** $x^2 - 11x + 24$

13. $x^2 - 11x + 28$ **14.** $y^2 - y - 20$

15. $s^2 + 13s + 30$ **16.** $b^2 + 14b + 33$

17. $a^2 - 2a - 48$ **18.** $x^2 - 17x + 60$

19. $x^2 - 8x + 7$ **20.** $x^2 + 7x - 18$

21. $m^2 + 3m - 28$ **22.** $a^2 + 10a + 25$

23. $x^2 - 6x - 40$ **24.** $x^2 - 11x + 10$

25. $x^2 - 14x + 49$ **26.** $s^2 - 4s - 32$

27. $p^2 - 10p - 24$ **28.** $x^2 - 11x - 60$

Name _____

Section _____ Date _____

ANSWERS

1. _____   2. _____

3. _____   4. _____

5. _____   6. _____

7. _____   8. _____

9. _____

10. _____

11. _____

12. _____

13. _____

14. _____

15. _____

16. _____

17. _____

18. _____

19. _____

20. _____

21. _____

22. _____

23. _____

24. _____

25. _____

26. _____

27. _____

28. _____

29.	
30.	
31.	
32.	
33.	
34.	
35.	
36.	
37.	
38.	
39.	
40.	
41.	
42.	
43.	
44.	
45.	
46.	
47.	
48.	
49.	
50.	
51.	
52.	
53.	
54.	
55.	
56.	

29. $x^2 + 5x - 66$ **30.** $a^2 + 2a - 80$

31. $c^2 + 19c + 60$ **32.** $t^2 - 4t - 60$

33. $n^2 + 5n - 50$ **34.** $x^2 - 16x + 63$

35. $x^2 + 7xy + 10y^2$ **36.** $x^2 - 8xy + 12y^2$

37. $a^2 - ab - 42b^2$ **38.** $m^2 - 8mn + 16n^2$

39. $x^2 - 13xy + 40y^2$ **40.** $r^2 - 9rs - 36s^2$

41. $b^2 + 6ab + 9a^2$ **42.** $x^2 + 3xy - 10y^2$

43. $x^2 - 2xy - 8y^2$ **44.** $u^2 + 6uv - 55v^2$

45. $25m^2 + 10mn + n^2$ **46.** $64m^2 - 16mn + n^2$

Factor each of the following trinomials completely.

47. $3a^2 - 3a - 126$ **48.** $2c^2 + 2c - 60$

49. $r^3 + 7r^2 - 18r$ **50.** $m^3 + 5m^2 - 14m$

51. $2x^3 - 20x^2 - 48x$ **52.** $3p^3 + 48p^2 - 108p$

53. $x^2y - 9xy^2 - 36y^3$ **54.** $4s^4 - 20s^3t - 96s^2t^2$

55. $m^3 - 29m^2n + 120mn^2$ **56.** $2a^3 - 52a^2b + 96ab^2$

Find all positive values for k for which each of the following can be factored.

57. $x^2 + kx + 8$

58. $x^2 + kx + 9$

59. $x^2 - kx + 16$

60. $x^2 - kx + 17$

61. $x^2 - kx - 5$

62. $x^2 - kx - 7$

63. $x^2 + 3x + k$

64. $x^2 + 5x + k$

65. $x^2 + 2x - k$

66. $x^2 + x - k$

Getting Ready for Section 4.3 [Section 3.4]

Multiply.

(a) $(2x - 1)(2x + 3)$

(b) $(3a - 1)(a + 4)$

(c) $(x - 4)(2x - 3)$

(d) $(2w - 11)(w + 2)$

(e) $(y + 5)(2y + 9)$

(f) $(2x + 1)(x - 12)$

(g) $(p + 9)(2p + 5)$

(h) $(3a - 5)(2a + 4)$

Answers

1. $x - 5$ **3.** $m + 6$ **5.** $p - 10$ **7.** $x - 8$ **9.** $x - 5y$

11. $(x + 3)(x + 5)$ **13.** $(x - 4)(x - 7)$ **15.** $(s + 3)(s + 10)$

17. $(a - 8)(a + 6)$ **19.** $(x - 1)(x - 7)$ **21.** $(m + 7)(m - 4)$

23. $(x + 4)(x - 10)$ **25.** $(x - 7)(x - 7)$ **27.** $(p - 12)(p + 2)$

29. $(x + 11)(x - 6)$ **31.** $(c + 4)(c + 15)$ **33.** $(n - 5)(n + 10)$

35. $(x + 2y)(x + 5y)$ **37.** $(a + 6b)(a - 7b)$ **39.** $(x - 5y)(x - 8y)$

41. $(b + 3a)(b + 3a)$ **43.** $(x + 2y)(x - 4y)$ **45.** $(5m + n)(5m + n)$

47. $3(a + 6)(a - 7)$ **49.** $r(r - 2)(r + 9)$ **51.** $2x(x - 12)(x + 2)$

53. $y(x + 3y)(x - 12y)$ **55.** $m(m - 5n)(m - 24n)$ **57.** 6 or 9

59. 8, 10, or 17   **61.** 4   **63.** 2   **65.** 3, 8, 15, 24, . . .   **a.** $4x^2 + 4x - 3$

b. $3a^2 + 11a - 4$ **c.** $2x^2 - 11x + 12$ **d.** $2w^2 - 7w - 22$

e. $2y^2 + 19y + 45$ **f.** $2x^2 - 23x - 12$ **g.** $2p^2 + 23p + 45$

h. $6a^2 + 2a - 20$

ANSWERS

57.

58.

59.

60.

61.

62.

63.

64.

65.

66.

a.

b.

c.

d.

e.

f.

g.

h.

4.3 Factoring Trinomials of the Form $ax^2 + bx + c$

4.3 OBJECTIVES

1. Factor a trinomial of the form $ax^2 + bx + c$
2. Completely factor a trinomial

Factoring trinomials is more time-consuming when the coefficient of the first term is not 1. Look at the following multiplication.

$$(5x + 2)(2x + 3) = 10x^2 + 19x + 6$$

Factors of $10x^2$ Factors of 6

Do you see the additional problem? We must consider all possible factors of the first coefficient (10 in our example) as well as those of the third term (6 in our example).

There is no easy way out! You need to form all possible combinations of factors and then check the middle term until the proper pair is found. If this seems a bit like guesswork, it is. In fact some call this process factoring by *trial and error*.

We can simplify the work a bit by reviewing the sign patterns found in Section 4.2.

Rules and Properties: Sign Patterns for Factoring Trinomials

NOTE Any time the leading coefficient is negative, factor out a negative one from the trinomial. This will leave one of these cases.

1. If all terms of a trinomial are positive, the signs between the terms in the binomial factors are both plus signs.
2. If the third term of the trinomial is positive and the middle term is negative, the signs between the terms in the binomial factors are both minus signs.
3. If the third term of the trinomial is negative, the signs between the terms in the binomial factors are opposite (one is + and one is −).

OBJECTIVE 1

Example 1   Factoring a Trinomial

Factor $3x^2 + 14x + 15$.

First, list the possible factors of 3, the coefficient of the first term.

$$3 = 1 \cdot 3$$

Now list the factors of 15, the last term.

$$15 = 1 \cdot 15$$
$$= 3 \cdot 5$$

Because the signs of the trinomial are all positive, we know any factors will have the form

The product of the numbers in the last blanks must be 15.

$$(\_x + \_)(\_x + \_)$$

The product of the numbers in the first blanks must be 3.

So the following are the possible factors and the corresponding middle terms:

Possible Factorizations	Middle Terms
$(x + 1)(3x + 15)$	$18x$
$(x + 15)(3x + 1)$	$46x$
$(3x + 3)(x + 5)$	$18x$
$(3x + 5)(x + 3)$	$14x$

The correct middle term

NOTE Take the time to multiply the binomial factors. This habit will ensure that you have an expression equivalent to the original problem.

So

$$3x^2 + 14x + 15 = (3x + 5)(x + 3)$$

 CHECK YOURSELF 1

Factor.

(a) $5x^2 + 14x + 8$ **(b)** $3x^2 + 20x + 12$

Example 2   Factoring a Trinomial

Factor $4x^2 - 11x + 6$.
    Because only the middle term is negative, we know the factors have the form

$$(\_x - \_)(\_x - \_)$$

Both signs are negative.

Now look at the factors of the first coefficient and the last term.

$$4 = 1 \cdot 4 \qquad 6 = 1 \cdot 6$$
$$= 2 \cdot 2 \qquad = 2 \cdot 3$$

This gives us the possible factors:

Possible Factorizations	Middle Terms
$(x - 1)(4x - 6)$	$-10x$
$(x - 6)(4x - 1)$	$-25x$
$(x - 2)(4x - 3)$	$-11x$

The correct middle term

RECALL Again, at least mentally, check your work by multiplying the factors.

Note that, in this example, we *stopped* as soon as the correct pair of factors was found. So
$$4x^2 - 11x + 6 = (x - 2)(4x - 3)$$

 CHECK YOURSELF 2

Factor.

(a) $2x^2 - 9x + 9$ **(b)** $6x^2 - 17x + 10$

Next, we will factor a trinomial whose last term is negative.

Example 3 **Factoring a Trinomial**

Factor $5x^2 + 6x - 8$.

Because the last term is negative, the factors have the form

$$(\_x + \_)(\_x - \_)$$

Consider the factors of the first coefficient and the last term.

$$5 = 1 \cdot 5 \qquad 8 = 1 \cdot 8$$
$$\qquad\qquad\qquad = 2 \cdot 4$$

The possible factors are then

Possible Factorizations	Middle Terms
$(x + 1)(5x - 8)$	$-3x$
$(x + 8)(5x - 1)$	$39x$
$(5x + 1)(x - 8)$	$-39x$
$(5x + 8)(x - 1)$	$3x$
$(x + 2)(5x - 4)$	$6x$

Again we stop as soon as the correct pair of factors is found.

$$5x^2 + 6x - 8 = (x + 2)(5x - 4)$$

 CHECK YOURSELF 3

Factor $4x^2 + 5x - 6$.

The same process is used to factor a trinomial with more than one variable.

Example 4 **Factoring a Trinomial**

Factor $6x^2 + 7xy - 10y^2$.

The form of the factors must be

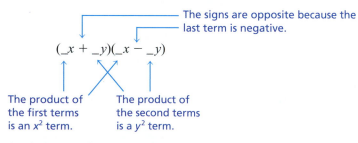

The signs are opposite because the last term is negative.

$$(\_x + \_y)(\_x - \_y)$$

The product of the first terms is an x^2 term.

The product of the second terms is a y^2 term.

Again look at the factors of the first and last coefficients.

$$6 = 1 \cdot 6 \qquad 10 = 1 \cdot 10$$
$$\ = 2 \cdot 3 \qquad\quad = 2 \cdot 5$$

Possible Factorizations	Middle Terms
$(x + y)(6x - 10y)$	$-4xy$
$(x + 10y)(6x - y)$	$59xy$
$(6x + y)(x - 10y)$	$-59xy$
$(6x + 10y)(x - y)$	$4xy$
$(x + 2y)(6x - 5y)$	$7xy$

NOTE Be certain that you have a pattern that matches up every possible pair of coefficients.

Once more, we stop as soon as the correct factors are found.

$$6x^2 + 7xy - 10y^2 = (x + 2y)(6x - 5y)$$

 CHECK YOURSELF 4

Factor $15x^2 - 4xy - 4y^2$.

Example 5 illustrates a special kind of trinomial called a *perfect square trinomial.*

Example 5   Factoring a Trinomial

Factor $9x^2 + 12xy + 4y^2$.

Because all terms are positive, the form of the factors must be

$$(\_x + \_y)(\_x + \_y)$$

Consider the factors of the first and last coefficients.

$$9 = 9 \cdot 1 \qquad 4 = 4 \cdot 1$$
$$ = 3 \cdot 3 \qquad = 2 \cdot 2$$

Possible Factorizations	Middle Terms
$(x + y)(9x + 4y)$	$13xy$
$(x + 4y)(9x + y)$	$37xy$
$(3x + 2y)(3x + 2y)$	$12xy$

So

NOTE Perfect square trinomials can be factored by using previous methods. Recognizing the special pattern simply saves time.

$$9x^2 + 12xy + 4y^2 = (3x + 2y)(3x + 2y)$$
$$= (3x + 2y)^2$$

Square   2(3x)(2y)   Square
of 3x   of 2y

This trinomial is the result of squaring a binomial, thus the special name of perfect square trinomial.

 CHECK YOURSELF 5

Factor.

(a) $4x^2 + 28x + 49$ **(b)** $16x^2 - 40xy + 25y^2$

Before looking at Example 6, review one important point from Section 4.2. Recall that when you factor trinomials, you should not forget to look for a common factor as the first step. If there is a common factor, factor it out and then factor the remaining trinomial as before.

OBJECTIVE 2

Example 6 Factoring a Trinomial

Factor $18x^2 - 18x + 4$.

First look for a common factor in all three terms. Here that factor is 2, so write

$$18x^2 - 18x + 4 = 2(9x^2 - 9x + 2)$$

By our earlier methods, we can factor the remaining trinomial as

NOTE If you don't see why this is true, you need to use your pencil to work it out before you move on!

$$9x^2 - 9x + 2 = (3x - 1)(3x - 2)$$

So

$$18x^2 - 18x + 4 = 2(3x - 1)(3x - 2)$$

Don't forget the 2 that was factored out!

 CHECK YOURSELF 6

Factor $16x^2 + 44x - 12$.

Now look at an example in which the common factor includes a variable.

Example 7 Factoring a Trinomial

Factor

$$6x^3 + 10x^2 - 4x$$

The common factor is $2x$.

So

$$6x^3 + 10x^2 - 4x = 2x(3x^2 + 5x - 2)$$

Because

$$3x^2 + 5x - 2 = (3x - 1)(x + 2)$$

we have

RECALL Remember to include the monomial factor.

$$6x^3 + 10x^2 - 4x = 2x(3x - 1)(x + 2)$$

 CHECK YOURSELF 7

Factor $6x^3 - 27x^2 + 30x$.

You have now had a chance to work with a variety of factoring techniques. Your success in factoring polynomials depends on your ability to recognize when to use which technique. Here are some guidelines to help you apply the factoring methods you have studied in this chapter.

Step by Step: Factoring Polynomials

Step 1   Look for a greatest common factor other than 1. If such a factor exists, factor out the GCF.

Step 2   If the polynomial that remains is a *trinomial*, try to factor the trinomial by the trial-and-error methods of Sections 4.2 and 4.3.

Example 8 illustrates the use of this strategy.

Example 8   Factoring a Trinomial

(a) Factor $5m^2n + 20n$.

First, we see that the GCF is $5n$. Factoring it out gives

$$5m^2n + 20n = 5n(m^2 + 4)$$

(b) Factor $3x^3 - 24x^2 + 48x$.

First, we see that the GCF is $3x$. Factoring out $3x$ yields

$$3x^3 - 24x^2 + 48x = 3x(x^2 - 8x + 16)$$
$$= 3x(x - 4)(x - 4) \quad \text{or} \quad 3x(x - 4)^2$$

(c) Factor $8r^2s + 20rs^2 - 12s^3$.

First, the GCF is $4s$, and we can write the original polynomial as

$$8r^2s + 20rs^2 - 12s^3 = 4s(2r^2 + 5rs - 3s^2)$$

Because the remaining polynomial is a trinomial, we can use the trial-and-error method to complete the factoring.

$$8r^2s + 20rs^2 - 12s^3 = 4s(2r - s)(r + 3s)$$

 CHECK YOURSELF 8 _____

Factor the following polynomials.

(a) $8a^3 + 32a^2b + 32ab^2$ **(b)** $7x^3 + 7x^2y - 42xy^2$ **(c)** $5m^4 + 15m^3 + 5m^2$

CHECK YOURSELF ANSWERS _____

1. (a) $(5x + 4)(x + 2)$; **(b)** $(3x + 2)(x + 6)$ **2. (a)** $(2x - 3)(x - 3)$;
(b) $(6x - 5)(x - 2)$ **3.** $(4x - 3)(x + 2)$ **4.** $(3x - 2y)(5x + 2y)$
5. (a) $(2x + 7)^2$; **(b)** $(4x - 5y)^2$ **6.** $4(4x - 1)(x + 3)$ **7.** $3x(2x - 5)(x - 2)$
8. (a) $8a(a + 2b)(a + 2b)$; **(b)** $7x(x + 3y)(x - 2y)$; **(c)** $5m^2(m^2 + 3m + 1)$

4.3 Exercises

Complete each of the following statements.

1. $4x^2 - 4x - 3 = (2x + 1)(\quad)$ 　　　 **2.** $3w^2 + 11w - 4 = (w + 4)(\quad)$

3. $6a^2 + 13a + 6 = (2a + 3)(\quad)$ 　　　 **4.** $25y^2 - 10y + 1 = (5y - 1)(\quad)$

5. $15x^2 - 16x + 4 = (3x - 2)(\quad)$ 　　　 **6.** $6m^2 + 5m - 4 = (3m + 4)(\quad)$

7. $16a^2 + 8ab + b^2 = (4a + b)(\quad)$ 　　　 **8.** $6x^2 + 5xy - 4y^2 = (3x + 4y)(\quad)$

9. $4m^2 + 5mn - 6n^2 = (m + 2n)(\quad)$ 　　 **10.** $10p^2 - pq - 3q^2 = (5p - 3q)(\quad)$

Factor each of the following polynomials completely.

11. $3x^2 + 7x + 2$ 　　　　　 **12.** $5y^2 + 8y + 3$

13. $2w^2 + 13w + 15$ 　　　　 **14.** $3x^2 - 16x + 21$

15. $5x^2 - 16x + 3$ 　　　　　 **16.** $2a^2 + 7a + 5$

17. $4x^2 - 12x + 5$ 　　　　　 **18.** $2x^2 + 11x + 12$

19. $3x^2 - 5x - 2$ 　　　　　 **20.** $4m^2 - 23m + 15$

21. $4p^2 + 19p - 5$ 　　　　　 **22.** $5x^2 - 36x + 7$

23. $6x^2 + 19x + 10$ 　　　　 **24.** $6x^2 - 7x - 3$

25. $15x^2 + x - 6$ 　　　　　 **26.** $12w^2 + 19w + 4$

27. $6m^2 + 25m - 25$ 　　　　 **28.** $8x^2 - 6x - 9$

29. $9x^2 - 12x + 4$ 　　　　　 **30.** $20x^2 - 23x + 6$

31. $12x^2 - 8x - 15$

32. $16a^2 + 40a + 25$

33. $3y^2 + 7y - 6$

34. $12x^2 + 11x - 15$

35. $8x^2 - 27x - 20$

36. $24v^2 + 5v - 36$

37. $2x^2 + 3xy + y^2$

38. $3x^2 - 5xy + 2y^2$

39. $5a^2 - 8ab - 4b^2$

40. $5x^2 + 7xy - 6y^2$

41. $9x^2 + 4xy - 5y^2$

42. $16x^2 + 32xy + 15y^2$

43. $6m^2 - 17mn + 12n^2$

44. $15x^2 - xy - 6y^2$

45. $36a^2 - 3ab - 5b^2$

46. $3q^2 - 17qr - 6r^2$

47. $x^2 + 4xy + 4y^2$

48. $25b^2 - 80bc + 64c^2$

Factor each of the following polynomials completely.

49. $20x^2 - 20x - 15$

50. $24x^2 - 18x - 6$

51. $8m^2 + 12m + 4$

52. $14x^2 - 20x + 6$

53. $15r^2 - 21rs + 6s^2$

54. $10x^2 + 5xy - 30y^2$

55. $2x^3 - 2x^2 - 4x$

56. $2y^3 + y^2 - 3y$

57. $2y^4 + 5y^3 + 3y^2$

58. $4z^3 - 18z^2 - 10z$

31. _____

32. _____

33. _____

34. _____

35. _____

36. _____

37. _____

38. _____

39. _____

40. _____

41. _____

42. _____

43. _____

44. _____

45. _____

46. _____

47. _____

48. _____

49. _____

50. _____

51. _____

52. _____

53. _____

54. _____

55. _____

56. _____

57. _____

58. _____

59. $36a^3 - 66a^2 + 18a$ **60.** $20n^4 - 22n^3 - 12n^2$

61. $9p^2 + 30pq + 21q^2$ **62.** $12x^2 + 2xy - 24y^2$

Factor each of the following polynomials completely.

63. $10(x + y)^2 - 11(x + y) - 6$ **64.** $8(a - b)^2 + 14(a - b) - 15$

65. $5(x - 1)^2 - 15(x - 1) - 350$ **66.** $3(x + 1)^2 - 6(x + 1) - 45$

67. $15 + 29x - 48x^2$ **68.** $12 + 4a - 21a^2$

69. $-6x^2 + 19x - 15$ **70.** $-3s^2 - 10s + 8$

 Getting Ready for Section 4.4 [Section 3.5]

Multiply.

   (a) $(x - 1)(x + 1)$ (b) $(a + 7)(a - 7)$

   (c) $(x - y)(x + y)$ (d) $(2x - 5)(2x + 5)$

   (e) $(3a - b)(3a + b)$ (f) $(5a - 4b)(5a + 4b)$

Answers

1. $2x - 3$ **3.** $3a + 2$ **5.** $5x - 2$ **7.** $4a + b$ **9.** $4m - 3n$
11. $(3x + 1)(x + 2)$ **13.** $(2w + 3)(w + 5)$ **15.** $(5x - 1)(x - 3)$
17. $(2x - 5)(2x - 1)$ **19.** $(3x + 1)(x - 2)$ **21.** $(4p - 1)(p + 5)$
23. $(3x + 2)(2x + 5)$ **25.** $(5x - 3)(3x + 2)$ **27.** $(6m - 5)(m + 5)$
29. $(3x - 2)(3x - 2) = (3x - 2)^2$ **31.** $(6x + 5)(2x - 3)$ **33.** $(3y - 2)(y + 3)$
35. $(8x + 5)(x - 4)$ **37.** $(2x + y)(x + y)$ **39.** $(5a + 2b)(a - 2b)$
41. $(9x - 5y)(x + y)$ **43.** $(3m - 4n)(2m - 3n)$ **45.** $(12a - 5b)(3a + b)$
47. $(x + 2y)(x + 2y) = (x + 2y)^2$ **49.** $5(2x - 3)(2x + 1)$
51. $4(2m + 1)(m + 1)$ **53.** $3(5r - 2s)(r - s)$ **55.** $2x(x - 2)(x + 1)$
57. $y^2(2y + 3)(y + 1)$ **59.** $6a(3a - 1)(2a - 3)$ **61.** $3(p + q)(3p + 7q)$
63. $(5x + 5y + 2)(2x + 2y - 3)$ **65.** $5(x - 11)(x + 6)$
67. $(1 + 3x)(15 - 16x)$ **69.** $(3x - 5)(-2x + 3)$ **a.** $x^2 - 1$
b. $a^2 - 49$ **c.** $x^2 - y^2$ **d.** $4x^2 - 25$ **e.** $9a^2 - b^2$ **f.** $25a^2 - 16b^2$

4.4 Difference of Squares and Perfect Square Trinomials

4.4 OBJECTIVES

1. Factor a binomial that is the difference of two squares
2. Factor a perfect square trinomial

In Section 3.5, we introduced some special products. Recall the following formula for the product of a sum and difference of two terms:

$$(a + b)(a - b) = a^2 - b^2$$

This also means that a binomial of the form $a^2 - b^2$, called a **difference of two squares**, has as its factors $a + b$ and $a - b$.

To use this idea for factoring, we can write

$$a^2 - b^2 = (a + b)(a - b)$$

A **perfect square** term has a coefficient that is a square (1, 4, 9, 16, 25, 36, etc.), and any variables have exponents that are multiples of 2 (x^2, y^4, z^6, etc.).

OBJECTIVE 1

Example 1 Identifying Perfect Square Terms

For each of the following, decide whether it is a perfect square term. If it is, rewrite the expression as an expression squared (called the *root*).

(a) $36x$

(b) $24x^6$

(c) $9x^4$

(d) $64x^6$

(e) $16x^9$

Only parts (c) and (d) are perfect square terms.

$9x^4 = (3x^2)^2$
$64x^6 = (8x^3)^2$

 CHECK YOURSELF 1 _____

For each of the following, decide whether it is a perfect square term. If it is, rewrite the expression as an expression squared.

(a) $36x^{12}$ **(b)** $4x^6$

(c) $9x^7$ **(d)** $25x^8$

(e) $16x^{25}$

We will now use the given equation to factor the difference between two perfect square terms.

Example 2   Factoring the Difference of Two Squares

Factor $x^2 - 16$.

Think $x^2 - 4^2$.

NOTE You could also write $(x - 4)(x + 4)$. The order doesn't matter because multiplication is commutative.

Because $x^2 - 16$ is a difference of squares, we have

$$x^2 - 16 = (x + 4)(x - 4)$$

 CHECK YOURSELF 2

Factor $m^2 - 49$.

Any time an expression is a difference of two squares, it can be factored.

Example 3   Factoring the Difference of Two Squares

Factor $4a^2 - 9$.

Think $(2a)^2 - 3^2$.

So

$$4a^2 - 9 = (2a)^2 - (3)^2$$
$$= (2a + 3)(2a - 3)$$

 CHECK YOURSELF 3

Factor $9b^2 - 25$.

The process for factoring a difference of squares does not change when more than one variable is involved.

Example 4   Factoring the Difference of Two Squares

NOTE Think $(5a)^2 - (4b^2)^2$.

Factor $25a^2 - 16b^4$.

$$25a^2 - 16b^4 = (5a + 4b^2)(5a - 4b^2)$$

 CHECK YOURSELF 4

Factor $49c^4 - 9d^2$.

We will now consider an example that combines common-term factoring with difference-of-squares factoring. Note that the common factor is always factored out as the *first step*.

Example 5   Removing the GCF First

NOTE Step 1
Factor out the GCF.
Step 2
Factor the remaining binomial.

Factor $32x^2y - 18y^3$.
    Note that $2y$ is a common factor, so

$$32x^2y - 18y^3 = 2y(\underline{16x^2 - 9y^2})$$

Difference of squares

$$= 2y(4x + 3y)(4x - 3y)$$

 **CHECK YOURSELF 5**

Factor $50a^3 - 8ab^2$.

Recall the following multiplication pattern.

⚠ **CAUTION**

Note that this is different from the sum of two squares (like $x^2 + y^2$), which never has integer factors.

$$(a + b)^2 = a^2 + 2ab + b^2$$

For example,

$$(x + 2)^2 = x^2 + 4x + 4$$
$$(x + 5)^2 = x^2 + 10x + 25$$
$$(2x + 1)^2 = 4x^2 + 4x + 1$$

Recognizing this pattern can simplify the process of factoring perfect square trinomials.

OBJECTIVE 2   ### Example 6   Factoring a Perfect Square Trinomial

Factor the trinomial $4x^2 + 12xy + 9y^2$.
    Note that this is a perfect square trinomial in which

$$a = 2x \quad \text{and} \quad b = 3y.$$

In factored form, we have

$$4x^2 + 12xy + 9y^2 = (2x + 3y)^2$$

 CHECK YOURSELF 6

Factor the trinomial $16u^2 + 24uv + 9v^2$.

Recognizing the same pattern can simplify the process of factoring perfect square trinomials in which the second term is negative.

Example 7   Factoring a Perfect Square Trinomial

Factor the trinomial $25x^2 - 10xy + y^2$.
    This is also a perfect square trinomial, in which

$a = 5x$ and $b = -y$.

In factored form, we have

$$25x^2 - 10xy + y^2 = (5x + (-y))^2 = (5x - y)^2$$

CHECK YOURSELF 7 _____

Factor the trinomial $4u^2 - 12uv + 9v^2$.

CHECK YOURSELF ANSWERS _____

1. **(a)** $(6x^6)^2$; **(b)** $(2x^3)^2$; **(d)** $(5x^4)^2$ **2.** $(m + 7)(m - 7)$ **3.** $(3b + 5)(3b - 5)$
4. $(7c^2 + 3d)(7c^2 - 3d)$ **5.** $2a(5a + 2b)(5a - 2b)$ **6.** $(4u + 3v)^2$
7. $(2u - 3v)^2$

4.4 Exercises

For each of the following binomials, is the binomial a difference of squares?

1. $3x^2 + 2y^2$

2. $5x^2 - 7y^2$

3. $16a^2 - 25b^2$

4. $9n^2 - 16m^2$

5. $16r^2 + 4$

6. $p^2 - 45$

7. $16a^2 - 12b^3$

8. $9a^2b^2 - 16c^2d^2$

9. $a^2b^2 - 25$

10. $4a^3 - b^3$

Factor the following binomials.

11. $m^2 - n^2$

12. $r^2 - 9$

13. $x^2 - 49$

14. $c^2 - d^2$ $(c+d)(c-d)$

15. $49 - y^2$

16. $81 - b^2$ $(9-b)(9+b)$ $36x^{12}$

17. $9b^2 - 16$

18. $36 - x^2$ $(6-x)(6+x)$ $6x$

19. $16w^2 - 49$

20. $4x^2 - 25$ $(2x-5)(2x+5)$

21. $4s^2 - 9r^2$

22. $64y^2 - x^2$ $(8y-x)(8y+x)$

23. $9w^2 - 49z^2$

24. $25x^2 - 81y^2$ $(5x-9y)(5x+9y)$

25. $16a^2 - 49b^2$

26. $64m^2 - 9n^2$ $(8m-3n)(8m+3n)$

27. $x^4 - 36$

28. $y^6 - 49$ $8y^3$

29. $x^2y^2 - 16$

30. $m^2n^2 - 64$ $(mn-8)(mn+8)$

ANSWERS

1. _____ 2. _____
3. _____ 4. _____
5. _____ 6. _____
7. _____ 8. _____
9. _____ 10. _____
11. _____
12. _____
13. _____
14. _____
15. _____
16. _____
17. _____
18. _____
19. _____
20. _____
21. _____
22. _____
23. _____
24. _____
25. _____
26. _____
27. _____
28. _____
29. _____
30. _____

31. _____

32. _____

33. _____

34. _____

35. _____

36. _____

37. _____

38. _____

39. _____

40. _____

41. _____

42. _____

43. _____

44. _____ 45. _____

46. _____

47. _____

48. _____ 49. _____

50. _____ 51. _____

52. _____

53. _____

54. _____

55. _____

56. _____

57. _____

58. _____

59. _____

60. _____

61. _____

62. _____

31. $25 - a^2b^2$

32. $49 - w^2z^2$

33. $r^4 - 4s^2$

34. $p^2 - 9q^4$

35. $81a^2 - 100b^6$

36. $64x^4 - 25y^4$

37. $18x^3 - 2xy^2$

38. $50a^2b - 2b^3$

39. $12m^3n - 75mn^3$

40. $63p^4 - 7p^2q^2$

41. $48a^2b^2 - 27b^4$

42. $20w^5 - 45w^3z^4$

Determine whether each of the following trinomials is a perfect square. If it is, factor the trinomial.

43. $x^2 - 14x + 49$

44. $x^2 + 9x + 16$

45. $x^2 - 18x - 81$

46. $x^2 + 10x + 25$

47. $x^2 - 18x + 81$

48. $x^2 - 24x + 48$

Factor the following trinomials.

49. $x^2 + 4x + 4$

50. $x^2 + 6x + 9$

51. $x^2 - 10x + 25$

52. $x^2 - 8x + 16$

53. $4x^2 + 12xy + 9y^2$

54. $16x^2 + 40xy + 25y^2$

55. $9x^2 - 24xy + 16y^2$

56. $9w^2 - 30wv + 25v^2$

57. $y^3 - 10y^2 + 25y$

58. $12b^3 - 12b^2 + 3b$

Factor each expression.

59. $x^2(x + y) - y^2(x + y)$

60. $a^2(b - c) - 16b^2(b - c)$

61. $2m^2(m - 2n) - 18n^2(m - 2n)$

62. $3a^3(2a + b) - 27ab^2(2a + b)$

63. Find a value for k so that $kx^2 - 25$ will have the factors $2x + 5$ and $2x - 5$.

64. Find a value for k so that $9m^2 - kn^2$ will have the factors $3m + 7n$ and $3m - 7n$.

65. Find a value for k so that $2x^3 - kxy^2$ will have the factors $2x$, $x - 3y$, and $x + 3y$.

66. Find a value for k so that $20a^3b - kab^3$ will have the factors $5ab$, $2a - 3b$, and $2a + 3b$.

67. Complete the following statement in complete sentences: "To factor a number you… ."

68. Complete this statement: To factor an algebraic expression into prime factors means… .

Getting Ready for Section 4.5 [Section 4.1]

Factor.

(a) $2x(3x + 2) - 5(3x + 2)$

(b) $3y(y - 4) + 5(y - 4)$

(c) $3x(x + 2y) + y(x + 2y)$

(d) $5x(2x - y) - 3(2x - y)$

(e) $4x(2x - 5y) - 3y(2x - 5y)$

Answers

1. No   **3.** Yes   **5.** No   **7.** No   **9.** Yes   **11.** $(m + n)(m - n)$
13. $(x + 7)(x - 7)$ **15.** $(7 + y)(7 - y)$ **17.** $(3b + 4)(3b - 4)$
19. $(4w + 7)(4w - 7)$ **21.** $(2s + 3r)(2s - 3r)$ **23.** $(3w + 7z)(3w - 7z)$
25. $(4a + 7b)(4a - 7b)$ **27.** $(x^2 + 6)(x^2 - 6)$ **29.** $(xy + 4)(xy - 4)$
31. $(5 + ab)(5 - ab)$ **33.** $(r^2 + 2s)(r^2 - 2s)$ **35.** $(9a + 10b^3)(9a - 10b^3)$
37. $2x(3x + y)(3x - y)$ **39.** $3mn(2m + 5n)(2m - 5n)$
41. $3b^2(4a + 3b)(4a - 3b)$ **43.** Yes; $(x - 7)^2$ **45.** No
47. Yes; $(x - 9)^2$ **49.** $(x + 2)^2$ **51.** $(x - 5)^2$ **53.** $(2x + 3y)^2$
55. $(3x - 4y)^2$ **57.** $y(y - 5)^2$ **59.** $(x + y)^2(x - y)$
61. $2(m - 2n)(m + 3n)(m - 3n)$ **63.** 4   **65.** 18   **67.**
a. $(3x + 2)(2x - 5)$ **b.** $(y - 4)(3y + 5)$
c. $(x + 2y)(3x + y)$ **d.** $(2x - y)(5x - 3)$
e. $(2x - 5y)(4x - 3y)$

4.5 Factoring by Grouping

4.5 OBJECTIVES

1. Factor a polynomial by grouping terms
2. Rewrite a polynomial so that it can be factored by the method of grouping terms

Some polynomials can be factored by grouping the terms and finding common factors within each group. Such a process is called factoring by grouping and will be explored in this section.

Recall that in Section 4.1, we looked at the expression

$$3x(x + y) + 2(x + y)$$

and found that we could factor out the common binomial, $(x + y)$, giving us

$$(x + y)(3x + 2)$$

That technique will be used in Example 1.

OBJECTIVE 1

Example 1   Factoring by Grouping Terms

Suppose we want to factor the polynomial

$$ax - ay + bx - by$$

As you can see, the polynomial has no common factors. However, look at what happens if we separate the polynomial into *two groups of two terms*.

NOTE Note that our example has *four* terms. That is a clue for trying the factoring by grouping method.

$$ax - ay + bx - by$$
$$= \underbrace{ax - ay}_{(1)} + \underbrace{bx - by}_{(2)}$$

Now *each* group has a common factor, and we can write the polynomial as

$$a(x - y) + b(x - y)$$

In this form, we can see that $x - y$ is the GCF. Factoring out $x - y$, we get

$$a(x - y) + b(x - y) = (x - y)(a + b)$$

 CHECK YOURSELF 1

Use the factoring by grouping method.

$$x^2 - 2xy + 3x - 6y$$

Be particularly careful of your treatment of algebraic signs when applying the factoring by grouping method. Consider Example 2.

Example 2   Factoring by Grouping Terms

Factor $2x^3 - 3x^2 - 6x + 9$.

We group the polynomial as follows.

$$\underbrace{2x^3 - 3x^2}_{(1)} - \underbrace{6x + 9}_{(2)}$$ Factor out the common factor of -3 from the second two terms.

NOTE Notice that $9 = (-3)(-3)$.

$$= x^2(2x - 3) - 3(2x - 3)$$
$$= (2x - 3)(x^2 - 3)$$

CHECK YOURSELF 2

Factor by grouping.

$3y^3 + 2y^2 - 6y - 4$

It may also be necessary to change the order of the terms as they are grouped. Look at Example 3.

OBJECTIVE 2   **Example 3   Factoring by Grouping Terms**

Factor $x^2 - 6yz + 2xy - 3xz$.

Grouping the terms as before, we have

$$\underbrace{x^2 - 6yz}_{(1)} + \underbrace{2xy - 3xz}_{(2)}$$

Do you see that we have accomplished nothing because there are no common factors in the first group?

We can, however, rearrange the terms to write the original polynomial as

$$\underbrace{x^2 + 2xy}_{(1)} - \underbrace{3xz - 6yz}_{(2)}$$

$$= x(x + 2y) - 3z(x + 2y)$$ We can now factor out the common factor of $x + 2y$ in group (1) and group (2).

$$= (x + 2y)(x - 3z)$$

Note: It is often true that the grouping can be done in more than one way. The factored form will be the same.

CHECK YOURSELF 3

We can write the polynomial of Example 3 as

$x^2 - 3xz + 2xy - 6yz$

Factor and verify that the factored form is the same in either case.

CHECK YOURSELF ANSWERS

1. $(x - 2y)(x + 3)$ **2.** $(3y + 2)(y^2 - 2)$ **3.** $(x - 3z)(x + 2y)$

4.5 Exercises

Factor each polynomial by grouping the first two terms and the last two terms.

1. $x^3 - 4x^2 + 3x - 12$

2. $x^3 - 6x^2 + 2x - 12$

3. $a^3 - 3a^2 + 5a - 15$

4. $6x^3 - 2x^2 + 9x - 3$

5. $10x^3 + 5x^2 - 2x - 1$

6. $x^5 + x^3 - 2x^2 - 2$

7. $x^4 - 2x^3 + 3x - 6$

8. $x^3 - 4x^2 + 2x - 8$

Factor each polynomial completely by factoring out any common factors and then factor by grouping. Do not combine like terms.

9. $3x - 6 + xy - 2y$

10. $2x - 10 + xy - 5y$

11. $ab - ac + b^2 - bc$

12. $ax + 2a + bx + 2b$

13. $3x^2 - 2xy + 3x - 2y$

14. $xy - 5y^2 - x + 5y$

15. $5s^2 + 15st - 2st - 6t^2$

16. $3a^3 + 3ab^2 + 2a^2b + 2b^3$

17. $3x^3 + 6x^2y - x^2y - 2xy^2$

18. $2p^4 + 3p^3q - 2p^3q - 3p^2q^2$

19. $x^4 + 5x^3 - 2x^2 - 10x$

20. $x^4y - 2x^3y + x^4 - 2x^3$

21. $2x^3 - 2x^2 + 3x^2 - 3x$

22. $3b^4 - 3b^3c + 2b^3c - 2b^2c^2$

Getting Ready for Section 4.6 [Section 3.4]

Multiply.

(a) $(2x - 1)(2x + 3)$ (b) $(3a - 1)(a + 4)$

(c) $(x - 4)(2x - 3)$ (d) $(2w - 11)(w + 2)$

(e) $(y + 5)(2y + 9)$ (f) $(2x + 1)(x - 12)$

(g) $(p + 9)(2p + 5)$ (h) $(3a - 5)(2a + 4)$

Answers

1. $(x - 4)(x^2 + 3)$ **3.** $(a - 3)(a^2 + 5)$ **5.** $(2x + 1)(5x^2 - 1)$

7. $(x - 2)(x^3 + 3)$ **9.** $(x - 2)(3 + y)$ **11.** $(b - c)(a + b)$

13. $(x + 1)(3x - 2y)$ **15.** $(s + 3t)(5s - 2t)$ **17.** $x(x + 2y)(3x - y)$

19. $x(x + 5)(x^2 - 2)$ **21.** $x(x - 1)(2x + 3)$ **a.** $4x^2 + 4x - 3$

b. $3a^2 + 11a - 4$ **c.** $2x^2 - 11x + 12$ **d.** $2w^2 - 7w - 22$

e. $2y^2 + 19y + 45$ **f.** $2x^2 - 23x - 12$ **g.** $2p^2 + 23p + 45$

h. $6a^2 + 2a - 20$

4.6 Using the *ac* Method to Factor

4.6 OBJECTIVES

1. Verify that a trinomial is correctly factored
2. Use the *ac* test to determine factorability
3. Use the results of the *ac* test to factor a trinomial
4. Completely factor a trinomial

In Sections 4.2 and 4.3 we used the trial-and-error method to factor trinomials. We also learned that not all trinomials can be factored. In this section we will look at the same kinds of trinomials, but in a slightly different context. We first determine whether a trinomial is factorable. We then use the results of that analysis to factor the trinomial.

Some students prefer the trial-and-error method for factoring because it is generally faster and more intuitive. Other students prefer the method of this section (called the *ac* method) because it yields the answer in a systematic way. We will let you determine which method you prefer.

We will begin by looking at some factored trinomials.

OBJECTIVE 1

Example 1   Matching Trinomials and Their Factors

Determine which of the following are true statements.

(a) $x^2 - 2x - 8 = (x - 4)(x + 2)$

This is a true statement. Using the FOIL method, we see that

$$(x - 4)(x + 2) = x^2 + 2x - 4x - 8$$
$$= x^2 - 2x - 8$$

(b) $x^2 + 6x + 5 = (x + 2)(x + 3)$

This is not a true statement.

$$(x + 2)(x + 3) = x^2 + 3x + 2x + 6 = x^2 + 5x + 6$$

(c) $x^2 + 5x - 14 = (x - 2)(x + 7)$

This is true: $(x - 2)(x + 7) = x^2 + 7x - 2x - 14 = x^2 + 5x - 14$

(d) $x^2 - 8x - 15 = (x - 5)(x - 3)$

This is false: $(x - 5)(x - 3) = x^2 - 3x - 5x + 15 = x^2 - 8x + 15$

CHECK YOURSELF 1

Determine which of the following are true statements.

(a) $2x^2 - 2x - 3 = (2x - 3)(x + 1)$
(b) $3x^2 + 11x - 4 = (3x - 1)(x + 4)$
(c) $2x^2 - 7x + 3 = (x - 3)(2x - 1)$

The first step in learning to factor a trinomial is to identify its coefficients. So that we are consistent, we first write the trinomial in standard $ax^2 + bx + c$ form and then label the three coefficients as a, b, and c.

Example 2   Identifying the Coefficients of $ax^2 + bx + c$

First, when necessary, rewrite the trinomial in $ax^2 + bx + c$ form. Then give the values for a, b, and c, in which a is the coefficient of the x^2 term, b is the coefficient of the x term, and c is the constant.

(a) $x^2 - 3x - 18$

$a = 1 \qquad b = -3 \qquad c = -18$

NOTE The negative sign is attached to the coefficients.

(b) $x^2 - 24x + 23$

$a = 1 \qquad b = -24 \qquad c = 23$

(c) $x^2 + 8 - 11x$

First rewrite the trinomial in descending order:

$x^2 - 11x + 8$

$a = 1 \qquad b = -11 \qquad c = 8$

CHECK YOURSELF 2

First, when necessary, rewrite the trinomials in $ax^2 + bx + c$ form. Then label a, b, and c, in which a is the coefficient of the x^2 term, b is the coefficient of the x term, and c is the constant.

(a) $x^2 + 5x - 14$ **(b)** $x^2 - 18x + 17$ **(c)** $x - 6 + 2x^2$

Not all trinomials can be factored. To discover if a trinomial is factorable, we try the *ac* **test.**

Definition: The *ac* Test

A trinomial of the form $ax^2 + bx + c$ is factorable if (and only if) there are two integers, m and n, such that

$$ac = mn \qquad \text{and} \qquad b = m + n$$

In Example 3 we will look for m and n to determine whether each trinomial is factorable.

OBJECTIVE 2   **Example 3   Using the *ac* Test**

Use the *ac* test to determine which of the following trinomials can be factored. Find the values of *m* and *n* for each trinomial that can be factored.

(a) $x^2 - 3x - 18$

First, we find the values of *a*, *b*, and *c*, so that we can find *ac*.

$$a = 1 \qquad b = -3 \qquad c = -18$$

$$ac = 1(-18) = -18 \qquad \text{and} \qquad b = -3$$

Then, we look for two numbers, *m* and *n*, such that their product is *ac*, and their sum is *b*. In this case, that means

$$mn = -18 \qquad \text{and} \qquad m + n = -3$$

We now look at all pairs of integers with a product of -18. We then look at the sum of each pair of integers, looking for a sum of -3.

mn	*m + n*	
$1(-18) = -18$	$1 + (-18) = -17$	
$2(-9) = -18$	$2 + (-9) = -7$	We need to look no
$3(-6) = -18$	$3 + (-6) = -3$	further than 3 and -6.
$6(-3) = -18$		
$9(-2) = -18$		
$18(-1) = -18$		

NOTE We could have chosen $m = -6$ and $n = 3$ as well.

3 and -6 are the two integers with a product of *ac* and a sum of *b*. We can say that

$$m = 3 \qquad \text{and} \qquad n = -6$$

Because we found values for *m* and *n*, we know that $x^2 - 3x - 18$ is factorable.

(b) $x^2 - 24x + 23$

We find that

$$a = 1 \qquad b = -24 \qquad c = 23$$

$$ac = 1(23) = 23 \qquad \text{and} \qquad b = -24$$

So

$$mn = 23 \qquad \text{and} \qquad m + n = -24$$

We now calculate integer pairs, looking for two numbers with a product of 23 and a sum of -24.

mn	*m + n*
$1(23) = 23$	$1 + 23 = 24$
$-1(-23) = 23$	$-1 + (-23) = -24$

$$m = -1 \qquad \text{and} \qquad n = -23$$

So, $x^2 - 24x + 23$ is factorable.

(c) $x^2 - 11x + 8$

We find that $a = 1$, $b = -11$, and $c = 8$. Therefore, $ac = 8$ and $b = -11$. Thus $mn = 8$ and $m + n = -11$. We calculate integer pairs:

mn	*m + n*
$1(8) = 8$	$1 + 8 = 9$
$2(4) = 8$	$2 + 4 = 6$
$-1(-8) = 8$	$-1 + (-8) = -9$
$-2(-4) = 8$	$-2 + (-4) = -6$

There are no other pairs of integers with a product of 8, and none of these pairs has a sum of -11. The trinomial $x^2 - 11x + 8$ is not factorable.

(d) $2x^2 + 7x - 15$

We find that $a = 2$, $b = 7$, and $c = -15$. Therefore, $ac = 2(-15) = -30$ and $b = 7$. Thus $mn = -30$ and $m + n = 7$. We calculate integer pairs:

mn	*m + n*
$1(-30) = -30$	$1 + (-30) = -29$
$2(-15) = -30$	$2 + (-15) = -13$
$3(-10) = -30$	$3 + (-10) = -7$
$5(-6) = -30$	$5 + (-6) = -1$
$6(-5) = -30$	$6 + (-5) = 1$
$10(-3) = -30$	$10 + (-3) = 7$

There is no need to go any further. We see that 10 and -3 have a product of -30 and a sum of 7, so

$$m = 10 \quad \text{and} \quad n = -3$$

Therefore, $2x^2 + 7x - 15$ is factorable.

It is not always necessary to evaluate all the products and sums to determine whether a trinomial is factorable. You may have noticed patterns and shortcuts that make it easier to find m and n. By all means, use them to help you find m and n. This is essential in mathematical thinking. You are taught a mathematical process that will always work for solving a problem. Such a process is called an **algorithm.** It is very easy to teach a computer to use an algorithm. It is very difficult (some would say impossible) for a computer to have insight. Shortcuts that you discover are *insights*. They may be the most important part of your mathematical education.

CHECK YOURSELF 3

Use the ac test to determine which of the following trinomials can be factored. Find the values of m and n for each trinomial that can be factored.

(a) $x^2 - 7x + 12$ **(b)** $x^2 + 5x - 14$

(c) $3x^2 - 6x + 7$ **(d)** $2x^2 + x - 6$

So far we have used the results of the *ac* test only to determine whether a trinomial is factorable. The results can also be used to help factor the trinomial.

OBJECTIVE 3

Example 4 **Using the Results of the *ac* Test to Factor**

Rewrite the middle term as the sum of two terms and then factor by grouping.

(a) $x^2 - 3x - 18$

We find that $a = 1$, $b = -3$, and $c = -18$, so $ac = -18$ and $b = -3$. We are looking for two numbers, m and n, where $mn = -18$ and $m + n = -3$. In Example 3, part (a), we looked at every pair of integers whose product (mn) was -18, to find a pair that had a sum ($m + n$) of -3. We found the two integers to be 3 and -6, because $3(-6) = -18$ and $3 + (-6) = -3$, so $m = 3$ and $n = -6$. We now use that result to rewrite the middle term as the sum of $3x$ and $-6x$.

$$x^2 + 3x - 6x - 18$$

We then factor by grouping:

$$x^2 + 3x - 6x - 18 = x(x + 3) - 6(x + 3)$$
$$= (x + 3)(x - 6)$$

(b) $x^2 - 24x + 23$

We use the results from Example 3, part (b), in which we found $m = -1$ and $n = -23$, to rewrite the middle term of the equation.

$$x^2 - 24x + 23 = x^2 - x - 23x + 23$$

Then we factor by grouping:

$$x^2 - x - 23x + 23 = (x^2 - x) - (23x - 23)$$
$$= x(x - 1) - 23(x - 1)$$
$$= (x - 1)(x - 23)$$

(c) $2x^2 + 7x - 15$

From Example 3, part (d), we know that this trinomial is factorable, and $m = 10$ and $n = -3$. We use that result to rewrite the middle term of the trinomial.

$$2x^2 + 7x - 15 = 2x^2 + 10x - 3x - 15$$
$$= (2x^2 + 10x) - (3x + 15)$$
$$= 2x(x + 5) - 3(x + 5)$$
$$= (x + 5)(2x - 3)$$

Careful readers will note that we did not ask you to factor Example 3, part (c), $x^2 - 11x + 8$. Recall that, by the *ac* method, we determined that this trinomial was not factorable.

CHECK YOURSELF 4

Use the results of Check Yourself 3 to rewrite the middle term as the sum of two terms and then factor by grouping.

(a) $x^2 - 7x + 12$ **(b)** $x^2 + 5x - 14$ **(c)** $2x^2 + x - 6$

Next, we look at some examples that require us to first find *m* and *n* and then factor the trinomial.

Example 5   Rewriting Middle Terms to Factor

Rewrite the middle term as the sum of two terms and then factor by grouping.

(a) $2x^2 - 13x - 7$

We find that $a = 2$, $b = -13$, and $c = -7$, so $mn = ac = -14$ and $m + n = b = -13$. Therefore,

mn	*m + n*
$1(-14) = -14$	$1 + (-14) = -13$

So, $m = 1$ and $n = -14$. We rewrite the middle term of the trinomial as follows:

$$2x^2 - 13x - 7 = 2x^2 + x - 14x - 7$$
$$= (2x^2 + x) - (14x + 7)$$
$$= x(2x + 1) - 7(2x + 1)$$
$$= (2x + 1)(x - 7)$$

(b) $6x^2 - 5x - 6$

We find that $a = 6$, $b = -5$, and $c = -6$, so $mn = ac = -36$ and $m + n = b = -5$.

mn	*m + n*
$1(-36) = -36$	$1 + (-36) = -35$
$2(-18) = -36$	$2 + (-18) = -16$
$3(-12) = -36$	$3 + (-12) = -9$
$4(-9)\ \ = -36$	$4 + (-9)\ \ = -5$

So, $m = 4$ and $n = -9$. We rewrite the middle term of the trinomial:

$$6x^2 - 5x - 6 = 6x^2 + 4x - 9x - 6$$
$$= (6x^2 + 4x) - (9x + 6)$$
$$= 2x(3x + 2) - 3(3x + 2)$$
$$= (3x + 2)(2x - 3)$$

✔ **CHECK YOURSELF 5**

Rewrite the middle term as the sum of two terms and then factor by grouping.

(a) $2x^2 - 7x - 15$

$$\frac{21}{-32}\qquad \frac{30}{-5}\quad 3$$

$$2x^2 - 2x - 5x - 15$$

(b) $6x^2 - 5x - 4$

$$\frac{12}{43}\qquad \frac{48}{14}2$$

$$6x^2 - 6x + 1x - 4$$

Be certain to check trinomials and binomial factors for any common monomial factor. (There is no common factor in the binomial unless it is also a common factor in the original trinomial.) Example 6 shows the factoring out of monomial factors.

OBJECTIVE 4   **Example 6   Factoring Out Common Factors**

Completely factor the trinomial.

$3x^2 + 12x - 15$

We could first factor out the common factor of 3:

$3x^2 + 12x - 15 = 3(x^2 + 4x - 5)$

Finding m and n for the trinomial $x^2 + 4x - 5$ yields $mn = -5$ and $m + n = 4$.

mn	$m + n$
$1(-5) = -5$	$1 + (-5) = -4$
$5(-1) = -5$	$-1 + (5) = 4$

So, $m = 5$ and $n = -1$. This gives us

$$
\begin{aligned}
3x^2 + 12x - 15 &= 3(x^2 + 4x - 5) \\
&= 3(x^2 + 5x - x - 5) \\
&= 3[(x^2 + 5x) - (x + 5)] \\
&= 3[x(x + 5) - (x + 5)] \\
&= 3[(x + 5)(x - 1)] \\
&= 3(x + 5)(x - 1)
\end{aligned}
$$

 CHECK YOURSELF 6

Completely factor the trinomial.

$6x^3 + 3x^2 - 18x$

Not all possible product pairs need to be tried to find m and n. A look at the sign pattern of the trinomial will eliminate many of the possibilities. Assuming the leading coefficient is positive, there are four possible sign patterns.

Pattern	Example	Conclusion
1. b and c are both positive.	$2x^2 + 13x + 15$	m and n must both be positive.
2. b is negative and c is positive.	$x^2 - 7x + 12$	m and n must both be negative.
3. b is positive and c is negative.	$x^2 + 3x - 10$	m and n are of opposite signs. (The value with the larger absolute value is positive.)
4. b and c are both negative.	$x^2 - 3x - 10$	m and n are of opposite signs. (The value with the larger absolute value is negative.)

CHECK YOURSELF ANSWERS

1. **(a)** False; **(b)** true; **(c)** true   2. **(a)** $a = 1, b = 5, c = -14$;
(b) $a = 1, b = -18, c = 17$; **(c)** $a = 2, b = 1, c = -6$
3. **(a)** Factorable, $m = -3, n = -4$; **(b)** factorable, $m = 7, n = -2$;
(c) not factorable; **(d)** factorable, $m = 4, n = -3$
4. **(a)** $x^2 - 3x - 4x + 12 = (x - 3)(x - 4)$;
(b) $x^2 + 7x - 2x - 14 = (x + 7)(x - 2)$;
(c) $2x^2 + 4x - 3x - 6 = (x + 2)(2x - 3)$
5. **(a)** $2x^2 - 10x + 3x - 15 = (x - 5)(2x + 3)$;
(b) $6x^2 - 8x + 3x - 4 = (3x - 4)(2x + 1)$ 6. $3x(2x - 3)(x + 2)$

4.6 Exercises

State whether each of the following is true or false.

1. $x^2 + 2x - 3 = (x + 3)(x - 1)$ **2.** $y^2 - 3y - 18 = (y - 6)(y + 3)$

3. $x^2 - 10x - 24 = (x - 6)(x + 4)$ **4.** $a^2 + 9a - 36 = (a - 12)(a + 4)$

5. $x^2 - 16x + 64 = (x - 8)(x - 8)$ **6.** $w^2 - 12w - 45 = (w - 9)(w - 5)$

7. $25y^2 - 10y + 1 = (5y - 1)(5y + 1)$

8. $6x^2 + 5xy - 4y^2 = (6x - 2y)(x + 2y)$

9. $10p^2 - pq - 3q^2 = (5p - 3q)(2p + q)$

10. $6a^2 + 13a + 6 = (2a + 3)(3a + 2)$

For each of the following trinomials, label a, b, and c.

11. $x^2 + 4x - 9$ **12.** $x^2 + 5x + 11$

13. $x^2 - 3x + 8$ **14.** $x^2 + 7x - 15$

15. $3x^2 + 5x - 8$ **16.** $2x^2 + 7x - 9$

17. $4x^2 + 11 + 8x$ **18.** $5x^2 - 9 + 7x$

19. $5x - 3x^2 - 10$ **20.** $9x - 7x^2 - 18$

Use the *ac* test to determine which of the following trinomials can be factored. Find the values of *m* and *n* for each trinomial that can be factored.

21. $x^2 + x - 6$ **22.** $x^2 + 2x - 15$

23. $x^2 + x + 2$ **24.** $x^2 - 3x + 7$

25. $x^2 - 5x + 6$ **26.** $x^2 - x + 2$

27. $2x^2 + 5x - 3$ **28.** $3x^2 - 14x - 5$

29. $6x^2 - 19x + 10$ **30.** $4x^2 + 5x + 6$

Rewrite the middle term as the sum of two terms and then factor by grouping.

31. $x^2 + 6x + 8$ **32.** $x^2 + 3x - 10$

33. $x^2 - 9x + 20$ **34.** $x^2 - 8x + 15$

35. $x^2 - 2x - 63$ **36.** $x^2 + 6x - 55$

Rewrite the middle term as the sum of two terms and then factor completely.

37. $x^2 + 8x + 15$ **38.** $x^2 - 11x + 24$

39. $x^2 - 11x + 28$ **40.** $y^2 - y - 20$

41. $s^2 + 13s + 30$ **42.** $b^2 + 14b + 33$

43. $a^2 - 2a - 48$ **44.** $x^2 - 17x + 60$

45. $x^2 - 8x + 7$ **46.** $x^2 + 7x - 18$

47. $x^2 - 6x - 40$ **48.** $x^2 - 11x + 10$

49. $x^2 - 14x + 49$ **50.** $s^2 - 4s - 32$

51. $p^2 - 10p - 24$ **52.** $x^2 - 11x - 60$

ANSWERS

27.

28.

29.

30.

31.

32.

33.

34.

35.

36.

37.

38.

39.

40.

41.

42.

43.

44.

45.

46.

47.

48.

49.

50.

51.

52.

53. _____

54. _____

55. _____

56. _____

57. _____

58. _____

59. _____

60. _____

61. _____

62. _____

63. _____

64. _____

65. _____

66. _____

67. _____

68. _____

69. _____

70. _____

71. _____

72. _____

73. _____

74. _____

75. _____

76. _____

77. _____

78. _____

79. _____

80. _____

53. $x^2 + 5x - 66$

54. $a^2 + 2a - 80$

55. $c^2 + 19c + 60$

56. $t^2 - 4t - 60$

57. $n^2 + 5n - 50$

58. $x^2 - 16x + 63$

59. $x^2 + 7xy + 10y^2$

60. $x^2 - 8xy + 12y^2$

61. $a^2 - ab - 42b^2$

62. $m^2 - 8mn + 16n^2$

63. $x^2 - 13xy + 40y^2$

64. $r^2 - 9rs - 36s^2$

65. $6x^2 + 19x + 10$

66. $6x^2 - 7x - 3$

67. $15x^2 + x - 6$

68. $12w^2 + 19w + 4$

69. $6m^2 + 25m - 25$

70. $8x^2 - 6x - 9$

71. $9x^2 - 12x + 4$

72. $20x^2 - 23x + 6$

73. $12x^2 - 8x - 15$

74. $16a^2 + 40a + 25$

75. $3y^2 + 7y - 6$

76. $12x^2 + 11x - 15$

77. $8x^2 - 27x - 20$

78. $24v^2 + 5v - 36$

79. $2x^2 + 3xy + y^2$

80. $3x^2 - 5xy + 2y^2$

81. $5a^2 - 8ab - 4b^2$

82. $5x^2 + 7xy - 6y^2$

83. $9x^2 + 4xy - 5y^2$

84. $16x^2 + 32xy + 15y^2$

85. $6m^2 - 17mn + 12n^2$

86. $15x^2 - xy - 6y^2$

87. $36a^2 - 3ab - 5b^2$

88. $3q^2 - 17qr - 6r^2$

89. $x^2 + 4xy + 4y^2$

90. $25b^2 - 80bc + 64c^2$

91. $20x^2 - 20x - 15$

92. $24x^2 - 18x - 6$

93. $8m^2 + 12m + 4$

94. $14x^2 - 20x + 6$

95. $15r^2 - 21rs + 6s^2$

96. $10x^2 + 5xy - 30y^2$

97. $2x^3 - 2x^2 - 4x$

98. $2y^3 + y^2 - 3y$

99. $2y^4 + 5y^3 + 3y^2$

100. $4z^3 - 18z^2 - 10z$

101. $36a^3 - 66a^2 + 18a$

102. $20n^4 - 22n^3 - 12n^2$

103. $9p^2 + 30pq + 21q^2$

104. $12x^2 + 2xy - 24y^2$

Find a positive value for k for which each of the following can be factored.

105. $x^2 + kx + 8$

106. $x^2 + kx + 9$

107. $x^2 - kx + 16$

108. $x^2 - kx + 17$

ANSWERS

81. _____

82. _____

83. _____

84. _____

85. _____

86. _____

87. _____

88. _____

89. _____

90. _____

91. _____

92. _____

93. _____

94. _____

95. _____

96. _____

97. _____

98. _____

99. _____

100. _____

101. _____

102. _____

103. _____

104. _____

105. _____

106. _____

107. _____

108. _____

109. _____

110. _____

111. _____

112. _____

113. _____

114. _____

a. _____

b. _____

c. _____

d. _____

e. _____

f. _____

g. _____

h. _____

109. $x^2 - kx - 5$

110. $x^2 - kx - 7$

111. $x^2 + 3x + k$

112. $x^2 + 5x + k$

113. $x^2 + 2x - k$

114. $x^2 + x - k$

 Getting Ready for Section 4.7 [Section 3.4]

Multiply.

   (a) $2x^2(x^2 + 3x - 5)$ (b) $(5a - 3)(2a + 4)$

   (c) $(5m - 3n)(5m + 3n)$ (d) $(x - 2y)(x^2 + 2xy + 4y^2)$

   (e) $(2w - 3z)(5w - z)$ (f) $x(x + 5y)(x - 5y)$

   (g) $(a + 3b)(a^2 - 3ab + 9b^2)$ (h) $2s(3s - r)(2s + r)$

Answers

1. True   **3.** False   **5.** True   **7.** False   **9.** True

11. $a = 1, b = 4, c = -9$ **13.** $a = 1, b = -3, c = 8$

15. $a = 3, b = 5, c = -8$ **17.** $a = 4, b = 8, c = 11$

19. $a = -3, b = 5, c = -10$ **21.** Factorable; $3, -2$

23. Not factorable   **25.** Factorable; $-3, -2$ **27.** Factorable; $6, -1$

29. Factorable; $-15, -4$ **31.** $x^2 + 2x + 4x + 8; (x + 2)(x + 4)$

33. $x^2 - 5x - 4x + 20; (x - 5)(x - 4)$ **35.** $x^2 - 9x + 7x - 63; (x - 9)(x + 7)$

37. $(x + 3)(x + 5)$ **39.** $(x - 4)(x - 7)$ **41.** $(s + 10)(s + 3)$

43. $(a - 8)(a + 6)$ **45.** $(x - 1)(x - 7)$ **47.** $(x - 10)(x + 4)$

49. $(x - 7)(x - 7) = (x - 7)^2$ **51.** $(p - 12)(p + 2)$ **53.** $(x + 11)(x - 6)$

55. $(c + 4)(c + 15)$ **57.** $(n + 10)(n - 5)$ **59.** $(x + 2y)(x + 5y)$

61. $(a - 7b)(a + 6b)$ **63.** $(x - 5y)(x - 8y)$ **65.** $(3x + 2)(2x + 5)$

67. $(5x - 3)(3x + 2)$ **69.** $(6m - 5)(m + 5)$

71. $(3x - 2)(3x - 2) = (3x - 2)^2$ **73.** $(6x + 5)(2x - 3)$

75. $(3y - 2)(y + 3)$ **77.** $(8x + 5)(x - 4)$ **79.** $(2x + y)(x + y)$

81. $(5a + 2b)(a - 2b)$ **83.** $(9x - 5y)(x + y)$ **85.** $(3m - 4n)(2m - 3n)$

87. $(12a - 5b)(3a + b)$ **89.** $(x + 2y)^2$ **91.** $5(2x - 3)(2x + 1)$

93. $4(2m + 1)(m + 1)$ **95.** $3(5r - 2s)(r - s)$ **97.** $2x(x - 2)(x + 1)$

99. $y^2(2y + 3)(y + 1)$ **101.** $6a(3a - 1)(2a - 3)$ **103.** $3(p + q)(3p + 7q)$

105. 6 or 9   **107.** 8 or 10 or 17   **109.** 4   **111.** 2   **113.** $3, 8, 15, 24, \ldots$

a. $2x^4 + 6x^3 - 10x^2$ **b.** $10a^2 + 14a - 12$ **c.** $25m^2 - 9n^2$ **d.** $x^3 - 8y^3$

e. $10w^2 - 17wz + 3z^2$ **f.** $x^3 - 25xy^2$ **g.** $a^3 + 27b^3$

h. $12s^3 + 2s^2r - 2sr^2$

4.7 Strategies in Factoring

4.7 OBJECTIVES

1. Recognize factoring patterns
2. Apply appropriate factoring strategies

In Sections 4.1 to 4.6 you have seen a variety of techniques for factoring polynomials. This section reviews those techniques and presents some guidelines for choosing an appropriate strategy or a combination of strategies.

1. Always look for a greatest common factor. If you find a GCF (other than 1), factor out the GCF as your first step.

   To factor $5x^2y - 10xy + 25xy^2$, the GCF is $5xy$, so

   $$5x^2y - 10xy + 25xy^2 = 5xy(x - 2 + 5y)$$

2. Now look at the number of terms in the polynomial you are trying to factor.

   a. If the polynomial is a *binomial,* consider the formula for the difference of two squares. Recall that a sum of squares will not, in general, factor over the real numbers.

      (i) To factor $x^2 - 49y^2$, recognize the difference of squares, so

      $$x^2 - 49y^2 = (x + 7y)(x - 7y)$$

      (ii) The binomial

      $$x^2 + 64$$

      cannot be further factored.

   b. If the polynomial is a *trinomial,* try to factor the trinomial as a product of two binomials, using trial and error.

      To factor $2x^2 - x - 6$, a consideration of possible factors of the first and last terms of the trinomial will lead to

      $$2x^2 - x - 6 = (2x + 3)(x - 2)$$

   > **NOTE** You may prefer to use the *ac* method shown in Section 4.6

   c. If the polynomial has *more than three terms,* try factoring by grouping.

      To factor $2x^2 - 3xy + 10x - 15y$, group the first two terms, and then the last two, and factor out common factors.

      $$2x^2 - 3xy + 10x - 15y = x(2x - 3y) + 5(2x - 3y)$$

      Now factor out the common factor $(2x - 3y)$.

      $$2x^2 - 3xy + 10x - 15y = (2x - 3y)(x + 5)$$

3. You should always factor the given polynomial completely. So after you apply one of the techniques given in part 2, another one may be necessary

   a. To factor

      $$6x^3 + 22x^2 - 40x$$

      first factor out the common factor of $2x$. So

      $$6x^3 + 22x^2 - 40x = 2x(3x^2 + 11x - 20)$$

      Now continue to factor the trinomial as before and

      $$6x^3 + 22x^2 - 40x = 2x(3x - 4)(x + 5)$$

b. To factor

$$x^3 - x^2y - 4x + 4y$$

first we proceed by grouping:

$$x^3 - x^2y - 4x + 4y = x^2(x - y) - 4(x - y)$$
$$= (x - y)(x^2 - 4)$$

Now because $x^2 - 4$ is a difference of two squares, we continue to factor and obtain

$$x^3 - x^2y - 4x + 4y = (x - y)(x + 2)(x - 2)$$

OBJECTIVE 1   **Example 1   Recognizing Factoring Patterns**

For each of the following expressions, state the appropriate first step for factoring the polynomial.

(a) $9x^2 - 18x - 72$

Find the GCF.

(b) $x^2 - 3x + 2xy - 6y$

Group the terms.

(c) $x^4 - 81y^4$

Factor the difference of squares.

(d) $3x^2 + 7x + 2$

Use the *ac* method (or trial and error).

 CHECK YOURSELF 1

For each of the following expressions, state the appropriate first step for factoring the polynomial.

(a) $5x^2 + 2x - 3$

(b) $a^4b^4 - 16$

(c) $3x^2 + 3x - 60$

(d) $2a^2 - 5a + 4ab - 10b$

OBJECTIVE 2   **Example 2   Factoring Polynomials**

For each of the following expressions, completely factor the polynomial.

(a) $9x^2 - 18x - 72$

The GCF is 9.

$$9x^2 - 18x - 72 = 9(x^2 - 2x - 8)$$
$$= 9(x - 4)(x + 2)$$

(b) $x^2 - 3x + 2xy - 6y$

Grouping the terms, we have

$$x^2 - 3x + 2xy - 6y = (x^2 - 3x) + (2xy - 6y)$$
$$= x(x - 3) + 2y(x - 3)$$
$$= (x - 3)(x + 2y)$$

(c) $x^4 - 81y^4$

Factoring the difference of squares, we find

$$x^4 - 81y^4 = (x^2 + 9y^2)(x^2 - 9y^2)$$
$$= (x^2 + 9y^2)(x - 3y)(x + 3y)$$

(d) $3x^2 + 7x + 2$

Using the *ac* method, we find $m = 1$ and $n = 6$.

$$3x^2 + 7x + 2 = 3x^2 + x + 6x + 2$$
$$= (3x^2 + x) + (6x + 2)$$
$$= x(3x + 1) + 2(3x + 1)$$
$$= (3x + 1)(x + 2)$$

CHECK YOURSELF 2

For each of the following expressions, completely factor the polynomial.

(a) $5x^2 + 2x - 3$

(b) $a^4b^4 - 16$

(c) $3x^2 + 3x - 60$

(d) $2a^2 - 5a + 4ab - 10b$

CHECK YOURSELF ANSWERS

1. (a) *ac* method (or trial and error); **(b)** factor the difference of squares;
(c) find the GCF; **(d)** group the terms
2. (a) $(5x - 3)(x + 1)$; **(b)** $(a^2b^2 + 4)(ab - 2)(ab + 2)$; **(c)** $3(x + 5)(x - 4)$;
(d) $(2a - 5)(a + 2b)$

4.7   Exercises

Factor each polynomial completely. To begin, state which method should be applied as the first step, given the guidelines of this section. Then continue the exercise and factor each polynomial completely.

1. $x^2 - 3x$

2. $4y^2 - 9$

3. $x^2 - 5x - 24$

4. $8x^3 + 10x$

5. $x(x - y) + 2(x - y)$

6. $5a^2 - 10a + 25$

7. $2x^2y - 6xy + 8y^2$

8. $2p - 6q + pq - 3q^2$

9. $y^2 - 13y + 40$

10. $m^3 + 27m^2n$

11. $3b^2 + 17b - 28$

12. $3x^2 + 6x - 5xy - 10y$

13. $3x^2 - 14xy - 24y^2$

14. $16c^2 - 49d^2$

15. $2a^2 + 11a + 12$

16. $m^3n^3 - mn$

17. $125r^3 + r^2$

18. $(x - y)^2 - 16$

19. $3x^2 - 30x + 63$

20. $3a^2 - 108$

21. $40a^2 + 5$

22. $4p^2 - 8p - 60$

23. $2w^2 - 14w - 36$

24. $xy^3 - 9xy$

25. $3a^2b - 48b^3$

26. $12b^3 - 86b^2 + 14b$

27. $x^4 - 3x^2 - 10$

28. $m^4 - 9n^4$

Factor completely.

29. $(x - 5)^2 - 169$

30. $(x - 7)^2 - 81$

31. $x^2 + 4xy + 4y^2 - 16$

32. $9x^2 + 12xy + 4y^2 - 25$

33. $6(x - 2)^2 + 7(x - 2) - 5$

34. $12(x + 1)^2 - 17(x + 1) + 6$

Getting Ready for Section 4.8 [Section 2.3]

Solve.

(a) $x - 5 = 0$
(b) $2x - 1 = 0$
(c) $3x + 2 = 0$
(d) $x + 4 = 0$
(e) $7 - x = 0$
(f) $9 - 4x = 0$

Answers

1. GCF, $x(x - 3)$ **3.** Trial and error, $(x - 8)(x + 3)$
5. GCF, $(x - y)(x + 2)$ **7.** GCF, $2y(x^2 - 3x + 4y)$
9. Trial and error, $(y - 5)(y - 8)$ **11.** Trial and error, $(b + 7)(3b - 4)$
13. Trial and error, $(3x + 4y)(x - 6y)$ **15.** Trial and error, $(2a + 3)(a + 4)$
17. GCF, $r^2(125r + 1)$ **19.** GCF, then trial and error, $3(x - 3)(x - 7)$
21. GCF, $5(8a^2 + 1)$ **23.** GCF, then trial and error, $2(w - 9)(w + 2)$
25. GCF, then difference of squares, $3b(a + 4b)(a - 4b)$
27. Trial and error, $(x^2 - 5)(x^2 + 2)$ **29.** $(x + 8)(x - 18)$
31. $(x + 2y + 4)(x + 2y - 4)$ **33.** $(2x - 5)(3x - 1)$

a. $x = 5$ **b.** $x = \dfrac{1}{2}$ **c.** $x = -\dfrac{2}{3}$ **d.** $x = -4$ **e.** $x = 7$

f. $x = \dfrac{9}{4}$

23. _____

24. _____

25. _____

26. _____

27. _____

28. _____

29. _____

30. _____

31. _____

32. _____

33. _____

34. _____

a. _____

b. _____

c. _____

d. _____

e. _____

f. _____

4.8 Solving Quadratic Equations by Factoring

4.8 OBJECTIVES

1. Solve quadratic equations by factoring
2. Solve applications of quadratic equations

The factoring techniques you have learned provide us with tools for solving equations that can be written in the form

$$ax^2 + bx + c = 0 \qquad a \neq 0$$

This is a quadratic equation in one variable, here x. You can recognize such a quadratic equation by the fact that the highest power of the variable x is the second power.

in which a, b, and c are constants.

An equation written in the form $ax^2 + bx + c = 0$ is called a **quadratic equation in standard form.** Using factoring to solve quadratic equations requires the **zero-product principle,** which says that if the product of two factors is 0, then one or both of the factors must be equal to 0. In symbols:

> **Definition:** Zero-Product Principle
>
> If $a \cdot b = 0$, then $a = 0$ or $b = 0$ or $a = b = 0$.

We can now apply this principle to solve quadratic equations.

OBJECTIVE 1

Example 1 Solving Equations by Factoring

Solve.

$$x^2 - 3x - 18 = 0$$

Factoring on the left, we have

$$(x - 6)(x + 3) = 0$$

NOTE To use the zero-product principle, 0 must be on one side of the equation.

By the zero-product principle, we know that one or both of the factors must be zero. We can then write

$$x - 6 = 0 \qquad \text{or} \qquad x + 3 = 0$$

Solving each equation gives

$$x = 6 \qquad \text{or} \qquad x = -3$$

The two solutions are 6 and -3.

Quadratic equations can be checked in the same way as linear equations were checked: by substitution. For instance, if $x = 6$, we have

$$6^2 - 3 \cdot 6 - 18 \stackrel{?}{=} 0$$
$$36 - 18 - 18 \stackrel{?}{=} 0$$
$$0 = 0$$

which is a true statement. We leave it to you to check the solution -3.

CHECK YOURSELF 1

Solve $x^2 - 9x + 20 = 0$.

$(x - 4)(x - 5) = 0$

Other factoring techniques are also used in solving quadratic equations. Example 2 illustrates this.

$x - 4 = 0 \quad x - 5 = 0$
$ +4 \quad\quad +5 +5$
$x = 4 \quad\quad x = 5$

| **Example 2** | **Solving Equations by Factoring** |

(a) Solve $x^2 - 5x = 0$.

Again, factor the left side of the equation and apply the zero-product principle.

$x(x - 5) = 0$

 C A U T I O N

A *common mistake* is to forget the statement $x = 0$ when you are solving equations of this type. Be sure to include the *two statements* obtained.

Now

$x = 0 \qquad \text{or} \qquad x - 5 = 0$
$\phantom{x = 0 \qquad \text{or} \qquad} x = 5$

The two solutions are 0 and 5.

(b) Solve $x^2 - 9 = 0$.

Factoring yields

$(x + 3)(x - 3) = 0$

$x + 3 = 0 \qquad \text{or} \qquad x - 3 = 0$
$ x = -3 \phantom{qquad \text{or} \qquad} x = 3$

NOTE The symbol $\pm$ is read "plus or minus."

The solutions may be written as $x = \pm 3$.

CHECK YOURSELF 2

Solve by factoring.

(a) $x^2 + 8x = 0$ 　　　　　　　　　　　　　**(b)** $x^2 - 16 = 0$

 C A U T I O N

Consider the equation
$x(2x - 1) = 3$

NOTE Students are sometimes tempted to write

$x = 3 \qquad \text{or} \qquad 2x - 1 = 3$

This is *not correct.* Instead, subtract 3 from both sides of the equation *as the first step* to write

$x(2x - 1) - 3 = 0$

Then proceed to write the equation in standard form. Only then can you factor and proceed as before.

Example 3 illustrates a crucial point. Our solution technique depends on the zero-product principle, which means that the product of factors *must be equal to 0*. The importance of this is shown now.

| **Example 3** | **Solving Equations by Factoring** |

Solve $2x^2 - x = 3$.

The first step in the solution is to write the equation in standard form (that is, write it so that one side of the equation is 0). So start by adding -3 to both sides of the equation. Then,

$2x^2 - x - 3 = 0$ 　　Make sure all terms are on one side of
　　　　　　　　　　the equation. The other side will be 0.

You can now factor and solve by using the zero-product principle.

$$(2x - 3)(x + 1) = 0$$

$$2x - 3 = 0 \quad \text{or} \quad x + 1 = 0$$

$$2x = 3 \qquad\qquad x = -1$$

$$x = \frac{3}{2}$$

The solutions are $\frac{3}{2}$ and -1.

 CHECK YOURSELF 3

Solve $3x^2 = 5x + 2$.

In all the previous examples, the quadratic equations had two distinct real number solutions. That may not always be the case, as we shall see.

Example 4   Solving Equations by Factoring

Solve $x^2 - 6x + 9 = 0$.

Factoring, we have

$$(x - 3)(x - 3) = 0$$

and

$$x - 3 = 0 \quad \text{or} \quad x - 3 = 0$$

$$x = 3 \qquad\qquad x = 3$$

The solution is 3.

A quadratic (or second-degree) equation always has *two* solutions. When an equation such as this one has two solutions that are the same number, we call 3 the **repeated** (or **double**) **solution** of the equation.

Although a quadratic equation will always have two solutions, they may not always be real numbers. You will learn more about this in a later course.

 CHECK YOURSELF 4

Solve $x^2 + 6x + 9 = 0$.

Always examine the quadratic member of an equation for common factors. It will make your work much easier, as Example 5 illustrates.

Example 5   Solving Equations by Factoring

Solve $3x^2 - 3x - 60 = 0$.

First, note the common factor 3 in the quadratic member of the equation. Factoring out the 3, we have

$$3(x^2 - x - 20) = 0$$

Now divide both sides of the equation by 3.

$$\frac{3(x^2 - x - 20)}{3} = \frac{0}{3}$$

or

$$x^2 - x - 20 = 0$$

We can now factor and solve as before.

$$(x - 5)(x + 4) = 0$$

$$x - 5 = 0 \quad \text{or} \quad x + 4 = 0$$
$$x = 5 \qquad\qquad x = -4$$

CHECK YOURSELF 5

Solve $2x^2 - 10x - 48 = 0$.

Many applications can be solved with quadratic equations.

OBJECTIVE 2   **Example 6   Solving an Application**

The Microhard corporation has found that the equation

$$P = x^2 - 7x - 94$$

describes the profit P, in thousands of dollars, for every x hundred computers sold. How many computers were sold if the profit was \$50,000?

If the profit was \$50,000, then $P = 50$. We now set up and solve the equation.

NOTE P is expressed in thousands so the value 50 is substituted for P, not 50,000.

$$50 = x^2 - 7x - 94$$
$$0 = x^2 - 7x - 144$$
$$0 = (x + 9)(x - 16)$$
$$x = -9 \quad \text{or} \quad x = 16$$

They cannot sell a negative number of computers, so $x = 16$. They sold 1600 computers.

CHECK YOURSELF 6

The Gerbil Babyfood corporation has found that the equation

$$P = x^2 - 6x - 7$$

describes the profit P, in hundreds of dollars, for every x thousand jars sold. How many jars were sold if the profit was \$2000?

CHECK YOURSELF ANSWERS

1. 4, 5   **2. (a)** 0, −8; **(b)** 4, −4   **3.** $-\frac{1}{3}$, 2   **4.** −3   **5.** −3, 8   **6.** 9000 jars

4.8 Exercises

Solve each of the following quadratic equations.

1. $(x - 3)(x - 4) = 0$

2. $(x - 7)(x + 1) = 0$

3. $(3x + 1)(x - 6) = 0$

4. $(5x - 4)(x - 6) = 0$

5. $x^2 - 2x - 3 = 0$

6. $x^2 + 5x + 4 = 0$

7. $x^2 - 7x + 6 = 0$

8. $x^2 + 3x - 10 = 0$

9. $x^2 + 8x + 15 = 0$

10. $x^2 - 3x - 18 = 0$

11. $x^2 + 4x - 21 = 0$

12. $x^2 - 12x + 32 = 0$

13. $x^2 - 4x = 12$

14. $x^2 + 8x = -15$

15. $x^2 + 5x = 14$

16. $x^2 = 11x - 24$

17. $2x^2 + 5x - 3 = 0$

18. $3x^2 + 7x + 2 = 0$

19. $4x^2 - 24x + 35 = 0$

20. $6x^2 + 11x - 10 = 0$

21. $4x^2 + 11x = -6$

22. $5x^2 + 2x = 3$

23. $5x^2 + 13x = 6$

24. $4x^2 = 13x + 12$

25. $x^2 - 2x = 0$

26. $x^2 + 5x = 0$

27. $x^2 = -8x$

28. $x^2 = 7x$

29. $5x^2 - 15x = 0$

30. $4x^2 + 20x = 0$

31. $x^2 - 25 = 0$

32. $x^2 = 49$

33. $x^2 = 81$

34. $x^2 = 64$

35. $2x^2 - 18 = 0$

36. $3x^2 - 75 = 0$

37. $3x^2 + 24x + 45 = 0$

38. $4x^2 - 4x = 24$

39. $2x(3x + 14) = 10$

40. $3x(5x + 9) = 6$

41. $(x + 3)(x - 2) = 14$

42. $(x - 5)(x + 2) = 18$

Solve the following problems.

43. Number problem. The product of two consecutive integers is 132. Find the two integers.

44. Number problem. The product of two consecutive positive even integers is 120. Find the two integers.

45. Number problem. The sum of an integer and its square is 72. What is the integer?

46. Number problem. The square of an integer is 56 more than the integer. Find the integer.

47. Geometry. If the sides of a square are increased by 3 in., the area is increased by 39 in.^{2}. What were the dimensions of the original square?

48. Geometry. If the sides of a square are decreased by 2 cm, the area is decreased by 36 cm^2. What were the dimensions of the original square?

49. Business and finance. The profit on a small appliance is given by $P = x^2 - 3x - 60$, in which x is the number of appliances sold per day. How many appliances were sold on a day when there was a $20 loss?

50. Business and finance. The relationship between the number of calculators x that a company can sell per month and the price of each calculator p is given by $x = 1700 - 100p$. Find the price at which a calculator should be sold to produce a monthly revenue of $7000. (*Hint:* Revenue $= xp$.)

ANSWERS

35. _____

36. _____

37. _____

38. _____

39. _____

40. _____

41. _____

42. _____

43. _____

44. _____

45. _____

46. _____

47. _____

48. _____

49. _____

50. _____

51. Write a short comparison that explains the difference between $ax^2 + bx + c$ and $ax^2 + bx + c = 0$.

52. When solving quadratic equations, some people try to solve an equation in the manner shown below, but this does not work! Write a paragraph to explain what is wrong with this approach.

$$2x^2 + 7x + 3 = 52$$
$$(2x + 1)(x + 3) = 52$$
$$2x + 1 = 52 \quad \text{or} \quad x + 3 = 52$$
$$x = \frac{51}{2} \quad \text{or} \quad x = 49$$

Answers

1. 3, 4   **3.** $-\dfrac{1}{3}$, 6   **5.** -1, 3   **7.** 1, 6   **9.** $-3, -5$ **11.** $-7, 3$

13. $-2, 6$ **15.** $-7, 2$ **17.** $-3, \dfrac{1}{2}$ **19.** $\dfrac{5}{2}, \dfrac{7}{2}$ **21.** $-\dfrac{3}{4}, -2$

23. $-3, \dfrac{2}{5}$ **25.** 0, 2   **27.** $0, -8$ **29.** 0, 3   **31.** $-5, 5$ **33.** $-9, 9$

35. $-3, 3$ **37.** $-5, -3$ **39.** $-5, \dfrac{1}{3}$ **41.** $4, -5$

43. 11, 12 or $-12, -11$ **45.** -9 or 8   **47.** 5 in. by 5 in.   **49.** 8

51.

4 Summary

DEFINITION/PROCEDURE	EXAMPLE	REFERENCE
An Introduction to Factoring		**Section 4.1**
Common Monomial Factor A single term that is a factor of every term of the polynomial. The greatest common factor (GCF) of a polynomial is the factor that is a product of (a) the largest possible numerical coefficient and (b) each variable with the smallest exponent in any term.	$4x^2$ is the greatest common monomial factor of $8x^4 - 12x^3 + 16x^2$.	 p. 350
Factoring a Monomial from a Polynomial 1. Determine the GCF for all terms. 2. Use the GCF to factor each term and then apply the distributive law in the form $$ab + ac = a(b + c)$$ The greatest common factor 3. Mentally check by multiplication.	$8x^4 - 12x^3 + 16x^2$ $= 4x^2(2x^2 - 3x + 4)$	 p. 350
Factoring Trinomials		**Sections 4.2 and 4.3**
Trial and Error To factor a trinomial, find the appropriate sign pattern and then find integer values that yield the appropriate coefficients for the trinomial.	$x^2 - 5x - 24$ $= (x -)(x +)$ $= (x - 8)(x + 3)$	 p. 360/366
Difference of Squares and Perfect Square Trinomials		**Section 4.4**
Factoring a Difference of Squares Use the following form: $$a^2 - b^2 = (a + b)(a - b)$$	To factor: $16x^2 - 25y^2$: Think: $(4x)^2 - (5y)^2$ so $16x^2 - 25y^2$ $= (4x + 5y)(4x - 5y)$	 p. 375
Factoring a Perfect Square Trinomial Use the following form: $$a^2 + 2ab + b^2 = (a + b)^2$$	$4x^2 + 12xy + 9y^2$ $= (2x)^2 + 2(2x)(3y) + (3y)^2$ $= (2x + 3y)^2$	 p. 377
Factoring by Grouping		**Section 4.5**
When there are four terms of a polynomial, factor the first pair and factor the last pair. If these two pairs have a common binomial factor, factor that out. The result will be the product of two binomials.	$4x^2 - 6x + 10x - 15$ $= 2x(2x - 3) + 5(2x - 3)$ $= (2x - 3)(2x + 5)$	 p. 382

Continued

Using the *ac* Method to Factor		Section 4.6
Factoring Trinomials To factor a trinomial, first use the *ac* test to determine factorability. If the trinomial is factorable, the *ac* test will yield two terms (which have as their sum the middle term) that allow the factoring to be completed by using the grouping method.	$x^2 + 3x - 28$ $ac = -28; b = 3$ $mn = -28; m + n = 3$ $m = 7, n = -4$ $x^2 + 7x - 4x - 28$ $= x(x + 7) - 4(x + 7)$ $= (x - 4)(x + 7)$	*p. 387*

Strategies in Factoring		Section 4.7
When factoring a polynomial, **1.** Look for the GCF. **2.** Consider the number of terms. 　**a.** If it is a binomial, look for a difference of squares. 　**b.** If it is trinomial, use the *ac* method or trial and error. 　**c.** If there are four or more terms, try grouping terms. **3.** Be certain that the polynomial is completely factored.	Given $12x^3 - 86x^2 + 14x$, factor out $2x$. $2x(6x^2 - 43x + 7)$ $= 2x(6x - 1)(x - 7)$	*p. 399*

Solving Quadratic Equations by Factoring		Section 4.8
1. Add or subtract the necessary terms on both sides of the equation so that the equation is in standard form (set equal to 0). **2.** Factor the quadratic expression. **3.** Set each factor equal to 0. **4.** Solve the resulting equations to find the solutions. **5.** Check each solution by substituting in the original equation.	To solve $x^2 + 7x = 30$ $x^2 + 7x - 30 = 0$ $(x + 10)(x - 3) = 0$ $x + 10 = 0$ or $x - 3 = 0$ $x = -10$ and $x = 3$ are solutions.	*p. 404*

Summary Exercises

This summary exercise set is provided to give you practice with each of the objectives of this chapter. Each exercise is keyed to the appropriate chapter section. When you are finished, you can check your answers to the odd-numbered exercises against those presented in the back of the text. If you have difficulty with any of these questions, go back and reread the examples from that section. Your instructor will give you guidelines on how to best use these exercises in your instructional setting.

[4.1] Factor each of the following polynomials.

1. $18a + 24$

2. $9m^2 - 21m$

3. $24s^2t - 16s^2$

4. $18a^2b + 36ab^2$

5. $35s^3 - 28s^2$

6. $3x^3 - 6x^2 + 15x$

7. $18m^2n^2 - 27m^2n + 18m^2n^3$

8. $121x^8y^3 + 77x^6y^3$

9. $8a^2b + 24ab - 16ab^2$

10. $3x^2y - 6xy^3 + 9x^3y - 12xy^2$

11. $x(2x - y) + y(2x - y)$

12. $5(w - 3z) - w(w - 3z)$

[4.2] Factor each of the following trinomials completely.

13. $x^2 + 9x + 20$

14. $x^2 - 10x + 24$

15. $a^2 - a - 12$

16. $w^2 - 13w + 40$

17. $x^2 + 12x + 36$

18. $r^2 - 9r - 36$

19. $b^2 - 4bc - 21c^2$

20. $m^2n + 4mn - 32n$

21. $m^3 + 2m^2 - 35m$

22. $2x^2 - 2x - 40$

23. $3y^3 - 48y^2 + 189y$

24. $3b^3 - 15b^2 - 42b$

[4.3] Factor each of the following trinomials completely.

25. $3x^2 + 8x + 5$

26. $5w^2 + 13w - 6$

27. $2b^2 - 9b + 9$

28. $8x^2 + 2x - 3$

29. $10x^2 - 11x + 3$

30. $4a^2 + 7a - 15$

31. $9y^2 - 3yz - 20z^2$

32. $8x^2 + 14xy - 15y^2$

33. $8x^3 - 36x^2 - 20x$

34. $9x^2 - 15x - 6$

35. $6x^3 - 3x^2 - 9x$

36. $5w^2 - 25wz + 30z^2$

[4.4]   Factor each of the following completely.

37. $p^2 - 49$

38. $25a^2 - 16$

39. $m^2 - 9n^2$

40. $16r^2 - 49s^2$

41. $25 - z^2$

42. $a^4 - 16b^2$

43. $25a^2 - 36b^2$

44. $x^6 - 4y^2$

45. $3w^3 - 12wz^2$

46. $9a^4 - 49b^2$

47. $2m^2 - 72n^4$

48. $3w^3z - 12wz^3$

49. $x^2 + 8x + 16$

50. $x^2 - 18x + 81$

51. $4x^2 + 12x + 9$

52. $9x^2 - 12x + 4$

53. $16x^3 + 40x^2 + 25x$

54. $4x^3 - 4x^2 + x$

[4.5–4.7]   Factor the following polynomials completely.

55. $x^2 - 4x + 5x - 20$

56. $x^2 + 7x - 2x - 14$

57. $6x^2 + 4x - 15x - 10$

58. $12x^2 - 9x - 28x + 21$

59. $6x^3 + 9x^2 - 4x^2 - 6x$

60. $3x^4 + 6x^3 + 5x^3 + 10x^2$

[4.8]   Solve each of the following quadratic equations.

61. $(x - 1)(2x + 3) = 0$

62. $x^2 - 5x + 6 = 0$

63. $x^2 - 10x = 0$

64. $x^2 = 144$

65. $x^2 - 2x = 15$

66. $3x^2 - 5x - 2 = 0$

67. $4x^2 - 13x + 10 = 0$

68. $2x^2 - 3x = 5$

69. $3x^2 - 9x = 0$

70. $x^2 - 25 = 0$

71. $2x^2 - 32 = 0$

72. $2x^2 - x - 3 = 0$

Self-Test for Chapter 4

Name _____

Section _____ Date _____

The purpose of this self-test is to help you check your progress and to review for the next in-class exam. Allow yourself about an hour to take this test. At the end of that hour check your answers against those given in the back of the text. Section references accompany the answers. If you missed any questions, go back to those sections and reread the examples until you master the concepts.

ANSWERS

Factor each of the following polynomials.

1. $12b + 18$

2. $9p^3 - 12p^2$

3. $5x^2 - 10x + 20$

4. $6a^2b - 18ab + 12ab^2$

Factor each of the following polynomials completely.

5. $a^2 - 10a + 25$

6. $64m^2 - n^2$

7. $49x^2 - 16y^2$

8. $32a^2b - 50b^3$

Factor each of the following polynomials completely.

9. $a^2 - 5a - 14$

10. $b^2 + 8b + 15$

11. $x^2 - 11x + 28$

12. $y^2 + 12yz + 20z^2$

13. $x^2 + 2x - 5x - 10$

14. $6x^2 + 2x - 9x - 3$

Factor each of the following polynomials completely.

15. $2x^2 + 15x - 8$

16. $3w^2 + 10w + 7$

17. $8x^2 - 2xy - 3y^2$

18. $6x^3 + 3x^2 - 30x$

1. _____

2. _____

3. _____

4. _____

5. _____

6. _____

7. _____

8. _____

9. _____

10. _____

11. _____

12. _____

13. _____

14. _____

15. _____

16. _____

17. _____

18. _____

19. _____

20. _____

21. _____

22. _____

23. _____

24. _____

25. _____

Solve each of the following equations for x.

19. $x^2 - 8x + 15 = 0$

20. $x^2 - 3x = 4$

21. $3x^2 + x - 2 = 0$

22. $4x^2 - 12x = 0$

23. $x(x - 4) = 0$

24. $(x - 3)(x - 2) = 30$

25. $x^2 - 14x = -49$

Cumulative Review
Chapters 0–4

The following exercises are presented to help you review concepts from earlier chapters. This is meant as a review and not as a comprehensive exam. The answers are presented in the back of the text. Section references accompany the answers. If you have difficulty with any of these exercises, be certain to at least read through the summary related to those sections.

ANSWERS

1. 17

2. -2

3. $9x^2 - x - 5$

4. $-4a^2 - 2a - 5$

5. $6b^2 + 8b - 3$

6. $15r^3s^2 - 12r^2s^2 + 18rs^3$

7. $6a^3 - 5a^2b + 3ab^2 - b^3$

8. $-y^2 + 3xy - 2x^2$

9. $3a + 2$

10. $x^2 - 2x + \dfrac{5}{2x} + 4$

11. $x = 2$

12. $x \le 33/5$

13. $t = \dfrac{2S - na}{n}$

Perform the indicated operations.

1. $7 - (-10)$

2. $(-34) \div (17)$

Perform each of the indicated operations.

3. $(7x^2 + 5x - 4) + (2x^2 - 6x - 1)$

$9x^2 - x - 5$

4. $(3a^2 - 2a) - (7a^2 + 5)$

$-4a^2 - 2a - 5$

5. Subtract $4b^2 - 3b$ from the sum of $6b^2 + 5b$ and $4b^2 - 3$.

$$6b^2 + 5b$$
$$+ 4b^2 - 3$$

$$10b^2 + 5b - 3$$
$$4b^2 + 3b$$

$$6b^2 + 8b - 3$$

6. $3rs(5r^2s - 4rs + 6rs^2)$

$15r^3s^2 - 12r^2s^2 + 18r^2s^3$

7. $(2a - b)(3a^2 - ab + b^2)$

$6a^3 - 2a^2b + 2ab^2 - 3a^2b + ab^2 - b^3$

8. $\dfrac{7xy^3 - 21x^2y^2 + 14x^3y}{-7xy}$

$-y^2 + 3xy$

9. $\dfrac{3a^2 - 10a - 8}{a - 4}$

10. $\dfrac{2x^3 - 8x + 5}{2x + 4}$

Solve the following equation for x.

11. $2 - 4(3x + 1) = 8 - 7x$

$2 - 12x - 4 = 8 - 7x$
$+4 +4$

$2 - 12x = 12 - 7x$
$+12x \quad +12x$
$2 = 12 + 5x$
$-12 \; -12$
$\dfrac{10}{5} = \dfrac{5x}{5}$

$x \le +\dfrac{33}{5}$

Solve the following inequality.

12. $4(x - 7) \le -(x - 5)$

$4x - 28 \le -x + 5$
$+28 \qquad +28$

$4x \le -x + 33$
$+x \quad +x$
$5x \le +33/5$

Solve the following equation for the indicated variable.

13. $S = \dfrac{n}{2}(a + t)$ for t

417

14. x^{17}

15. $16x^5y^7$

16. $9x^4y^6$

17. $4xy^2$

18. $108x^7$

19. $12w^4(3w-4)$

20. $5xy(x-3+2y)$

21. $(5x+3y)^2$

22. $4p(p^2-36q^2)$

23. $(a+3)(a+1)$

24. $2w(w^2-2w-12)$

25. $(3x+2y)(x+3y)$

26. $3, 4$

27. $-4, 4$

28. $2/3, -1$

29. 6

30. 5 in $\times$ 21 in

Simplify the following expressions.

14. x^6x^{11} x^{17}

15. $(3x^2y^3)(2x^3y^4)$
$16x^5y^7$

16. $(3x^2y^3)^2(-4x^3y^2)^0$
$9x^4y^6$

17. $\dfrac{16x^2y^5}{4xy^3}$
$4xy^2$

18. $(3x^2)^3(2x)^2$
$(27x^5)(4x^2)$
$108x^7$

Factor each of the following polynomials completely.

19. $36w^5 - 48w^4$
$12w^4(3w-4)$

20. $5x^2y - 15xy + 10xy^2$
$5xy(x-3+2y)$

21. $25x^2 + 30xy + 9y^2$

22. $4p^3 - 144pq^2$
$4p(p^2-36q^2)$

23. $a^2 + 4a + 3$
$(a+3)(a+1)$

24. $2w^3 - 4w^2 - 24w$
$2w(w^2-2w-12)$

25. $3x^2 + 11xy + 6y^2$
$(3x+2y)(x+3y)$

Solve each of the following equations.

26. $a^2 - 7a + 12 = 0$
$(a-3)(a-4)$
$3, 4$

27. $3w^2 - 48 = 0$
$3, 4$

28. $15x^2 + 5x = 10$
$5x(3x+1)=10$

Solve the following problems.

29. Number problem. Twice the square of a positive integer is 12 more than 10 times that integer. What is the integer?

30. Geometry. The length of a rectangle is 1 in. more than 4 times its width. If the area of the rectangle is 105 in.^{2}, find the dimensions of the rectangle.

RATIONAL EXPRESSIONS

INTRODUCTION

In the United States, disorders of the heart and circulatory system kill more people than all other causes combined. The major risk factors for heart disease are smoking, high blood pressure, obesity, high cholesterol, and a family history of heart problems. Although nothing can be done about family history, everyone can affect the first four risk factors through diet and exercise.

One quick way to check your risk of heart problems is to compare your waist and hip measurements. Measure around your waist at the navel and around your hips at the largest point. These measures may be in inches or centimeters. Use the ratio w/h to assess your risk. For women, $w/h \geq 0.8$ indicates an increased health risk, and for men, $w/h \geq 0.95$ is the indicator of an increased risk.

The American Medical Association sponsored a study using Body Mass Index, or BMI, which used height and weight measurements:

$$\text{BMI} = \frac{705w}{h^2}$$

in which w = weight in pounds
h = height in inches

This study concluded that people with BMI ≤ 21 had the lowest rates of heart disease, and that an increase of only 2 points in the BMI dramatically raises the risk of heart problems.

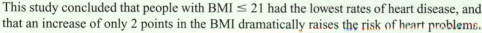

Medical professionals and researchers continue to disagree about how accurate these indicators are because each is a statistical average. One issue is how well the measures relate to the percentage of total body fat. A person may have a relatively low percentage of body fat and be in excellent health but have a BMI over 21 because of a very muscular build or large bone structure.

Pre-Test Chapter 5

This pre-test will provide a preview of the types of exercises you will encounter in each section of this chapter. The answers for these exercises can be found in the back of the text. If you are working on your own, or ahead of the class, this pre-test can help you identify the sections in which you should focus more of your time.

Write each fraction in simplest form.

[5.1] **1.** $\dfrac{-15a^4 b^7}{25a^6 b}$ **2.** $\dfrac{x^2 - 16}{2x - 8}$ **3.** $\dfrac{3x^2 - 2x - 1}{6x^2 + 2x}$

Add or subtract as indicated.

[5.3] **4.** $\dfrac{7a}{12} + \dfrac{19a}{12}$ **5.** $\dfrac{5x}{x+1} + \dfrac{5}{x+1}$ **6.** $\dfrac{x^2}{x-6} - \dfrac{36}{x-6}$

[5.4] **7.** $\dfrac{5}{2w} - \dfrac{3}{w^2}$ **8.** $\dfrac{4}{b-3} - \dfrac{1}{b}$ **9.** $\dfrac{2}{3x-3} - \dfrac{5}{2x-2}$

10. $\dfrac{4x}{x^2 - 8x + 15} + \dfrac{6}{x-3}$

Multiply or divide as indicated.

[5.2] **11.** $\dfrac{-4a^2}{6ab^3} \cdot \dfrac{3ab^2}{-4ab}$ **12.** $\dfrac{x^2 + 5x + 4}{2x^2 + 2x} \cdot \dfrac{x^2 - x - 12}{x^2 - 16}$

13. $\dfrac{8b^4}{5bc} \div \dfrac{12b^2 c^2}{15bc^3}$ **14.** $\dfrac{x^2 y + 2xy^2}{x^2 - 4y^2} \div \dfrac{4x^2 y}{x^2 - xy - 2y^2}$

Simplify the complex fractions.

[5.7] **15.** $\dfrac{\frac{x^3}{16}}{\frac{x^2}{24}}$ **16.** $\dfrac{2 - \frac{x}{y}}{4 - \frac{x^2}{y^2}}$

What values for x, if any, must be excluded in the following algebraic fractions?

[5.5] **17.** $\dfrac{5}{x-3}$ **18.** $\dfrac{4}{x^2 - 3x - 10}$

Solve the following equations for x.

[5.5] **19.** $\dfrac{x}{4} - \dfrac{x}{5} = 2$ **20.** $\dfrac{x}{x-2} + 1 = \dfrac{x+4}{x-2}$

21. $\dfrac{7}{x} - \dfrac{1}{x-3} = \dfrac{9}{x^2 - 3x}$ **22.** $\dfrac{x-3}{8} = \dfrac{x-2}{10}$

Solve the following applications.

[5.6] **23. Number problem.** One number is 4 times another. If the sum of their reciprocals is $\dfrac{1}{4}$, find the two numbers.

24. Science and medicine. Mark drove 240 mi to visit Sandra. Returning by a shorter route, he found that the trip was only 200 mi, but traffic slowed his speed by 8 mi/h. If the two trips took exactly the same time, what was his rate each way?

25. Number problem. A 55-ft cable is to be cut into two pieces whose lengths have the ratio 3 to 8. Find the lengths of the two pieces.

5.1 Simplifying Rational Expressions

5.1 OBJECTIVES

1. Find the GCF for two monomials and simplify a rational expression
2. Find the GCF for two polynomials and simplify a rational expression

Much of our work with rational expressions (also called algebraic fractions) will be similar to your work in arithmetic. For instance, in algebra, as in arithmetic, many fractions name the same number. You will remember from Chapter 0 that

$$\frac{1}{4} = \frac{1 \cdot 2}{4 \cdot 2} = \frac{2}{8}$$

and

$$\frac{1}{4} = \frac{1 \cdot 3}{4 \cdot 3} = \frac{3}{12}$$

So $\frac{1}{4}, \frac{2}{8}$, and $\frac{3}{12}$ all name the same number; they are called **equivalent fractions.** These examples illustrate what is called the **Fundamental Principle of Fractions.** In algebra it becomes

---

Rules and Properties: Fundamental Principle of Rational Expressions

For polynomials P, Q, and R,

$$\frac{P}{Q} = \frac{PR}{QR} \qquad \text{when } Q \neq 0 \text{ and } R \neq 0$$

NOTE A rational expression is sometimes called an algebraic fraction, or simply a fraction.

This principle allows us to multiply or divide the numerator and denominator of a fraction by the same nonzero polynomial. The result will be an expression that is equivalent to the original one.

Our objective in this section is to simplify rational expressions by using the fundamental principle. In algebra, as in arithmetic, to write a fraction in simplest form, you divide the numerator and denominator of the fraction by their greatest common factor (GCF). The numerator and denominator of the resulting fraction will have no common factors other than 1, and the fraction is then in **simplest form.** The following rule summarizes this procedure.

NOTE Step 2 uses the Fundamental Principle of Fractions. The GCF is R in the Fundamental Principle of Rational Expressions rule above.

Step by Step: To Write Rational Expressions in Simplest Form

Step 1 Factor the numerator and denominator.
Step 2 Divide the numerator and denominator by the GCF. The resulting fraction will be in lowest terms.

OBJECTIVE 1 | Example 1 | Writing Fractions in Simplest Form

(a) Write $\dfrac{18}{30}$ in simplest form.

RECALL This is the same as dividing both the numerator and denominator of $\dfrac{18}{30}$ by 6.

$$\frac{18}{30} = \frac{2 \cdot 3 \cdot 3}{2 \cdot 3 \cdot 5} = \frac{\overset{1}{\cancel{2}} \cdot \overset{1}{\cancel{3}} \cdot 3}{\underset{1}{\cancel{2}} \cdot \underset{1}{\cancel{3}} \cdot 5} = \frac{3}{5}$$

Divide by the GCF. The slash lines indicate that we have divided the numerator and denominator by 2 and by 3.

(b) Write $\dfrac{4x^3}{6x}$ in simplest form.

$$\frac{4x^3}{6x} = \frac{\overset{1}{\cancel{2}} \cdot 2 \cdot \overset{1}{\cancel{x}} \cdot x \cdot x}{\underset{1}{\cancel{2}} \cdot 3 \cdot \underset{1}{\cancel{x}}} = \frac{2x^2}{3}$$

(c) Write $\dfrac{15x^3y^2}{20xy^4}$ in simplest form.

$$\frac{15x^3y^2}{20xy^4} = \frac{3 \cdot \overset{1}{\cancel{5}} \cdot \overset{1}{\cancel{x}} \cdot x \cdot x \cdot \overset{1}{\cancel{y}} \cdot \overset{1}{\cancel{y}}}{2 \cdot 2 \cdot \underset{1}{\cancel{5}} \cdot \underset{1}{\cancel{x}} \cdot \underset{1}{\cancel{y}} \cdot \underset{1}{\cancel{y}} \cdot y \cdot y} = \frac{3x^2}{4y^2}$$

(d) Write $\dfrac{3a^2b}{9a^3b^2}$ in simplest form.

NOTE With practice you will be able to simplify these terms without writing out the factorizations.

$$\frac{3a^2b}{9a^3b^2} = \frac{\overset{1}{\cancel{3}} \cdot \overset{1}{\cancel{a}} \cdot \overset{1}{\cancel{a}} \cdot \overset{1}{\cancel{b}}}{\underset{1}{\cancel{3}} \cdot 3 \cdot \underset{1}{\cancel{a}} \cdot \underset{1}{\cancel{a}} \cdot a \cdot \underset{1}{\cancel{b}} \cdot b} = \frac{1}{3ab}$$

(e) Write $\dfrac{10a^5b^4}{2a^2b^3}$ in simplest form.

$$\frac{10a^5b^4}{2a^2b^3} = \frac{5 \cdot \overset{1}{\cancel{2}} \cdot \overset{1}{\cancel{a}} \cdot \overset{1}{\cancel{a}} \cdot a \cdot a \cdot a \cdot \overset{1}{\cancel{b}} \cdot \overset{1}{\cancel{b}} \cdot \overset{1}{\cancel{b}} \cdot b}{\underset{1}{\cancel{2}} \cdot \underset{1}{\cancel{a}} \cdot \underset{1}{\cancel{a}} \cdot \underset{1}{\cancel{b}} \cdot \underset{1}{\cancel{b}} \cdot \underset{1}{\cancel{b}}} = \frac{5a^3b}{1} = 5a^3b$$

CHECK YOURSELF 1

Write each fraction in simplest form.

NOTE Most of the methods of this chapter build on our factoring work of the last chapter.

(a) $\dfrac{30}{66}$ (b) $\dfrac{5x^4}{15x}$ (c) $\dfrac{12xy^4}{18x^3y^2}$ (d) $\dfrac{5m^2n}{10m^3n^3}$ (e) $\dfrac{12a^4b^6}{2a^3b^4}$

In simplifying arithmetic fractions, common factors are generally easy to recognize. With rational expressions the factoring techniques you studied in Chapter 4 will have to be used as the *first step* in determining those factors.

OBJECTIVE 2   **Example 2   Writing Fractions in Simplest Form**

Write each fraction in simplest form.

(a) $\dfrac{2x - 4}{x^2 - 4} = \dfrac{2(x - 2)}{(x + 2)(x - 2)}$ Factor the numerator and denominator.

$$= \dfrac{2(x \!\!\!\!\diagdown\!\!\!\! - 2)}{(x + 2)(x \!\!\!\!\diagdown\!\!\!\! - 2)}$$ Divide by the GCF $x - 2$. The slash lines indicate that we have divided by that common factor.

$$= \dfrac{2}{x + 2}$$

NOTE

$3x^2 - 3$
$= 3(x^2 - 1)$
$= 3(x - 1)(x + 1)$

(b) $\dfrac{3x^2 - 3}{x^2 - 2x - 3} = \dfrac{3(x - 1)(x + 1)}{(x - 3)(x + 1)}$

$$= \dfrac{3(x - 1)}{x - 3}$$

(c) $\dfrac{2x^2 + x - 6}{2x^2 - x - 3} = \dfrac{(x + 2)(2x - 3)}{(x + 1)(2x - 3)}$

$$= \dfrac{x + 2}{x + 1}$$

C A U T I O N

Pick any value, other than 0, for x and substitute. You will quickly see that

$\dfrac{x + 2}{x + 1} \neq \dfrac{2}{1}$

For example, if $x = 4$,

$\dfrac{4 + 2}{4 + 1} = \dfrac{6}{5}$

Be Careful! The expression $\dfrac{x + 2}{x + 1}$ is already in simplest form. Students are often tempted to divide as follows:

$$\dfrac{\diagup\!\!\!\!x + 2}{\diagup\!\!\!\!x + 1} \quad \text{is } not \text{ equal to} \quad \dfrac{2}{1}$$

The x's are *terms* in the numerator and denominator. They *cannot* be divided out. Only *factors* can be divided. The fraction

$$\dfrac{x + 2}{x + 1}$$

is in its simplest form.

CHECK YOURSELF 2

Write each fraction in simplest form.

(a) $\dfrac{5x - 15}{x^2 - 9}$ (b) $\dfrac{a^2 - 5a + 6}{3a^2 - 6a}$

(c) $\dfrac{3x^2 + 14x - 5}{3x^2 + 2x - 1}$ (d) $\dfrac{5p - 15}{p^2 - 4}$

Remember the rules for signs in division. The quotient of a positive number and a negative number is always negative. Thus there are three equivalent ways to write such a quotient. For instance,

$$\frac{-2}{3} = \frac{2}{-3} = -\frac{2}{3}$$

The quotient of two positive numbers or two negative numbers is always positive. For example,

$$\frac{-2}{-3} = \frac{2}{3}$$

Example 3 Writing Fractions in Simplest Form

Write each fraction in simplest form.

NOTE In part (a), the final quotient is written in the most common way with the minus sign in the numerator.

(a) $\dfrac{6x^2}{-3xy} = \dfrac{2 \cdot \overset{1}{\cancel{3}} \cdot \overset{1}{\cancel{x}} \cdot x}{(-1) \cdot \underset{1}{\cancel{3}} \cdot \underset{1}{\cancel{x}} \cdot y} = \dfrac{2x}{-y} = \dfrac{-2x}{y}$

(b) $\dfrac{-5a^2b}{-10b^2} = \dfrac{\overset{1}{\cancel{(-1)}} \cdot \overset{1}{\cancel{5}} \cdot a \cdot a \cdot \overset{1}{\cancel{b}}}{\underset{1}{\cancel{(-1)}} \cdot 2 \cdot \underset{1}{\cancel{5}} \cdot \underset{1}{\cancel{b}} \cdot b} = \dfrac{a^2}{2b}$

CHECK YOURSELF 3

Write each fraction in simplest form.

(a) $\dfrac{8x^3y}{-4xy^2}$ (b) $\dfrac{-16a^4b^2}{-12a^2b^5}$

It is sometimes necessary to factor out a monomial before simplifying the fraction.

Example 4 Writing Fractions in Simplest Form

Write each fraction in simplest form.

(a) $\dfrac{6x^2 + 2x}{2x^2 + 12x} = \dfrac{2x(3x + 1)}{2x(x + 6)} = \dfrac{3x + 1}{x + 6}$

(b) $\dfrac{x^2 - 4}{x^2 + 6x + 8} = \dfrac{(x + 2)(x - 2)}{(x + 2)(x + 4)} = \dfrac{x - 2}{x + 4}$

(c) $\dfrac{x + 3}{x^2 + 4x + 12} = \dfrac{x + 3}{(x + 3)(x + 4)} = \dfrac{1}{x + 4}$

CHECK YOURSELF 4

Simplify each fraction.

(a) $\dfrac{3x^3 - 6x^2}{9x^4 - 3x^2}$

(b) $\dfrac{x^2 - 9}{x^2 - 12x + 27}$

Simplifying certain rational expressions will be easier with the following result. First, verify for yourself that

$$5 - 8 = -(8 - 5)$$

In general, it is true that

$$a - b = -(b - a)$$

or, by dividing both sides of the equation by $b - a$,

$$\frac{a - b}{b - a} = \frac{-(b - a)}{b - a}$$

So dividing by $b - a$ on the right, we have

NOTE Remember that a and b cannot be divided out because they are not factors.

$$\frac{a - b}{b - a} = -1$$

Let's look at some applications of this result in Example 5.

Example 5   Writing Rational Expressions in Simplest Form

Write each fraction in simplest form.

(a) $\dfrac{2x - 4}{4 - x^2} = \dfrac{2(x - 2)}{(2 + x)(2 - x)}$ This is equal to −1.

$$= \frac{2(-1)}{2 + x}$$

$$= \frac{-2}{2 + x}$$

(b) $\dfrac{9 - x^2}{x^2 + 2x - 15} = \dfrac{(3 + x)(3 - x)}{(x + 5)(x - 3)}$ This is equal to −1.

$$= \frac{(3 + x)(-1)}{x + 5}$$

$$= \frac{-x - 3}{x + 5}$$

CHECK YOURSELF 5

Write each fraction in simplest form.

(a) $\dfrac{3x - 9}{9 - x^2}$

(b) $\dfrac{x^2 - 6x - 27}{81 - x^2}$

CHECK YOURSELF ANSWERS

1. (a) $\dfrac{5}{11}$; (b) $\dfrac{x^3}{3}$; (c) $\dfrac{2y^2}{3x^2}$; (d) $\dfrac{1}{2mn^2}$; (e) $6ab^2$ 2. (a) $\dfrac{5}{x + 3}$; (b) $\dfrac{a - 3}{3a}$; (c) $\dfrac{x + 5}{x + 1}$;

(d) $\dfrac{5(p - 3)}{(p + 2)(p - 2)}$ 3. (a) $\dfrac{-2x^2}{y}$; (b) $\dfrac{4a^2}{3b^3}$ 4. (a) $\dfrac{x - 2}{3x^2 - 1}$; (b) $\dfrac{x + 3}{x - 9}$

5. (a) $\dfrac{-3}{x + 3}$; (b) $\dfrac{-x - 3}{x + 9}$

Name _____

Section _____   Date _____

Write each fraction in simplest form.

1. $\dfrac{16}{24}$

2. $\dfrac{56}{64}$

3. $\dfrac{80}{180}$

4. $\dfrac{18}{30}$

5. $\dfrac{4x^5}{6x^2}$

6. $\dfrac{10x^2}{15x^4}$

7. $\dfrac{9x^3}{27x^6}$

8. $\dfrac{25w^6}{20w^2}$

9. $\dfrac{10a^2b^5}{25ab^2}$

10. $\dfrac{18x^4y^3}{24x^2y^3}$

11. $\dfrac{42x^3y}{14xy^3}$

12. $\dfrac{18pq}{45p^2q^2}$

13. $\dfrac{2xyw^2}{6x^2y^3w^3}$

14. $\dfrac{3c^2d^2}{6bc^3d^3}$

15. $\dfrac{10x^5y^5}{2x^3y^4}$

16. $\dfrac{3bc^6d^3}{bc^3d}$

17. $\dfrac{-4m^3n}{6mn^2}$

18. $\dfrac{-15x^3y^3}{-20xy^4}$

19. $\dfrac{-8ab^3}{-16a^3b}$

20. $\dfrac{14x^2y}{-21xy^4}$

21. $\dfrac{8r^2s^3t}{-16rs^4t^3}$

22. $\dfrac{-10a^3b^2c^3}{15ab^4c}$

ANSWERS

1. _____   2. _____

3. _____   4. _____

5. _____

6. _____

7. _____

8. _____

9. _____

10. _____

11. _____

12. _____

13. _____

14. _____

15. _____

16. _____

17. _____

18. _____

19. _____

20. _____

21. _____

22. _____

Write each expression in simplest form.

23. $\dfrac{3x + 18}{5x + 30}$
  24. $\dfrac{4x - 28}{5x - 35}$

25. $\dfrac{3x - 6}{5x - 15}$
  26. $\dfrac{x^2 - 25}{3x - 15}$

27. $\dfrac{6a - 24}{a^2 - 16}$
  28. $\dfrac{5x - 5}{x^2 - 4}$

29. $\dfrac{x^2 + 3x + 2}{5x + 10}$
  30. $\dfrac{4w^2 - 20w}{w^2 - 2w - 15}$

31. $\dfrac{x^2 - 6x - 16}{x^2 - 64}$
  32. $\dfrac{y^2 - 25}{y^2 - y - 20}$

33. $\dfrac{2m^2 + 3m - 5}{2m^2 + 11m + 15}$
  34. $\dfrac{6x^2 - x - 2}{3x^2 - 5x + 2}$

35. $\dfrac{p^2 + 2pq - 15q^2}{p^2 - 25q^2}$
  36. $\dfrac{4r^2 - 25s^2}{2r^2 + 3rs - 20s^2}$

37. $\dfrac{y - 7}{7 - y}$
  38. $\dfrac{5 - y}{y - 5}$

39. $\dfrac{2x - 10}{25 - x^2}$
  40. $\dfrac{3a - 12}{16 - a^2}$

41. $\dfrac{25 - a^2}{a^2 + a - 30}$
  42. $\dfrac{2x^2 - 7x + 3}{9 - x^2}$

43. $\dfrac{x^2 + xy - 6y^2}{4y^2 - x^2}$
  44. $\dfrac{16z^2 - w^2}{2w^2 - 5wz - 12z^2}$

45. $\dfrac{x^2 + 4x + 4}{x + 2}$
  46. $\dfrac{4x^2 + 12x + 9}{2x + 3}$

47. $\dfrac{xy - 2y + 4x - 8}{2y + 6 - xy - 3x}$
  48. $\dfrac{ab - 3a + 5b - 15}{15 + 3a^2 - 5b - a^2b}$

49. Geometry. The area of the rectangle is represented by $6x^2 + 19x + 10$. What is the length?

3x + 2

50. Geometry. The volume of the box is represented by $(x^2 + 5x + 6)(x + 5)$. Find the polynomial that represents the area of the bottom of the box.

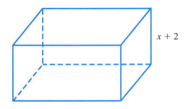

x + 2

51. To work with rational expressions correctly, it is important to understand the difference between a *factor* and a *term* of an expression. In your own words, write definitions for both, explaining the difference between the two.

52. Give some examples of terms and factors in rational expressions and explain how both are affected when a fraction is simplified.

53. Show how the following rational expression can be simplified:

$$\frac{x^2 - 9}{4x + 12}$$

Note that your simplified fraction is equivalent to the given fraction. Are there other rational expressions equivalent to this one? Write another rational expression that you think is equivalent to this one. Exchange papers with another student. Do you agree that the other student's fraction is equivalent to yours? Why or why not?

54. Explain the reasoning involved in each step when simplifying the fraction $\dfrac{42}{56}$.

55. Describe why $\dfrac{3}{5}$ and $\dfrac{27}{45}$ are *equivalent fractions*.

49. _____

50. _____

51. _____

52. _____

53. _____

54. _____

55. _____

Getting Ready for Section 5.2 [Section 0.2]

Perform the indicated operations.

(a) $\dfrac{2}{3} \cdot \dfrac{4}{5}$ (b) $\dfrac{5}{6} \cdot \dfrac{4}{11}$

(c) $\dfrac{4}{7} \div \dfrac{8}{5}$ (d) $\dfrac{1}{6} \div \dfrac{7}{9}$

(e) $\dfrac{5}{8} \cdot \dfrac{16}{15}$ (f) $\dfrac{15}{21} \div \dfrac{10}{7}$

(g) $\dfrac{15}{8} \cdot \dfrac{24}{25}$ (h) $\dfrac{28}{16} \div \dfrac{21}{20}$

Answers

1. $\dfrac{2}{3}$ 3. $\dfrac{4}{9}$ 5. $\dfrac{2x^3}{3}$ 7. $\dfrac{1}{3x^3}$ 9. $\dfrac{2ab^3}{5}$ 11. $\dfrac{3x^2}{y^2}$ 13. $\dfrac{1}{3xy^2w}$

15. $5x^2y$ 17. $\dfrac{-2m^2}{3n}$ 19. $\dfrac{b^2}{2a^2}$ 21. $\dfrac{-r}{2st^2}$ 23. $\dfrac{3}{5}$ 25. $\dfrac{3(x-2)}{5(x-3)}$

27. $\dfrac{6}{a+4}$ 29. $\dfrac{x+1}{5}$ 31. $\dfrac{x+2}{x+8}$ 33. $\dfrac{m-1}{m+3}$ 35. $\dfrac{p-3q}{p-5q}$

37. -1 39. $\dfrac{-2}{x+5}$ 41. $\dfrac{-a-5}{a+6}$ 43. $\dfrac{-x-3y}{2y+x}$ 45. $x+2$

47. $\dfrac{-(y+4)}{y+3}$ 49. $2x+5$ 51.   53.   55.

a. $\dfrac{8}{15}$ b. $\dfrac{10}{33}$ c. $\dfrac{5}{14}$ d. $\dfrac{3}{14}$ e. $\dfrac{2}{3}$ f. $\dfrac{1}{2}$ g. $\dfrac{9}{5}$ h. $\dfrac{5}{3}$

5.2 Multiplying and Dividing Rational Expressions

5.2 OBJECTIVES

1. Write the product of two rational expressions in simplest form
2. Write the quotient of two rational expressions in simplest form

In arithmetic, you found the product of two fractions by multiplying the numerators and the denominators. For example,

$$\frac{2}{5} \cdot \frac{3}{7} = \frac{2 \cdot 3}{5 \cdot 7} = \frac{6}{35}$$

In symbols, we have

NOTE P, Q, R, and S again represent polynomials.

Rules and Properties: Multiplying Rational Expressions

$$\frac{P}{Q} \cdot \frac{R}{S} = \frac{PR}{QS} \qquad \text{when } Q \neq 0 \text{ and } S \neq 0$$

It is easier to divide the numerator and denominator by any common factors *before* multiplying. Consider the following.

NOTE Divide by the common factors of 3 and 4. The alternative is to multiply *first:*

$$\frac{3}{8} \cdot \frac{4}{9} = \frac{12}{72}$$

and then use the GCF to reduce to lowest terms

$$\frac{12}{72} = \frac{1}{6}$$

$$\frac{3}{8} \cdot \frac{4}{9} = \frac{\overset{1}{\cancel{3}} \cdot \overset{1}{\cancel{4}}}{\underset{2}{\cancel{8}} \cdot \underset{3}{\cancel{9}}} = \frac{1}{6}$$

In algebra, we multiply fractions in exactly the same way.

Step by Step: To Multiply Rational Expressions

Step 1 Factor the numerators and denominators.
Step 2 Divide the numerator and denominator by any common factors.
Step 3 Write the product of the remaining factors in the numerator over the product of the remaining factors in the denominator.

Example 1 illustrates this process.

OBJECTIVE 1

Example 1 Multiplying Rational Expressions

Multiply the following fractions.

NOTE Divide by the common factors of 5, x^2, and y.

(a) $\dfrac{2x^3}{5y^2} \cdot \dfrac{10y}{3x^2} = \dfrac{2x^3 \cdot 10y}{5y^2 \cdot 3x^2} = \dfrac{20x^3 y}{15x^2 y^2} = \dfrac{4x}{3y}$

(b) $\dfrac{x}{x^2 - 3x} \cdot \dfrac{6x - 18}{9x} = \dfrac{x}{x(x-3)} \cdot \dfrac{6(x-3)}{9x}$ ⟵ Factor

NOTE Divide by the common factors of 3, x, and $x - 3$.

$$= \frac{\overset{1}{\cancel{x}} \cdot \overset{2}{\cancel{6}}(x \overset{1}{\cancel{-3}})}{\underset{1}{\cancel{x}}(x \underset{1}{\cancel{-3}}) \cdot \underset{3}{\cancel{9}}x} = \frac{2}{3x}$$

(c) $\dfrac{4}{x^2 - 2x} \cdot \dfrac{10 - 5x}{8} = \dfrac{4}{x(x - 2)} \cdot \dfrac{5(2 - x)}{8}$

RECALL

$\dfrac{2 - x}{x - 2} = \dfrac{-(x - 2)}{x - 2} = -1$

$$= \dfrac{\overset{1}{\cancel{4}} \cdot 5(2 \overset{-1}{\cancel{- x}})}{x(x \cancel{- 2}) \cdot \underset{2}{\cancel{8}}} = \dfrac{-5}{2x}$$

NOTE Divide by the common factors of $x - 4$, x, and 3.

(d) $\dfrac{x^2 - 2x - 8}{3x^2} \cdot \dfrac{6x}{3x - 12} = \dfrac{\overset{1}{(\cancel{x - 4})}(x + 2)}{\underset{x}{3x^2}} \cdot \dfrac{\overset{2}{\cancel{6x}}}{\underset{1}{3(\cancel{x - 4})}}$

$$= \dfrac{2(x + 2)}{3x}$$

(e) $\dfrac{x^2 - y^2}{5x - 5y} \cdot \dfrac{10xy}{x^2 + 2xy + y^2} = \dfrac{\overset{1}{(\cancel{x - y})}\overset{1}{(\cancel{x + y})}}{\underset{1}{5(\cancel{x - y})}} \cdot \dfrac{\overset{2}{\cancel{10xy}}}{\underset{1}{(\cancel{x + y})(x + y)}}$

$$= \dfrac{2xy}{x + y}$$

✔ CHECK YOURSELF 1

Multiply the following fractions.

(a) $\dfrac{3x^2}{5y^2} \cdot \dfrac{10y^5}{15x^3}$ **(b)** $\dfrac{5x + 15}{x} \cdot \dfrac{2x^2}{x^2 + 3x}$ **(c)** $\dfrac{x}{2x - 6} \cdot \dfrac{3x - x^2}{2}$

(d) $\dfrac{3x - 15}{6x^2} \cdot \dfrac{2x}{x^2 - 25}$ **(e)** $\dfrac{x^2 - 5x - 14}{4x^2} \cdot \dfrac{8x}{x^2 - 49}$

You can also use your experience from arithmetic in dividing fractions. Recall that, to divide fractions, we *invert the divisor* (the *second* fraction) and multiply. For example,

RECALL $\dfrac{6}{5}$ is the reciprocal of $\dfrac{5}{6}$.

$$\dfrac{2}{3} \div \dfrac{5}{6} = \dfrac{2}{3} \cdot \dfrac{6}{5} = \dfrac{2 \cdot 6}{3 \cdot 5} = \dfrac{12}{15} = \dfrac{4}{5}$$

In symbols, we have

Rules and Properties: Dividing Rational Expressions

NOTE Once more P, Q, R, and S are polynomials.

$$\dfrac{P}{Q} \div \dfrac{R}{S} = \dfrac{P}{Q} \cdot \dfrac{S}{R} = \dfrac{PS}{QR}$$

when $Q \neq 0$, $R \neq 0$, and $S \neq 0$.

We divide rational expressions in exactly the same way.

Step by Step: To Divide Rational Expressions

Step 1   Invert the divisor and change the operation to multiplication.
Step 2   Proceed, using the steps for multiplying rational expressions.

Example 2 illustrates this approach.

OBJECTIVE 2   **Example 2   Dividing Rational Expressions**

Divide the following fractions.

(a) $\dfrac{6}{x^2} \div \dfrac{9}{x^3} = \dfrac{6}{x^2} \cdot \dfrac{x^3}{9}$ Invert the divisor and multiply.

$= \dfrac{\cancel{6}x^{\cancel{3}}}{\cancel{9}x^{\cancel{2}}}$ No simplification can be done until the divisor is inverted. Then divide by the common factors of 3 and x^2.

$= \dfrac{2x}{3}$

(b) $\dfrac{3x^2 y}{8xy^3} \div \dfrac{9x^3}{4y^4} = \dfrac{3x^2 y}{8xy^3} \cdot \dfrac{4y^4}{9x^3}$

$= \dfrac{y^2}{6x^2}$

(c) $\dfrac{2x + 4y}{9x - 18y} \div \dfrac{4x + 8y}{3x - 6y} = \dfrac{2x + 4y}{9x - 18y} \cdot \dfrac{3x - 6y}{4x + 8y}$

NOTE Factor all numerators and denominators *before* dividing out any common factors.

$= \dfrac{\overset{1}{\cancel{2}}\overset{1}{\cancel{(x + 2y)}} \cdot \overset{1}{\cancel{3}}\overset{1}{\cancel{(x - 2y)}}}{\underset{3}{\cancel{9}}\underset{1}{\cancel{(x - 2y)}} \cdot \underset{2}{\cancel{4}}\underset{1}{\cancel{(x + 2y)}}}$

$= \dfrac{1}{6}$

(d) $\dfrac{x^2 - x - 6}{2x - 6} \div \dfrac{x^2 - 4}{4x^2} = \dfrac{x^2 - x - 6}{2x - 6} \cdot \dfrac{4x^2}{x^2 - 4}$

$= \dfrac{\overset{1}{\cancel{(x - 3)}}\overset{1}{\cancel{(x + 2)}} \cdot \overset{2}{\cancel{4}}x^2}{\underset{1}{\cancel{2}}\underset{1}{\cancel{(x - 3)}} \cdot \underset{1}{\cancel{(x + 2)}}(x - 2)}$

$= \dfrac{2x^2}{x - 2}$

CHECK YOURSELF 2

Divide the following fractions.

(a) $\dfrac{4}{x^5} \div \dfrac{12}{x^3}$

(b) $\dfrac{5xy^2}{7x^3 y} \div \dfrac{10y^2}{14x^3}$

(c) $\dfrac{3x - 9y}{2x + 10y} \div \dfrac{x^2 - 3xy}{4x + 20y}$

(d) $\dfrac{x^2 - 9}{4x} \div \dfrac{x^2 - 2x - 15}{2x - 10}$

CHECK YOURSELF ANSWERS

1. (a) $\dfrac{2y^3}{5x}$; (b) 10; (c) $\dfrac{-x^2}{4}$; (d) $\dfrac{1}{x(x + 5)}$; (e) $\dfrac{2(x + 2)}{x(x + 7)}$

2. (a) $\dfrac{1}{3x^2}$; (b) $\dfrac{x}{y}$; (c) $\dfrac{6}{x}$; (d) $\dfrac{x - 3}{2x}$

5.2 Exercises

Multiply the following fractions.

1. $\dfrac{3}{7} \cdot \dfrac{14}{27}$

2. $\dfrac{9}{20} \cdot \dfrac{5}{36}$

3. $\dfrac{x}{2} \cdot \dfrac{y}{6}$

4. $\dfrac{w}{2} \cdot \dfrac{5}{14}$

5. $\dfrac{3a}{2} \cdot \dfrac{4}{a^2}$

6. $\dfrac{5x^3}{3x} \cdot \dfrac{9}{20x}$

7. $\dfrac{3x^3y}{10xy^3} \cdot \dfrac{5xy^2}{9xy^2}$

8. $\dfrac{8xy^5}{5x^3y^2} \cdot \dfrac{15y^2}{16xy^3}$

9. $\dfrac{-4ab^2}{15a^3} \cdot \dfrac{25ab}{-16b^3}$

10. $\dfrac{-7xy^2}{12x^2y} \cdot \dfrac{24x^3y^5}{-21x^2y^7}$

11. $\dfrac{-3m^3n}{10mn^3} \cdot \dfrac{5mn^2}{-9mn^3}$

12. $\dfrac{3x}{2x-6} \cdot \dfrac{x^2-3x}{6}$

13. $\dfrac{x^2+5x}{3x^2} \cdot \dfrac{10x}{5x+25}$

14. $\dfrac{x^2-3x-10}{5x} \cdot \dfrac{15x^2}{3x-15}$

15. $\dfrac{p^2-8p}{4p} \cdot \dfrac{12p^2}{p^2-64}$

16. $\dfrac{a^2-81}{a^2+9a} \cdot \dfrac{5a^2}{a^2-7a-18}$

17. $\dfrac{m^2-4m-21}{3m^2} \cdot \dfrac{m^2+7m}{m^2-49}$

18. $\dfrac{2x^2-x-3}{3x^2+7x+4} \cdot \dfrac{3x^2-11x-20}{4x^2-9}$

19. $\dfrac{4r^2-1}{2r^2-9r-5} \cdot \dfrac{3r^2-13r-10}{9r^2-4}$

20. $\dfrac{a^2+ab}{2a^2-ab-3b^2} \cdot \dfrac{4a^2-9b^2}{5a^2-4ab}$

21. $\dfrac{x^2-4y^2}{x^2-xy-6y^2} \cdot \dfrac{7x^2-21xy}{5x-10y}$

22. $\dfrac{a^2-9b^2}{a^2+ab-6b^2} \cdot \dfrac{6a^2-12ab}{7a-21b}$

23. $\dfrac{2x-6}{x^2+2x} \cdot \dfrac{3x}{3-x}$

24. $\dfrac{3x-15}{x^2+3x} \cdot \dfrac{4x}{5-x}$

Divide the following fractions.

25. $\dfrac{5}{8} \div \dfrac{15}{16}$

26. $\dfrac{4}{9} \div \dfrac{12}{18}$

27. $\dfrac{5}{x^2} \div \dfrac{10}{x}$

28. $\dfrac{w^2}{3} \div \dfrac{w}{9}$

29. $\dfrac{4x^2y^2}{9x^3} \div \dfrac{8y^2}{27xy}$

30. $\dfrac{8x^3y}{27xy^3} \div \dfrac{16x^3y}{45y}$

31. $\dfrac{3x + 6}{8} \div \dfrac{5x + 10}{6}$

32. $\dfrac{x^2 - 2x}{4x} \div \dfrac{6x - 12}{8}$

33. $\dfrac{4a - 12}{5a + 15} \div \dfrac{8a^2}{a^2 + 3a}$

34. $\dfrac{6p - 18}{9p} \div \dfrac{3p - 9}{p^2 + 2p}$

35. $\dfrac{x^2 + 2x - 8}{9x^2} \div \dfrac{x^2 - 16}{3x - 12}$

36. $\dfrac{16x}{4x^2 - 16} \div \dfrac{4x - 24}{x^2 - 4x - 12}$

37. $\dfrac{x^2 - 9}{2x^2 - 6x} \div \dfrac{2x^2 + 5x - 3}{4x^2 - 1}$

38. $\dfrac{2m^2 - 5m - 7}{4m^2 - 9} \div \dfrac{5m^2 + 5m}{2m^2 + 3m}$

39. $\dfrac{a^2 - 9b^2}{4a^2 + 12ab} \div \dfrac{a^2 - ab - 6b^2}{12ab}$

40. $\dfrac{r^2 + 2rs - 15s^2}{r^3 + 5r^2s} \div \dfrac{r^2 - 9s^2}{5r^3}$

41. $\dfrac{x^2 - 16y^2}{3x^2 - 12xy} \div (x^2 + 4xy)$

42. $\dfrac{p^2 - 4pq - 21q^2}{4p - 28q} \div (2p^2 + 6pq)$

43. $\dfrac{x - 7}{2x + 6} \div \dfrac{21 - 3x}{x^2 + 3x}$

44. $\dfrac{x - 4}{x^2 + 2x} \div \dfrac{16 - 4x}{3x + 6}$

Perform the indicated operations.

45. $\dfrac{x^2 + 5x}{3x - 6} \cdot \dfrac{x^2 - 4}{3x^2 + 15x} \cdot \dfrac{6x}{x^2 + 6x + 8}$

46. $\dfrac{m^2 - n^2}{m^2 - mn} \cdot \dfrac{6m}{2m^2 + mn - n^2} \cdot \dfrac{8m - 4n}{12m^2 + 12mn}$

47. $\dfrac{x^2 - 2x - 8}{2x - 8} \cdot \dfrac{x^2 + 5x}{x^2 + 5x + 6} \div \dfrac{x^2 + 2x - 15}{x^2 - 9}$

48. $\dfrac{14x - 7}{x^2 + 3x - 4} \cdot \dfrac{x^2 + 6x + 8}{2x^2 + 5x - 3} \div \dfrac{x^2 + 2x}{x^2 + 2x - 3}$

27. _____

28. _____

29. _____

30. _____

31. _____

32. _____

33. _____

34. _____

35. _____

36. _____

37. _____

38. _____

39. _____

40. _____

41. _____

42. _____

43. _____

44. _____

45. _____

46. _____

47. _____

48. _____

49. _____

50. _____

51. _____

52. _____

a. _____

b. _____

c. _____

d. _____

e. _____

f. _____

g. _____

h. _____

49. Science and technology. Herbicides constitute $\frac{2}{3}$ of all pesticides used in the United States. Insecticides are $\frac{1}{4}$ of all pesticides used in the United States. The ratio of herbicides to insecticides used in the United States can be written $\frac{2}{3} \div \frac{1}{4}$. Write this ratio in simplest form.

50. Science and technology. Fungicides account for $\frac{1}{10}$ of the pesticides used in the United States. Insecticides account for $\frac{1}{4}$ of all the pesticides used in the United States. The ratio of fungicides to insecticides used in the United States can be written $\frac{1}{10} \div \frac{1}{4}$. Write this ratio in simplest form.

51. Science and technology. The ratio of insecticides to herbicides applied to wheat, soybeans, corn, and cotton can be expressed as $\frac{7}{10} \div \frac{4}{5}$. Simplify this ratio.

52. Geometry. Find the area of the rectangle shown.

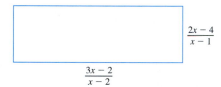

$\frac{2x-4}{x-1}$

$\frac{3x-2}{x-2}$

 Getting Ready for Section 5.3 [Section 0.2]

Perform the indicated operations.

(a) $\frac{3}{10} + \frac{4}{10}$ (b) $\frac{5}{8} - \frac{4}{8}$ (c) $\frac{5}{12} - \frac{1}{12}$ (d) $\frac{7}{16} + \frac{3}{16}$

(e) $\frac{7}{20} + \frac{9}{20}$ (f) $\frac{13}{8} - \frac{5}{8}$ (g) $\frac{11}{6} - \frac{2}{6}$ (h) $\frac{5}{9} + \frac{7}{9}$

Answers

1. $\frac{2}{9}$ 3. $\frac{xy}{12}$ 5. $\frac{6}{a}$ 7. $\frac{x^2}{6y^2}$ 9. $\frac{5}{12a}$ 11. $\frac{m^2}{6n^3}$ 13. $\frac{2}{3}$

15. $\frac{3p^2}{p+8}$ 17. $\frac{m+3}{3m}$ 19. $\frac{2r-1}{3r-2}$ 21. $\frac{7x}{5}$ 23. $\frac{-6}{x+2}$

25. $\frac{2}{3}$ 27. $\frac{1}{2x}$ 29. $\frac{3y}{2}$ 31. $\frac{9}{20}$ 33. $\frac{a-3}{10a}$ 35. $\frac{x-2}{3x^2}$

37. $\frac{2x+1}{2x}$ 39. $\frac{3b}{a+2b}$ 41. $\frac{1}{3x^2}$ 43. $\frac{-x}{6}$ 45. $\frac{2x}{3(x+4)}$

47. $\frac{x}{2}$ 49. $\frac{8}{3}$ 51. $\frac{7}{8}$ a. $\frac{7}{10}$ b. $\frac{1}{8}$ c. $\frac{1}{3}$

d. $\frac{5}{8}$ e. $\frac{4}{5}$ f. 1   g. $\frac{3}{2}$ h. $\frac{4}{3}$

5.3 Adding and Subtracting Like Rational Expressions

5.3 OBJECTIVES

1. Write the sum or difference of two rational expressions whose numerator and denominator are monomials
2. Write the sum or difference of two rational expressions whose numerator and denominator are polynomials

You probably remember from arithmetic that **like fractions** are fractions that have the same denominator. The same is true in algebra.

$\dfrac{2}{5}$, $\dfrac{12}{5}$, and $\dfrac{4}{5}$ are like fractions.

$\dfrac{x}{3}$, $\dfrac{y}{3}$, and $\dfrac{z-5}{3}$ are like fractions.

$\dfrac{3x}{2}$, $\dfrac{x}{4}$, and $\dfrac{3x}{8}$ are unlike fractions.

$\dfrac{3}{x}$, $\dfrac{2}{x^2}$, and $\dfrac{x+1}{x^3}$ are unlike fractions.

NOTE The fractions have different denominators.

In arithmetic, the sum or difference of like fractions was found by adding or subtracting the numerators and writing the result over the common denominator. For example,

$$\frac{3}{11} + \frac{5}{11} = \frac{3+5}{11} = \frac{8}{11}$$

In symbols, we have

Rules and Properties: To Add or Subtract Like Rational Expressions

$$\frac{P}{R} + \frac{Q}{R} = \frac{P+Q}{R} \qquad R \neq 0$$

$$\frac{P}{R} - \frac{Q}{R} = \frac{P-Q}{R} \qquad R \neq 0$$

Adding or subtracting like rational expressions is just as straightforward. You can use the following steps.

Step by Step: To Add or Subtract Like Rational Expressions

Step 1 Add or subtract the numerators.
Step 2 Write the sum or difference over the common denominator.
Step 3 Write the resulting fraction in simplest form.

437

OBJECTIVE 1

Example 1   Adding and Subtracting Rational Expressions

Add or subtract as indicated. Express your results in simplest form.

Add the numerators.

(a) $\dfrac{2x}{15} + \dfrac{x}{15} = \dfrac{\overbrace{2x + x}}{15}$

$= \dfrac{3x}{15} = \dfrac{x}{5}$

Simplify.

Subtract the numerators.

(b) $\dfrac{5y}{6} - \dfrac{y}{6} = \dfrac{\overbrace{5y - y}}{6}$

$= \dfrac{4y}{6} = \dfrac{2y}{3}$

Simplify.

(c) $\dfrac{3}{x} + \dfrac{5}{x} = \dfrac{3 + 5}{x} = \dfrac{8}{x}$

(d) $\dfrac{9}{a^2} - \dfrac{7}{a^2} = \dfrac{9 - 7}{a^2} = \dfrac{2}{a^2}$

(e) $\dfrac{7}{2ab} - \dfrac{5}{2ab} = \dfrac{7 - 5}{2ab}$

$= \dfrac{2}{2ab}$

$= \dfrac{1}{ab}$

✔ **CHECK YOURSELF 1**

Add or subtract as indicated.

(a) $\dfrac{3a}{10} + \dfrac{2a}{10}$ **(b)** $\dfrac{7b}{8} - \dfrac{3b}{8}$

(c) $\dfrac{4}{x} + \dfrac{3}{x}$ **(d)** $\dfrac{5}{3xy} - \dfrac{2}{3xy}$

If polynomials are involved in the numerators or denominators, the process is exactly the same.

OBJECTIVE 2

Example 2   Adding and Subtracting Rational Expressions

Add or subtract as indicated. Express your results in simplest form.

(a) $\dfrac{5}{x + 3} + \dfrac{2}{x + 3} = \dfrac{5 + 2}{x + 3} = \dfrac{7}{x + 3}$

(b) $\dfrac{4x}{x-4} - \dfrac{16}{x-4} = \dfrac{4x-16}{x-4}$

Factor and simplify.

NOTE The final answer is always written in simplest form.

$$= \dfrac{4(\overset{1}{\cancel{x-4}})}{\underset{1}{\cancel{x-4}}} = 4$$

(c) $\dfrac{a-b}{3} + \dfrac{2a+b}{3} = \dfrac{(a-b)+(2a+b)}{3}$

$$= \dfrac{a-b+2a+b}{3}$$

$$= \dfrac{\overset{1}{\cancel{3}}a}{\underset{1}{\cancel{3}}} = a$$

Be sure to enclose the second numerator in parentheses!

(d) $\dfrac{3x+y}{2x} - \dfrac{x-3y}{2x} = \dfrac{(3x+y)-(x-3y)}{2x}$

Change both signs.

$$= \dfrac{3x+y-x+3y}{2x}$$

$$= \dfrac{2x+4y}{2x}$$

$$= \dfrac{\overset{1}{\cancel{2}}(x+2y)}{\underset{1}{\cancel{2}}x}$$ Factor and divide by the common factor of 2.

$$= \dfrac{x+2y}{x}$$

(e) $\dfrac{3x-5}{x^2+x-2} - \dfrac{2x-4}{x^2+x-2} = \dfrac{(3x-5)-(2x-4)}{x^2+x-2}$

Put the second numerator in parentheses.

Change both signs.

$$= \dfrac{3x-5-2x+4}{x^2+x-2}$$

$$= \dfrac{x-1}{x^2+x-2}$$

$$= \dfrac{(\overset{1}{\cancel{x-1}})}{(x+2)(\underset{1}{\cancel{x-1}})}$$ Factor and divide by the common factor of $x-1$.

$$= \dfrac{1}{x+2}$$

(f) $\dfrac{2x + 7y}{x + 3y} - \dfrac{x + 4y}{x + 3y} = \dfrac{(2x + 7y) - (x + 4y)}{x + 3y}$

Change both signs.

$= \dfrac{2x + 7y - x - 4y}{x + 3y}$

$= \dfrac{x + 3y}{x + 3y} = 1$

CHECK YOURSELF 2

Add or subtract as indicated.

(a) $\dfrac{4}{x - 5} - \dfrac{2}{x - 5}$

(b) $\dfrac{3x}{x + 3} + \dfrac{9}{x + 3}$

(c) $\dfrac{5x - y}{3y} - \dfrac{2x - 4y}{3y}$

(d) $\dfrac{5x + 8}{x^2 - 2x - 15} - \dfrac{4x + 5}{x^2 - 2x - 15}$

CHECK YOURSELF ANSWERS

1. (a) $\dfrac{a}{2}$; (b) $\dfrac{b}{2}$; (c) $\dfrac{7}{x}$; (d) $\dfrac{1}{xy}$ 2. (a) $\dfrac{2}{x - 5}$; (b) 3; (c) $\dfrac{x + y}{y}$; (d) $\dfrac{1}{x - 5}$

Add or subtract as indicated. Express your results in simplest form.

1. $\dfrac{7}{18} + \dfrac{5}{18}$

2. $\dfrac{5}{18} - \dfrac{2}{18}$

3. $\dfrac{13}{16} - \dfrac{9}{16}$

4. $\dfrac{5}{12} + \dfrac{11}{12}$

5. $\dfrac{x}{8} + \dfrac{3x}{8}$

6. $\dfrac{5y}{16} + \dfrac{7y}{16}$

7. $\dfrac{7a}{10} - \dfrac{3a}{10}$

8. $\dfrac{5x}{12} - \dfrac{x}{12}$

9. $\dfrac{5}{x} + \dfrac{3}{x}$

10. $\dfrac{9}{y} - \dfrac{3}{y}$

11. $\dfrac{8}{w} - \dfrac{2}{w}$

12. $\dfrac{7}{z} + \dfrac{9}{z}$

13. $\dfrac{2}{xy} + \dfrac{3}{xy}$

14. $\dfrac{8}{ab} + \dfrac{4}{ab}$

15. $\dfrac{2}{3cd} + \dfrac{4}{3cd}$

16. $\dfrac{5}{4cd} + \dfrac{11}{4cd}$

17. $\dfrac{7}{x-5} + \dfrac{9}{x-5}$

18. $\dfrac{11}{x+7} - \dfrac{4}{x+7}$

19. $\dfrac{2x}{x-2} - \dfrac{4}{x-2}$

20. $\dfrac{7w}{w+3} + \dfrac{21}{w+3}$

21. $\dfrac{8p}{p+4} + \dfrac{32}{p+4}$

22. $\dfrac{5a}{a-3} - \dfrac{15}{a-3}$

23. $\dfrac{x^2}{x+4} + \dfrac{3x-4}{x+4}$

24. $\dfrac{x^2}{x-3} - \dfrac{9}{x-3}$

25. $\dfrac{m^2}{m-5} - \dfrac{25}{m-5}$

26. $\dfrac{s^2}{s+3} + \dfrac{2s-3}{s+3}$

ANSWERS

1.
2.
3.
4.
5.
6.
7.
8.
9.
10.
11.
12.
13.
14.
15.
16.
17.   18.
19.   20.
21.   22.
23.   24.
25.   26.

27. _____

28. _____

29. _____

30. _____

31. _____

32. _____

33. _____

34. _____

35. _____

36. _____

37. _____

38. _____

39. _____

40. _____

41. _____

42. _____

43. _____

44. _____

27. $\dfrac{a-1}{3} + \dfrac{2a-5}{3}$

28. $\dfrac{y+2}{3} - \dfrac{4y+8}{3}$

29. $\dfrac{3x-1}{4} - \dfrac{x+7}{4}$

30. $\dfrac{4x+2}{3} - \dfrac{x-1}{3}$

31. $\dfrac{4m+7}{6m} - \dfrac{2m+5}{6m}$

32. $\dfrac{6x-y}{4y} - \dfrac{2x+3y}{4y}$

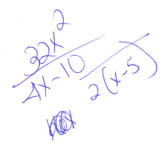

33. $\dfrac{4w-7}{w-5} - \dfrac{2w+3}{w-5}$

34. $\dfrac{3b-8}{b-6} + \dfrac{b-16}{b-6}$

35. $\dfrac{x-7}{x^2-x-6} + \dfrac{2x-2}{x^2-x-6}$

36. $\dfrac{5a-12}{a^2-8a+15} - \dfrac{3a-2}{a^2-8a+15}$

37. $\dfrac{y^2}{2y+8} + \dfrac{3y-4}{2y+8}$

38. $\dfrac{x^2}{4x-12} - \dfrac{9}{4x-12}$

39. $\dfrac{7w}{w+3} + \dfrac{21}{w+3}$

40. $\dfrac{2x}{x-3} - \dfrac{6}{x-3}$

41. $\dfrac{x^2}{x^2+x-6} - \dfrac{6}{(x+3)(x-2)} + \dfrac{x}{(x^2+x-6)}$

42. $\dfrac{-12}{x^2+x-12} + \dfrac{x^2}{(x+4)(x-3)} + \dfrac{x}{x^2+x-12}$

43. **Geometry.** Find the perimeter of the given figure.

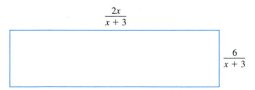

44. **Geometry.** Find the perimeter of the given figure.

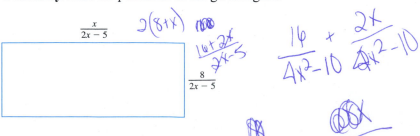

Getting Ready for Section 5.4 [Section 0.2]

(a) $\dfrac{3}{4} + \dfrac{1}{2}$

(b) $\dfrac{5}{6} - \dfrac{2}{3}$

(c) $\dfrac{7}{10} - \dfrac{3}{5}$

(d) $\dfrac{5}{8} + \dfrac{3}{4}$

(e) $\dfrac{5}{6} + \dfrac{3}{8}$

(f) $\dfrac{7}{8} - \dfrac{3}{5}$

(g) $\dfrac{9}{10} - \dfrac{2}{15}$

(h) $\dfrac{5}{12} + \dfrac{7}{18}$

Answers

1. $\dfrac{2}{3}$ **3.** $\dfrac{1}{4}$ **5.** $\dfrac{x}{2}$ **7.** $\dfrac{2a}{5}$ **9.** $\dfrac{8}{x}$ **11.** $\dfrac{6}{w}$ **13.** $\dfrac{5}{xy}$

15. $\dfrac{2}{cd}$ **17.** $\dfrac{16}{x-5}$ **19.** 2 **21.** 8 **23.** $x-1$ **25.** $m+5$

27. $a-2$ **29.** $\dfrac{x-4}{2}$ **31.** $\dfrac{m+1}{3m}$ **33.** 2 **35.** $\dfrac{3}{x+2}$

37. $\dfrac{y-1}{2}$ **39.** 7 **41.** 1 **43.** 4 **a.** $\dfrac{5}{4}$ **b.** $\dfrac{1}{6}$ **c.** $\dfrac{1}{10}$

d. $\dfrac{11}{8}$ **e.** $\dfrac{29}{24}$ **f.** $\dfrac{11}{40}$ **g.** $\dfrac{23}{30}$ **h.** $\dfrac{29}{36}$

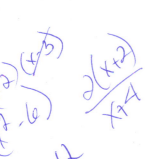

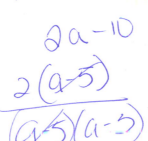

5.4 Adding and Subtracting Unlike Rational Expressions

5.4 OBJECTIVES

1. Write the sum of two unlike rational expressions in simplest form
2. Write the difference of two unlike rational expressions in simplest form

Adding or subtracting **unlike rational expressions** (fractions that do not have the same denominator) requires a bit more work than adding or subtracting the like rational expressions of Section 5.3. When the denominators are not the same, we must use the idea of the *least common denominator* (LCD). Each fraction is "built up" to an equivalent fraction having the LCD as a denominator. You can then add or subtract as before.

OBJECTIVE 1

Example 1   Finding the LCD and Adding Fractions

Add $\dfrac{5}{9} + \dfrac{1}{6}$.

Step 1   To find the LCD, factor each denominator.

$9 = 3 \cdot 3$ ⟵ 3 appears twice.

$6 = 2 \cdot 3$

To form the LCD, include each factor the greatest number of times it appears in any single denominator. In this example, use one 2, because 2 appears only once in the factorization of 6. Use two 3s, because 3 appears twice in the factorization of 9. Thus the LCD for the fractions is $2 \cdot 3 \cdot 3 = 18$.

Step 2   "Build up" each fraction to an equivalent fraction with the LCD as the denominator. Do this by multiplying the numerator and denominator of the given fractions by the same number.

NOTE Do you see that this uses the fundamental principle in the following form?

$\dfrac{P}{Q} = \dfrac{PR}{QR}$

$\dfrac{5}{9} = \dfrac{5 \cdot 2}{9 \cdot 2} = \dfrac{10}{18}$

$\dfrac{1}{6} = \dfrac{1 \cdot 3}{6 \cdot 3} = \dfrac{3}{18}$

Step 3   Add the fractions.

$\dfrac{5}{9} + \dfrac{1}{6} = \dfrac{10}{18} + \dfrac{3}{18} = \dfrac{13}{18}$

$\dfrac{13}{18}$ is in simplest form and so we are done!

✔ **CHECK YOURSELF 1** _____

Add the fractions.

(a) $\dfrac{1}{6} + \dfrac{3}{8}$

(b) $\dfrac{3}{10} + \dfrac{4}{15}$

The process of finding the sum or difference is exactly the same in algebra as it is in arithmetic. We can summarize the steps with the following rule:

Step by Step: To Add or Subtract Unlike Rational Expressions

Step 1 Find the least common denominator of all the fractions.
Step 2 Convert each fraction to an equivalent fraction with the LCD as a denominator.
Step 3 Add or subtract the like fractions formed in step 2.
Step 4 Write the sum or difference in simplest form.

OBJECTIVES 1–2   **Example 2   Adding and Subtracting Unlike Rational Expressions**

(a) Add $\dfrac{3}{2x} + \dfrac{4}{x^2}$.

Step 1 Factor the denominators.

$2x = 2 \cdot x$
$x^2 = x \cdot x$

NOTE Although the product of the denominators will be a common denominator, it is not necessarily the *least* common denominator (LCD).

The LCD must contain the factors 2 and x. The factor x must appear *twice* because it appears twice as a factor in the second denominator.

The LCD is $2 \cdot x \cdot x$, or $2x^2$.

Step 2

$\dfrac{3}{2x} = \dfrac{3 \cdot x}{2x \cdot x} = \dfrac{3x}{2x^2}$

$\dfrac{4}{x^2} = \dfrac{4 \cdot 2}{x^2 \cdot 2} = \dfrac{8}{2x^2}$

$$\frac{9}{30} + \frac{8}{30} = \frac{17}{30}$$

Step 3

$\dfrac{3}{2x} + \dfrac{4}{x^2} = \dfrac{3x}{2x^2} + \dfrac{8}{2x^2}$

$= \dfrac{3x + 8}{2x^2}$

The sum is in simplest form.

(b) Subtract $\dfrac{4}{3x^2} - \dfrac{3}{2x^3}$.

Step 1 Factor the denominators.

$3x^2 = 3 \cdot x \cdot x$
$2x^3 = 2 \cdot x \cdot x \cdot x$

The LCD must contain the factors 2, 3, and x. The LCD is

$2 \cdot 3 \cdot x \cdot x \cdot x$ or $6x^3$ The factor x must appear 3 times. Do you see why?

Step 2

RECALL Both the numerator and the denominator must be multiplied by the same quantity.

$\dfrac{4}{3x^2} = \dfrac{4 \cdot 2x}{3x^2 \cdot 2x} = \dfrac{8x}{6x^3}$

$\dfrac{3}{2x^3} = \dfrac{3 \cdot 3}{2x^3 \cdot 3} = \dfrac{9}{6x^3}$

Step 3

$\dfrac{4}{3x^2} - \dfrac{3}{2x^3} = \dfrac{8x}{6x^3} - \dfrac{9}{6x^3}$

$= \dfrac{8x - 9}{6x^3}$

The difference is in simplest form.

✔ **CHECK YOURSELF 2**

Add or subtract as indicated.

(a) $\dfrac{5}{x^2} + \dfrac{3}{x^3}$

(b) $\dfrac{3}{5x} - \dfrac{1}{4x^2}$

We can also add fractions with more than one variable in the denominator. Example 3 illustrates this type of sum.

Example 3 Adding Unlike Rational Expressions

Add $\dfrac{2}{3x^2y} + \dfrac{3}{4x^3}$.

Step 1 Factor the denominators.

$3x^2y = 3 \cdot x \cdot x \cdot y$
$4x^3 = 2 \cdot 2 \cdot x \cdot x \cdot x$

The LCD is $12x^3y$. Do you see why?

Step 2

$$\frac{2}{3x^2y} = \frac{2 \cdot 4x}{3x^2y \cdot 4x} = \frac{8x}{12x^3y}$$

$$\frac{3}{4x^3} = \frac{3 \cdot 3y}{4x^3 \cdot 3y} = \frac{9y}{12x^3y}$$

Step 3

NOTE The y in the numerator and that in the denominator cannot be divided out because the form is not a factor.

$$\frac{2}{3x^2y} + \frac{3}{4x^3} = \frac{8x}{12x^3y} + \frac{9y}{12x^3y}$$

$$= \frac{8x + 9y}{12x^3y}$$

 CHECK YOURSELF 3

Add.

$$\frac{2}{3x^2y} + \frac{1}{6xy^2}$$

Rational expressions with binomials in the denominator can also be added by taking the approach shown in Example 3. Example 4 illustrates this approach with binomials.

Example 4   Adding and Subtracting Unlike Rational Expressions

(a) Add $\dfrac{5}{x} + \dfrac{2}{x - 1}$.

Step 1   The LCD must have factors of x and $x - 1$. The LCD is $x(x - 1)$.

Step 2

$$\frac{5}{x} = \frac{5(x - 1)}{x(x - 1)}$$

$$\frac{2}{x - 1} = \frac{2x}{(x - 1)x} = \frac{2x}{x(x - 1)}$$

Step 3

$$\frac{5}{x} + \frac{2}{x - 1} = \frac{5(x - 1)}{x(x - 1)} + \frac{2x}{x(x - 1)}$$

$$= \frac{5x - 5 + 2x}{x(x - 1)}$$

$$= \frac{7x - 5}{x(x - 1)}$$

(b) Subtract $\dfrac{3}{x-2} - \dfrac{4}{x+2}$.

Step 1 The LCD must have factors of $x-2$ and $x+2$. The LCD is $(x-2)(x+2)$.

Step 2

NOTE Multiply numerator and denominator by $x+2$.

$$\frac{3}{x-2} = \frac{3(x+2)}{(x-2)(x+2)}$$

NOTE Multiply numerator and denominator by $x-2$.

$$\frac{4}{x+2} = \frac{4(x-2)}{(x+2)(x-2)}$$

Step 3

$$\frac{3}{x-2} - \frac{4}{x+2} = \frac{3(x+2) - 4(x-2)}{(x+2)(x-2)}$$

Note that the x-term becomes negative and the constant term becomes positive.

$$= \frac{3x + 6 - 4x + 8}{(x+2)(x-2)}$$

$$= \frac{-x + 14}{(x+2)(x-2)}$$

CHECK YOURSELF 4

Add or subtract as indicated.

(a) $\dfrac{3}{x+2} + \dfrac{5}{x}$

(b) $\dfrac{4}{x+3} - \dfrac{2}{x-3}$

Example 5 will show how factoring must sometimes be used in forming the LCD.

Example 5 **Adding and Subtracting Unlike Rational Expressions**

(a) Add $\dfrac{3}{2x-2} + \dfrac{5}{3x-3}$.

Step 1 Factor the denominators.

$$2x - 2 = 2(x-1)$$
$$3x - 3 = 3(x-1)$$

CAUTION

$x - 1$ is not used twice in forming the LCD.

The LCD must have factors of 2, 3, and $x-1$. The LCD is $2 \cdot 3(x-1)$, or $6(x-1)$.

Step 2

$$\frac{3}{2x - 2} = \frac{3}{2(x - 1)} = \frac{3 \cdot 3}{2(x - 1) \cdot 3} = \frac{9}{6(x - 1)}$$

$$\frac{5}{3x - 3} = \frac{5}{3(x - 1)} = \frac{5 \cdot 2}{3(x - 1) \cdot 2} = \frac{10}{6(x - 1)}$$

Step 3

$$\frac{3}{2x - 2} + \frac{5}{3x - 3} = \frac{9}{6(x - 1)} + \frac{10}{6(x - 1)}$$

$$= \frac{9 + 10}{6(x - 1)}$$

$$= \frac{19}{6(x - 1)}$$

(b) Subtract $\dfrac{3}{2x - 4} - \dfrac{6}{x^2 - 4}$.

Step 1 Factor the denominators.

$$2x - 4 = 2(x - 2)$$
$$x^2 - 4 = (x + 2)(x - 2)$$

The LCD must have factors of 2, $x - 2$, and $x + 2$. The LCD is $2(x - 2)(x + 2)$.

Step 2

NOTE Multiply numerator and denominator by $x + 2$.

$$\frac{3}{2x - 4} = \frac{3}{2(x - 2)} = \frac{3(x + 2)}{2(x - 2)(x + 2)}$$

NOTE Multiply numerator and denominator by 2.

$$\frac{6}{x^2 - 4} = \frac{6}{(x + 2)(x - 2)} = \frac{6 \cdot 2}{2(x + 2)(x - 2)}$$

$$= \frac{12}{2(x + 2)(x - 2)}$$

Step 3

$$\frac{3}{2x - 4} - \frac{6}{x^2 - 4} = \frac{3(x + 2) - 12}{2(x - 2)(x + 2)}$$

NOTE Remove the parentheses and combine like terms in the numerator.

$$= \frac{3x + 6 - 12}{2(x - 2)(x + 2)}$$

$$= \frac{3x - 6}{2(x - 2)(x + 2)}$$

Step 4 Simplify the difference.

NOTE Factor the numerator and divide by the common factor, $x - 2$.

$$\frac{3x - 6}{2(x - 2)(x + 2)} = \frac{3\overset{1}{(x - 2)}}{2\underset{1}{(x - 2)}(x + 2)}$$

$$= \frac{3}{2(x + 2)}$$

(c) Subtract $\dfrac{5}{x^2 - 1} - \dfrac{2}{x^2 + 2x + 1}$.

Step 1   Factor the denominators.

$$x^2 - 1 = (x - 1)(x + 1)$$
$$x^2 + 2x + 1 = (x + 1)(x + 1)$$

The LCD is $(x - 1)(x + 1)(x + 1)$.

This factor is needed twice.

Step 2

$$\frac{5}{(x - 1)(x + 1)} = \frac{5(x + 1)}{(x - 1)(x + 1)(x + 1)}$$

$$\frac{2}{(x + 1)(x + 1)} = \frac{2(x - 1)}{(x + 1)(x + 1)(x - 1)}$$

Step 3

$$\frac{5}{x^2 - 1} - \frac{2}{x^2 + 2x + 1} = \frac{5(x + 1) - 2(x - 1)}{(x - 1)(x + 1)(x + 1)}$$

NOTE Remove the parentheses and simplify in the numerator.

$$= \frac{5x + 5 - 2x + 2}{(x - 1)(x + 1)(x + 1)}$$

$$= \frac{3x + 7}{(x - 1)(x + 1)(x + 1)}$$

CHECK YOURSELF 5

Add or subtract as indicated.

(a) $\dfrac{5}{2x + 2} + \dfrac{1}{5x + 5}$

(b) $\dfrac{3}{x^2 - 9} - \dfrac{1}{2x - 6}$

(c) $\dfrac{4}{x^2 - x - 2} - \dfrac{3}{x^2 + 4x + 3}$

Recall from Section 5.1 that

$$a - b = -(b - a)$$

Let's see how this can be used in adding or subtracting rational expressions.

Example 6   Adding Unlike Rational Expressions

Add $\dfrac{4}{x-5} + \dfrac{2}{5-x}$.

Rather than try a denominator of $(x-5)(5-x)$, let's rewrite one of the denominators.

NOTE Replace $5-x$ with $-(x-5)$. We now use the fact that

$$\frac{a}{-b} = \frac{-a}{b}$$

$$\frac{4}{x-5} + \frac{2}{5-x} = \frac{4}{x-5} + \frac{2}{-(x-5)}$$

$$= \frac{4}{x-5} + \frac{-2}{x-5}$$

The LCD is now $x-5$, and we can combine the rational expressions as

$$\frac{4-2}{x-5} = \frac{2}{x-5}$$

CHECK YOURSELF 6

Subtract the fractions.

$$\frac{3}{x-3} - \frac{1}{3-x}$$

CHECK YOURSELF ANSWERS

1. (a) $\dfrac{13}{24}$; (b) $\dfrac{17}{30}$ **2.** (a) $\dfrac{5x+3}{x^3}$; (b) $\dfrac{12x-5}{20x^2}$ **3.** $\dfrac{4y+x}{6x^2y^2}$

4. (a) $\dfrac{8x+10}{x(x+2)}$; (b) $\dfrac{2x-18}{(x+3)(x-3)}$ **5.** (a) $\dfrac{27}{10(x+1)}$; (b) $\dfrac{-1}{2(x+3)}$;

(c) $\dfrac{x+18}{(x+1)(x-2)(x+3)}$ **6.** $\dfrac{4}{x-3}$

5.4　Exercises

Add or subtract as indicated. Express your result in simplest form.

1. $\dfrac{3}{7} + \dfrac{5}{6}$

2. $\dfrac{7}{12} - \dfrac{4}{9}$

3. $\dfrac{13}{25} - \dfrac{7}{20}$

4. $\dfrac{3}{5} + \dfrac{7}{9}$

5. $\dfrac{y}{4} + \dfrac{3y}{5}$

6. $\dfrac{5x}{6} - \dfrac{2x}{3}$

7. $\dfrac{7a}{3} - \dfrac{a}{7}$

8. $\dfrac{3m}{4} + \dfrac{m}{9}$

9. $\dfrac{3}{x} - \dfrac{4}{5}$

10. $\dfrac{5}{x} + \dfrac{2}{3}$

11. $\dfrac{5}{a} + \dfrac{a}{5}$

12. $\dfrac{y}{3} - \dfrac{3}{y}$

13. $\dfrac{5}{m} + \dfrac{3}{m^2}$

14. $\dfrac{4}{x^2} - \dfrac{3}{x}$

15. $\dfrac{2}{x^2} - \dfrac{5}{7x}$

16. $\dfrac{7}{3w} + \dfrac{5}{w^3}$

17. $\dfrac{7}{9s} + \dfrac{5}{s^2}$

18. $\dfrac{11}{x^2} - \dfrac{5}{7x}$

19. $\dfrac{3}{4b^2} + \dfrac{5}{3b^3}$

20. $\dfrac{4}{5x^3} - \dfrac{3}{2x^2}$

21. $\dfrac{x}{x+2} + \dfrac{2}{5}$

22. $\dfrac{3}{4} - \dfrac{a}{a-1}$

23. $\dfrac{y}{y-4} - \dfrac{3}{4}$

24. $\dfrac{m}{m+3} + \dfrac{2}{3}$

25. $\dfrac{4}{x} + \dfrac{3}{x+1}$

26. $\dfrac{2}{x} - \dfrac{1}{x-2}$

27. $\dfrac{5}{a-1} - \dfrac{2}{a}$

28. $\dfrac{4}{x+2} + \dfrac{3}{x}$

29. $\dfrac{4}{2x-3} + \dfrac{2}{3x}$

30. $\dfrac{7}{2y-1} - \dfrac{3}{2y}$

31. $\dfrac{2}{x+1} + \dfrac{3}{x+3}$

32. $\dfrac{5}{x-1} + \dfrac{2}{x+2}$

33. $\dfrac{4}{y-2} - \dfrac{1}{y+1}$

34. $\dfrac{5}{x+4} - \dfrac{3}{x-1}$

35. $\dfrac{2}{b-3} + \dfrac{3}{2b-6}$

36. $\dfrac{4}{a+5} - \dfrac{3}{4a+20}$

37. $\dfrac{x}{x+4} - \dfrac{2}{3x+12}$

38. $\dfrac{x}{x-3} + \dfrac{5}{2x-6}$

39. $\dfrac{4}{3m+3} + \dfrac{1}{2m+2}$

40. $\dfrac{3}{5y-5} - \dfrac{2}{3y-3}$

41. $\dfrac{4}{5x-10} - \dfrac{1}{3x-6}$

42. $\dfrac{2}{3w+3} + \dfrac{5}{2w+2}$

43. $\dfrac{7}{3c+6} - \dfrac{2c}{7c+14}$

44. $\dfrac{5}{3c-12} + \dfrac{4c}{5c-20}$

45. $\dfrac{y-1}{y+1} - \dfrac{y}{3y+3}$

46. $\dfrac{x+2}{x-2} - \dfrac{x}{3x-6}$

ANSWERS

23.

24.

25.

26.

27.

28.

29.

30.

31.

32.

33.

34.

35.   36.

37.   38.

39.

40.

41.

42.

43.

44.

45.

46.

47. _____

48. _____

49. _____

50. _____

51. _____

52. _____

53. _____

54. _____

55. _____

56. _____

57. _____

58. _____

59. _____

60. _____

61. _____

62. _____

63. _____

64. _____

65. _____

66. _____

67. _____

68. _____

47. $\dfrac{3}{x^2 - 4} + \dfrac{2}{x + 2}$

48. $\dfrac{4}{x - 2} + \dfrac{3}{x^2 - x - 2}$

49. $\dfrac{3x}{x^2 - 3x + 2} - \dfrac{1}{x - 2}$

50. $\dfrac{a}{a^2 - 1} - \dfrac{4}{a + 1}$

51. $\dfrac{2x}{x^2 - 5x + 6} + \dfrac{4}{x - 2}$

52. $\dfrac{7a}{a^2 + a - 12} - \dfrac{4}{a + 4}$

53. $\dfrac{2}{3x - 3} - \dfrac{1}{4x + 4}$

54. $\dfrac{2}{5w + 10} - \dfrac{3}{2w - 4}$

55. $\dfrac{4}{3a - 9} - \dfrac{3}{2a + 4}$

56. $\dfrac{2}{3b - 6} + \dfrac{3}{4b + 8}$

57. $\dfrac{5}{x^2 - 16} - \dfrac{3}{x^2 - x - 12}$

58. $\dfrac{3}{x^2 + 4x + 3} - \dfrac{1}{x^2 - 9}$

59. $\dfrac{2}{y^2 + y - 6} + \dfrac{3y}{y^2 - 2y - 15}$

60. $\dfrac{2a}{a^2 - a - 12} - \dfrac{3}{a^2 - 2a - 8}$

61. $\dfrac{6x}{x^2 - 9} - \dfrac{5x}{x^2 + x - 6}$

62. $\dfrac{4y}{y^2 + 6y + 5} + \dfrac{2y}{y^2 - 1}$

63. $\dfrac{3}{a - 7} + \dfrac{2}{7 - a}$

64. $\dfrac{5}{x - 5} - \dfrac{3}{5 - x}$

65. $\dfrac{2x}{2x - 3} - \dfrac{1}{3 - 2x}$

66. $\dfrac{9m}{3m - 1} + \dfrac{3}{1 - 3m}$

Add or subtract, as indicated.

67. $\dfrac{1}{a - 3} - \dfrac{1}{a + 3} + \dfrac{2a}{a^2 - 9}$

68. $\dfrac{1}{p + 1} + \dfrac{1}{p - 3} - \dfrac{4}{p^2 - 2p - 3}$

69. $\dfrac{2x^2 + 3x}{x^2 - 2x - 63} + \dfrac{7 - x}{x^2 - 2x - 63} - \dfrac{x^2 - 3x + 21}{x^2 - 2x - 63}$

70. $-\dfrac{3 - 2x^2}{x^2 - 9x + 20} - \dfrac{4x^2 + 2x + 1}{x^2 - 9x + 20} + \dfrac{2x^2 + 3x}{x^2 - 9x + 20}$

71. Number problem. Use a rational expression to represent the sum of the reciprocals of two consecutive even integers.

72. Number problem. One number is two less than another. Use a rational expression to represent the sum of the reciprocals of the two numbers.

73. Geometry. Refer to the rectangle in the figure. Find an expression that represents its perimeter.

$\dfrac{2x + 1}{5}$

$\dfrac{4}{3x + 1}$

74. Geometry. Refer to the triangle in the figure. Find an expression that represents its perimeter.

$\dfrac{3}{4x}$

$\dfrac{1}{x^2}$

$\dfrac{5}{x^2}$

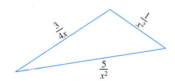

Getting Ready for Section 5.5 [Section 2.3]

Solve each of the following equations.

(a) $x + 7 = 10$

(b) $x - 8 = -3$

(c) $4x - 5 = 2$

(d) $5x + 8 = 4$

(e) $4(x - 2) - 5 = 7$

(f) $4(2x + 1) - 3 = -23$

(g) $3(5x - 2) - 4(4x + 1) = -20$

(h) $5(3x + 2) - 3(4x - 3) = -2$

ANSWERS

69. _____

70. _____

71. _____

72. _____

73. _____

74. _____

a. _____

b. _____

c. _____

d. _____

e. _____

f. _____

g. _____

h. _____

© 2005 McGraw-Hill Companies

Answers

1. $\dfrac{53}{42}$ **3.** $\dfrac{17}{100}$ **5.** $\dfrac{17y}{20}$ **7.** $\dfrac{46a}{21}$ **9.** $\dfrac{15 - 4x}{5x}$ **11.** $\dfrac{25 + a^2}{5a}$

13. $\dfrac{5m + 3}{m^2}$ **15.** $\dfrac{14 - 5x}{7x^2}$ **17.** $\dfrac{7s + 45}{9s^2}$ **19.** $\dfrac{9b + 20}{12b^3}$

21. $\dfrac{7x + 4}{5(x + 2)}$ **23.** $\dfrac{y + 12}{4(y - 4)}$ **25.** $\dfrac{7x + 4}{x(x + 1)}$ **27.** $\dfrac{3a + 2}{a(a - 1)}$

29. $\dfrac{2(8x - 3)}{3x(2x - 3)}$ **31.** $\dfrac{5x + 9}{(x + 1)(x + 3)}$ **33.** $\dfrac{3(y + 2)}{(y - 2)(y + 1)}$

35. $\dfrac{7}{2(b - 3)}$ **37.** $\dfrac{3x - 2}{3(x + 4)}$ **39.** $\dfrac{11}{6(m + 1)}$ **41.** $\dfrac{7}{15(x - 2)}$

43. $\dfrac{49 - 6c}{21(c + 2)}$ **45.** $\dfrac{2y - 3}{3(y + 1)}$ **47.** $\dfrac{2x - 1}{(x - 2)(x + 2)}$

49. $\dfrac{2x + 1}{(x - 1)(x - 2)}$ **51.** $\dfrac{6}{x - 3}$ **53.** $\dfrac{5x + 11}{12(x - 1)(x + 1)}$

55. $\dfrac{-a + 43}{6(a - 3)(a + 2)}$ **57.** $\dfrac{2x + 3}{(x + 4)(x - 4)(x + 3)}$

59. $\dfrac{3y^2 - 4y - 10}{(y + 3)(y - 2)(y - 5)}$ **61.** $\dfrac{x}{(x - 3)(x - 2)}$ **63.** $\dfrac{1}{a - 7}$

65. $\dfrac{2x + 1}{2x - 3}$ **67.** $\dfrac{2}{a - 3}$ **69.** $\dfrac{x - 2}{x - 9}$ **71.** $\dfrac{2x + 2}{x(x + 2)}$

73. $\dfrac{2(6x^2 + 5x + 21)}{5(3x + 1)}$ **a.** 3   **b.** 5   **c.** $\dfrac{7}{4}$ **d.** $\dfrac{-4}{5}$

e. 5   **f.** -3 **g.** 10   **h.** -7

5.5 Equations Involving Rational Expressions

5.5 OBJECTIVES

1. Solve an equation with integer denominators
2. Determine the excluded values for the variables of a rational expression
3. Solve a rational equation
4. Solve a proportion for an unknown

In Chapter 2, you learned how to solve a variety of equations. We now want to extend that work to the solution of **rational equations,** which are equations that involve rational expressions.

To solve a rational equation, we multiply each term of the equation by the LCD of any fractions. The resulting equation should be equivalent to the original equation and be cleared of all fractions.

NOTE The resulting equation *will* be equivalent unless a solution results that makes a denominator in the original equation zero. More about this later!

OBJECTIVE 1

Example 1   Solving Equations with Integer Denominators

Solve

NOTE This equation has three terms: $\frac{x}{2}$, $-\frac{1}{3}$, and $\frac{2x+3}{6}$. The sign of the term is not used to find the LCD.

$$\frac{x}{2} - \frac{1}{3} = \frac{2x+3}{6}$$

The LCD for $\frac{x}{2}$, $\frac{1}{3}$, and $\frac{2x+3}{6}$ is 6. Multiply both sides of the equation by 6. Using the distributive property, we multiply *each* term by 6.

NOTE By the multiplication property of equality, this equation is equivalent to the original equation.

$$6 \cdot \frac{x}{2} - 6 \cdot \frac{1}{3} = 6\left(\frac{2x+3}{6}\right)$$

or

$$3x - 2 = 2x + 3$$

Solving as before, we have

$$3x - 2x = 3 + 2 \qquad \text{or} \qquad x = 5$$

To check, substitute 5 for x in the *original* equation.

$$\frac{(5)}{2} - \frac{1}{3} \stackrel{?}{=} \frac{2(5)+3}{6}$$

$$\frac{13}{6} \stackrel{\checkmark}{=} \frac{13}{6} \qquad \text{(True)}$$

457

CAUTION

Be Careful! Many students have difficulty because they don't distinguish between adding and subtracting *expressions* (as we did in Sections 5.3 and 5.4) and solving equations (illustrated in the above example). In the **expression**

$$\frac{x+1}{2} + \frac{x}{3}$$

we want to add the two fractions to form a single fraction. In the **equation**

$$\frac{x+1}{2} = \frac{x}{3} + 1$$

we want to solve for x.

CHECK YOURSELF 1

Solve and check.

$$\frac{x}{4} - \frac{1}{6} = \frac{4x-5}{12}$$

In Example 1, all the denominators were integers. What happens when we allow variables in the denominator? Recall that, for any fraction, the denominator must not be equal to zero. When a fraction has a variable in the denominator, we must exclude any value for the variable that would result in division by zero.

OBJECTIVE 2   **Example 2   Finding Excluded Values for *x***

In the following rational expressions, what values for x must be excluded?

(a) $\frac{x}{5}$. Here x can have any value, so none need be excluded.

(b) $\frac{3}{x}$. If $x = 0$, then $\frac{3}{x}$ is undefined; 0 is the excluded value.

(c) $\frac{5}{x-2}$. If $x = 2$, then $\frac{5}{x-2} = \frac{5}{(2)-2} = \frac{5}{0}$, which is undefined, so 2 is the excluded value.

CHECK YOURSELF 2

What values for x, if any, must be excluded?

(a) $\dfrac{x}{7}$
(b) $\dfrac{5}{x}$
(c) $\dfrac{7}{x - 5}$

If the denominator of a rational expression contains a product of two or more variable factors, the zero-product principle must be used to determine the excluded values for the variable.

In some cases, you have to factor the denominator to see the restrictions on the values for the variable.

Example 3   Finding Excluded Values for *x*

What values for *x* must be excluded in each fraction?

(a) $\dfrac{3}{x^2 - 6x - 16}$

Factoring the denominator, we have

$$\frac{3}{x^2 - 6x - 16} = \frac{3}{(x - 8)(x + 2)}$$

Letting $x - 8 = 0$ or $x + 2 = 0$, we see that 8 and -2 make the denominator 0, so both 8 and -2 must be excluded.

(b) $\dfrac{3}{x^2 + 2x - 48}$

The denominator is zero when

$$x^2 + 2x - 48 = 0$$

Factoring, we find

$$(x - 6)(x + 8) = 0$$

The denominator is zero when

$$x = 6 \qquad \text{or} \qquad x = -8$$

The excluded values are 6 and -8.

CHECK YOURSELF 3

What values for x must be excluded in the following fractions?

(a) $\dfrac{5}{x^2 - 3x - 10}$
(b) $\dfrac{7}{x^2 + 5x - 14}$

The steps for solving an equation involving fractions are summarized in the following rule.

> **Step by Step:** To Solve a Rational Equation
>
> **Step 1** Remove the fractions in the equation by multiplying each term by the LCD of all the fractions.
> **Step 2** Solve the equation resulting from Step 1 as before.
> **Step 3** Check your solution in the *original equation*.

NOTE The equation that is formed in Step 2 can be solved by the methods of Sections 2.3 and 4.8.

We can also solve rational equations with variables in the denominator by using the above algorithm. Example 4 illustrates this approach.

OBJECTIVE 3 | **Example 4** **Solving Rational Equations**

Solve

$$\frac{7}{4x} - \frac{3}{x^2} = \frac{1}{2x^2}$$

NOTE The factor x appears twice in the LCD.

The LCD of the three terms in the equation is $4x^2$, so we multiply both sides of the equation by $4x^2$.

$$4x^2 \cdot \frac{7}{4x} - 4x^2 \cdot \frac{3}{x^2} = 4x^2 \cdot \frac{1}{2x^2}$$

Simplifying, we have

$$7x - 12 = 2$$
$$7x = 14$$
$$x = 2$$

We leave the check to you. Be sure to return to the original equation.

CHECK YOURSELF 4

Solve and check.

$$\frac{5}{2x} - \frac{4}{x^2} = \frac{7}{2x^2}$$

The process of solving rational equations is exactly the same when there are binomials in the denominators.

Example 5   Solving Rational Equations

(a) Solve

NOTE There are three terms.

$$\frac{x}{x-3} - 2 = \frac{1}{x-3}$$

The LCD is $x - 3$, so we multiply each side (every term) by $x - 3$.

NOTE Each of the terms is multiplied by $x - 3$.

$$(x-3)^{1} \cdot \left(\frac{x}{x-3}_{1}\right) - 2(x-3) = (x-3)^{1} \cdot \left(\frac{1}{x-3}_{1}\right)$$

C A U T I O N

Be careful of the signs!

Simplifying, we have

$$x - 2(x - 3) = 1$$
$$x - 2x + 6 = 1$$
$$-x = -5$$
$$x = 5$$

To check, substitute 5 for x in the original equation.

$$\frac{(5)}{(5) - 3} - 2 \stackrel{?}{=} \frac{1}{(5) - 3}$$

$$\frac{5}{2} - 2 \stackrel{?}{=} \frac{1}{2}$$

$$\frac{1}{2} \stackrel{\checkmark}{=} \frac{1}{2}$$

(b) Solve

RECALL Remember that $x^2 - 9 = (x - 3)(x + 3)$

$$\frac{3}{x-3} - \frac{7}{x+3} = \frac{2}{x^2 - 9}$$

In factored form, the three denominators are $x - 3$, $x + 3$, and $(x + 3)(x - 3)$. This means that the LCD is $(x + 3)(x - 3)$, and so we multiply:

$$(x-3)^{1}(x+3)\left(\frac{3}{x-3}_{1}\right) - (x+3)^{1}(x-3)\left(\frac{7}{x+3}_{1}\right) = (x+3)^{1}(x-3)^{1}\left(\frac{2}{x^2-9}_{1}\right)$$

Simplifying, we have

$$3(x + 3) - 7(x - 3) = 2$$
$$3x + 9 - 7x + 21 = 2$$
$$-4x + 30 = 2$$
$$-4x = -28$$
$$x = 7$$

CHECK YOURSELF 5 _____

Solve and check.

(a) $\dfrac{x}{x - 5} - 2 = \dfrac{2}{x - 5}$

(b) $\dfrac{4}{x - 4} - \dfrac{3}{x + 1} = \dfrac{5}{x^2 - 3x - 4}$

You should be aware that some rational equations have no solutions. Example 6 shows that possibility.

Example 6   Solving Rational Equations

Solve

$$\frac{x}{x - 2} - 7 = \frac{2}{x - 2}$$

The LCD is $x - 2$, and so we multiply each side (every term) by $x - 2$.

$$(x - 2)\left(\frac{x}{x - 2}\right) - 7(x - 2) = (x - 2)\left(\frac{2}{x - 2}\right)$$

Simplifying, we have

$$x - 7x + 14 = 2$$
$$-6x = -12$$
$$x = 2$$

Now, when we try to check our result, we have

NOTE 2 is substituted for x in the original equation.

$$\frac{(2)}{(2) - 2} - 7 \stackrel{?}{=} \frac{2}{(2) - 2} \qquad \text{or} \qquad \frac{2}{0} - 7 \stackrel{?}{=} \frac{2}{0}$$

These terms are undefined.

What went wrong? Remember that two of the terms in our original equation were $\dfrac{x}{x - 2}$ and $\dfrac{2}{x - 2}$. The variable x cannot have the value 2 because 2 is an excluded value (it makes the denominator 0). So our original equation has *no solution*.

CHECK YOURSELF 6 _____

Solve, if possible.

$$\frac{x}{x + 3} - 6 = \frac{-3}{x + 3}$$

Equations involving fractions may also lead to quadratic equations, as Example 7 illustrates.

Example 7   Solving Rational Equations

Solve

$$\frac{x}{x-4} = \frac{15}{x-3} - \frac{2x}{x^2 - 7x + 12}$$

The LCD is $(x - 4)(x - 3)$. Multiply each side (every term) by $(x - 4)(x - 3)$.

$$\frac{x}{(x-4)}\overset{1}{(x-4)}(x-3) = \frac{15}{(x-3)}(x-4)\overset{1}{(x-3)} - \frac{2x}{(x-4)(x-3)}\overset{1}{(x-4)}\overset{1}{(x-3)}$$

Simplifying, we have

$$x(x-3) = 15(x-4) - 2x$$

Multiply to remove the parentheses:

$$x^2 - 3x = 15x - 60 - 2x$$

NOTE This equation is *quadratic.* It can be solved by the methods of Section 4.4.

In standard form, the equation is

$$x^2 - 16x + 60 = 0$$

or

$$(x-6)(x-10) = 0$$

Setting the factors to 0, we have

$$x - 6 = 0 \qquad \text{or} \qquad x - 10 = 0$$
$$x = 6 \qquad\qquad\qquad x = 10$$

So $x = 6$ and $x = 10$ are possible solutions. We leave the check of *each* solution to you.

 ### CHECK YOURSELF 7

Solve and check.

$$\frac{3x}{x+2} - \frac{2}{x+3} = \frac{36}{x^2 + 5x + 6}$$

The following equation is a special kind of equation involving fractions:

$$\frac{135}{t} = \frac{180}{t + 1}$$

An equation of the form $\frac{a}{b} = \frac{c}{d}$ is said to be in **proportion form,** or, more simply, it is called a **proportion.** This type of equation occurs often enough in algebra that it is worth developing some special methods for its solution. First, we need some definitions.

A **ratio** is a means of comparing two quantities. A ratio can be written as a fraction. For instance, the ratio of 2 to 3 can be written as $\frac{2}{3}$. A statement that two ratios are equal is called a *proportion.* A proportion has the form

$$\frac{a}{b} = \frac{c}{d}$$ In this proportion, *a* and *d* are called the **extremes** of the proportion, and *b* and *c* are called the **means.**

A useful property of proportions is easily developed. If

NOTE *bd* is the LCD of the denominators.

$$\frac{a}{b} = \frac{c}{d}$$ we multiply both sides by $b \cdot d$

$$\left(\frac{a}{b}\right)bd = \left(\frac{c}{d}\right)bd \quad \text{or} \quad ad = bc$$

Rules and Properties: Proportions

If $\frac{a}{b} = \frac{c}{d}$, then $ad = bc$.

In words:

In any proportion, the product of the extremes (*ad*) is equal to the product of the means (*bc*).

Because a proportion is a special kind of rational equation, this rule gives us an alternative approach to solving equations that are in the proportion form.

OBJECTIVE 4   **Example 8   Solving a Proportion for an Unknown**

Solve each equation for *x*.

NOTE The extremes are *x* and 15. The means are 5 and 12.

(a) $\dfrac{x}{5} = \dfrac{12}{15}$

Set the product of the extremes equal to the product of the means.

$15x = 5 \cdot 12$

$15x = 60$

$x = 4$

Our solution is 4. You can check as before, by substituting in the original proportion.

(b) $\dfrac{x + 3}{10} = \dfrac{x}{7}$

Set the product of the extremes equal to the product of the means. Be certain to use parentheses with a numerator with more than one term.

$$7(x + 3) = 10x$$
$$7x + 21 = 10x$$
$$21 = 3x$$
$$7 = x$$

We leave it to you to check this result.

 CHECK YOURSELF 8

Solve each equation for x.

(a) $\dfrac{x}{8} = \dfrac{3}{4}$ 　　　　　　　　　　**(b)** $\dfrac{x - 1}{9} = \dfrac{x + 1}{12}$

As the examples of this section illustrated, *whenever* an equation involves rational expressions, the *first step* of the solution is to clear the equation of fractions by multiplication.

The following algorithm summarizes our work in solving equations that involve rational expressions.

Step by Step:　To Solve an Equation Involving Fractions

Step 1　Remove the fractions appearing in the equation by multiplying each side (every term) by the LCD of all the fractions.

Step 2　Solve the equation resulting from step 1. If the equation is linear, use the methods of Section 2.3 for the solution. If the equation is quadratic, use the methods of Section 4.4.

Step 3　Check all solutions by substitution in the *original equation*. Be sure to discard any *extraneous* solutions, that is, solutions that would result in a zero denominator in the original equation.

CHECK YOURSELF ANSWERS

1. $x = 3$　　**2. (a)** None; **(b)** 0; **(c)** 5　　**3. (a)** $-2, 5$; **(b)** $-7, 2$　　**4.** $x = 3$

5. (a) $x = 8$; **(b)** $x = -11$　　**6.** No solution　　**7.** $x = -5$ or $x = \dfrac{8}{3}$

8. (a) $x = 6$; **(b)** $x = 7$

5.5 Exercises

What values for x, if any, must be excluded in each of the following algebraic fractions?

1. $\dfrac{x}{15}$

2. $\dfrac{8}{x}$

3. $\dfrac{17}{x}$

4. $\dfrac{x}{8}$

5. $\dfrac{3}{x-2}$

6. $\dfrac{x-1}{5}$

7. $\dfrac{-5}{x+4}$

8. $\dfrac{4}{x+3}$

9. $\dfrac{x-5}{2}$

10. $\dfrac{x-1}{x-5}$

11. $\dfrac{3x}{(x+1)(x-2)}$

12. $\dfrac{5x}{(x-3)(x+7)}$

13. $\dfrac{x-1}{(2x-1)(x+3)}$

14. $\dfrac{x+3}{(3x+1)(x-2)}$

15. $\dfrac{7}{x^2-9}$

16. $\dfrac{5x}{x^2+x-2}$

17. $\dfrac{x+3}{x^2-7x+12}$

18. $\dfrac{3x-4}{x^2-49}$

19. $\dfrac{2x-1}{3x^2+x-2}$

20. $\dfrac{3x+1}{4x^2-11x+6}$

Solve and check each of the following equations for x.

21. $\dfrac{x}{2}+3=6$

22. $\dfrac{x}{3}-2=1$

23. $\dfrac{x}{2} - \dfrac{x}{3} = 2$

24. $\dfrac{x}{6} - \dfrac{x}{8} = 1$

25. $\dfrac{x}{5} - \dfrac{1}{3} = \dfrac{x-7}{3}$

26. $\dfrac{x}{6} + \dfrac{3}{4} = \dfrac{x-1}{4}$

27. $\dfrac{x}{4} - \dfrac{1}{5} = \dfrac{4x+3}{20}$

28. $\dfrac{x}{12} - \dfrac{1}{6} = \dfrac{2x-7}{12}$

29. $\dfrac{3}{x} + 2 = \dfrac{7}{x}$

30. $\dfrac{4}{x} - 3 = \dfrac{16}{x}$

31. $\dfrac{4}{x} + \dfrac{3}{4} = \dfrac{10}{x}$

32. $\dfrac{3}{x} = \dfrac{5}{3} - \dfrac{7}{x}$

33. $\dfrac{5}{2x} - \dfrac{1}{x} = \dfrac{9}{2x^2}$

34. $\dfrac{4}{3x} + \dfrac{1}{x} = \dfrac{14}{3x^2}$

35. $\dfrac{2}{x-3} + 1 = \dfrac{7}{x-3}$

36. $\dfrac{x}{x+1} + 2 = \dfrac{14}{x+1}$

37. $\dfrac{12}{x+3} = \dfrac{x}{x+3} + 2$

38. $\dfrac{5}{x-3} + 3 = \dfrac{x}{x-3}$

39. $\dfrac{3}{x-5} + 4 = \dfrac{2x+5}{x-5}$

40. $\dfrac{24}{x+5} - 2 = \dfrac{x+2}{x+5}$

41. $\dfrac{2}{x+3} + \dfrac{1}{2} = \dfrac{x+6}{x+3}$

42. $\dfrac{6}{x-5} - \dfrac{2}{3} = \dfrac{x-9}{x-5}$

43. $\dfrac{x}{3x+12} + \dfrac{x-1}{x+4} = \dfrac{5}{3}$

44. $\dfrac{x}{4x-12} - \dfrac{x-4}{x-3} = \dfrac{1}{8}$

45. $\dfrac{x}{x-3} - 2 = \dfrac{3}{x-3}$

46. $\dfrac{x}{x-5} + 2 = \dfrac{5}{x-5}$

47. $\dfrac{x-1}{x+3} - \dfrac{x-3}{x} = \dfrac{3}{x^2+3x}$

48. $\dfrac{x}{x-2} - \dfrac{x+1}{x} = \dfrac{8}{x^2-2x}$

ANSWERS

23.

24.

25.

26.

27.

28.

29.

30.

31.

32.

33.

34.

35.

36.

37.

38.

39.

40.

41.

42.

43.

44.

45.

46.

47.

48.

49. _____

50. _____

51. _____

52. _____

53. _____

54. _____

55. _____

56. _____

57. _____

58. _____

59. _____

60. _____

61. _____

62. _____

63. _____

64. _____

65. _____

66. _____

67. _____

68. _____

69. _____

70. _____

71. _____

72. _____

49. $\dfrac{1}{x-2} - \dfrac{2}{x+2} = \dfrac{2}{x^2-4}$

50. $\dfrac{1}{x+4} + \dfrac{1}{x-4} = \dfrac{12}{x^2-16}$

51. $\dfrac{5}{x-4} = \dfrac{1}{x+2} - \dfrac{2}{x^2-2x-8}$

52. $\dfrac{11}{x+2} = \dfrac{5}{x^2-x-6} + \dfrac{1}{x-3}$

53. $\dfrac{3}{x-1} - \dfrac{1}{x+9} = \dfrac{18}{x^2+8x-9}$

54. $\dfrac{2}{x+2} = \dfrac{3}{x+6} + \dfrac{9}{x^2+8x+12}$

55. $\dfrac{3}{x+3} + \dfrac{25}{x^2+x-6} = \dfrac{5}{x-2}$

56. $\dfrac{5}{x+6} + \dfrac{2}{x^2+7x+6} = \dfrac{3}{x+1}$

57. $\dfrac{7}{x-5} - \dfrac{3}{x+5} = \dfrac{40}{x^2-25}$

58. $\dfrac{3}{x-3} - \dfrac{18}{x^2-9} = \dfrac{5}{x+3}$

59. $\dfrac{2x}{x-3} + \dfrac{2}{x-5} = \dfrac{3x}{x^2-8x+15}$

60. $\dfrac{x}{x-4} = \dfrac{5x}{x^2-x-12} - \dfrac{3}{x+3}$

61. $\dfrac{2x}{x+2} = \dfrac{5}{x^2-x-6} - \dfrac{1}{x-3}$

62. $\dfrac{3x}{x-1} = \dfrac{2}{x-2} - \dfrac{2}{x^2-3x+2}$

63. $\dfrac{7}{x-2} + \dfrac{16}{x+3} = 3$

64. $\dfrac{5}{x-2} + \dfrac{6}{x+2} = 2$

65. $\dfrac{11}{x-3} - 1 = \dfrac{10}{x+3}$

66. $\dfrac{17}{x-4} - 2 = \dfrac{10}{x+2}$

Solve each of the following equations for x.

67. $\dfrac{x}{11} = \dfrac{12}{33}$

68. $\dfrac{4}{x} = \dfrac{16}{20}$

69. $\dfrac{5}{8} = \dfrac{20}{x}$

70. $\dfrac{x}{10} = \dfrac{9}{30}$

71. $\dfrac{x+1}{5} = \dfrac{20}{25}$

72. $\dfrac{2}{5} = \dfrac{x-2}{20}$

73. $\dfrac{3}{5} = \dfrac{x-1}{20}$

74. $\dfrac{5}{x-3} = \dfrac{15}{21}$

75. $\dfrac{x}{6} = \dfrac{x+5}{16}$

76. $\dfrac{x-2}{x+2} = \dfrac{12}{20}$

77. $\dfrac{x}{x+7} = \dfrac{10}{17}$

78. $\dfrac{x}{10} = \dfrac{x+6}{30}$

79. $\dfrac{2}{x-1} = \dfrac{6}{x+9}$

80. $\dfrac{3}{x-3} = \dfrac{4}{x-5}$

81. $\dfrac{1}{x+3} = \dfrac{7}{x^2-9}$

82. $\dfrac{1}{x+5} = \dfrac{4}{x^2+3x-10}$

73. _____
74. _____
75. _____
76. _____
77. _____
78. _____
79. _____
80. _____
81. _____
82. _____
a. _____
b. _____
c. _____
d. _____
e. _____
f. _____

Getting Ready for Section 5.6 [Section 1.1]

Write each of the following phrases using symbols. Use the variable x to represent the number in each case.

(a) One-fourth of a number added to four-fifths of the same number
(b) 6 times a number, decreased by 12
(c) The quotient when 5 more than a number is divided by 6
(d) Three times the length of a side of a rectangle decreased by 4
(e) A distance traveled divided by 5
(f) The speed of a truck that is 5 mi/h slower than a car

Answers

1. None   **3.** 0   **5.** 2   **7.** -4 **9.** None   **11.** $-1, 2$

13. $-3, \dfrac{1}{2}$ **15.** $-3, 3$ **17.** 3, 4   **19.** $-1, \dfrac{2}{3}$ **21.** 6

23. 12   **25.** 15   **27.** 7   **29.** 2   **31.** 8   **33.** 3
35. 8   **37.** 2   **39.** 11   **41.** -5 **43.** -23
45. No solution   **47.** 6   **49.** 4   **51.** -4 **53.** -5

55. No solution   **57.** $-\dfrac{5}{2}$ **59.** $-\dfrac{1}{2}, 6$ **61.** $-\dfrac{1}{2}$ **63.** $-\dfrac{1}{3}, 7$

65. $-8, 9$ **67.** 4   **69.** 32   **71.** 3   **73.** 13   **75.** 3   **77.** 10

79. 6   **81.** 10   **a.** $\dfrac{1}{4}x + \dfrac{4}{5}x$ **b.** $6x - 12$ **c.** $\dfrac{x+5}{6}$ **d.** $3x - 4$

e. $\dfrac{x}{5}$ **f.** $x - 5$

5.6 Applications of Rational Expressions

5.6 OBJECTIVES

1. Solve a word problem that leads to a rational equation
2. Use a proportion to solve a word problem

Many word problems will lead to rational equations that must be solved by using the methods of Section 5.5. The five steps in solving word problems are, of course, the same as you saw earlier.

OBJECTIVE 1

Example 1 Solving a Numerical Application

If one-third of a number is added to three-fourths of that same number, the sum is 26. Find the number.

Step 1 Read the problem carefully. You want to find the unknown number.

Step 2 Choose a letter to represent the unknown. Let x be the unknown number.

Step 3 Form an equation.

NOTE The equation expresses the relationship between the two parts of the number.

$$\frac{1}{3}x + \frac{3}{4}x = 26$$

One-third of number

Three-fourths of number

Step 4 Solve the equation. Multiply each side (every term) of the equation by 12, the LCD.

$$12 \cdot \frac{1}{3}x + 12 \cdot \frac{3}{4}x = 12 \cdot 26$$

Simplifying yields

$$4x + 9x = 312$$
$$13x = 312$$
$$x = 24$$

NOTE Be sure to answer the question raised in the problem.

The number is 24.

Step 5 Check your solution by returning to the *original problem*. If the number is 24, we have

$$\frac{1}{3}(24) + \frac{3}{4}(24) = 8 + 18 = 26$$

and the solution is verified.

 CHECK YOURSELF 1 _____

The sum of two-fifths of a number and one-half of that number is 18. *Find the number.*

470

Number problems that involve reciprocals can be solved by using rational equations. Example 2 illustrates this approach.

Example 2   Solving a Numerical Application

One number is twice another number. If the sum of their reciprocals is $\frac{3}{10}$, what are the two numbers?

Step 1   You want to find the two numbers.

Step 2   Let x be one number. Then $2x$ is the other number.

Twice the first

Step 3

NOTE The reciprocal of a fraction is the fraction obtained by switching the numerator and denominator.

$$\frac{1}{x} + \frac{1}{2x} = \frac{3}{10}$$

The reciprocal of the first number, x

The reciprocal of the second number, $2x$

Step 4   The LCD of the fractions is $10x$, so we multiply by $10x$.

$$10x\left(\frac{1}{x}\right) + 10x\left(\frac{1}{2x}\right) = 10x\left(\frac{3}{10}\right)$$

Simplifying, we have

$$10 + 5 = 3x$$
$$15 = 3x$$
$$5 = x$$

NOTE x was one number, and $2x$ was the other.

The numbers are 5 and 10.

Step 5

Again check the result by returning to the original problem. If the numbers are 5 and 10, we have

$$\frac{1}{(5)} + \frac{1}{2(5)} = \frac{2+1}{10} = \frac{3}{10}$$

The sum of the reciprocals is $\frac{3}{10}$.

CHECK YOURSELF 2

One number is 3 times another. If the sum of their reciprocals is $\frac{2}{9}$, find the two numbers.

The solution of many motion problems will also involve rational equations. Recall that the key equation for solving all motion problems relates the distance traveled, the speed or rate, and the time:

> **Definition: Motion Problem Relationships**
>
> $d = r \cdot t$
>
> Often we will use this equation in different forms by solving for r or for t.
>
> $r = \dfrac{d}{t}$ or $t = \dfrac{d}{r}$

Example 3   Solving an Application Involving $r = \dfrac{d}{t}$

Vince took 2 h longer to drive 225 mi than he did on a trip of 135 mi. If his speed was the same both times, how long did each trip take?

Step 1   You want to find the times taken for the 225-mi trip and for the 135-mi trip.

NOTE It is often helpful to choose your variable to "suggest" the unknown quantity—here t for time.

Step 2   Let t be the time for the 135-mi trip (in hours).

2 h longer

Then $\overbrace{t + 2}$ is the time for the 225-mi trip.
   It is often helpful to arrange the information in tabular form such as that shown.

RECALL Rate is distance divided by time. The rightmost column is formed by using that relationship.

	Distance	Time	Rate
135-mi trip	135	t	$\dfrac{135}{t}$
225-mi trip	225	$t + 2$	$\dfrac{225}{t + 2}$

Step 3   In forming the equation, remember that the speed (or rate) for each trip was the same. That is the *key* idea. We can equate the rates for the two trips that were found in step 2. The two rates are shown in the rightmost column of the table. Thus we can write

$$\frac{135}{t} = \frac{225}{t + 2}$$

NOTE The equation is in proportion form. So we could solve by setting the product of the means equal to the product of the extremes.

Step 4 To solve the above equation, multiply each side by $t(t + 2)$, the LCD of the fractions.

$$t(t + 2)\left(\frac{135}{t}\right) = t(t + 2)\left(\frac{225}{t + 2}\right)$$

Simplifying, we have

$$135(t + 2) = 225t$$
$$135t + 270 = 225t$$
$$270 = 90t$$
$$t = 3 \text{ h}$$

The time for the 135-mi trip was 3 h, and the time for the 225-mi trip was 5 h. We leave it to you to check this result.

CHECK YOURSELF 3 _____

Cynthia took 2 h longer to bicycle 75 mi than she did on a trip of 45 mi. If her speed was the same each time, find the time for each trip.

Example 4 uses the $d = r \cdot t$ relationship to find the speed.

Example 4   **Solving an Application Involving $d = r \cdot t$**

A train makes a trip of 300 mi in the same time that a bus can travel 250 mi. If the speed of the train is 10 mi/h faster than the speed of the bus, find the speed of each.

Step 1 You want to find the speeds of the train and of the bus.

Step 2 Let r be the speed (or rate) of the bus (in miles per hour).

Then $\underline{r + 10}$ is the rate of the train.
   10 mi/h faster

Again, form a chart of the information.

RECALL Time is distance divided by rate. Here the rightmost column is found by using that relationship.

	Distance	Rate	Time
Train	300	$r + 10$	$\dfrac{300}{r + 10}$
Bus	250	r	$\dfrac{250}{r}$

Step 3 To form an equation, remember that the times for the train and bus are the same. We can equate the expressions for time found in Step 2. Again, working from the rightmost column, we have

$$\frac{300}{r + 10} = \frac{250}{r}$$

Step 4 We multiply each term by $r(r + 10)$, the LCD of the fractions.

$$\overset{1}{\cancel{r}}(r + 10)\left(\frac{250}{\underset{1}{\cancel{r}}}\right) = r(\cancel{r + 10})\overset{1}{\left(\frac{300}{\underset{1}{\cancel{r + 10}}}\right)}$$

Simplifying, we have

$$250(r + 10) = 300r$$
$$250r + 2500 = 300r$$
$$2500 = 50r$$
$$r = 50 \text{ mi/h}$$

RECALL Remember to find the rates of both vehicles.

The rate of the bus is 50 mi/h, and the rate of the train is 60 mi/h. You can check this result.

CHECK YOURSELF 4

A car makes a trip of 280 mi in the same time that a truck travels 245 mi. If the speed of the truck is 5 mi/h slower than that of the car, find the speed of each.

Example 5 involves fractions in decimal form. Mixture problems often use percentages, and those percentages can be written as decimals. Example 5 illustrates this method.

Example 5   Solving an Application Involving Solutions

A solution of antifreeze is 20% alcohol. How much pure alcohol must be added to 12 quarts (qt) of the solution to make a 40% solution?

Step 1 You want to find the number of quarts of pure alcohol that must be added.

Step 2 Let x be the number of quarts of pure alcohol to be added.

Step 3 To form our equation, note that the amount of alcohol present before mixing *must be the same* as the amount in the combined solution.

A picture will help.

12 qt   +   x qt   =   12 + x qt
20%   100%   40%

So

NOTE Express the percentages as decimals in the equation.

$$\underbrace{12(0.20)}_{} + \underbrace{x(1.00)}_{} = \underbrace{(12 + x)(0.40)}_{}$$

The amount of alcohol in the first solution (20% is 0.20)

The amount of pure alcohol ("pure" is 100%, or 1.00)

The amount of alcohol in the mixture

Step 4 Most students prefer to clear the decimals at this stage. Multiplying by 100 will move the decimal point *two places to the right.* We then have

$$12(20) + x(100) = (12 + x)(40)$$
$$240 + 100x = 480 + 40x$$
$$60x = 240$$
$$x = 4 \text{ qt}$$

CHECK YOURSELF 5

How much pure alcohol must be added to 500 cm^3 of a 40% alcohol mixture to make a solution that is 80% alcohol?

There are many types of applications that lead to proportions in their solution. Typically these applications will involve a common ratio, such as miles to gallons or miles to hours, and they can be solved with three basic steps.

Step by Step: To Solve an Application Using a Proportion

Step 1   Assign a variable to represent the unknown quantity.
Step 2   Write a proportion, using the known and unknown quantities. Be sure each ratio involves the same units.
Step 3   Solve the proportion written in Step 2 for the unknown quantity.

Example 6 illustrates this approach.

OBJECTIVE 2   **Example 6   Solving an Application Using a Proportion**

A car uses 3 gal of gas to travel 105 mi. At that mileage rate, how many gallons will be used on a trip of 385 mi?

Step 1   Assign a variable to represent the unknown quantity. Let x be the number of gallons of gas that will be used on the 385-mi trip.

Step 2   Write a proportion. Note that the ratio of miles to gallons must stay the same.

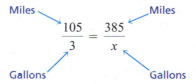

$$\frac{105}{3} = \frac{385}{x}$$

Miles   Miles

Gallons   Gallons

Step 3 Solve the proportion. The product of the extremes is equal to the product of the means.

$$105x = 3 \cdot 385$$
$$105x = 1155$$
$$\frac{105x}{105} = \frac{1155}{105}$$
$$x = 11 \text{ gal}$$

NOTE To verify your solution, return to the original problem and check that the two ratios are equivalent.

So 11 gal of gas will be used for the 385-mi trip.

CHECK YOURSELF 6

A car uses 8 L of gasoline in traveling 100 km. At that rate, how many liters of gas will be used on a trip of 250 km?

Proportions can also be used to solve problems in which a quantity is divided by using a specific ratio. Example 7 shows how.

Example 7   Solving an Application Using a Proportion

A 60 in. long piece of wire is to be cut into two pieces whose lengths have the ratio 5 to 7. Find the length of each piece.

Step 1 Let x represent the length of the shorter piece. Then $60 - x$ is the length of the longer piece.

RECALL A picture of the problem always helps.

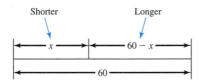

Step 2 The two pieces have the ratio $\frac{5}{7}$, so

NOTE On the left and right, we have the ratio of the length of the shorter piece to that of the longer piece.

$$\frac{x}{60 - x} = \frac{5}{7}$$

Step 3 Solving as before, we get

$$7x = (60 - x)5$$
$$7x = 300 - 5x$$
$$12x = 300$$
$$x = 25 \quad \text{(Shorter piece)}$$
$$60 - x = 35 \quad \text{(Longer piece)}$$

The pieces have lengths of 25 in. and 35 in.

CHECK YOURSELF 7

A 21 ft long board is to be cut into two pieces so that the ratio of their lengths is 3 to 4. Find the lengths of the two pieces.

CHECK YOURSELF ANSWERS

1. The number is 20.   **2.** The numbers are 6 and 18.
3. 75-mi trip: 5 h; 45-mi trip: 3 h   **4.** Car: 40 mi/h; truck: 35 mi/h   **5.** 1000 cm^3
6. 20 L   **7.** 9 ft; 12 ft

5.6 Exercises

Solve the following word problems.

1. **Number problem.** If two-thirds of a number is added to one-half of that number, the sum is 35. Find the number.

2. **Number problem.** If one-third of a number is subtracted from three-fourths of that number, the difference is 15. What is the number?

3. **Number problem.** If one-fourth of a number is subtracted from two-fifths of a number, the difference is 3. Find the number.

4. **Number problem.** If five-sixths of a number is added to one-fifth of the number, the sum is 31. What is the number?

5. **Number problem.** If one-third of an integer is added to one-half of the next consecutive integer, the sum is 13. What are the two integers?

6. **Number problem.** If one-half of one integer is subtracted from three-fifths of the next consecutive integer, the difference is 3. What are the two integers?

7. **Number problem.** One number is twice another number. If the sum of their reciprocals is $\frac{1}{4}$, find the two numbers.

8. **Number problem.** One number is 3 times another. If the sum of their reciprocals is $\frac{1}{6}$, find the two numbers.

9. **Number problem.** One number is 4 times another. If the sum of their reciprocals is $\frac{5}{12}$, find the two numbers.

10. **Number problem.** One number is 3 times another. If the sum of their reciprocals is $\frac{4}{15}$, what are the two numbers?

11. **Number problem.** One number is 5 times another number. If the sum of their reciprocals is $\frac{6}{35}$, what are the two numbers?

12. **Number problem.** One number is 4 times another. The sum of their reciprocals is $\frac{5}{24}$. What are the two numbers?

13. **Number problem.** If the reciprocal of 5 times a number is subtracted from the reciprocal of that number, the result is $\frac{4}{25}$. What is the number?

14. Number problem. If the reciprocal of a number is added to 4 times the reciprocal of that number, the result is $\frac{5}{9}$. Find the number.

15. Science and medicine. Lee can ride his bicycle 50 mi in the same time it takes him to drive 125 mi. If his driving rate is 30 mi/h faster than his rate bicycling, find each rate.

16. Science and medicine. Tina can run 12 mi in the same time it takes her to bicycle 72 mi. If her bicycling rate is 20 mi/h faster than her running rate, find each rate.

17. Science and medicine. An express bus can travel 275 mi in the same time that it takes a local bus to travel 225 mi. If the rate of the express bus is 10 mi/h faster than that of the local bus, find the rate for each bus.

18. Science and medicine. A passenger train can travel 325 mi in the same time a freight train takes to travel 200 mi. If the speed of the passenger train is 25 mi/h faster than the speed of the freight train, find the speed of each.

19. Science and medicine. A light plane took 1 h longer to travel 450 mi on the first portion of a trip than it took to fly 300 mi on the second. If the speed was the same for each portion, what was the flying time for each part of the trip?

20. Science and medicine. A small business jet took 1 h longer to fly 810 mi on the first part of a flight than to fly 540 mi on the second portion. If the jet's rate was the same for each leg of the flight, what was the flying time for each leg?

21. Science and medicine. Charles took 2 h longer to drive 240 mi on the first day of a vacation trip than to drive 144 mi on the second day. If his average driving rate was the same on both days, what was his driving time for each of the days?

22. Science and medicine. Ariana took 2 h longer to drive 360 mi on the first day of a trip than she took to drive 270 mi on the second day. If her speed was the same on both days, what was the driving time each day?

23. Science and medicine. An airplane took 3 h longer to fly 1200 mi than it took for a flight of 480 mi. If the plane's rate was the same on each trip, what was the time of each flight?

ANSWERS

14. _____

15. _____

16. _____

17. _____

18. _____

19. _____

20. _____

21. _____

22. _____

23. _____

24. **Science and medicine.** A train travels 80 mi in the same time that a light plane can travel 280 mi. If the speed of the plane is 100 mi/h faster than that of the train, find each of the rates.

25. **Science and medicine.** Jan and Tariq took a canoeing trip, traveling 6 mi upstream against a 2 mi/h current. They then returned to the same point downstream. If their entire trip took 4 h, how fast can they paddle in still water? [*Hint:* If r is their rate (in miles per hour) in still water, their rate upstream is $r - 2$ and their rate downstream is $r + 2$.]

26. **Science and medicine.** A plane flies 720 mi against a steady 30-mi/h headwind and then returns to the same point with the wind. If the entire trip takes 10 h, what is the plane's speed in still air?

27. **Science and medicine.** How much pure alcohol must be added to 40 oz of a 25% solution to produce a mixture that is 40% alcohol?

28. **Science and medicine.** How many centiliters (cL) of pure acid must be added to 200 cL of a 40% acid solution to produce a 50% solution?

29. **Science and medicine.** A speed of 60 mi/h corresponds to 88 ft/s. If a light plane's speed is 150 mi/h, what is its speed in feet per second?

30. **Business and finance.** If 342 cups of coffee can be made from 9 lb of coffee, how many cups can be made from 6 lb of coffee?

31. **Social science.** A car uses 5 gal of gasoline on a trip of 160 mi. At the same mileage rate, how much gasoline will a 384-mi trip require?

32. **Social science.** A car uses 12 L of gasoline in traveling 150 km. At that rate, how many liters of gasoline will be used in a trip of 400 km?

33. **Business and finance.** Sveta earns $13,500 commission in 20 weeks in her new sales position. At that rate, how much will she earn in 1 year (52 weeks)?

34. **Business and finance.** Kevin earned $165 interest for 1 year on an investment of $1500. At the same rate, what amount of interest would be earned by an investment of $2500?

35. **Social science.** A company is selling a natural insect control that mixes ladybug beetles and praying mantises in the ratio of 7 to 4. If there are a total of 110 insects per package, how many of each type of insect is in a package?

36. **Social science.** A woman casts a 4 ft–shadow. At the same time, a 72-ft building casts a 48 ft–shadow. How tall is the woman?

37. **Business and finance.** A brother and sister are to divide an inheritance of $12,000 in the ratio of 2 to 3. What amount will each receive?

38. **Business and finance.** In Bucks County, the property tax rate is $25.32 per $1000 of assessed value. If a house and property have a value of $128,000, find the tax the owner will have to pay.

Assessed
Value: $128,000

Tax Rate: $25.32
per $1000

a. _____

b. _____

c. _____

d. _____

e. _____

f. _____

 Getting Ready for Section 5.7 [Section 0.2]

Divide each of the following.

(a) $\dfrac{3}{5} \div 6$ (b) $\dfrac{2}{7} \div \dfrac{3}{7}$

(c) $\dfrac{11}{4} \div \dfrac{5}{6}$ (d) $\dfrac{5}{8} \div \dfrac{10}{3}$

(e) $\dfrac{15}{4} \div \dfrac{5}{8}$ (f) $\dfrac{9}{8} \div \dfrac{3}{4}$

Answers

1. 30   **3.** 20   **5.** 15, 16   **7.** 6, 12   **9.** 3, 12   **11.** 7, 35
13. 5   **15.** 20 mi/h bicycling, 50 mi/h driving
17. Express 55 mi/h, local 45 mi/h   **19.** 3 h, 2 h   **21.** 5 h, 3 h
23. 5 h, 2 h   **25.** 4 mi/h   **27.** 10 oz   **29.** 220 ft/s   **31.** 12 gal
33. $35,100   **35.** 70 ladybugs, 40 praying mantises
37. Brother $4800, sister $7200   **a.** 1/10   **b.** 2/3   **c.** 33/10
d. 3/16   **e.** 6   **f.** 3/2

5.7 Complex Rational Expressions

Recall, from arithmetic, the manner in which you were taught to divide fractions. The rule was referred to as invert-and-multiply. We will see why this rule works.

$$\frac{3}{5} \div \frac{2}{3}$$

We can write

$$\frac{3}{5} \div \frac{2}{3} = \frac{\dfrac{3}{5}}{\dfrac{2}{3}} = \frac{\dfrac{3}{5} \cdot \dfrac{3}{2}}{\dfrac{2}{3} \cdot \dfrac{3}{2}}$$ We are multiplying by 1.

Interpret the division as a fraction.

$$= \frac{\dfrac{3}{5} \cdot \dfrac{3}{2}}{1}$$

$$\frac{2}{3} \cdot \frac{3}{2} = 1$$

$$= \frac{3}{5} \cdot \frac{3}{2}$$

We then have

$$\frac{3}{5} \div \frac{2}{3} = \frac{3}{5} \cdot \frac{3}{2} = \frac{9}{10}$$

By comparing these expressions, you should see the rule for dividing fractions. Invert the fraction that follows the division symbol and multiply.

A fraction that has a fraction in its numerator, in its denominator, or in both is called a **complex fraction.** For example, the following are complex fractions.

$$\frac{\dfrac{5}{6}}{\dfrac{3}{4}}, \qquad \frac{\dfrac{4}{x}}{\dfrac{3}{x^2}}, \qquad \text{and} \qquad \frac{\dfrac{a+2}{3}}{\dfrac{a-2}{5}}$$

Remember that we can always multiply the numerator and the denominator of a fraction by the same nonzero term.

RECALL This is the Fundamental Principle of Fractions.

$$\frac{P}{Q} = \frac{P \cdot R}{Q \cdot R} \qquad \text{in which } Q \neq 0 \text{ and } R \neq 0$$

To simplify a complex fraction, multiply the numerator and denominator by the LCD of all fractions that appear within the complex fraction.

OBJECTIVE 1   **Example 1   Simplifying Complex Fractions**

Simplify $\dfrac{\dfrac{3}{4}}{\dfrac{5}{8}}$.

The LCD of $\dfrac{3}{4}$ and $\dfrac{5}{8}$ is 8. So multiply the numerator and denominator by 8.

$$\dfrac{\dfrac{3}{4}}{\dfrac{5}{8}} = \dfrac{\dfrac{3}{4} \cdot 8}{\dfrac{5}{8} \cdot 8} = \dfrac{3 \cdot 2}{5 \cdot 1} = \dfrac{6}{5}$$

 CHECK YOURSELF 1 _____

Simplify.

(a) $\dfrac{\dfrac{4}{7}}{\dfrac{3}{7}}$

(b) $\dfrac{\dfrac{3}{8}}{\dfrac{5}{6}}$

The same method can be used to simplify a complex fraction when variables are involved in the expression. Consider Example 2.

OBJECTIVE 2   **Example 2   Simplifying Complex Rational Expressions**

Simplify $\dfrac{\dfrac{5}{x}}{\dfrac{10}{x^2}}$.

The LCD of $\dfrac{5}{x}$ and $\dfrac{10}{x^2}$ is x^2, so multiply the numerator and denominator by x^2.

NOTE Be sure to write the result in simplest form.

$$\dfrac{\dfrac{5}{x}}{\dfrac{10}{x^2}} = \dfrac{\left(\dfrac{5}{x}\right)x^2}{\left(\dfrac{10}{x^2}\right)x^2} = \dfrac{5x}{10} = \dfrac{x}{2}$$

CHECK YOURSELF 2 _____

Simplify.

(a) $\dfrac{\dfrac{6}{x^3}}{\dfrac{9}{x^2}}$

(b) $\dfrac{\dfrac{m^4}{15}}{\dfrac{m^3}{20}}$

We may also have a sum or a difference in the numerator or denominator of a complex fraction. The simplification steps are exactly the same. Consider Example 3.

Example 3   Simplifying Complex Fractions

Simplify $\dfrac{1 + \dfrac{x}{y}}{1 - \dfrac{x}{y}}$.

The LCD of 1, $\dfrac{x}{y}$, 1, and $\dfrac{x}{y}$ is y, so multiply the numerator and denominator by y.

NOTE We use the distributive property to multiply *each term* in the numerator and in the denominator by y.

$$\frac{1 + \dfrac{x}{y}}{1 - \dfrac{x}{y}} = \frac{\left(1 + \dfrac{x}{y}\right)y}{\left(1 - \dfrac{x}{y}\right)y} = \frac{1 \cdot y + \dfrac{x}{y} \cdot y}{1 \cdot y - \dfrac{x}{y} \cdot y}$$

$$= \frac{y + x}{y - x}$$

CHECK YOURSELF 3

Simplify.

$$\frac{\dfrac{x}{y} - 2}{\dfrac{x}{y} + 2}$$

The following algorithm summarizes our work to this point with simplifying complex fractions.

Step by Step: To Simplify Complex Rational Expressions

Step 1   Multiply the numerator and denominator of the complex rational expression by the LCD of all the fractions that appear within the complex rational expression.

Step 2   Write the resulting fraction in simplest form.

A second method for simplifying complex fractions uses the fact that

RECALL To divide by a fraction, we invert the divisor (it *follows* the division sign) and multiply.

$$\frac{\dfrac{P}{Q}}{\dfrac{R}{S}} = \frac{P}{Q} \div \frac{R}{S} = \frac{P}{Q} \cdot \frac{S}{R}$$

To use this method, we must write the numerator and denominator of the complex fraction as single fractions. We can then divide the numerator by the denominator as before.

In Example 4, we use this method to simplify the complex rational expression we saw in Example 3.

Example 4   Simplifying Complex Fractions

Simplify $\dfrac{1 + \dfrac{x}{y}}{1 - \dfrac{x}{y}}$.

To use this method, we rewrite both the numerator and the denominator as single fractions.

$$\frac{1 + \dfrac{x}{y}}{1 - \dfrac{x}{y}} = \frac{\dfrac{y}{y} + \dfrac{x}{y}}{\dfrac{y}{y} - \dfrac{x}{y}} = \frac{\dfrac{y + x}{y}}{\dfrac{y - x}{y}}$$

Now we invert and multiply.

$$\frac{\dfrac{y + x}{y}}{\dfrac{y - x}{y}} = \frac{y + x}{y} \cdot \frac{y}{y - x} = \frac{y + x}{y - x}$$

Not surprisingly, we have the same result as we found in Example 3.

CHECK YOURSELF 4

Simplify $\dfrac{\dfrac{x}{y} - 2}{\dfrac{x}{y} + 2}$.

CHECK YOURSELF ANSWERS

1. (a) $\dfrac{4}{3}$; (b) $\dfrac{9}{20}$ 2. (a) $\dfrac{2}{3x}$; (b) $\dfrac{4m}{3}$ 3. $\dfrac{x - 2y}{x + 2y}$ 4. $\dfrac{x - 2y}{x + 2y}$

5.7 Exercises

Simplify each complex fraction.

1. $\dfrac{\dfrac{2}{3}}{\dfrac{6}{8}}$

2. $\dfrac{\dfrac{5}{6}}{\dfrac{10}{15}}$

3. $\dfrac{1 + \dfrac{1}{2}}{2 + \dfrac{1}{4}}$

4. $\dfrac{1 + \dfrac{3}{4}}{2 - \dfrac{1}{8}}$

5. $\dfrac{\dfrac{x}{8}}{\dfrac{x^2}{4}}$

6. $\dfrac{\dfrac{m^2}{10}}{\dfrac{m^3}{15}}$

7. $\dfrac{\dfrac{3}{a}}{\dfrac{2}{a^2}}$

8. $\dfrac{\dfrac{6}{x^2}}{\dfrac{9}{x^3}}$

9. $\dfrac{\dfrac{y + 1}{y}}{\dfrac{y - 1}{2y}}$

10. $\dfrac{\dfrac{w + 3}{4w}}{\dfrac{w - 3}{2w}}$

11. $\dfrac{2 - \dfrac{1}{x}}{2 + \dfrac{1}{x}}$

12. $\dfrac{3 + \dfrac{1}{a}}{3 - \dfrac{1}{a}}$

13. $\dfrac{3 - \dfrac{x}{y}}{\dfrac{6}{y}}$

14. $\dfrac{2 + \dfrac{x}{y}}{\dfrac{4}{y}}$

15. $\dfrac{\dfrac{x^2}{y^2} - 1}{\dfrac{x}{y} + 1}$

16. $\dfrac{\dfrac{a}{b} + 2}{\dfrac{a^2}{b^2} - 4}$

ANSWERS

1. _____
2. _____
3. _____
4. _____
5. _____
6. _____
7. _____
8. _____
9. _____
10. _____
11. _____
12. _____
13. _____
14. _____
15. _____
16. _____

17. $\dfrac{1 + \dfrac{3}{x} - \dfrac{4}{x^2}}{1 + \dfrac{2}{x} - \dfrac{3}{x^2}}$

18. $\dfrac{1 - \dfrac{2}{r} - \dfrac{8}{r^2}}{1 - \dfrac{1}{r} - \dfrac{6}{r^2}}$

19. $\dfrac{\dfrac{2}{x} - \dfrac{1}{xy}}{\dfrac{1}{xy} + \dfrac{2}{y}}$

20. $\dfrac{\dfrac{1}{xy} + \dfrac{2}{x}}{\dfrac{3}{y} - \dfrac{1}{xy}}$

21. $\dfrac{\dfrac{2}{x-1} + 1}{1 - \dfrac{3}{x-1}}$

22. $\dfrac{\dfrac{3}{a+2} - 1}{1 + \dfrac{2}{a+2}}$

23. $\dfrac{1 - \dfrac{1}{y-1}}{y - \dfrac{8}{y+2}}$

24. $\dfrac{1 + \dfrac{1}{x+2}}{x - \dfrac{18}{x-3}}$

25. $1 + \dfrac{1}{1 + \dfrac{1}{x}}$

26. $1 + \dfrac{1}{1 - \dfrac{1}{y}}$

27. Science and medicine. Herbicides constitute $\dfrac{2}{3}$ of all pesticides used in the United States. Insecticides are $\dfrac{1}{4}$ of all pesticides used in the United States. The ratio of herbicides to insecticides used in the United States can be written $\dfrac{2}{3} \div \dfrac{1}{4}$. Write this ratio in simplest form.

28. Science and medicine. Fungicides account for $\dfrac{1}{10}$ of the pesticides used in the United States. Insecticides account for $\dfrac{1}{4}$ of all the pesticides used in the United States. The ratio of fungicides to insecticides used in the United States can be written $\dfrac{1}{10} \div \dfrac{1}{4}$. Write this ratio in simplest form.

29. Science and medicine. The ratio of insecticides to herbicides applied to wheat, soybeans, corn, and cotton can be expressed as $\dfrac{7}{10} \div \dfrac{4}{5}$. Simplify this ratio.

30. Geometry. The area of the rectangle shown below is $\dfrac{2(3x-2)}{x-1}$. Find the width.

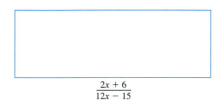

$$\frac{3x-2}{x-2}$$

31. Geometry. The area of the rectangle shown below is $\dfrac{2}{3}$. Find the width.

$$\frac{2x+6}{12x-15}$$

32. Science and medicine. The combined resistance of two resistors R_1 and R_2 in a parallel circuit is given by the formula

$$R_T = \frac{1}{\dfrac{1}{R_1} + \dfrac{1}{R_2}}$$

Simplify the formula.

33. Complex fractions have some interesting patterns. Work with a partner to evaluate each complex fraction in the sequence below. This is an interesting sequence of fractions because the numerators and denominators are a famous sequence of whole numbers, and the fractions get closer and closer to a number called "the golden mean."

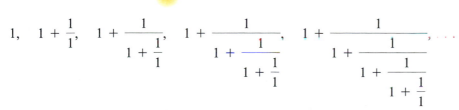

$$1, \quad 1+\frac{1}{1}, \quad 1+\frac{1}{1+\dfrac{1}{1}}, \quad 1+\frac{1}{1+\dfrac{1}{1+\dfrac{1}{1}}}, \quad 1+\frac{1}{1+\dfrac{1}{1+\dfrac{1}{1+\dfrac{1}{1}}}}, \ldots$$

After you have evaluated these first five, you no doubt will see a pattern in the resulting fractions that allows you to go on indefinitely without having to evaluate more complex fractions. Write each of these fractions as decimals. Write your observations about the sequence of fractions and about the sequence of decimal fractions.

Answers

1. $\dfrac{8}{9}$ 3. $\dfrac{2}{3}$ 5. $\dfrac{1}{2x}$ 7. $\dfrac{3a}{2}$ 9. $\dfrac{2(y+1)}{y-1}$ 11. $\dfrac{2x-1}{2x+1}$

13. $\dfrac{3y-x}{6}$ 15. $\dfrac{x-y}{y}$ 17. $\dfrac{x+4}{x+3}$ 19. $\dfrac{2y-1}{1+2x}$

21. $\dfrac{x+1}{x-4}$ 23. $\dfrac{y+2}{(y-1)(y+4)}$ 25. $\dfrac{2x+1}{x+1}$ 27. $\dfrac{8}{3}$

29. $\dfrac{7}{8}$ 31. $\dfrac{4x-5}{x+3}$ 33.

5 Summary

DEFINITION/PROCEDURE	EXAMPLE	REFERENCE
Simplifying Rational Expressions		**Section 5.1**
Rational Expressions These have the form Numerator Fraction bar → $\dfrac{P}{Q}$ Denominator in which P and Q are polynomials and Q cannot have the value 0.	$\dfrac{x^2 - 3x}{x - 2}$ is a rational expression. The variable x cannot have the value 2.	p. 421
Writing in Simplest Form A fraction is in simplest form if its numerator and denominator have no common factors other than 1. To write in simplest form: **1.** Factor the numerator and denominator. **2.** Divide the numerator and denominator by all common factors. The resulting fraction will be in simplest form.	$\dfrac{x + 2}{x - 1}$ is in simplest form. $\dfrac{x^2 - 4}{x^2 - 2x - 8}$ $= \dfrac{(x - 2)(x + 2)}{(x - 4)(x + 2)}$ $= \dfrac{(x - 2)(x + 2)^{1}}{(x - 4)(x + 2)_{1}}$ $= \dfrac{x - 2}{x - 4}$	p. 421
Multiplying and Dividing Rational Expressions		**Section 5.2**
Multiplying Rational Expressions $\dfrac{P}{Q} \cdot \dfrac{R}{S} = \dfrac{PR}{QS}$ in which $Q \neq 0$ and $S \neq 0$.	$\dfrac{2}{3} \cdot \dfrac{4}{5} = \dfrac{2 \cdot 4}{3 \cdot 5} = \dfrac{8}{15}$	p. 431
Multiplying Rational Expressions **1.** Factor the numerators and denominators. **2.** Divide the numerator and denominator by any common factors. **3.** Write the product of the remaining factors in the numerator over the product of the remaining factors in the denominator.	$\dfrac{2x - 4}{x^2 - 4} \cdot \dfrac{x^2 + 2x}{6x + 18}$ $= \dfrac{2(x - 2)}{(x - 2)(x + 2)} \cdot \dfrac{x(x + 2)}{6(x + 3)}$ $= \dfrac{2(x - 2)}{(x - 2)(x + 2)} \cdot \dfrac{x(x + 2)}{6(x + 3)}$ $= \dfrac{x}{3(x + 3)}$	p. 431
Dividing Rational Expressions $\dfrac{P}{Q} \div \dfrac{R}{S} = \dfrac{P}{Q} \cdot \dfrac{S}{R}$ in which $Q \neq 0$, $R \neq 0$, and $S \neq 0$. In words, invert the divisor (the second fraction) and multiply.	$\dfrac{4}{9} \div \dfrac{8}{12}$ $= \dfrac{4}{9} \cdot \dfrac{12}{8} = \dfrac{2}{3}$ $\dfrac{3x}{2x - 6} \div \dfrac{9x^2}{x^2 - 9}$ $= \dfrac{3x}{2x - 6} \cdot \dfrac{x^2 - 9}{9x^2}$ $= \dfrac{3x}{2(x - 3)} \cdot \dfrac{(x + 3)(x - 3)^{1}}{9x^2}$ $= \dfrac{x + 3}{6x}$	p. 432

Continued

491

Adding and Subtracting Like Rational Expressions		Section 5.3
Like Rational Expressions 1. Add or subtract the numerators. 2. Write the sum or difference over the common denominator. 3. Write the resulting fraction in simplest form.	$$\frac{2x}{x^2 + 3x} + \frac{6}{x^2 + 3x}$$ $$= \frac{2x + 6}{x^2 + 3x}$$ $$= \frac{2(\cancel{x + 3})}{x(\cancel{x + 3})} = \frac{2}{x}$$	p. 437

Adding and Subtracting Unlike Rational Expressions		Section 5.4
The Least Common Denominator Finding the LCD: 1. Factor each denominator. 2. Write each factor the greatest number of times it appears in any single denominator. 3. The LCD is the product of the factors found in Step 2.	For $\dfrac{2}{x^2 + 2x + 1}$ and $\dfrac{3}{x^2 + x}$ Factor: $x^2 + 2x + 1 = (x + 1)(x + 1)$ $x^2 + x = x(x + 1)$ The LCD is $x(x + 1)(x + 1)$	p. 444
Unlike Rational Expressions To add or subtract unlike rational expressions: 1. Find the LCD. 2. Convert each rational expression to an equivalent rational expression with the LCD as a common denominator. 3. Add or subtract the like rational expressions formed. 4. Write the sum or difference in simplest form.	$$\frac{2}{x^2 + 2x + 1} - \frac{3}{x^2 + x}$$ $$= \frac{2x}{x(x + 1)(x + 1)}$$ $$- \frac{3(x + 1)}{x(x + 1)(x + 1)}$$ $$= \frac{2x - 3x - 3}{x(x + 1)(x + 1)}$$ $$= \frac{-x - 3}{x(x + 1)(x + 1)}$$	p. 445

Equations Involving Rational Expressions		Section 5.5
1. Remove the fractions in the equation by multiplying each term by the LCD of all the fractions. 2. Solve the equation resulting from Step 1. 3. Check your solution using the original equation. Discard any extraneous solutions.	$$2 - \frac{4}{x} = \frac{2}{3}$$ LCD: $3x$ $$2(3x) - \frac{4}{x}(3x) = \frac{2}{3}(3x)$$ $$6x - 12 = 2x$$ $$4x = 12$$ $$x = 3$$	p. 460

Complex Rational Expressions		Section 5.7
Simplifying Complex Fractions $$\frac{\dfrac{a}{b}}{\dfrac{c}{d}} = \frac{a}{b} \div \frac{c}{d} = \frac{a}{b} \cdot \frac{d}{c}$$	$$\frac{\dfrac{3}{8}}{\dfrac{5}{6}} = \frac{3}{8} \div \frac{5}{6}$$ $$= \frac{3}{\cancel{8}} \cdot \frac{\cancel{6}}{5}$$ $$= \frac{9}{20}$$	p. 485

Summary Exercises

This summary exercise set is provided to give you practice with each of the objectives of this chapter. Each exercise is keyed to the appropriate chapter section. When you are finished, you can check your answers to the odd-numbered exercises against those presented in the back of the text. If you have difficulty with any of these questions, go back and reread the examples from that section. Your instructor will give you guidelines on how to best use these exercises in your instructional setting.

[5.1] Write each fraction in simplest form.

1. $\dfrac{6a^2}{9a^3}$

2. $\dfrac{-12x^4y^3}{18x^2y^2}$

3. $\dfrac{w^2 - 25}{2w - 8}$

4. $\dfrac{3x^2 + 11x - 4}{2x^2 + 11x + 12}$

5. $\dfrac{m^2 - 2m - 3}{9 - m^2}$

6. $\dfrac{3c^2 - 2cd - d^2}{6c^2 + 2cd}$

[5.2] Multiply or divide as indicated.

7. $\dfrac{6x}{5} \cdot \dfrac{10}{18x^2}$

8. $\dfrac{-2a^2}{ab^3} \cdot \dfrac{3ab^2}{-4ab}$

9. $\dfrac{2x + 6}{x^2 - 9} \cdot \dfrac{x^2 - 3x}{4}$

10. $\dfrac{a^2 + 5a + 4}{2a^2 + 2a} \cdot \dfrac{a^2 - a - 12}{a^2 - 16}$

11. $\dfrac{3p}{5} \div \dfrac{9p^2}{10}$

12. $\dfrac{8m^3}{5mn} \div \dfrac{12m^2n^2}{15mn^3}$

13. $\dfrac{x^2 + 7x + 10}{x^2 + 5x} \div \dfrac{x^2 - 4}{2x^2 - 7x + 6}$

14. $\dfrac{2w^2 + 11w - 21}{w^2 - 49} \div (4w - 6)$

15. $\dfrac{a^2b + 2ab^2}{a^2 - 4b^2} \div \dfrac{4a^2b}{a^2 - ab - 2b^2}$

16. $\dfrac{2x^2 + 6x}{4x} \cdot \dfrac{6x + 12}{x^2 + 2x - 3} \div \dfrac{x^2 - 4}{x^2 - 3x + 2}$

[5.3] Add or subtract as indicated.

17. $\dfrac{x}{9} + \dfrac{2x}{9}$

18. $\dfrac{7a}{15} - \dfrac{2a}{15}$

19. $\dfrac{8}{x + 2} + \dfrac{3}{x + 2}$

20. $\dfrac{y - 2}{5} - \dfrac{2y + 3}{5}$

21. $\dfrac{7r - 3s}{4r} + \dfrac{r - s}{4r}$

22. $\dfrac{x^2}{x - 4} - \dfrac{16}{x - 4}$

23. $\dfrac{5w - 6}{w - 4} - \dfrac{3w + 2}{w - 4}$

24. $\dfrac{x + 3}{x^2 - 2x - 8} + \dfrac{2x + 3}{x^2 - 2x - 8}$

[5.4] Add or subtract as indicated.

25. $\dfrac{5x}{6} + \dfrac{x}{3}$

26. $\dfrac{3y}{10} - \dfrac{2y}{5}$

27. $\dfrac{5}{2m} - \dfrac{3}{m^2}$

28. $\dfrac{x}{x - 3} - \dfrac{2}{3}$

29. $\dfrac{4}{x - 3} - \dfrac{1}{x}$

30. $\dfrac{2}{s + 5} + \dfrac{3}{s + 1}$

31. $\dfrac{5}{w - 5} - \dfrac{2}{w - 3}$

32. $\dfrac{4x}{2x - 1} + \dfrac{2}{1 - 2x}$

33. $\dfrac{2}{3x - 3} - \dfrac{5}{2x - 2}$

34. $\dfrac{4y}{y^2 - 8y + 15} + \dfrac{6}{y - 3}$

35. $\dfrac{3a}{a^2 + 5a + 4} + \dfrac{2a}{a^2 - 1}$

36. $\dfrac{3x}{x^2 + 2x - 8} - \dfrac{1}{x - 2} + \dfrac{1}{x + 4}$

[5.7] Simplify the complex fractions.

37. $\dfrac{\dfrac{x^2}{12}}{\dfrac{x^3}{8}}$

38. $\dfrac{3 + \dfrac{1}{a}}{3 - \dfrac{1}{a}}$

39. $\dfrac{1 + \dfrac{x}{y}}{1 - \dfrac{x}{y}}$

40. $\dfrac{1 + \dfrac{1}{p}}{p^2 - 1}$

41. $\dfrac{\dfrac{1}{m} - \dfrac{1}{n}}{\dfrac{1}{m} + \dfrac{1}{n}}$

42. $\dfrac{2 - \dfrac{x}{y}}{4 - \dfrac{x^2}{y^2}}$

43. $\dfrac{\dfrac{2}{a+1}+1}{1-\dfrac{4}{a+1}}$

44. $\dfrac{\dfrac{a}{b}-1-\dfrac{2b}{a}}{\dfrac{1}{b^2}-\dfrac{1}{a^2}}$

[5.5] What values for x, if any, must be excluded in the following rational expressions?

45. $\dfrac{x}{5}$

46. $\dfrac{3}{x-4}$

47. $\dfrac{2}{(x+1)(x-2)}$

48. $\dfrac{7}{x^2-16}$

49. $\dfrac{x-1}{x^2+3x+2}$

50. $\dfrac{2x+3}{3x^2+x-2}$

Solve the following proportions for x.

51. $\dfrac{x-3}{8}=\dfrac{x-2}{10}$

52. $\dfrac{1}{x-3}=\dfrac{7}{x^2-x-6}$

Solve the following equations for x.

53. $\dfrac{x}{4}-\dfrac{x}{5}=2$

54. $\dfrac{13}{4x}+\dfrac{3}{x^2}=\dfrac{5}{2x}$

55. $\dfrac{x}{x-2}+1=\dfrac{x+4}{x-2}$

56. $\dfrac{x}{x-4}-3=\dfrac{4}{x-4}$

57. $\dfrac{x}{2x-6}-\dfrac{x-4}{x-3}=\dfrac{1}{8}$

58. $\dfrac{7}{x}-\dfrac{1}{x-3}=\dfrac{9}{x^2-3x}$

59. $\dfrac{x}{x-5}=\dfrac{3x}{x^2-7x+10}+\dfrac{8}{x-2}$

60. $\dfrac{6}{x+5}+1=\dfrac{3}{x-5}$

61. $\dfrac{24}{x+2}-2=\dfrac{2}{x-3}$

[5.6] Solve the following applications.

62. Number problem. If two-fifths of a number is added to one-half of that number, the sum is 27. Find the number.

63. Number problem. One number is 3 times another. If the sum of their reciprocals is $\dfrac{1}{3}$, what are the two numbers?

64. **Number problem.** If the reciprocal of 4 times a number is subtracted from the reciprocal of that number, the result is $\frac{1}{8}$. What is the number?

65. **Science and medicine.** Robert made a trip of 240 mi. Returning by a different route, he found that the distance was only 200 mi, but traffic slowed his speed by 8 mi/h. If the trip took the same amount of time in both directions, what was Robert's rate each way?

66. **Science and medicine.** On the first day of a vacation trip, Jovita drove 225 mi. On the second day it took her 1 h longer to drive 270 mi. If her average speed was the same on both days, how long did she drive each day?

67. **Science and medicine.** A light plane flies 700 mi against a steady 20-mi/h headwind and then returns, with the wind, to the same point. If the entire trip took 12 h, what was the speed of the plane in still air?

68. **Science and medicine.** How much pure alcohol should be added to 300 mL of a 30% solution to obtain a 40% solution?

69. **Science and medicine.** A chemist has a 10% acid solution and a 40% solution. How much of the 40% solution should be added to 300 mL of the 10% solution to produce a mixture with a concentration of 20%?

70. **Business and finance.** Melina wants to invest a total of $10,800 in two types of savings accounts. If she wants the ratio of the amounts deposited in the two accounts to be 4 to 5, what amount should she invest in each account?

Self-Test for Chapter 5

The purpose of this self-test is to help you check your progress and to review for the next in-class exam. Allow yourself about an hour to take this test. At the end of that hour check your answers against those given in the back of the text. Section references accompany the answers. If you missed any questions, go back to those sections and reread the examples until you master the concepts.

Write each fraction in simplest form.

1. $\dfrac{-21x^5y^3}{28xy^5}$ **2.** $\dfrac{4a - 24}{a^2 - 6a}$ **3.** $\dfrac{3x^2 + x - 2}{3x^2 - 8x + 4}$

Add or subtract as indicated.

4. $\dfrac{3a}{8} + \dfrac{5a}{8}$ **5.** $\dfrac{2x}{x + 3} + \dfrac{6}{x + 3}$ **6.** $\dfrac{7x - 3}{x - 2} - \dfrac{2x + 7}{x - 2}$

7. $\dfrac{x}{3} + \dfrac{4x}{5}$ **8.** $\dfrac{3}{s} - \dfrac{2}{s^2}$

9. $\dfrac{5}{x - 2} - \dfrac{1}{x + 3}$ **10.** $\dfrac{6}{w - 2} + \dfrac{9w}{w^2 - 7w + 10}$

Multiply or divide as indicated.

11. $\dfrac{3pq^2}{5pq^3} \cdot \dfrac{20p^2q}{21q}$ **12.** $\dfrac{x^2 - 3x}{5x^2} \cdot \dfrac{10x}{x^2 - 4x + 3}$

13. $\dfrac{2x^2}{3xy} \div \dfrac{8x^2y}{9xy}$ **14.** $\dfrac{3m - 9}{m^2 - 2m} \div \dfrac{m^2 - m - 6}{m^2 - 4}$

Simplify the complex fractions.

15. $\dfrac{\dfrac{x^2}{18}}{\dfrac{x^3}{12}}$ **16.** $\dfrac{2 - \dfrac{m}{n}}{4 - \dfrac{m^2}{n^2}}$

What values for x, if any, must be excluded in the following rational expressions?

17. $\dfrac{8}{x - 4}$ **18.** $\dfrac{3}{x^2 - 9}$

19. _____

20. _____

21. _____

22. _____

23. _____

24. _____

25. _____

Solve the following equations for x.

19. $\dfrac{x}{3} - \dfrac{x}{4} = 3$

20. $\dfrac{5}{x} - \dfrac{x-3}{x+2} = \dfrac{22}{x^2+2x}$

Solve the following proportions.

21. $\dfrac{x-1}{5} = \dfrac{x+2}{8}$

22. $\dfrac{2x-1}{7} = \dfrac{x}{4}$

Solve the following applications.

23. Number problem. One number is 3 times another. If the sum of their reciprocals is $\dfrac{1}{3}$, find the two numbers.

24. Science and medicine. Mark drove 250 mi to visit Sandra. Returning by a shorter route, he found that the trip was only 225 mi, but traffic slowed his speed by 5 mi/h. If the two trips took exactly the same time, what was his rate each way?

25. Construction. A cable that is 55 ft long is to be cut into two pieces whose lengths have the ratio 4 to 7. Find the lengths of the two pieces.

The following questions are presented to help you review concepts from earlier chapters. This is meant as a review and not as a comprehensive exam. The answers are presented in the back of the text. Section references accompany the answers. If you have difficulty with any of these questions, be certain to at least read through the summary related to those sections.

Perform the indicated operation.

1. $x^2y - 4xy - x^2y + 2xy$

2. $\dfrac{12a^3b}{9ab}$

3. $(5x^2 - 2x + 1) - (3x^2 + 3x - 5)$

4. $(5a^2 + 6a) - (2a^2 - 1)$

5. $4 + 3(7 - 4)^2$

6. $|3 - 5| - |-4 + 3|$

Multiply.

7. $(x - 2y)(2x + 3y)$

8. $(x + 7)(x + 4)$

Divide.

9. $(2x^2 + 3x - 1) \div (x + 2)$

10. $(x^2 - 5) \div (x - 1)$

Solve each equation and check your results.

11. $4x - 3 = 2x + 5$

12. $2 - 3(2x + 1) = 11$

Factor each polynomial completely.

13. $x^2 - 5x - 14$

14. $3m^2n - 6mn^2 + 9mn$

15. $a^2 - 9b^2$

16. $2x^3 - 28x^2 + 96x$

Solve the following word problems. Show the equation used for each solution.

17. Number problem. 2 more than 4 times a number is 30. Find the number.

ANSWERS

1. _____
2. _____
3. _____
4. _____
5. _____
6. _____
7. _____
8. _____
9. _____
10. _____
11. _____
12. _____
13. _____
14. _____
15. _____
16. _____
17. _____

18. _____

19. _____

20. _____

21. _____

22. _____

23. _____

24. _____

25. _____

26. _____

27. _____

28. _____

29. _____

30. _____

18. **Number problem.** If the reciprocal of 4 times a number is subtracted from the reciprocal of that number, the result is $\frac{3}{16}$. What is the number?

19. **Science and medicine.** A speed of 60 mi/h corresponds to 88 ft/s. If a race car is traveling at 180 mi/h, what is its speed in feet per second?

20. **Geometry.** The length of a rectangle is 3 in. less than twice its width. If the area of the rectangle is 35 in.^{2}, find the dimensions of the rectangle.

Write each rational expression in simplest form.

21. $\dfrac{m^2 - 4m}{3m - 12}$

22. $\dfrac{a^2 - 49}{3a^2 + 22a + 7}$

Perform the indicated operations.

23. $\dfrac{4}{3r} + \dfrac{1}{2r^2}$

24. $\dfrac{2}{x - 3} - \dfrac{5}{3x + 9}$

25. $\dfrac{3x^2 + 9x}{x^2 - 9} \cdot \dfrac{2x^2 - 9x + 9}{2x^3 - 3x^2}$

26. $\dfrac{4w^2 - 25}{2w^2 - 5w} \div (6w + 15)$

Simplify the complex rational expressions.

27. $\dfrac{1 - \dfrac{1}{x}}{2 + \dfrac{1}{x}}$

28. $\dfrac{3 - \dfrac{m}{n}}{9 - \dfrac{m^2}{n^2}}$

Solve the following equations for x.

29. $\dfrac{5}{3x} + \dfrac{1}{x^2} = \dfrac{5}{2x}$

30. $\dfrac{10}{x - 3} - 2 = \dfrac{5}{x + 3}$

AN INTRODUCTION TO GRAPHING

INTRODUCTION

Graphs are used to discern patterns and trends that may be difficult to see when looking at a list of numbers or other kinds of data. The word *graph* comes from Latin and Greek roots and means "to draw a picture." This is just what a graph is in mathematics: It is a picture of a relationship between two or more variables. But, as in art, graphs can be difficult to interpret without a little practice and training. This chapter is the beginning of that training, which is important because graphs are used in every field in which numbers are used.

In the field of pediatric medicine, there has been controversy about the use of somatotropin (human growth hormone) to help children whose growth has been impeded by various health problems. The reason for the controversy is that many doctors are giving this expensive drug therapy to children who are simply shorter than average or shorter than their parents want them to be. The determination of which children are healthy but small in stature and which children have health defects that keep them from growing is an issue that has been vigorously argued by professionals here and in Europe, where the therapy is being used.

Some of the measures used to distinguish between the two groups are blood tests and age and height measurements. The age and height measurements are graphed and monitored over several years of a child's life to gauge the rate of growth. If, during a certain period, the child's rate of growth slows to below 4.5 centimeters per year, this indicates that something may be seriously wrong. The graph can also indicate if the child's size fits within a range considered normal at each age of the child's life.

Pre-Test Chapter 6

This pre-test provides a preview of the types of exercises you will encounter in each section of this chapter. The answers for these exercises can be found in the back of the text. If you are working on your own, or ahead of the class, this pre-test can help you find the sections in which you should focus more of your time.

Determine which of the ordered pairs are solutions for the given equations.

[6.1]   **1.** $x - y = 12$ $(15, 3), (9, 6), (18, 6)$ **2.** $3x + 2y = 6$ $(1, 2), (0, 3), (2, 0)$

3. Complete the ordered pairs so that each is a solution for the given equation.

$2x + y = 5$ $(1,\ ), (0,\ ), (\ , 11)$

4. Find three solutions for each of the following equations.

$2x - 3y = 8$ $6x + y = 11$

[6.2]   **5.** Give the coordinates of the points graphed below.

6. Plot the points with the given coordinates: $S(-1, 2)$, $T(3, 0)$.

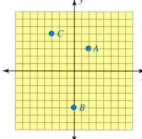

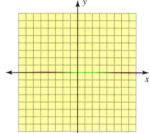

Graph each of the following equations.

[6.3]   **7.** $x + y = 5$

8. $y = \dfrac{1}{2}x - 1$

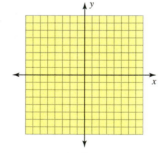

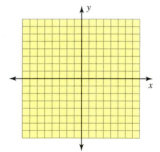

9. $y = -2$

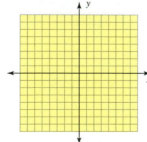

Find the slope of the line through the following pairs of points.

[6.4] **10.** $(-2, 5)$ and $(5, 12)$ **11.** $(-1, -3)$ and $(7, -11)$

12. Find the constant of variation k if y varies directly with x, and $y = 56$ when $x = 7$.

[6.5] The total commercial fishery landings at the Cameron, LA port (in millions of pounds, rounded) for each of the years 1998 to 2002 is shown in the table below. Use this table to answer exercises 13 and 14.

Commercial Fishery Landings; Cameron, LA

Year	1997	1998	1999	2000	2001
Landings	380	257	406	415	324

Source: National Marine Fisheries Service, Fisheries Statistics and Economics Division.

13. Construct a line graph displaying the given data.

14. Predict the commercial fishery landings at the Cameron, LA port for 2002.

[6.4] **15.** Pete's commission varies directly with the number of appliances he sells. If his commission last month was $800 and he sold 20 appliances, what would his salary be if he sold 25 appliances?

ANSWERS

10. _____

11. _____

12. _____

13. _____

14. _____

15. _____

6.1 Solutions of Equations in Two Variables

6.1 OBJECTIVES

1. Find solutions for an equation in two variables
2. Use ordered-pair notation to write solutions for equations in two variables

We discussed finding solutions for equations in Chapter 2. Recall that a solution is a value for the variable that "satisfies" the equation or makes the equation a true statement. For instance, we know that 4 is a solution of the equation

$$2x + 5 = 13$$

We know this is true because, when we replace x with 4, we have

$$2(4) + 5 \overset{?}{=} 13$$
$$8 + 5 \overset{?}{=} 13$$
$$13 = 13 \qquad \text{A true statement}$$

RECALL An equation is two expressions connected by an equal-sign.

We now want to consider **equations in two variables.** An example is

$$x + y = 5$$

What will a solution look like? It is not going to be a single number, because there are two variables. Here a solution will be a pair of numbers—one value for each of the variables, x and y. Suppose that x has the value 3. In the equation $x + y = 5$, you can substitute 3 for x.

$$(3) + y = 5$$

Solving for y gives

$$y = 2$$

NOTE An equation in two variables "pairs" two numbers, one for x and one for y.

So the pair of values $x = 3$ and $y = 2$ satisfies the equation because

$$3 + 2 = 5$$

The pair of numbers that satisfy an equation is called a **solution** for the equation in two variables.

How many such pairs are there? Choose any value for x (or for y). You can always find the other *paired* or *corresponding* value in an equation of this form. We say that there are an *infinite* number of pairs that will satisfy the equation. Each of these pairs is a solution. We will find some other solutions for the equation $x + y = 5$ in Example 1.

OBJECTIVE 1 | **Example 1 Solving for Corresponding Values**

For the equation $x + y = 5$, find (a) y if $x = 5$ and (b) x if $y = 4$.

(a) If $x = 5$

$$(5) + y = 5 \qquad \text{or} \qquad y = 0$$

(b) If $y = 4$,

$$x + (4) = 5 \qquad \text{or} \qquad x = 1$$

So the pairs $x = 5$, $y = 0$ and $x = 1$, $y = 4$ are both solutions.

CHECK YOURSELF 1

For the equation $2x + 3y = 26$,

(a) If $x = 4$, $y = ?$ **(b)** If $y = 0$, $x = ?$

To simplify writing the pairs that satisfy an equation, we use **ordered-pair notation.** The numbers are written in parentheses and are separated by a comma. For example, we know that the values $x = 3$ and $y = 2$ satisfy the equation $x + y = 5$. So we write the pair as

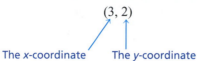

$(3, 2)$

The *x*-coordinate   The *y*-coordinate

CAUTION

$(3, 2)$ means $x = 3$ and $y = 2$.
$(2, 3)$ means $x = 2$ and $y = 3$.
$(3, 2)$ and $(2, 3)$ are different, which is why we call them *ordered pairs.*

The first number of the pair is *always* the value for x and is called the **x-coordinate.** The second number of the pair is *always* the value for y and is the **y-coordinate.**

Using this ordered-pair notation, we can say that $(3, 2)$, $(5, 0)$, and $(1, 4)$ are all *solutions* for the equation $x + y = 5$. Each pair gives values for x and y that will satisfy the equation.

OBJECTIVE 2   **Example 2**   **Identifying Solutions of Two-Variable Equations**

Which of the ordered pairs (a) $(2, 5)$, (b) $(5, -1)$, and (c) $(3, 4)$ are solutions for the equation $2x + y = 9$?

(a) To check whether $(2, 5)$ is a solution, let $x = 2$ and $y = 5$ and see if the equation is satisfied.

$2x + y = 9$ The original equation.

x y

$2(2) + (5) \stackrel{?}{=} 9$ Substitute 2 for x and 5 for y.

$4 + 5 \stackrel{?}{=} 9$

$9 = 9$ A true statement

NOTE $(2, 5)$ is a solution because a *true statement* results.

$(2, 5)$ is a solution for the equation.

(b) For $(5, -1)$, let $x = 5$ and $y = -1$.

$2(5) + (-1) \stackrel{?}{=} 9$

$10 - 1 \stackrel{?}{=} 9$

$9 = 9$ A true statement

So $(5, -1)$ is a solution.

(c) For $(3, 4)$, let $x = 3$ and $y = 4$. Then

$2(3) + (4) \stackrel{?}{=} 9$

$6 + 4 \stackrel{?}{=} 9$

$10 \stackrel{?}{=} 9$ *Not* a true statement

So $(3, 4)$ is *not* a solution for the equation.

CHECK YOURSELF 2 _____

Which of the ordered pairs (3, 4), (4, 3), (1, −2), and (0, −5) are solutions for the following equation?

$3x − y = 5$

If the equation contains only one variable, then the missing variable can take on any value.

Example 3   Identifying Solutions of One-Variable Equations

Which of the ordered pairs, (2, 0), (0, 2), (5, 2), (2, 5), and (2, −1) are solutions for the equation $x = 2$?

A solution is any ordered pair in which the *x*-coordinate is 2. That makes (2, 0), (2, 5), and (2, −1) solutions for the given equation.

CHECK YOURSELF 3 _____

Which of the ordered pairs (3, 0), (0, 3), (3, 3), (−1, 3), and (3, −1) are solutions for the equation $y = 3$?

Remember that, when an ordered pair is presented, the first number is always the *x*-coordinate and the second number is always the *y*-coordinate.

Example 4   Completing Ordered-Pair Solutions

Complete the ordered pairs (a) (9,   ), (b) (, −1), (c) (0,   ), and (d) (, 0) for the equation $x − 3y = 6$.

NOTE The *x*-coordinate is sometimes called the **abscissa** and the *y*-coordinate the **ordinate**.

(a) The first number, 9, appearing in (9,   ) represents the *x*-value. To complete the pair (9,   ), substitute 9 for *x* and then solve for *y*.

$$(9) − 3y = 6$$
$$−3y = −3$$
$$y = 1$$

(9, 1) is a solution.

(b) To complete the pair (, −1), let *y* be −1 and solve for *x*.

$$x − 3(−1) = 6$$
$$x + 3 = 6$$
$$x = 3$$

(3, −1) is a solution.

(c) To complete the pair (0,   ), let *x* be 0.

$$(0) − 3y = 6$$
$$−3y = 6$$
$$y = −2$$

(0, −2) is a solution.

(d) To complete the pair (, 0), let y be 0.

$$x - 3(0) = 6$$
$$x - 0 = 6$$
$$x = 6$$

$(6, 0)$ is a solution.

 CHECK YOURSELF 4 _____

Complete the given ordered pairs so that each is a solution for the equation $2x + 5y = 10$.

(10,), (, 4), (0,), and (, 0)

Example 5   Finding Some Solutions of a Two-Variable Equation

Find four solutions for the equation

$$2x + y = 8$$

NOTE Generally, you want to pick values for x (or for y) so that the resulting equation in one variable is easy to solve.

In this case the values used to form the solutions are *up to you*. You can assign any value for x (or for y). We demonstrate with some possible choices.

Solution with $x = 2$:

$$2(2) + y = 8$$
$$4 + y = 8$$
$$y = 4$$

$(2, 4)$ is a solution.

Solution with $y = 6$:

$$2x + (6) = 8$$
$$2x = 2$$
$$x = 1$$

$(1, 6)$ is a solution.

Solution with $x = 0$:

$$2(0) + y = 8$$
$$y = 8$$

NOTE The solutions (0, 8) and (4, 0) will have special significance later in graphing. They are also easy to find!

$(0, 8)$ is a solution.

Solution with $y = 0$:

$$2x + (0) = 8$$
$$2x = 8$$
$$x = 4$$

$(4, 0)$ is a solution.

CHECK YOURSELF 5

Find four solutions for x − 3y = 12.

Applications involving two-variable equations are fairly common.

Example 6   Applications of Two-Variable Equations

NOTE We will look at variable and fixed costs in more detail in Section 7.1.

Suppose that it costs the manufacturer $1.25 for each stapler that is produced. In addition, fixed costs (related to staplers) are $110 per day.

(a) Write an equation relating the total daily costs C to the number of staplers produced x in a day.

Since each stapler costs $1.25 to produce, the cost of producing staplers is $1.25x$. Adding the fixed cost to this gives us an equation for the total daily costs.

$$C = 1.25x + 110$$

(b) What is the total cost of producing 500 staplers in a day?

We substitute 500 for x in the equation from part (a) and calculate the total cost.

$$C = 1.25(500) + 110$$
$$= 625 + 110$$
$$= 735$$

It costs the manufacturer a total of $735 to produce 500 staplers in one day.

(c) How many staplers can be produced for $1110?

In this case, we substitute 1110 for C in the equation from part (a) and solve for x.

$$(1110) = 1.25x + 110 \qquad \text{Subtract 110 from both sides.}$$
$$1000 = 1.25x \qquad \qquad \text{Divide both sides by 1.25.}$$
$$800 = x$$

800 staplers can be produced at a cost of $1110.

CHECK YOURSELF 6

Suppose that the stapler manufacturer earns a profit of $1.80 on each stapler shipped. However, it costs $120 to operate each day.

(a) Write an equation relating the daily profit P to the number of staplers shipped x in a day.
(b) What is the total profit of shipping 500 staplers in a day?
(c) How many staplers need to be shipped to produce a profit of $1500?

CHECK YOURSELF ANSWERS

1. (a) $y = 6$; **(b)** $x = 13$ **2.** $(3, 4)$, $(1, -2)$, and $(0, -5)$ are solutions
3. $(0, 3)$, $(3, 3)$, and $(-1, 3)$ are solutions   **4.** $(10, -2)$, $(-5, 4)$, $(0, 2)$, and $(5, 0)$
5. $(6, -2)$, $(3, -3)$, $(0, -4)$, and $(12, 0)$ are four possibilities
6. (a) $P = 1.80x - 120$; **(b)** $780; **(c)** 900

6.1 Exercises

Determine which of the ordered pairs are solutions for the given equation.

1. $x + y = 6$ $(4, 2), (-2, 4), (0, 6), (-3, 9)$

1. _____

2. $x - y = 12$ $(13, 1), (13, -1), (12, 0), (6, 6)$

2. _____

3. $2x - y = 8$ $(5, 2), (4, 0), (0, 8), (6, 4)$

3. _____

4. _____

4. $x + 5y = 20$ $(10, -2), (10, 2), (20, 0), (25, -1)$

5. _____

5. $3x + y = 6$ $(2, 0), (2, 3), (0, 2), (1, 3)$

6. _____

7. _____

6. $x - 2y = 8$ $(8, 0), (0, 4), (5, -1), (10, -1)$

8. _____

7. $2x - 3y = 6$ $(0, 2), (3, 0), (6, 2), (0, -2)$

9. _____

8. $8x + 4y = 16$ $(2, 0), (6, -8), (0, 4), (6, -6)$

10. _____

11. _____

9. $3x - 2y = 12$ $(4, 0), \left(\dfrac{2}{3}, -5\right), (0, 6), \left(5, \dfrac{3}{2}\right)$

12. _____

13. _____

10. $3x + 4y = 12$ $(-4, 0), \left(\dfrac{2}{3}, \dfrac{5}{2}\right), (0, 3), \left(\dfrac{2}{3}, 2\right)$

14. _____

15. _____

11. $y = 4x$ $(0, 0), (1, 3), (2, 8), (8, 2)$

16. _____

12. $y = 2x - 1$ $(0, -2), (0, -1), \left(\dfrac{1}{2}, 0\right), (3, -5)$

13. $x = 3$ $(3, 5), (0, 3), (3, 0), (3, 7)$

14. $y = 5$ $(0, 5), (3, 5), (-2, -5), (5, 5)$

Complete the ordered pairs so that each is a solution for the given equation.

15. $x + y = 12$ $(4, \ \ ), (\ \ , 5), (0, \ \ ), (\ \ , 0)$

16. $x - y = 7$ $(\ \ , 4), (15, \ \ ), (0, \ \ ), (\ \ , 0)$

ANSWERS

17. _____

18. _____

19. _____

20. _____

21. _____

22. _____

23. _____

24. _____

25. _____

26. _____

27. _____

28. _____

29. _____

30. _____

31. _____

32. _____

33. _____

34. _____

35. _____

36. _____

37. _____

38. _____

17. $3x + y = 9$ $(3,\ ), (\ , 9), (\ , -3), (0,\ )$

18. $x + 5y = 20$ $(0,\ ), (\ , 2), (10,\ ), (\ , 0)$

19. $5x - y = 15$ $(\ , 0), (2,\ ), (4,\ ), (\ , -5)$

20. $x - 3y = 9$ $(0,\ ), (12,\ ), (\ , 0), (\ , -2)$

21. $3x - 2y = 12$ $(\ , 0), (\ , -6), (2,\ ), (\ , 3)$

22. $2x + 5y = 20$ $(0,\ ), (5,\ ), (\ , 0), (\ , 6)$

23. $y = 3x + 9$ $(\ , 0), \left(\dfrac{2}{3},\ \right), (0,\ ), \left(-\dfrac{2}{3},\ \right)$

24. $3x + 4y = 12$ $(0,\ ), \left(\ , \dfrac{3}{4}\right), (\ , 0), \left(\dfrac{8}{3},\ \right)$

25. $y = 3x - 4$ $(0,\ ), (\ , 5), (\ , 0), \left(\dfrac{5}{3},\ \right)$

26. $y = -2x + 5$ $(0,\ ), (\ , 5), \left(\dfrac{3}{2},\ \right), (\ , 1)$

Find four solutions for each of the following equations. **Note:** Your answers may vary from those shown in the answer section.

27. $x - y = 7$

28. $x + y = 18$

29. $2x - y = 6$

30. $3x - y = 12$

31. $x + 4y = 8$

32. $x + 3y = 12$

33. $2x - 5y = 10$

34. $2x + 7y = 14$

35. $y = 2x + 3$

36. $y = 8x - 5$

37. $x = -5$

38. $y = 8$

An equation in three variables has an ordered triple as a solution. For example, $(1, 2, 2)$ is a solution to the equation $x + 2y - z = 3$. Complete the ordered-triple solutions for each equation.

39. $x + y + z = 0$ $(2, -3, \ )$

40. $2x + y + z = 2$ $(\ , -1, 3)$

41. $x + y + z = 0$ $(1, \ , 5)$

42. $x + y - z = 1$ $(4, \ , 3)$

43. $2x + y + z = 2$ $(-2, \ , 1)$

44. $x + y - z = 1$ $(-2, 1, \ )$

45. Business and finance. When an employee produces x units per hour, the hourly wage in dollars is given by $y = 0.75x + 8$. What are the hourly wages for the following number of units: 2, 5, 10, 15, and 20?

46. Science and medicine. Celsius temperature readings can be converted to Fahrenheit readings using the formula $F = \dfrac{9}{5}C + 32$. What is the Fahrenheit temperature that corresponds to each of the following Celsius temperatures: -10, 0, 15, 100?

47. Geometry. The area of a square is given by $A = s^2$. What is the area of the squares whose sides are 5, 10, 12, and 15 cm?

48. Business and finance. When x number of units are sold, the price of each unit (in dollars) is given by $p = \dfrac{-x}{2} + 75$. Find the unit price when the following quantities are sold: 2, 7, 9, 11.

49. Statistics. The number of programs for the disabled in the United States from 1993 to 1997 is approximated by the equation $y = 162x + 4365$ in which x is the number of years after 1993. Complete the following table.

x	1	2	3	4	6
y					

50. Statistics. Your monthly pay as a car salesperson is determined by using the equation $S = 200x + 1500$ in which x is the number of cars you can sell each month.

(a) Complete the following table.

x	12	15	17	18
S				

ANSWERS

39. _____

40. _____

41. _____

42. _____

43. _____

44. _____

45. _____

46. _____

47. _____

48. _____

49. _____

50. _____

50. _____

51. _____

52. _____

a. _____

b. _____

c. _____

d. _____

e. _____

f. _____

g. _____

h. _____

i. _____

j. _____

(b) You are offered a job at a salary of $56,400 per year. How many cars would you have to sell per month to equal this salary?

51. You now have had practice solving equations with one variable and equations with two variables. Compare equations with one variable to equations with two variables. How are they alike? How are they different?

52. Each of the following sentences describes pairs of numbers that are related. After completing the sentences in parts (a) to (g), write two of your own sentences in (h) and (i).

(a) The *number of hours you work* determines the *amount you are* _____.

(b) The *number of gallons of gasoline* you put in your car determines *the amount you* _____.

(c) The *amount of the* _____ in a restaurant is related to *the amount of the tip.*

(d) The *sales amount of a purchase in a store* determines _____.

(e) The *age of an automobile* is related to _____.

(f) The *amount of electricity you use in a month* determines _____.

(g) The *cost of food for a family of four* and _____.

Think of two more:

(h) .

(i) 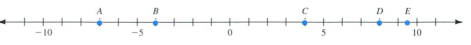.

![runner icon] **Getting Ready for Section 6.2 [Section 0.5]**

Plot points with the following coordinates on the number line shown.

(a) −3   (b) 7   (c) 0   (d) −8   (e) $\dfrac{3}{2}$

```
    |-------|-------|-------|-------|
   −10     −5       0       5       10
```

Give the coordinate of each of the following points.

(f) *A*   (g) *B*   (h) *C*   (i) *D*   (j) *E*

```
           A       B              C      D  E
    |-------|-------|-------|-------|------|--|
   −10     −5       0       5       10
```

Answers

1. $(4, 2), (0, 6), (-3, 9)$ **3.** $(5, 2), (4, 0), (6, 4)$ **5.** $(2, 0), (1, 3)$

7. $(3, 0), (6, 2), (0, -2)$ **9.** $(4, 0), \left(\frac{2}{3}, -5\right), \left(5, \frac{3}{2}\right)$ **11.** $(0, 0), (2, 8)$

13. $(3, 5), (3, 0), (3, 7)$ **15.** $(4, 8), (7, 5), (0, 12), (12, 0)$

17. $(3, 0), (0, 9), (4, -3), (0, 9)$ **19.** $(3, 0), (-5, 2), (10, 5), (2, 0)$

21. $(4, 0), (0, -6), (2, -3), (6, 3)$ **23.** $(-3, 0), \left(\frac{2}{3}, 11\right), (0, 9), \left(-\frac{2}{3}, 7\right)$

25. $(0, -4), (3, 5), \left(\frac{4}{3}, 0\right), \left(\frac{5}{3}, 1\right)$ **27.** $(0, -7), (2, -5), (4, -3), (6, -1)$

29. $(0, -6), (3, 0), (6, 6), (9, 12)$ **31.** $(8, 0), (-4, 3), (0, 2), (4, 1)$

33. $(-5, -4), (0, -2), (5, 0), (10, 2)$ **35.** $(0, 3), (1, 5), (2, 7), (3, 9)$

37. $(-5, 0), (-5, 1), (-5, 2), (-5, 3)$ **39.** $(2, -3, 1)$ **41.** $(1, -6, 5)$

43. $(-2, 5, 1)$ **45.** \$9.50, \$11.75, \$15.50, \$19.25, \$23

47. 25 cm², 100 cm², 144 cm², 225 cm²   **49.** 4527, 4689, 4851, 5013, 5337

51.

a–e.

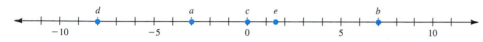

f. -7 **g.** -4 **h.** 4 **i.** 8 **j.** $\dfrac{19}{2}$

6.2 The Rectangular Coordinate System

In Section 6.1, we saw that ordered pairs could be used to write the solutions of equations in two variables. The next step is to graph those ordered pairs as points in a plane.

Because there are two numbers (one for x and one for y), we need two number lines. One line is drawn horizontally, and the other is drawn vertically; their point of intersection (at their respective zero points) is called the *origin*. The horizontal line is called the **x-axis,** and the vertical line is called the **y-axis.** Together the lines form the **rectangular coordinate system.**

The axes divide the plane into four regions called **quadrants,** which are numbered (usually by Roman numerals) counterclockwise from the upper right.

NOTE This system is also called the **Cartesian coordinate system,** named in honor of its inventor, René Descartes (1596–1650), a French mathematician and philosopher.

y-axis

Quadrant II   Quadrant I

Origin   *x*-axis

The origin is the point with coordinates (0, 0).

Quadrant III   Quadrant IV

We now want to establish correspondences between ordered pairs of numbers (x, y) and points in the plane.

For any ordered pair

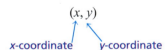

$$(x, y)$$

x-coordinate   *y*-coordinate

the following are true:

1. If the x-coordinate is

   Positive, the point corresponding to that pair is located x units to the *right* of the y-axis.
   Negative, the point is x units to the *left* of the y-axis.
   Zero, the point is on the y-axis.

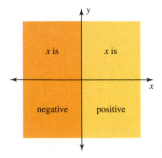

2. If the *y*-coordinate is

Positive, the point is *y* units *above* the *x*-axis.
Negative, the point is *y* units *below* the *x*-axis.
Zero, the point is on the *x*-axis.

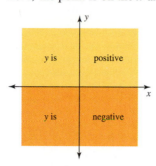

Example 1 illustrates how to use these guidelines to give coordinates to points in the plane.

OBJECTIVE 1   **Example 1   Identifying the Coordinates for a Given Point**

Give the coordinates for the given point.

RECALL The *x*-coordinate gives the *horizontal* distance from the *y*-axis. The *y*-coordinate gives the *vertical* distance from the *x*-axis.

(a)

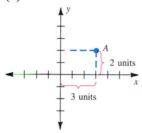

Point *A* is 3 units to the *right* of the *y*-axis and 2 units *above* the *x*-axis. Point *A* has coordinates $(3, 2)$.

(b)

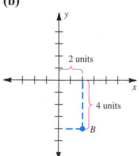

Point *B* is 2 units to the *right* of the *y*-axis and 4 units *below* the *x*-axis. Point *B* has coordinates $(2, -4)$.

(c)

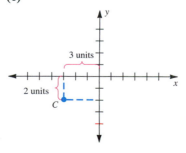

Point *C* is 3 units to the *left* of the *y*-axis and 2 units *below* the *x*-axis. *C* has coordinates $(-3, -2)$.

(d)

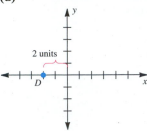

Point *D* is 2 units to the *left* of the *y*-axis and *on* the *x*-axis. Point *D* has coordinates $(-2, 0)$.

 CHECK YOURSELF 1

Give the coordinates of points P, Q, R, and S.

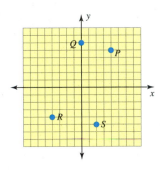

P _____

Q _____

R _____

S _____

Reversing the process above will allow us to graph (or plot) a point in the plane given the coordinates of the point. You can use the following steps.

NOTE The graphing of individual points is sometimes called **point plotting.**

Step by Step: To Graph a Point in the Plane

Step 1   Start at the origin.
Step 2   Move right or left according to the value of the *x*-coordinate.
Step 3   Move up or down according to the value of the *y*-coordinate.

OBJECTIVE 2   **Example 2   Graphing Points**

(a) Graph the point corresponding to the ordered pair $(4, 3)$.

Move 4 units to the right on the *x*-axis. Then move 3 units up from the point you stopped at on the *x*-axis. This locates the point corresponding to $(4, 3)$.

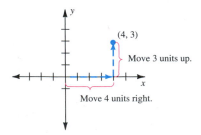

(b) Graph the point corresponding to the ordered pair $(-5, 2)$.

In this case move 5 units *left* (because the x-coordinate is negative) and then 2 units *up*.

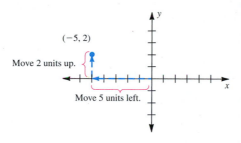

(c) Graph the point corresponding to $(-4, -2)$.

Here move 4 units *left* and then 2 units *down* (the y-coordinate is negative).

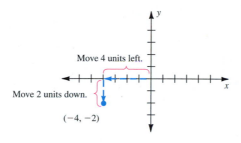

NOTE Any point on an axis will have 0 for one of its coordinates.

(d) Graph the point corresponding to $(0, -3)$.

There is *no* horizontal movement because the x-coordinate is 0. Move 3 units *down*.

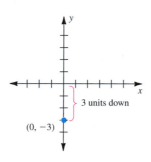

(e) Graph the point corresponding to $(5, 0)$.

Move 5 units *right*. The desired point is on the x-axis because the y-coordinate is 0.

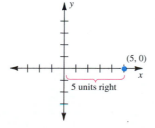

CHECK YOURSELF 2

Graph the points corresponding to M(4, 3), N(−2, 4), P(−5, −3), and Q(0, −3).

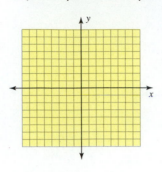

CHECK YOURSELF ANSWERS

1. $P(4, 5)$, $Q(0, 6)$, $R(−4, −4)$, and $S(2, −5)$

2.

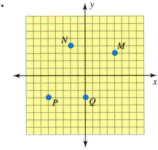

6.2 Exercises

Give the coordinates of the points graphed below.

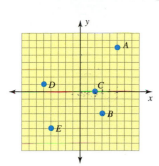

1. A $\left(5, 6\right)$

2. B $(3, 3)$

3. C $(2, 0)$

4. D $(-5, 1)$

5. E $(-4, -5)$

Give the coordinates of the points graphed below.

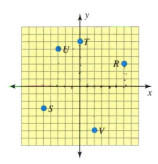

6. R

7. S

8. T

9. U

10. V

Plot points with the following coordinates on the graph below.

11. $M(5, 3)$

12. $N(0, -3)$

13. $P(-2, 6)$

14. $Q(5, 0)$

15. $R(-4, -6)$

16. $S(-3, -4)$

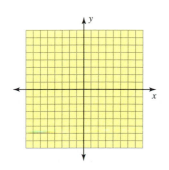

Plot points with the following coordinates on the given graph.

17. $F(-3, -1)$

18. $G(4, 3)$

19. $H(5, -2)$

20. $I(-3, 0)$

1. _____
2. _____
3. _____
4. _____
5. _____
6. _____
7. _____
8. _____
9. _____
10. _____
11. _____
12. _____
13. _____
14. _____
15. _____
16. _____
17. _____
18. _____
19. _____
20. _____

21. _____

22. _____

23. _____

24. _____

25. _____

26. _____

21. $J(-5, 3)$ **22.** $K(0, 6)$

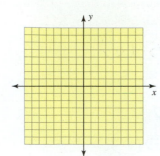

23. Plot points with coordinates $(2, 3), (3, 4),$ and $(4, 5)$ on the given graph. What do you observe? Can you give the coordinates of another point with the same property?

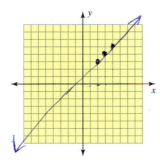

24. Plot points with coordinates $(-1, 4), (0, 3),$ and $(1, 2)$ on the given graph. What do you observe? Can you give the coordinates of another point with the same property?

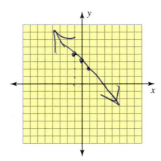

25. In each of the following statements, fill in the blank with *always, sometimes,* or *never*.

    (a) In the plane, a point on an axis _____ has a coordinate equal to zero.

    (b) The ordered pair (a, b) is _____ equal to the ordered pair (b, a).

26. In each of the following statements, fill in the blank with *always, sometimes,* or *never*.

    (a) If, in the ordered pair (a, b), a and b have different signs, then the point (a, b) is _____ in the second quadrant.

    (b) If $a \neq b$, then the ordered pair (a, b) is _____ equal to the ordered pair (b, a).

27. **Science and medicine.** A local plastics company is sponsoring a plastics recycling contest for each of the local community. The focus of the contest is collecting plastic milk, juice, and water jugs. The company will award $200 plus the current market price of the jugs collected to the group that collects the most jugs in a single month. The number of jugs collected and the amount of money won can be represented as an ordered pair.

(a) In April, group *A* collected 1500 lb of jugs to win first place. The prize for the month was $350. On the following graph, *x* represents the pounds of jugs and *y* represents the amount of money that the group won. Graph the point that represents the winner for April.

(b) In May, group *B* collected 2300 lb of jugs to win first place. The prize for the month was $430. Graph the point that represents the May winner on the same axis you used in part (a).

(c) In June, group *C* collected 1200 lb of jugs to win the contest. The prize for the month was $320. Graph the point that represents the June winner on the same axis as used before.

RECYCLES
PLAST

28. **Statistics.** The table gives the hours *x* that Damien studied for five different math exams and the resulting grades *y*. Plot the data given in the table.

x	4	5	5	2	6
y	83	89	93	75	95

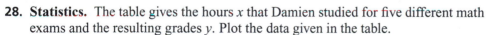

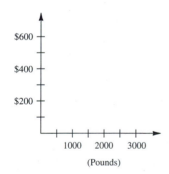

x	y
2	75
4	83
5	89
5	93
6	95

Recursive math

29. **Science and medicine.** The table gives the average temperature y (in degrees Fahrenheit) for each of the first 6 months of the year, x. The months are numbered 1 through 6, with 1 corresponding to January. Plot the data given in the table.

x	1	2	3	4	5	6
y	4	14	26	33	42	51

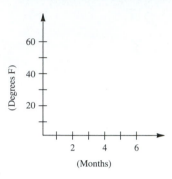

30. **Business and finance.** The table gives the total salary of a salesperson, y, for each of the four quarters of the year, x. Plot the data given in the table.

x	1	2	3	4
y	$6000	$5000	$8000	$9000

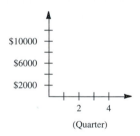

31. **Statistics.** The table shows the number of runs scored by the Anaheim Angels in each game of the 2002 World Series.

Game	1	2	3	4	5	6	7
Runs	3	11	10	3	4	6	4

Source: Major League Baseball.

Plot the data given in the table.

32. **Statistics.** The following table shows the number of wins and total points for the five teams in the Atlantic Division of the National Hockey League in the early part of the 1999–2000 season.

Team	Wins	Points
New Jersey Devils	5	12
Philadelphia Flyers	4	10
New York Rangers	4	9
Pittsburgh Penguins	2	6
New York Islanders	2	5

Plot the data given in the table.

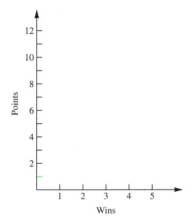

33. How would you describe a rectangular coordinate system? Explain what information is needed to locate a point in a coordinate system.

34. Some newspapers have a special day that they devote to automobile ads. Use this special section or the Sunday classified ads from your local newspaper to find all the want ads for a particular automobile model. Make a list of the model year and asking price for 10 ads, being sure to get a variety of ages for this model. After collecting the information, make a scatter plot of the age and the asking price for the car.

   Describe your graph, including an explanation of how you decided which variable to put on the vertical axis and which on the horizontal axis. What trends or other information are given by the graph?

35. _____

a. _____

b. _____

c. _____

d. _____

e. _____

f. _____

35. The map shown below uses letters and numbers to label a grid that helps to locate a city. For instance, Salem is located at E-4.

(a) Find the coordinates for the following: White Swan, Newport, and Wheeler.

(b) What cities correspond to the following coordinates: A2, F4, and A5?

 Getting Ready for Section 6.3 [Section 2.3]

Solve each of the following equations.

(a) $2x - 2 = 6$

(b) $2 - 5x = 12$

(c) $7y + 10 = -11$

(d) $-3 + 5x = 1$

(e) $6 - 3x = 8$

(f) $-4y + 6 = 3$

Answers

1. (5, 6)   **3.** (2, 0)   **5.** (−4, −5)   **7.** (−5, −3)   **9.** (−3, 5)

11–21.   **23.** The points lie on a line; (1, 2)

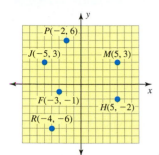

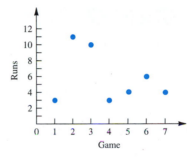

25. (a) Always; **(b)** sometimes

27. (a) (1500, 350); **(b)** (2300, 430); **(c)** (1200, 320)

29.

```
(Degrees F)
60
40
20
        2    4    6
      (Months)
```

31.

```
12
10 Runs
 8
 6
 4
 2
   0  1  2  3  4  5  6  7
            Game
```

33.   **35. (a)** A7, F2, C2; **(b)** Oysterville, Sweet Home, Mineral

a. 4   **b.** −2   **c.** −3   **d.** $\dfrac{4}{5}$ **e.** $-\dfrac{2}{3}$ **f.** $\dfrac{3}{4}$

6.3 Graphing Linear Equations

We are now ready to combine our work of Sections 6.1 and 6.2. In Section 6.1 you learned to write the solutions of equations in two variables as ordered pairs. Then, in Section 6.2, these ordered pairs were graphed in the plane. Putting these ideas together will let us graph certain equations. Example 1 illustrates this approach.

OBJECTIVE 1

Example 1   Graphing a Linear Equation

Graph $x + 2y = 4$.

NOTE We are going to find *three* solutions for the equation. We'll point out why shortly.

Step 1   Find some solutions for $x + 2y = 4$. To find solutions, we choose any convenient values for x, say $x = 0$, $x = 2$, and $x = 4$. Given these values for x, we can substitute and then solve for the corresponding value for y. So

If $x = 0$, then $y = 2$, so $(0, 2)$ is a solution.
If $x = 2$, then $y = 1$, so $(2, 1)$ is a solution.
If $x = 4$, then $y = 0$, so $(4, 0)$ is a solution.

A handy way to show this information is in a table such as this:

NOTE The table is just a convenient way to display the information. It is the same as writing $(0, 2)$, $(2, 1)$, and $(4, 0)$.

x	y
0	2
2	1
4	0

Step 2   We now graph the solutions found in step 1.

$x + 2y = 4$

x	y
0	2
2	1
4	0

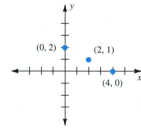

What pattern do you see? It appears that the three points lie on a straight line, and that is in fact the case.

Step 3 Draw a straight line through the three points graphed in Step 2.

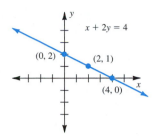

NOTE The arrows on the end of the line mean that the line extends indefinitely in either direction.

NOTE The graph is a "picture" of the solutions for the given equation.

The line shown is the **graph** of the equation $x + 2y = 4$. It represents *all* of the ordered pairs that are solutions (an infinite number) for that equation.

Every ordered pair that is a solution will have its graph on this line. Any point on the line will have coordinates that are a solution for the equation.

Note: Why did we suggest finding *three* solutions in step 1? Two points determine a line, so technically you need only two. The third point that we find is a check to catch any possible errors.

 CHECK YOURSELF 1

Graph $2x - y = 6$, using the steps shown in Example 1.

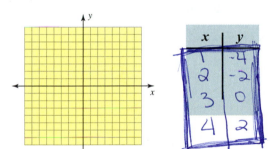

An equation that can be written in the form

$$Ax + By = C$$

in which A, B, and C are real numbers and A and B are not both 0 is called a **linear equation in two variables.** The graph of this equation is a *straight line*.

The steps of graphing follow.

Step by Step: To Graph a Linear Equation

Step 1 Find at least three solutions for the equation and put your results in tabular form.

Step 2 Graph the solutions found in Step 1.

Step 3 Draw a straight line through the points determined in Step 2 to form the graph of the equation.

Example 2   **Graphing a Linear Equation**

Graph $y = 3x$.

NOTE Let $x = 0$, 1, and 2, and substitute to determine the corresponding y-values. Again the choices for x are simply convenient. Other values for x would serve the same purpose.

Step 1   Some solutions are

x	y
0	0
1	3
2	6

Step 2   Graph the points.

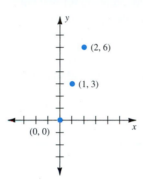

NOTE Connecting any pair of these points produces the same line.

Step 3   Draw a line through the points.

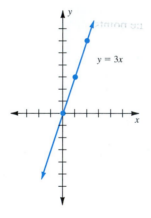

 CHECK YOURSELF 2

Graph the equation $y = -2x$ after completing the table of values.

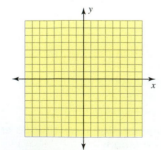

x	y
0	
1	
2	

Let's work through another example of graphing a line from its equation.

Example 3   Graphing a Linear Equation

Graph $y = 2x + 3$.

Step 1   Some solutions are

x	y
0	3
1	5
2	7

Step 2   Graph the points corresponding to these values.

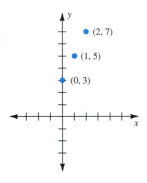

Step 3   Draw a line through the points.

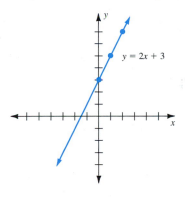

 CHECK YOURSELF 3 _____

Graph the equation $y = 3x - 2$ after completing the table of values.

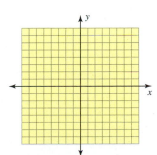

x	y
0	
1	
2	

When graphing equations, particularly if fractions are involved, a careful choice of values for x can simplify the process. Consider Example 4.

Example 4   Graphing a Linear Equation

Graph

$$y = \frac{3}{2}x - 2$$

As before, we want to find solutions for the given equation by picking convenient values for x. Note that in this case, choosing *multiples of 2* will avoid fractional values for y and make the plotting of those solutions much easier. For instance, here we might choose values of -2, 0, and 2 for x.

Step 1   If $x = -2$:

$$y = \frac{3}{2}(-2) - 2$$

$$= -3 - 2 = -5$$

If $x = 0$:

$$y = \frac{3}{2}(0) - 2$$

$$= 0 - 2 = -2$$

If $x = 2$:

$$y = \frac{3}{2}(2) - 2$$

$$= 3 - 2 = 1$$

NOTE Suppose we do *not* choose a multiple of 2, say, $x = 3$. Then

$$y = \frac{3}{2}(3) - 2$$

$$= \frac{9}{2} - 2$$

$$= \frac{5}{2}$$

$\left(3, \dfrac{5}{2}\right)$ is still a valid solution, but we must graph a point with fractions as coordinates.

In tabular form, the solutions are

x	y
-2	-5
0	-2
2	1

Step 2   Graph the points determined above.

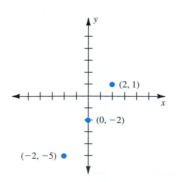

Step 3 Draw a line through the points.

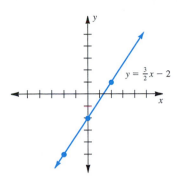

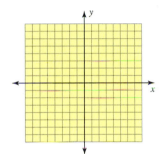

CHECK YOURSELF 4

Graph the equation $y = -\dfrac{1}{3}x + 3$ after completing the table of values.

x	y
−3	
0	
3	

Some special cases of linear equations are illustrated in Examples 5 and 6.

OBJECTIVE 2   **Example 5   Graphing an Equation That Results in a Vertical Line**

Graph $x = 3$.

The equation $x = 3$ is equivalent to $x + 0(y) = 3$. Some solutions follow.

If $y = 1$:   If $y = 4$:   If $y = -2$:

$x + 0(1) = 3$ $x + 0(4) = 3$ $x + 0(-2) = 3$

$x = 3$ $x = 3$ $x = 3$

In tabular form,

x	y
3	1
3	4
3	−2

What do you observe? The variable x has the value 3, regardless of the value of y. Look at the following graph.

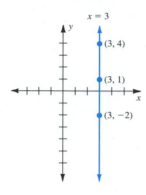

The graph of $x = 3$ is a vertical line crossing the x-axis at $(3, 0)$.

Note that graphing (or plotting) points in this case is not really necessary. Simply recognize that the graph of $x = 3$ *must* be a vertical line (parallel to the y-axis) that intercepts the x-axis at $(3, 0)$.

 CHECK YOURSELF 5

Graph the equation $x = -2$.

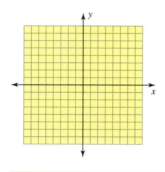

Example 6 is a related example involving a horizontal line.

Example 6   Graphing an Equation That Results in a Horizontal Line

Graph $y = 4$.

Because $y = 4$ is equivalent to $0(x) + y = 4$, any value for x paired with 4 for y will form a solution. A table of values might be

x	y
-2	4
0	4
2	4

Here is the graph.

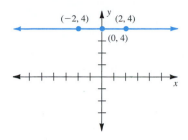

This time the graph is a horizontal line that crosses the y-axis at $(0, 4)$. Again the graphing of points is not required. The graph of $y = 4$ *must* be horizontal (parallel to the x-axis) and intercepts the y-axis at $(0, 4)$.

 CHECK YOURSELF 6 _____

Graph the equation $y = -3$.

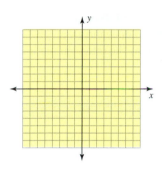

The following box summarizes our work in Examples 5 and 6.

> **Definition: Vertical and Horizontal Lines**
>
> **1.** The graph of $x = a$ is a *vertical line* crossing the x-axis at $(a, 0)$.
> **2.** The graph of $y = b$ is a *horizontal line* crossing the y-axis at $(0, b)$.

To simplify the graphing of certain linear equations, some students prefer the **intercept method** of graphing. This method makes use of the fact that the solutions that are easiest to find are those with an x-coordinate or a y-coordinate of 0. For instance, to graph the equation

$$4x + 3y = 12$$

NOTE With practice, all this can be done mentally, which is the big advantage of this method.

first, let $x = 0$ and then solve for y.

$$4(0) + 3y = 12$$
$$3y = 12$$
$$y = 4$$

So $(0, 4)$ is one solution. Now we let $y = 0$ and solve for x.

$$4x + 3(0) = 12$$
$$4x = 12$$
$$x = 3$$

A second solution is (3, 0).

The two points corresponding to these solutions can now be used to graph the equation.

RECALL Only two points are needed to graph a line. A third point is used only as a check.

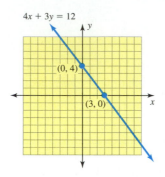

NOTE The intercepts are the points where the line crosses the *x*- and *y*-axes.

The ordered pair (3, 0) is called the ***x*-intercept,** and the ordered pair (0, 4) is the ***y*-intercept** of the graph. Using these points to draw the graph gives the name to this method. Let's look at a second example of graphing by the intercept method.

OBJECTIVE 3   **Example 7   Using the Intercept Method to Graph a Line**

Graph $3x - 5y = 15$, using the intercept method.

To find the *x*-intercept, let $y = 0$.

$$3x - 5(0) = 15$$
$$x = 5$$

The *x*-intercept is (5, 0).

To find the *y*-intercept, let $x = 0$.

$$3(0) - 5y = 15$$
$$y = -3$$

The *y*-intercept is (0, −3).

So (5, 0) and (0, −3) are solutions for the equation, and we can use the corresponding points to graph the equation.

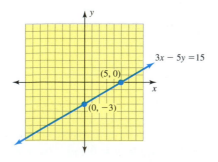

CHECK YOURSELF 7 _____

Graph $4x + 5y = 20$, *using the intercept method.*

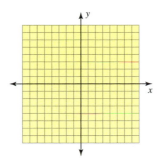

NOTE Finding a third "checkpoint" is always a good idea.

This all looks quite easy, and for many equations it is. What are the drawbacks? For one, you don't have a third checkpoint, and it is possible for errors to occur. You can, of course, still find a third point (other than the two intercepts) to be sure your graph is correct. A second difficulty arises when the *x*- and *y*-intercepts are very close to one another (or are actually the same point—the origin). For instance, if we have the equation

$$3x + 2y = 1$$

the intercepts are $\left(\frac{1}{3}, 0\right)$ and $\left(0, \frac{1}{2}\right)$. It is difficult to draw a line accurately through these intercepts, so choose other solutions farther away from the origin for your points.

We summarize the steps of graphing by the intercept method for appropriate equations.

Step by Step: Graphing a Line by the Intercept Method

Step 1   To find the *x*-intercept: Let $y = 0$, then solve for *x*.
Step 2   To find the *y*-intercept: Let $x = 0$, then solve for *y*.
Step 3   Graph the *x*- and *y*-intercepts.
Step 4   Draw a straight line through the intercepts.

A third method of graphing linear equations involves **solving the equation for *y*.** The reason we use this extra step is that it often will make finding solutions for the equation much easier.

OBJECTIVE 4   **Example 8   Graphing a Linear Equation by Solving for *y***

Graph $2x + 3y = 6$.

Rather than finding solutions for the equation in this form, we solve for *y*.

RECALL Solving for *y* means that we want to leave *y* isolated on the left.

$$2x + 3y = 6$$
$$3y = 6 - 2x \qquad \text{Subtract } 2x.$$
$$y = \frac{6 - 2x}{3} \qquad \text{Divide by 3.}$$
or $\quad y = 2 - \frac{2}{3}x$

Now find your solutions by picking convenient values for x.

NOTE Again, to choose convenient values for x, we suggest you look at the equation carefully. Here, for instance, choosing multiples of 3 for x will make the work much easier.

If $x = -3$:

$$y = 2 - \frac{2}{3}(-3)$$

$$= 2 + 2 = 4$$

So $(-3, 4)$ is a solution.

If $x = 0$:

$$y = 2 - \frac{2}{3}(0)$$

$$= 2$$

So $(0, 2)$ is a solution.

If $x = 3$:

$$y = 2 - \frac{2}{3}(3)$$

$$= 2 - 2 = 0$$

So $(3, 0)$ is a solution.

We can now plot the points that correspond to these solutions and form the graph of the equation as before.

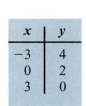

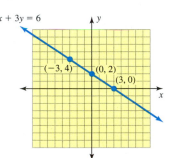

 CHECK YOURSELF 8

Graph the equation $5x + 2y = 10$. Solve for y to determine solutions.

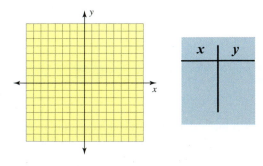

Often, in applications, we have graphs that do not fit nicely into a standard 8-by-8 grid. In these cases, we should show the portion of the graph that displays the interesting properties of the graph. We may need to scale the x- or y-axis to meet our needs or even set the axes so that it seems they do not intersect at the origin.

When we scale the axes, it is important to include numbers on the axes at convenient grid lines. If we set the axes so that they do not seem to intersect at the origin, we include a mark to indicate this. Both of these situations are illustrated in Example 9.

Example 9   Graphing in Nonstandard Windows

NOTE In business, the constant, 2500, is called the **fixed cost**. The slope, 45, is referred to as the **variable cost**.

The cost, y, to produce x CD players is given by the equation $y = 45x + 2500$. Graph the cost equation, with appropriately scaled and set axes.

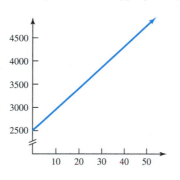

The y-intercept is (0, 2500). The slope is 45. Beginning at the intercept, we can find a second point by moving 10 units in the x direction and $10 \cdot 45$ or 450 units in the y-direction.

 CHECK YOURSELF 9 _____

Graph the cost equation given by $y = 60x + 1200$, with appropriately scaled and set axes.

CHECK YOURSELF ANSWERS _____

1.

x	y
1	−4
2	−2
3	0

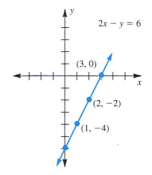

2.

x	y
0	0
1	−2
2	−4

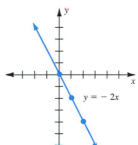

3.

x	y
0	−2
1	1
2	4

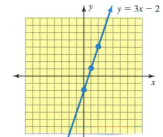

4.

x	y
−3	4
0	3
3	2

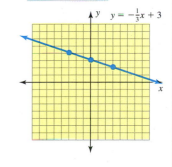

5.

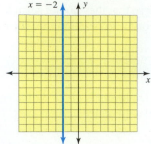

$x = -2$

6.

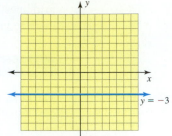

$y = -3$

7.

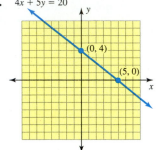

$4x + 5y = 20$

$(0, 4)$

$(5, 0)$

8.

x	y
0	5
2	0
4	-5

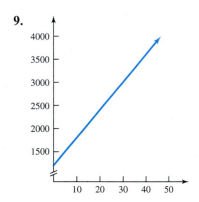

$y = -\frac{5}{2}x + 5$

9.

4000
3500
3000
2500
2000
1500

10   20   30   40   50

Graph each of the following equations.

ANSWERS

1. $x + y = 6$

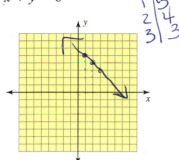

2. $x - y = 5$

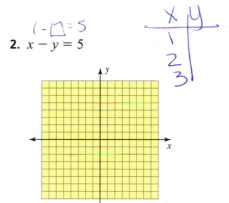

1. _____

2. _____

3. _____

4. _____

5. _____

6. _____

7. _____

8. _____

3. $x - y = -3$

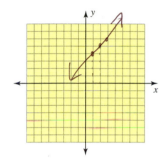

4. $x + y = -3$

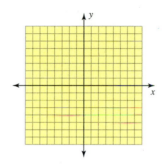

5. $2x + y = 2$

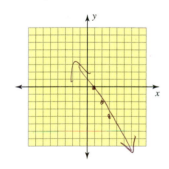

6. $x - 2y = 6$

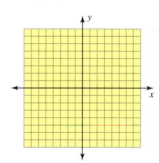

7. $3x + y = 0$

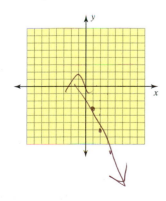

8. $3x - y = 6$

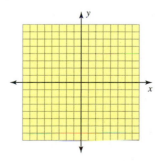

9. _____

10. _____

11. _____

12. _____

13. _____

14. _____

15. _____

16. _____

9. $x + 4y = 8$

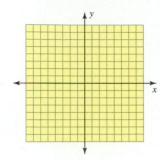

10. $2x - 3y = 6$

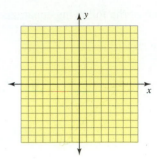

11. $y = 5x$

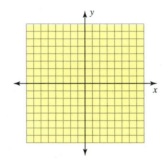

12. $y = -4x$

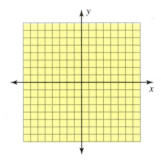

13. $y = 2x - 1$

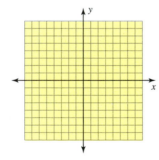

14. $y = 4x + 3$

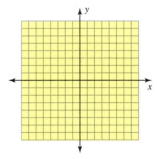

15. $y = -3x + 1$

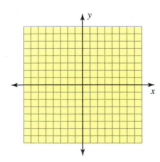

16. $y = -3x - 3$

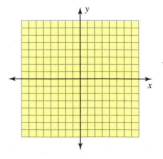

17. $y = \frac{1}{3}x$

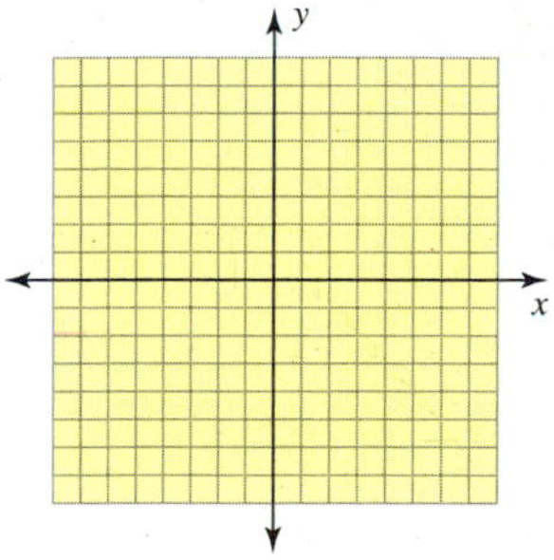

18. $y = -\frac{1}{4}x$

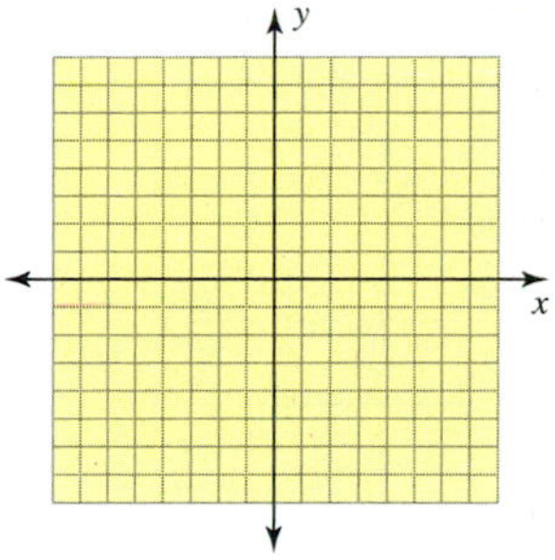

19. $y = \frac{2}{3}x - 3$

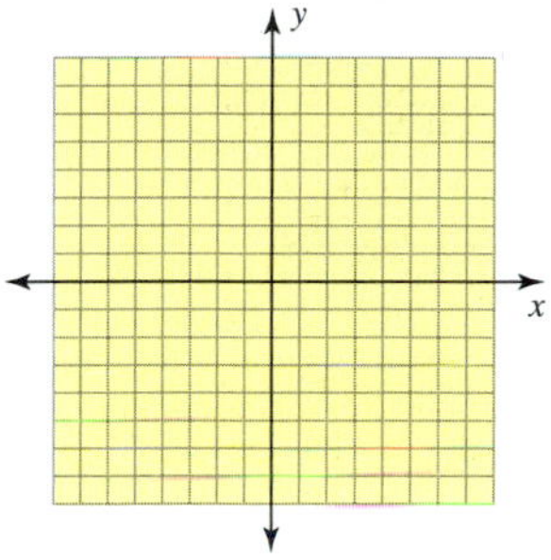

20. $y = \frac{3}{4}x + 2$

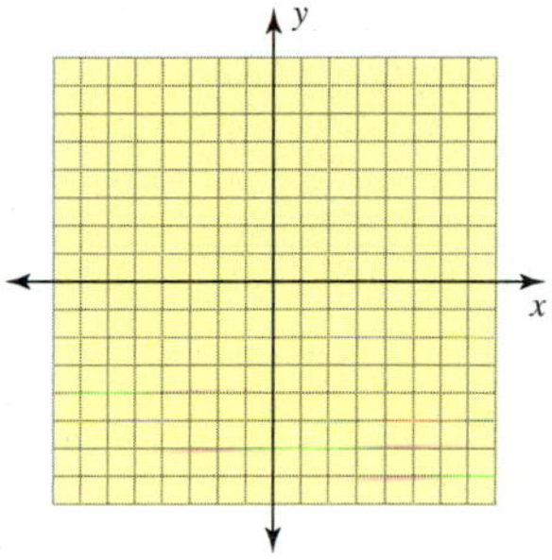

21. $x = 5$

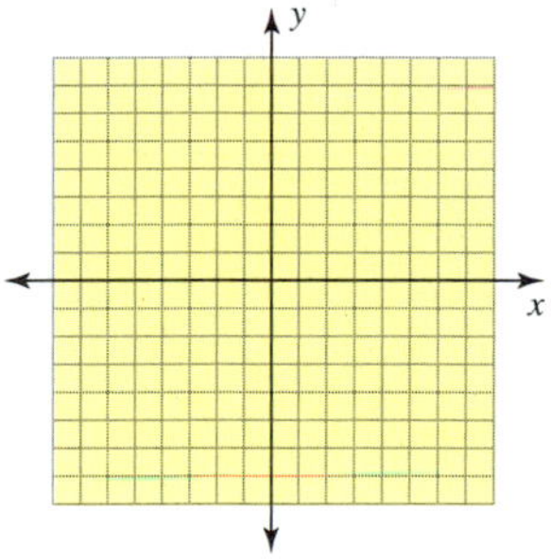

22. $y = -3$

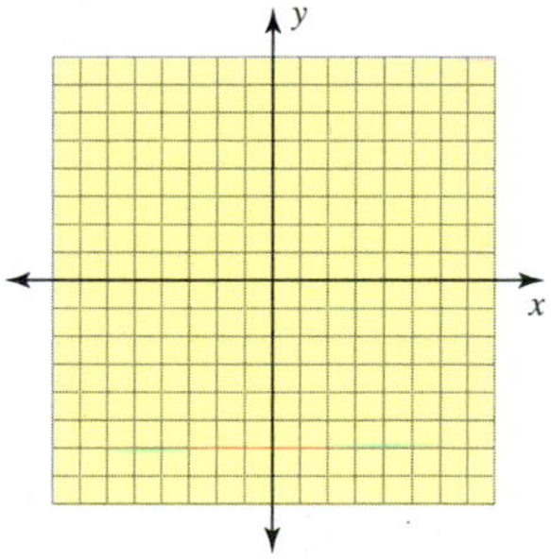

23. $y = 1$

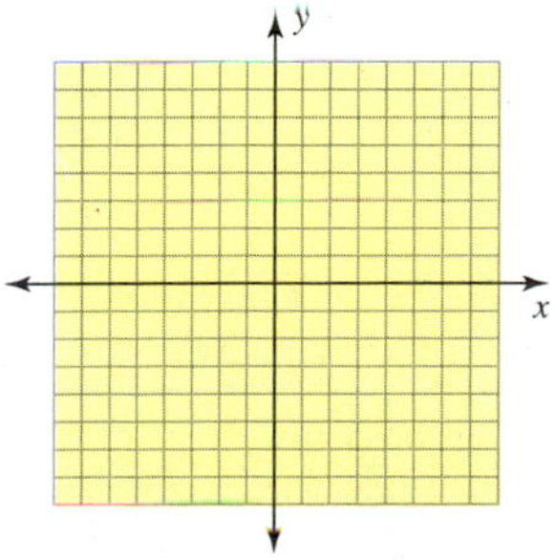

24. $x = -2$

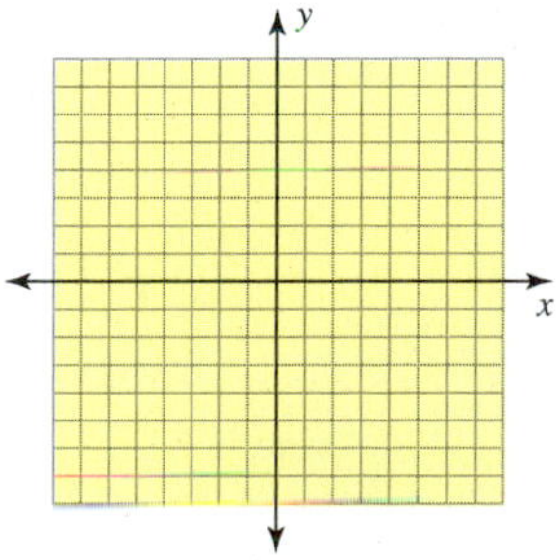

ANSWERS

17. ____________

18. ____________

19. ____________

20. ____________

21. ____________

22. ____________

23. ____________

24. ____________

Graph each of the following equations, using the intercept method.

25. $x - 2y = 4$

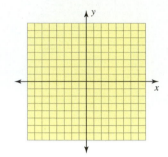

26. $6x + y = 6$

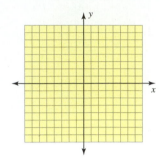

27. $5x + 2y = 10$

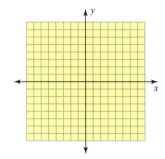

28. $2x + 3y = 6$

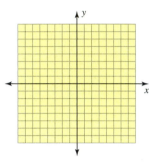

29. $3x + 5y = 15$

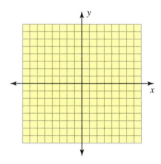

30. $4x + 3y = 12$

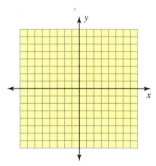

Graph each of the following equations by first solving for y.

31. $x + 3y = 6$

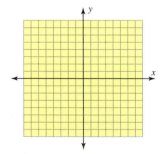

32. $x - 2y = 6$

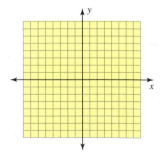

33. $3x + 4y = 12$

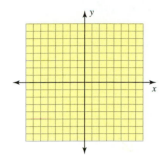

34. $2x - 3y = 12$

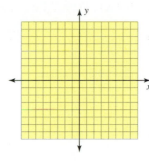

35. $5x - 4y = 20$

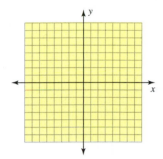

36. $7x + 3y = 21$

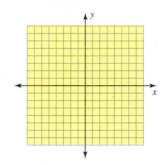

37. In each of the following statements, fill in the blank with *always, sometimes,* or *never*.

    (a) If the ordered pair (x, y) is a solution to an equation in two variables, then the point (x, y) is _____ on the graph of the equation.

    (b) If the graph of a linear equation $Ax + By = C$ passes through the origin, then C _____ equals zero.

38. In each of the following statements, fill in the blank with *always, sometimes,* or *never*.

    (a) If the ordered pair (x, y) is *not* a solution to an equation in two variables, then the point (x, y) is _____ on the graph of the equation.

    (b) The graph of a horizontal line _____ passes through the origin.

Write an equation that describes the following relationships between x and y. Then graph each relationship.

39. y is twice x.

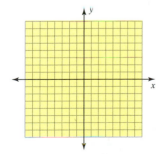

40. y is 2 less than x.

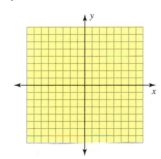

ANSWERS

33. _____

34. _____

35. _____

36. _____

37. _____

38. _____

39. _____

40. _____

41. y is 3 less than 3 times x.

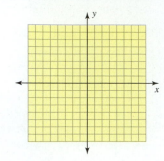

42. y is 4 more than twice x.

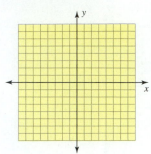

43. The difference of x and the product of 4 and y is 12.

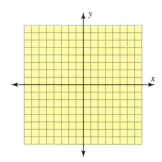

44. The difference of twice x and y is 6.

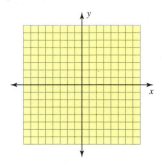

Graph each pair of equations on the same axes. Give the coordinates of the point where the lines intersect.

45. $x + y = 4$
$x - y = 2$

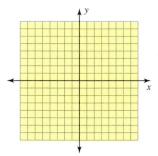

46. $x - y = 3$
$x + y = 5$

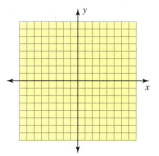

47. Science and medicine. The equation $y = 0.10x + 200$ describes the amount of winnings a group earns for collecting plastic jugs in the recycling contest described in Exercise 27 at the end of Section 6.2. Sketch the graph of the line on the given coordinate system.

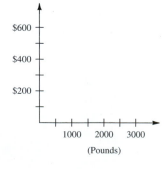

48. Science and medicine. The contest sponsor will award a prize only if the winning group in the contest collects 100 lb of jugs or more. Use your graph obtained in Exercise 47 to determine the minimum prize possible.

49. Business and finance. A high school class wants to raise some money by recycling newspapers. The class decides to rent a truck for a weekend and to collect the newspapers from homes in the neighborhood. The market price for recycled newsprint is currently $11 per ton. The equation $y = 11x - 100$ describes the amount of money the class will make, in which y is the amount of money made in dollars, x is the number of tons of newsprint collected, and 100 is the cost in dollars to rent the truck.

(a) Using the given axes, draw a graph that represents the relationship between newsprint collected and money earned.

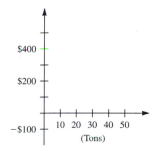

(b) The truck costs the class $100. How many tons of newspapers must the class collect to break even on this project?

(c) If the class members collect 16 tons of newsprint, how much money will they earn?

(d) Six months later the price of newsprint is $17 a ton, and the cost to rent the truck has risen to $125. Write the equation that describes the amount of money the class might make at that time.

50. Business and finance. The cost of producing a number of items x is given by $C = mx + b$, in which b is the fixed cost and m is the variable cost (the cost of producing one more item).

(a) If the fixed cost is $40 and the variable cost is $10, write the cost equation.

(b) Graph the cost equation.

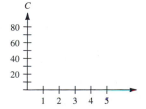

_____

51. _____

52. _____

  a. _____

  b. _____

  c. _____

  d. _____

(c) The revenue generated from the sale of x items is given by $R = 50x$. Graph the revenue equation on the same set of axes as the cost equation.

(d) How many items must be produced for the revenue to equal the cost (the break-even point)?

Graph each set of equations on the same coordinate system. Do the lines intersect? What are the y-intercepts?

51. $y = 3x$

    $y = 3x + 4$

    $y = 3x - 5$

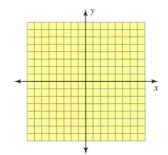

52. $y = -2x$

    $y = -2x + 3$

    $y = -2x - 5$

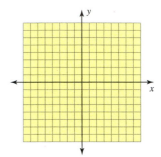

 Getting Ready for Section 6.4 [Section 1.4]

Evaluate the following expressions.

(a) $\dfrac{7 - 3}{8 - 4}$
    (b) $\dfrac{-9 - 5}{-4 - 3}$
    (c) $\dfrac{4 - (-2)}{6 - 2}$
    (d) $\dfrac{-4 - (-4)}{8 - 2}$

Answers

1. $x + y = 6$

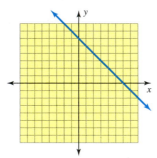

3. $x - y = -3$

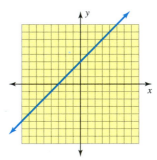

5. $2x + y = 2$

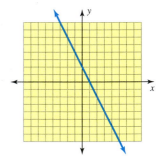

7. $3x + y = 0$

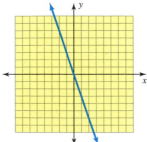

9. $x + 4y = 8$

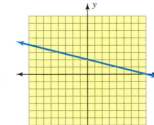

11. $y = 5x$

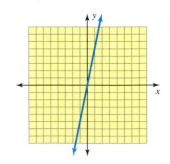

13. $y = 2x - 1$

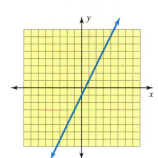

15. $y = -3x + 1$

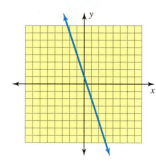

17. $y = \dfrac{1}{3}x$

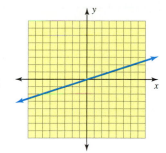

19. $y = \dfrac{2}{3}x - 3$

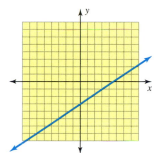

21. $x = 5$

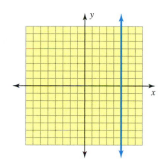

23. $y = 1$

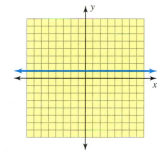

25. $x - 2y = 4$

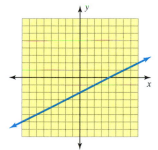

27. $5x + 2y = 10$

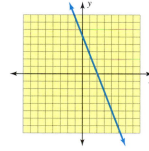

29. $3x + 5y = 15$

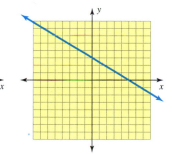

31. $y = 2 - \dfrac{x}{3}$

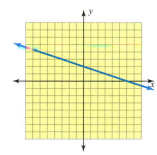

33. $y = 3 - \dfrac{3}{4}x$

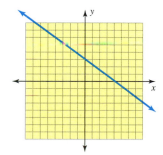

35. $y = -5 + \dfrac{5}{4}x$

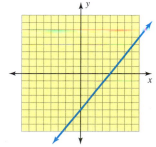

37. (a) Always; **(b)** always

39. $y = 2x$

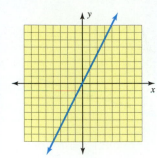

41. $y = 3x - 3$

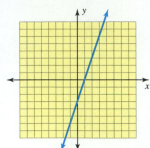

43. $x - 4y = 12$

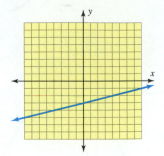

45. (3, 1)

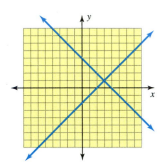

47. Graph

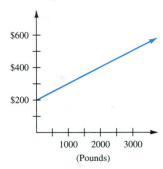

(Pounds)

49. (a) See graph below;

(b) $\dfrac{100}{11}$ or ≈ 9 tons; **(c)** \$76;

(d) $y = 17x - 125$

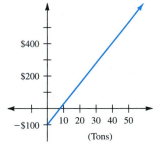

(Tons)

a. 1   **b.** 2   **c.** $\dfrac{3}{2}$ **d.** 0

51. The lines do not intersect.
The y-intercepts are (0, 0), (0, 4), and (0, −5).

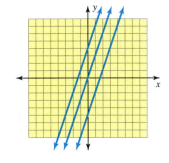

6.4 The Slope of a Line

6.4 OBJECTIVES

1. Find the slope of a line through two given points
2. Find the slopes of vertical and horizontal lines
3. Find the slope of a line from its graph
4. Write and graph the equation for a direct-variation relationship

In Section 6.3 we saw that the graph of an equation such as

RECALL An equation such as $y = 2x + 3$ is a *linear equation in two variables.* Its graph is always a straight line.

$$y = 2x + 3$$

is a straight line. In this section we want to develop an important idea related to the equation of a line and its graph, called the **slope** of a line. Finding the slope of a line gives us a numerical measure of the "steepness," or inclination, of that line.

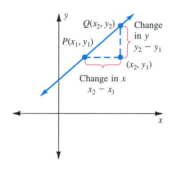

To find the slope of a line, we first let $P(x_1, y_1)$ and $Q(x_2, y_2)$ be any two distinct points on that line. The **horizontal change** (or the change in x) between the points is $x_2 - x_1$. The **vertical change** (or the change in y) between the points is $y_2 - y_1$.

NOTE x_1 is read "x sub 1," x_2 is read "x sub 2," and so on. The 1 in x_1 and the 2 in x_2 are called **subscripts.**

We call the ratio of the vertical change, $y_2 - y_1$, to the horizontal change, $x_2 - x_1$, the *slope* of the line as we move along the line from P to Q. That ratio is usually denoted by the letter m, and so we have the following formula:

NOTE The difference $x_2 - x_1$ is sometimes called the **run** between points P and Q. The difference $y_2 - y_1$ is called the **rise**. So the slope may be thought of as "rise over run."

Definition: The Slope of a Line

If $P(x_1, y_1)$ and $Q(x_2, y_2)$ are any two points on a line, then m, the slope of the line, is given by

$$m = \frac{\text{vertical change}}{\text{horizontal change}} = \frac{y_2 - y_1}{x_2 - x_1} \quad \text{when } x_2 \neq x_1$$

This definition provides exactly the numerical measure of "steepness" that we want. If a line "rises" as we move from left to right, the slope will be positive—the steeper the line, the larger the numerical value of the slope. If the line "falls" from left to right, the slope will be negative.

Let's proceed to some examples.

OBJECTIVE 1 **Example 1 Finding the Slope Given Two Points**

Find the slope of the line containing points with coordinates (1, 2) and (5, 4).

Let $P(x_1, y_1) = (1, 2)$ and $Q(x_2, y_2) = (5, 4)$. By the definition of slope, we have

$$m = \frac{y_2 - y_1}{x_2 - x_1} = \frac{(4) - (2)}{(5) - (1)} = \frac{2}{4} = \frac{1}{2}$$

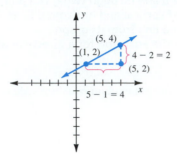

Note: We would have found the same slope if we had reversed P and Q and subtracted in the other order. In that case, $P(x_1, y_1) = (5, 4)$ and $Q(x_2, y_2) = (1, 2)$, so

$$m = \frac{(2) - (4)}{(1) - (5)} = \frac{-2}{-4} = \frac{1}{2}$$

It makes no difference which point is labeled (x_1, y_1) and which is (x_2, y_2). The resulting slope will be the same. You must simply stay with your choice once it is made and *not* reverse the order of the subtraction in your calculations.

 CHECK YOURSELF 1

Find the slope of the line containing points with coordinates (2, 3) and (5, 5).

By now you should be comfortable subtracting negative numbers. We apply that skill to finding a slope.

Example 2 Finding the Slope

Find the slope of the line containing points with the coordinates $(-1, -2)$ and $(3, 6)$.
Again, applying the definition, we have

$$m = \frac{(6) - (-2)}{(3) - (-1)} = \frac{6 + 2}{3 + 1} = \frac{8}{4} = 2$$

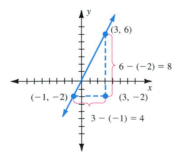

The next figure compares the slopes found in Example 1 and this example. Line l_1, from Example 1, has slope $\frac{1}{2}$. Line l_2, from this example, has slope 2. Do you see the idea of slope measuring steepness? The greater the slope, the more steeply the line is inclined upward.

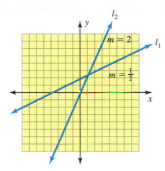

 CHECK YOURSELF 2 _____

Find the slope of the line containing points with coordinates (−1, 2) and (2, 7). Draw a sketch of this line and the line of Check Yourself 1. Compare the lines and the two slopes.

Next, we look at lines with a negative slope.

Example 3   Finding a Negative Slope

Find the slope of the line containing points with coordinates $(-2, 3)$ and $(1, -3)$.
    By the definition,

$$m = \frac{(-3) - (3)}{(1) - (-2)} = \frac{-6}{3} = -2$$

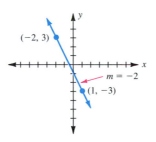

This line has a *negative* slope. The line *falls* as we move from left to right.

 CHECK YOURSELF 3 _____

Find the slope of the line containing points with coordinates (−1, 3) and (1, −3).

We have seen that lines with positive slope rise from left to right and lines with negative slope fall from left to right. What about lines with a slope of zero? A line with a slope of 0 is especially important in mathematics.

OBJECTIVE 2   **Example 4   Finding the Slope of a Horizontal Line**

Find the slope of the line containing points with coordinates $(-5, 2)$ and $(3, 2)$.
By the definition,

$$m = \frac{(2) - (2)}{(3) - (-5)} = \frac{0}{8} = 0$$

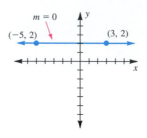

The slope of the line is 0. In fact, that will be the case for any horizontal line. Because any two points on the line have the same y-coordinate, the vertical change $y_2 - y_1$ must always be 0, and so the resulting slope is 0.

 CHECK YOURSELF 4 —————

Find the slope of the line containing points with coordinates $(-2, -4)$ and $(3, -4)$.

Because division by 0 is undefined, it is possible to have a line with an undefined slope.

Example 5   Finding the Slope of a Vertical Line

Find the slope of the line containing points with coordinates $(2, -5)$ and $(2, 5)$.
By the definition,

$$m = \frac{(5) - (-5)}{(2) - (2)} = \frac{10}{0} \qquad \text{Remember that division by 0 is undefined.}$$

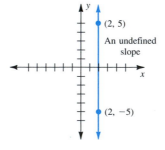

We say that the vertical line has an undefined slope. On a vertical line, any two points have the same x-coordinate. This means that the horizontal change $x_2 - x_1$ must always be 0, and because division by 0 is undefined, the slope of a vertical line will always be undefined.

CHECK YOURSELF 5

Find the slope of the line containing points with the coordinates (−3, −5) and (−3, 2).

Given the graph of a line, we can find the slope of that line. Example 6 illustrates this.

OBJECTIVE 3   **Example 6   Finding the Slope from the Graph**

Find the slope of the graphed line.

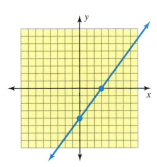

We can find the slope by identifying any two points. It is usually easiest to use the x- and y-intercepts. In this case, those intercepts are (3, 0) and (0, −4).
Using the definition of slope, we find

$$m = \frac{(0) - (-4)}{(3) - (0)} = \frac{4}{3}$$

The slope of the line is $\frac{4}{3}$.

 CHECK YOURSELF 6

Find the slope of the graphed line.

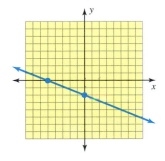

In Section 6.3, we saw that a line could be drawn from two ordered pairs. Given equations of the form $y = kx$, it is fairly easy to find two ordered pairs. In Example 7, we will use those ordered pairs to find the graph of the equation.

Lisa Marie
Jennifer Calabrese

Example 7   Graphing an Equation of the Form y = kx

(a) Find the graph of the equation $y = -2x$.

From the following table, we know that the ordered pairs $(0, 0)$ and $(1, -2)$ are solutions to the equation.

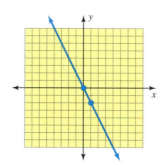

The graph is displayed here.

$\dfrac{2}{1} = -2$ $y = -2x$

Note that the slope of the line that passes through the points $(0, 0)$ and $(1, -2)$ is

$$m = \frac{(0) - (-2)}{(0) - (1)} = \frac{2}{-1} = -2$$

(b) Find the graph of the equation $y = \dfrac{1}{3}x$.

From the following table, we know that the ordered pairs $(0, 0)$ and $(3, 1)$ are solutions to the equation.

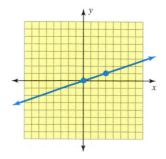

The graph is displayed here.

Note that the slope of the line that passes through the points $(0, 0)$ and $(3, 1)$ is

$$m = \frac{(0) - (1)}{(0) - (3)} = \frac{1}{3}$$

✓ CHECK YOURSELF 7

Find the graph of the equation $y = -\frac{1}{2}x$.

In Example 7, we noted that the slope of the line for the equation $y = -2x$ is -2, and the slope of the line for the equation $y = \frac{1}{3}x$ is $\frac{1}{3}$. This leads us to the following observation.

The slope of a line for an equation of the form $y = kx$ will always be k. Because k is the slope, we generally write the form as

$y = mx$

Note that $(0, 0)$ will be a solution for any equation of this form. As a result, the line for an equation of the form $y = mx$ will always pass through the origin.

The following sketch summarizes the results of our previous examples.

NOTE As the slope gets closer to 0, the line gets "flatter."

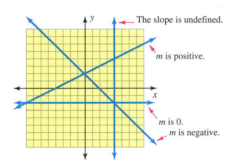

The slope is undefined.

m is positive.

m is 0.
m is negative.

Four lines are illustrated in the figure. Note that

1. The slope of a line that rises from left to right is positive.
2. The slope of a line that falls from left to right is negative.
3. The slope of a horizontal line is 0.
4. A vertical line has an undefined slope.

There are numerous applications involving lines that pass through the origin. Consider the following scenario. Pedro makes \$25 an hour as an electrician. If he works for 1 h, he makes \$25; if he works for 2 h, he makes \$50; and so on. We say his total pay **varies directly** with the number of hours worked. This type of situation occurs so frequently that we use special terminology to describe it.

Definition: Direct Variation

If *y* is a constant multiple of *x*, we write

$y = kx$ in which *k* is a constant.

We say that *y varies directly* with *x*, or that *y* is *directly proportional* to *x*. The constant *k* is called the **constant of variation.**

OBJECTIVE 4 | **Example 8 Writing an Equation for Direct Variation**

Marina earns $9 an hour as a tutor. Write the equation that describes the relationship between the number of hours she works and her pay.

Her pay (*P*) is equal to the rate of pay (*r*) times the number of hours worked (*h*), so

$$P = r \cdot h \quad \text{or} \quad P = 9h$$

 CHECK YOURSELF 8 _____

Sorina is driving at a constant rate of 50 mi/h. Write the equation that shows the distance she travels (d) in h hours.

RECALL *k* is the constant of variation.

If two things vary directly and values are given for *x* and *y*, we can find *k*. This property is illustrated in Example 9.

Example 9 Finding the Constant of Variation

If *y* varies directly with *x*, and *y* = 30 when *x* = 6, find *k*.
Because *y* varies directly with *x*, we know from the definition that

NOTE The direct variation equation is *y* = *kx*. (6, 30) is one ordered pair that satisfies the equation.

$$y = kx$$

We need to find *k*. We do this by substituting 30 for *y* and 6 for *x*.

$$(30) = k(6) \quad \text{or} \quad k = 5$$

 CHECK YOURSELF 9 _____

If y varies directly with x and y = 100 when x = 25, find the constant of variation.

The graph for a linear equation of direct variation will always pass through the origin. Example 10 illustrates this.

Example 10   Graphing an Equation of Direct Variation

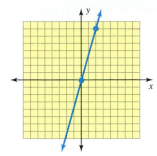

Let y vary directly with x, with a constant of variation $k = 3.5$. Graph the equation of variation.

The equation of variation is $y = 3.5x$, so the graph will have a slope of 3.5. Two points that satisfy the relationship are $(0, 0)$ and $(3.5, 7)$.

CHECK YOURSELF 10

Let y vary directly with x, with a constant of variation $k = \dfrac{7}{3}$. Graph the equation of variation.

CHECK YOURSELF ANSWERS

1. $m = \dfrac{2}{3}$ **2.** $m = \dfrac{5}{3}$

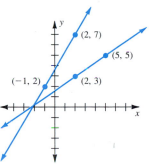

$(2, 7)$
$(5, 5)$
$(-1, 2)$ $(2, 3)$

3. $m = -3$ **4.** $m = 0$ **5.** m is undefined   **6.** $m = -\dfrac{2}{5}$

7.

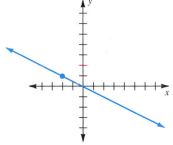

[Note: Your second point could have been $(-2, 1)$ or $(2, -1)$.]

8. $d = 50h$ **9.** $k = 4$

10.

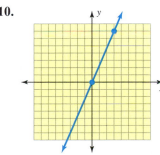

6.4 Exercises

Find the slope of the line through the following pairs of points.

1. (5, 7) and (9, 11)

2. (4, 9) and (8, 17)

3. (−2, −5) and (2, 15)

4. (−3, 2) and (0, 17)

5. (−2, 3) and (3, 7)

6. (−3, −4) and (3, −2)

7. (−3, 2) and (2, −8)

8. (−6, 1) and (2, −7)

9. (3, 3) and (5, 0)

10. (−2, 4) and (3, 1)

11. (5, −4) and (5, 2)

12. (−5, 4) and (2, 4)

13. (−4, −2) and (3, 3)

14. (−5, −3) and (−5, 2)

15. (−3, −4) and (2, −4)

16. (−5, 7) and (2, −2)

17. (−1, 7) and (2, 3)

18. (−4, −2) and (6, 4)

In Exercises 19 to 24, two points are shown. Find the slope of the line through the given points.

19.

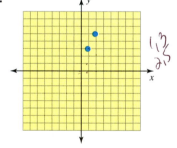

20.

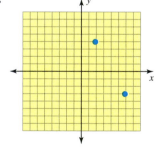

21.

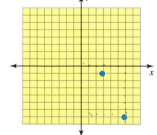

22.

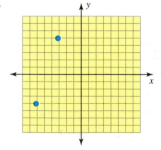

23.

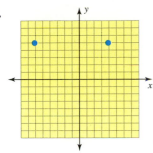

24.

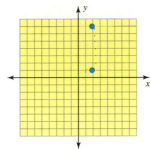

ANSWERS

23. _____

24. _____

25. _____

26. _____

27. _____

28. _____

29. _____

30. _____

In Exercises 25 to 30, find the slope of the lines graphed.

25.

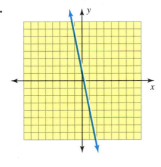

s

26.

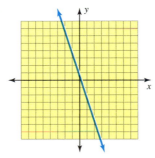

27.

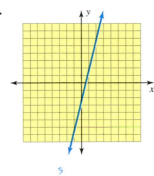

28.

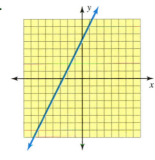

29.

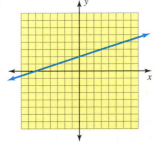

30.

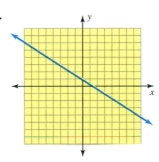

Find the graph of the following equations.

31. $y = -4x$

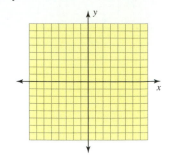

32. $y = 3x$

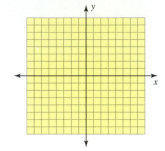

33. $y = \dfrac{2}{3}x$

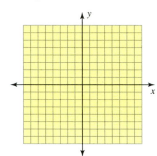

34. $y = -\dfrac{3}{4}x$

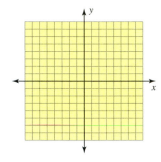

35. $y = \dfrac{5}{4}x$

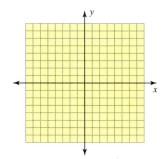

36. $y = -\dfrac{4}{5}x$

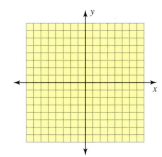

37. Consider the equation $y = 2x - 5$.

 (a) Complete the following table:

x	y
3	
4	

 (b) Use the ordered pairs found in part (a) to calculate the slope of the line.

 (c) What do you observe concerning the slope found in part (b) and the given equation?

38. Repeat exercise 37 for $y = \dfrac{3}{2}x + 5$ and

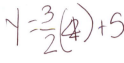

x	y
2	16
4	11

$y = \dfrac{3}{2}(4) + 5$ $\dfrac{rise}{run}$ $\dfrac{3}{2}$

$\dfrac{16-11}{4-2}$ $\dfrac{5}{2}$

$\cdot 3 - 3$

$\cdot 1 - (-1)$

$\dfrac{-6}{2}$

$(\cdot 3)$

_____

39. Repeat exercise 37 for $y = -\dfrac{1}{3}x + 2$ and

x	y
3	
6	

_____

40. Repeat exercise 37 for $y = -4x - 6$ and

x	y
-1	
-3	

_____

41. Consider the equation: $y = 2x + 3$.

(a) Complete the following table of values.

Point	x	y
A	5	
B	6	
C	7	
D	8	
E	9	

_____

(b) As the x-coordinate changes by 1 (for example, as you move from point A to point B), how much do the corresponding y-coordinates change?

(c) Is your answer to part (b) the same if you move from B to C? from C to D? from D to E?

(d) Describe the "growth rate" of the line using these observations. Complete the following statement: When the x-value grows by 1 unit, the y-value _____.

_____

42. Repeat Exercise 41 using $y = 2x + 5$.

_____

43. Repeat Exercise 41 using $y = -4x + 50$.

44. Repeat Exercise 41 using $y = -4x + 40$.

45. _____

46. _____

47. _____

48. _____

49. _____

50. _____

51. _____

52. _____

53. _____

54. _____

In the following exercises, (a) plot the given point; (b) using the given slope, move from the point plotted in (a) to plot a new point; (c) draw the line that passes through the points plotted in (a) and (b).

45. $(3, 1)$, $m = 2$

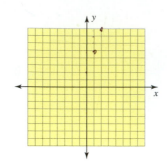

46. $(-1, 4)$, $m = -2$

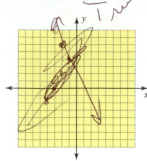

47. $(-2, -1)$, $m = -4$

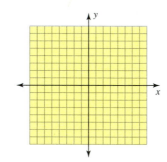

48. $(-3, 5)$, $m = 2$

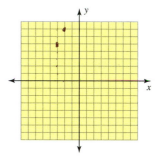

49. Business and finance. Robin earns $12 per hour. Write an equation that shows how much she makes (S) in h hours.

50. Business and finance. Kwang earns $11.50 per hour. Write an equation that shows how much he earns (S) in h hours.

In exercises 51 to 54, find the constant of variation k.

51. y varies directly with x; $y = 54$ when $x = 6$.

52. m varies directly with n; $m = 144$ when $n = 8$.

53. y varies directly with x; $y = 2100$ when $x = 600$.

54. y varies directly with x; $y = 400$ when $x = 1000$.

In Exercises 55 to 60, y varies directly with x and the value of k is given. Graph the equation of variation.

ANSWERS

55. _____

56. _____

57. _____

58. _____

59. _____

60. _____

61. _____

55. $k = 2$

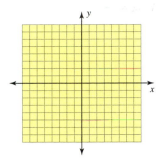

56. $k = 4$

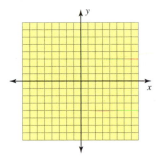

57. $k = 2.5$

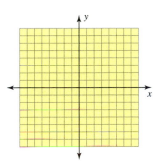

58. $k = \dfrac{11}{5}$

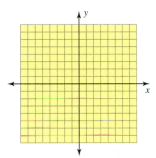

59. $k = 100$

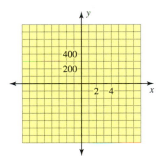

60. $k = 300$

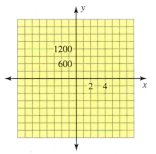

61. Business and finance. At a factory that makes grinding wheels, Kalila makes $0.20 for each wheel completed. Sketch the equation of direct variation.

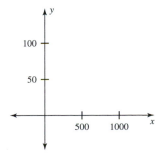

62. _____

63. _____

64. _____

a. _____

b. _____

c. _____

d. _____

e. _____

f. _____

62. Business and finance. Palmer makes $1.25 per page for each page that he types. Sketch the equation of direct variation.

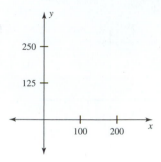

63. Business and finance. Josephine works part-time in a local video store. Her salary varies directly as the number of hours worked. Last week she earned $43.20 for working 8 hours. This week she earned $118.80. How many hours did she work this week?

64. Business and finance. The revenue for a sandwich shop is directly proportional to its advertising budget. When the owner spent $2000 a month on advertising, the revenue was $120,000. If the revenue is now $180,000, how much is the owner spending on advertising?

 Getting Ready for Section 6.5 [Section 0.3]

Write each number as a percent.

(a) 0.06

(b) 0.375

(c) 2.4

(d) $\dfrac{43}{100}$

(e) $\dfrac{2}{5}$

(f) $2\dfrac{2}{3}$

Answers

1. 1   **3.** 5   **5.** $\dfrac{4}{5}$ **7.** -2 **9.** $-\dfrac{3}{2}$ **11.** Undefined   **13.** $\dfrac{5}{7}$

15. 0   **17.** $-\dfrac{4}{3}$ **19.** 2   **21.** -2 **23.** 0   **25.** 4   **27.** -5 **29.** $\dfrac{1}{3}$

31. $y = -4x$

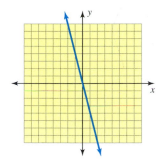

33. $y = \dfrac{2}{3}x$

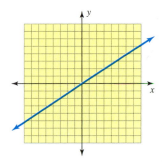

35. $y = \dfrac{5}{4}x$

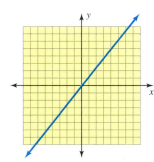

37. **(a)** (3, 1), (4, 3); **(b)** 2; **(c)** Slope equals coefficient of x, 0

39. **(a)** (3, 1), (6, 0); **(b)** $-\dfrac{1}{3}$; **(c)** slope equals coefficient of x

41. **(a)** (5, 13), (6, 15), (7, 17), (8, 19), (9, 21); **(b)** 2; **(c)** yes; **(d)** increases by 2

43. **(a)** (5, 30), (6, 26), (7, 22), (8, 18), (9, 14); **(b)** 4; **(c)** yes; **(d)** decreases by 4

45.

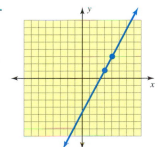

47.

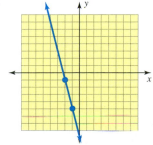

49. $S = 12h$ **51.** 9   **53.** 3.5

55.
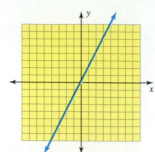

57.
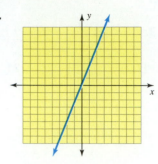

59.
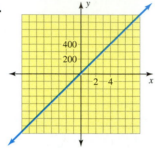

61.
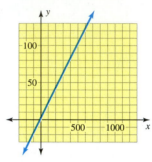

63. 22 h   **a.** 6%   **b.** 37.5%   **c.** 240%   **d.** 43%   **e.** 40%

f. $266\frac{2}{3}\%$

6.5 Reading Graphs*

6.5 OBJECTIVES

1. Read and interpret a table
2. Read and interpret a pie chart
3. Read different types of bar graphs
4. Read and interpret a line graph
5. Create and make a prediction from a line graph

NOTE Spreadsheets and databases are commonly used software tools for working with tables of data.

A **table** is a display of information in parallel rows or columns. Tables can be used anywhere that information is to be summarized.

Following is a table describing land area and world population. Each entry in the table is sometimes called a **cell.** This table will be used for Examples 1 and 2.

NOTE Antarctica has been omitted from this table.

Continent or Region	Land Area (1000 mi^2)	Population (in millions)		
		1900	1950	2000
North America	9,400	106	221	305
South America	6,900	38	111	515
Europe	3,800	400	392	510
Asia (including Russia)	17,400	932	1,591	4,178
Africa	11,700	118	229	889
Oceana (including Australia)	3,300	6	12	32
World totals	52,500	1,600	2,556	6,429

Source: Bureau of the Census; U.S. Dept. of Commerce.

OBJECTIVE 1

Example 1   Reading a Table

From the land area and world population table given previously, find each of the following.

(a) What was the population of Africa in 1950?

Continent or Region	Land Area (1000 mi^2)	Population (in millions)		
		1900	1950	2000
North America	9,400	106	221	305
South America	6,900	38	111	515
Europe	3,800	400	392	510
Asia (including Russia)	17,400	932	1,591	4,178
Africa	11,700	118	229	889
Oceana (including Australia)	3,300	6	12	32
World totals	52,500	1,600	2,556	6,429

Source: Bureau of the Census; U.S. Dept. of Commerce.

Looking at the cell in the row labeled Africa and the column labeled 1950, we find the number 229. Because we are told the population is given in millions, we conclude that the population in Africa in 1950 was 229,000,000.

* This section is included for those instructors who have it in their curricula. It is optional in the sense that it is not required for the remainder of the text.

(b) What is the land area of Asia, in square miles?

The cell in the row Asia and column Land Area reads 17,400. The land area is given in 1000 mi^2 units, so the actual land area of Asia is 17,400,000 mi^2.

CHECK YOURSELF 1

Use the land area and world population table to answer each question.

(a) What was the population of South America in 1900?
(b) What is the land area of Europe?

We frequently can use a table to find information not explicitly given in the table. We will use the land area and world population table again to illustrate this point.

Example 2 Interpreting a Table

Use the land area and world population table to answer each of the following questions.

(a) By what percentage did the population of Africa change between 1950 and 2000?

In Example 1, we found that the 1950 population of Africa was 229 million. The 2000 population of Africa was 889 million, or 660 million more people than in 1950. The percentage change is

$$\frac{660,000,000}{229,000,000} \approx 2.882 = 288.2\%$$

So, the population of Africa increased by about 288.2% between 1950 and 2000.

(b) What is the land area of Asia as a percentage of the Earth's total land area (excluding Antarctica)?

In Example 1 we concluded that Asia is 17,400,000 mi^2. We find that there are 52,500,000 mi^2 of land area (excluding Antarctica) from the row labeled World Totals. Therefore, the percentage of the Earth's total area made up by Asia is

$$\frac{17,400,000}{52,500,000} \approx 33.1\%$$

(c) What was the density of the population (people per square mile) in North America in 2000?

North America had a population of 305 million in 2000. With a land area of 9,400,000 mi², the population density of North America is given by

$$\frac{305,000,000}{9,400,000} \approx 32.4 \text{ people per square mile}$$

 CHECK YOURSELF 2

Use the land area and world population table to answer each of the following questions.

(a) By what percentage did the population of South America change between 1900 and 1950?
(b) What is the land area of Europe as a percentage of the Earth's total land area (excluding Antarctica)?
(c) Did the world population increase by a greater percentage between 1900 and 1950 or between 1950 and 2000?

If we need to present the relationship between two sets of data in only a column or two of a table, it is often easier to use pictures. A **graph** is a diagram that represents the connection between two or more quantities.

NOTE Pie charts are sometimes called *circle graphs*.

The first graph we will look at is called a **pie chart.** We use a circle to represent some total that interests us. Wedges (or sectors) are drawn in the circle to show how much of the whole each part makes up.

OBJECTIVE 2   **Example 3   Reading a Pie Chart**

This pie chart represents the results of a survey that asked students how they get to school most often.

NOTE The total of the percentages of the wedges is always 100%.

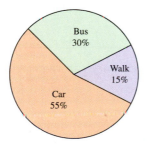

(a) What percentage of the students walk to school?

We see that 15% walk to school.

(b) What percentage of the students do not arrive by car?

Because 55% arrive by car, there are 100% − 55%, or 45%, who do not.

CHECK YOURSELF 3 _____

This pie chart represents the results of a survey that asked students whether they bought lunch, brought it, or skipped lunch altogether.

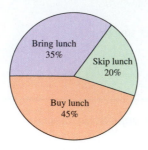

(a) What percentage of the students skipped lunch?

(b) What percentage of the students did not buy lunch?

If we know what the whole pie represents, we can also find out more about what each wedge represents, as illustrated by Example 4.

Example 4   Interpreting a Pie Chart

This pie chart shows how Sarah spent her $12,000 college scholarship.

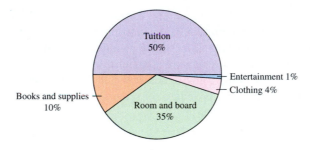

(a) How much did she spend on tuition?

50% of her $12,000 scholarship, or $6000.

(b) How much did she spend on clothing and entertainment?

Together, 5% of the money was spent on clothing and entertainment, and 0.05 × 12,000 = 600. Therefore, $600 was spent on clothing and entertainment.

CHECK YOURSELF 4 _____

This pie chart shows how Rebecca spends an average 24-hour school day.

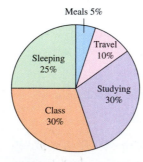

(a) How many hours does she spend sleeping each day?

(b) How many hours does she spend altogether studying and in class?

A **bar graph** provides yet another way to present information. It shows the relationship between two sets of data.

OBJECTIVE 3   **Example 5   Reading a Bar Graph**

The following bar graph represents the number of cars sold in the United States in each year listed (sales are in thousands).

U. S. Car Sales

NOTE Remember that sales numbers are in thousands.

(a) How many cars were sold in the United States in 1990?

We frequently have to estimate our answer when reading a bar graph. In this case, there were approximately 9,500,000 cars sold in 1990.

(b) In which year listed did the most car sales occur?

The tallest bar occurs in 1985; therefore, more cars were sold in 1985 than in any other year represented on the graph.

CHECK YOURSELF 5 _____

The following bar graph represents the response to a 1995 Gallup poll that asked people what their favorite spectator sport was.

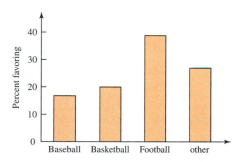

(a) Find the percentage of people for whom football is their favorite spectator sport.

(b) Of the three major sports listed, which was named as a favorite by the fewest people?

When we use bar graphs to display additional information, we often use different colors for different bars. With such graphs, it is necessary to include a legend. A **legend** is a key describing what each color or shade of bar represents.

Example 6   Reading a Bar Graph

The following bar graph compares the number of cars sold in the world and in the United States in each year listed (sales are in thousands).

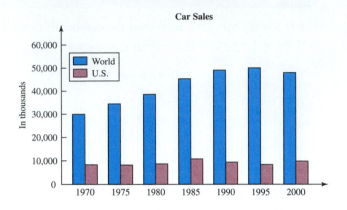

(a) How many cars were sold in the world in 1985? In the United States in 1985?

The legend tells us that the blue bar represents worldwide sales and the maroon bar represents U.S. sales.

It would appear that there were about 45,000,000 cars sold in the world in 1985. Approximately 11,000,000 of these were sold in the United States.

(b) What percentage of global sales did the United States account for in 1985?

$$\frac{11}{45} \approx 0.244 = 24.4\%$$

CHECK YOURSELF 6

The following bar graph represents the average student age at Berndt Community College.

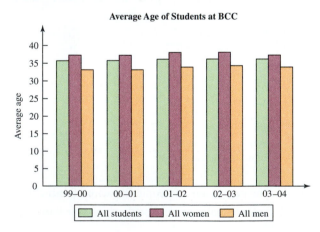

(a) What was the average age of female students in 2003–2004?

(b) Did the average age of female students increase or decrease between 2002–2003 and 2003–2004?

(c) Who tends to be older, male students or female students?

Another useful type of graph is called a **line graph.** In a line graph, one set of data is usually related to time. Line graphs give us a way of visualizing changes to something over time.

OBJECTIVE 4 | **Example 7   Reading and Interpreting a Line Graph**

The following line graph represents the number of social security beneficiaries in each of the years listed.

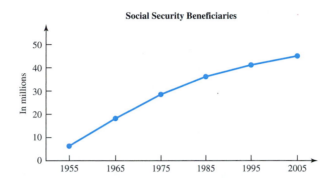

(a) How many social security beneficiaries were there in 1975?

We find the point on the line graph that lies above 1975. We then determine that this point lies at about 28 on the vertical axis.

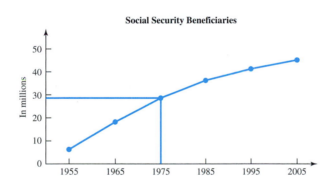

Therefore, we conclude that there were approximately 28 million social security beneficiaries in 1975.

(b) How many beneficiaries were there in 1990?

The line connecting 1985 and 1995 passes through 1990 at about 39 million.

(c) What was the mean number of annual beneficiaries between 1955 and 2005?

RECALL The mean is computed by adding all of the values in a set and dividing by the number of values in the set.

We estimate the number of beneficiaries in each of the years 1955, 1965, 1975, 1985, and 2005 to be 6 million, 18 million, 28 million, 36 million, 41 million, and 45 million respectively.

Therefore, we calculate the mean number of beneficiaries as

$$\frac{6 + 18 + 28 + 36 + 41 + 45}{6} = \frac{174}{6} = 29$$

So, the mean number of annual beneficiaries is 29 million.

✓ CHECK YOURSELF 7

The following graph indicates the high temperatures in Baltimore, Maryland, for a week in September.

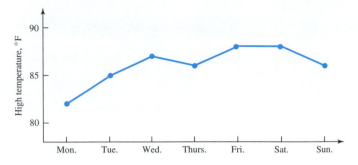

(a) What was the high temperature on Friday?

(b) Find the mean high temperature for that week.

We use many of the techniques learned in Section 6.2 to create a line graph. We start with the rectangular coordinate system. Generally, we make the *x*-axis represent the time quantity and scale the *y*-axis accordingly.

For each time given, we plot the point corresponding to whatever it is we are measuring. Finally, we "connect the dots." That is, we draw a line segment from each point to the next point (immediately to the right).

OBJECTIVE 5   **Example 8**   **Creating a Line Graph**

Use the following data to create a line graph.

FBI: Larceny-Theft Cases (in hundred thousands)	
Year	**Thefts**
1985	69
1990	79
1995	80
2000	70

Source: Bureau of Justice Statistics.

We set and scale our axes appropriately (and label them). Then we plot the points given in the table, with *x* given by the year, and *y* given by the number of thefts.

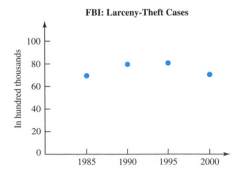

Finally, we connect the dots.

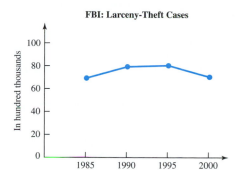

FBI: Larceny-Theft Cases

CHECK YOURSELF 8

Create a line graph based on the following table of data that describes the cost of a first-class postage stamp by year.

First-Class Postage Stamps	
Year	**Cost (in cents)**
1960	4
1965	5
1970	6
1975	10
1980	15
1985	22
1990	25
1995	32
2000	37

An important feature of line graphs is that they allow us to (cautiously) make predictions beyond the given data. To do this, we extend the sketched curve as needed.

Example 9   Making a Prediction from a Line Graph

Use the line graph in Example 8 to predict the number of larceny-theft cases reported to the FBI in 2005.

We extend the line segment connecting the points at 1995 and 2000 to 2005.

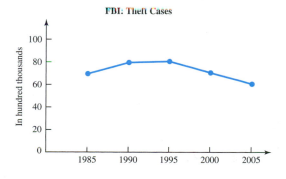

FBI: Theft Cases

We predict 60 hundred thousand, or 6,000,000, larceny-theft cases being reported to the FBI in 2005.

CHECK YOURSELF 9

Use the postage stamp line graph created in Check Yourself 8 to predict the cost of a postage stamp in the year 2005.

You should not use a line graph to predict occurrences too far out of range of the original data (at least not if you are looking for an accurate prediction). To illustrate this, consider the following.

Between the ages of 3 and 10 months most kittens gain about 1 lb per month. Therefore, if a kitten weighs 2 lb at 3 months, it would be reasonable to predict that the kitten will weigh 9 lb at 10 months. We could even guess that the kitten might weigh around 11 lb after 1 year.

But, if we extrapolate further, we would then predict that this same kitten will weigh 59 lb when it is 5 years old. This is, of course, ridiculous. The problem is that our initial data describes only a small segment (within the first year). When we try to move too far beyond the original data, we run into difficulties because the original data no longer applies to our prediction.

CHECK YOURSELF ANSWERS

1. **(a)** 38,000,000; **(b)** 3,800,000 mi^2 2. **(a)** 192.1%; **(b)** 7.2%;
(c) 1950–2000 (152% vs. 60%)   3. **(a)** 20%; **(b)** 55%   4. **(a)** 6 h; **(b)** 14.4 h
5. **(a)** 38%; **(b)** baseball   6. **(a)** 37 years; **(b)** it decreased; **(c)** female students
7. **(a)** 88°F; **(b)** 86°F

8.
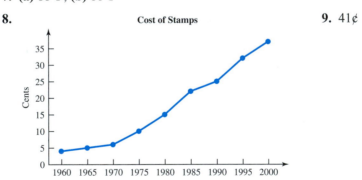
Cost of Stamps

9. 41¢

Name _____

Section _____ Date _____

Use the following table for exercises 1 to 8.

World Motor Vehicle Production, 1950–1997

Year	United States	Canada	Europe	Japan	Other	World Total
			Production (in thousands)			
1997	12,119	2,571	17,773	10,975	10,024	53,463
1996	11,799	2,397	17,550	10,346	9,241	51,332
1995	11,985	2,408	17,045	10,196	8,349	49,983
1994	12,263	2,321	16,195	10,554	8,167	49,500
1993	10,898	2,246	15,208	11,228	7,205	46,785
1992	9,729	1,961	17,628	12,499	6,269	48,088
1991	8,811	1,888	17,804	13,245	5,180	46,928
1990	9,783	1,928	18,866	13,487	4,496	48,554
1985	11,653	1,933	16,113	12,271	2,939	44,909
1980	8,010	1,324	15,496	11,043	2,692	38,565
1970	8,284	1,160	13,049	5,289	1,637	29,419
1960	7,905	398	6,837	482	866	16,488
1950	8,006	388	1,991	32	160	10,577

Note: As far as can be determined, production refers to vehicles locally manufactured.
Source: American Automobile Manufacturers Assn.

ANSWERS

1. _____

2. _____

3. _____

4. _____

5. _____

6. _____

7. _____

8. _____

1. What was the motor vehicle production in Japan in 1950? 1997?

2. What was the motor vehicle production in countries outside the United States in 1950? 1997?

3. What was the percentage increase in motor vehicle production in the United States from 1950 to 1997?

4. What was the percentage increase in motor vehicle production in countries outside the United States from 1950 to 1997?

5. What percentage of world motor vehicle production occurred in Japan in 1997?

6. What percentage of world motor vehicle production occurred in the United States in 1997?

7. What percentage of world motor vehicle production occurred outside the United States and Japan in 1997?

8. Between 1950 and 1997, did the production of motor vehicles increase by a greater percentage in Canada or Europe?

9. _____

10. _____

11. _____

12. _____

13. _____

14. _____

15. _____

16. _____

The following pie chart represents the way a new company ships its goods. Use the information presented to complete exercises 9 to 12.

9. What percentage was shipped by air freight?

10. What percentage was shipped by truck?

11. What percentage was shipped by truck or second-day air freight?

12. If the company shipped a total of 550 items last month, how many were shipped using second-day air freight?

In exercises 13 to 18, use the given bar graph representing the number of bankruptcy filings during a recent 5-year period.

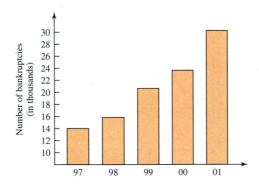

13. How many people filed for bankruptcy in 1998?

14. How many people filed for bankruptcy in 2001?

15. What was the increase in filings from 1999 to 2001?

16. What was the increase in filings from 1997 to 2001?

17. Which year had the greatest increase in filings?

18. In which year did the greatest percent increase in filings occur?

Use the line graph showing ticket sales for the last 6 months of the year to answer exercises 19 and 20.

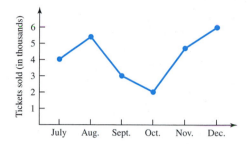

19. What month had the greatest number of ticket sales?

20. Between what two months did the greatest decrease in ticket sales occur?

Use the following table to complete exercises 21 and 22.

Year	Population of United States (in millions)
1950	151
1960	179
1970	203
1980	227
1990	249
2000	281

21. Create a line graph to present the data given in the table.

22. Use the line graph to predict the U.S. population in the year 2010.

ANSWERS

17. _____

18. _____

19. _____

20. _____

21. _____

22. _____

Answers

1. 32,000; 10,975,000 **3.** 51.4% **5.** 20.5% **7.** 56.8% **9.** 55%
11. 85% **13.** 16,000 **15.** 9000 **17.** 2001 **19.** December
21.

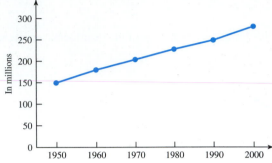

6 Summary

DEFINITION/PROCEDURE	EXAMPLE	REFERENCE
Solutions of Equations in Two Variables		**Section 6.1**
Solutions of Equations A pair of values that satisfies the equation. Solutions for equations in two variables are written as *ordered pairs*. An ordered pair has the form (x, y) *x*-coordinate *y*-coordinate	If $2x - y = 10$, $(6, 2)$ is a solution for the equation, because substituting 6 for x and 2 for y gives a true statement.	p. 504
The Rectangular Coordinate System		**Section 6.2**
The Rectangular Coordinate System A system formed by two perpendicular axes that intersect at a point called the **origin**. The horizontal line is called the **x-axis.** The vertical line is called the **y-axis.** *Graphing Points from Ordered Pairs* The coordinates of an ordered pair allow you to associate a point in the plane with the ordered pair. To graph a point in the plane, **1.** Start at the origin. **2.** Move right or left according to the value of the x-coordinate: to the right if x is positive or to the left if x is negative. **3.** Then move up or down according to the value of the y-coordinate: up if y is positive and down if y is negative.	To graph the point corresponding to $(2, 3)$: 	p. 516
Graphing Linear Equations		**Section 6.3**
Linear Equation An equation that can be written in the form $$Ax + By = C$$ in which A and B are not both 0.	$2x - 3y = 4$ is a linear equation.	p. 527
Graphing Linear Equations **1.** Find at least three solutions for the equation, and put your results in tabular form. **2.** Graph the solutions found in step 1. **3.** Draw a straight line through the points determined in step 2 to form the graph of the equation.		p. 527

Continued

The Slope of a Line		Section 6.4
Slope The slope of a line gives a numerical measure of the steepness of the line. The slope m of a line containing the distinct points in the plane $P(x_1, y_1)$ and $Q(x_2, y_2)$ is given by $$m = \frac{\text{vert change}}{\text{horiz change}} = \frac{y_2 - y_1}{x_2 - x_1} \quad \text{when } x_2 \neq x_1$$	To find the slope of the line through $(-2, -3)$ and $(4, 6)$: $$m = \frac{(6) - (-3)}{(4) - (-2)}$$ $$= \frac{6 + 3}{4 + 2} = \frac{9}{6} = \frac{3}{2}$$	*p. 549*
The Graph of y = mx A line passing through the origin with slope m.	To graph $y = -4x$: **1.** Plot $(0, 0)$ **2.** $\dfrac{\text{Rise}}{\text{Run}} = -4 = \dfrac{-4}{1}$	*p. 555*
Direct Variation If y is a constant multiple of x, we write $$y = kx$$ and say y varies directly as x	If $y = 20$ when $x = 5$, and y varies directly as x, $$(20) = k(5)$$ $$k = 4 \quad y = 4x$$	*p. 555*

Reading Graphs		Section 6.5
Tables A **table** is a rectangular display of information or data.	**(Federal Highway Admin., U.S. Dept. of Transportation)** Year — Cars Pictured 1980 — 12 1/6 cars 1985 — 13 1/6 cars 1990 — 14 1/3 cars 1995 — 13 2/3 cars 2000 — 13 cars	*p. 567*
Graphs **Graph.** A diagram that relates two different pieces of information. **Pie chart.** A graph that shows the component parts of a whole. **Bar graph.** One of the most common types of graph. It relates the amounts of items to each other. **Line graph.** A graph in which one of the axes is usually related to time. We can use line graphs to make **predictions** about events.		*pp. 569–573*

Summary Exercises

This summary exercise set is provided to give you practice with each of the objectives of this chapter. Each exercise is keyed to the appropriate chapter section. When you are finished, you can check your answers to the odd-numbered exercises against those presented in the back of the text. If you have difficulty with any of these questions, go back and reread the examples from that section. Your instructor will give you guidelines on how to best use these exercises in your instructional setting.

[6.1] Tell whether the number shown in parentheses is a solution for the given equation.

1. $7x + 2 = 16$ (2)

2. $5x - 8 = 3x + 2$ (4)

3. $7x - 2 = 2x + 8$ (2)

4. $4x + 3 = 2x - 11$ (-7)

5. $x + 5 + 3x = 2 + x + 23$ (6)

6. $\frac{2}{3}x - 2 = 10$ (21)

Determine which of the ordered pairs are solutions for the given equations.

7. $x - y = 6$ $(6, 0), (3, 3), (3, -3), (0, -6)$

8. $2x + y = 8$ $(4, 0), (2, 2), (2, 4), (4, 2)$

9. $2x + 3y = 6$ $(3, 0), (6, 2), (-3, 4), (0, 2)$

10. $2x - 5y = 10$ $(5, 0), \left(\frac{5}{2}, -1\right), \left(2, \frac{2}{5}\right), (0, -2)$

Complete the ordered pairs so that each is a solution for the given equation.

11. $x + y = 8$ $(4, \ ), (\ , 8), (8, \ ), (6, \ )$

12. $x - 2y = 10$ $(0, \ ), (12, \ ), (\ , -2), (8, \ )$

13. $2x + 3y = 6$ $(3, \ ), (6, \ ), (\ , -4), (-3, \ )$

14. $y = 3x + 4$ $(2, \ ), (\ , 7), \left(\frac{1}{3}, \ \right), \left(\frac{4}{3}, \ \right)$

Find four solutions for each of the following equations.

15. $x + y = 10$

16. $2x + y = 8$

17. $2x - 3y = 6$

18. $y = -\frac{3}{2}x + 2$

583

[6.2]   Give the coordinates of the points in the following graph.

19. *A*

20. *B*

21. *E*

22. *F*

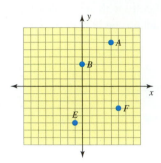

Plot points with the coordinates shown.

23. $P(6, 0)$

24. $Q(5, 4)$

25. $T(-2, 4)$

26. $U(4, -2)$

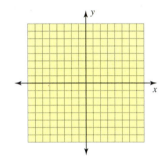

[6.3]   Graph each of the following equations.

27. $x + y = 5$

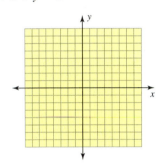

28. $x - y = 6$

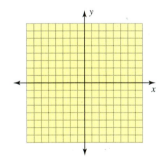

29. $y = 2x$

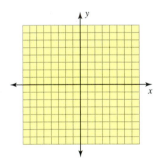

30. $y = -3x$

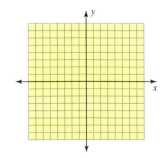

31. $y = \dfrac{3}{2}x$

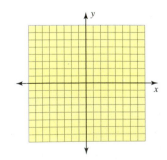

32. $y = 3x + 2$

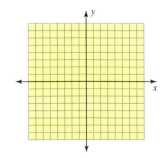

33. $y = 2x - 3$

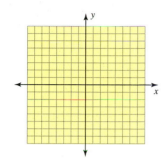

34. $y = -3x + 4$

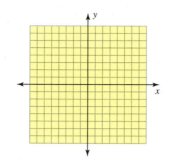

35. $y = \dfrac{2}{3}x + 2$

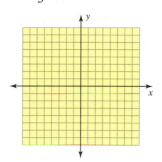

36. $3x - y = 3$

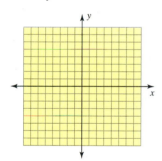

37. $2x + y = 6$

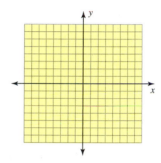

38. $3x + 2y = 12$

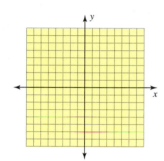

39. $3x - 4y = 12$

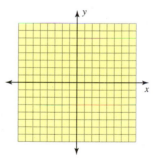

40. $x = 3$

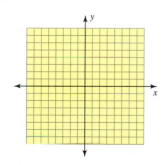

41. $y = -2$

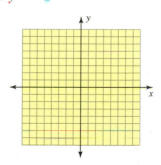

Graph each of the following equations.

42. $5x - 3y = 15$

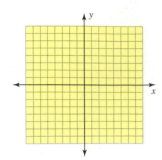

43. $4x + 3y = 12$

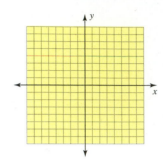

Graph each equation by first solving for y.

44. $2x + y = 6$

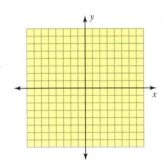

45. $3x + 2y = 6$

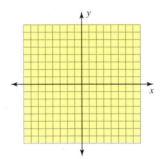

[6.4]   Find the slope of the line through each of the following pairs of points.

46. $(3, 4)$ and $(5, 8)$

47. $(-2, 3)$ and $(1, -6)$

48. $(-2, 5)$ and $(2, 3)$

49. $(-5, -2)$ and $(1, 2)$

50. $(-2, 6)$ and $(5, 6)$

51. $(-3, 2)$ and $(-1, -3)$

52. $(-3, -6)$ and $(5, -2)$

53. $(-6, -2)$ and $(-6, 3)$

In exercises 54 to 57, find the slope of the lines graphed.

54.

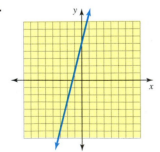

55.

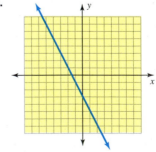

56.

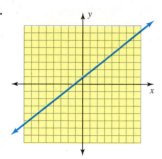

57.

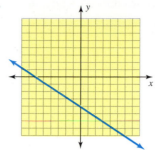

Graph each of the following equations.

58. $y = 6x$

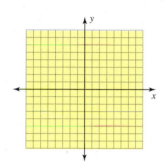

59. $y = -6x$

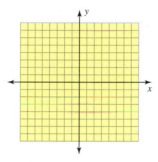

60. $y = \dfrac{2}{5}x$

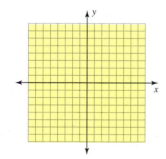

61. $y = -\dfrac{3}{4}x$

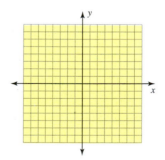

Solve for k, the constant of variation.

62. y varies directly as x; $y = 20$ when $x = 40$

63. y varies directly as x; $y = 5$ when $x = 3$

In Exercises 64 and 65, y varies directly with x and the value of k is given. Graph the equation of variation.

64. $k = 4$

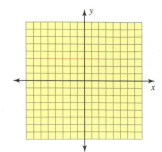

65. $k = -3.5$

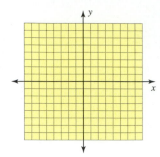

[6.5]   Use the following table describing technology available in public schools to answer Exercises 66 to 69.

Technology in U.S. Public Schools, 1995–1998				
	Number of Schools			
Technology	**1995**	**1996**	**1997**	**1998**
Schools with modems[1]	30,768	37,889	40,876	61,930
Elementary	16,010	20,250	22,234	35,066
Junior high	5,652	6,929	7,417	10,996
Senior high[2]	8,790	10,277	10,781	14,540
Schools with networks[1]	24,604	29,875	32,299	49,178
Elementary	11,693	14,868	16,441	26,422
Junior high	4,599	5,590	6,035	9,003
Senior high[2]	8,159	9,166	9,565	12,853
Schools with CD-ROMs[1]	34,480	43,499	46,388	64,200
Elementary	18,343	24,353	26,377	37,908
Junior high	6,510	7,952	8,410	11,023
Senior high[2]	9,327	10,756	11,140	13,985
Schools with Internet access[1]	NA	14,211	35,762	60,224
Elementary	NA	7,608	21,026	34,195
Junior high	NA	2,707	5,752	10,888
Senior high[2]	NA	3,736	8,984	13,829

NA = Not applicable. (1) Includes schools for special and adult education, not shown separate with grade spans of K-3, K-5, K-6, K-8, and K-12. (2) Includes schools with grade spans of technical and alternative high schools and schools with grade spans of 7–12, 9–12, and 10.
Source: Quality Education Data, Inc., Denver, CO.

66. What is the increase in the number of schools with modems from 1995 to 1998?

67. How many senior high schools had either modems, networks, or CD-ROMs in 1998?

68. What is the percent increase in public schools with Internet access from 1996 to 1998?

69. What is the percent increase in elementary schools who have either modems, networks, or Internet access from 1996 to 1998?

[6.5]   Use the following pie charts describing car sales by size in 1983 and 1993 to answer questions 70 to 73.

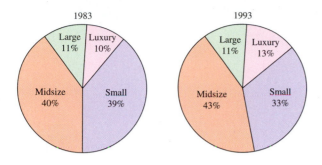

70. What was the percentage of midsize and small cars sold in 1993?

71. What was the percentage of large and luxury cars sold in 1993?

72. Sales of which type of car increased the most between 1983 and 1993?

73. If 21,303,000 U.S. cars were sold in 1993, how many small cars were sold?

In Exercises 74 and 75, use the following graph showing enrollment at Berndt Community College.

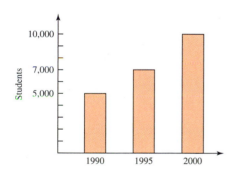

74. How many more students were enrolled in 2000 than in 1990?

75. What was the percent increase from 1990 to 1995?

[6.5]   Use the following line graph to complete Exercises 76 to 78.

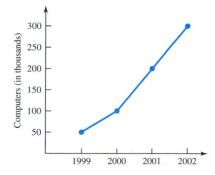

76. How many more personal computers were sold in 2000 than in 1999?

77. What was the percent increase in sales from 1999 to 2002?

78. Predict the sales of personal computers in the year 2003.

79. Use the following table to create a line graph comparing annual gross income (in thousands) of 30-year-olds with their years of formal education.

Years of Education	Gross Income
8	16
10	21
12	23
14	28
16	31

The purpose of this self-test is to help you check your progress and to review for the next in-class exam. Allow yourself about an hour to take this test. At the end of that hour check your answers against those given in the back of the text. Section references accompany the answers. If you missed any questions, go back to those sections and reread the examples until you master the concepts.

ANSWERS

Determine which of the ordered pairs are solutions for the given equations.

1. $x + y = 9$ $(3, 6), (9, 0), (3, 2)$

2. $4x - y = 16$ $(4, 0), (3, -1), (5, 4)$

Complete the ordered pairs so that each is a solution for the given equation.

3. $x + 3y = 12$ $(3, \ ), (\ , 2), (9, \ )$

4. $4x + 3y = 12$ $(3, \ ), (\ , 4), (\ , 3)$

Find four solutions for each of the following equations.

5. $x - y = 7$ **6.** $5x - 6y = 30$

Give the coordinates of the points in the following graph.

7. A

8. B

9. C

Plot points with the coordinates shown.

10. $S(1, -2)$

11. $T(0, 3)$

12. $U(-2, -3)$

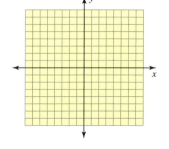

ANSWERS

1. _____

2. _____

3. _____

4. _____

5. _____

6. _____

7. _____

8. _____

9. _____

10. _____

11. _____

12. _____

ANSWERS

13. _____

14. _____

15. _____

16. _____

17. _____

18. _____

Graph each of the following equations.

13. $x + y = 4$

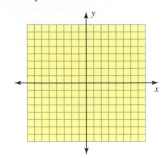

14. $y = 3x$

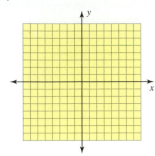

15. $y = \dfrac{3}{4}x - 4$

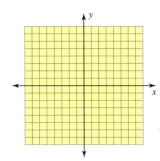

16. $x + 3y = 6$

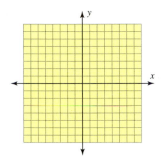

17. $2x + 5y = 10$

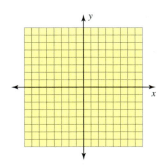

18. $y = -4$

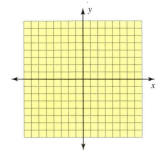

Find the slope of the line through the following pairs of points.

19. $(-3, 5)$ and $(2, 10)$

20. $(-2, 6)$ and $(2, 9)$

21. $(4, 6)$ and $(4, 8)$

22. $(7, 9)$ and $(3, 9)$

23. Find the slope of the line graphed.

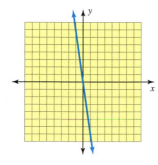

24. Graph the equation $y = 2x$.

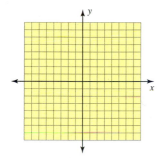

25. Solve for the constant of variation if y varies directly with x and $y = 35$ when $x = 7$.

The pie chart below represents the portion of the $40 million tourism industry spent by tourists from each country. Use the chart to complete Exercises 26 and 27.

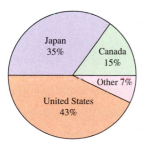

26. What percentage of the total tourism dollars is accounted for by Canada?

27. How many dollars does the United States account for?

ANSWERS

19. _____

20. _____

21. _____

22. _____

23. _____

24. _____

25. _____

26. _____

27. _____

Use the following line graph to answer Exercises 28 to 30.

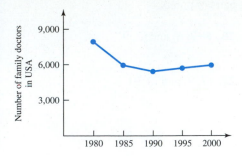

28. How many fewer family doctors were there in the United States in 1990 than in 1980?

29. What was the total change in the number of family doctors between 1980 and 2000?

30. In what 5-year period was the decrease in family doctors the greatest? What was the decrease?

ACTIVITY 4: GRAPHING WITH A CALCULATOR

Each activity in this text is designed to either enhance your understanding of the topics of the preceding chapter, provide you with a mathematical extension of those topics, or both. The activities can be undertaken by one student, but they are better suited for a small group project. Occasionally, it is only through discussion that different facets of the activity become apparent. For material related to this activity, visit the text website at www.mhhe.com/streeter.

The graphing calculator is a tool that can be used to help you solve many different kinds of problems. This activity will walk you through several features of the TI-83 Plus. By the time you complete this activity, you will be able to graph equations, change the viewing window to better accommodate a graph, or look at a table of values that represent some of the solutions for an equation. The first portion of this activity will demonstrate how you can create the graph of an equation. The features described here can be found on most graphing calculators. See your calculator manual to learn how to get your particular calculator model to perform this activity.

Menus and Graphing

1. To graph the equation $y = 2x + 3$ on a graphing calculator, follow these steps.

    a. Press the $\boxed{\text{Y} =}$ key.

    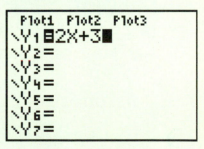

    b. Type $2x + 3$ at the Y_1 prompt. (This represents the first equation. You can type up to 10 separate equations.) Use the $\boxed{\text{X, T, }\theta\text{, }n}$ key for the variable.

    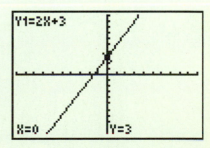

    c. Press the $\boxed{\text{GRAPH}}$ key to see the graph.
    d. Press the $\boxed{\text{TRACE}}$ key to display the equation. Once you have selected the $\boxed{\text{TRACE}}$ key, you can use the left and right arrows of the calculator to move the cursor along the line. Experiment with this movement. Look at the coordinates at the bottom of the display screen as you move along the line.

NOTE Be sure the window is the standard window to see the same graph displayed.

Frequently, we can learn more about an equation if we look at a different section of the graph than the one offered on the display screen. The portion of the graph displayed is called the **window.** The second portion of the activity explains how this window can be changed.

2. WINDOW. Press the WINDOW key. The **standard** graphing screen is shown.

Xmin = left edge of screen
Xmax = right edge of screen
Xscl = scale given by each tick on *x*-axis
Ymin = bottom edge of screen
Ymax = top edge of screen
Yscl = scale given by each tick on *y*-axis
Xres = resolution (do not alter this)

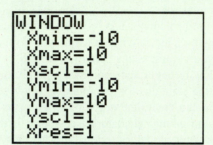

Note: To turn the scales off, enter a 0 for Xscl or Yscl. Do this when the intervals used are very large.

By changing the values for Xmin, Xmax, Ymin, and Ymax, you can adjust the viewing window. Change the viewing window so that Xmin = 0, Xmax = 40, Ymin = 0, and Ymax = 10. Again, press GRAPH. Notice that the tick marks along the *x*-axis are now much closer together. Changing Xscl from 1 to 5 will improve the display. Try it.

Sometimes we can learn something important about a graph by zooming in or zooming out. The third portion of this activity discusses this feature of the TI-83 Plus.

3. ZOOM

   a. Press the ZOOM key. There are 10 options. Use the ▼ key to scroll down.

   b. Selecting the first option, ZBox, allows the user to enlarge the graph within a specified rectangle.

      i. Graph the equation $y = x^2 + x - 1$ in the standard window. *Note:* To type in the exponent, use the x^2 key or the ∧ key.

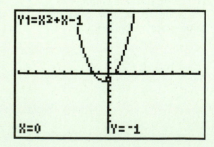

ii.   When ZBox is selected, a blinking "+" cursor will appear in the graph window. Use the arrow keys to move the cursor to where you would like a corner of the screen to be; then press the ENTER key.

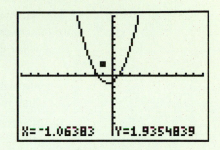

iii.   Use the arrow keys to trace out the box containing the desired portion of the graph. Do not press the ENTER key until you have reached the diagonal corner and a full box is on your screen.

After using the down arrow   *After using the right arrow*

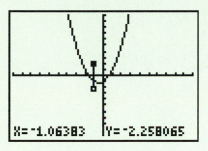   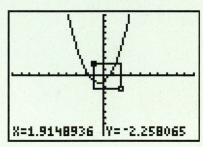

After pressing the ENTER *key a second time*

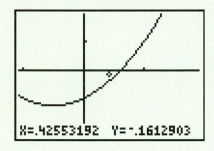

Now the desired portion of a graph can be seen more clearly.

The Zbox feature is especially useful when analyzing the roots (*x*-intercepts) of an equation.

c.   Another feature that allows us to focus is ZoomIn. Press the ZoomIn button in the Zoom menu. Place the cursor in the center of the portion of the graph you are interested in and press the ENTER key. The window will reset with the cursor at the center of a zoomed-in view.

d.   ZoomOut works like ZoomIn, except that it sets the view larger (that is, it zooms out) to enable you to see a larger portion of the graph.

e.   ZStandard sets the window to the standard window. This is a quick and convenient way to reset the viewing window.

f.   ZSquare recalculates the view so that one horizontal unit is the same length as one vertical unit. This is sometimes necessary to get an accurate view of a graph since the width of the calculator screen is greater than its height.

Home Screen: This is where all the basic computations take place. To get to the home screen from any other screen, press 2nd , Mode . This accesses the QUIT feature. To clear the home screen of calculations, press the CLEAR key (once or twice).

Tables: The final feature that we will look at here is Table. Enter the function $y = 2x + 3$ into the Y = menu. Then press 2nd , WINDOW to access the TBLSET menu. Set the table as shown here and press 2nd , GRAPH to access the TABLE feature. You will see the screen shown here.

The following exercises are presented to help you review concepts from earlier chapters. This is meant as a review and not as a comprehensive exam. The answers are presented in the back of the text. Section references accompany the answers. If you have difficulty with any of these exercises, be certain to at least read through the summary related to those sections.

ANSWERS

1. 3
2. 5
3. 37
4. -53
5. 69
6. -120
7. -5
8. 3
9. 108
10. 3
11. -23 69
12. 9
13. $-4/3$
14. $-\frac{8}{3}$
15. 10
16. 1
17. $C = \frac{5}{9}(F-32)$
18. $x < 4$
19. $x \le 3$

Perform the indicated operations.

1. $9 + (-6)$

2. $-4 - (-9)$

3. $25 - (-12)$

4. $-32 + (-21)$

5. $(-23)(-3)$

6. $(12)(-10)$

7. $30 \div (-6)$

8. $(-24) \div (-8)$

Evaluate the expressions if $x = -3$, $y = 4$, and $z = -5$.

9. $3x^2y$

$3(-3^2)4$

10. $-3z - 3y$

$15 - 7 =$

11. $-3(-2y + 3z)$

$-3(-8 + -15)$

12. $\dfrac{3y - 2x}{5y + 6x}$

$12 - -6 \quad \dfrac{18}{2}$

$20 + -18$

Solve the following equations and check your results.

13. $5x - 2 = 2x - 6$

$ +2 +2 \quad \frac{4}{3}$

$5x = 2x + 4$

$-2x \quad -2x$

$3x = 4$

14. $3(x - 2) = 2(3x + 1)$

$3x - 6 = 6x + 2$

$+6 +6$

$3x = 6x + 8$

15. $\dfrac{5}{6}x - 3 = 2 + \dfrac{1}{3}x$

$\frac{3}{6} \quad \frac{5}{6}x = 5 + \frac{1}{3}x$

$-\frac{1}{3} \quad -\frac{1}{3}$

$\frac{3}{6}x$

16. $4(2 - x) + 9 = 7 + 6x$

17. Solve the equation $F = \dfrac{9}{5}C + 32$ for C.

$F - \frac{9}{5} = C + 32$

$-32 \quad -32$

$\frac{9}{5} \quad \frac{9}{5}$

Solve the following inequalities.

18. $4x - 9 < 7$

$+9 \quad +9$

$4x < 16$

$\frac{4x}{4} < \frac{16}{4}$

$x < 4$

19. $-5x + 15 \ge 2x - 6$

$+6 \quad +6$

$-5x + 21 \ge 2x$

$+5x \quad +5x$

$21 \ge 7x$

$3 \ge x$

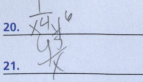

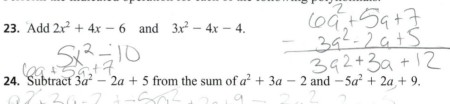

ANSWERS

20. $\frac{1}{x^4 y^6}$

21. $\frac{y^5}{x}$

22. 1

23. $5x^2 - 10$

24. $7a^2 + 7a + 2$

25. 19

26. 26

27. $6x^2 - 2yx - 20y$

28. $6x^3 - 3x^2 - 45$

29. $4a^2 - 49b^2$

30. $4pn^2(3pn + 5n - 4n)$

31. $(y-3)(y-5)$

32. $(3x+2)(x-1)$

33. $2(x+2)(x-1)$

34. $(2a+3b)(3a-b)$

35. $-3, 11$

36. $4/5, 2/7$

37. $\frac{-5a^3}{3b^2}$

38. $\frac{w-2}{w+3}$

Use the properties of exponents to simplify the following expressions and write the results with positive exponents.

20. $(x^2 y^3)^{-2}$

21. $\dfrac{x^3 y^2}{x^4 y^{-3}}$

22. $(x^6 y^{-3})^0$

Perform the indicated operation for each of the following polynomials.

23. Add $2x^2 + 4x - 6$ and $3x^2 - 4x - 4$.

24. Subtract $3a^2 - 2a + 5$ from the sum of $a^2 + 3a - 2$ and $-5a^2 + 2a + 9$.

Evaluate each polynomial for the indicated variable value.

25. $2x^2 - 5x + 7$ for $x = 4$

26. $x^3 + 3x^2 - 7x + 8$ for $x = -2$

Multiply the following polynomials.

27. $(3x - 5y)(2x + 4y)$

28. $3x(x - 3)(2x + 5)$

29. $(2a + 7b)(2a - 7b)$

Completely factor each polynomial.

30. $12p^2 n^2 + 20pn^2 - 16pn^3$

31. $y^3 - 3y^2 - 5y + 15$

32. $9a^2 b - 49b$

33. $6x^2 - 2x - 4$

34. $6a^2 + 7ab - 3b^2$

Solve each of the following by factoring.

35. $x^2 - 8x - 33 = 0$

36. $35x^2 - 38x = -8$

Simplify each of the following rational expressions.

37. $\dfrac{-35a^4 b^5}{21ab^7}$

38. $\dfrac{2w^2 - w - 6}{2w^2 + 9w + 9}$

Add or subtract as indicated. Simplify your answer.

39. $\dfrac{2}{a-5} - \dfrac{1}{a}$

40. $\dfrac{2w}{w^2-9w+20} + \dfrac{8}{w-4}$

Multiply or divide as indicated.

41. $\dfrac{4xy^3}{5xy^2} \cdot \dfrac{15x^3y}{16y^2}$

42. $\dfrac{m^2-3m}{m^2-9} \div \dfrac{4m^2}{m^2-m-12}$

Solve each of the following equations.

43. $\dfrac{w}{w-2} + 1 = \dfrac{w+4}{w-2}$

44. $\dfrac{7}{x} - \dfrac{1}{x-3} = \dfrac{9}{x^2-3x}$

Graph each of the following equations.

45. $3x + 4y = 12$

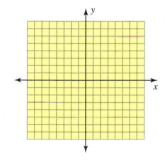

46. $y = -7$

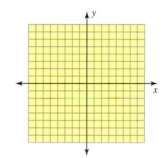

47. $x = 2y$

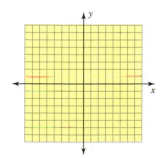

48. Find the slope of the line passing through the points $(-3, 5)$ and $(1, 13)$.

ANSWERS

39. _____

40. _____

41. _____

42. _____

43. _____

44. _____

45. _____

46. _____

47. _____

48. _____

ANSWERS

49. _____

50. _____

51. $l=7$ $w=5$

52. $41, 43, 45$

53. 112.50

49. Graph the equation of variation if $k = -4$.

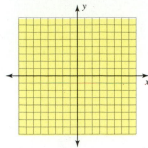

50. Find the constant of variation if y varies directly with x, and if $y = 450$ when $x = 15$.

Solve the following problems.

51. The length of a rectangle is 3 in. less than twice its width. If the perimeter is 24 in., find the dimensions of the rectangle.

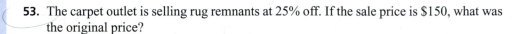

$P=24$
$2n-3=l$ $P=2w+2l$ $24=2n+2(2w-3)$

52. The sum of three consecutive odd integers is 129. Find the three integers.

53. The carpet outlet is selling rug remnants at 25% off. If the sale price is \$150, what was the original price?

602

GRAPHING AND INEQUALITIES

INTRODUCTION

In the pharmaceutical-making process, great caution must be exercised to ensure that the medicines and drugs are pure and contain precisely what is indicated on the label. Guaranteeing such purity is a task for which the quality control division of the pharmaceutical company has responsibility.

A lab technician working in quality control must run a series of tests on samples of every ingredient, even simple ingredients such as salt (NaCl). One such test is a measure of how much weight is lost as a sample is dried. The technician must set up a 3-hour procedure that involves cleaning and drying bottles and stoppers and then weighing them while they are empty and again when they contain samples of the substance to be heated and dried. At the end of the procedure, to compute the percentage of weight loss from drying, the technician uses the formula

$$L = \frac{W_g - W_f}{W_g - T} \cdot 100$$

in which L = percentage loss in drying
W_g = weight of container and sample
W_f = weight of container and sample after drying process has been completed
T = weight of empty container

The pharmaceutical company may have a standard of acceptability for this substance. For instance, the substance may not be acceptable if the loss of weight from drying is greater than 10%. The technician would then use the following inequality to calculate acceptable weight loss:

$$10 \geq \frac{W_g - W_f}{W_g - T} \cdot 100$$

We will further examine inequalities in this chapter.

Pre-Test Chapter 7

ANSWERS

1. _____

2. _____

3. _____

4. _____

5. _____

6. _____

7. _____

8. _____

9. _____

10. _____

11. _____

12. _____

This pre-test provides a preview of the types of exercises you will encounter in each section of this chapter. The answers for these exercises can be found in the back of the text. If you are working on your own, or ahead of the class, this pre-test can help you find the sections in which you should focus more of your time.

[7.1] **1.** Find the slope and y-intercept of the line given by $y = -2x + 3$.

2. Find the slope and y-intercept from the equation $5x - 4y = 20$.

Write the equation of the line with the given slope and y-intercept. Then graph each line.

3. Slope 5 and y-intercept $(0, -2)$ **4.** Slope -4 and y-intercept $(0, 6)$

[7.2] Determine if the following pairs of lines are parallel, perpendicular, or neither.

5. L_1 through $(3, 7)$ and $(1, 11)$ L_1 with equation $y = 3x + 9$

6. L_2 through $(-5, -1)$ and $(6, -23)$ L_2 with equation $3y + x = 9$

[7.3] **7.** Find the equation of the line through $(-5, 8)$ and parallel to the line $4x + y = 8$.

Write the equation of the line L satisfying each of the following sets of geometric conditions.

8. L passes through $(5, -6)$ and is perpendicular to $3x - 5y = 15$.

9. L has y-intercept $(0, -3)$ and is parallel to $-3x + 5y = -15$.

[7.4] Graph each of the following inequalities.

10. $2x + y < 5$ **11.** $-3x + 5y \geq 15$

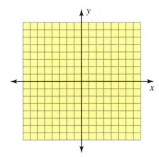   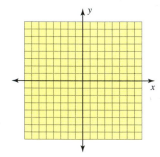

[7.5] Evaluate $f(x)$ as indicated.

12. $f(x) = -2x^3 - 6x - 4x^2 + 9$; find $f(-1)$ and $f(1)$.

7.1 The Slope-Intercept Form

7.1 OBJECTIVES

1. Find the slope and *y*-intercept from the equation of a line
2. Given the slope and *y*-intercept, write the equation of a line
3. Use the slope and *y*-intercept to graph a line
4. Solve an application of slope-intercept equations

In Chapter 6, we used two points to find the slope of a line. In this chapter we use the slope and *y*-intercept to find the graph of an equation.

First, we want to consider finding the equation of a line when its slope and *y*-intercept are known.

Suppose that the *y*-intercept of a line is $(0, b)$. Then the point at which the line crosses the *y*-axis has coordinates $(0, b)$, as shown in the sketch at left.

Now, using any other point (x, y) on the line and using our definition of slope, we can write

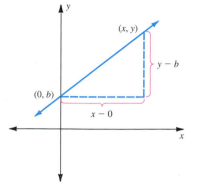

$$m = \frac{y - b}{x - 0}$$

Change in *y*.

Change in *x*.

or

$$m = \frac{y - b}{x}$$

Multiplying both sides of this equation by *x*, we have

$$mx = y - b$$

Finally, adding *b* to both sides of this equation gives

$$mx + b = y$$

or

$$y = mx + b$$

We can summarize this discussion as follows:

NOTE In this form, the equation is *solved for y*. The coefficient of *x* gives the slope of the line, and the constant term gives the *y*-intercept.

Definition: The Slope-Intercept Form for a Line

An equation of the line with slope *m* and *y*-intercept $(0, b)$ is

$$y = mx + b$$

OBJECTIVE 1

Example 1 Finding the Slope and *y*-Intercept

Find the slope and *y*-intercept of the equation

$$y = -\frac{2}{3}x - 5$$

    m   *b*

The slope of the line is $-\frac{2}{3}$; the *y*-intercept is $(0, -5)$.

CHECK YOURSELF 1 _____

Find the slope and y-intercept for the graph of each of the following equations.

(a) $y = -3x - 7$ 　　　　　　　　 **(b)** $y = \dfrac{3}{4}x + 5$

As Example 2 illustrates, we may have to solve for y as the first step in determining the slope and the y-intercept for the graph of an equation.

> **Example 2   Finding the Slope and y-Intercept**

Find the slope and y-intercept of the equation

$$3x + 2y = 6$$

First, we must solve the equation for y.

NOTE If we write the equation as

$$y = \dfrac{-3x + 6}{2}$$

it is more difficult to identify the slope and the intercept.

$$3x + 2y = 6$$
$$2y = -3x + 6 \qquad \text{Add } (-3x) \text{ to both sides.}$$
$$y = -\dfrac{3}{2}x + 3 \qquad \text{Divide each term by 2.}$$

The equation is now in slope-intercept form. The slope is $-\dfrac{3}{2}$, and the y-intercept is $(0, 3)$.

CHECK YOURSELF 2 _____

Find the slope and y-intercept for the graph of the equation

$$2x - 5y = 10$$

As we mentioned earlier, knowing certain properties of a line (namely, its slope and y-intercept) will also allow us to write the equation of the line by using the slope-intercept form. Example 3 illustrates this approach.

OBJECTIVE 2

> **Example 3   Writing the Equation of a Line**

Write the equation of a line with slope $-\dfrac{3}{4}$ and y-intercept $(0, -3)$.

We know that $m = -\dfrac{3}{4}$ and $b = -3$. In this case,

$$y = \overset{m}{-\dfrac{3}{4}}x + \overset{b}{(-3)}$$

or

$$y = -\dfrac{3}{4}x - 3$$

which is the desired equation.

✔ **CHECK YOURSELF 3**

Write the equation of a line with the following:

(a) slope -2 and *y*-intercept $(0, 7)$

(b) slope $\frac{2}{3}$ and *y*-intercept $(0, -3)$

We can also use the slope and *y*-intercept of a line in drawing its graph. Consider Example 4.

OBJECTIVE 3   **Example 4   Graphing a Line Using the Slope and *y*-intercept**

Graph the line with slope $\frac{2}{3}$ and *y*-intercept $(0, 2)$.

Because the *y*-intercept is $(0, 2)$, we begin by plotting the point $(0, 2)$. Because the horizontal change (or run) is 3, we move 3 units to the right *from that y-intercept*. Then because the vertical change (or rise) is 2, we move 2 units up to locate another point on the desired graph. Note that we will have located that second point at $(3, 4)$. The final step is to simply draw a line through that point and the *y*-intercept.

NOTE

$m = \dfrac{2}{3} = \dfrac{\text{rise}}{\text{run}}$

The line rises from left to right because the slope is positive.

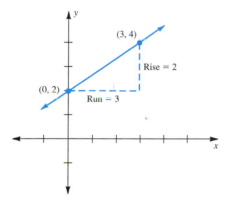

The equation of this line is $y = \frac{2}{3}x + 2$.

✔ **CHECK YOURSELF 4**

Graph the equation of a line with slope $\frac{3}{5}$ and y-intercept $(0, -2)$.

Step by Step: Graphing by Using the Slope-Intercept Form

Step 1   Write the original equation of the line in slope-intercept form.

Step 2   Determine the slope *m* and the *y*-intercept $(0, b)$.

Step 3   Plot the *y*-intercept at $(0, b)$.

Step 4   Use *m* (the change in *y* over the change in *x*) to determine a second point on the desired line.

Step 5   Draw a line through the two points determined in steps 1 to 4 to complete the graph.

You have now seen two methods for graphing lines: the slope-intercept method (this section) and the intercept method (Section 6.3). When you graph a linear equation, you first should decide which is the appropriate method.

Example 5 Selecting an Appropriate Graphing Method

Decide which of the two methods for graphing lines—the intercept method or the slope-intercept method—is more appropriate for graphing equations (a), (b), and (c).

(a) $2x - 5y = 10$

Because both intercepts are easy to find, you should choose the intercept method to graph this equation.

(b) $2x + y = 6$

This equation can be quickly graphed by either method. As it is written, you might choose the intercept method. It can, however, be rewritten as $y = -2x + 6$. In that case the slope-intercept method is more appropriate.

(c) $y = \dfrac{1}{4}x - 4$

Because the equation is in slope-intercept form, that is the more appropriate method to choose.

CHECK YOURSELF 5

Which would be more appropriate for graphing each equation, the intercept method or the slope-intercept method?

(a) $x + y = -2$ **(b)** $3x - 2y = 12$ **(c)** $y = -\dfrac{1}{2}x - 6$

The slope-intercept form lends itself easily to applications involving linear equations. Consider the cost of manufacturing a product; there are really two costs involved.

Fixed costs are those costs that are independent of the number of units produced. That is, no matter how many (or how few) items are produced, fixed costs remain the same. Fixed costs include the cost of leasing property, insurance costs, and labor costs for salaried personnel. If we write an equation relating total costs to the number of units produced, then the fixed costs are represented by the constant term.

Variable costs are those costs that change depending on the number of units produced. These include the costs of materials, labor costs for hourly employees, many utility costs, and service costs. In the slope-intercept equation, variable costs correspond to the slope m because the dollar amount increases each time another unit is produced.

NOTE The y-intercept b in the equation $y = mx + b$ is the constant term.

NOTE Variable costs are sometimes called *marginal costs*.

OBJECTIVE 4

Example 6 An Application of Slope-Intercept Equations

S-Bar Electronics determines that the cost to produce each stereo is $26. In addition, the cost to keep its factory open each month is $3500.

(a) Write an equation relating the total monthly cost of producing stereos to the number of stereos produced.

The y-intercept represents the fixed costs, so we have $b = 3500$. The slope m is given by the variable costs, which are \$26 per stereo. Putting this together gives us the equation

$$y = 26x + 3500$$

(b) Use the cost equation to determine the total cost of producing 320 stereos in a month.

In the cost equation, we substitute 320 for x and calculate y:

$$y = 26(320) + 3500$$
$$= 11{,}820$$

So it costs S-Bar Electronics \$11,820 to produce 320 stereos in a month.

CHECK YOURSELF 6

A manager at the chic new restaurant Sweet Eats determines that the average dinner costs the restaurant \$18 to produce. In addition, it costs the restaurant \$620 to stay open each evening.

(a) Write an equation relating the total nightly cost of operation to the number of dinners served.

(b) How much does it cost the restaurant to serve 50 dinners in an evening?

CHECK YOURSELF ANSWERS

1. (a) Slope is -3, y-intercept is $(0, -7)$; **(b)** Slope is $\dfrac{3}{4}$, y-intercept is $(0, 5)$

2. The slope is $\dfrac{2}{5}$; the y-intercept is $(0, -2)$

3. (a) $y = -2x + 7$; **(b)** $y = \dfrac{2}{3}x - 3$

4.

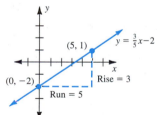

5. (a) Either; **(b)** intercept; **(c)** slope-intercept
6. (a) $y = 18x + 620$; **(b)** \$1520

7.1   Exercises

Find the slope and *y*-intercept of the line represented by each of the following equations.

ANSWERS

1. _____

2. _____

3. _____

4. _____

5. _____

6. _____

7. _____

8. _____

9. _____

10. _____

11. _____

12. _____

13. _____

14. _____

15. _____

16. _____

17. _____

18. _____

1. $y = 3x + 5$

2. $y = -7x + 3$

3. $y = -2x - 5$

4. $y = 5x - 2$

5. $y = \dfrac{3}{4}x + 1$

6. $y = -4x$

7. $y = \dfrac{2}{3}x$

8. $y = -\dfrac{3}{5}x - 2$

9. $4x + 3y = 12$

10. $2x + 5y = 10$

11. $y = 9$

12. $2x - 3y = 6$

13. $3x - 2y = 8$

14. $x = 5$

Write the equation of the line with given slope and *y*-intercept. Then graph each line using the slope and *y*-intercept.

15. Slope: 3; *y*-intercept: (0, 5)

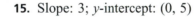
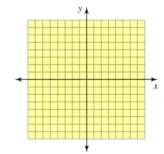

16. Slope: −2; *y*-intercept: (0, 4)

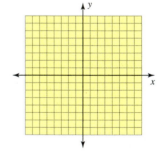

17. Slope: −3; *y*-intercept: (0, 4)

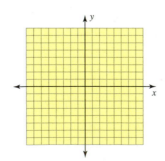

18. Slope: 5; *y*-intercept: (0, −2)

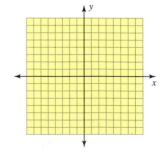

19. Slope: $\frac{1}{2}$; y-intercept: $(0, -2)$

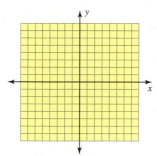

20. Slope: $-\frac{3}{4}$; y-intercept: $(0, 8)$

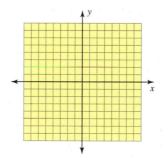

21. Slope: $-\frac{2}{3}$; y-intercept: $(0, 0)$

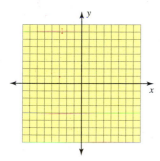

22. Slope: $\frac{2}{3}$; y-intercept: $(0, -2)$

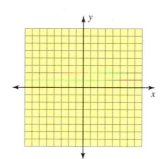

23. Slope: $\frac{3}{4}$; y-intercept: $(0, 3)$

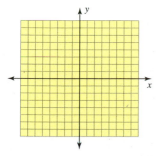

24. Slope: -3; y-intercept: $(0, 0)$

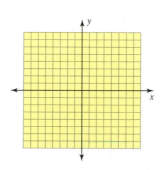

ANSWERS

19. _____

20. _____

21. _____

22. _____

23. _____

24. _____

25. _____

26. _____

27. _____

28. _____

29. _____

30. _____

31. _____

32. _____

In exercises 25 to 32, match the graph with one of the following equations.

(a) $y = 2x$ **(b)** $y = x + 1$ **(c)** $y = -x + 3$ **(d)** $y = 2x + 1$

(e) $y = -3x - 2$ **(f)** $y = \frac{2}{3}x + 1$ **(g)** $y = -\frac{3}{4}x + 1$ **(h)** $y = -4x$

25.

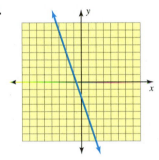

26.

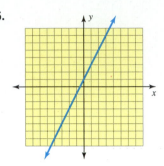

27.

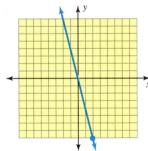

28.

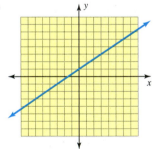

29.

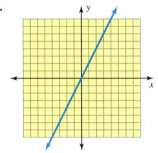

30.

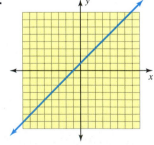

31.

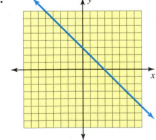

32.

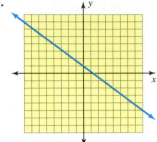

33. In each of the following statements, fill in the blank with *always, sometimes,* or *never.*

    (a) The slope of a line through the origin is _____ zero.

    (b) A line with an undefined slope is _____ the same as a line with a slope of zero.

34. In each of the following statements fill in the blank with *always, sometimes,* or *never.*

    (a) Lines _____ have exactly one *x*-intercept.

    (b) The *y*-intercept of a line through the origin is _____ zero.

In which quadrant(s) are there no solutions for each line?

35. $y = -x + 1$

36. $y = 3x + 2$

37. $y = -2x - 5$

38. $y = -5x - 7$

39. $y = 3$

40. $x = -2$

41. Social science. The equation $y = 0.10x + 200$ describes the award money in a recycling contest. What are the slope and the *y*-intercept for this equation?

42. Business and finance. The equation $y = 15x - 100$ describes the amount of money a high school class might earn from a paper drive. What are the slope and *y*-intercept for this equation?

43. Science and medicine. On a certain February day in Philadelphia, the temperature at 6:00 A.M. was 10°F. By 2:00 P.M. the temperature was up to 26°F. What was the hourly rate of temperature change?

44. Construction. A roof rises 8.75 ft over a horizontal distance of 15.09 ft. Find the slope of the roof to the nearest hundredth.

33.	
34.	
35.	
36.	
37.	
38.	
39.	
40.	
41.	
42.	
43.	
44.	

45. _____

46. _____

47.  _____

48. _____

49. _____

50. _____

45. **Science and medicine.** An airplane covered 15 mi of its route while decreasing its altitude by 24,000 ft. Find the slope of the line of descent that was followed. (1 mi = 5280 ft) Round to the nearest hundredth.

46. **Technology.** Driving down a mountain, Tom finds that he has descended 1800 ft in elevation by the time he is 3.25 mi horizontally away from the top of the mountain. Find the slope of his descent to the nearest hundredth.

47. Complete the following statement: "The difference between undefined slope and zero slope is . . . ."

48. Complete the following: "The slope of a line tells you . . . ."

49. **Statistics.** In a study on nutrition conducted in 1984, 18 normal adults aged 23 to 61 years old were measured for body fat, which is given as percentage of weight. The mean (average) body fat percentage for women 40 years old was 28.6% and for women 53 years old was 38.4%. Work with a partner to decide how to show this information by plotting points on a graph. Try to find a linear equation that will tell you percentage of body fat based on a woman's age. What does your equation give for 20 years of age? For 60? Do you think a linear model works well for predicting body fat percentage in women as they age?

50. **Business and finance.** On two occasions last month, Sam Johnson rented a car on a business trip. Both times it was the same model from the same company, and both times it was in San Francisco. Sam now has to fill out an expense account form and needs to know how much he was charged per mile and the base rate. On both occasions he dropped the car at the airport booth and just got the total charge, not the details. All Sam knows is that he was charged $210 for 625 mi on the first occasion and $133.50 for 370 mi on the second trip. Sam has called accounting to ask for help. Plot these two points on a graph and draw the line that goes through them. What question does the slope of the line answer for Sam? How does the y-intercept help? Write a memo to Sam explaining the answers to his question and how a knowledge of algebra and graphing has helped you find the answers.

51. On the same graph, sketch the following lines:

$$y = 2x - 1 \quad \text{and} \quad y = 2x + 3$$

What do you observe about these graphs? Will the lines intersect?

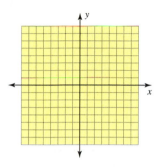

52. Repeat exercise 51 using

$$y = -2x + 4 \quad \text{and} \quad y = -2x + 1$$

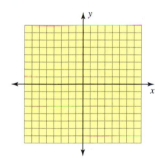

53. On the same graph, sketch the following lines:

$$y = \frac{2}{3}x \quad \text{and} \quad y = -\frac{3}{2}x$$

What do you observe concerning these graphs? Find the product of the slopes of these two lines.

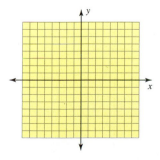

54. Repeat exercise 53 using

$$y = \frac{4}{3}x \quad \text{and} \quad y = -\frac{3}{4}x$$

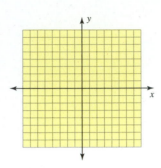

55. Based on exercises 53 and 54, write the equation of a line that is perpendicular to

$$y = \frac{3}{5}x$$

 Getting Ready for Section 7.2 [Section 6.4]

Find the slope of the line connecting the given points.

    (a) $(-4, 6)$ and $(3, 20)$ (b) $(2, 8)$ and $(-6, -8)$

    (c) $(5, -7)$ and $(-5, 3)$ (d) $(2, 8)$ and $(2, 5)$

    (e) $(6, 9)$ and $(3, 9)$ (f) $(4, 6)$ and $(-4, -2)$

Answers

1. Slope 3, y-intercept $(0, 5)$ **3.** Slope -2, y-intercept $(0, -5)$

5. Slope $\frac{3}{4}$, y-intercept $(0, 1)$ **7.** Slope $\frac{2}{3}$, y-intercept $(0, 0)$

9. Slope $-\frac{4}{3}$, y-intercept $(0, 4)$ **11.** Slope 0, y-intercept $(0, 9)$

13. Slope $\frac{3}{2}$, y-intercept $(0, -4)$

15. $y = 3x + 5$ **17.** $y = -3x + 4$

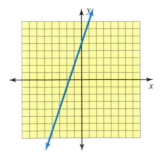

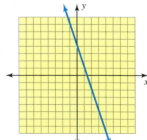

19. $y = \dfrac{1}{2}x - 2$

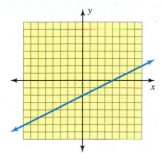

21. $y = -\dfrac{2}{3}x$

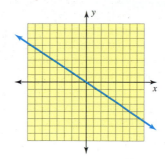

23. $y = \dfrac{3}{4}x + 3$

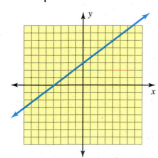

25. g   **27.** e   **29.** h   **31.** c   **33.** (a) Sometimes; (b) never
35. III   **37.** I   **39.** III and IV   **41.** Slope $= 0.10$; y-intercept $= (0, 200)$
43. 2°F/h   **45.** -0.30 **47.**   **49.**

51. Parallel lines; no

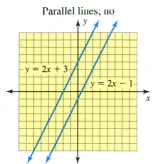

53. Perpendicular lines; -1

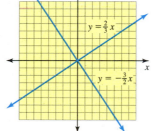

55. $y = -\dfrac{5}{3}x$ **a.** 2   **b.** 2   **c.** -1 **d.** Undefined   **e.** 0   **f.** 1

7.2 Parallel and Perpendicular Lines

7.2 OBJECTIVES

1. Determine whether two lines are parallel
2. Determine whether two lines are perpendicular
3. Find the slope of a line parallel or perpendicular to a given line

For most inexperienced drivers, the most difficult driving maneuver to master is parallel parking. What is parallel parking? It is the act of backing into a curbside space so that the car's tires are parallel to the curb.

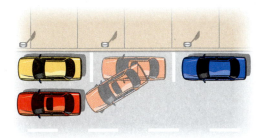

How can you tell that you've done a good job of parallel parking? Most people check to see that both the front tires and the back tires are the same distance (8 in. or so) from the curb. This is checking to be certain that the car is parallel to the curb.

How can we tell that two equations represent parallel lines? Look at the following sketch.

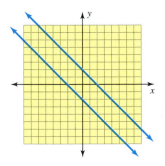

If two lines are parallel, they have the same slope. If their equations are in slope-intercept form, you simply compare the slopes.

OBJECTIVE 1   **Example 1   Determining That Two Lines Are Parallel**

Which two equations represent parallel lines?

(a) $y = 2x + 3$

(b) $y = -\dfrac{1}{2}x - 5$

(c) $y = -2x + \dfrac{1}{2}$

(d) $y = -2x - 9$

Because (c) and (d) both have a slope of -2, the lines are parallel.

 CHECK YOURSELF 1

Which two equations represent parallel lines?

(a) $y = -5x + 5$

(b) $y = \dfrac{1}{5}x - 5$

(c) $y = -5x + \dfrac{1}{2}$

(d) $y = 5x - 9$

More formally, we can state the following about parallel lines.

> **Rules and Properties: Slopes of Parallel Lines**
>
> For nonvertical lines L_1 and L_2, if line L_1 has slope m_1 and line L_2 has slope m_2, then
>
> L_1 is parallel to L_2 if, and only if, $m_1 = m_2$
>
> **Note:** All vertical lines are parallel to each other.

NOTE This means that if the lines are parallel, then their slopes are equal. Conversely, if the slopes are equal, then the lines are parallel.

As we discovered in Chapter 6, we can find the slope of a line from any two points on the line.

> **Example 2   Determining If Two Lines Are Parallel**

Are lines L_1 through $(2, 3)$ and $(4, 6)$ and L_2 through $(-4, 2)$ and $(0, 8)$ parallel, or do they intersect?

$$m_1 = \frac{6 - 3}{4 - 2} = \frac{3}{2}$$

NOTE Unless, of course, L_1 and L_2 are actually the *same line*. In this case a quick sketch will show that the lines are distinct.

$$m_2 = \frac{8 - 2}{0 - (-4)} = \frac{6}{4} = \frac{3}{2}$$

Because the slopes of the lines are equal, the lines are parallel. They do *not* intersect.

 CHECK YOURSELF 2

Are lines L_1 through $(-2, -1)$ and $(1, 4)$ and L_2 through $(-3, 4)$ and $(0, 8)$ parallel, or do they intersect?

Many important characteristics of lines are evident from a city map.

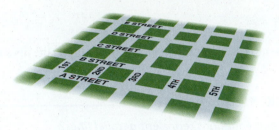

Note that Fourth Street and Fifth Street are parallel. Just as these streets never meet, it is true that two distinct parallel lines will never meet.

Recall that the point at which two lines meet is called their **intersection.** This is also true with two streets. We call the common area of the two streets the intersection.

In this case, the two streets meet at right angles. When two lines meet at right angles, we say that they are **perpendicular.**

Rules and Properties: Slopes of Perpendicular Lines

For nonvertical lines L_1 and L_2, if line L_1 has slope m_1 and line L_2 has slope m_2, then

L_1 is perpendicular to L_2 if, and only if, $m_1 = -\dfrac{1}{m_2}$

or equivalently

$m_1 \cdot m_2 = -1$

Note: Horizontal lines are perpendicular to vertical lines.

OBJECTIVE 2   **Example 3   Determining That Two Lines Are Perpendicular**

Which two equations represent perpendicular lines?

(a) $y = 2x + 3$

(b) $y = -\dfrac{1}{2}x - 5$

(c) $y = -2x + \dfrac{1}{2}$

(d) $y = -2x - 9$

Because the product of the slopes for (a) and (b) is

$$2\left(-\dfrac{1}{2}\right) = -1$$

these two lines are perpendicular. Note that none of the other pairs of slopes have a product of -1.

 CHECK YOURSELF 3

Which two equations represent perpendicular lines?

(a) $y = -5x + 5$ **(b)** $y = -\dfrac{1}{5}x - 5$

(c) $y = -5x + \dfrac{1}{2}$ **(d)** $y = 5x - 9$

Example 4   Determining If Two Lines Are Perpendicular

Are lines L_1 through points $(-2, 3)$ and $(1, 7)$ and L_2 through points $(2, 4)$ and $(6, 1)$ perpendicular?

NOTE $\left(\dfrac{4}{3}\right)\left(-\dfrac{3}{4}\right) = -1$

$$m_1 = \dfrac{7 - 3}{1 - (-2)} = \dfrac{4}{3}$$

$$m_2 = \dfrac{1 - 4}{6 - 2} = -\dfrac{3}{4}$$

Because the slopes are negative reciprocals, the lines are perpendicular.

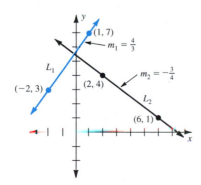

 CHECK YOURSELF 4

Are lines L_1 through points (1, 3) and (4, 1) and L_2 through points (−2, 4) and (2, 10) perpendicular?

If we already have a line, we can use its slope to determine the slope of other lines that are parallel or perpendicular to our line.

OBJECTIVE 3

Example 5   Finding Parallel and Perpendicular Slopes

(a) Find the slope of all lines parallel to the line given by $5x + y = 1$.

RECALL The slope-intercept equation of a line is given by $y = mx + b$, in which m represents the slope and b gives the y-intercept.

We begin by writing the equation for the given line in slope-intercept form. We do this by isolating y.

$$5x + y = 1$$
$$y = -5x + 1$$

In this form, we see that the slope of this line is -5. Therefore, all lines parallel to the given line have a slope of -5.

(b) Find the slope of all lines perpendicular to the line given by $5x + y = 1$.

From part (a), we know the slope of the given line is -5. Using our rules and properties of perpendicular lines, we have

$$m_1 = -5$$
$$m_2 = -\frac{1}{m_1} = -\frac{1}{(-5)} = \frac{1}{5}$$

All lines perpendicular to the given line have a slope of $\frac{1}{5}$.

CHECK YOURSELF 5

(a) Find the slope of all lines parallel to the line given by $x - 5y = 3$.
(b) Find the slope of all lines perpendicular to the line given by $x - 5y = 3$.

We can also use the slope-intercept form to determine whether the graphs of given equations will be parallel, intersecting, or perpendicular lines.

Example 6   Verifying That Two Lines Are Parallel

Show that the graphs of $3x + 2y = 4$ and $6x + 4y = 12$ are parallel lines.
First, we solve each equation for y:

$$3x + 2y = 4$$
$$2y = -3x + 4$$
$$y = -\frac{3}{2}x + 2$$

NOTE The slopes are the same, but the y-intercepts are different. Therefore, the lines are distinct.

$$6x + 4y = 12$$
$$4y = -6x + 12$$
$$y = -\frac{3}{2}x + 3$$

Because the two lines have the same slope, here $-\frac{3}{2}$, the lines are parallel.

CHECK YOURSELF 6

Show that the graphs of the equations

$$-3x + 2y = 4 \qquad and \qquad 2x + 3y = 9$$

are perpendicular lines.

Many professions require people to sketch plans of one sort or another. In particular, architects are often required to sketch their designs on a rectangular coordinate system. This makes determining the relationship of adjacent objects such as walls and ceilings a fairly straightforward process as seen in Example 7.

Example 7   An Application of Perpendicular Lines

Design plans for a project need to be checked by the architect Nicolas. On the sketch, one line passes through the points (3, 6) and (7, 3). A second line also passes through (7, 3), as well as through the point (13, 11). Should Nicolas approve these plans if the two lines are supposed to be perpendicular?

To determine whether the two lines are perpendicular, we compute the slope of each. The slope of the first line is given by

$$m_1 = \frac{\text{change in } y}{\text{change in } x} = \frac{y_2 - y_1}{x_2 - x_1}$$

$$= \frac{3 - 6}{7 - 3} = \frac{-3}{4}$$

$$= -\frac{3}{4}$$

The slope of the second line is found similarly.

$$m_2 = \frac{11 - 3}{13 - 7} = \frac{8}{6}$$

$$= \frac{4}{3}$$

The product of the two slopes is

$$m_1 \cdot m_2 = \left(-\frac{3}{4}\right)\left(\frac{4}{3}\right) = -1$$

so the lines are perpendicular. Nicholas should approve the plans.

CHECK YOURSELF 7

On another portion of Nicolas's plans is a line through the points (4, 9) and (6, 8). Is this line parallel to the first line [through the points (3, 6) and (7, 3)]?

CHECK YOURSELF ANSWERS

1. (a) and (c)   **2.** The lines intersect

3. (b) and (d)   **4.** The lines are perpendicular

5. (a) $\dfrac{1}{5}$; (b) -5

6. $y = \dfrac{3}{2}x + 2$ **7.** The lines are not parallel

$$y = -\dfrac{2}{3}x + 3$$

$$\left(\dfrac{3}{2}\right)\left(-\dfrac{2}{3}\right) = -1$$

7.2   Exercises

Section ___   Date ___

In exercises 1 to 4, determine which two equations represent parallel lines.

1. (a) $y = -4x + 5$ (b) $y = 4x + 5$ (c) $y = \frac{1}{4}x + 5$ (d) $y = -4x + 9$

2. (a) $y = 3x - 5$ (b) $y = -3x + 5$ (c) $y = 3x + 2$ (d) $y = -\frac{1}{3}x - 5$

3. (a) $y = \frac{2}{3}x + 3$ (b) $y = -\frac{3}{2}x - 6$ (c) $y = 4x + 12$ (d) $y = 4x - 3$

4. (a) $y = \frac{9}{4}x - 3$ (b) $y = \frac{4}{9}x + 7$ (c) $y = \frac{4}{9}x - 7$ (d) $y = -\frac{9}{4}x + 7$

In exercises 5 to 8, determine which two equations represent perpendicular lines.

5. (a) $y = 6x - 3$ (b) $y = \frac{1}{6}x + 3$ (c) $y = -\frac{1}{6}x + 3$ (d) $y = \frac{1}{6}x - 3$

6. (a) $y = \frac{2}{3}x - 8$ (b) $y = -\frac{2}{3}x - 6$ (c) $y = \frac{2}{3}x - 6$ (d) $y = \frac{3}{2}x - 6$

7. (a) $y = \frac{1}{3}x - 9$ (b) $y = 3x - 9$ (c) $y = \frac{1}{3}x + 9$ (d) $y = -\frac{1}{3}x + 9$

8. (a) $y = \frac{5}{9}x - 6$ (b) $y = 6x - \frac{5}{9}$ (c) $y = -\frac{1}{6}x + \frac{5}{9}$ (d) $y = \frac{1}{6}x - \frac{5}{9}$

Are the following pairs of lines parallel, perpendicular, or neither?

9. L_1 through $(-2, -3)$ and $(4, 3)$
L_2 through $(3, 5)$ and $(5, 7)$

10. L_1 through $(2, 4)$ and $(1, 8)$
L_2 through $(-1, -1)$ and $(-5, 2)$

11. L_1 through $(8, 5)$ and $(3, -2)$
L_2 through $(-2, 4)$ and $(4, -1)$

12. L_1 through $(-2, -3)$ and $(3, -1)$
L_2 through $(-3, 1)$ and $(7, 5)$

ANSWERS
1. 2. 3. 4. 5. 6. 7. 8. 9. 10. 11. 12.

13. _____

14. _____

15. _____

16. _____

17. _____

18. _____

19. _____

20. _____

21. _____

13. L_1 with equation $x - 3y = 6$
L_2 with equation $3x + y = 3$

14. L_1 with equation $x + 2y = 4$
L_2 with equation $2x + 4y = 5$

15. Find the slope of any line parallel to the line through points $(-2, 3)$ and $(4, 5)$.

16. Find the slope of any line perpendicular to the line through points $(0, 5)$ and $(-3, -4)$.

17. A line passing through $(-1, 2)$ and $(4, y)$ is parallel to a line with slope 2. What is the value of y?

18. A line passing through $(2, 3)$ and $(5, y)$ is perpendicular to a line with slope $\frac{3}{4}$. What is the value of y?

In exercises 19 to 21, use the concept of slope to determine if the given figure is a parallelogram or a rectangle.

19.

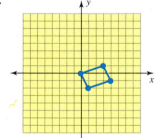

20.

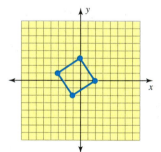

21.

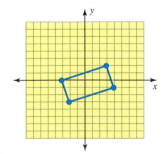

In exercises 22 to 24, use the concept of slope to determine whether the given figure is a right triangle (i.e., does the triangle contain a right angle?).

22.

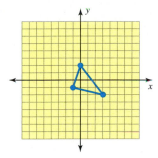

23.

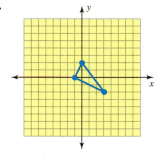

24.

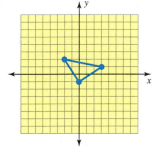

In exercises 25 and 26, use the concept of slope to draw a line perpendicular to the given line segment, passing through the marked point.

25.

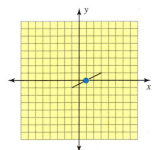

26.

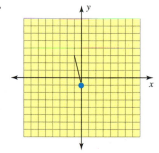

27. Geometry. Floor plans for a building have the four corners of a room located at the points $(2, 3)$, $(11, 6)$, $(-3, 18)$, and $(8, 21)$. Determine whether the side through the points $(2, 3)$ and $(11, 6)$ is parallel to the side through the points $(-3, 18)$ and $(8, 21)$.

28. Geometry. For the floor plans given in exercise 27, determine whether the side through the points $(2, 3)$ and $(11, 6)$ is perpendicular to the side through the points $(2, 3)$ and $(-3, 18)$.

29. Geometry. Determine whether or not the room described in exercise 27 is a rectangle.

ANSWERS

22. _____

23. _____

24. _____

25. _____

26. _____

27. _____

28. _____

29. _____

a. _____

b. _____

c. _____

d. _____

e. _____

f. _____

Getting Ready for Section 7.3 [Section 2.4]

Solve the following equations for y.

(a) $3x + 4y = 12$ (b) $x - y = -5$ (c) $2x + 6 = 4y$

(d) $y + 5 = x$ (e) $5x - 2y = 10$ (f) $-y + 3x = -5$

Answers

1. a and d   **3.** c and d   **5.** a and c   **7.** b and d   **9.** Parallel

11. Neither   **13.** Perpendicular   **15.** $\dfrac{1}{3}$ **17.** 12   **19.** Parallelogram

21. Rectangle   **23.** Yes

25.   **27.** Not parallel   **29.** Not a rectangle

a. $y = -\dfrac{3}{4}x + 3$ **b.** $y = x + 5$ **c.** $y = \dfrac{1}{2}x + \dfrac{3}{2}$ **d.** $y = x - 5$

e. $y = \dfrac{5}{2}x - 5$ **f.** $y = 3x + 5$

7.3 The Point-Slope Form

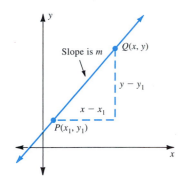

1. Given a point and the slope, find the equation of a line
2. Given two points, find the equation of a line
3. Find the equation of a line from given geometric conditions.

Often in mathematics it is useful to be able to write the equation of a line, given its slope and *any* point on the line. In this section, we will derive a third form for the equation of a line for this purpose.

Suppose that a line has slope m and that it passes through the known point $P(x_1, y_1)$. Let $Q(x, y)$ be any other point on the line. Once again we can use the definition of slope and write

$$m = \frac{y - y_1}{x - x_1}$$

NOTE We use subscripts (x_1, y_1) to indicate a fixed point on the line.

Multiplying both sides by $x - x_1$ gives

$$m(x - x_1) = y - y_1$$

or

$$y - y_1 = m(x - x_1)$$

This equation is called the *point-slope form* for the equation of a line, and all points lying on the line [including (x_1, y_1)] will satisfy this equation. We can state the following general result.

Rules and Properties: Point-Slope Form for the Equation of a Line

RECALL The equation of a line with undefined slope passing through the point (x_1, y_1) is given by $x = x_1$.

The equation of a line with slope m that passes through point (x_1, y_1) is given by

$$y - y_1 = m(x - x_1)$$

OBJECTIVE 1 | **Example 1   Finding the Equation of a Line**

Write the equation for the line that passes through point $(3, -1)$ with a slope of 3.
 Letting $(x_1, y_1) = (3, -1)$ and $m = 3$ in point-slope form, we have

$$y - (-1) = 3(x - 3)$$

or

$$y + 1 = 3x - 9$$

We can write the final result in slope-intercept form as

$$y = 3x - 10$$

 CHECK YOURSELF 1 _____

Write the equation of the line that passes through point $(-2, -4)$ with a slope of $\frac{3}{2}$.

Write your result in slope-intercept form.

Because we know that two points determine a line, it is natural that we should be able to write the equation of a line passing through two given points. Using the point-slope form together with the slope formula will allow us to write such an equation.

OBJECTIVE 2

Example 2   Finding the Equation of a Line

Write the equation of the line passing through (2, 4) and (4, 7).

First, we find m, the slope of the line. Here

$$m = \frac{7 - 4}{4 - 2} = \frac{3}{2}$$

NOTE We could just as well have chosen to let

$(x_1, y_1) = (4, 7)$

The resulting equation will be the same in either case. Take time to verify this for yourself.

Now we apply the point-slope form with $m = \dfrac{3}{2}$ and $(x_1, y_1) = (2, 4)$:

$$y - 4 = \frac{3}{2}(x - 2)$$

$$y - 4 = \frac{3}{2}x - 3$$

$$y = \frac{3}{2}x + 1$$

CHECK YOURSELF 2

Write the equation of the line passing through (−2, 5) and (1, 3). Write your result in slope-intercept form.

A line with slope zero is a horizontal line. A line with an undefined slope is vertical. Example 3 illustrates the equations of such lines.

OBJECTIVE 3

Example 3   Finding the Equations of Horizontal and Vertical Lines

(a) Find the equation of a line passing through (7, −2) with a slope of zero.

We could find the equation by letting $m = 0$. Substituting the ordered pair (7, −2) into the slope-intercept form, we can solve for b.

$$y = mx + b$$
$$-2 = 0(7) + b$$
$$-2 = b$$

So,

$$y = 0 \cdot x - 2 \qquad \text{or} \qquad y = -2$$

It is far easier to remember that any line with a zero slope is a horizontal line and has the form

$$y = b$$

The value for b will always be the y-coordinate for the given point.

(b) Find the equation of a line with undefined slope passing through (4, −5).

A line with undefined slope is vertical. It will always be of the form $x = a$, in which a is the x-coordinate for the given point. The equation is

$$x = 4$$

CHECK YOURSELF 3 _____

(a) Find the equation of a line with zero slope that passes through point $(-3, 5)$.
(b) Find the equation of a line passing through $(-3, -6)$ with undefined slope.

Alternate methods for finding the equation of a line through two points exist and have particular significance in other fields of mathematics, such as statistics. Example 4 shows such an alternate approach.

Example 4   Finding the Equation of a Line

Write the equation of the line through points $(-2, 3)$ and $(4, 5)$.
First, we find m, as before:

$$m = \frac{5 - 3}{4 - (-2)} = \frac{2}{6} = \frac{1}{3}$$

We now make use of the slope-intercept equation, but in a slightly different form.
Because $y = mx + b$, we can write

$$b = y - mx$$

NOTE We substitute these values because the line must pass through $(-2, 3)$.

Now letting $x = -2$, $y = 3$, and $m = \frac{1}{3}$, we can calculate b:

$$b = 3 - \left(\frac{1}{3}\right)(-2)$$

$$= 3 + \frac{2}{3} = \frac{11}{3}$$

With $m = \frac{1}{3}$ and $b = \frac{11}{3}$, we can apply the slope-intercept form to write the equation of the desired line. We have

$$y = \frac{1}{3}x + \frac{11}{3}$$

CHECK YOURSELF 4 _____

Repeat the Check Yourself 2 exercise, using the technique illustrated in Example 4.

We now know that we can write the equation of a line once we have been given appropriate geometric conditions, such as a point on the line and the slope of that line. In some applications the slope may be given not directly but through specified parallel or perpendicular lines.

Example 5   Finding the Equation of a Line

Find the equation of the line passing through $(-4, -3)$ and parallel to the line determined by $3x + 4y = 12$.
First, we find the slope of the given parallel line, as before:

$$3x + 4y = 12$$

$$4y = -3x + 12$$

NOTE The slope of the given line is $-\frac{3}{4}$.

$$y = -\frac{3}{4}x + 3$$

Now because the slope of the desired line must also be $-\dfrac{3}{4}$, we can use the point-slope form to write the required equation:

NOTE The line must pass through $(-4, -3)$, so let $(x_1, y_1) = (-4, -3)$.

$$y - (-3) = -\frac{3}{4}[x - (-4)]$$

This simplifies to

$$y = -\frac{3}{4}x - 6$$

and we have our equation in slope-intercept form.

 CHECK YOURSELF 5 ⎯⎯⎯⎯⎯⎯⎯⎯⎯⎯⎯⎯

Find the equation of the line passing through (5, 4) and perpendicular to the line with equation 2x − 5y = 10.

Hint: Recall that the slopes of perpendicular lines are negative reciprocals of each other.

The following chart summarizes the various forms of the equation of a line.

Form	Equation for Line L	Conditions
Standard	$Ax + By = C$	Constants A and B cannot both be zero.
Slope-intercept	$y = mx + b$	Line L has y-intercept $(0, b)$ with slope m.
Point-slope	$y - y_1 = m(x - x_1)$	Line L passes through point (x_1, y_1) with slope m.
Horizontal	$y = a$	Slope is zero.
Vertical	$x = b$	Slope is undefined.

In real-world applications, we rarely begin with the equation of a line. Rather, we usually have points relating two variables based on actual data. If we believe that the two variables are linearly related, we can provide a line through the data points. This line can then be used to determine other points and make predictions.

Example 6 **An Application of the Point-Slope Form**

A marketing firm spent $12,400 in advertisements in January for Alexa's Used Car Emporium. February sales at Alexa's were $341,000. In August, the firm spent $8600 on advertisements, and sales in the following month were $265,000.

(a) Write a linear equation relating the amount spent on advertisements x with sales in the following month y. Write your answer in slope-intercept form.

We identify the two points (12,400, 341,000) and (8600, 265,000). The slope of the line through these two points is

$$m = \frac{265{,}000 - 341{,}000}{8600 - 12{,}400}$$

$$= \frac{-76{,}000}{-3800}$$

$$= 20$$

We may choose either point to use with the point-slope formula. Here, we will let

$(x_1, y_1) = (12,400, 341,000)$

which gives

$y - 341,000 = 20(x - 12,400)$

We solve this equation for y and simplify to write it in slope-intercept form.

$y - 341,000 = 20(x - 12,400)$
$y - 341,000 = 20x - 20 \cdot 12,400$
$y - 341,000 = 20x - 248,000$
$\qquad\qquad y = 20x - 248,000 + 341,000$
$\qquad\qquad y = 20x + 93,000$

(b) Use the equation found in part (a) to predict the sales amount if $10,000 is spent on advertisements in the previous month.

We evaluate the slope-intercept equation found in part (a) with $x = 10,000$.

$y = 20(10,000) + 93,000$
$ = 200,000 + 93,000$
$ = 293,000$

So Alexa can assume that if she spends $10,000 on advertisements one month, she will record $293,000 in sales in the following month.

CHECK YOURSELF 6

The average cost of tuition and fees at public 4-year colleges was $2159 in 1991. By 1996, the average cost had risen to $3151.

(a) Assuming that the cost y is related to the year x linearly, determine an equation relating the cost to the year. Write your answer in slope-intercept form. *Hint:* If you let $x = 0$ correspond to 1990, then x represents the number of years after 1990; so 1996 corresponds to $x = 6$.
(b) Use the equation found in part (a) to predict the average cost of tuition and fees at public 4-year colleges in 2005.

CHECK YOURSELF ANSWERS

1. $y = \dfrac{3}{2}x - 1$ **2.** $y = -\dfrac{2}{3}x + \dfrac{11}{3}$ **3. (a)** $y = 5$; **(b)** $x = -3$

4. $y = -\dfrac{2}{3}x + \dfrac{11}{3}$ **5.** $y = -\dfrac{5}{2}x + \dfrac{33}{2}$

6. (a) $y = 198.4x + 1960.60$; **(b)** $4936.60

7.3 Exercises

Write the equation of the line passing through each of the given points with the indicated slope. Give your results in slope-intercept form, where possible.

1. _____

2. _____

3. _____

4. _____

5. _____

6. _____

7. _____

8. _____

9. _____

10. _____

11. _____

12. _____

13. _____

14. _____

15. _____

16. _____

17. _____

18. _____

19. _____

20. _____

21. _____

22. _____

1. $(0, 2)$, $m = 3$

2. $(0, -4)$, $m = -2$

3. $(0, 2)$, $m = \dfrac{3}{2}$

4. $(0, -3)$, $m = -2$

5. $(0, 4)$, $m = 0$

6. $(0, 5)$, $m = -\dfrac{3}{5}$

7. $(0, -5)$, $m = \dfrac{5}{4}$

8. $(0, -4)$, $m = -\dfrac{3}{4}$

9. $(1, 2)$, $m = 3$

10. $(-1, 2)$, $m = 3$

11. $(-2, -3)$, $m = -3$

12. $(1, -4)$, $m = -4$

13. $(5, -3)$, $m = \dfrac{2}{5}$

14. $(4, 3)$, $m = 0$

15. $(2, -3)$, m is undefined

16. $(2, -5)$, $m = \dfrac{1}{4}$

Write the equation of the line passing through each of the given pairs of points. Write your result in slope-intercept form, where possible.

17. $(2, 3)$ and $(5, 6)$

18. $(3, -2)$ and $(6, 4)$

19. $(-2, -3)$ and $(2, 0)$

20. $(-1, 3)$ and $(4, -2)$

21. $(-3, 2)$ and $(4, 2)$

22. $(-5, 3)$ and $(4, 1)$

23. $(2, 0)$ and $(0, -3)$

24. $(2, -3)$ and $(2, 4)$

25. $(0, 4)$ and $(-2, -1)$

26. $(-4, 1)$ and $(3, 1)$

Write the equation of the line L satisfying the given geometric conditions.

27. L has slope 4 and y-intercept $(0, -2)$.

28. L has slope $-\dfrac{2}{3}$ and y-intercept $(0, 4)$.

29. L has x-intercept $(4, 0)$ and y-intercept $(0, 2)$.

30. L has x-intercept $(-2, 0)$ and slope $\dfrac{3}{4}$.

31. L has y-intercept $(0, 4)$ and a 0 slope.

32. L has x-intercept $(-2, 0)$ and an undefined slope.

33. L passes through point $(3, 2)$ with a slope of 5.

34. L passes through point $(-2, -4)$ with a slope of $-\dfrac{3}{2}$.

35. L has y-intercept $(0, 3)$ and is parallel to the line with equation $y = 3x - 5$.

36. L has y-intercept $(0, -3)$ and is parallel to the line with equation $y = \dfrac{2}{3}x + 1$.

37. L has y-intercept $(0, 4)$ and is perpendicular to the line with equation $y = -2x + 1$.

38. L has y-intercept $(0, 2)$ and is parallel to the line with equation $y = -1$.

39. L has y-intercept $(0, 3)$ and is parallel to the line with equation $y = 2$.

40. L has y-intercept $(0, 2)$ and is perpendicular to the line with equation $2x - 3y = 6$.

23. _____

24. _____

25. _____

26. _____

27. _____

28. _____

29. _____

30. _____

31. _____

32. _____

33. _____

34. _____

35. _____

36. _____

37. _____

38. _____

39. _____

40. _____

41. _____

42. _____

43. _____

44. _____

45. _____

46. _____

47. _____

48. _____

49. _____

50. _____

51. _____

52. _____

53. _____

41. *L* passes through point $(-3, 2)$ and is parallel to the line with equation $y = 2x - 3$.

42. *L* passes through point $(-4, 3)$ and is parallel to the line with equation $y = -2x + 1$.

43. *L* passes through point $(3, 2)$ and is parallel to the line with equation $y = \frac{4}{3}x + 4$.

44. *L* passes through point $(-2, -1)$ and is perpendicular to the line with equation $y = 3x + 1$.

45. *L* passes through point $(5, -2)$ and is perpendicular to the line with equation $y = -3x - 2$.

46. *L* passes through point $(3, 4)$ and is perpendicular to the line with equation $y = -\frac{3}{5}x + 2$.

47. *L* passes through $(-2, 1)$ and is parallel to the line with equation $x + 2y = 4$.

48. *L* passes through $(-3, 5)$ and is parallel to the *x*-axis.

49. Describe the process for finding the equation of a line if you are given two points on the line.

50. How would you find the equation of a line if you were given the slope and the *x-intercept*?

51. Science and medicine. A temperature of 10°C corresponds to a temperature of 50°F. Also 40°C corresponds to 104°F. Find the linear equation relating *F* and *C*.

52. Business and finance. In planning for a new item, a manufacturer assumes that the number of items produced *x* and the cost in dollars *C* of producing these items are related by a linear equation. Projections are that 100 items will cost $10,000 to produce and that 300 items will cost $22,000 to produce. Find the equation that relates *C* and *x*.

53. Technology. A word processing station was purchased by a company for $10,000. After 4 years it is estimated that the value of the station will be $4000. If the value in dollars *V* and the time the station has been in use *t* are related by a linear equation, find the equation that relates *V* and *t*.

54. Business and finance. Two years after an expansion, a company had sales of $42,000. Four years later the sales were $102,000. Assuming that the sales in dollars S and the time in years t are related by a linear equation, find the equation relating S and t.

Getting Ready for Section 7.4 [Section 2.7]

Graph each of the following inequalities on a number line.

(a) $x < 3$

(b) $x \geq -2$

(c) $2x \leq 8$

(d) $3x \geq -9$

(e) $-3x < 12$

(f) $-2x \leq 10$

(g) $\dfrac{2}{3}x \leq 4$

(h) $-\dfrac{3}{4}x \geq 6$

Answers

1. $y = 3x + 2$ **3.** $y = \dfrac{3}{2}x + 2$ **5.** $y = 4$ **7.** $y = \dfrac{5}{4}x - 5$

9. $y = 3x - 1$ **11.** $y = -3x - 9$ **13.** $y = \dfrac{2}{5}x - 5$ **15.** $x = 2$

17. $y = x + 1$ **19.** $y = \dfrac{3}{4}x - \dfrac{3}{2}$ **21.** $y = 2$ **23.** $y = \dfrac{3}{2}x - 3$

25. $y = \dfrac{5}{2}x + 4$ **27.** $y = 4x - 2$ **29.** $y = -\dfrac{1}{2}x + 2$

31. $y = 4$ **33.** $y = 5x - 13$ **35.** $y = 3x + 3$ **37.** $y = \dfrac{1}{2}x + 4$

39. $y = 3$ **41.** $y = 2x + 8$ **43.** $y = \dfrac{4}{3}x - 2$ **45.** $y = \dfrac{1}{3}x - \dfrac{11}{3}$

47. $y = -\dfrac{1}{2}x$ **49.** **51.** $F = \dfrac{9}{5}C + 32$

53. $V = -1500t + 10{,}000$

a.

b.

c.

d.

e.

f.

g.

h.

7.4 Graphing Linear Inequalities

7.4 OBJECTIVE

1. Graph a linear inequality in two variables

In Section 2.7 you learned to graph inequalities in one variable on a number line. We now want to extend our work with graphing to include linear inequalities in two variables. We begin with a definition.

> ### Definition: Linear Inequality in Two Variables
>
> An inequality that can be written in the form
>
> $Ax + By < C$
>
> in which A and B are not both 0, is called a **linear inequality in two variables.**

NOTE The inequality symbols ≤, >, and ≥ can also be used.

Some examples of linear inequalities in two variables are

$$x + 3y > 6 \qquad y \leq 3x + 1 \qquad 2x - y \geq 3$$

The *graph* of a linear inequality is always a region (actually a half–plane) of the plane whose boundary is a straight line. Let's look at an example of graphing such an inequality.

OBJECTIVE 1

Example 1 Graphing a Linear Inequality

Graph $2x + y < 4$.

First, replace the inequality symbol ($<$) with an equal–sign. We then have $2x + y = 4$. This equation forms the **boundary line** of the graph of the original inequality. You can graph the line by any of the methods discussed earlier.

The boundary line for our inequality is shown at left. We see that the boundary line separates the plane into two regions, each of which is called a **half–plane.**

We now need to choose the correct half–plane. Choose any convenient test point not on the boundary line. The origin (0, 0) is a good choice because it results in an easy calculation.

Substitute $x = 0$ and $y = 0$ into the inequality.

$$2 \cdot 0 + 0 \overset{?}{<} 4$$
$$0 + 0 \overset{?}{<} 4$$
$$0 < 4 \qquad \text{A true statement}$$

NOTE The dotted line indicates that the points on the line $2x + y = 4$ are *not* part of the solution to the inequality $2x + y < 4$.

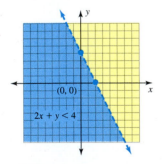

NOTE You can always use the origin for a test point unless the boundary line passes through the origin.

Because the inequality is *true* for the test point, we shade the half–plane containing that test point (the origin). The origin and all other points *below* the boundary line then represent solutions for our original inequality.

 CHECK YOURSELF 1

Graph the inequality $x + 3y < 3$.

638

The process is similar when the boundary line is included in the solution.

Example 2 Graphing a Linear Inequality

Graph $4x - 3y \geq 12$.

NOTE Again, we replace the inequality symbol ($\geq$) with an equal–sign to write the equation for our boundary line.

First, graph the boundary line $4x - 3y = 12$.

Note: When equality *is included* ($\leq$ or $\geq$), use a *solid line* for the graph of the boundary line. This means the line is included in the graph of the linear inequality.

The graph of our boundary line (a solid line here) is shown on the figure.

NOTE Although any of our graphing methods can be used here, the intercept method is probably the most efficient.

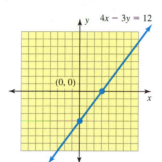

Again, we use $(0, 0)$ as a convenient test point. Substituting 0 for x and for y in the original inequality, we have

$$4 \cdot 0 - 3 \cdot 0 \geq 12$$

$$0 \geq 12 \qquad \text{A false statement}$$

Because the inequality is *false* for the test point, we shade the half–plane that does *not* contain that test point, here $(0, 0)$.

NOTE All points *on and below* the boundary line represent solutions for our original inequality.

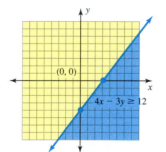

✔ CHECK YOURSELF 2

Graph the inequality $3x + 2y \geq 6$.

Example 3 Graphing a Linear Inequality with One Variable

Graph $x \leq 5$.

The boundary line is $x = 5$. Its graph is a solid line because equality is included. Using $(0, 0)$ as a test point, we substitute 0 for x with the result

$$0 \leq 5 \qquad \text{A true statement}$$

Because the inequality is *true* for the test point, we shade the half–plane containing the origin.

NOTE If the correct half–plane is obvious, you may not need to use a test point. Did you know without testing which half–plane to shade in this example?

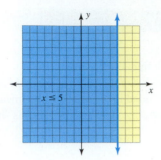

CHECK YOURSELF 3 _____

Graph the inequality y < 2.

As we mentioned earlier, we may have to use a point other than the origin as our test point. Example 4 illustrates this approach.

Example 4 **Graphing a Linear Inequality through the Origin**

Graph $2x + 3y < 0$.

The boundary line is $2x + 3y = 0$. Its graph is shown on the figure.

NOTE We use a dotted line for our boundary line because equality is not included.

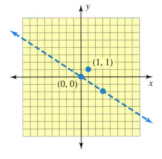

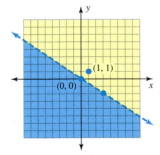

We cannot use (0, 0) as our test point in this case. Do you see why?

Choose any other point *not* on the line. We chose (1, 1) as a test point. Substituting 1 for *x* and 1 for *y* gives

$$2 \cdot 1 + 3 \cdot 1 \overset{?}{<} 0$$
$$2 + 3 \overset{?}{<} 0$$
$$5 < 0 \qquad \text{A false statement}$$

Because the inequality is *false* at our test point, we shade the half–plane *not* containing (1, 1). This is shown in the graph in the margin.

CHECK YOURSELF 4 _____

Graph the inequality x − 2y < 0.

The following steps summarize our work in graphing linear inequalities in two variables.

> **Step by Step: To Graph a Linear Inequality**
>
> **Step 1** Replace the inequality symbol with an equal–sign to form the equation of the boundary line of the graph.
> **Step 2** Graph the boundary line. Use a dotted line if equality is not included ($<$ or $>$). Use a solid line if equality is included ($\leq$ or $\geq$).
> **Step 3** Choose any convenient test point *not* on the line.
> **Step 4** If the inequality is *true* at the checkpoint, shade the half–plane including the test point. If the inequality is *false* at the checkpoint, shade the half–plane not including the test point.

NOTE We will develop the idea of graphing "bounded" regions more fully in Chapter 8.

Linear inequalities and their graphs may be used to represent *feasible regions* in applications. These regions include all points that satisfy some set of conditions determined by the application.

> ## Example 5   Applications of Linear Inequalities

A hospital food service can serve at most 1000 meals per day. Patients on a normal diet receive three meals per day and patients on a special diet receive four meals per day.

(a) Write a linear inequality that describes the number of patients that can be served in a day and sketch its graph.

RECALL "At most" means less than or equal to.

Let x be the number of people served a normal diet. Then $3x$ represents the number of meals served to the people on a normal diet. Let y be the number of people served a special diet. Then $4y$ represents the number of meals served to the people on the special diet. We know that the total number of meals served is at most 1000. Writing this as an inequality gives

$$3x + 4y \leq 1000$$

or

$$y \leq -\frac{3}{4}x + 250 \qquad \text{We solved the inequality for } y.$$

We only need to graph this inequality in the first quadrant. Do you see why?

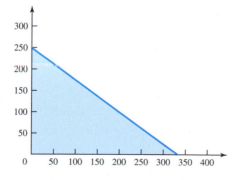

(b) Can the hospital food service serve 100 patients on a normal diet and 100 patients on the special diet in a day?

We can substitute 100 for both x and y in the inequality found in part (a) and see if a true statement results.

$$y \leq -\frac{3}{4}x + 250$$

$$100 \stackrel{?}{\leq} -\frac{3}{4}(100) + 250$$

$$100 \stackrel{?}{\leq} -75 + 250$$

$$100 \leq 175 \qquad\qquad \text{True!}$$

Since this final inequality is true, we conclude that the hospital food service can serve 100 people on each type of diet in a single day. Graphically, we see that the point (100, 100) is in the solution region.

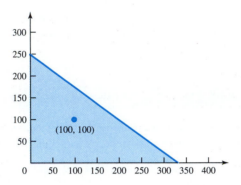

(c) If the hospital serves 200 people on a normal diet, then what is the maximum number of people who can be served the special diet?

Substitute 200 for x and calculate y.

$$y \leq -\frac{3}{4}x + 250$$

$$y \leq -\frac{3}{4}(200) + 250$$

$$y \leq -150 + 250$$

$$y \leq 100$$

We conclude that if the hospital serves 200 meals to patients on a normal diet, they can serve, at most, 100 patients on the special diet.

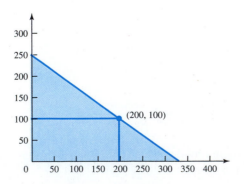

CHECK YOURSELF 5

A manufacturer produces a standard and deluxe model of 13-in. television sets. The standard model requires 12 h to produce, while the deluxe model requires 18 h to produce. The manufacturer has a total of 360 h of labor available in a week.

(a) Write a linear inequality to represent the number of each type of television set the manufacturer can produce in a week and graph the inequality. *Hint:* You only need to graph the inequality in the first quadrant.

(b) If the manufacturer needs to produce 16 standard models one week, what is the largest number of deluxe models that can be produced in that week?

CHECK YOURSELF ANSWERS

1.

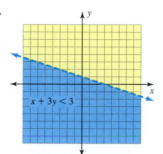

2.

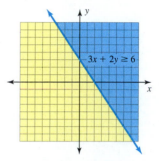

3.

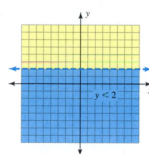

4.

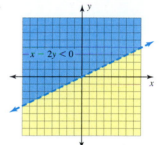

5. (a) $12x + 18y \leq 360$ or $y \leq -\dfrac{2}{3}x + 20$; **(b)** nine deluxe models

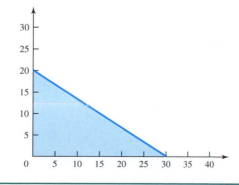

7.4 Exercises

In exercises 1 to 8, we have graphed the boundary line for the linear inequality. Determine the correct half–plane in each case, and complete the graph.

1. $x + y < 5$

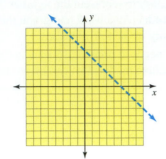

2. $x - y \geq 4$

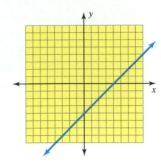

3. $x - 2y \geq 4$

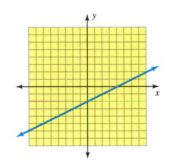

4. $2x + y < 6$

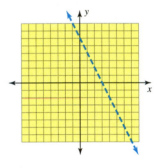

5. $x \leq -3$

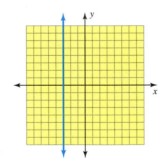

6. $y \geq 2x$

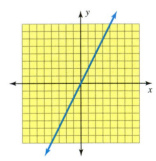

7. $y < 2x - 6$

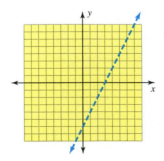

8. $y > 3$

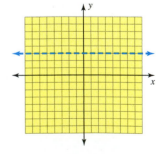

Graph each of the following inequalities.

9. $x + y < 3$

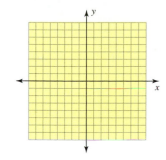

10. $x - y \geq 4$

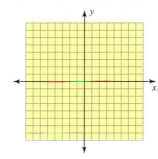

11. $x - y \leq 5$

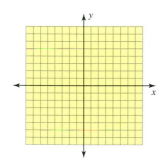

12. $x + y > 5$

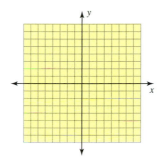

13. $2x + y < 6$

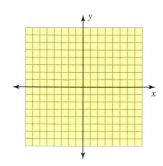

14. $3x + y \geq 6$

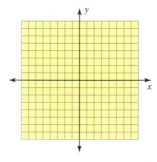

15. $x \leq 3$

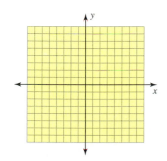

16. $4x + y \geq 4$

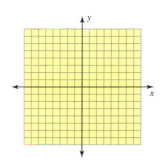

17.

18.

19.

20.

21.

22.

23.

24.

17. $x - 5y < 5$

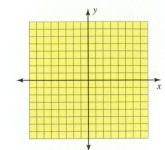

18. $y > 3$

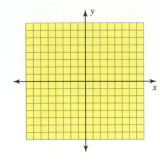

19. $y < -4$

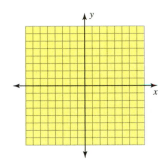

20. $4x + 3y > 12$

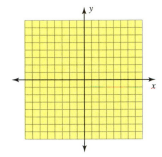

21. $2x - 3y \geq 6$

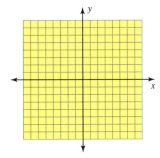

22. $x \geq -2$

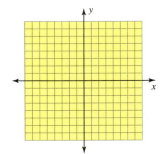

23. $3x + 2y \geq 0$

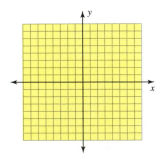

24. $3x + 5y < 15$

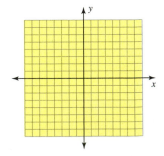

25. $5x + 2y > 10$

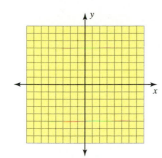

26. $x - 3y \geq 0$

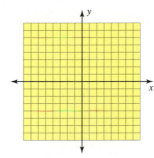

27. $y \leq 2x$

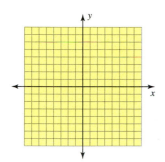

28. $3x - 4y < 12$

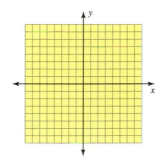

29. $y > 2x - 3$

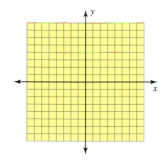

30. $y \geq -2x$

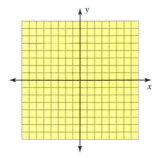

31. $y < -2x - 3$

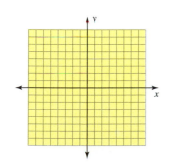

32. $y \leq 3x + 4$

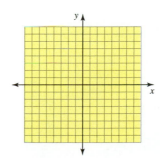

ANSWERS

25. _____

26. _____

27. _____

28. _____

29. _____

30. _____

31. _____

32. _____

33. _____

34. _____

35. _____

36. _____

37. _____

38. _____

Graph each of the following inequalities.

33. $2(x + y) - x > 6$

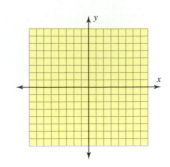

34. $3(x + y) - 2y < 3$

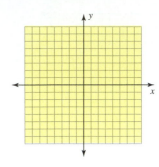

35. $4(x + y) - 3(x + y) \leq 5$

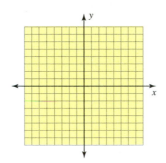

36. $5(2x + y) - 4(2x + y) \geq 4$

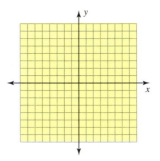

37. Business and finance. Suppose you have two part-time jobs. One is at a video store that pays $9 per hour and the other is at a convenience store that pays $8 per hour. Between the two jobs, you want to earn at least $240 per week. Write an inequality that shows the various number of hours you can work at each job.

38. Number problems. You have at least $30 in change in your drawer, consisting of dimes and quarters. Write an inequality that shows the different number of coins in your drawer.

39. Business and finance. Linda Williams has just begun a nursery business and seeks your advice. She has limited funds to spend and wants to stock two kinds of fruit-bearing plants. She lives in the northeastern part of Texas and thinks that blueberry bushes and peach trees would sell well there. Linda can buy blueberry bushes from a supplier for $2.50 each and young peach trees for $5.50 each. She wants to know what combination she should buy and keep her outlay to $500 or less. Write an inequality and draw a graph to depict what combinations of blueberry bushes and peach trees she can buy for the amount of money she has. Explain the graph and her options.

40. Statistics. After reading an article on the front page of *The New York Times* titled "You Have to Be Good at Algebra to Figure Out the Best Deal for Long Distance," Rafaella De La Cruz decided to apply her skills in algebra to try to decide between two competing long-distance companies. It was difficult at first to get the companies to explain their charge policies. They both kept repeating that they were 25% cheaper than their competition. Finally, Rafaella found someone who explained that the charge depended on when she called, where she called, how long she talked, and how often she called. "Too many variables!" she exclaimed. So she decided to ask one company what they charged as a base amount, just for using the service.

Company A said that they charged $5 for the privilege of using their long-distance service, whether or not she made any phone calls, and that because of this fee they were able to allow her to call anywhere in the United States after 6 P.M. for only $0.15 a minute. Complete this table of charges based on this company's plan:

Total Minutes Long Distance in 1 Month (After 6 P.M.)	Total Charge
0 minutes	
10 minutes	
30 minutes	
60 minutes	
120 minutes	

Use this table to make a graph of the monthly charges from Company A based on the number of minutes of long distance.

Rafaella wanted to compare this offer to Company B, which she was currently using. She looked at her phone bill and saw that one month she had been charged $7.50 for 30 minutes and another month she had been charged $11.25 for 45 minutes of long-distance calling. These calls were made after 6 P.M. to her relatives in Indiana and Arizona. Draw a graph on the same set of axes you made for Company A's figures. Use your graph and what you know about linear inequalities to advise Rafaella about which company is best.

Getting Ready for Section 7.5 [Section 1.5]

Evaluate each expression for the given variable value.

(a) $2x + 1$ $(x = 2)$

(b) $2x + 1$ $(x = -2)$

(c) $3 - 2x$ $(x = 1)$

(d) $3 - 2x$ $(x = -1)$

(e) $x^2 - 2\ (x = 2)$ (f) $x^2 - 2\ (x = -2)$

(g) $x^2 + 5\ (x = 1)$ (h) $x^2 + 5\ (x = -1)$

Answers

1. $x + y < 5$

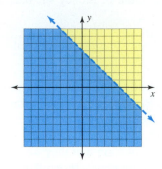

3. $x - 2y \geq 4$

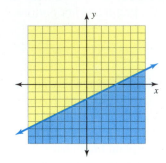

5. $x \leq -3$

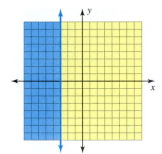

7. $y < 2x - 6$

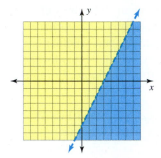

9. $x + y < 3$

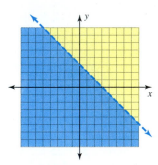

11. $x - y \leq 5$

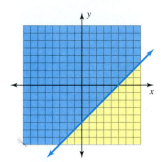

13. $2x + y < 6$

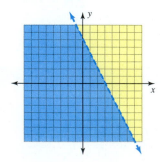

15. $x \leq 3$

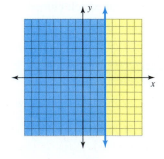

17. $x - 5y < 5$

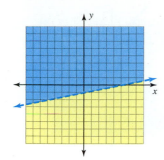

19. $y < -4$

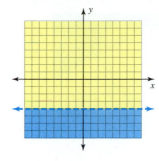

21. $2x - 3y \geq 6$

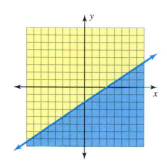

23. $3x + 2y \geq 0$

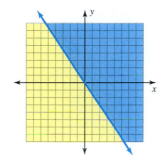

25. $5x + 2y > 10$

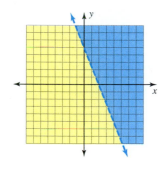

27. $y \leq 2x$

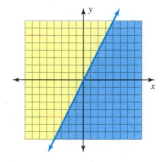

29. $y > 2x - 3$

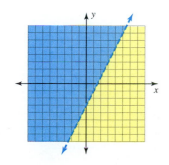

31. $y < -2x - 3$

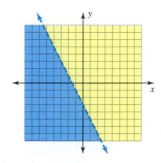

33. $x + 2y > 6$

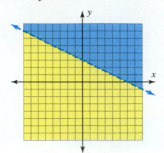

35. $x + y \leq 5$

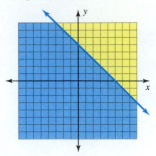

37. $9x + 8y \geq 240$ **39.**   **a.** 5   **b.** -3 **c.** 1   **d.** 5

e. 2   **f.** 2   **g.** 6   **h.** 6

7.5 An Introduction to Functions

7.5 OBJECTIVES

1. Evaluate expressions
2. Evaluate functions
3. Express the equation of a line as a linear function
4. Graph a linear function

Variables can be used to represent unknown real numbers. Together with the operations of addition, subtraction, multiplication, division, and exponentiation, numbers and variables form expressions such as

$$3x + 5 \qquad 7x - 4 \qquad x^2 - 3x - 10 \qquad x^4 - 2x^2 + 3x + 4$$

Four different actions can be taken with expressions. We can

1. Substitute values for the variable(s) and **evaluate the expression.**
2. Rewrite an expression as some simpler equivalent expression. This rewriting is called **simplifying the expression.**
3. Set two expressions equal to each other and **solve for the stated variable.**
4. Set two expressions equal to each other and **graph the equation.**

Throughout this book, everything we do involves one of these four actions. We now return our focus to the first item, evaluating expressions. As we saw in Section 1.5, expressions can be evaluated for an indicated value of the variable(s). Example 1 illustrates.

OBJECTIVE 1

Example 1   Evaluating Expressions

Evaluate the expression $x^4 - 2x^2 + 3x + 4$ for the indicated value of x.

(a) $x = 0$

Substituting 0 for x in the expression yields

$$(0)^4 - 2(0)^2 + 3(0) + 4 = 0 - 0 + 0 + 4$$
$$= 4$$

(b) $x = 2$

Substituting 2 for x in the expression yields

$$(2)^4 - 2(2)^2 + 3(2) + 4 = 16 - 8 + 6 + 4$$
$$= 18$$

(c) $x = -1$

Substituting -1 for x in the expression yields

$$(-1)^4 - 2(-1)^2 + 3(-1) + 4 = 1 - 2 - 3 + 4$$
$$= 0$$

CHECK YOURSELF 1

Evaluate the expression $2x^3 - 3x^2 + 3x + 1$ for the indicated value of x.

(a) $x = 0$ **(b)** $x = 1$ **(c)** $x = -2$

We could design a machine whose function would be to crank out the value of an expression for each given value of x. We could call this machine something simple such as f. Our *function* machine might look like this.

For example, when we put -1 into the machine, the machine would substitute -1 for x in the expression, and 5 would come out the other end because

$$2(-1)^3 + 3(-1)^2 - 5(-1) - 1 = -2 + 3 + 5 - 1 = 5$$

In fact, the idea of the function machine is very useful in mathematics. Your graphing calculator can be used as a function machine. You can enter the expression into the calculator as Y_1 and then evaluate Y_1 for different values of x.

Generally, in mathematics, we do not write $Y_1 = 2x^3 + 3x^2 - 5x - 1$. Instead, we write $f(x) = 2x^3 + 3x^2 - 5x - 1$, which is read as "$f$ of x is equal to. . . ." Instead of calling f a function machine, we say that f is a function of x. The greatest benefit of this notation is that it lets us easily note the input value of x along with the output of the function. Instead of "Evaluate y for $x = 4$" we say "Find $f(4)$." This means that, given the function f, $f(c)$ designates the value of the function when the variable is equal to c.

NOTE $f(x)$ does not mean f times x.

OBJECTIVE 2

Example 2   Evaluating Expressions with Function Notation

Given $f(x) = x^3 - 3x^2 + x + 5$, find the following:

(a) $f(0)$

Substituting 0 for x in the expression, we get

$$f(0) = (0)^3 - 3(0)^2 + (0) + 5$$
$$= 5$$

(b) $f(-3)$

Substituting -3 for x in the expression, we get

$$f(-3) = (-3)^3 - 3(-3)^2 + (-3) + 5$$
$$= -27 - 27 - 3 + 5$$
$$= -52$$

(c) $f\left(\dfrac{1}{2}\right)$

Substituting $\dfrac{1}{2}$ for x in the expression, we get

$$f\left(\frac{1}{2}\right) = \left(\frac{1}{2}\right)^3 - 3\left(\frac{1}{2}\right)^2 + \left(\frac{1}{2}\right) + 5$$
$$= \frac{1}{8} - 3\left(\frac{1}{4}\right) + \frac{1}{2} + 5$$
$$= \frac{1}{8} - \frac{3}{4} + \frac{1}{2} + 5$$

$$= \frac{1}{8} - \frac{6}{8} + \frac{4}{8} + 5$$

$$= -\frac{1}{8} + 5$$

$$= \frac{39}{8}$$

CHECK YOURSELF 2

Given f(x) = 2x³ − x² + 3x − 2, find the following.

(a) $f(0)$ **(b)** $f(3)$ **(c)** $f\left(-\frac{1}{2}\right)$

Given a function f, the pair of numbers $(x, f(x))$ is very significant. We always write them in that order. In part (a) of Example 2 we saw that given $f(x) = x^3 - 3x^2 + x + 5$, $f(0) = 5$, which meant that the ordered pair $(0, 5)$ was associated with the function. The ordered pair consists of the x-value first and the function value at that x (the $f(x)$) second.

Example 3   Finding Ordered Pairs

Given the function $f(x) = 2x^2 - 3x + 5$, find the ordered pair $(x, f(x))$ associated with each given value for x.

(a) $x = 0$

$f(0) = 5$

so the ordered pair is $(0, 5)$.

(b) $x = -1$

$f(-1) = 2(-1)^2 - 3(-1) + 5 = 10$

The ordered pair is $(-1, 10)$.

(c) $x = \frac{1}{4}$

$$f\left(\frac{1}{4}\right) = 2\left(\frac{1}{16}\right) - 3\left(\frac{1}{4}\right) + 5 = \frac{35}{8}$$

The ordered pair is $\left(\frac{1}{4}, \frac{35}{8}\right)$.

CHECK YOURSELF 3

Given f(x) = 2x³ − x² + 3x − 2, find the ordered pair associated with each given value of x.

(a) $x = 0$ **(b)** $x = 3$ **(c)** $x = -\frac{1}{2}$

Any linear equation $Ax + By = C$ in which $B \neq 0$ can be written using function notation. Because the process for finding a value for y given a particular value for x is the same as finding $f(x)$ given some value for x, we know that both y and $f(x)$ will always yield the same number given some value for x.

Consider the equation $y = x + 1$. If $x = 1$, then we determine y by substituting 1 for x in the equation.

$$y = (1) + 1 = 2$$

We then express this using ordered-pair notation $(x, y) = (1, 2)$ as a solution to the equation $y = x + 1$.

If we write the function $f(x) = x + 1$ and evaluate $f(1)$, we get

$$f(1) = (1) + 1 = 2$$

So, $(x, f(x)) = (1, 2)$ is an ordered pair associated with the function.

This leads us to the conclusion that $y = f(x)$ represents the relationship between function notation and equations in two variables.

To write an equation in two variables (x, y) as a function, follow these two steps.

1. Solve the equation for y, if possible.
2. Replace y with $f(x)$.

If we are unable to complete step 1, then we cannot write y as a function of x. With linear equations, $Ax + By = C$, this is equivalent to the condition $B = 0$. That is, the equation for a vertical line $x = a$ does not represent a function.

On the other hand, *all* nonvertical lines may be written as functions.

OBJECTIVE 3   **Example 4   Writing Equations as Functions**

Rewrite each linear equation as a function of x.

(a) $y = 3x - 4$

Because the equation is already solved for y, we simply replace y with $f(x)$.

$$f(x) = 3x - 4$$

(b) $2x - 3y = 6$

This equation must first be solved for y.

$$-3y = -2x + 6$$
$$y = \frac{2}{3}x - 2$$

We then rewrite the equation as

$$f(x) = \frac{2}{3}x - 2$$

 CHECK YOURSELF 4

Rewrite each equation as a function of x.

(a) $y = -2x + 5$ **(b)** $3x + 5y = 15$

The process of finding the graph of a linear function is identical to the process of finding the graph of a linear equation.

OBJECTIVE 4   **Example 5   Graphing a Linear Function**

Graph the function

$$f(x) = 3x - 5$$

We could use the slope and y-intercept to graph the line, or we can find three points (the third is a checkpoint) and draw the line through them. We will do the latter.

$$f(0) = -5 \qquad f(1) = -2 \qquad f(2) = 1$$

We use the three points $(0, -5)$, $(1, -2)$, and $(2, 1)$ to graph the line.

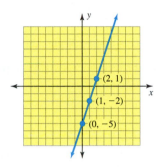

CHECK YOURSELF 5

Graph the function

$$f(x) = 5x - 3$$

One benefit of having a function written in $f(x)$ form is that it makes it fairly easy to substitute values for x. In Example 5, we substituted the values 0, 1, and 2. Sometimes it is useful to substitute nonnumeric values for x.

Example 6   Substituting Nonnumeric Values for x

Let $f(x) = 2x + 3$. Evaluate f as indicated.

(a) $f(a)$

Substituting a for x in our equation, we see that

$$f(a) = 2a + 3$$

(b) $f(2 + h)$

Substituting $2 + h$ for x in our equation, we get

$$f(2 + h) = 2(2 + h) + 3$$

Distributing the 2 and then simplifying, we have

$$f(2 + h) = 4 + 2h + 3$$
$$= 2h + 7$$

CHECK YOURSELF 6

Let $f(x) = 4x - 2$. Evaluate f as indicated.

(a) $f(b)$ (b) $f(4 + h)$

Functions and function notation are used in many applications, as illustrated by Example 7.

Example 7   Applications of Functions

The profit function for stereos sold by S-Bar Electronics is given by

NOTE f is one name choice for a function. In applications, functions are often named in some way that relates to their usage.

$$P(x) = 34x - 3500$$

in which x is the number of stereos sold and $P(x)$ is the profit produced by selling x stereos.

(a) Determine the profit if S-Bar Electronics sells 75 stereos.

$$P(75) = 34(75) - 3500$$
$$= 2550 - 3500$$
$$= -950$$

Because the profit is negative, we determine that S-Bar Electronics suffers a *loss* of $950 if it only sells 75 stereos.

(b) Determine the profit from the sale of 150 stereos.

$$P(150) = 34(150) - 3500$$
$$= 5100 - 3500$$
$$= 1600$$

S-Bar Electronics earns a profit of $1600 when it sells 150 stereos.

(c) Find the break-even point (that is, the number of stereos that must be sold for the profit to equal zero).

We set $P(x) = 0$ and solve for x. That is, we are trying to find x so that the profit is zero.

NOTE If they sell 103 stereos, they earn a profit of $2; however, if they sell 102 stereos, they lose $32. Since the number of stereos sold must be a whole number, we set the break-even point to 103.

$$0 = 34x - 3500$$
$$3500 = 34x$$
$$x = \frac{3500}{34} \approx 103$$

The break-even point for the profit equation is 103 stereos.

CHECK YOURSELF 7

The manager at Sweet Eats determines that the profit produced by serving x dinners is given by the function

$P(x) = 42x - 620$

(a) Determine the profit if the restaurant serves 20 dinners.
(b) How many dinners must be served to reach the break-even point?

CHECK YOURSELF ANSWERS

1. **(a)** 1; **(b)** 3; **(c)** -33 2. **(a)** -2; **(b)** 52; **(c)** -4

3. **(a)** $(0, -2)$; **(b)** $(3, 52)$; **(c)** $\left(-\dfrac{1}{2}, -4\right)$

4. **(a)** $f(x) = -2x + 5$; **(b)** $f(x) = -\dfrac{3}{5}x + 3$ **5.**

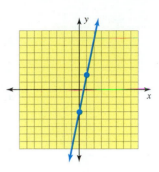

6. **(a)** $4b - 2$; **(b)** $4h + 14$ 7. **(a)** \$220; **(b)** 15

7.5 Exercises

1. ————————
2. ————————
3. ————————
4. ————————
5. ————————
6. ————————
7. ————————
8. ————————
9. ————————
10. ————————
11. ————————
12. ————————
13. ————————
14. ————————
15. ————————
16. ————————
17. ————————
18. ————————
19. ————————
20. ————————

In exercises 1 to 10, evaluate each function for the value specified.

1. $f(x) = x^2 - x - 2$; find (a) $f(0)$, (b) $f(-2)$, and (c) $f(1)$.

2. $f(x) = x^2 - 7x + 10$; find (a) $f(0)$, (b) $f(5)$, and (c) $f(-2)$.

3. $f(x) = 3x^2 + x - 1$; find (a) $f(-2)$, (b) $f(0)$, and (c) $f(1)$.

4. $f(x) = -x^2 - x - 2$; find (a) $f(-1)$, (b) $f(0)$, and (c) $f(2)$.

5. $f(x) = x^3 - 2x^2 + 5x - 2$; find (a) $f(-3)$, (b) $f(0)$, and (c) $f(1)$.

6. $f(x) = -2x^3 + 5x^2 - x - 1$; find (a) $f(-1)$, (b) $f(0)$, and (c) $f(2)$.

7. $f(x) = -3x^3 + 2x^2 - 5x + 3$; find (a) $f(-2)$, (b) $f(0)$, and (c) $f(3)$.

8. $f(x) = -x^3 + 5x^2 - 7x - 8$; find (a) $f(-3)$, (b) $f(0)$, and (c) $f(2)$.

9. $f(x) = 2x^3 + 4x^2 + 5x + 2$; find (a) $f(-1)$, (b) $f(0)$, and (c) $f(1)$.

10. $f(x) = -x^3 + 2x^2 - 7x + 9$; find (a) $f(-2)$, (b) $f(0)$, and (c) $f(2)$.

In exercises 11 to 20, rewrite each equation as a function of x.

11. $y = -3x + 2$ **12.** $y = 5x + 7$

13. $y = 4x - 8$ **14.** $y = -7x - 9$

15. $3x + 2y = 6$ **16.** $4x + 3y = 12$

17. $-2x + 6y = 9$ **18.** $-3x + 4y = 11$

19. $-5x - 8y = -9$ **20.** $4x - 7y = -10$

ANSWERS

21. _____

22. _____

23. _____

24. _____

25. _____

26. _____

27. _____

28. _____

29. _____

30. _____

31. _____

32. _____

In exercises 21 to 26, graph the functions.

21. $f(x) = 3x + 7$

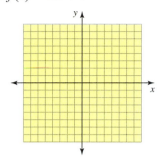

22. $f(x) = -2x - 5$

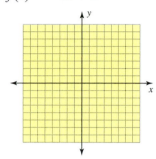

23. $f(x) = -2x + 7$

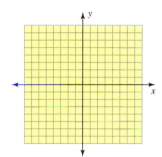

24. $f(x) = -3x + 8$

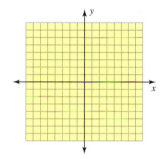

25. $f(x) = -x - 1$

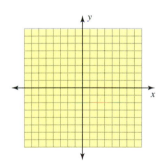

26. $f(x) = -2x - 5$

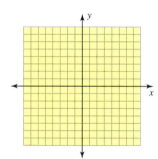

In exercises 27 to 32, if $f(x) = 4x - 3$, find the following:

27. $f(5)$

28. $f(0)$

29. $f(4)$

30. $f(-1)$

31. $f(-4)$

32. $f\left(\dfrac{1}{2}\right)$

33. _____

34. _____

35. _____

36. _____

37. _____

38. _____

39. _____

40. _____

41. _____

42. _____

43. _____

44. _____

45. _____

46. _____

47. _____

48. _____

49. _____

In exercises 33 to 38, if $f(x) = 5x - 1$, find the following:

33. $f(a)$ **34.** $f(2r)$

35. $f(x + 1)$ **36.** $f(a - 2)$

37. $f(x + h)$ **38.** $\dfrac{f(x + h) - f(x)}{h}$

In exercises 39 to 42, if $g(x) = -3x + 2$, find the following:

39. $g(m)$ **40.** $g(5n)$

41. $g(x + 2)$ **42.** $g(s - 1)$

In exercises 43 to 46, let $f(x) = 2x + 3$.

43. Find $f(1)$.   **44.** Find $f(3)$.

45. Form the ordered pairs $(1, f(1))$ and $(3, f(3))$.

46. Write the equation of the line passing through the points determined by the ordered pairs in exercise 45.

47. In each of the following statements, fill in the blank with *always, sometimes,* or *never*.

    (a) When evaluating a function f, we can _____ write $f(0) = 0$.

    (b) The break-even point for a product _____ occurs at the number of units necessary for the cost and revenue functions to be equal.

48. In each of the following statements, fill in the blank with *always, sometimes,* or *never*.

    (a) The graph of a function _____ has more than one y-intercept.

    (b) Horizontal lines _____ represent the graphs of functions.

49. Let $f(x) = 5x - 2$. Find (a) $f(4) - f(3)$; (b) $f(9) - f(8)$; (c) $f(12) - f(11)$. (d) How do the results of (a) through (c) compare to the slope of the line that is the graph of f?

50. Repeat exercise 49 with $f(x) = 7x + 1$.

51. Repeat exercise 49 with $f(x) = mx + b$.

52. Based on your work in exercises 49 to 51, write a paragraph discussing how slope relates to function notation.

Solve the following applications involving functions.

53. Business and finance. The inventor of a new product believes that the cost of producing the product is given by the function

$$C(x) = 1.75x + 7000$$

How much does it cost to produce 2000 units of her invention?

54. Business and finance. If the inventor in exercise 53 charges \$4 per unit, then her profit for producing and selling x units is given by the function

$$P(x) = 2.25x - 7000$$

(a) What is her profit if she sells 2000 units?
(b) What is her profit if she sells 5000 units?
(c) What is the break-even point for sales?

Answers

1. (a) -2; **(b)** 4; **(c)** -2 **3. (a)** 9; **(b)** -1; **(c)** 3
5. (a) -62; **(b)** -2; **(c)** 2 **7. (a)** 45; **(b)** 3; **(c)** -75
9. (a) -1; **(b)** 2; **(c)** 13 **11.** $f(x) = -3x + 2$ **13.** $f(x) = 4x - 8$
15. $f(x) = -\dfrac{3}{2}x + 3$ **17.** $f(x) = \dfrac{1}{3}x + \dfrac{3}{2}$ **19.** $f(x) = -\dfrac{5}{8}x + \dfrac{9}{8}$
21. $f(x) = 3x + 7$ **23.** $f(x) = -2x + 7$

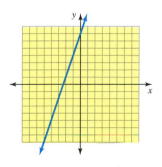

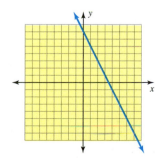

25. $f(x) = -x - 1$

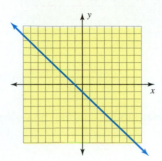

27. 17   **29.** 13   **31.** -19
33. $5a - 1$ **35.** $5x + 4$ **37.** $5x + 5h - 1$ **39.** $-3m + 2$
41. $-3x - 4$ **43.** 5   **45.** $(1, 5), (3, 9)$ **47.** **(a)** Sometimes; **(b)** always
49. **(a)** 5; **(b)** 5; **(c)** 5; **(d)** same   **51.** **(a)** m; **(b)** m; **(c)** m; **(d)** same
53. $10,500

7 Summary

DEFINITION/PROCEDURE	EXAMPLE	REFERENCE
The Slope-Intercept Form		Section 7.1
The slope-intercept form for the equation of a line is $$y = mx + b$$ in which the line has slope m and y-intercept $(0, b)$.	For the equation $$y = \frac{2}{3}x - 3$$ the slope m is $\frac{2}{3}$ and b, which determines the y-intercept, is -3.	p. 605
Parallel and Perpendicular Lines		Section 7.2
Two lines are parallel if and only if they have the same slope, so $$m_1 = m_2$$	$y = 3x - 5$ and $y = 3x + 2$ are parallel Parallel lines	p. 619
Two lines are perpendicular if and only if their slopes are negative reciprocals, that is, when $$m_1 \cdot m_2 = -1$$	$y = 5x + 2$ and $y = -\frac{1}{5}x - 3$ are perpendicular Perpendicular lines	p. 620

Continued

The Point-Slope Form		Section 7.3
The equation of a line with slope m that passes through the point (x_1, y_1) is $$y - y_1 = m(x - x_1)$$	The line with slope $\frac{1}{3}$ passing through $(4, 3)$ has the equation $$y - 3 = \frac{1}{3}(x - 4)$$	
		p. 629

Graphing Linear Inequalities — Section 7.4

The Graphing Steps
1. Replace the inequality symbol with an equal–sign to form the equation of the boundary line of the graph.
2. Graph the boundary line. Use a dotted line if equality is not included ($<$ or $>$). Use a solid line if equality is included ($\leq$ or $\geq$).
3. Choose any convenient test point *not* on the line.
4. If the inequality is *true* at the checkpoint, shade the half–plane including the test point. If the inequality is *false* at the checkpoint, shade the half–plane that does not include the checkpoint.

To graph $x - 2y < 4$:
$x - 2y = 4$ is the boundary line. Using $(0, 0)$ as the checkpoint, we have

$$(0) - 2(0) \overset{?}{<} 4$$
$$0 < 4 \quad \text{(True)}$$

Shade the half–plane that includes $(0, 0)$.

p. 641

An Introduction to Functions — Section 7.5

Given a function f, $f(c)$ designates the value of the function when the variable is equal to c.

$$f(x) = 2x^3 - x^2 + 1$$
$$f(-2) = 2(-2)^3 - (-2)^2 + 1$$
$$= 2(-8) - (4) + 1$$
$$= -19$$

p. 654

Summary Exercises

This summary exercise set is provided to give you practice with each of the objectives of this chapter. Each exercise is keyed to the appropriate chapter section. When you are finished, you can check your answers to the odd-numbered exercises against those presented in the back of the text. If you have difficulty with any of these questions, go back and reread the examples from that section. Your instructor will give you guidelines on how to best use these exercises in your instructional setting.

[7.1] Find the slope of the line through each of the following pairs of points.

1. $(3, 4)$ and $(5, 8)$

2. $(-2, 3)$ and $(1, -6)$

3. $(-2, 5)$ and $(2, 3)$

4. $(-5, -2)$ and $(1, 2)$

5. $(-2, 6)$ and $(5, 6)$

6. $(-3, 2)$ and $(-1, -3)$

7. $(-3, -6)$ and $(5, -2)$

8. $(-6, -2)$ and $(-6, 3)$

Find the slope and y-intercept of the line represented by each of the following equations.

9. $y = 2x + 5$

10. $y = -4x - 3$

11. $y = -\dfrac{3}{4}x$

12. $y = \dfrac{2}{3}x + 3$

13. $2x + 3y = 6$

14. $5x - 2y = 10$

15. $y = -3$

16. $x = 2$

Write the equation of the line with the given slope and y-intercept. Then graph each line, *using* the slope and y-intercept.

17. Slope $= 2$; y-intercept: $(0, 3)$

18. Slope $= \dfrac{3}{4}$; y-intercept: $(0, -2)$

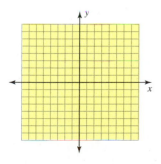

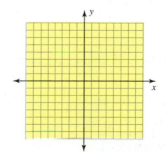

19. Slope: $-\dfrac{2}{3}$; y-intercept: (0, 2)

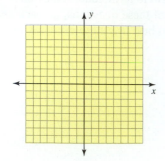

[7.2]   Are the following pairs of lines parallel, perpendicular, or neither?

20. L_1 through $(-3, -2)$ and $(1, 3)$
L_2 through $(0, 3)$ and $(4, 8)$

21. L_1 through $(-4, 1)$ and $(2, -3)$
L_2 through $(0, -3)$ and $(2, 0)$

22. L_1 with equation $x + 2y = 6$
L_2 with equation $x + 3y = 9$

23. L_1 with equation $4x - 6y = 18$
L_2 with equation $2x - 3y = 6$

[7.3]   Write the equation of the line passing through each of the following points with the indicated slope. Give your results in slope-intercept form, where possible.

24. $(0, -5)$, $m = \dfrac{2}{3}$

25. $(0, -3)$, $m = 0$

26. $(2, 3)$, $m = 3$

27. $(4, 3)$, m is undefined

28. $(3, -2)$, $m = \dfrac{5}{3}$

29. $(-2, -3)$, $m = 0$

30. $(-2, -4)$, $m = -\dfrac{5}{2}$

31. $(-3, 2)$, $m = -\dfrac{4}{3}$

32. $\left(\dfrac{2}{3}, -5\right)$, $m = 0$

33. $\left(-\dfrac{5}{2}, -1\right)$, m is undefined

Write the equation of the line L satisfying each of the following sets of geometric conditions.

34. L passes through $(-3, -1)$ and $(3, 3)$.

35. L passes through $(0, 4)$ and $(5, 3)$.

36. L has slope $\dfrac{3}{4}$ and y-intercept $(0, 3)$.

37. L passes through $(4, -3)$ with a slope of $-\dfrac{5}{4}$.

38. L has y-intercept $(0, -4)$ and is parallel to the line with equation $3x - y = 6$.

39. L passes through $(3, -2)$ and is perpendicular to the line with equation $3x - 5y = 15$.

40. L passes through $(2, -1)$ and is perpendicular to the line with the equation $3x - 2y = 5$.

41. L passes through the point $(-5, -2)$ and is parallel to the line with the equation $4x - 3y = 9$.

[7.4] Graph each of the following inequalities.

42. $x + y \leq 4$

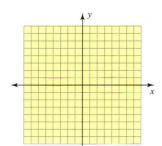

43. $x - y > 5$

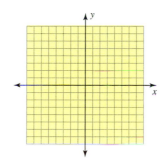

44. $2x + y < 6$

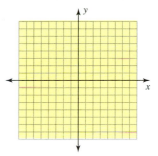

45. $2x - y \geq 6$

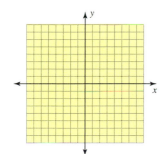

46. $x > 3$

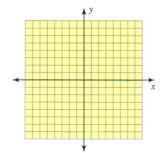

47. $y \leq 2$

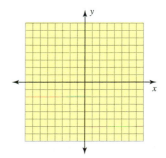

[7.5] In exercises 48 to 53, evaluate $f(x)$ for the value specified.

48. $f(x) = x^2 - 3x + 5$; find (a) $f(0)$, (b) $f(-1)$, and (c) $f(1)$.

49. $f(x) = -2x^2 + x - 7$; find (a) $f(0)$, (b) $f(2)$, and (c) $f(-2)$.

50. $f(x) = x^3 - x^2 - 2x + 5$; find (a) $f(-1)$, (b) $f(0)$, and (c) $f(2)$.

51. $f(x) = -x^2 + 7x - 9$; find (a) $f(-3)$, (b) $f(0)$, and (c) $f(1)$.

52. $f(x) = 3x^2 - 5x + 1$; find (a) $f(-1)$, (b) $f(0)$, and (c) $f(2)$.

53. $f(x) = x^3 + 3x - 5$; find (a) $f(2)$, (b) $f(0)$, and (c) $f(1)$.

In exercises 54 to 57, rewrite each equation as a function of x.

54. $y = 4x + 7$

55. $y = -7x - 3$

56. $4x + 5y = 40$

57. $-3x - 2y = 12$

In exercises 58 to 63, graph the function.

58. $f(x) = 2x + 3$

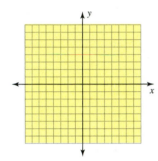

59. $f(x) = 3x - 6$

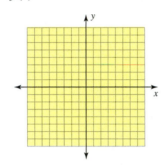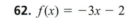

60. $f(x) = -5x + 6$

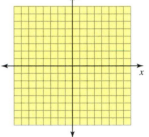

61. $f(x) = -x + 3$

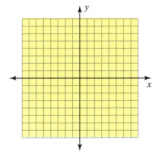

62. $f(x) = -3x - 2$

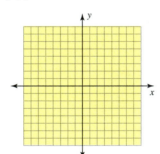

63. $f(x) = -2x + 6$

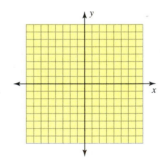

In exercises 64 to 69, evaluate each function as indicated.

64. $f(x) = 5x + 3$; find $f(2)$ and $f(0)$.

65. $f(x) = -3x + 5$; find $f(0)$ and $f(1)$.

66. $f(x) = 7x - 5$; find $f\left(\dfrac{5}{4}\right)$ and $f(-1)$.

67. $f(x) = -2x + 5$; find $f(0)$ and $f(-2)$.

68. $f(x) = -5x + 3$; find $f(a)$, $f(2b)$, and $f(x + 2)$.

69. $f(x) = 7x - 1$; find $f(a)$, $f(3b)$, and $f(x - 1)$.

Self-Test for Chapter 7

Name _____

Section _____ Date _____

ANSWERS

1. _____

2. _____

3. _____

4. _____

5. _____

6. _____

7. _____

8. _____

9. _____

10. _____

11. _____

The purpose of this self-test is to help you check your progress and to review for the next in-class exam. Allow yourself about an hour to take this test. At the end of that hour check your answers against those given in the back of the text. Section references accompany the answers. If you missed any questions, go back to those sections and reread the examples until you master the concepts.

Find the slope of the line through the following pairs of points.

1. $(-3, 5)$ and $(2, 10)$

2. $(-2, 6)$ and $(2, 9)$

Find the slope and y-intercept of the line represented by each of the following equations.

3. $4x - 5y = 10$

4. $y = -\dfrac{2}{3}x - 9$

Write the equation of the line with the given slope and y-intercept. Then graph each line.

5. Slope -3 and y-intercept $(0, 6)$

6. Slope 5 and y-intercept $(0, -3)$

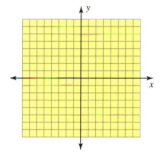

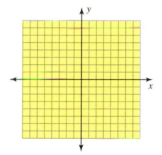

Determine if the following pairs of lines are parallel, perpendicular, or neither.

7. L_1 through $(2, 5)$ and $(4, 9)$ L_2 through $(-7, 1)$ and $(-2, 11)$

8. L_1 with equation $y = 5x - 8$ L_2 with equation $5y + x = 3$

9. Find the equation of the line through $(-5, 8)$ and perpendicular to the line $4x + 2y = 8$.

Write the equation of the line L satisfying each of the following sets of geometric conditions.

10. L passes through the points $(-4, 3)$ and $(-1, 7)$.

11. L passes through $(-3, 7)$ and is perpendicular to $2x - 3y = 7$.

12. _____

13. _____

14. _____

15. _____

16. _____

17. _____

18. _____

19. _____

20. _____

12. L has y-intercept $(0, 5)$ and is parallel to $4x + 6y = 12$.

13. L has y-intercept $(0, -8)$ and is parallel to the x-axis.

Graph each of the following inequalities.

14. $x + y < 3$

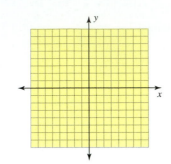

15. $3x + y \geq 9$

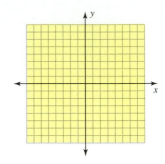

16. $x \leq 7$

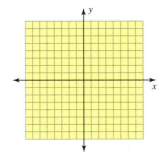

In each of the following, evaluate $f(x)$ for the value given.

17. $f(x) = x^2 - 4x - 5$; find $f(0)$ and $f(-2)$.

18. $f(x) = -x^3 + 5x - 3x^2 - 8$; find $f(-1)$ and $f(1)$.

19. $f(x) = -7x - 15$; find $f(0)$ and $f(-3)$.

20. $f(x) = 3x - 25$; find $f(a)$ and $f(x - 1)$.

ACTIVITY 5: GRAPHING WITH THE INTERNET

Each activity in this text is designed to either enhance your understanding of the topics of the preceding chapter, provide you with a mathematical extension of those topics, or both. The activities can be undertaken by one student, but they are better suited for a small group project. Occasionally, it is only through discussion that different facets of the activity become apparent. For material related to this activity, visit the text website at www.mhhe.com/streeter.

I. Find a Graphing Tutorial on the Internet

Search the Internet to find a website that allows you to create a graph from an equation. You may use one of the major search engines or go through an algebra tutorial website that you are familiar with.

II. Use the Graphing Tutorial

When you find a website that allows you to enter an equation or function to be graphed, enter the following function.

$$y = 2x \quad (\textit{Hint:} \text{ You may need to use } f(x) = 2x \text{ instead of } y).$$

The "viewing window" or "range" describes the limits of the coordinate system (that is, the minimum and maximum values for the variables).

1. What range was shown?
2. Does the website allow you to change the range?

To your existing graph, add the graph of the equation $y = -3x + 5$.

3. Find the point at which the lines intersect. Does the website provide you with a way to do this or do you need to estimate the point?
4. Briefly describe how the plot distinguishes between your first and second equations.

III. Evaluate the Website

5. Describe any shortcomings to the graphing capability. What improvements might you recommend?
6. Describe other algebra tutorial content the website makes available.
7. Consider some topic from your algebra course that you found difficult. Does the website provide tutorial information for this topic? How useful is the tutorial provided?

Cumulative Review
Chapters 0–7

ANSWERS

1. _____

2. _____

3. _____

4. _____

5. _____

6. _____

7. _____

8. _____

9. _____

10. _____

11. _____

12. _____

13. _____

14. _____

15. _____

16. _____

The following questions are presented to help you review concepts from earlier chapters. This is meant as a review and not as a comprehensive exam. The answers are presented in the back of the text. Section references accompany the answers. If you have difficulty with any of these questions, be certain to at least read through the summary related to those sections.

Perform the indicated operation.

1. $3x^2y^2 - 5xy - 2x^2y^2 + 2xy$

2. $\dfrac{36m^5n^2}{27m^2n}$

3. $(x^2 - 3x + 5) - (x^2 - 2x - 4)$

4. $(5z^2 - 3z) - (2z^2 - 5)$

Multiply.

5. $(2x - 3)(x + 7)$

6. $(2a - 2b)(a + 4b)$

Divide.

7. $(x^2 + 3x + 2) \div (x - 3)$

8. $(x^4 - 2x) \div (x + 2)$

Solve each equation and check your results.

9. $5x - 2 = 2x - 6$

10. $3(x - 2) = 2(3x + 1) - 2$

Factor each polynomial completely.

11. $x^2 - x - 56$

12. $4x^3y - 2x^2y^2 + 8x^4y$

13. $8a^3 - 18ab^2$

14. $15x^2 - 21xy + 6y^2$

Find the slope of the line through the following pairs of points.

15. $(2, -4)$ and $(-3, -9)$

16. $(-1, 7)$ and $(3, -2)$

Perform the indicated operations.

17. $\dfrac{x^2 + 7x + 10}{x^2 + 5x} \cdot \dfrac{2x^2 - 7x + 6}{x^2 - 4}$

18. $\dfrac{2a^2 + 11a - 21}{a^2 - 49} \div (2a - 3)$

19. $\dfrac{5}{2m} + \dfrac{3}{m^2}$

20. $\dfrac{4}{x - 3} - \dfrac{2}{x}$

21. $\dfrac{3y}{y^2 + 5y + 4} + \dfrac{2y}{y^2 - 1}$

Solve the following equations for x.

22. $\dfrac{13}{4x} + \dfrac{3}{x^2} = \dfrac{5}{2x}$

23. $\dfrac{6}{x + 5} + 1 = \dfrac{3}{x - 5}$

Solve the following applications.

24. If the reciprocal of 4 times a number is subtracted from the reciprocal of that number, the result is $\dfrac{1}{12}$. What is the number?

25. Kyoko drove 280 mi to attend a business conference. In returning from the conference along a different route, the trip was only 240 mi, but traffic slowed her speed by 7 mi/h. If her driving time was the same both ways, what was her speed each way?

26. A laser printer can print 400 form letters in 30 min. At that rate, how long will it take the printer to complete a job requiring 1680 letters?

27. Write the equation of the line perpendicular to the line $7x - y = 15$ with y-intercept of $(0, 2)$.

17. _____

18. _____

19. _____

20. _____

21. _____

22. _____

23. _____

24. _____

25. _____

26. _____

27. _____

28. _____

29. _____

30. _____

28. Write the equation of the line with slope of -5 and y-intercept $(0, 3)$.

29. Graph the inequality $4x - 2y \geq 8$.

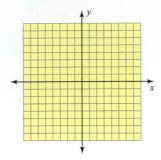

30. If $f(x) = x^2 + 3x$, find $f(-1)$.

SYSTEMS OF LINEAR EQUATIONS

INTRODUCTION

In the United States, almost all electricity is generated by the burning of fossil fuels (coal, oil, and gas); by nuclear fission; or by water-powered turbines in hydroelectric dams. About 65% of the electric power we use comes from burning fossil fuels. Because of this dependence on a nonrenewable resource and concern over pollution caused by burning fossil fuels, there has been some urgency in developing ways to utilize other power sources. Some of the most promising projects have been in solar- and wind-generated energy.

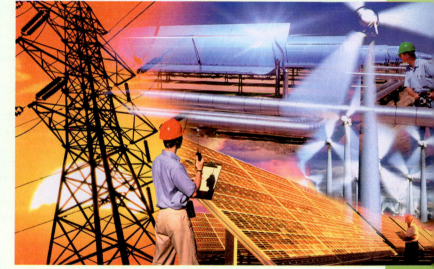

Alternative sources of energy are expensive compared to the cost of the traditional methods of generating electricity described above. As the average price per kilowatt hour (kWh) of electric power has increased (costs to residential users have increased by about $0.0028 per kWh per year since 1970), alternative energy sources look more promising. Additionally, the costs of manufacturing and installing banks of wind turbines in windy locations have declined.

When will the cost of generating electricity using wind power be equal to or less than the cost of using the traditional energy mix? Economists use equations such as the following to make projections and then advise about the feasibility of investing in wind power plants for large cities:

$C_1 = \$0.054 + 0.0028t$

$C_2 = \$0.25 - 0.0035t$

in which C_1 and C_2 represent cost per kWh in 2000 and t is the time in years since 1970

C_1 = cost of present mix of energy sources*

C_2 = cost of wind-powered electricity

* Of course, the true cost of burning fossil fuels also includes the damage to the environment and people's health.

Pre-Test Chapter 8

ANSWERS

1. _____

2. _____

3. _____

4. _____

5. _____

6. _____

7. _____

8. _____

9. _____

10. _____

11. _____

12. _____

This pre-test will provide a preview of the types of exercises you will encounter in each section of this chapter. The answers for these exercises can be found in the back of the text. If you are working on your own, or ahead of the class, this pre-test can help you find the sections in which you should focus more of your time.

[8.1] Solve each of the following systems by graphing.

1. $-4x + y = 4$
$4x - y = 4$

2. $2x + y = -2$
$x - 3y = 6$

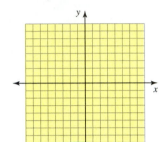
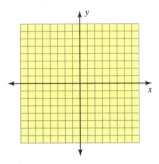

[8.2–8.3] Solve each of the following systems.

3. $x - y = 9$
$x + y = 3$

4. $3x - 4y = 2$
$-6x + 8y = -4$

5. $x + y = 5$
$x = y + 3$

6. $x - 2y = 5$
$3x + y = 8$

Solve the following problems. Be sure to show the equations used.

7. The sum of two numbers is 40, and their difference is 8. Find the two numbers.

8. A rope 30 m long is cut into two pieces so that one piece is 4 m longer than the other. How long is each piece?

9. Nila has 45 coins with a value of $8.40. If the coins are all dimes and quarters, how many of each coin does he have?

10. Jackson was able to travel 36 mi downstream in 4 h. In returning upstream, it took 6 h to make the trip. How fast can his canoe travel in still water? What was the rate of the river current?

[8.4] Solve each of the following linear inequalities graphically.

11. $x + 2y < 4$
$x - y < 5$

12. $2x - 6y \leq 12$
$y \leq 3$

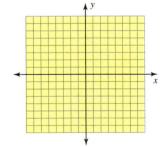
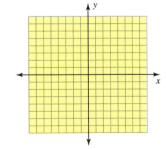

8.1 Systems of Linear Equations: Solving by Graphing

From our work in Section 6.1, we know that an equation of the form $x + y = 3$ is a linear equation. Remember that its graph is a straight line. Often we will want to consider two equations together. They then form a **system of linear equations.** An example of such a system is

$$x + y = 3$$
$$3x - y = 5$$

A solution for a linear equation in two variables is any ordered pair that satisfies the equation. Often there is just one ordered pair that satisfies both equations of a system. It is called the **solution for the system.** For instance, there is one solution for the system above, and it is (2, 1) because, replacing x with 2 and y with 1, we have

$x + y = 3$	$3x - y = 5$
$(2) + (1) \stackrel{?}{=} 3$	$3 \cdot (2) - (1) \stackrel{?}{=} 5$
$3 = 3$	$6 - 1 \stackrel{?}{=} 5$
	$5 = 5$

NOTE There is no other ordered pair that satisfies both equations.

Because both statements are true, the ordered pair (2, 1) satisfies both equations.

One approach to finding the solution for a system of linear equations is the **graphical method.** To use this, we graph the two lines on the same coordinate system. The coordinates of the point where the lines intersect is the solution for the system.

OBJECTIVE 1 | **Example 1 Solving by Graphing**

Solve the system by graphing.

$$x + y = 6$$
$$x - y = 4$$

NOTE Use the intercept method to graph each equation.

First, we determine solutions for the equations of our system. For $x + y = 6$, two solutions are (6, 0) and (0, 6). For $x - y = 4$, two solutions are (4, 0) and (0, −4). Using these intercepts, we graph the two equations. The lines intersect at the point (5, 1).

NOTE By substituting 5 for x and 1 for y into the two original equations, we can check that (5, 1) is indeed the solution for our system.

$x + y = 6$	$x - y = 4$
$(5) + (1) \stackrel{?}{=} 6$	$(5) - (1) \stackrel{?}{=} 4$
$6 = 6$	$4 = 4$

Both statements must be true for (5, 1) to be a solution for the system.

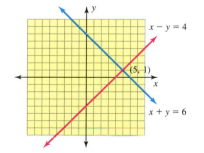

(5, 1) is the solution of the system.
It is the only point that lies on both lines.

✔ **CHECK YOURSELF 1**

Solve the system by graphing.

$2x - y = 4$

$x + y = 5$

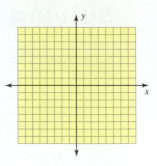

Example 2 shows how to graph a system when one of the equations represents a horizontal line.

Example 2   Solving by Graphing

Solve the system by graphing.

$3x + 2y = 6$

$y = 6$

For $3x + 2y = 6$, two solutions are $(2, 0)$ and $(0, 3)$. These represent the x- and y-intercepts of the graph of the equation. The equation $y = 6$ represents a horizontal line that crosses the y-axis at the point $(0, 6)$. Using these intercepts, we graph the two equations. The lines will intersect at the point $(-2, 6)$. So this is the solution to our system.

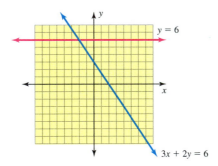

✔ **CHECK YOURSELF 2**

Solve the system by graphing.

$4x + 5y = 20$

$y = 8$

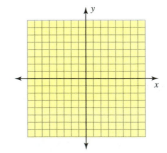

The systems in Examples 1 and 2 both had exactly one solution. A system with one solution is called a **consistent system.** It is possible for a system of equations to have no solution. Such a system is called an **inconsistent system.** We present such a system here.

OBJECTIVE 2   **Example 3**   **Solving an Inconsistent System**

Solve the system by graphing.

$2x + y = 2$

$2x + y = 4$

We can graph the two lines as before. For $2x + y = 2$, two solutions are $(0, 2)$ and $(1, 0)$. For $2x + y = 4$, two solutions are $(0, 4)$ and $(2, 0)$. Using these intercepts, we graph the two equations.

NOTE In slope-intercept form, our equations are

$y = -2x + 2$

and

$y = -2x + 4$

Both lines have slope -2.

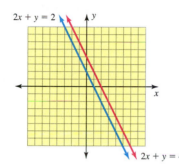

Notice that the slope for each of these lines is -2, but they have different y-intercepts. This means that the lines are parallel (they will never intersect). Because the lines have no points in common, there is no ordered pair that will satisfy both equations. The system has no solution. It is *inconsistent.*

CHECK YOURSELF 3

Solve the system by graphing.

$x - 3y = 3$

$x - 3y = 6$

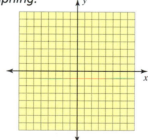

There is one more possibility for linear systems, as Example 4 illustrates.

OBJECTIVE 3   **Example 4**   **Solving a Dependent System**

Solve the system by graphing.

NOTE Multiplying the first equation by 2 results in the second equation.

$x - 2y = 4$

$2x - 4y = 8$

Graphing as before and using the intercept method, we find

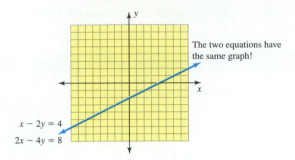

$x - 2y = 4$
$2x - 4y = 8$

The two equations have the same graph!

Because the graphs coincide, there are *infinitely many* solutions for this system. Every point on the graph of $x - 2y = 4$ is also on the graph of $2x - 4y = 8$, so any ordered pair satisfying $x - 2y = 4$ also satisfies $2x - 4y = 8$. This is called a *dependent* system, and any point on the line is a solution.

 CHECK YOURSELF 4

Solve the system by graphing.

$x + \ y = 4$
$2x + 2y = 8$

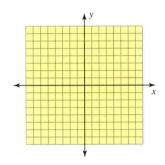

The following summarizes our work in this section.

Step by Step: To Solve a System of Equations by Graphing

Step 1 Graph both equations on the same coordinate system.
Step 2 Determine the solution to the system as follows.
   a. If the lines intersect at one point, the solution is the ordered pair corresponding to that point. This is called a **consistent system.**

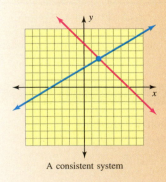

A consistent system

NOTE There is no ordered pair that lies on both lines.

b. If the lines are parallel, there are no solutions. This is called an **inconsistent system.**

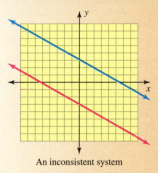

An inconsistent system

NOTE Any ordered pair that corresponds to a point on the line is a solution.

c. If the two equations have the same graph, then the system has infinitely many solutions. This is called a **dependent system.**

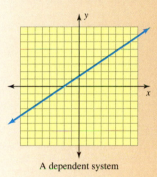

A dependent system

Step 3 Check the solution in both equations, if necessary.

CHECK YOURSELF ANSWERS

1.

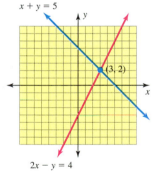

$x + y = 5$

$(3, 2)$

$2x - y = 4$

2.

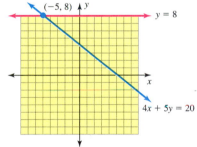

$(-5, 8)$ $y = 8$

$4x + 5y = 20$

3. There is no solution. The lines are parallel, so the system is inconsistent.

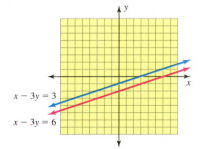

$x - 3y = 3$

$x - 3y = 6$

4. There are infinitely many solutions.

$x + y = 4$
$2x + 2y = 8$

A dependent system

8.1 Exercises

Solve each of the following systems by graphing.

1. $2x + 2y = 12$
    $x - y = 4$

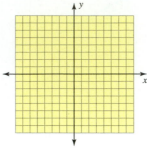

2. $x - y = 8$
    $x + y = 2$

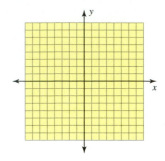

3. $-x + y = 3$
    $x + y = 5$

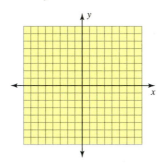

4. $x + y = 7$
    $-x + y = 3$

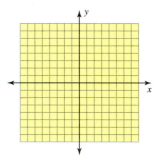

5. $x + 2y = 4$
    $x - y = 1$

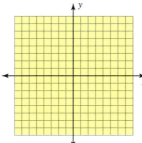

6. $3x + y = 6$
    $x + y = 4$

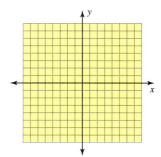

7. $2x + y = 8$
    $2x - y = 0$

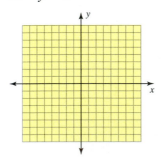

8. $x - 2y = -2$
    $x + 2y = 6$

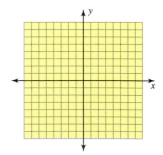

9. $x + 3y = 12$
　　$2x - 3y = 6$

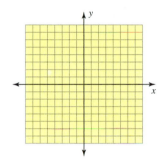

10. $2x - y = 4$
　　$2x - y = 6$

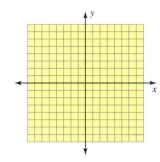

9. _____

10. _____

11. _____

12. _____

13. _____

14. _____

15. _____

16. _____

11. $3x + 2y = 12$
　　　$y = 3$

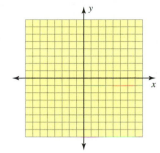

12. $x - 2y = 8$
　　$3x - 2y = 12$

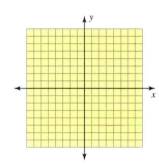

13. $x - y = 4$
　　$2x - 2y = 8$

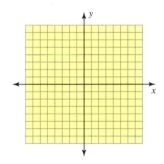

14. $2x - y = 8$
　　　$x = 2$

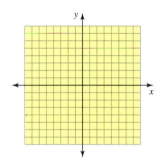

15. $x - 4y = -4$
　　$x + 2y = 8$

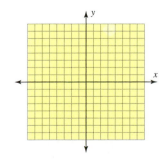

16. $x - 6y = 6$
　　$-x + y = 4$

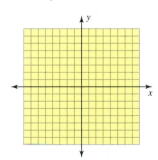

ANSWERS

17. _____

18. _____

19. _____

20. _____

21. _____

22. _____

23. _____

24. _____

17. $3x - 2y = 6$
$\quad\;\; 2x - \;\;y = 5$

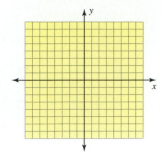

18. $4x + 3y = 12$
$\qquad x + \;\;y = \;\;2$

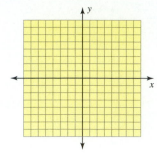

19. $3x - y = 3$
$\quad\;\; 3x - y = 6$

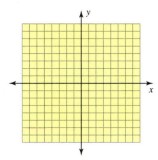

20. $3x - 6y = 9$
$\qquad x - 2y = 3$

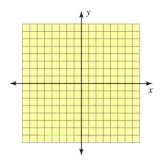

21. $\qquad 2y = \;\;\;3$
$\quad\; x - 2y = -3$

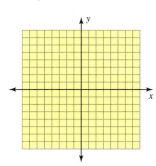

22. $\quad\;\; x + \;\;y = -6$
$\quad\; -x + 2y = \;\;\;6$

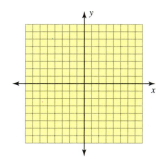

23. $x = \;\;\;4$
$\quad\; y = -6$

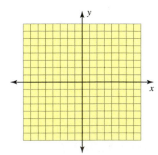

24. $x = -3$
$\quad\; y = \;\;\;5$

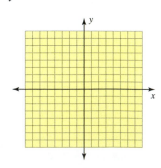

25. Find values for m and b in the following system so that the solution to the system is $(1, 2)$.

$$mx + 3y = 8$$
$$-3x + 4y = b$$

26. Find values for m and b in the following system so that the solution to the system is $(-3, 4)$.

$$5x + 7y = b$$
$$mx + \ y = 22$$

27. Complete the following statements in your own words:

"To solve an equation means to. . . ."
"To solve a system of equations means to. . . ."

28. A system of equations such as the one that follows is sometimes called a "2-by-2 system of linear equations."

$$3x + 4y = 1$$
$$x - 2y = 6$$

Explain this term.

29. Complete this statement in your own words: "All the points on the graph of the equation $2x + 3y = 6$. . . ." Exchange statements with other students. Do you agree with other students' statements?

For exercises 30 and 31, fill in the blank with *always, sometimes,* or *never.*

30. A linear system _____ has at least one solution.

31. If the graphs of two linear equations in a system have different slopes, the system _____ has exactly one solution.

Getting Ready for Section 8.2 [Section 1.6]

Simplify each of the following expressions.

(a) $(2x + y) + (x - y)$

(b) $(x + y) + (-x + y)$

(c) $(3x + 2y) + (-3x - 3y)$

(d) $(x - 5y) + (2x + 5y)$

(e) $2(x + y) + (3x - 2y)$

(f) $2(2x - y) + (-4x - 3y)$

(g) $3(2x + y) + 2(-3x + y)$

(h) $3(2x - 4y) + 4(x + 3y)$

25. _____

26. _____

27. _____

28. _____

29. _____

30. _____

31. _____

a. _____

b. _____

c. _____

d. _____

e. _____

f. _____

g. _____

h. _____

Answers

1. $\left. \begin{array}{l} x + y = 6 \\ x - y = 4 \end{array} \right\}$ (5, 1)

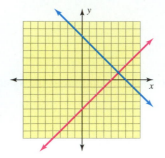

3. $\left. \begin{array}{l} -x + y = 3 \\ x + y = 5 \end{array} \right\}$ (1, 4)

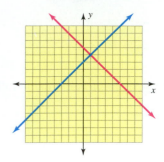

5. $\left. \begin{array}{l} x + 2y = 4 \\ x - y = 1 \end{array} \right\}$ (2, 1)

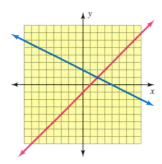

7. $\left. \begin{array}{l} 2x + y = 8 \\ 2x - y = 0 \end{array} \right\}$ (2, 4)

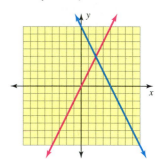

9. $\left. \begin{array}{l} x + 3y = 12 \\ 2x - 3y = 6 \end{array} \right\}$ (6, 2)

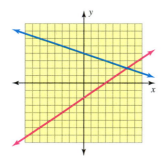

11. $\left. \begin{array}{l} 3x + 2y = 12 \\ y = 3 \end{array} \right\}$ (2, 3)

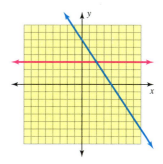

13. $\left. \begin{array}{l} x - y = 4 \\ 2x - 2y = 8 \end{array} \right\}$ Dependent

Infinitely many solutions

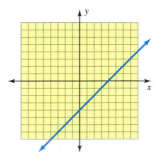

15. $\left. \begin{array}{l} x - 4y = -4 \\ x + 2y = 8 \end{array} \right\}$ (4, 2)

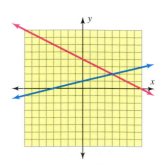

17. $\left.\begin{array}{r} 3x - 2y = 6 \\ 2x - y = 5 \end{array}\right\}$ $(4, 3)$

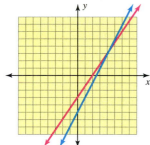

19. $\left.\begin{array}{r} 3x - y = 3 \\ 3x - y = 6 \end{array}\right\}$ Inconsistent

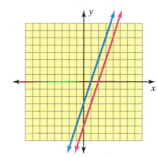

21. $\left.\begin{array}{r} 2y = 3 \\ x - 2y = -3 \end{array}\right\}$ $\left(0, \dfrac{3}{2}\right)$

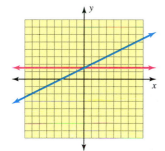

23. $\left.\begin{array}{r} x = 4 \\ y = -6 \end{array}\right\}$ $(4, -6)$

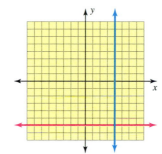

25. $m = 2, b = 5$ **27.**   **29.** 🔆   **31.** Always   **a.** $3x$

b. $2y$ **c.** $-y$ **d.** $3x$ **e.** $5x$ **f.** $-5y$ **g.** $5y$ **h.** $10x$

8.2 Systems of Linear Equations: Solving by Adding

8.2 OBJECTIVES

1. Solve systems of linear equations using the addition method
2. Solve applications of systems of linear equations

The graphical method of solving equations, shown in Section 8.1, has two definite disadvantages. First, it is time-consuming to graph each system that you want to solve. More importantly, the graphical method is not precise. For instance, look at the graph of the system

$$x - 2y = 4$$
$$3x + 2y = 6$$

which follows,

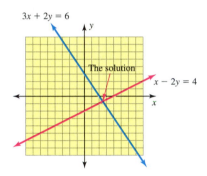

The exact solution for the system is $\left(\dfrac{5}{2}, -\dfrac{3}{4}\right)$, but that is difficult to read from the graph.

Fortunately, there are algebraic methods that do not have this disadvantage and will allow you to find exact solutions for a system of equations.

One algebraic method of finding a solution is called the **addition method.**

OBJECTIVE 1

> **Example 1   Solving a System by the Addition Method**

Solve the system.

$$x + y = 8$$
$$x - y = 2$$

NOTE This method uses the fact that if

$a = b$ and $c = d$

then

$a + c = b + d$

This is the **additive property of** equality. Note that by the additive property, if equals are added to equals, the resulting sums are equal.

Note that the coefficients of the y-terms are the *additive inverses* of one another (1 and -1) and that adding the two equations will "eliminate" the variable y. That addition step is shown here.

$$
\begin{array}{rcl}
x + y &=& 8 \\
x - y &=& 2 \\
\hline
2x &=& 10 \\
x &=& 5
\end{array}
$$

By adding, we eliminate the variable y. The resulting equation contains *only* the variable x.

NOTE This is also called **solution by elimination** for this reason.

690

We now know that 5 is the x-coordinate of our solution. Substitute 5 for x into *either* of the original equations.

$$x + y = 8$$
$$(5) + y = 8$$
$$y = 3$$

So $(5, 3)$ is the solution.

To check, replace x and y with these values in *both* of the original equations.

$$\begin{array}{ll} x + y = 8 & \qquad x - y = 2 \\ \hline (5) + (3) = 8 & \qquad (5) - (3) = 2 \\ \quad\;\; 8 = 8 \quad \text{(True)} & \qquad\quad\;\; 2 = 2 \;\; \text{(True)} \end{array}$$

Because $(5, 3)$ satisfies both equations, it is the solution.

CHECK YOURSELF 1

Solve the system by adding.

$$x - y = -2$$
$$x + y = 6$$

Example 2   Solving a System by the Addition Method

Solve the system.

$$-3x + 2y = 12$$
$$3x - y = -9$$

In this case, adding will eliminate the x-terms.

NOTE It does not matter which variable is eliminated. Choose the one that requires the least work.

$$\begin{array}{r} -3x + 2y = 12 \\ 3x - y = -9 \\ \hline y = 3 \end{array}$$

Now substitute 3 for y in either equation. From the first equation

$$-3x + 2(3) = 12$$
$$-3x = 6$$
$$x = -2$$

and $(-2, 3)$ is the solution.

Show that you get the same x-coordinate by substituting 3 for y in the second equation rather than in the first. Then check the solution.

CHECK YOURSELF 2

Solve the system by adding.

$$5x - 2y = 9$$
$$-5x + 3y = -11$$

Note that in both Examples 1 and 2 we found an equation in a single variable by adding. We could do this because the coefficients of one of the variables were opposites. This gave 0 as a coefficient for one of the variables after we added the two equations. In some systems, you will not be able to directly eliminate either variable by adding. However, an equivalent system can always be written by multiplying one or both of the equations by a nonzero constant so that the coefficients of x (or of y) are opposites. Example 3 illustrates this approach.

> **Example 3 Solving a System by the Addition Method**

Solve the system.

$$2x + y = 13$$
$$3x + y = 18$$

NOTE Remember that multiplying both sides of an equation by some nonzero number does not change the solutions. So even though we have "altered" the equations, they are equivalent and will have the same solutions.

Note that adding the equations in this form will not eliminate either variable. You will still have terms in x and in y. However, look at what happens if we multiply both sides of the second equation by -1 as the first step.

$$2x + y = 13 \qquad \longrightarrow \qquad 2x + y = 13$$

$$3x + y = 18 \quad \xrightarrow[\text{by } -1]{\text{Multiply}} \quad -3x - y = -18$$

Now we can add.

$$
\begin{array}{r}
2x + y = 13 \\
-3x - y = -18 \\
\hline
-x \quad\;\; = -5 \\
x = \quad 5
\end{array}
$$

Substitute 5 for x in the first equation.

$$2(5) + y = 13$$
$$ y = \;\; 3$$

$(5, 3)$ is the solution. We leave it to the reader to check this solution.

CHECK YOURSELF 3 _____

Solve the system by adding.

$$x - 2y = \;\; 9$$
$$x + 3y = -1$$

To summarize, multiplying both sides of one of the equations by a nonzero constant can yield an equivalent system in which the coefficients of the x-terms or the y-terms are opposites. This means that a variable can be eliminated by adding.

Example 4   Solving a System by the Addition Method

Solve the system.

$$x + 4y = 2$$
$$3x - 2y = -22$$

One approach is to multiply both sides of the second equation by 2. Do you see that the coefficients of the y-terms will then be opposites?

$$x + 4y = 2 \longrightarrow x + 4y = 2$$

$$3x - 2y = -22 \xrightarrow[\text{by 2}]{\text{Multiply}} 6x - 4y = -44$$

If we add the resulting equations, the variable y will be eliminated and we can solve for x.

NOTE Now the coefficients of the y-terms are opposites.

$$
\begin{array}{r}
x + 4y = 2 \\
6x - 4y = -44 \\
\hline
7x = -42 \\
x = -6
\end{array}
$$

Now substitute -6 for x in the first equation of this example to find y.

NOTE We could substitute -6 for x in the second equation to find y.

$$(-6) + 4y = 2$$
$$4y = 8$$
$$y = 2$$

So $(-6, 2)$ is the solution.

Again you should check this result. As is often the case, there are several ways to solve the system. For example, what if we multiply both sides of our original equation by -3? The coefficients of the x-terms will then be opposites and adding will eliminate the variable x so that we can solve for y. Try that for yourself in the Check Yourself 4 exercise.

CHECK YOURSELF 4

Solve the system by eliminating x.

$$x + 4y = 2$$
$$3x - 2y = -22$$

It may be necessary to multiply each equation separately so that one of the variables will be eliminated when the equations are added. Example 5 illustrates this approach.

Example 5   Solving a System by the Addition Method

Solve the system.

$$4x + 3y = 11$$
$$3x - 2y = 4$$

Do you see that, if we want to have integer coefficients, multiplying in one equation will not help in this case? We will have to multiply in both equations.

NOTE The minus sign is used with the 4 so that the coefficients of the *x*-term are opposites.

To eliminate *x*, we can multiply both sides of the first equation by 3 and both sides of the second equation by -4. The coefficients of the *x*-terms will then be opposites.

$$4x + 3y = 11 \xrightarrow[\text{by 3}]{\text{Multiply}} 12x + 9y = 33$$

$$3x - 2y = 4 \xrightarrow[\text{by } -4]{\text{Multiply}} -12x + 8y = -16$$

Adding the resulting equations gives

$$17y = 17$$
$$y = 1$$

Now substituting 1 for *y* in the first equation, we have

$$4x + 3(1) = 11$$
$$4x = 8$$
$$x = 2$$

NOTE Check (2, 1) in both equations of the original system.

and (2, 1) is the solution.

 CHECK YOURSELF 5 _____

Solve the system by eliminating y.

$4x + 3y = 11$
$3x - 2y = 4$

The following summarizes the solution steps we have illustrated.

Step by Step: To Solve a System of Linear Equations by Adding

Step 1   If necessary, multiply both sides of one or both equations by nonzero numbers to form an equivalent system in which the coefficients of one of the variables are opposites.
Step 2   Add the equations of the new system.
Step 3   Solve the resulting equation for the remaining variable.
Step 4   Substitute the value found in Step 3 into either of the original equations to find the value of the second variable.
Step 5   Check your solution in both of the original equations.

In Section 8.1 we saw that some systems had *infinitely* many solutions. Example 6 shows how this is indicated when we are using the addition method of solving equations.

Example 6   Solving a Dependent System

Solve the system.

$x + 3y = -2$
$3x + 9y = -6$

We multiply both sides of the first equation by -3.

$$x + 3y = -2 \xrightarrow[\text{by } -3]{\text{Multiply}} -3x - 9y = 6$$

$$3x + 9y = -6 \xrightarrow{} 3x + 9y = -6$$

$$\overline{0 = 0}$$

Adding, we see that both variables have been eliminated, and we have the true statement $0 = 0$.

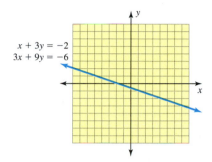

NOTE The lines coincide. That will be the case whenever *adding eliminates both variables* and a true statement results.

Look at the graph of the system.

$x + 3y = -2$
$3x + 9y = -6$

As we see, the two equations have the *same* graph. This means that the system is *dependent,* and there are *infinitely many solutions.* Any (x, y) that satisfies $x + 3y = -2$ will also satisfy $3x + 9y = -6$.

CHECK YOURSELF 6

Solve the system by adding.

$$x - 2y = 3$$
$$-2x + 4y = -6$$

Earlier we encountered systems that had *no* solutions. Example 7 illustrates what happens when we try to solve such a system with the addition method.

Example 7   Solving an Inconsistent System

Solve the system.

$$3x - y = 4$$
$$-6x + 2y = -5$$

We multiply both sides of the first equation by 2.

NOTE Be sure to multiply the 4 by 2.

$$3x - y = 4 \xrightarrow[\text{by 2}]{\text{Multiply}} 6x - 2y = 8$$

$$-6x + 2y = -5 \xrightarrow{} -6x + 2y = -5$$

$$\overline{0 = 3}$$

We now add the two equations.

Again both variables have been eliminated by addition. But this time we have the *false* statement $0 = 3$ because we tried to solve a system whose graph consists of two parallel lines, as we see in the following graph. Because the two lines do not intersect, there is *no* solution for the system. It is *inconsistent.*

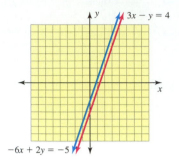

CHECK YOURSELF 7

Solve the system by adding.

$5x + 15y = 20$

$x + 3y = 3$

NOTE Remember that, in Chapter 2, all the unknowns in the problem had to be expressed in terms of that single variable.

In Chapter 2 we solved word problems by using equations in a single variable. Now you have the background to use two equations in two variables to solve word problems. The five steps for solving word problems stay the same (in fact, we give them again for reference in our first application example). Many students find that using two equations and two variables makes writing the necessary equation much easier, as Example 8 illustrates.

OBJECTIVE 2 | **Example 8 Solving an Application with Two Linear Equations**

Ryan bought 8 pens and 7 pencils and paid a total of $14.80. Ashleigh purchased 2 pens and 10 pencils and paid $7. Find the cost for a single pen and a single pencil.

NOTE Here are the steps for using a single variable:

1. Read the problem carefully. What do you want to find?

Step 1 You want to find the cost of a single pen and the cost of a single pencil.

2. Assign variables to the unknown quantities.

Step 2   Let x be the cost of a pen and y be the cost of a pencil.

3. Translate the problem to the language of algebra to form a system of equations.

Step 3   Write the two necessary equations.

$$8x + 7y = 14.80$$
$$2x + 10y = 7.00$$

In the first equation, $8x$ is the total cost of the pens Ryan bought and $7y$ is the total cost of the pencils Ryan bought. The second equation is formed in a similar fashion.

4. Solve the system.

Step 4   Solve the system formed in Step 3. We multiply the second equation by -4. Then adding will eliminate the variable x.

$$8x + 7y = 14.80$$
$$-8x - 40y = -28.00$$

Now adding the equations, we have

$$-33y = -13.20$$
$$y = 0.40$$

Substituting 0.40 for y in the first equation, we have

$$8x + 7(0.40) = 14.80$$
$$8x + 2.80 = 14.80$$
$$8x = 12.00$$
$$x = 1.50$$

5. Verify your result by returning to the original problem.

Step 5   From the results of Step 4 we see that the pens are $1.50 each and the pencils are 40¢ each.

To check these solutions, replace x with $1.50 and y with 0.40 in the first equation.

$$8(1.50) + 7(0.40) = 14.80$$
$$12.00 + 2.80 = 14.80$$
$$14.80 = 14.80 \quad \text{(True)}$$

We leave it to you to check these values in the second equation.

 CHECK YOURSELF 8

Alana bought three digital tapes and two compact disks on sale for $66. At the same sale, Chen bought three digital tapes and four compact disks for $96. Find the individual price for a tape and a disk.

Example 9 shows how sketches can be helpful in setting up a problem.

Example 9   Using a Sketch to Help Solve an Application

An 18-ft board is cut into two pieces, one of which is 4 ft longer than the other. How long is each piece?

NOTE You should always draw a sketch of the problem when it is appropriate.

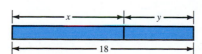

Step 1   You want to find the two lengths.

Step 2   Let x be the length of the longer piece and y the length of the shorter piece.

Step 3   Write the equations for the solution.

$x + y = 18$ ⟵ The total length is 18.

$x - y = \;\; 4$ ⟵ The difference in lengths is 4.

NOTE Our second equation could also be written as

$x = y + 4$

Step 4   To solve the system, add:

$$
\begin{array}{r}
x + y = 18 \\
x - y = \;\; 4 \\
\hline
2x \quad\;\; = 22 \\
x = 11
\end{array}
$$

Replace x with 11 in the first equation.

$(11) + y = 18$

$y = 7$

The longer piece has length 11 ft, and the shorter piece has length 7 ft.

Step 5   We leave it to you to check this result in the original problem.

CHECK YOURSELF 9 _____

A 20-ft board is cut into two pieces, one of which is 6 ft longer than the other. How long is each piece?

Using two equations in two variables also helps in solving **mixture problems.**

Example 10   Solving a Mixture Problem Involving Coins

Winnifred has collected $4.50 in nickels and dimes. If she has 55 coins, how many of each kind of coin does she have?

Step 1   You want to find the number of nickels and the number of dimes.

Step 2   Let

NOTE We choose appropriate variables—*n* for nickels, *d* for dimes.

n = number of nickels

d = number of dimes

Step 3   Write the equations for the solution.

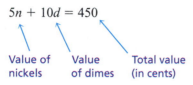

$n + d = 55$ ⟵ There are 55 coins in all.

RECALL The value of a number of coins is the value per coin times the number of coins: 5*n*, 10*d*, etc.

$5n + 10d = 450$

Value of nickels   Value of dimes   Total value (in cents)

Step 4   We now have the system

$$n + \ \ d = 55$$
$$5n + 10d = 450$$

We choose to solve this system by addition. Multiply the first equation by -5. We then add the equations to eliminate the variable n.

$$
\begin{array}{rl}
-5n - \ \ 5d = & -275 \\
5n + 10d = & \ \ \ 450 \\
\hline
5d = & \ \ \ 175 \\
d = & \ \ \ \ \ 35
\end{array}
$$

We now substitute 35 for d in the first equation.

$$n + (35) = 55$$
$$n = 20$$

There are 20 nickels and 35 dimes.

Step 5   We leave it to you to check this result. Just verify that the value of these coins is $4.50.

CHECK YOURSELF 10

Tickets for a play cost $8 or $6. If 350 tickets were sold in all and receipts were $2500, how many tickets of each price were sold?

We can also solve mixture problems that involve percentages by using two equations in two unknowns. Look at Example 11.

Example 11   Solving a Mixture Problem Involving Chemicals

There are two solutions in a chemistry lab: a 20% acid solution and a 60% acid solution. How many milliliters of each should be mixed to produce 200 mL of a 44% acid solution?

20%   60%   44%
x mL   y mL   200 mL

Step 1 You need to know the amount of each solution to use.

Step 2 Let

x = amount of 20% acid solution

y = amount of 60% acid solution

Step 3 Note that a 20% acid solution is 20% acid and 80% water.

We can write equations from the total amount of the solution, here 200 mL, and from the amount of acid in that solution. Many students find a chart helpful in organizing the information at this point. Here, for example, we might have

NOTE The amount of acid is the amount of solution times the percentage of acid (as a decimal). That is the key to forming the third column of our table.

	Amount of Solution	% Acid	Amount of Acid
	x	0.20	$0.20x$
	y	0.60	$0.60y$
Final solution	200	0.44	$(0.44)(200)$

NOTE The first equation is the total amount of the solution from the first column of our table.

Now we are ready to form our system.

$$x + y = 200$$
$$0.20x + 0.60y = \underline{0.44(200)}$$

NOTE The second equation is the amount of acid from the third column of our table. The sum of the acid in the two solutions equals the acid in the mixture.

Acid in 20%   Acid in 60%   Acid in
solution   solution   mixture

Step 4 If we multiply the second equation by 100 to clear it of decimals, we have

$$x + y = 200 \xrightarrow[\text{by } -20]{\text{Multiply}} -20x - 20y = -4000$$

$$20x + 60y = 8800 \longrightarrow \underline{20x + 60y = 8800}$$

$$40y = 4800$$
$$y = 120$$

Substituting 120 for y in the first equation, we have

$$x + (120) = 200$$
$$x = 80$$

The amounts to be mixed are 80 mL (20% acid solution) and 120 mL (60% acid solution).

Step 5 You can check this solution by verifying that the amount of acid from the 20% solution added to the amount from the 60% solution is equal to the amount of acid in the mixture.

CHECK YOURSELF 11

You have a 30% alcohol solution and a 50% alcohol solution. How much of each solution should be combined to make 400 mL of a 45% alcohol solution?

A related kind of application involves interest. The key equation involves the *principal* (the amount invested), the annual *interest rate,* the *time* (in years) that the money is invested, and the amount of *interest* you receive.

$$I = P \cdot r \cdot t$$

Interest   Principal   Rate   Time

For 1 year we have

$$I = P \cdot r \qquad \text{because } t = 1$$

Example 12   Solving an Investment Application

Jeremy inherits $20,000 and invests part of the money in bonds with an interest rate of 11%. The remainder of the money is in savings at a 9% rate. What amount has he invested at each rate if he receives $2040 in interest for 1 year?

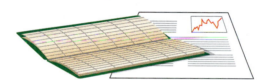

Step 1   You want to find the amounts invested at 11% and at 9%.

NOTE The amount invested at 11% could have been represented by y and the amount invested at 9% by x.

Step 2   Let x = the amount invested at 11% and y = the amount invested at 9%. Once again you may find a chart helpful at this point.

NOTE The formula $I = P \cdot r$ (interest equals principal times rate) is the key to forming the third column of our table.

	Principal	Rate	Interest
	x	11%	$0.11x$
	y	9%	$0.09y$
Totals	20,000		2040

Step 3   Form the equations for the solution, using the first and third columns of the table.

$$x + y = 20,000 \quad \longleftarrow \text{He has \$20,000 invested in all.}$$

NOTE The decimal form of 11% and 9% is used in the equation.

$$0.11x + 0.09y = 2040$$

The interest   The interest   The total
at 11%   at 9%   interest
(rate · principal)

Step 4   To solve the following system, use addition.

$$x + \qquad y = 20{,}000$$
$$0.11x + 0.09y = 2{,}040$$

To do this, multiply both sides of the first equation by -9. Multiplying both sides of the second equation by 100 will clear decimals. Adding the resulting equations will eliminate y.

$$-9x - 9y = -180{,}000$$
$$\underline{11x + 9y = 204{,}000}$$
$$2x = 24{,}000$$
$$x = 12{,}000$$

Now, substitute 12,000 for x in the first equation and solve for y.

$$(12{,}000) + y = 20{,}000$$
$$y = 8{,}000$$

NOTE Be sure to answer the question asked in the problem.

Jeremy has $12,000 invested at 11% and $8000 invested at 9%.

Step 5   To check, the interest at 11% is ($12,000)(0.11), or $1320. The interest at 9% is ($8000)(0.09), or $720. The total interest is $2040, and the solution is verified.

 CHECK YOURSELF 12 _____

Jan has $2000 more invested in a stock that pays 9% interest than in a savings account paying 8%. If her total interest for 1 year is $860, how much does she have invested at each rate?

RECALL Distance, rate, and time of travel are related by the equation

Distance Rate Time

    In Chapters 2 and 5, we solved **motion problems;** they involved a distance traveled, the rate, and the time of travel. Example 13 shows the use of $d = r \cdot t$ in forming a system of equations to solve a motion problem.

Example 13   Solving a Motion Problem

A boat can travel 36 mi downstream in 2 h. Coming back upstream, the trip takes 3 h. Find the rate of the boat in still water and the rate of the current.

Step 1   You want to find the two rates (of the boat and the current).

Step 2   Let

x = rate of boat in still water

y = rate of current

Step 3   To write the equations, think about the following: What is the effect of the current? Suppose the boat's rate in still water is 10 mi/h and the current is 2 mi/h.

The current *increases* the rate *downstream* to 12 mi/h (10 + 2). The current *decreases* the rate *upstream* to 8 mi/h (10 − 2). So here the rate downstream will be $x + y$ and the rate upstream will be $x - y$. At this point a chart of information is helpful.

	Distance	Rate	Time
Downstream	36	$x + y$	2
Upstream	36	$x - y$	3

From the relationship $d = r \cdot t$ we can now use our table to write the system

$36 = 2(x + y)$ (From line 1 of our table)

$36 = 3(x - y)$ (From line 2 of our table)

Step 4   Removing the parentheses in the equations of Step 3, we have

$2x + 2y = 36$

$3x - 3y = 36$

By either of our earlier methods, this system gives values of 15 for x and 3 for y. The rate in still water is 15 mi/h, and the rate of the current is 3 mi/h. We leave the check to you.

CHECK YOURSELF 13

A plane flies 480 mi with the wind in 4 h. In returning against the wind, the trip takes 6 h. What is the rate of the plane in still air? What was the rate of the wind?

CHECK YOURSELF ANSWERS

1. (2, 4)   **2.** (1, −2)   **3.** (5, −2)   **4.** (−6, 2)   **5.** (2, 1)
6. There are infinitely many solutions. It is a dependent system.
7. There is no solution. The system is inconsistent.   **8.** Tape $12, disk $15
9. 7 ft, 13 ft   **10.** 150 $6 tickets, 200 $8 tickets
11. 100 mL (30%), 300 mL (50%)   **12.** $4000 at 8%, $6000 at 9%
13. Plane's rate in still air, 100 mi/h; wind's rate, 20 mi/h

Name _____

Section _____ Date _____

ANSWERS

1. _____
2. _____
3. _____
4. _____
5. _____
6. _____
7. _____
8. _____
9. _____
10. _____
11. _____
12. _____
13. _____
14. _____
15. _____
16. _____
17. _____
18. _____
19. _____
20. _____
21. _____
22. _____
23. _____
24. _____

8.2 Exercises

Solve each of the following systems by addition. If a unique solution does not exist, state whether the system is inconsistent or dependent.

1. $x + y = 6$
   $x - y = 4$

2. $x - y = 8$
   $x + y = 2$

3. $-x + y = 3$
   $x + y = 5$

4. $x + y = 7$
   $-x + y = 3$

5. $2x - y = 1$
   $-2x + 3y = 5$

6. $x - 2y = 2$
   $x + 2y = -14$

7. $x + 3y = 12$
   $2x - 3y = 6$

8. $-3x + y = 8$
   $3x - 2y = -10$

9. $x + 2y = -2$
   $3x + 2y = -12$

10. $4x - 3y = 22$
    $4x + 5y = 6$

11. $4x - 3y = 6$
    $4x + 5y = 22$

12. $2x + 3y = 1$
    $5x + 3y = 16$

13. $2x + y = 8$
    $2x + y = 2$

14. $5x + 4y = 7$
    $5x - 2y = 19$

15. $3x - 5y = 2$
    $2x - 5y = -2$

16. $2x - y = 4$
    $2x - y = 6$

17. $x + y = 3$
    $3x - 2y = 4$

18. $x - y = -2$
    $2x + 3y = 21$

19. $-5x + 2y = -3$
    $x - 3y = -15$

20. $x + 5y = 10$
    $-2x - 10y = -20$

21. $7x + y = 10$
    $2x + 3y = -8$

22. $3x - 4y = 2$
    $4x - y = 20$

23. $5x + 2y = 28$
    $x - 4y = -23$

24. $7x + 2y = 17$
    $x - 5y = 13$

704   SECTION 8.2

25. $3x - 4y = 2$
$-6x + 8y = -4$

26. $-x + 5y = 19$
$4x + 3y = -7$

27. $5x - 2y = 31$
$4x + 3y = 11$

28. $7x + 3y = -13$
$5x + 2y = -8$

29. $3x - 2y = 12$
$5x - 3y = 21$

30. $-4x + 5y = -6$
$5x - 2y = 16$

31. $-2x + 7y = 2$
$3x - 5y = -14$

32. $3x + 4y = 0$
$5x - 3y = -29$

33. $7x + 4y = 20$
$5x + 6y = 19$

34. $5x + 4y = 5$
$7x - 6y = 36$

35. $2x - 7y = 6$
$-4x + 3y = -12$

36. $3x + 2y = -18$
$7x - 6y = -42$

37. $5x - y = 20$
$4x + 3y = 16$

38. $3x + y = -5$
$5x - 4y = 20$

39. $3x + y = 1$
$5x + y = 2$

40. $2x - y = 2$
$2x + 5y = -1$

41. $3x + 4y = 3$
$6x - 2y = 1$

42. $3x + 3y = 1$
$2x + 4y = 2$

43. $5x - 2y = \dfrac{9}{5}$
$3x + 4y = -1$

44. $2x + 3y = -\dfrac{1}{12}$
$5x + 4y = \dfrac{2}{3}$

Solve the following systems by adding. If a unique solution does not exist, state whether the system is inconsistent or dependent.

45. $\dfrac{x}{3} - \dfrac{y}{4} = -\dfrac{1}{2}$

$\dfrac{x}{2} - \dfrac{y}{5} = \dfrac{3}{10}$

46. $\dfrac{1}{3}x - \dfrac{1}{2}y = \dfrac{5}{6}$

$\dfrac{1}{2}x - \dfrac{2}{5}y = \dfrac{9}{10}$

ANSWERS

25. _____
26. _____
27. _____
28. _____
29. _____
30. _____
31. _____
32. _____
33. _____
34. _____
35. _____
36. _____
37. _____
38. _____
39. _____
40. _____
41. _____
42. _____
43. _____
44. _____
45. _____
46. _____

47. _____

48. _____

49. _____

50. _____

51. _____

52. _____

53. _____

54. _____

55. _____

56. _____

57. _____

47. $0.4x - 0.2y = 0.6$
$0.5x - 0.6y = 9.5$

48. $0.2x + 0.37y = 0.8$
$-0.6x + 1.4y = 2.62$

Solve each of the following problems. Be sure to show the equations used for the solution.

49. Number problems. The sum of two numbers is 40. Their difference is 8. Find the two numbers.

50. Number problems. Eight eagle stamps and two raccoon stamps cost $2.80. Three eagle stamps and four raccoon stamps cost $2.35. Find the cost of each kind of stamp.

51. Number problems. Robin bought four chocolate bars and a pack of gum and paid $2.75. Meg bought two chocolate bars and three packs of gum and paid $2.25. Find the cost of each.

52. Number problems. Xavier bought five red delicious apples and four Granny Smith apples at a cost of $4.81. Dean bought one of each of the two types at a cost of $1.08. Find the cost for each kind of apple.

53. Number problems. Eight disks and five zip disks cost a total of $27.50. Two disks and four zip disks cost $16.50. Find the unit cost for each.

54. Crafts. A 30-m rope is cut into two pieces so that one piece is 6 m longer than the other. How long is each piece?

55. Crafts. An 18-ft board is cut into two pieces, one of which is twice as long as the other. How long is each piece?

56. Number problems. Jill has $3.50 in nickels and dimes. If she has 50 coins, how many of each type of coin does she have?

57. Number problems. Richard has 22 coins with a total value of $4. If the coins are all quarters and dimes, how many of each type of coin does he have?

58. Number problems. Theater tickets are $8 for general admission and $5 for students. During one evening 240 tickets were sold, and the receipts were $1680. How many of each type of ticket was sold?

59. Number problems. Four hundred tickets were sold for a concert. The receipts from ticket sales were $3100, and the ticket prices were $7 and $9. How many of each type of ticket was sold?

60. Business and finance. A coffee merchant has coffee beans that sell for $9 per pound and $12 per pound. The two types are to be mixed to create 100 lb of a mixture that will sell for $11.25 per pound. How much of each type of bean should be used in the mixture?

61. Business and finance. Peanuts are selling for $2 per pound, and cashews are selling for $5 per pound. How much of each type of nut would be needed to create 20 lb of a mixture that would sell for $2.75 per pound?

62. Science and medicine. A chemist has a 25% and a 50% acid solution. How much of each solution should be used to form 200 mL of a 35% acid solution?

25% acid

50% acid

63. Science and medicine. A pharmacist wishes to prepare 150 mL of a 20% alcohol solution. She has a 30% solution and a 15% solution in her stock. How much of each should be used in forming the desired mixture?

64. Science and medicine. You have two alcohol solutions, one a 15% solution and one a 45% solution. How much of each solution should be used to obtain 300 mL of a 25% solution?

65. Business and finance. Otis has a total of $12,000 invested in two accounts. One account pays 8% and the other 9%. If his interest for 1 year is $1010, how much does he have invested at each rate?

66. Business and finance. Amy invests a part of $8000 in bonds paying 12% interest. The remainder is in a savings account at 8%. If she receives $840 in interest for 1 year, how much does she have invested at each rate?

ANSWERS

58. _____

59. _____

60. _____

61. _____

62. _____

63. _____

64. _____

65. _____

66. _____

67. _____

68. _____

69. _____

70. _____

71. _____

72. _____

67. Science and medicine. A plane flies 450 mi with the wind in 3 h. Flying back against the wind, the plane takes 5 h to make the trip. What was the rate of the plane in still air? What was the rate of the wind?

68. Science and medicine. An airliner made a trip of 1800 mi in 3 h, flying east across the country with the jetstream directly behind it. The return trip, against the jetstream, took 4 h. Find the speed of the plane in still air and the speed of the jetstream.

Each of the following applications (exercises 69 to 76) can be solved by the use of a system of linear equations. Match the application with the system on the right that could be used for its solution.

69. Number problems. One number is 4 less than 3 times another. If the sum of the numbers is 36, what are the two numbers?

(a) $12x + 5y = 116$
$\quad\ \ 8x + 12y = 112$

70. Number problems. Suppose that a movie theater sold 300 adult and student tickets for a showing with a revenue of $1440. If the adult tickets were $6 and the student tickets $4, how many of each type of ticket were sold?

(b) $\quad x + \quad\ y = 8000$
$\quad 0.06x + 0.09y = \quad 600$

71. Geometry. The length of a rectangle is 3 cm more than twice its width. If the perimeter of the rectangle is 36 cm, find the dimensions of the rectangle.

(c) $\quad x + \quad\ y = 200$
$\quad 0.20x + 0.60y = \quad 90$

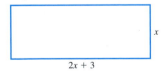

2x + 3

x

72. Number problems. An order of 12 dozen roller-ball pens and 5 dozen ballpoint pens cost $116. A later order for 8 dozen roller-ball pens and 12 dozen ballpoint pens cost $112. What was the cost of 1 dozen of each of the pens?

(d) $x + y = 36$
$\quad\ \ y = 3x - 4$

73. Business and finance. A candy merchant wishes to mix peanuts selling at $2/lb with cashews selling for $5.50/lb to form 140 lb of a mixed-nut blend that will sell for $3/lb. What amount of each type of nut should be used?

(e) $5(x - y) = 80$
$4(x + y) = 80$

74. Business and finance. Rolando has investments totaling $8000 in two accounts, one a savings account paying 6% interest and the other a bond paying 9%. If the annual interest from the two investments was $600, how much did he have invested at each rate?

(f) $x + y = 300$
$6x + 4y = 1440$

75. Science and medicine. A chemist wants to combine a 20% alcohol solution with a 60% solution to form 200 mL of a 45% solution. How much of each of the solutions should be used to form the mixture?

(g) $L = 2W + 3$
$2L + 2W = 36$

76. Science and medicine. A boat traveled 80 mi upstream in 5 h. Returning downstream with the current, the boat took 4 h to make the trip. What was the boat's speed in still water? What was the speed of the river's current?

(h) $x + y = 140$
$2x + 5.5y = 420$

For exercises 77 to 79, fill in the blank with *always, sometimes,* or *never.*

77. Both variables are _____ eliminated when the equations of a linear system are added.

78. A system is _____ both inconsistent and dependent.

79. It is _____ possible to use the addition method to solve a linear system.

ANSWERS

73. _____

74. _____

75. _____

76. _____

77. _____

78. _____

79. _____

80.

a. _____

b. _____

c. _____

d. _____

80. Work with a partner to solve the following problems.

Your friend Valerie has contacted you about going into business with her. She wants to start a small manufacturing business making and selling sweaters to specialty boutiques. She explains that the initial investment for each of you will be $1500 for a knitting machine. She has worked hard to come up with an estimate for expenses and thinks that they will be close to $1600 a month for overhead. She says that each sweater manufactured will cost $28 to produce and that the sweaters will sell for at least $70. She wants to know if you are willing to invest the money you have saved for college costs. You have faith in Valerie's ability to carry out her plan. But, you have worked hard to save this money. Use graphs and equations to help you decide if this is a good opportunity. Think about whether you need more information from Valerie. Write a letter summarizing your thoughts.

 Getting Ready for Section 8.3 [Section 2.3]

Solve each of the following equations.

(a) $2x + 3(x + 1) = 13$ 　　　　　　(b) $3(y - 1) + 4y = 18$

(c) $x + 2(3x - 5) = 25$ 　　　　　　(d) $3x - 2(x - 7) = 12$

Answers

1. $(5, 1)$　　**3.** $(1, 4)$　　**5.** $(2, 3)$　　**7.** $(6, 2)$　　**9.** $\left(-5, \dfrac{3}{2}\right)$　　**11.** $(3, 2)$

13. Inconsistent system　　**15.** $(4, 2)$　　**17.** $(2, 1)$　　**19.** $(3, 6)$

21. $(2, -4)$　　**23.** $\left(3, \dfrac{13}{2}\right)$　　**25.** Dependent system　　**27.** $(5, -3)$

29. $(6, 3)$　　**31.** $(-8, -2)$　　**33.** $\left(2, \dfrac{3}{2}\right)$　　**35.** $(3, 0)$　　**37.** $(4, 0)$

39. $\left(\dfrac{1}{2}, -\dfrac{1}{2}\right)$　　**41.** $\left(\dfrac{1}{3}, \dfrac{1}{2}\right)$　　**43.** $\left(\dfrac{1}{5}, -\dfrac{2}{5}\right)$　　**45.** $(3, 6)$　　**47.** $(-11, -25)$

49. 24, 16　　**51.** Chocolate 60¢, gum 35¢　　**53.** Disk $1.25, zip disk $3.50

55. 12 ft, 6 ft　　**57.** 10 dimes, 12 quarters　　**59.** 250 at $7, 150 at $9

61. 15 lb peanuts, 5 lb cashews　　**63.** 50 mL of 30%, 100 mL of 15%

65. $7000 at 8%, $5000 at 9%　　**67.** 120 mi/h, 30 mi/h　　**69.** d　　**71.** g

73. h　　**75.** c　　**77.** Sometimes　　**79.** Always　　**a.** 2　　**b.** 3

c. 5　　**d.** -2

8.3 Systems of Linear Equations: Solving by Substitution

8.3 OBJECTIVES

1. Solve systems using the substitution method
2. Choose an appropriate method for solving a system
3. Solve applications of systems of equations

In Sections 8.1 and 8.2, we looked at graphing and addition as methods of solving linear systems. A third method is called **solving by substitution.**

OBJECTIVE 1

Example 1   Solving a System by Substitution

Solve by substitution.

$$x + y = 12$$
$$y = 3x$$

Notice that the second equation says that y and $3x$ name the same quantity. So we may substitute $3x$ for y in the first equation. We then have

Replace y with $3x$ in the first equation.

NOTE The resulting equation contains only the variable x, so substitution is just another way of eliminating one of the variables from our system.

$$x + 3x = 12$$
$$4x = 12$$
$$x = 3$$

We can now substitute 3 for x in the first equation to find the corresponding y-coordinate of the solution.

$$(3) + y = 12$$
$$y = 9$$

NOTE The solution for a system is written as an ordered pair.

So $(3, 9)$ is the solution.

This last step is identical to the one you saw in Section 8.2. As before, you can substitute the known coordinate value back into either of the original equations to find the value of the remaining variable. The check is also identical.

CHECK YOURSELF 1

Solve by substitution.

$x - y = 9$
$y = 4x$

The same technique can be readily used any time one of the equations is *already solved* for x or for y, as Example 2 illustrates.

Example 2 **Solving a System by Substitution**

Solve by substitution.

$$2x + 3y = 3$$
$$y = 2x - 7$$

Because the second equation tells us that y is $2x - 7$, we can replace y with $2x - 7$ in the first equation. This gives

NOTE Now y is eliminated from the equation, and we can proceed to solve for x.

$$2x + 3\overbrace{(2x - 7)}^{y} = 3$$
$$2x + 6x - 21 = 3$$
$$8x = 24$$
$$x = 3$$

We now know that 3 is the x-coordinate for the solution. So substituting 3 for x in the second equation, we have

$$y = 2(3) - 7$$
$$= 6 - 7$$
$$= -1$$

The solution is $(3, -1)$. Once again you should verify this result by letting $x = 3$ and $y = -1$ in the original system.

CHECK YOURSELF 2

Solve by substitution.

$2x - 3y = 6$
$\quad x = 4y - 2$

As we have seen, the substitution method works very well when one of the given equations is already solved for x or y. It is also useful if you can readily solve for x or for y in one of the equations.

Example 3 **Solving a System by Substitution**

Solve by substitution.

$\quad x - 2y = 5$
$3x + \quad y = 8$

Neither equation is solved for a variable. That is easily handled in this case. Solving for x in the first equation, we have

$$x = 2y + 5$$

NOTE The second equation could have been solved for y with the result substituted into the first equation.

Now substitute $2y + 5$ for x in the second equation.

$$3(2y + 5) + y = 8$$
$$6y + 15 + y = 8$$
$$7y = -7$$
$$y = -1$$

Substituting -1 for y in the second equation yields

$$3x + (-1) = 8$$
$$3x = 9$$
$$x = 3$$

So $(3, -1)$ is the solution. You should check this result by substituting 3 for x and -1 for y in the equations of the original system.

✔ CHECK YOURSELF 3

Solve by substitution.

$$3x - y = 5$$
$$x + 4y = 6$$

Inconsistent systems and dependent systems will show up in a fashion similar to that which we saw in Section 8.2. Example 4 illustrates this approach.

Example 4   Solving an Inconsistent or Dependent System

Solve the following systems by substitution.

(a) $4x - 2y = 6$
$\quad\quad\quad y = 2x - 3$

From the second equation we can substitute $2x - 3$ for y in the first equation.

NOTE Be sure to change both signs in the parentheses.

$$4x - 2(2x - 3) = 6$$
$$4x - 4x + 6 = 6$$
$$6 = 6$$

Both variables have been eliminated, and we have the true statement $6 = 6$.

Recall from Section 8.2 that a true statement tells us that the graphs of the two equations are lines which coincide. We call this system dependent. There are an infinite number of solutions.

(b) $3x - 6y = 9$
$\quad\quad\quad x = 2y + 2$

Substitute $2y + 2$ for x in the first equation.

$$3(2y + 2) - 6y = 9$$
$$6y + 6 - 6y = 9 \quad\quad \text{This time we have}$$
$$6 = 9 \quad\quad\quad\quad \text{a false statement.}$$

This means that the system is *inconsistent* and that the graphs of the two equations are parallel lines. There is no solution.

CHECK YOURSELF 4 _____

Indicate whether the given system is inconsistent (no solution) or dependent (an infinite number of solutions).

(a) $5x + 15y = 10$

$x = -3y + 1$

(b) $12x - 4y = 8$

$y = 3x - 2$

The following summarizes our work in this section.

Step by Step: To Solve a System of Linear Equations by Substitution

Step 1   Solve one of the given equations for x or y. If this is already done, go on to Step 2.

Step 2   Substitute this expression for x or for y into the other equation.

Step 3   Solve the resulting equation for the remaining variable.

Step 4   Substitute the known value into either of the original equations to find the value of the second variable.

Step 5   Check your solution in both of the original equations.

You have now seen three different ways to solve systems of linear equations: by graphing, adding, and substitution. The natural question is, Which method should I use in a given situation?

Graphing is the least exact of the methods, and solutions may have to be estimated.

The algebraic methods—addition and substitution—give exact solutions, and both will work for any system of linear equations. In fact, you may have noticed that several examples in this section could just as easily have been solved by adding (Example 3, for instance).

The choice of which algebraic method (substitution or addition) to use is yours and depends largely on the given system. Here are some guidelines designed to help you choose an appropriate method for solving a linear system.

Rules and Properties: Choosing an Appropriate Method for Solving a System

1. If one of the equations is already solved for x (or for y), then substitution is the preferred method.

2. If the coefficients of x (or of y) are the same, or opposites, in the two equations, then addition is the preferred method.

3. If solving for x (or for y) in either of the given equations will result in fractional coefficients, then addition is the preferred method.

OBJECTIVE 2   **Example 5   Choosing an Appropriate Method for Solving a System**

Select the most appropriate method for solving each of the following systems.

(a) $5x + 3y = 9$
$2x - 7y = 8$

Addition is the most appropriate method because solving for a variable will result in fractional coefficients.

(b) $7x + 26 = 8$
$x = 3y - 5$

Substitution is the most appropriate method because the second equation is already solved for x.

(c) $8x - 9y = 11$
$4x + 9y = 15$

Addition is the most appropriate method because the coefficients of y are opposites.

 CHECK YOURSELF 5

Select the most appropriate method for solving each of the following systems.

(a) $2x + 5y = 3$
$8x - 5y = -13$

(b) $4x - 3y = 2$
$y = 3x - 4$

(c) $3x - 5y = 2$
$x = 3y - 2$

(d) $5x - 2y = 19$
$4x + 6y = 38$

Number problems, such as those presented in Chapter 2, are sometimes more easily solved by the methods presented in this section. Example 6 illustrates this approach.

OBJECTIVE 3   **Example 6   Solving a Number Problem by Substitution**

The sum of two numbers is 25. If the second number is 5 less than twice the first number, what are the two numbers?

NOTE

1. What do you want to find?
2. Assign variables. This time we use two letters, x and y.
3. Write equations for the solution. Here two equations are needed because we have introduced two variables.

Step 1   You want to find the two unknown numbers.

Step 2   Let $x =$ the first number and $y =$ the second number.

Step 3

$x + y = 25$

The sum   is 25.

$y = 2x - 5$

The second   is 5 less than
number   twice the first.

4. Solve the system of equations.

Step 4

$$x + y = 25$$
$$y = 2x - 5$$

NOTE We use the substitution method because the second equation is already solved for y.

Substitute $2x - 5$ for y in the first equation.

$$x + (2x - 5) = 25$$
$$3x - 5 = 25$$
$$x = 10$$

From the first equation,

$$(10) + y = 25$$
$$y = 15$$

The two numbers are 10 and 15.

5. Check the result.

Step 5 The sum of the numbers is 25. The second number, 15, is 5 less than twice the first number, 10. The solution checks.

CHECK YOURSELF 6 _____

The sum of two numbers is 28. The second number is 4 more than twice the first number. What are the numbers?

Sketches are always helpful in solving applications from geometry. Let's look at such an example.

Example 7   Solving an Application from Geometry

The length of a rectangle is 3 m more than twice its width. If the perimeter of the rectangle is 42 m, find the dimensions of the rectangle.

Step 1 You want to find the dimensions (length and width) of the rectangle.

NOTE We used x and y as our two variables in the previous examples. Use whatever letters you want. The process is the same, and sometimes it helps you remember what letter stands for what. Here $L =$ length and $W =$ width.

Step 2 Let L be the length of the rectangle and W the width. Now draw a sketch of the problem.

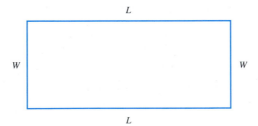

Step 3 Write the equations for the solution.

$$L = \underline{2W + 3}$$

    3 more than twice
    the width

$$\underline{2L + 2W} = 42$$

the perimeter

Step 4 Solve the system.

$$L = 2W + 3$$
$$2L + 2W = 42$$

NOTE Substitution is used because one equation is already solved for a variable.

From this first equation we can substitute $2W + 3$ for L in this last equation.

$$2(2W + 3) + 2W = 42$$
$$4W + 6 + 2W = 42$$
$$6W = 36$$
$$W = 6$$

Replace W with 6 to find L.

$$L = 2(6) + 3$$
$$= 12 + 3$$
$$= 15$$

The length is 15 m, the width is 6 m.

Step 5 Check these results. The perimeter is $2L + 2W$, which should give us 42 m.

$$2(15) + 2(6) \stackrel{?}{=} 42$$
$$30 + 12 \stackrel{\checkmark}{=} 42 \qquad \text{True}$$

 CHECK YOURSELF 7 _____

The length of each of the two equal legs of an isosceles triangle is 5 in. less than the length of the base. If the perimeter of the triangle is 50 in., find the lengths of the legs and the base.

CHECK YOURSELF ANSWERS _____

1. $(-3, -12)$ **2.** $(6, 2)$ **3.** $(2, 1)$

4. **(a)** Inconsistent system; **(b)** dependent system

5. **(a)** Addition; **(b)** substitution; **(c)** substitution; **(d)** addition

6. The numbers are 8 and 20.   **7.** The legs have length 15 in.; the base is 20 in.

8.3 Exercises

Solve each of the following systems by substitution.

1. $x + y = 10$
$\quad\ y = 4x$

2. $x - y = 4$
$\quad\ x = 3y$

3. $2x - y = 10$
$\quad\ x = -2y$

4. $x + 3y = 10$
$\quad\ 3x = y$

5. $3x + 2y = 12$
$\quad\ y = 3x$

6. $4x - 3y = 24$
$\quad\ y = -4x$

7. $x + y = 5$
$\quad\ y = x - 3$

8. $x + y = 9$
$\quad\ x = y + 3$

9. $x - y = 4$
$\quad\ x = 2y - 2$

10. $x - y = 7$
$\quad\ y = 2x - 12$

11. $2x + y = \ \ 7$
$\quad\ y - x = -8$

12. $3x - y = -15$
$\quad\ x = y - 7$

13. $2x - 5y = 10$
$\quad\ x - y = 8$

14. $4x - 3y = 0$
$\quad\ y = x + 1$

15. $3x + 4y = 9$
$\quad\ y - 3x = 1$

16. $5x - 2y = -5$
$\quad\ y - 5x = \ \ 3$

17. $3x - 18y = 4$
$\quad\ x = 6y + 2$

18. $4x + 5y = 6$
$\quad\ y = 2x - 10$

19. $5x - 3y = 6$
$\quad\ y = 3x - 6$

20. $8x - 4y = 16$
$\quad\ y = 2x - 4$

21. $8x - 5y = 16$
$\quad\ y = 4x - 5$

22. $6x - 5y = 27$
$\quad\ x = 5y + 2$

23. $x + 3y = 7$
$\quad\ x - \ y = 3$

24. $2x - y = -4$
$\quad\ x + y = -5$

25. $\ \ 6x - 3y = \ \ 9$
$\quad -2x + \ y = -3$

26. $5x - 6y = 21$
$\quad\ x - 2y = \ \ 5$

27. $x - 7y = 3$
    $2x - 5y = 15$

28. $4x - 12y = 5$
    $-x + 3y = -1$

For exercises 29 and 30, fill in the blank with *always, sometimes,* or *never.*

29. It is _____ possible to use the substitution method to solve a linear system.

30. The substitution method is _____ easier to use than the addition method.

Solve each of the following systems by using either addition or substitution. If a unique solution does not exist, state whether the system is dependent or inconsistent.

31. $2x + 3y = -6$
    $x = 3y + 6$

32. $7x + 3y = 31$
    $y = -2x + 9$

33. $2x - y = 1$
    $-2x + 3y = 5$

34. $x + 3y = 12$
    $2x - 3y = 6$

35. $6x + 2y = 4$
    $y = -3x + 2$

36. $3x - 2y = 15$
    $-x + 5y = -5$

37. $x + 2y = -2$
    $3x + 2y = -12$

38. $10x + 2y = 7$
    $y = -5x + 3$

39. $2x - 3y = 14$
    $4x + 5y = -5$

40. $2x + 3y = 1$
    $5x + 3y = 16$

41. $4x - 2y = 0$
    $x = \dfrac{3}{2}$

42. $4x - 3y = \dfrac{11}{2}$
    $y = -\dfrac{3}{2}$

Solve each system.

43. $\dfrac{1}{3}x + \dfrac{1}{2}y = 5$

    $\dfrac{x}{4} - \dfrac{y}{5} = -2$

44. $\dfrac{5x}{2} - y = \dfrac{9}{10}$

    $\dfrac{3x}{4} + \dfrac{5y}{6} = \dfrac{2}{3}$

45. $0.4x - 0.2y = 0.6$
    $2.5x - 0.3y = 4.7$

46. $0.4x - 0.1y = 5$
    $6.4x + 0.4y = 60$

ANSWERS

27. _____

28. _____

29. _____

30. _____

31. _____

32. _____

33. _____

34. _____

35. _____

36. _____

37. _____

38. _____

39. _____

40. _____

41. _____

42. _____

43. _____

44. _____

45. _____

46. _____

47. _____

48. _____

49. _____

50. _____

51. _____

52. _____

53. _____

54. _____

55. _____

56. _____

Solve each of the following problems. Be sure to show the equation used for the solution.

47. **Number problems.** The sum of two numbers is 100. The second is 3 times the first. Find the two numbers.

48. **Number problems.** The sum of two numbers is 70. The second is 10 more than 3 times the first. Find the numbers.

49. **Number problems.** The sum of two numbers is 56. The second is 4 less than twice the first. What are the two numbers?

50. **Number problems.** The difference of two numbers is 4. The larger is 8 less than twice the smaller. What are the two numbers?

51. **Number problems.** The difference of two numbers is 22. The larger is 2 more than 3 times the smaller. Find the two numbers.

52. **Number problems.** One number is 18 more than another, and the sum of the smaller number and twice the larger number is 45. Find the two numbers.

53. **Number problems.** One number is 5 times another. The larger number is 9 more than twice the smaller. Find the two numbers.

54. **Business and finance.** Two packages together weigh 32 kilograms (kg). The smaller package weighs 6 kg less than the larger. How much does each package weigh?

55. **Business and finance.** A washer-dryer combination costs $1200. If the washer costs $220 more than the dryer, what does each appliance cost separately?

56. **Social science.** In a town election, the winning candidate had 220 more votes than the loser. If 810 votes were cast in all, how many votes did each candidate receive?

57. **Business and finance.** An office desk and chair together cost $850. If the desk costs $50 less than twice as much as the chair, what did each cost?

58. **Geometry.** The length of a rectangle is 2 in. more than twice its width. If the perimeter of the rectangle is 34 in., find the dimensions of the rectangle.

59. **Geometry.** The perimeter of an isosceles triangle is 37 in. The lengths of the two equal legs are 6 in. less than 3 times the length of the base. Find the lengths of the three sides.

60. You have a part-time job writing the *Consumer Concerns* column for your local newspaper. Your topic for this week is clothes dryers, and you are planning to compare the Helpmate and the Whirlgarb dryers, both readily available in stores in your area. The information you have is that the Helpmate dryer is listed at $520, and it costs 22.5¢ to dry an average–sized load at the utility rates in your city. The Whirlgarb dryer is listed at $735, and it costs 15.8¢ to run for each normal load. The maintenance costs for both dryers are about the same. Working with a partner, write a short article giving your readers helpful advice about these appliances. What should they consider when buying one of these clothes dryers?

 Getting Ready for Section 8.4 [Section 2.7]

Graph the solution sets for the following linear inequalities.

  (a) $x + y > 8$ (b) $2x - y \le 6$

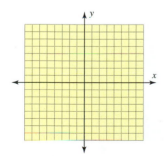

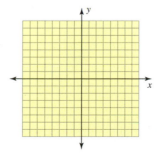

(c) $3x + 4y \geq 12$

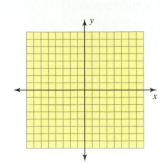

(d) $y > 2x$

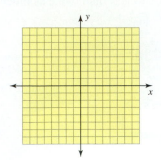

(e) $y \leq -3$

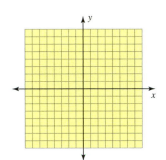

(f) $x > 5$

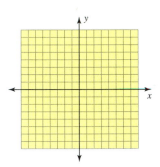

Answers

1. $(2, 8)$ **3.** $(4, -2)$ **5.** $\left(\frac{4}{3}, 4\right)$ **7.** $(4, 1)$ **9.** $(10, 6)$

11. $(5, -3)$ **13.** $(10, 2)$ **15.** $\left(\frac{1}{3}, 2\right)$ **17.** No solution   **19.** $(3, 3)$

21. $\left(\frac{3}{4}, -2\right)$ **23.** $(4, 1)$ **25.** Infinite number of solutions   **27.** $(10, 1)$

29. Always   **31.** $(0, -2)$ **33.** $(2, 3)$ **35.** Dependent system

37. $\left(-5, \frac{3}{2}\right)$ **39.** $\left(\frac{5}{2}, -3\right)$ **41.** $\left(\frac{3}{2}, 3\right)$ **43.** $(0, 10)$ **45.** $(2, 1)$

47. 25, 75   **49.** 20, 36   **51.** 32, 10   **53.** 3, 15

55. Washer \$710, dryer \$490   **57.** Desk \$550, chair \$300

59. 7 in., 15 in., 15 in.

a. $x + y > 8$ **b.** $2x - y \leq 6$

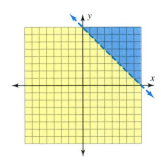

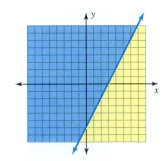

c. $3x + 4y \geq 12$

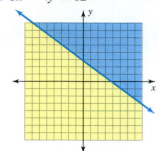

d. $y > 2x$

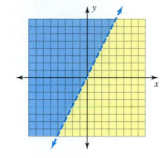

e. $y \leq -3$

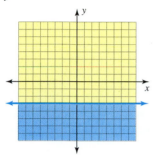

f. $x > 5$

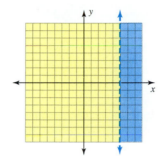

8.4 Systems of Linear Inequalities

8.4 OBJECTIVES

1. Graph a system of linear inequalities
2. Solve an application of linear inequalities

Our previous work in this chapter dealt with finding the solution set of a system of linear equations. That solution set represented the points of intersection of the graphs of the equations in the system. In this section, we extend that idea to include systems of linear inequalities.

NOTE You might want to review graphing linear inequalities in Section 7.4 at this point.

In this case, the solution set for each inequality is all ordered pairs that satisfy that inequality. *The graph of the solution set of a system of linear inequalities* is then the intersection of the graphs of the individual inequalities.

OBJECTIVE 1

Example 1   Solving a System by Graphing

Solve the following system of linear inequalities by graphing.

$x + y > 4$

$x - y < 2$

We start by graphing each inequality separately. The boundary line is drawn, and using $(0, 0)$ as a test point, we see that we should shade the half plane above the line in both graphs.

NOTE The boundary line is dashed to indicate it is *not* included in the graph.

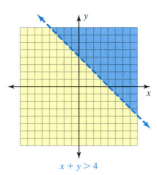
$x + y > 4$

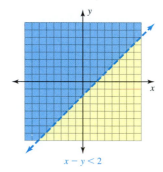
$x - y < 2$

In practice, the graphs of the two inequalities are combined on the same set of axes, as shown in the following graph. This graph of the solution set of the original system is the intersection of the two original graphs.

NOTE Points on the lines are not included in the solution.

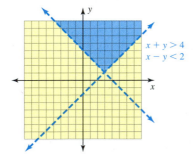
$x + y > 4$
$x - y < 2$

CHECK YOURSELF 1

Solve the following system of linear inequalities by graphing.

$2x - y < 4$

$x + y < 3$

Most applications of systems of linear inequalities lead to bounded regions. This requires a system of three or more inequalities, as shown in Example 2.

Example 2   Solving a System by Graphing

Solve the following system of linear inequalities by graphing.

$$x + 2y \le 6$$
$$x + y \le 5$$
$$x \ge 2$$
$$y \ge 0$$

On the same set of axes, we graph the boundary line of each of the inequalities. We then choose the appropriate half planes, indicating each with an arrow. The set of solutions is the intersection of those regions.

NOTE The vertices of the shaded region are given because they have particular significance in later applications of this concept. Can you see how the coordinates of the vertices were determined?

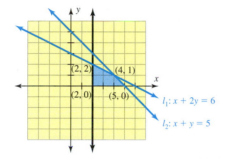

$l_1: x + 2y = 6$

$l_2: x + y = 5$

(2, 2)   (4, 1)

(2, 0)   (5, 0)

CHECK YOURSELF 2

Solve the following system of linear inequalities by graphing.

$$2x - y \le 8 \qquad x \ge 0$$
$$x + y \le 7 \qquad y \ge 0$$

Next, we look at an application of our work with systems of linear inequalities.

OBJECTIVE 2

Example 3   Solving a Business-Based Application

A manufacturer produces a standard model and a deluxe model of a 13-in. television set. The standard model requires 12 h of labor to produce, whereas the deluxe model requires 18 h. The labor available is limited to 360 h per week. Also, the plant capacity is limited to producing a total of 25 sets per week. Write a system of inequalities representing this situation. Then, draw a graph of the region representing the number of sets that can be produced, given these conditions.

We let x represent the number of standard-model sets produced and y the number of deluxe-model sets. Because the labor is limited to 360 h, we have

NOTE The total labor is limited to (or less than or equal to) 360 h.

$$12x \quad + \quad 18y \le 360$$

12 h per   18 h per
standard set   deluxe set

The total production, here $x + y$ sets, is limited to 25, so we can write

$$x + y \le 25$$

For convenience in graphing, we divide both expressions in the first inequality by 6, to write the equivalent system:

NOTE We have $x \geq 0$ and $y \geq 0$ because the number of sets produced cannot be negative.

$$2x + 3y \leq 60$$
$$x + y \leq 25$$
$$x \geq 0$$
$$y \geq 0$$

We now graph the system of inequalities as before. The shaded area represents all possibilities in terms of the number of sets that can be produced.

NOTE The shaded area is called the *feasible region*. All points in the region meet the given conditions of the problem and represent possible production options.

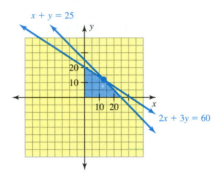

CHECK YOURSELF 3

A manufacturer produces DVD players and compact disk players. The DVD players require 10 h of labor to produce and the disk players require 20 h. The labor hours available are limited to 300 h per week. Existing orders require that at least 10 DVD players and at least 5 disk players be produced per week.

Write a system of inequalities representing this situation. Then, draw a graph of the region representing the possible production options.

CHECK YOURSELF ANSWERS

1. $2x - y < 4$
$x + y < 3$

2. $2x - y \leq 8$
$x + y \leq 7$
$x \geq 0$
$y \geq 0$

3. Let x be the number of DVD players and y be the number of CD players. The system is

$$10x + 20y \leq 300$$
$$x \geq 10$$
$$y \geq 5$$

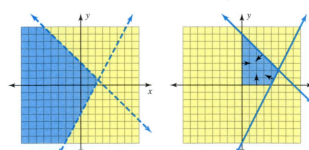
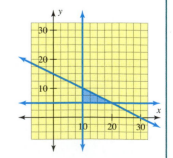

8.4 Exercises

Solve each of the following systems of linear inequalities graphically.

1. $x + 2y \leq 4$
$\quad x - y \geq 1$

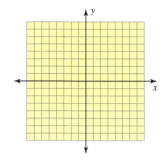

2. $3x - y > 6$
$\quad x + y < 6$

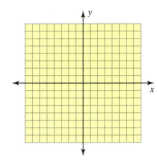

3. $3x + y < 6$
$\quad x + y > 4$

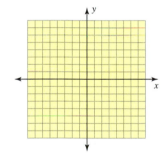

4. $2x + y \geq 8$
$\quad x + y \geq 4$

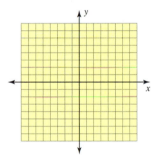

5. $x + 3y \leq 12$
$\quad 2x - 3y \leq 6$

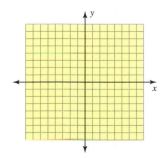

6. $x - 2y > 8$
$\quad 3x - 2y > 12$

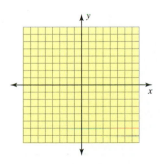

ANSWERS

1. _____

2. _____

3. _____

4. _____

5. _____

6. _____

7. $3x + 2y \le 12$

$x \ge\ \ 2$

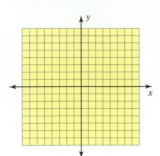

8. $2x + y \le 6$

$y \ge 1$

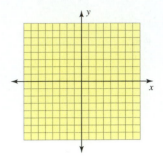

9. $2x + y \le 8$

$x > 1$

$y > 2$

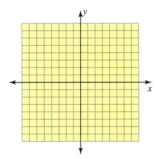

10. $3x - y \le 6$

$x \ge 1$

$y \le 3$

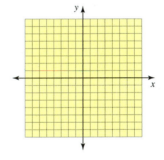

11. $x + 2y \le 8$

$2 \le x \le 6$

$y \ge 0$

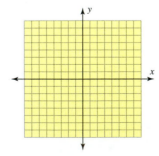

12. $x + y < 6$

$0 \le y \le 3$

$x \ge 1$

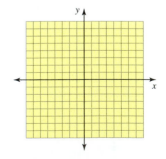

13. $3x + y \leq 6$
$x + y \leq 4$
$x \geq 0$
$y \geq 0$

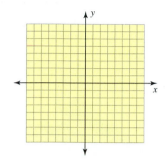

14. $x - 2y \geq -2$
$x + 2y \leq 6$
$x \geq 0$
$y \geq 0$

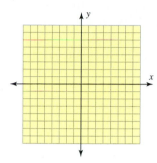

15. $4x + 3y \leq 12$
$x + 4y \leq 8$
$x \geq 0$
$y \geq 0$

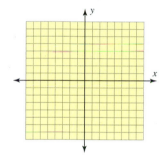

16. $2x + y \leq 8$
$x + y \geq 3$
$x \geq 0$
$y \geq 0$

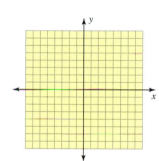

17. $x - 4y \leq -4$
$x + 2y \leq 8$
$x \geq 2$

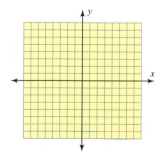

18. $x - 3y \geq -6$
$x + 2y \geq 4$
$x \leq 4$

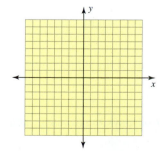

ANSWERS

13. _____

14. _____

15. _____

16. _____

17. _____

18. _____

ANSWERS

19. _____

20. _____

21. _____

22. _____

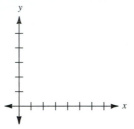

23. _____

Draw the appropriate graphs in each of the following.

19. A manufacturer produces both two-slice and four-slice toasters. The two-slice toaster takes 6 h of labor to produce and the four-slice toaster 10 h. The labor available is limited to 300 h per week, and the total production capacity is 40 toasters per week. Write a system of inequalities representing this situation. Then, draw a graph of the feasible region, given these conditions, in which x is the number of two-slice toasters and y is the number of four-slice toasters.

20. A small firm produces both AM and AM/FM car radios. The AM radios take 15 h to produce, and the AM/FM radios take 20 h. The number of production hours is limited to 300 h per week. The plant's capacity is limited to a total of 18 radios per week, and existing orders require that at least 4 AM radios and at least 3 AM/FM radios be produced per week. Write a system of inequalities representing this situation. Then, draw a graph of the feasible region given these conditions, in which x is the number of AM radios and y the number of AM/FM radios.

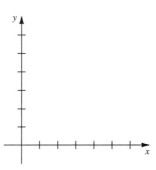

For exercises 21 and 22, fill in the blanks with *always, sometimes,* or *never*.

21. The graph of the solution set of a system of two linear inequalities _____ includes the origin.

22. The graph of the solution set of a system of two linear inequalities is _____ bounded.

23. Describe a system of linear inequalities for which there is no solution.

24. Write the system of inequalities whose graph is the shaded region.

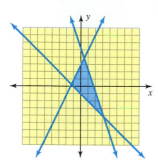

25. Write the system of inequalities whose graph is the shaded region.

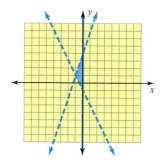

Answers

1. $x + 2y \le 4$
    $x - y \ge 1$

3. $3x + y < 6$
    $x + y > 4$

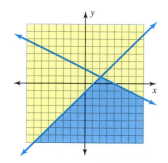

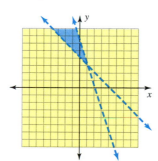

5. $x + 3y \le 12$
    $2x - 3y \le 6$

7. $3x + 2y \le 12$
    $x \ge 2$

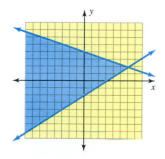

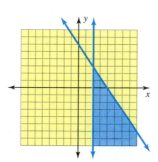

9. $2x + y \leq 8$
$\quad x > 1$
$\quad y > 2$

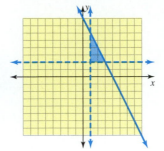

11. $x + 2y \leq 8$
$\quad 2 \leq x \leq 6$
$\quad y \geq 0$

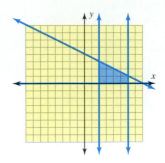

13. $3x + y \leq 6$
$\quad x + y \leq 4$
$\quad x \geq 0$
$\quad y \geq 0$

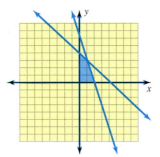

15. $4x + 3y \leq 12$
$\quad x + 4y \leq 8$
$\quad x \geq 0$
$\quad y \geq 0$

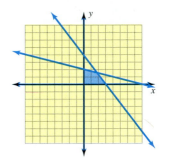

17. $x - 4y \leq -4$
$\quad x + 2y \leq 8$
$\quad x \geq 2$

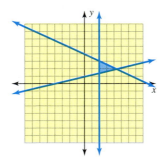

19. $6x + 10y \leq 300$
$\quad x + y \leq 40$
$\quad x \geq 0$
$\quad y \geq 0$

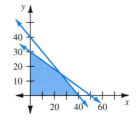

21. Sometimes

23.

25. $x \leq 0$
$\quad y < 3x + 4$
$\quad y > -2x - 1$

8   Summary

DEFINITION/PROCEDURE	EXAMPLE	REFERENCE
Systems of Linear Equations: Solving by Graphing		**Section 8.1**
A System of Equations Two or more equations considered together.	$x + y = 4$ $2x - y = 5$	*p. 679*
Solution The solution of a system of two equations in two unknowns is an ordered pair that satisfies each equation of the system.	$(x, y) = (3, 1)$	*p. 679*

Solving by Graphing
 1. Graph both equations on the same coordinate system.
 2. The system may have
    a. *One solution.* The lines intersect at one point (a consistent system). The solution is the ordered pair corresponding to that point.
    b. *No solution.* The lines are parallel (an inconsistent system).
    c. *Infinitely many solutions.* The two equations have the same graph (a dependent system). Any ordered pair corresponding to a point on the line is a solution.
	
	 A consistent system
	
	 An inconsistent system
	
	 A dependent system
		p. 682

Systems of Linear Equations: Solving by Adding		**Section 8.2**
Solving by Adding 1. If necessary, multiply both sides of one or both equations by nonzero numbers to form an equivalent system in which the coefficients of one of the variables are opposites. 2. Add the equations of the new system.	$2x - y = 4$ $3x + 2y = 13$ Multiply the first equation by 2. $4x - 2y = 8$ $3x + 2y = 13$	

Continued

Systems of Linear Equations: Solving by Adding		**Section 8.2**
3. Solve the resulting equation for the remaining variable. 4. Substitute the value found in Step 3 into either of the original equations to find the value of the second variable. 5. Check your solution in both of the original equations.	Add. $$7x = 21$$ $$x = 3$$ In the original equation, $$2(3) - y = 4$$ $$y = 2$$ (3, 2) is the solution.	*p. 694*
Applying Systems of Equations Often word problems can be solved by using two variables and two equations to represent the unknowns and the given relationships in the problem. *The Solution Steps* 1. Read the problem carefully. Then reread it to decide what you are asked to find. 2. Assign variables to the unknown quantities. 3. Translate the problem to the language of algebra to form a system of equations. 4. Solve the system. 5. Verify your solution in the original problem.		*p. 696*

Systems of Linear Equations: Solving by Substitution		**Section 8.3**
Solving by Substitution 1. Solve one of the given equations for x or for y. If this is already done, go on to Step 2. 2. Substitute this expression for x or for y into the other equation. 3. Solve the resulting equation for the remaining variable. Steps 4 and 5 are the same as in Section 8.2.	$$x - 2y = 3$$ $$2x + 3y = 13$$ From the first equation, $$x = 2y + 3$$ Substitute in the second equation: $$2(2y + 3) + 3y = 13$$ $$4y + 6 + 3y = 13$$ $$7y + 6 = 13$$ $$7y = 7$$ $$y = 1$$	*p. 714*

Systems of Linear Inequalities		**Section 8.4**
A *system of linear inequalities* is two or more linear inequalities considered together. The *graph of the solution set* of a system of linear inequalities is the intersection of the graphs of the individual inequalities. *Solving Systems of Linear Inequalities Graphically* 1. Graph each inequality, shading the appropriate half plane, on the same set of coordinate axes. 2. The graph of the system is the intersection of the regions shaded in Step 1.	To solve $$x + 2y \leq 8$$ $$x + y \leq 6$$ $$x \geq 0$$ $$y \geq 0$$ graphically 	*p. 724*

Summary Exercises

This summary exercise set is provided to give you practice with each of the objectives of this chapter. Each exercise is keyed to the appropriate chapter section. When you are finished, you can check your answers to the odd-numbered exercises against those presented in the back of the text. If you have difficulty with any of these questions, go back and reread the examples from that section. Your instructor will give you guidelines on how to best use these exercises in your instructional setting.

[8.1] Solve each of the following systems by graphing.

1. $x + y = 6$
$x - y = 2$

2. $x - y = 8$
$2x + y = 7$

3. $x + 2y = 4$
$x + 2y = 6$

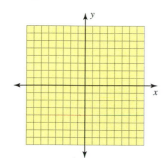
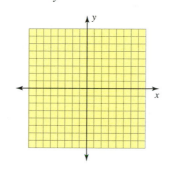
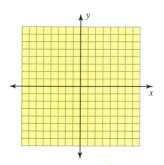

4. $2x - y = 8$
$y = 2$

5. $2x - 4y = 8$
$x - 2y = 4$

6. $3x + 2y = 6$
$4x - y = 8$

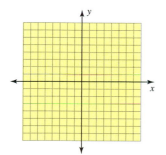
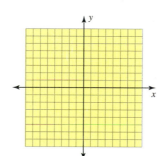
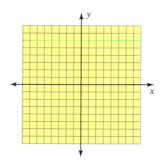

[8.2] Solve each of the following systems by addition. If a unique solution does not exist, state whether the system is inconsistent or dependent.

7. $x + y = 8$
$x - y = 2$

8. $-x - y = 4$
$x - y = -8$

9. $2x - 3y = 16$
$5x + 3y = 19$

10. $2x + y = 7$
$3x - y = 3$

11. $3x - 5y = 14$
$3x + 2y = 7$

12. $2x - 4y = 8$
$x - 2y = 4$

13. $4x - 3y = -22$
$4x + 5y = -6$

14. $5x - 2y = 17$
$3x - 2y = 9$

15. $4x - 3y = 10$
$2x - 3y = 6$

16. $2x + 3y = -10$
$-2x + 5y = 10$

17. $3x + 2y = 3$
$6x + 4y = 5$

18. $3x - 2y = 23$
$x + 5y = -15$

19. $5x - 2y = -1$
$10x + 3y = 12$

20. $x - 3y = 9$
$5x - 15y = 45$

21. $2x - 3y = 18$
$5x - 6y = 42$

22. $3x + 7y = 1$
$4x - 5y = 30$

23. $5x - 4y = 12$
$3x + 5y = 22$

24. $6x + 5y = -6$
$9x - 2y = 10$

25. $4x - 3y = 7$
$-8x + 6y = -10$

26. $3x + 2y = 8$
$-x - 5y = -20$

27. $3x - 5y = -14$
$6x + 3y = -2$

[8.3] Solve each of the following systems by substitution. If a unique solution does not exist, state whether the system is inconsistent or dependent.

28. $x + 2y = 10$
$y = 2x$

29. $x - y = 10$
$x = -4y$

30. $2x - y = 10$
$x = 3y$

31. $2x + 3y = 2$
$y = x - 6$

32. $4x + 2y = 4$
$y = 2 - 2x$

33. $x + 5y = 20$
$x = y + 2$

34. $6x + y = 2$
$y = 3x - 4$

35. $2x + 6y = 10$
$x = 6 - 3y$

36. $2x + y = 9$
$x - 3y = 22$

37. $x - 3y = 17$
$2x + y = 6$

38. $2x + 3y = 4$
$y = 2$

39. $4x - 5y = -2$
$x = -3$

40. $-6x + 3y = -4$
$y = -\dfrac{2}{3}$

41. $5x - 2y = -15$
$y = 2x + 6$

42. $3x + y = 15$
$x = 2y + 5$

Solve each of the following systems by either addition or substitution. If a unique solution does not exist, state whether the system is inconsistent or dependent.

43. $x - 4y = 0$
$4x + y = 34$

44. $2x + y = 2$
$y = -x$

45. $3x - 3y = 30$
$x = -2y - 8$

46. $5x + 4y = 40$
$x + 2y = 11$

47. $x - 6y = -8$
$2x + 3y = 4$

48. $4x - 3y = 9$
$2x + y = 12$

49. $9x + y = 9$
$x + 3y = 14$

50. $3x - 2y = 8$
$-6x + 4y = -16$

51. $3x - 2y = 8$
$2x - 3y = 7$

Solve the following problems. Be sure to show the equations used.

52. **Number problems.** The sum of two numbers is 40. If their difference is 10, find the two numbers.

53. **Number problems.** The sum of two numbers is 17. If the larger number is 1 more than 3 times the smaller, what are the two numbers?

54. **Number problems.** The difference of two numbers is 8. The larger number is 2 less than twice the smaller. Find the numbers.

55. **Business and finance.** Five writing tablets and three pencils cost $8.25. Two tablets and two pencils cost $3.50. Find the cost for each item.

56. **Construction.** A cable 200 ft long is cut into two pieces so that one piece is 12 ft longer than the other. How long is each piece?

57. **Business and finance.** An amplifier and a pair of speakers cost $925. If the amplifier costs $75 more than the speakers, what does each cost?

58. **Business and finance.** A sofa and chair cost $850 as a set. If the sofa costs $100 more than twice as much as the chair, what is the cost of each?

59. **Geometry.** The length of a rectangle is 4 cm more than its width. If the perimeter of the rectangle is 64 cm, find the dimensions of the rectangle.

60. **Geometry.** The perimeter of an isosceles triangle is 29 in. The lengths of the two equal legs are 2 in. more than twice the length of the base. Find the lengths of the three sides.

61. **Number problems.** Darryl has 30 coins with a value of $5.50. If they are all nickels and quarters, how many of each kind of coin does he have?

62. **Business and finance.** Tickets for a concert sold for $11 and $8. If 600 tickets were sold for one evening and the receipts were $5550, how many of each kind of ticket were sold?

63. **Science and medicine.** A laboratory has a 20% acid solution and a 50% acid solution. How much of each should be used to produce 600 mL of a 40% acid solution?

64. **Science and medicine.** A service station wishes to mix 40 L of a 78% antifreeze solution. How many liters of a 75% solution and a 90% solution should be used in forming the mixture?

65. **Business and finance.** Martha has $18,000 invested. Part of the money is invested in a bond that yields 11% interest. The remainder is in her savings account, which pays 7%. If she earns $1660 in interest for 1 year, how much does she have invested at each rate?

66. **Science and medicine.** A boat travels 24 mi upstream in 3 h. It then takes 3 h to go 36 mi downstream. Find the speed of the boat in still water and the speed of the current.

67. **Science and medicine.** A plane flying with the wind makes a trip of 2200 mi in 4 h. Returning against the wind, it can travel only 1800 mi in 4 h. What is the plane's rate in still air? What is the wind speed?

[8.4]   Solve these systems of linear inequalities.

68. $x - y < 7$
$x + y > 3$

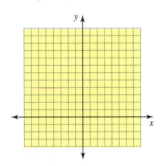

69. $x - 2y \leq -2$
$x + 2y \leq 6$

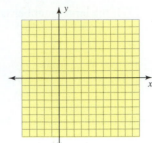

70. $x - 6y < 6$
$-x + y < 4$

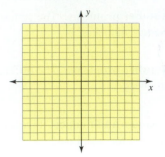

71. $2x + y \leq 8$
$x \geq 1$
$y \geq 0$

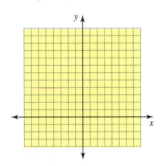

72. $2x + y \leq 6$
$x \geq 1$
$y \geq 0$

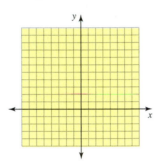

73. $4x + y \leq 8$
$x \geq 0$
$y \geq 2$

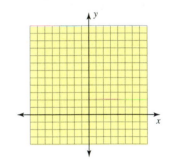

74. $4x + 2y \leq 8$
$x + y \leq 3$
$x \geq 0$
$y \geq 0$

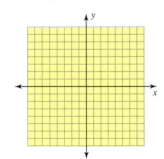

75. $3x + y \leq 6$
$x + y \leq 4$
$x \geq 0$
$y \geq 0$

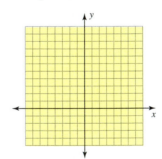

Self-Test for Chapter 8

Name _____

Section _____ Date _____

The purpose of this self-test is to help you check your progress and to review for the next in-class exam. Allow yourself about an hour to take this test. At the end of that hour check your answers against those given in the back of the text. Section references accompany the answers. If you missed any questions, go back to those sections and reread the examples until you master the concepts.

Solve each of the following systems by graphing. If a unique solution does not exist, state whether the system is inconsistent or dependent.

1. $x + y = 5$
$x - y = 3$

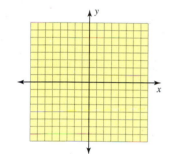

2. $x + 2y = 8$
$x - y = 2$

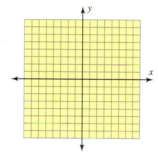

3. $x - 3y = 3$
$x - 3y = 6$

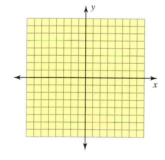

4. $4x - y = 4$
$x - 2y = -6$

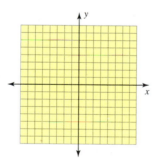

ANSWERS

1. _____
2. _____
3. _____
4. _____
5. _____
6. _____
7. _____
8. _____
9. _____
10. _____
11. _____
12. _____

Solve each of the following systems by addition. If a unique solution does not exist, state whether the system is inconsistent or dependent.

5. $x + y = 5$
$x - y = 3$

6. $x + 2y = 8$
$x - y = 2$

7. $3x + y = 6$
$-3x + 2y = 3$

8. $3x + 2y = 11$
$5x + 2y = 15$

9. $3x - 6y = 12$
$x - 2y = 4$

10. $4x + y = 2$
$8x - 3y = 9$

11. $2x - 5y = 2$
$3x + 4y = 26$

12. $x + 3y = 6$
$3x + 9y = 9$

Solve each of the following systems by substitution. If a unique solution does not exist, state whether the system is inconsistent or dependent.

13. $x + y = 8$
$y = 3x$

14. $x - y = 9$
$x = -2y$

15. $2x - y = 10$
$x = y + 4$

16. $x - 3y = -7$
$y = x - 1$

17. $3x + y = -6$
$y = 2x + 9$

18. $4x + 2y = 8$
$y = 3 - 2x$

19. $5x + y = 10$
$x + 2y = -7$

20. $3x - 2y = 5$
$2x + y = 8$

Solve each of the following problems. Be sure to show the equations used.

21. Number problems. The sum of two numbers is 30, and their difference is 6. Find the two numbers.

22. Construction. A rope 50 m long is cut into two pieces so that one piece is 8 m longer than the other. How long is each piece?

23. Geometry. The length of a rectangle is 4 in. less than twice its width. If the perimeter of the rectangle is 64 in., what are the dimensions of the rectangle?

24. Number problems. Murray has 30 coins with a value of $5.70. If the coins are all dimes and quarters, how many of each coin does he have?

25. Science and medicine. Jackson was able to travel 36 mi downstream in 2 h. In returning upstream, it took 3 h to make the trip. How fast can his boat travel in still water? What was the rate of the river current?

Solve the following systems of inequalities.

26. $x + y < 3$
$x - 2y < 6$

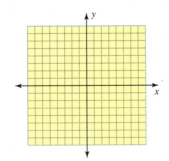

27. $4y + 3x \geq 12$
$x \geq 1$

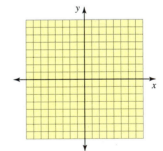

28. $2y + x \leq 8$
$y + x \leq 6$
$x \geq 0$
$y \geq 0$

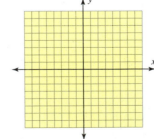

ACTIVITY 6: GROWTH OF CHILDREN—FITTING A LINEAR MODEL TO DATA

Each activity in this text is designed to either enhance your understanding of the topics of the preceding chapter, provide you with a mathematical extension of those topics, or both. The activities can be undertaken by one student, but they are better suited for a small group project. Occasionally it is only through discussion that different facets of the activity become apparent. For material related to this activity, visit the text website at www.mhhe.com/streeter.

When you walk into a home where children have lived, you will often find a wall with notches showing the heights of the children at various points in their lives. Nearly every time you bring a child to see the doctor, the child's height and weight are recorded, regardless of the reason for the office visit.

The National Institute of Health (NIH) through the Centers for Disease Control and Prevention (CDC) publishes and updates data detailing heights and weights of children for the populace as a whole and for what they consider to be healthy children. These data were most recently updated in November 2000.

In this activity, we explore the graphing of trend data and its predictive capability. We use a child's height and weight at various ages for our data.

I.

1. Locate the medical records of a child between 2–5 years old whom you know.
2. Create a table with three columns and seven rows. Label the table "Height and Weight of [name]; Year 2."
3. Label the first column "Age (months)," the second column, "Height (in.)," and the third column, "Weight (lb)."
4. In the first column, write the numbers (one to each row): 14, 16, 18, 20, 22, 24.
5. In the second column, list the child's height at each of the months listed in the first column.
6. Do likewise in the third column with the child's weight.

II.

1. Create a scatterplot of the data in your table.
2. Describe the trend of the data. Fit a line to the data and give the equation of the line.
3. Use the equation of the line to predict the child's height and weight at 10 months of age.
4. Use the equation of the line to predict the child's height and weight at 28 months of age.
5. Use the equation of the line to predict the child's height and weight at 240 months (20 years) of age.
6. According to the model, how tall will the child be when he or she is 50 years old (600 months)?
7. Discuss the accuracy of the predictions made in Steps 3 to 6. What can you discern about using scatterplot data to make predictions in general?

Sample Data Set and Solutions

I.

Median Heights and Weights for Children: Year 2		
Age (months)	Median Height for Girls (in.)	Median Weight for Boys (lb)
14	30	24
16	30.75	25
18	31.5	25.75
20	32.5	26.25
22	33	27.25
24	34	28

Source: Centers for Disease Control and Prevention.

II. 1.

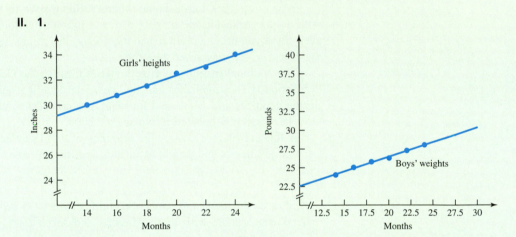

2. Both sets of data appear linear. The lines shown have the following equations:

Girls' heights: $y = 0.4x + 24.4$

Boys' weights: $y = 0.39x + 18.6$

3–6.

Predictions				
	Age (months)			
	10	28	240	600
Girls' heights (in.)	28.4	35.6	120.4	264.4
Boys' weights (lb)	22.5	29.52	112.2	252.6

While the predictions for 10 and 28 months are reasonably accurate, 20-year-old women tend to be shorter than the 10 ft given by the prediction and 20-year-old men tend to be heavier than the predicted 112 lb. Likewise, women do not grow to be 22 ft tall by the age of 50.

7. We need to be careful that we stay close to the original data source when using data to predict. That is, we can reasonably predict a person's height at 10 months from the data of their second year of age, but we cannot extrapolate that to their height 50 years later.

Cumulative Review
Chapters 0–8

Name _____

Section _____ Date _____

The following exercises are presented to help you review concepts from earlier chapters. This is meant as a review and not as a comprehensive exam. The answers are presented in the back of the text. Section references accompany the answers. If you have difficulty with any of these exercises, be certain to at least read through the summary related to those sections.

Perform each of the indicated operations.

1. $(5x^2 - 9x + 3) + (3x^2 + 2x - 7)$

2. Subtract $9w^2 + 5w$ from the sum of $8w^2 - 3w$ and $2w^2 - 4$.

3. $7xy(4x^2y - 2xy + 3xy^2)$

4. $(3s - 7)(5s + 4)$

5. $\dfrac{5x^3y - 10x^2y^2 + 15xy^2}{-5xy}$

6. $\dfrac{4x^2 + 6x - 4}{2x - 1}$

Solve the following equation for x.

7. $5 - 3(2x - 7) = 8 - 4x$

Factor each of the following polynomials completely.

8. $24a^3 - 16a^2$

9. $7m^2n - 21mn - 49mn^2$

10. $a^2 - 64b^2$

11. $5p^3 - 80pq^2$

12. $a^2 - 14a + 48$

13. $2w^3 - 8w^2 - 42w$

Solve each of the following equations.

14. $x^2 - 9x + 20 = 0$

15. $2x^2 - 32 = 0$

Solve the following applications.

16. Twice the square of a positive integer is 35 more than 9 times that integer. What is the integer?

ANSWERS

1. _____

2. _____

3. _____

4. _____

5. _____

6. _____

7. _____

8. _____

9. _____

10. _____

11. _____

12. _____

13. _____

14. _____

15. _____

16. _____

743

17. _____

18. _____

19. _____

20. _____

21. _____

22. _____

23. _____

24. _____

25. _____

26. _____

17. The length of a rectangle is 2 in. more than 3 times its width. If the area of the rectangle is 85 in.^{2}, find the dimensions of the rectangle.

Write each fraction in simplest form.

18. $\dfrac{m^2 - 4m}{3m - 12}$

19. $\dfrac{a^2 - 49}{3a^2 + 22a + 7}$

Perform the indicated operations.

20. $\dfrac{3x^2 + 9x}{x^2 - 9} \cdot \dfrac{2x^2 - 9x + 9}{2x^3 - 3x^2}$

21. $\dfrac{4w^2 - 25}{2w^2 - 5w} \div (6w + 15)$

Graph each of the following equations.

22. $x - y = 5$

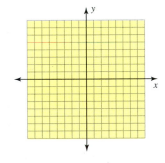

23. $y = \dfrac{2}{3}x + 3$

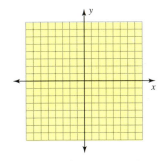

24. $2x - 5y = 10$

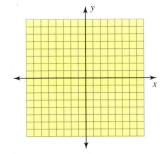

25. $y = -5$

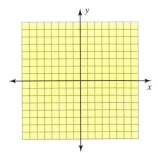

26. Find the slope of the line through the pair of points $(-2, -3)$ and $(5, 7)$.

27. Find the slope and *y*-intercept of the line described by the equation $5x - 3y = 15$.

28. Given the slope and *y*-intercept for the following line, write the equation of the line. Then graph the line.

Slope $= 2$; *y*-intercept: $(0, -5)$

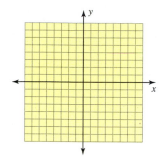

Graph each of the following inequalities.

29. $x + 2y < 6$ **30.** $3x - 4y \geq 12$

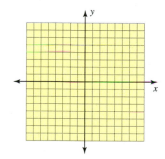   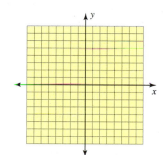

Solve the following system by graphing.

31. $3x + 2y = 6$
$x + 2y = -2$

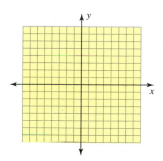

32. _____

33. _____

34. _____

35. _____

36. _____

37. _____

38. _____

39. _____

40. _____

Solve each of the following systems. If a unique solution does not exist, state whether the system is inconsistent or dependent.

32. $5x + 2y = 30$
$\quad\ x - 4y = 17$

33. $2x - 6y = 8$
$\qquad\quad x = 3y + 4$

34. $4x - 5y = 20$
$\quad\ 2x + 3y = 10$

35. $4x + 2y = 11$
$\quad\ 2x + \ y = \ 5$

36. $4x - 3y = 7$
$\quad\ 6x + 6y = 7$

Solve each of the following applications.

37. One number is 4 less than 5 times another. If the sum of the numbers is 26, what are the two numbers?

38. Cynthia bought five blank VHS tapes and four cassette tapes for $28.50. Charlie bought four VHS tapes and two cassette tapes for $21.00. Find the cost of each type of tape.

39. Receipts for a concert, attended by 450 people, were $2775. If reserved-seat tickets were $7 and general-admission tickets were $4, how many of each type of ticket were sold?

40. Anthony invested part of his $12,000 inheritance in a bond paying 9% and the other part in a savings account paying 6%. If his interest from the two investments was $930 in the first year, how much did he have invested at each rate?

EXPONENTS AND RADICALS

INTRODUCTION

In designing a public building, an engineer or an architect must include a plan for safety. The Uniform Building Code states size and location requirements for exits: "If two exits are required in a building, they must be placed apart a distance not less than one-half the length of the maximum overall diagonal dimension of the building. . . ." Stated in algebraic terms, if the building is rectangular and if d is the distance between exits, l is the length of the building, and w is the width of the building, then

$$d \geq \frac{1}{2}\sqrt{l^2 + w^2}$$

So, for example, if a rectangular building is 50 ft by 40 ft, the diagonal dimension is $\sqrt{50^2 + 40^2}$, and the distance d between the exits must be equal to or more than half of this value. Thus,

$$d \geq \frac{1}{2}\sqrt{50^2 + 40^2}$$

In this case, the distance between the exits must be 32 ft or more.

The use of a radical sign often shows up in the measurement of distances and is based on the Pythagorean theorem, which describes the relationship between the sides of a right triangle: $a^2 + b^2 = c^2$. Using algebra to interpret statements such as the one in the building code quoted above is an example of how algebra can make complicated statements clearer and easier to understand.

Pre-Test Chapter 9

ANSWERS

1. _____

2. _____

3. _____

4. _____

5. _____

6. _____

7. _____

8. _____

9. _____

10. _____

11. _____

12. _____

13. _____

14. _____

15. _____

16. _____

17. _____

18. _____

19. _____

20. _____

21. _____

22. _____

23. _____

24. _____ 25. _____

This pre-test will provide a preview of the types of exercises you will encounter in each section of this chapter. The answers for these exercises can be found in the back of the text. If you are working on your own, or ahead of the class, this pre-test can help you identify the sections in which you should focus more of your time.

Evaluate, if possible.

[9.1] **1.** $\sqrt{144}$ **2.** $\sqrt[3]{64}$ **3.** $\sqrt{-121}$ **4.** $-\sqrt[3]{-125}$

Simplify each of the following radical expressions.

[9.2] **5.** $\sqrt{72}$ **6.** $\sqrt{27x^3}$ **7.** $\sqrt{\dfrac{9}{36}}$ **8.** $\sqrt{\dfrac{7}{25}}$

Simplify by combining like terms.

[9.3] **9.** $3\sqrt{7} - 2\sqrt{7} + 6\sqrt{7}$ **10.** $\sqrt{5} + \sqrt{20}$

      11. $3\sqrt{12} - \sqrt{27}$ **12.** $\sqrt{18} - \sqrt{98} + \sqrt{50}$

Simplify each of the following radical expressions.

[9.4] **13.** $\sqrt{2x} \cdot \sqrt{5x}$ **14.** $\sqrt{2}(\sqrt{5} - 3\sqrt{2})$

      15. $(\sqrt{3} + 2)(\sqrt{3} + 3)$ **16.** $\dfrac{\sqrt{5}}{\sqrt{3}}$ **17.** $\dfrac{44 - \sqrt{605}}{11}$

Find length x in each triangle. Express your answer to the nearest hundredth.

[9.6] **18.**

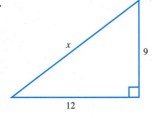

19.

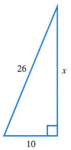

20.

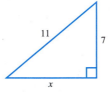

21.

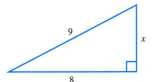

Solve the following word problem.

22. If the diagonal of a rectangle is 25 cm and the width of the rectangle is 16 cm, what is the length of the rectangle? Express your answer to the nearest hundredth of a centimeter.

Find the distance between the given two points. Express your answer to the nearest hundredth.

23. $(-4, 9)$ and $(-18, 9)$ **24.** $(-3, 4)$ and $(-6, -2)$

[9.5] **25.** Solve $\sqrt{x - 3} = 9$ for x.

9.1 Roots and Radicals

1. Use the radical notation to represent roots
2. Approximate a square root
3. Evaluate cube and fourth roots
4. Distinguish between rational and irrational numbers

In Chapter 3, we discussed the properties of exponents. Over the next four sections, we will work with a new notation that "reverses" the process of raising to a power.

From our work in Chapter 0, we know that when we have a statement such as

$$x^2 = 9$$

it is read as "x squared equals 9."

Here we are concerned with the relationship between the variable x and the number 9. We call that relationship the **square root** and say, equivalently, that "x is the square root of 9."

We know from experience that x must be 3 (because $3^2 = 9$) or -3 [because $(-3)^2 = 9$]. We see that 9 has two square roots, 3 and -3. In fact, every positive number will have *two* square roots. In general, if $x^2 = a$, we call x a *square root of a*.

We are now ready for our new notation. The symbol $\sqrt{}$ is called a **radical sign.** We just saw that 3 was the positive square root of 9. We also call 3 the **principal square root** of 9 and write

$$\sqrt{9} = 3$$

to indicate that 3 is the principal square root of 9.

NOTE The symbol $\sqrt{}$ first appeared in print in 1525. In Latin, "radix" means **root,** and this was contracted to a small r. The present symbol may have evolved from the manuscript form of that small r.

> **Definition:** Square Root
>
> $\sqrt{a}$ is the *positive* (or *principal*) square root of a. It is the positive number whose square is a.

OBJECTIVE 1

Example 1 Finding Principal Square Roots

Find the following square roots.

(a) $\sqrt{49} = 7$ Because 7 is the positive number we must square to get 49.

(b) $\sqrt{\dfrac{4}{9}} = \dfrac{2}{3}$ Because $\dfrac{2}{3}$ is the positive number we must square to get $\dfrac{4}{9}$.

CHECK YOURSELF 1

Find the following square roots.

(a) $\sqrt{64}$ (b) $\sqrt{144}$ (c) $\sqrt{\dfrac{16}{25}}$

NOTE When you use the radical sign, you are referring to the *positive square root:*
$\sqrt{25} = 5$

Each non-negative number has two square roots. For instance, 25 has square roots of 5 and -5 because

$$5^2 = 25 \qquad \text{and} \qquad (-5)^2 = 25$$

If you want to indicate the negative square root, you must use a minus sign in front of the radical.

NOTE $-\sqrt{x}$ is the negative square root of x.

$$-\sqrt{25} = -5$$

Example 2   Finding Square Roots

Find the following square roots.

(a) $\sqrt{100} = 10$ The principal root

(b) $-\sqrt{100} = -10$ The negative square root

(c) $-\sqrt{\dfrac{9}{16}} = -\dfrac{3}{4}$

 CHECK YOURSELF 2 _____

Find the following square roots.

(a) $\sqrt{16}$ (b) $-\sqrt{16}$ (c) $-\sqrt{\dfrac{16}{25}}$

 **CAUTION**

Be Careful! Do not confuse $-\sqrt{9}$ with $\sqrt{-9}$
The expression $-\sqrt{9}$ is -3, whereas $\sqrt{-9}$ is not a real number.

Every number that we have encountered in this text is a **real number.** The square roots of negative numbers are *not* real numbers. For instance, $\sqrt{-9}$ is *not* a real number because there is *no* real number x such that

$$x^2 = -9$$

Example 3 summarizes our discussion thus far.

Example 3   Finding Square Roots

Evaluate each of the following square roots.

(a) $\sqrt{36} = 6$ (b) $\sqrt{121} = 11$

(c) $-\sqrt{64} = -8$ (d) $\sqrt{-64}$ is not a real number.

(e) $\sqrt{0} = 0$ (Because $0 \cdot 0 = 0$)

✔ **CHECK YOURSELF 3** _____

Evaluate, if possible.

(a) $\sqrt{81}$ (b) $\sqrt{49}$ (c) $-\sqrt{49}$ (d) $\sqrt{-49}$

All calculators have square root keys, but the only integers for which the calculator gives the exact value of the square root are perfect square integers. For all other positive integers, *a calculator gives only an approximation of the correct answer.* In Example 4 you will use your calculator to approximate square roots.

OBJECTIVE 2

 Example 4   Approximating Square Roots

Use your calculator to approximate each square root to the nearest hundredth.

NOTE The ≈ sign means "is approximately equal to."

(a) $\sqrt{45} \approx 6.708203932 \approx 6.71$ **(b)** $\sqrt{8} \approx 2.83$

(c) $\sqrt{20} \approx 4.47$ **(d)** $\sqrt{273} \approx 16.52$

✔ **CHECK YOURSELF 4**

Use your calculator to approximate each square root to the nearest hundredth.

(a) $\sqrt{3}$ **(b)** $\sqrt{14}$ **(c)** $\sqrt{91}$ **(d)** $\sqrt{756}$

As we mentioned earlier, finding the square root of a number is the reverse of squaring a number. We can extend that idea to work with other roots of numbers. For instance, the **cube root** of a number is the number we must cube (or raise to the third power) to get that number. For example, the cube root of 8 is 2 because $2^3 = 8$, and we write

NOTE $\sqrt[3]{8}$ is read "the cube root of 8."

$$\sqrt[3]{8} = 2$$

The parts of a radical expression are summarized as follows.

Definition: Parts of a Radical Expression

NOTE The index for $\sqrt[3]{a}$ is 3.

Every radical expression contains three parts as shown here. The principal *n*th root of *a* is written as

NOTE The index of 2 for square roots is generally not written. We understand that $\sqrt{a}$ is the principal square root of *a*.

Index

$$\sqrt[n]{a}$$

Radical sign   Radicand

To illustrate, the *cube root* of 64 is written

Index⟶ $\sqrt[3]{64} = 4$
of 3

because $4^3 = 64$. And

Index⟶ $\sqrt[4]{81} = 3$
of 4

is the *fourth root* of 81 because $3^4 = 81$.

We can find roots of negative numbers as long as the index is *odd* (3, 5, etc.). For example,

$$\sqrt[3]{-64} = -4$$

because $(-4)^3 = -64$.

If the index is *even* (2, 4, etc.), roots of negative numbers are *not* real numbers. For example,

$$\sqrt[4]{-16}$$

NOTE The *even power* of a real number is always *positive* or *zero*.

is not a real number because there is no real number x such that $x^4 = -16$.

The following table shows the most common roots.

NOTE It would be helpful for your work here and in future mathematics classes to memorize these roots.

Square Roots		Cube Roots	Fourth Roots
$\sqrt{1} = 1$ $\sqrt{49} = 7$		$\sqrt[3]{1} = 1$	$\sqrt[4]{1} = 1$
$\sqrt{4} = 2$ $\sqrt{64} = 8$		$\sqrt[3]{8} = 2$	$\sqrt[4]{16} = 2$
$\sqrt{9} = 3$ $\sqrt{81} = 9$		$\sqrt[3]{27} = 3$	$\sqrt[4]{81} = 3$
$\sqrt{16} = 4$ $\sqrt{100} = 10$		$\sqrt[3]{64} = 4$	$\sqrt[4]{256} = 4$
$\sqrt{25} = 5$ $\sqrt{121} = 11$		$\sqrt[3]{125} = 5$	$\sqrt[4]{625} = 5$
$\sqrt{36} = 6$ $\sqrt{144} = 12$			

You can use the table in Example 5, which summarizes the discussion so far.

OBJECTIVE 3

Example 5   Evaluating Roots

Evaluate each of the following.

NOTE The cube root of a negative number will be negative.

(a) $\sqrt[5]{32} = 2$ because $2^5 = 32$.

(b) $\sqrt[3]{-125} = -5$ because $(-5)^3 = -125$.

NOTE The fourth root of a negative number is not a real number.

(c) $\sqrt[4]{-81}$ is not a real number.

CHECK YOURSELF 5

Evaluate, if possible.

(a) $\sqrt[3]{64}$ (b) $\sqrt[4]{16}$ (c) $\sqrt[4]{-256}$ (d) $\sqrt[3]{-8}$

The radical notation helps us to distinguish between two important types of numbers: rational numbers and irrational numbers.

A **rational number** can be represented by a fraction whose numerator and denominator are integers and whose denominator is not zero. The form of a rational number is

$$\frac{a}{b} \qquad a \text{ and } b \text{ are integers, } b \neq 0$$

NOTE Notice that each radicand is a **perfect-square integer** (that is, an integer that is the square of another integer).

Certain square roots are rational numbers also. For example,

$$\sqrt{4} \qquad \sqrt{25} \qquad \text{and} \qquad \sqrt{64}$$

NOTE The fact that the square root of 2 is irrational will be proved in later mathematics courses and was known to Greek mathematicians over 2000 years ago.

represent the rational numbers 2, 5, and 8, respectively.

An **irrational number** is a number that *cannot* be written as the ratio of two integers. For example, the square root of any positive number that is not itself a perfect square is an irrational number. Because the radicands are *not* perfect squares, the expressions $\sqrt{2}$, $\sqrt{3}$, and $\sqrt{5}$ represent irrational numbers.

OBJECTIVE 4   Example 6   **Identifying Rational Numbers**

Which of the following numbers are rational and which are irrational?

$$\sqrt{\frac{2}{3}} \qquad \sqrt{\frac{4}{9}} \qquad \sqrt{7} \qquad \sqrt{16} \qquad \sqrt{25}$$

Here $\sqrt{7}$ and $\sqrt{\frac{2}{3}}$ are irrational numbers. The numbers $\sqrt{16}$ and $\sqrt{25}$ are rational because 16 and 25 are perfect squares. Also $\sqrt{\frac{4}{9}}$ is rational because $\sqrt{\frac{4}{9}} = \frac{2}{3}$.

 CHECK YOURSELF 6

Determine whether each root is rational or irrational.

(a) $\sqrt{26}$ **(b)** $\sqrt{49}$ **(c)** $\sqrt{\frac{6}{7}}$

(d) $\sqrt{105}$ **(e)** $\sqrt{\frac{16}{9}}$

NOTE The decimal representation of a rational number always terminates or repeats. For instance,

$$\frac{3}{8} = 0.375$$

$$\frac{5}{11} = 0.454545\dots$$

An important fact about the irrational numbers is that their decimal representations are always *nonterminating* and *nonrepeating*. We can therefore only approximate irrational numbers with a decimal that has been rounded. A calculator can be used to find roots. However, note that the values found for the irrational roots are only approximations. For instance, $\sqrt{2}$ is approximately 1.414 (to three decimal places), and we can write

$$\sqrt{2} \approx 1.414$$

With a calculator we find that

NOTE 1.414 is an approximation to the number whose square is 2.

$$(1.414)^2 = 1.999396$$

The set of all rational numbers and the set of all irrational numbers together form the set of *real numbers*. The real numbers will represent every point that can be pictured on the number line. Some examples are shown.

NOTE For this reason we refer to the number line as the **real number line**.

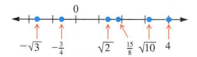

The following diagram summarizes the relationships among the various numeric sets.

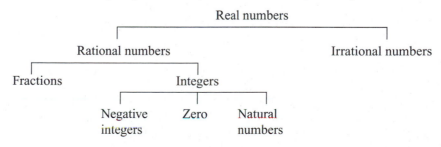

We conclude our work in this section by developing a general result that we will need later. Start by looking at two numerical examples.

$$\sqrt{2^2} = \sqrt{4} = 2$$
$$\sqrt{(-2)^2} = \sqrt{4} = 2 \qquad \text{because } (-2)^2 = 4$$

Consider the value of $\sqrt{x^2}$ when x is positive or negative.

NOTE This is because the principal square root of a number is always positive or zero.

In the first equation when $x = 2$:

$$\sqrt{2^2} = 2$$

In the second equation when $x = -2$:

$$\sqrt{(-2)^2} \neq -2$$
$$\sqrt{(-2)^2} = -(-2) = 2$$

Comparing the results of these two equations, we see that $\sqrt{x^2}$ is x if x is positive (or 0) and $\sqrt{x^2}$ is $-x$ if x is negative. We can write

$$\sqrt{x^2} = \begin{cases} x & \text{when } x \geq 0 \\ -x & \text{when } x < 0 \end{cases}$$

From your earlier work with absolute values you will remember that

$$|x| = \begin{cases} x & \text{when } x \geq 0 \\ -x & \text{when } x < 0 \end{cases}$$

and we can summarize the discussion by writing

$$\sqrt{x^2} = |x| \qquad \text{for any real number } x$$

Example 7   Evaluating Radical Expressions

NOTE Alternatively in part (b), we could write
$$\sqrt{(-4)^2} = \sqrt{16} = 4$$

Evaluate each of the following.

(a) $\sqrt{5^2} = 5$

(b) $\sqrt{(-4)^2} = |-4| = 4$

CHECK YOURSELF 7

Evaluate.

(a) $\sqrt{6^2}$

(b) $\sqrt{(-6)^2}$

CHECK YOURSELF ANSWERS

1. (a) 8; **(b)** 12; **(c)** $\dfrac{4}{5}$ **2. (a)** 4; **(b)** -4; **(c)** $-\dfrac{4}{5}$ **3. (a)** 9; **(b)** 7; **(c)** -7;

(d) not a real number   **4. (a)** ≈ 1.73; **(b)** ≈ 3.74; **(c)** ≈ 9.54; **(d)** ≈ 27.50

5. (a) 4; **(b)** 2; **(c)** not a real number; **(d)** -2 **6. (a)** Irrational;

(b) rational (because $\sqrt{49} = 7$); **(c)** irrational; **(d)** irrational

(e) rational $\left(\text{because } \sqrt{\dfrac{16}{9}} = \dfrac{4}{3}\right)$ **7. (a)** 6; **(b)** 6

9.1 Exercises

Name _____

Section _____ Date _____

Evaluate, if possible.

1. $\sqrt{25}$

2. $\sqrt{121}$

3. $\sqrt{400}$

4. $\sqrt{64}$

5. $-\sqrt{100}$

6. $\sqrt{-121}$

7. $\sqrt{-81}$

8. $-\sqrt{81}$

9. $\sqrt{\dfrac{36}{25}}$

10. $-\sqrt{\dfrac{1}{25}}$

11. $\sqrt{-\dfrac{4}{25}}$

12. $\sqrt{\dfrac{4}{25}}$

13. $\sqrt{\dfrac{121}{100}}$

14. $\sqrt{\dfrac{256}{25}}$

15. $\sqrt[3]{-125}$

16. $\sqrt[4]{-16}$

17. $\sqrt[4]{-81}$

18. $-\sqrt[3]{64}$

19. $-\sqrt[3]{27}$

20. $\sqrt[3]{-27}$

21. $\sqrt[4]{1296}$

22. $\sqrt[3]{1000}$

23. $\sqrt[3]{\dfrac{1}{27}}$

24. $\sqrt[3]{-\dfrac{8}{27}}$

State whether each of the following roots is rational or irrational.

25. $\sqrt{21}$

26. $\sqrt{36}$

27. $\sqrt{100}$

28. $\sqrt{7}$

ANSWERS

1. _____   2. _____

3. _____   4. _____

5. _____

6. _____

7. _____

8. _____

9. _____   10. _____

11. _____

12. _____

13. _____

14. _____

15. _____

16. _____

17. _____

18. _____   19. _____

20. _____   21. _____

22. _____

23. _____

24. _____

25. _____

26. _____

27. _____

28. _____

29. _____

30. _____

31. _____

32. _____

33. _____

34. _____

35. _____

36. _____

37. _____

38. _____

39. _____

40. _____

41. _____

42. _____

43. _____

44. _____

45. _____

46. _____

47. _____

48. _____

49. _____

50. _____

51. _____

52. _____

53. _____

54. _____

55. _____

56. _____

29. $\sqrt[3]{9}$ 30. $\sqrt[3]{27}$

31. $\sqrt[4]{16}$ 32. $\sqrt{\dfrac{4}{9}}$

33. $\sqrt{\dfrac{9}{15}}$ 34. $\sqrt[3]{5}$

35. $\sqrt[3]{-27}$ 36. $-\sqrt[4]{81}$

Use your calculator to approximate the square root to the nearest hundredth.

37. $\sqrt{11}$ 38. $\sqrt{14}$

39. $\sqrt{7}$ 40. $\sqrt{23}$

41. $\sqrt{65}$ 42. $\sqrt{78}$

43. $\sqrt{\dfrac{2}{5}}$ 44. $\sqrt{\dfrac{4}{3}}$

45. $\sqrt{\dfrac{8}{9}}$ 46. $\sqrt{\dfrac{7}{15}}$

47. $-\sqrt{18}$ 48. $-\sqrt{31}$

49. $-\sqrt{27}$ 50. $-\sqrt{65}$

For exercises 51 to 56, find the two expressions that are equivalent.

51. $\sqrt{-16}, \; -\sqrt{16}, \; -4$ 52. $-\sqrt{25}, \; -5, \; \sqrt{-25}$

53. $\sqrt[3]{-125}, \; -\sqrt[3]{125}, \; |-5|$ 54. $\sqrt[5]{-32}, \; -\sqrt[5]{32}, \; |-2|$

55. $\sqrt[4]{10,000}, \; 100, \; \sqrt[3]{1000}$ 56. $10^2, \; \sqrt{10,000}, \; \sqrt[3]{100,000}$

In exercises 57 to 62, label the statement as true or false.

57. $\sqrt{16x^{16}} = 4x^4$ **58.** $\sqrt{(x-4)^2} = x - 4$

59. $\sqrt{16x^{-4}y^{-4}}$ is a real number   **60.** $\sqrt{x^2 + y^2} = x + y$

61. $\dfrac{\sqrt{x^2 - 25}}{x - 5} = \sqrt{x + 5}$ **62.** $\sqrt{2} + \sqrt{6} = \sqrt{8}$

63. Geometry. The area of a square is 32 ft^2. Find the length of a side to the nearest hundredth.

64. Geometry. The area of a square is 83 ft^2. Find the length of the side to the nearest hundredth.

65. Geometry. The area of a circle is 147 ft^2. Find the radius to the nearest hundredth.

66. Geometry. If the area of a circle is 72 cm^2, find the radius to the nearest hundredth.

67. Science and medicine. The time in seconds that it takes for an object to fall from rest is given by $t = \dfrac{1}{4}\sqrt{s}$, in which s is the distance fallen. Find the time required for an object to fall to the ground from a building that is 800 ft high.

68. Science and medicine. Find the time required for an object to fall to the ground from a building that is 1400 ft high. (Use the formula found in exercise 67.)

In exercises 69 to 71, the area is given in square feet. Find the length of a side of the square. Round your answer to the nearest hundredth of a foot.

69.

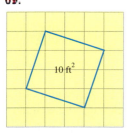

10 ft^2

70.

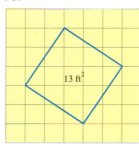

13 ft^2

71.

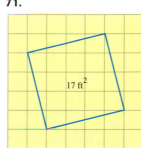

17 ft^2

ANSWERS

57. _____
58. _____
59. _____
60. _____
61. _____
62. _____
63. _____
64. _____
65. _____
66. _____
67. _____
68. _____
69. _____
70. _____
71. _____

72. _____

73. _____

74. _____

75. _____

76. _____

77. _____

78. _____

72. Is there any prime number whose square root is an integer? Explain your answer.

73. Explain the difference between the conjugate, in which the middle sign is changed, of a binomial and the opposite of a binomial. To illustrate, use $4 - \sqrt{7}$.

74. Determine two consecutive integers whose square roots are also consecutive integers.

75. Determine the missing binomial in the following: $(\sqrt{3} - 2)(\quad) = -1$.

76. Try the following using your calculator.

   (a) Choose a number greater than 1 and find its square root. Then find the square root of the result and continue in this manner, observing the successive square roots. Do these numbers seem to be approaching a certain value? If so, what?

   (b) Choose a number greater than 0 but less than 1 and find its square root. Then find the square root of the result, and continue in this manner, observing successive square roots. Do these numbers seem to be approaching a certain value? If so, what?

77. (a) Can a number be equal to its own square root?

   (b) Other than the number(s) found in part (a), is a number always greater than its square root? Investigate.

78. Let a and b be positive numbers. If a is greater than b, is it always true that the square root of a is greater than the square root of b? Investigate.

79. Suppose that a weight is attached to a string of length L, and the other end of the string is held fixed. If we pull the weight and then release it, allowing the weight to swing back and forth, we can observe the behavior of a simple pendulum. The period T is the time required for the weight to complete a full cycle, swinging

forward and then back. The following formula may be used to describe the relationship between T and L.

$$T = 2\pi\sqrt{\frac{L}{g}}$$

If L is expressed in centimeters, then $g = 980$ cm/s². For each of the following string lengths, calculate the corresponding period. Round to the nearest tenth of a second.

(a) 30 cm   (b) 50 cm   (c) 70 cm   (d) 90 cm   (e) 110 cm

Getting Ready for Section 9.2 [Section 3.4]

Find each of the following products.

(a) $(4x^2)(2x)$ (b) $(9a^4)(5a)$ (c) $(16m^2)(3m)$ (d) $(8b^3)(2b)$
(e) $(27p^6)(3p)$ (f) $(81s^4)(s^3)$ (g) $(100y^4)(2y)$ (h) $(49m^6)(2m)$

Answers

1. 5   **3.** 20   **5.** -10 **7.** Not a real number   **9.** $\frac{6}{5}$

11. Not a real number   **13.** $\frac{11}{10}$ **15.** -5 **17.** Not a real number

19. -3 **21.** 6   **23.** $\frac{1}{3}$ **25.** Irrational   **27.** Rational

29. Irrational   **31.** Rational   **33.** Irrational   **35.** Rational
37. 3.32   **39.** 2.65   **41.** 8.06   **43.** 0.63   **45.** 0.94   **47.** -4.24
49. -5.20 **51.** $-\sqrt{16}, -4$ **53.** $\sqrt[3]{-125}, -\sqrt[3]{125}$
55. $\sqrt[4]{10,000}, \sqrt[3]{1000}$ **57.** False   **59.** True   **61.** False   **63.** 5.66 ft
65. 6.84 ft   **67.** 7.07 s   **69.** 3.16 ft   **71.** 4.12 ft
73. Conjugate: $4 + \sqrt{7}$; opposite: $-4 + \sqrt{7}$ **75.** $\sqrt{3} + 2$ **77.**

79. **(a)** 1.1 s; **(b)** 1.4 s; **(c)** 1.7 s; **(d)** 1.9 s; **(e)** 2.1 s   **a.** $8x^3$ **b.** $45a^5$
c. $48m^3$ **d.** $16b^4$ **e.** $81p^7$ **f.** $81s^7$ **g.** $200y^5$ **h.** $98m^7$

79.
a.
b.
c.
d.
e.
f.
g.
h.

9.2 Simplifying Radical Expressions

9.2 OBJECTIVES

1. Simplify expressions involving numeric radicals
2. Simplify expressions involving algebraic radicals

In Section 9.1, we introduced radical notation. For most applications, we will want to make sure that all radical expressions are in *simplest form*. To accomplish this, the following three conditions must be satisfied.

> **Rules and Properties:** Square Root Expressions in Simplest Form
>
> An expression involving square roots is in *simplest form* if
>
> 1. There are no perfect-square factors in a radical.
> 2. No fraction appears inside a radical.
> 3. No radical appears in the denominator of a fraction.

For instance, considering condition 1,

$\sqrt{17}$ is in simplest form because 17 has *no* perfect-square factors

whereas

$\sqrt{12}$ is *not* in simplest form

because it does contain a perfect-square factor.

$$\sqrt{12} = \sqrt{4 \cdot 3}$$

A perfect square

To simplify radical expressions, we need to develop two important properties. First, look at the following expressions:

$$\sqrt{4 \cdot 9} = \sqrt{36} = 6$$
$$\sqrt{4} \cdot \sqrt{9} = 2 \cdot 3 = 6$$

Because this tells us that $\sqrt{4 \cdot 9} = \sqrt{4} \cdot \sqrt{9}$, the following general rule for radicals is suggested.

> **Rules and Properties:** Property 1 of Radicals
>
> For any positive real numbers a and b,
>
> $$\sqrt{ab} = \sqrt{a} \cdot \sqrt{b}$$
>
> In words, the square root of a product is the product of the square roots.

   Example 1 illustrates how this property is applied in simplifying expressions when radicals are involved.

OBJECTIVE 1

Example 1   Simplifying Radical Expressions

NOTE Perfect-square factors are 1, 4, 9, 16, 25, 36, 49, 64, 81, 100, and so on.

Simplify each expression.

(a) $\sqrt{12} = \sqrt{4 \cdot 3}$

A perfect square

NOTE Notice that we have removed the perfect-square factor from inside the radical, so the expression is in simplest form.

$= \sqrt{4} \cdot \sqrt{3}$ Apply Property 1.

$= 2\sqrt{3}$

(b) $\sqrt{45} = \sqrt{9 \cdot 5}$

A perfect square

$= \sqrt{9} \cdot \sqrt{5}$

$= 3\sqrt{5}$

NOTE We look for the *largest* perfect-square factor, here 36. Then apply Property 1.

(c) $\sqrt{72} = \sqrt{36 \cdot 2}$

A perfect square

$= \sqrt{36} \cdot \sqrt{2}$

$= 6\sqrt{2}$

(d) $5\sqrt{18} = 5\sqrt{9 \cdot 2}$

A perfect square

$= 5 \cdot \sqrt{9} \cdot \sqrt{2} = 5 \cdot 3 \cdot \sqrt{2} = 15\sqrt{2}$

 **C A U T I O N**

Be Careful!   Even though

$\sqrt{a \cdot b} = \sqrt{a} \cdot \sqrt{b}$

The expression $\sqrt{a + b}$ is *not the same* as $\sqrt{a} + \sqrt{b}$

Let $a = 4$ and $b = 9$, and substitute.

$\sqrt{a + b} = \sqrt{4 + 9} = \sqrt{13}$

$\sqrt{a} + \sqrt{b} = \sqrt{4} + \sqrt{9} = 2 + 3 = 5$

Because $\sqrt{13} \neq 5$, we see that the expressions $\sqrt{a + b}$ and $\sqrt{a} + \sqrt{b}$ are not, in general, the same.

 CHECK YOURSELF 1

Simplify.

(a) $\sqrt{20}$ **(b)** $\sqrt{75}$ **(c)** $\sqrt{98}$ **(d)** $\sqrt{48}$

The process is the same if variables are involved in a radical expression. In our remaining work with radicals, we will assume that all variables represent positive real numbers.

OBJECTIVE 2

Example 2   Simplifying Algebraic Radical Expressions

Simplify each of the following radicals.

(a) $\sqrt{x^3} = \sqrt{x^2 \cdot x}$

            A perfect square

$= \sqrt{x^2} \cdot \sqrt{x}$

NOTE $\sqrt{x^2} = x$ (as long as x is positive).

$= x\sqrt{x}$

(b) $\sqrt{4b^3} = \sqrt{4 \cdot b^2 \cdot b}$

            Perfect squares

$= \sqrt{4b^2} \cdot \sqrt{b}$

$= 2b\sqrt{b}$

RECALL Notice that we want the perfect-square factor to have the largest possible even exponent, here 4. Keep in mind that

$a^2 \cdot a^2 = a^4$

(c) $\sqrt{18a^5} = \sqrt{9 \cdot a^4 \cdot 2a}$

          Perfect squares

$= \sqrt{9a^4} \cdot \sqrt{2a}$

$= 3a^2\sqrt{2a}$

CHECK YOURSELF 2

Simplify.

(a) $\sqrt{9x^3}$

(b) $\sqrt{27m^3}$

(c) $\sqrt{50b^5}$

To develop a second property for radicals, look at the following expressions:

$$\sqrt{\frac{16}{4}} = \sqrt{4} = 2$$

$$\frac{\sqrt{16}}{\sqrt{4}} = \frac{4}{2} = 2$$

Because $\sqrt{\dfrac{16}{4}} = \dfrac{\sqrt{16}}{\sqrt{4}}$, a second general rule for radicals is suggested.

Rules and Properties: Property 2 of Radicals

For any positive real numbers a and b,

$$\sqrt{\frac{a}{b}} = \frac{\sqrt{a}}{\sqrt{b}}$$

In words, the square root of a quotient is the quotient of the square roots.

This property is used in a fashion similar to Property 1 in simplifying radical expressions. Remember that our second condition for a radical expression to be in simplest form states that no fraction should appear inside a radical. Example 3 illustrates how expressions that violate that condition are simplified.

Example 3 Simplifying Radical Expressions

Write each expression in simplest form.

NOTE Apply Property 2 to write the numerator and denominator as separate radicals.

(a) $\sqrt{\dfrac{9}{4}} = \dfrac{\sqrt{9}}{\sqrt{4}}$ Remove any perfect squares from the radical.

$= \dfrac{3}{2}$

(b) $\sqrt{\dfrac{2}{25}} = \dfrac{\sqrt{2}}{\sqrt{25}}$ Apply Property 2.

$= \dfrac{\sqrt{2}}{5}$

(c) $\sqrt{\dfrac{8x^2}{9}} = \dfrac{\sqrt{8x^2}}{\sqrt{9}}$ Apply Property 2.

NOTE Factor $8x^2$ as $4x^2 \cdot 2$.

$= \dfrac{\sqrt{4x^2 \cdot 2}}{3}$

$= \dfrac{\sqrt{4x^2} \cdot \sqrt{2}}{3}$ Apply Property 1 in the numerator.

$= \dfrac{2x\sqrt{2}}{3}$

CHECK YOURSELF 3

Simplify.

(a) $\sqrt{\dfrac{25}{16}}$

(b) $\sqrt{\dfrac{7}{9}}$

(c) $\sqrt{\dfrac{12x^2}{49}}$

In the three expressions in Example 3, the denominator of the fraction appearing in the radical was a perfect square, and we were able to write each expression in simplest radical form by removing that perfect square from the denominator.

If the denominator of the fraction in the radical is *not* a perfect square, we can still apply Property 2 of radicals. As we show in Example 4, the third condition for a radical to be in simplest form is then violated, and a new technique is necessary.

Example 4 Simplifying Radical Expressions

Write each expression in simplest form.

NOTE We begin by applying Property 2.

(a) $\sqrt{\dfrac{1}{3}} = \dfrac{\sqrt{1}}{\sqrt{3}} = \dfrac{1}{\sqrt{3}}$

Do you see that $\dfrac{1}{\sqrt{3}}$ is still not in simplest form because of the radical in the denominator? To solve this problem, we multiply the numerator and denominator by $\sqrt{3}$. Note that the denominator will become

$$\sqrt{3} \cdot \sqrt{3} = \sqrt{9} = 3$$

We then have

RECALL We can do this because we are multiplying the fraction by $\dfrac{\sqrt{3}}{\sqrt{3}}$ or 1, which does not change its value.

$$\frac{1}{\sqrt{3}} = \frac{1 \cdot \sqrt{3}}{\sqrt{3} \cdot \sqrt{3}} = \frac{\sqrt{3}}{3}$$

The expression $\dfrac{\sqrt{3}}{3}$ is now in simplest form because all three of our conditions are satisfied.

(b) $\sqrt{\dfrac{2}{5}} = \dfrac{\sqrt{2}}{\sqrt{5}}$

NOTE
$\sqrt{2} \cdot \sqrt{5} = \sqrt{2 \cdot 5} = \sqrt{10}$
$\sqrt{5} \cdot \sqrt{5} = 5$

$$= \frac{\sqrt{2} \cdot \sqrt{5}}{\sqrt{5} \cdot \sqrt{5}}$$

$$= \frac{\sqrt{10}}{5}$$

and the expression is in simplest form because again our three conditions are satisfied.

(c) $\sqrt{\dfrac{3x}{7}} = \dfrac{\sqrt{3x}}{\sqrt{7}}$

NOTE We multiply numerator and denominator by $\sqrt{7}$ to "clear" the denominator of the radical. This is also known as "rationalizing" the denominator.

$$= \frac{\sqrt{3x} \cdot \sqrt{7}}{\sqrt{7} \cdot \sqrt{7}}$$

$$= \frac{\sqrt{21x}}{7}$$

The expression is in simplest form.

CHECK YOURSELF 4

Simplify.

(a) $\sqrt{\dfrac{1}{2}}$ **(b)** $\sqrt{\dfrac{2}{3}}$ **(c)** $\sqrt{\dfrac{2y}{5}}$

Both of the properties of radicals given in this section are true for cube roots, fourth roots, and so on. Here we have limited ourselves to simplifying expressions involving square roots.

CHECK YOURSELF ANSWERS

1. (a) $2\sqrt{5}$; **(b)** $5\sqrt{3}$; **(c)** $7\sqrt{2}$; **(d)** $4\sqrt{3}$ **2. (a)** $3x\sqrt{x}$; **(b)** $3m\sqrt{3m}$;
(c) $5b^2\sqrt{2b}$ **3. (a)** $\dfrac{5}{4}$; **(b)** $\dfrac{\sqrt{7}}{3}$; **(c)** $\dfrac{2x\sqrt{3}}{7}$ **4. (a)** $\dfrac{\sqrt{2}}{2}$; **(b)** $\dfrac{\sqrt{6}}{3}$; **(c)** $\dfrac{\sqrt{10y}}{5}$

9.2 Exercises

Use Property 1 to simplify each of the following radical expressions. Assume that all variables represent positive real numbers.

1. $\sqrt{48}$

2. $\sqrt{50}$

3. $\sqrt{28}$

4. $\sqrt{108}$

5. $\sqrt{45}$

6. $\sqrt{80}$

7. $\sqrt{54}$

8. $\sqrt{180}$

9. $\sqrt{200}$

10. $\sqrt{96}$

11. $\sqrt{567}$

12. $\sqrt{300}$

13. $3\sqrt{12}$

14. $5\sqrt{24}$

15. $\sqrt{3x^2}$

16. $\sqrt{7a^2}$

17. $\sqrt{3y^4}$

18. $\sqrt{10x^6}$

19. $\sqrt{2r^3}$

20. $\sqrt{7a^7}$

21. $\sqrt{125b^2}$

22. $\sqrt{98m^4}$

23. $\sqrt{24x^4}$

24. $\sqrt{72x^3}$

25. $\sqrt{54a^5}$

26. $\sqrt{200y^6}$

27. $\sqrt{x^3y^2}$

28. $\sqrt{a^2b^5}$

29. $\sqrt{a^6b^4c}$

30. $\sqrt{x^3y^4z^2}$

1. _____	2. _____	
3. _____	4. _____	
5. _____	6. _____	
7. _____	8. _____	
9. _____	10. _____	
11. _____		
12. _____		
13. _____		
14. _____		
15. _____		
16. _____		
17. _____		
18. _____		
19. _____		
20. _____		
21. _____		
22. _____		
23. _____		
24. _____		
25. _____		
26. _____		
27. _____		
28. _____		
29. _____		
30. _____		

Use Property 2 to simplify each of the following radical expressions.

31. $\sqrt{\dfrac{9}{16}}$ $\qquad\qquad$ **32.** $\sqrt{\dfrac{49}{25}}$

33. $\sqrt{\dfrac{3}{4}}$ $\qquad\qquad$ **34.** $\sqrt{\dfrac{7}{16}}$

35. $\sqrt{\dfrac{3}{49}}$ $\qquad\qquad$ **36.** $\sqrt{\dfrac{10}{49}}$

Use the properties for radicals to simplify each of the following expressions. Assume that all variables represent positive real numbers.

37. $\sqrt{\dfrac{8a^2}{25}}$ $\qquad\qquad$ **38.** $\sqrt{\dfrac{12y^2}{49}}$

39. $\sqrt{\dfrac{1}{5}}$ $\qquad\qquad$ **40.** $\sqrt{\dfrac{1}{7}}$

41. $\sqrt{\dfrac{5}{2}}$ $\qquad\qquad$ **42.** $\sqrt{\dfrac{5}{3}}$

43. $\sqrt{\dfrac{3a}{5}}$ $\qquad\qquad$ **44.** $\sqrt{\dfrac{2x}{5}}$

45. $\sqrt{\dfrac{3x^4}{7}}$ $\qquad\qquad$ **46.** $\sqrt{\dfrac{5m^2}{2}}$

47. $\sqrt{\dfrac{8s^3}{7}}$ $\qquad\qquad$ **48.** $\sqrt{\dfrac{12x^3}{5}}$

Decide whether each of the following is already written in simplest form. If it is not, explain what needs to be done.

49. $\sqrt{10mn}$ $\qquad\qquad$ **50.** $\sqrt{18ab}$

51. $\sqrt{\dfrac{98x^2y}{7x}}$ $\qquad\qquad$ **52.** $\dfrac{\sqrt{6xy}}{3x}$

53. Find the area and perimeter of this square:

$\sqrt{3}$

$\sqrt{3}$

One of these measures, the area, is a rational number, and the other, the perimeter, is an irrational number. Explain how this happened. Will the area always be a rational number? Explain.

54. **(a)** Evaluate the three expressions $\dfrac{n^2 - 1}{2}, n, \dfrac{n^2 + 1}{2}$ using odd values of n: 1, 3, 5, 7, and so forth. Make a chart like the one that follows and complete it.

n	$a = \dfrac{n^2 - 1}{2}$	$b = n$	$c = \dfrac{n^2 + 1}{2}$	a^2	b^2	c^2
1						
3						
5						
7						
9						
11						
13						
15						

(b) Check for each of these sets of three numbers to see if this statement is true: $\sqrt{a^2 + b^2} = \sqrt{c^2}$. For how many of your sets of three did this work? Sets of three numbers for which this statement is true are called *Pythagorean triples* because $a^2 + b^2 = c^2$. Can the radical equation be written in this way: $\sqrt{a^2 + b^2} = a + b$? Explain your answer.

Getting Ready for Section 9.3 [Section 1.6]

Use the distributive property to combine the like terms in each of the following expressions.

(a) $5x + 6x$ (b) $8a - 3a$

(c) $10y - 12y$ (d) $7m + 10m$

(e) $9a + 7a - 12a$ (f) $5s - 8s + 4s$

(g) $12m + 3n - 6m$ (h) $8x + 5y - 4x$

Answers

1. $4\sqrt{3}$ **3.** $2\sqrt{7}$ **5.** $3\sqrt{5}$ **7.** $3\sqrt{6}$ **9.** $10\sqrt{2}$ **11.** $9\sqrt{7}$

13. $6\sqrt{3}$ **15.** $x\sqrt{3}$ **17.** $y^2\sqrt{3}$ **19.** $r\sqrt{2r}$ **21.** $5b\sqrt{5}$

23. $2x^2\sqrt{6}$ **25.** $3a^2\sqrt{6a}$ **27.** $xy\sqrt{x}$ **29.** $a^3b^2\sqrt{c}$ **31.** $\dfrac{3}{4}$

33. $\dfrac{\sqrt{3}}{2}$ **35.** $\dfrac{\sqrt{3}}{7}$ **37.** $\dfrac{2a\sqrt{2}}{5}$ **39.** $\dfrac{\sqrt{5}}{5}$ **41.** $\dfrac{\sqrt{10}}{2}$

43. $\dfrac{\sqrt{15a}}{5}$ **45.** $\dfrac{x^2\sqrt{21}}{7}$ **47.** $\dfrac{2s\sqrt{14s}}{7}$ **49.** Simplest form

51. Remove the perfect-square factors from the radical and simplify.

53.   **a.** $11x$ **b.** $5a$ **c.** $-2y$ **d.** $17m$ **e.** $4a$

f. s **g.** $6m + 3n$ **h.** $4x + 5y$

9.3 Adding and Subtracting Radicals

9.3 OBJECTIVES

1. Add and subtract expressions involving numeric radicals
2. Add and subtract expressions involving algebraic radicals

Two radicals that have the same index and the same radicand (the expression inside the radical) are called **like radicals.** For example,

$2\sqrt{3}$ and $5\sqrt{3}$ are like radicals.

$\sqrt{2}$ and $\sqrt{5}$ are not like radicals—they have different radicands.

NOTE *Indices* is the plural of *index*.

$\sqrt{2}$ and $\sqrt[3]{2}$ are not like radicals—they have different indices (2 and 3, representing a square root and a cube root).

Like radicals can be added (or subtracted) in the same way as like terms. We apply the distributive property and then combine the coefficients:

$$2\sqrt{5} + 3\sqrt{5} = (2 + 3)\sqrt{5} = 5\sqrt{5}$$

OBJECTIVE 1

Example 1 Adding and Subtracting Like Numeric Radicals

Simplify each expression.

NOTE Apply the distributive property and then combine the coefficients.

(a) $5\sqrt{2} + 3\sqrt{2} = (5 + 3)\sqrt{2} = 8\sqrt{2}$

(b) $7\sqrt{5} - 2\sqrt{5} = (7 - 2)\sqrt{5} = 5\sqrt{5}$

(c) $8\sqrt{7} - \sqrt{7} + 2\sqrt{7} = (8 - 1 + 2)\sqrt{7} = 9\sqrt{7}$

CHECK YOURSELF 1 —————————

Simplify.

(a) $2\sqrt{5} + 7\sqrt{5}$
(c) $5\sqrt{3} - 2\sqrt{3} + \sqrt{3}$

(b) $9\sqrt{7} - \sqrt{7}$

If a sum or difference involves terms that are *not* like radicals, we may be able to combine terms after simplifying the radicals according to our earlier rules.

Example 2 Adding and Subtracting Numeric Radicals

Simplify each expression.

(a) $3\sqrt{2} + \sqrt{8}$

We do not have like radicals, but we can simplify $\sqrt{8}$. Remember that

$$\sqrt{8} = \sqrt{4 \cdot 2} = 2\sqrt{2}$$

so

$$3\sqrt{2} + \overset{\sqrt{8}}{\overline{\sqrt{8}}} = 3\sqrt{2} + 2\sqrt{2}$$
$$= (3 + 2)\sqrt{2} = 5\sqrt{2}$$

(b) $5\sqrt{3} - \sqrt{12} = 5\sqrt{3} - \sqrt{4 \cdot 3}$
$$= 5\sqrt{3} - \sqrt{4} \cdot \sqrt{3}$$
$$= 5\sqrt{3} - 2\sqrt{3}$$
$$= 3\sqrt{3}$$

NOTE The radicals can now be combined. Do you see why?

CHECK YOURSELF 2

Simplify.

(a) $\sqrt{2} + \sqrt{18}$ **(b)** $5\sqrt{3} - \sqrt{27}$

Example 3 illustrates that we may need to apply our earlier methods for adding fractions when working with radical expressions.

Example 3   Adding Radical Expressions

Add $\dfrac{\sqrt{5}}{3} + \dfrac{2}{\sqrt{5}}$.

Our first step will be to rationalize the denominator of the second fraction, to write the sum as

$$\frac{\sqrt{5}}{3} + \frac{2\sqrt{5}}{\sqrt{5} \cdot \sqrt{5}}$$

or

$$\frac{\sqrt{5}}{3} + \frac{2\sqrt{5}}{5}$$

The LCD of the fractions is 15 and rewriting each fraction with that denominator, we have

$$\frac{\sqrt{5} \cdot 5}{3 \cdot 5} + \frac{2\sqrt{5} \cdot 3}{5 \cdot 3} = \frac{5\sqrt{5} + 6\sqrt{5}}{15}$$
$$= \frac{11\sqrt{5}}{15}$$

CHECK YOURSELF 3

Subtract $\dfrac{3}{\sqrt{10}} - \dfrac{\sqrt{10}}{5}$.

If variables are involved in radical expressions, the process of combining terms proceeds in a fashion similar to that shown in Examples 1 and 2. Consider Example 4. We again assume that all variables represent positive real numbers.

OBJECTIVE 2 | **Example 4   Simplifying Expressions Involving Variables**

Simplify each expression.

NOTE Because like radicals are involved, we apply the distributive property and combine terms as before.

(a) $5\sqrt{3x} - 2\sqrt{3x} = 3\sqrt{3x}$

(b) $2\sqrt{3a^3} + 5a\sqrt{3a}$

NOTE Simplify the first term.

$= 2\sqrt{a^2 \cdot 3a} + 5a\sqrt{3a}$

$= 2\sqrt{a^2} \cdot \sqrt{3a} + 5a\sqrt{3a}$

$= 2a\sqrt{3a} + 5a\sqrt{3a}$

NOTE The radicals can now be combined.

$= 7a\sqrt{3a}$

CHECK YOURSELF 4

Simplify each expression.

(a) $2\sqrt{7y} + 3\sqrt{7y}$

(b) $\sqrt{20a^2} - a\sqrt{45}$

Example 5   Adding or Subtracting Algebraic Radical Expressions

Add or subtract as indicated.

(a) $2\sqrt{6} + \sqrt{\dfrac{2}{3}}$

We apply the quotient property to the *second term* and rationalize the denominator.

NOTE Multiply by $\dfrac{\sqrt{3}}{\sqrt{3}}$, or 1.

$$\sqrt{\dfrac{2}{3}} = \dfrac{\sqrt{2}}{\sqrt{3}} = \dfrac{\sqrt{2} \cdot \sqrt{3}}{\sqrt{3} \cdot \sqrt{3}} = \dfrac{\sqrt{6}}{3}$$

So

NOTE $\dfrac{\sqrt{6}}{3}$ and $\dfrac{1}{3}\sqrt{6}$ are equivalent.

$$2\sqrt{6} + \sqrt{\dfrac{2}{3}} = 2\sqrt{6} + \dfrac{\sqrt{6}}{3}$$

$$= \left(2 + \dfrac{1}{3}\right)\sqrt{6} = \dfrac{7}{3}\sqrt{6}$$

(b) $\sqrt{20x} - \sqrt{\dfrac{x}{5}}$

Again we first simplify the two expressions. So

$$\sqrt{20x} - \sqrt{\dfrac{x}{5}} = 2\sqrt{5x} - \dfrac{\sqrt{x} \cdot \sqrt{5}}{\sqrt{5} \cdot \sqrt{5}}$$

$$= 2\sqrt{5x} - \dfrac{\sqrt{5x}}{5}$$

$$= \left(2 - \dfrac{1}{5}\right)\sqrt{5x} = \dfrac{9}{5}\sqrt{5x}$$

CHECK YOURSELF 5

Add or subtract as indicated.

(a) $3\sqrt{7} + \sqrt{\dfrac{1}{7}}$

(b) $\sqrt{40x} - \sqrt{\dfrac{2x}{5}}$

CHECK YOURSELF ANSWERS

1. **(a)** $9\sqrt{5}$; **(b)** $8\sqrt{7}$; **(c)** $4\sqrt{3}$ 2. **(a)** $4\sqrt{2}$; **(b)** $2\sqrt{3}$ 3. $\dfrac{\sqrt{10}}{10}$

4. **(a)** $5\sqrt{7y}$; **(b)** $-a\sqrt{5}$ 5. **(a)** $\dfrac{22}{7}\sqrt{7}$; **(b)** $\dfrac{9}{5}\sqrt{10x}$

Simplify by combining like terms.

1. $3\sqrt{2} + 5\sqrt{2}$

2. $\sqrt{3} + 5\sqrt{3}$

3. $11\sqrt{7} - 4\sqrt{7}$

4. $7\sqrt{3} - 5\sqrt{2}$

5. $5\sqrt{7} + 3\sqrt{6}$

6. $3\sqrt{5} - 5\sqrt{5}$

7. $3\sqrt{5} - 7\sqrt{5}$

8. $2\sqrt{11} + 5\sqrt{11}$

9. $2\sqrt{3x} + 5\sqrt{3x}$

10. $7\sqrt{2a} - 3\sqrt{2a}$

11. $2\sqrt{3} + \sqrt{3} + 3\sqrt{3}$

12. $3\sqrt{5} + 2\sqrt{5} + \sqrt{5}$

13. $6\sqrt{11} - 5\sqrt{11} + 3\sqrt{11}$

14. $3\sqrt{10} - 2\sqrt{10} + \sqrt{10}$

15. $2\sqrt{5x} + 5\sqrt{5x} - \sqrt{5x}$

16. $8\sqrt{3b} - 2\sqrt{3b} + \sqrt{3b}$

17. $2\sqrt{3} + \sqrt{12}$

18. $7\sqrt{3} + 2\sqrt{27}$

19. $\sqrt{20} - \sqrt{5}$

20. $\sqrt{98} - 3\sqrt{2}$

21. $2\sqrt{6} - \sqrt{54}$

22. $2\sqrt{3} - \sqrt{27}$

23. $\sqrt{72} + \sqrt{50}$

24. $\sqrt{27} - \sqrt{12}$

25. $3\sqrt{12} - \sqrt{48}$

26. $5\sqrt{8} + 2\sqrt{18}$

27. $\sqrt{3} + \sqrt{\dfrac{1}{3}}$

28. $\sqrt{6} - \sqrt{\dfrac{1}{6}}$

29. $\sqrt{12} + \sqrt{27} - \sqrt{3}$

30. $\sqrt{50} + \sqrt{32} - \sqrt{8}$

31. $3\sqrt{24} - \sqrt{54} + \sqrt{6}$

32. $\sqrt{63} - 2\sqrt{28} + 5\sqrt{7}$

33. $\dfrac{\sqrt{12}}{3} - \dfrac{1}{\sqrt{3}}$

34. $\dfrac{\sqrt{20}}{5} + \dfrac{2}{\sqrt{5}}$

ANSWERS

1. _____
2. _____
3. _____
4. _____
5. _____
6. _____ 7. _____
8. _____ 9. _____
10. _____ 11. _____
12. _____ 13. _____
14. _____ 15. _____
16. _____ 17. _____
18. _____ 19. _____
20. _____ 21. _____
22. _____ 23. _____
24. _____ 25. _____
26. _____
27. _____
28. _____
29. _____
30. _____
31. _____
32. _____
33. _____
34. _____

35. _____

36. _____

37. _____

38. _____

39. _____

40. _____

41. _____

42. _____

43. _____

44. _____

45. _____

46. _____

47. _____

48. _____

49. _____

Simplify by combining like terms.

35. $a\sqrt{27} - 2\sqrt{3a^2}$

36. $5\sqrt{2y^2} - 3y\sqrt{8}$

37. $5\sqrt{3x^3} + 2\sqrt{27x}$

38. $7\sqrt{2a^3} - \sqrt{8a}$

Use a calculator to find a decimal approximation for each of the following. Round your answer to the nearest hundredth.

39. $\sqrt{3} - \sqrt{2}$

40. $\sqrt{7} + \sqrt{11}$

41. $\sqrt{5} + \sqrt{3}$

42. $\sqrt{17} - \sqrt{13}$

43. $4\sqrt{3} - 7\sqrt{5}$

44. $8\sqrt{2} + 3\sqrt{7}$

45. $5\sqrt{7} + 8\sqrt{13}$

46. $7\sqrt{2} - 4\sqrt{11}$

47. Geometry. Find the perimeter of the rectangle shown in the figure.

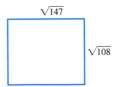

$\sqrt{36}$

$\sqrt{49}$

48. Geometry. Find the perimeter of the rectangle shown in the figure. Write your answer in radical form.

$\sqrt{147}$

$\sqrt{108}$

49. Geometry. Find the perimeter of the triangle shown in the figure.

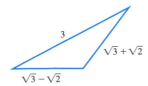

3

$\sqrt{3} + \sqrt{2}$

$\sqrt{3} - \sqrt{2}$

50. Geometry. Find the perimeter of the triangle shown in the figure.

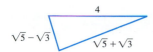

Getting Ready for Section 9.4 [Section 3.4]

Perform the indicated multiplication.

(a) $2(x + 5)$

(b) $3(a - 3)$

(c) $m(m - 8)$

(d) $y(y + 7)$

(e) $(w + 2)(w - 2)$

(f) $(x - 3)(x + 3)$

(g) $(x + y)(x + y)$

(h) $(b - 7)(b - 7)$

Answers

1. $8\sqrt{2}$ **3.** $7\sqrt{7}$ **5.** Cannot be simplified **7.** $-4\sqrt{5}$ **9.** $7\sqrt{3x}$
11. $6\sqrt{3}$ **13.** $4\sqrt{11}$ **15.** $6\sqrt{5x}$ **17.** $4\sqrt{3}$ **19.** $\sqrt{5}$

21. $-\sqrt{6}$ **23.** $11\sqrt{2}$ **25.** $2\sqrt{3}$ **27.** $\frac{4}{3}\sqrt{3}$ **29.** $4\sqrt{3}$

31. $4\sqrt{6}$ **33.** $\frac{1}{3}\sqrt{3}$ **35.** $a\sqrt{3}$ **37.** $(5x + 6)\sqrt{3x}$

39. 0.32 **41.** 3.97 **43.** -8.72 **45.** 42.07 **47.** 26
49. $2\sqrt{3} + 3$ **a.** $2x + 10$ **b.** $3a - 9$ **c.** $m^2 - 8m$
d. $y^2 + 7y$ **e.** $w^2 - 4$ **f.** $x^2 - 9$ **g.** $x^2 + 2xy + y^2$
h. $b^2 - 14b + 49$

9.4 Multiplying and Dividing Radicals

9.4 OBJECTIVES

1. Multiply expressions involving radical expressions
2. Divide expressions involving radical expressions

In Section 9.2 we stated the first property for radicals:

$$\sqrt{ab} = \sqrt{a} \cdot \sqrt{b} \quad \text{when } a \text{ and } b \text{ are any positive real numbers}$$

That property has been used to simplify radical expressions up to this point. Suppose we want to find a product, such as $\sqrt{3} \cdot \sqrt{5}$.

We can use our first radical rule in the opposite manner.

NOTE The product of square roots is equal to the square root of the product of the radicands.

$$\sqrt{a} \cdot \sqrt{b} = \sqrt{ab}$$

so

$$\sqrt{3} \cdot \sqrt{5} = \sqrt{3 \cdot 5} = \sqrt{15}$$

We may have to simplify after multiplying, as Example 1 illustrates.

OBJECTIVE 1

Example 1 Multiplying Radical Expressions

Multiply and then simplify each expression.

(a) $\sqrt{5} \cdot \sqrt{10} = \sqrt{5 \cdot 10} = \sqrt{50}$
$$= \sqrt{25 \cdot 2} = 5\sqrt{2}$$

(b) $\sqrt{12} \cdot \sqrt{6} = \sqrt{12 \cdot 6} = \sqrt{72}$
$$= \sqrt{36 \cdot 2} = \sqrt{36} \cdot \sqrt{2} = 6\sqrt{2}$$

An alternative approach would be to simplify $\sqrt{12}$ first.

$$\sqrt{12} \cdot \sqrt{6} = 2\sqrt{3}\,\sqrt{6} = 2\sqrt{18}$$
$$= 2\sqrt{9 \cdot 2} = 2\sqrt{9}\,\sqrt{2}$$
$$= 2 \cdot 3\sqrt{2} = 6\sqrt{2}$$

(c) $\sqrt{10x} \cdot \sqrt{2x} = \sqrt{20x^2} = \sqrt{4x^2 \cdot 5}$
$$= \sqrt{4x^2} \cdot \sqrt{5} = 2x\sqrt{5}$$

CHECK YOURSELF 1

Simplify.

(a) $\sqrt{3} \cdot \sqrt{6}$ **(b)** $\sqrt{3} \cdot \sqrt{18}$ **(c)** $\sqrt{8a} \cdot \sqrt{3a}$

If coefficients are involved in a product, we can use the commutative and associative properties to change the order and grouping of the factors. This is illustrated in Example 2.

> **Example 2   Multiplying Radical Expressions**

Multiply.

NOTE In practice, it is not necessary to show the intermediate steps.

$$(2\sqrt{5})(3\sqrt{6}) = (2 \cdot 3)(\sqrt{5} \cdot \sqrt{6})$$
$$= 6\sqrt{5 \cdot 6}$$
$$= 6\sqrt{30}$$

CHECK YOURSELF 2

Multiply $(3\sqrt{7})(5\sqrt{3})$.

The distributive property can also be applied in multiplying radical expressions. Consider the following.

> **Example 3   Multiplying Radical Expressions**

Multiply.

(a) $\sqrt{3}(\sqrt{2} + \sqrt{3})$

$$= \sqrt{3} \cdot \sqrt{2} + \sqrt{3} \cdot \sqrt{3}$$ The distributive property
$$= \sqrt{6} + 3$$ Multiply the radicals.

(b) $\sqrt{5}(2\sqrt{6} + 3\sqrt{3})$

$$= \sqrt{5} \cdot 2\sqrt{6} + \sqrt{5} \cdot 3\sqrt{3}$$ The distributive property
$$= 2 \cdot \sqrt{5} \cdot \sqrt{6} + 3 \cdot \sqrt{5} \cdot \sqrt{3}$$ The commutative property
$$= 2\sqrt{30} + 3\sqrt{15}$$

CHECK YOURSELF 3

Multiply.

(a) $\sqrt{5}(\sqrt{6} + \sqrt{5})$ **(b)** $\sqrt{3}(2\sqrt{5} + 3\sqrt{2})$

The FOIL pattern we used for multiplying binomials in Section 3.4 can also be applied in multiplying radical expressions. This is shown in Example 4.

> **Example 4   Multiplying Radical Expressions**

Multiply.

(a) $(\sqrt{3} + 2)(\sqrt{3} + 5)$

$$= \sqrt{3} \cdot \sqrt{3} + 5\sqrt{3} + 2\sqrt{3} + 2 \cdot 5$$
$$= 3 + 5\sqrt{3} + 2\sqrt{3} + 10$$ Combine like terms.
$$= 13 + 7\sqrt{3}$$

Be Careful! This result *cannot* be further simplified: 13 and $7\sqrt{3}$ are *not* like terms.

C A U T I O N

NOTE You can use the pattern $(a + b)(a - b) = a^2 - b^2$, where $a = \sqrt{7}$ and $b = 2$, for the same result. $\sqrt{7} + 2$ and $\sqrt{7} - 2$ are called **conjugates** of each other. Note that their product is the rational number 3. The product of conjugates will *always be rational*.

(b) $(\sqrt{7} + 2)(\sqrt{7} - 2) = \sqrt{7} \cdot \sqrt{7} - 2\sqrt{7} + 2\sqrt{7} - 4$

$$= 7 - 4 = 3$$

(c) $(\sqrt{3} + 5)^2 = (\sqrt{3} + 5)(\sqrt{3} + 5)$

$$= \sqrt{3} \cdot \sqrt{3} + 5\sqrt{3} + 5\sqrt{3} + 5 \cdot 5$$

$$= 3 + 5\sqrt{3} + 5\sqrt{3} + 25$$

$$= 28 + 10\sqrt{3}$$

CHECK YOURSELF 4

Multiply.

(a) $(\sqrt{5} + 3)(\sqrt{5} - 2)$ **(b)** $(\sqrt{3} + 4)(\sqrt{3} - 4)$ **(c)** $(\sqrt{2} - 3)^2$

We can also use our second property for radicals in the opposite manner.

NOTE The quotient of square roots is equal to the square root of the quotient of the radicands.

$$\frac{\sqrt{a}}{\sqrt{b}} = \sqrt{\frac{a}{b}}$$

One use of this property to divide radical expressions is illustrated in Example 5.

OBJECTIVE 2   **Example 5   Dividing Radical Expressions**

Simplify.

NOTE The clue to recognizing when to use this approach is in noting that 48 is divisible by 3.

(a) $\dfrac{\sqrt{48}}{\sqrt{3}} = \sqrt{\dfrac{48}{3}} = \sqrt{16} = 4$

(b) $\dfrac{\sqrt{200}}{\sqrt{2}} = \sqrt{\dfrac{200}{2}} = \sqrt{100} = 10$

(c) $\dfrac{\sqrt{125x^2}}{\sqrt{5}} = \sqrt{\dfrac{125x^2}{5}} = \sqrt{25x^2} = 5x$

CHECK YOURSELF 5

Simplify.

(a) $\dfrac{\sqrt{75}}{\sqrt{3}}$ **(b)** $\dfrac{\sqrt{81s^2}}{\sqrt{9}}$

There is one final quotient form that you may encounter in simplifying expressions, and it will be extremely important in our work with quadratic equations in Chapter 10. This form is shown in Example 6.

Example 6   Simplifying Radical Expressions

Simplify the expression

$$\frac{3 + \sqrt{72}}{3}$$

CAUTION

Be Careful! Students are sometimes tempted to write

$$\frac{\cancel{3} + 6\sqrt{2}}{\cancel{3}} = 1 + 6\sqrt{2}$$

This is *not* correct. We must divide *both terms* of the numerator by the common factor.

First, we must simplify the radical in the numerator.

$$\frac{3 + \sqrt{72}}{3} = \frac{3 + \sqrt{36 \cdot 2}}{3}$$ Use Property 1 to simplify $\sqrt{72}$.

$$= \frac{3 + \sqrt{36} \cdot \sqrt{2}}{3} = \frac{3 + 6\sqrt{2}}{3}$$

$$= \frac{3(1 + 2\sqrt{2})}{3} = 1 + 2\sqrt{2}$$ *Factor* the numerator and then divide by the *common* factor 3.

✔ **CHECK YOURSELF 6**

Simplify $\dfrac{15 + \sqrt{75}}{5}$.

CHECK YOURSELF ANSWERS

1. (a) $3\sqrt{2}$; **(b)** $3\sqrt{6}$; **(c)** $2a\sqrt{6}$ **2.** $15\sqrt{21}$ **3. (a)** $\sqrt{30} + 5$;
(b) $2\sqrt{15} + 3\sqrt{6}$ **4. (a)** $-1 + \sqrt{5}$; **(b)** -13; **(c)** $11 - 6\sqrt{2}$
5. (a) 5; **(b)** $3s$ **6.** $3 + \sqrt{3}$

Name _____

Section _____ Date _____

9.4   Exercises

Perform the indicated multiplication. Then simplify each radical expression.

ANSWERS

1. _____ 2. _____

3. _____ 4. _____

5. _____ 6. _____

7. _____ 8. _____

9. _____ 10. _____

11. _____ 12. _____

13. _____ 14. _____

15. _____ 16. _____

17. _____ 18. _____

19. _____

20. _____

21. _____

22. _____

23. _____

24. _____

25. _____

26. _____

27. _____

28. _____

29. _____

30. _____

31. _____

32. _____

33. _____

34. _____

1. $\sqrt{7} \cdot \sqrt{5}$

2. $\sqrt{3} \cdot \sqrt{7}$

3. $\sqrt{5} \cdot \sqrt{11}$

4. $\sqrt{13} \cdot \sqrt{5}$

5. $\sqrt{3} \cdot \sqrt{10m}$

6. $\sqrt{7a} \cdot \sqrt{13}$

7. $\sqrt{2x} \cdot \sqrt{15}$

8. $\sqrt{15} \cdot \sqrt{2b}$

9. $\sqrt{3} \cdot \sqrt{7} \cdot \sqrt{2}$

10. $\sqrt{5} \cdot \sqrt{7} \cdot \sqrt{3}$

11. $\sqrt{3} \cdot \sqrt{12}$

12. $\sqrt{8} \cdot \sqrt{8}$

13. $\sqrt{10x} \cdot \sqrt{10x}$

14. $\sqrt{5a} \cdot \sqrt{15a}$

15. $\sqrt{27} \cdot \sqrt{2}$

16. $\sqrt{8} \cdot \sqrt{10}$

17. $\sqrt{2x} \cdot \sqrt{6x}$

18. $\sqrt{3a} \cdot \sqrt{15a}$

19. $3\sqrt{2x} \cdot \sqrt{6x}$

20. $3\sqrt{2a} \cdot \sqrt{5a}$

21. $(3a\sqrt{3a})(5\sqrt{7a})$

22. $(2x\sqrt{5x})(3\sqrt{11x})$

23. $(3\sqrt{5})(2\sqrt{10})$

24. $(4\sqrt{3})(3\sqrt{6})$

25. $\sqrt{3}(\sqrt{2} + \sqrt{3})$

26. $\sqrt{3}(\sqrt{5} - \sqrt{3})$

27. $\sqrt{3}(2\sqrt{5} - 3\sqrt{3})$

28. $\sqrt{7}(2\sqrt{3} + 3\sqrt{7})$

29. $(\sqrt{3} + 5)(\sqrt{3} + 3)$

30. $(\sqrt{3} - 5)(\sqrt{3} + 2)$

31. $(\sqrt{5} - 1)(\sqrt{5} + 3)$

32. $(\sqrt{2} + 3)(\sqrt{2} - 7)$

33. $(\sqrt{5} - 2)(\sqrt{5} + 2)$

34. $(\sqrt{7} + 5)(\sqrt{7} - 5)$

35. $(\sqrt{7} + 3)(\sqrt{7} - 3)$

36. $(\sqrt{11} - 3)(\sqrt{11} + 3)$

37. $(\sqrt{x} + 3)(\sqrt{x} - 3)$

38. $(\sqrt{a} - 4)(\sqrt{a} + 4)$

39. $(\sqrt{3} + 2)^2$

40. $(\sqrt{5} - 3)^2$

41. $(\sqrt{y} - 5)^2$

42. $(\sqrt{x} + 4)^2$

Perform the indicated division. Rationalize the denominator if necessary. Then simplify each radical expression.

43. $\dfrac{\sqrt{98}}{\sqrt{2}}$

44. $\dfrac{\sqrt{108}}{\sqrt{3}}$

45. $\dfrac{\sqrt{72a^2}}{\sqrt{2}}$

46. $\dfrac{\sqrt{48m^2}}{\sqrt{3}}$

47. $\dfrac{4 + \sqrt{48}}{4}$

48. $\dfrac{12 + \sqrt{108}}{6}$

49. $\dfrac{5 + \sqrt{175}}{5}$

50. $\dfrac{18 + \sqrt{567}}{9}$

51. $\dfrac{-8 - \sqrt{512}}{4}$

52. $\dfrac{-9 - \sqrt{108}}{3}$

53. $\dfrac{6 + \sqrt{18}}{3}$

54. $\dfrac{6 - \sqrt{20}}{2}$

55. $\dfrac{15 - \sqrt{75}}{5}$

56. $\dfrac{8 + \sqrt{48}}{4}$

57. Geometry. Find the area of the rectangle shown.

$\sqrt{3}$

$\sqrt{11}$

35.

36.

37.

38.

39.

40.

41.

42.

43.

44.

45.

46.

47.

48.

49.

50.

51.

52.

53.

54.

55.

56.

57.

58. _____

59. _____

60. _____

61. _____

a. _____

b. _____

c. _____

d. _____

e. _____

f. _____

58. Geometry. Find the area of the rectangle shown in the figure.

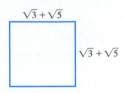

$\sqrt{3} + \sqrt{5}$

$\sqrt{3} + \sqrt{5}$

59. Complete this statement: "$\sqrt{2} \cdot \sqrt{5} = \sqrt{10}$ because. . . ."

60. Explain why $2\sqrt{3} + 5\sqrt{3} = 7\sqrt{3}$ but $7\sqrt{3} + 3\sqrt{5} \neq 10\sqrt{8}$.

61. When you look out over an unobstructed landscape or seascape, the distance to the visible horizon depends on your height above the ground. The equation

$$d = \sqrt{\frac{3}{2}h}$$

is a good estimate of this, in which d = distance to horizon in miles and h = height of viewer above the ground. Work with a partner to make a chart of distances to the horizon given different elevations. Use the actual heights of tall buildings or prominent landmarks in your area. The local library should have a list of these. Be sure to consider the view to the horizon you get when flying in a plane. What would your elevation have to be to see from one side of your city or town to the other? From one side of your state or county to the other?

Getting Ready for Section 9.5 [Section 0.3]

Solve the following equations.

(a) $x^2 = 16$ (b) $x^2 = 25$ (c) $2x^2 = 72$

(d) $5x^2 = 125$ (e) $(x - 1)^2 = 9$ (f) $(x + 2)^2 = 16$

Answers

1. $\sqrt{35}$ **3.** $\sqrt{55}$ **5.** $\sqrt{30m}$ **7.** $\sqrt{30x}$ **9.** $\sqrt{42}$ **11.** 6
13. $10x$ **15.** $3\sqrt{6}$ **17.** $2x\sqrt{3}$ **19.** $6x\sqrt{3}$ **21.** $15a^2\sqrt{21}$
23. $30\sqrt{2}$ **25.** $\sqrt{6} + 3$ **27.** $2\sqrt{15} - 9$ **29.** $18 + 8\sqrt{3}$
31. $2 + 2\sqrt{5}$ **33.** 1   **35.** -2 **37.** $x - 9$ **39.** $7 + 4\sqrt{3}$
41. $y - 10\sqrt{y} + 25$ **43.** 7   **45.** $6a$ **47.** $1 + \sqrt{3}$ **49.** $1 + \sqrt{7}$
51. $-2 - 4\sqrt{2}$ **53.** $2 + \sqrt{2}$ **55.** $3 - \sqrt{3}$ **57.** $\sqrt{33}$
59.   **61.**   **a.** $-4, 4$ **b.** $-5, 5$ **c.** $-6, 6$

d. $-5, 5$ **e.** $-2, 4$ **f.** $-6, 2$

9.5 Solving Radical Equations

9.5 OBJECTIVE

1. Solve an equation containing a radical expression

In this section, we wish to establish procedures for solving equations involving radical expressions. The basic technique we will use involves raising both sides of an equation to some power. However, doing so requires some caution.

For example, we will begin with the equation $x = 1$. Squaring both sides gives us $x^2 = 1$, which has two solutions, 1 and -1. Clearly -1 is not a solution to the original equation. We refer to -1 as an *extraneous solution.*

We must be aware of the possibility of extraneous solutions any time we raise both sides of an equation to any *even power.* Having said that, we are now prepared to introduce the power property of equality.

NOTE
$$x^2 = 1$$
$$x^2 - 1 = 0$$
$$(x + 1)(x - 1) = 0$$
So the solutions are 1 and -1.

Rules and Properties: The Power Property of Equality

Given any two expressions a and b and any positive integer n,

If $a = b$, then $a^n = b^n$.

Although you will never lose a solution when applying the power property, you will often find an extraneous one as a result of raising both sides of the equation to some power. Because of this, it is very important that you *check all solutions.*

OBJECTIVE 1

Example 1 Solving a Radical Equation

NOTE Notice that
$(\sqrt{x + 2})^2 = x + 2$

That is why squaring both sides of the equation removes the radical.

Solve $\sqrt{x + 2} = 3$.

Squaring each side, we have

$$(\sqrt{x + 2})^2 = 3^2$$
$$x + 2 = 9$$
$$x = 7$$

Substituting 7 into the original equation, we find

$$\sqrt{7 + 2} \overset{?}{=} 3$$
$$\sqrt{9} \overset{?}{=} 3$$
$$3 = 3$$

Because 7 is the only value that makes this a true statement, the solution for the equation is 7.

✔ **CHECK YOURSELF 1**

Solve the equation $\sqrt{x - 5} = 4$.

Example 2   Solving a Radical Equation

NOTE Applying the power property will only remove the radical if that radical is isolated on one side of the equation.

NOTE Notice that on the right $(-1)^2 = 1$.

Solve $\sqrt{4x + 5} + 1 = 0$.

We must *first isolate the radical* on the left side:

$$\sqrt{4x + 5} = -1$$

Then, squaring both sides, we have

$$(\sqrt{4x + 5})^2 = (-1)^2$$
$$4x + 5 = 1$$

and solving for x, we find that

$$x = -1$$

Now check the solution by substituting -1 for x in the original equation:

NOTE $\sqrt{1} = 1$, the principal root.

NOTE 2 is never equal to 0, so -1 is *not* a solution for the original equation.

$$\sqrt{4(-1) + 5} + 1 \overset{?}{=} 0$$
$$\sqrt{1} + 1 \overset{?}{=} 0$$
and $$2 \neq 0$$

Because -1 is an extraneous solution, there are *no solutions* to the original equation.

CHECK YOURSELF 2

Solve $\sqrt{3x - 2} + 2 = 0$.

Next, consider an example in which the procedure involves squaring a binomial.

Example 3   Solving a Radical Equation

Solve $\sqrt{x + 3} = x + 1$.

We can square each side, as before.

$$(\sqrt{x + 3})^2 = (x + 1)^2$$
$$x + 3 = x^2 + 2x + 1$$

Simplifying this gives us the quadratic equation

$$x^2 + x - 2 = 0$$

Factoring, we have

$$(x - 1)(x + 2) = 0$$

which gives us the possible solutions

NOTE Verify this for yourself by substituting 1 and then -2 for x in the original equation.

$$x = 1 \qquad \text{or} \qquad x = -2$$

Now we check for extraneous solutions and find that $x = 1$ is a valid solution, but that $x = -2$ does not yield a true statement.

CAUTION

Be Careful! Sometimes (as in this example), one side of the equation contains a binomial. In that case, we must remember the middle term when we square the binomial. The square of a binomial *is always a trinomial.*

✔ CHECK YOURSELF 3

Solve $\sqrt{x - 5} = x - 7$.

It is not always the case that one of the solutions is extraneous. We may have zero, one, or two valid solutions when we generate a quadratic from a radical equation.

In Example 4 we see a case in which both of the solutions derived will satisfy the equation.

Example 4   Solving a Radical Equation

Solve $\sqrt{7x + 1} - 1 = 2x$.

First, *we must isolate the term involving the radical.*

$$\sqrt{7x + 1} = 2x + 1$$

We can now square both sides of the equation.

$$7x + 1 = 4x^2 + 4x + 1$$

Now we write the quadratic equation in standard form.

$$4x^2 - 3x = 0$$

Factoring, we have

$$x(4x - 3) = 0$$

which yields two possible solutions

$$x = 0 \qquad \text{or} \qquad x = \frac{3}{4}$$

Checking the solutions by substitution, we find that both values for x give true statements, as follows.

Letting x be 0, we have

$$\sqrt{7(0) + 1} - 1 \stackrel{?}{=} 2(0)$$
$$\sqrt{1} - 1 \stackrel{?}{=} 0$$

or $\qquad\qquad 0 = 0$ A true statement.

Letting x be $\frac{3}{4}$, we have

$$\sqrt{7\left(\frac{3}{4}\right) + 1} - 1 \stackrel{?}{=} 2\left(\frac{3}{4}\right)$$

$$\sqrt{\frac{25}{4}} - 1 \stackrel{?}{=} \frac{3}{2}$$

$$\frac{5}{2} - 1 \stackrel{?}{=} \frac{3}{2}$$

$$\frac{3}{2} = \frac{3}{2}$$ Again a true statement. The set of solutions is $\left\{0, \frac{3}{4}\right\}$.

CHECK YOURSELF 4 _____

Solve $\sqrt{5x + 1} - 1 = 3x$.

We summarize our work in this section in the following algorithm for solving equations involving radicals.

Step by Step: Solving Equations Involving Radicals

Step 1   Isolate a radical on one side of the equation.

Step 2   Raise each side of the equation to the smallest power that will eliminate the isolated radical.

Step 3   If any radicals remain in the equation derived in Step 2, return to Step 1 and continue until no radical remains.

Step 4   Solve the resulting equation to determine any possible solutions.

Step 5   Check all solutions to determine whether extraneous solutions may have resulted from Step 2.

CHECK YOURSELF ANSWERS _____

1. 21   **2.** No solution   **3.** 9   **4.** $0, -\dfrac{1}{9}$

Name _____

Section _____ Date _____

Solve each of the equations. Be sure to check your solutions.

1. $\sqrt{x} = 2$

2. $\sqrt{x} - 3 = 0$

3. $2\sqrt{y} - 1 = 0$

4. $3\sqrt{2z} = 9$

5. $\sqrt{m + 5} = 3$

6. $\sqrt{y + 7} = 5$

7. $\sqrt{x - 3} = 4$

8. $\sqrt{x + 4} = 3$

9. $\sqrt{2x - 1} = 3$

10. $\sqrt{3x + 1} = 4$

11. $\sqrt{2x + 4} - 4 = 0$

12. $\sqrt{3x + 3} - 6 = 0$

13. $\sqrt{3x - 2} + 2 = 0$

14. $\sqrt{4x + 1} + 3 = 0$

15. $x + \sqrt{x - 1} = 7$

16. $x - \sqrt{x + 2} = 10$

17. $x + \sqrt{2x - 5} = 10$

18. $2x - \sqrt{3x - 2} = 8$

19. $\sqrt{2x - 3} + 1 = 3$

20. $\sqrt{3x + 1} - 2 = -1$

21. $2\sqrt{3z + 2} - 1 = 5$

22. $3\sqrt{4q - 1} - 2 = 7$

23. $\sqrt{15 - 2x} = x$

24. $\sqrt{48 - 2y} = y$

25. $\sqrt{x + 5} = x - 1$

26. $\sqrt{2x - 1} = x - 8$

27. $\sqrt{3m - 2} + m = 10$

28. $\sqrt{2x + 1} + x = 7$

29. $\sqrt{t + 9} + 3 = t$

30. $\sqrt{2y + 7} + 4 = y$

ANSWERS

1. _____ 2. _____

3. _____ 4. _____

5. _____ 6. _____

7. _____ 8. _____

9. _____ 10. _____

11. _____ 12. _____

13. _____

14. _____

15. _____

16. _____

17. _____

18. _____

19. _____

20. _____

21. _____

22. _____

23. _____

24. _____

25. _____

26. _____

27. _____

28. _____

29. _____

30. _____

31. _____

32. _____

33. _____

34. _____

35. _____

36. _____

a. _____

b. _____

c. _____

d. _____

e. _____

f. _____

Solve for the indicated variable.

31. $h = \sqrt{pq}$ for q

32. $c = \sqrt{a^2 + b^2}$ for a

33. $v = \sqrt{2gR}$ for R

34. $v = \sqrt{2gR}$ for g

35. $r = \sqrt{\dfrac{S}{2\pi}}$ for S

36. $r = \sqrt{\dfrac{3V}{4\pi}}$ for V

 Getting Ready for Section 9.6 [Section 9.1]

Evaluate the following. Round your answer to the nearest thousandth.

(a) $\sqrt{16}$

(b) $\sqrt{49}$

(c) $\sqrt{121}$

(d) $\sqrt{12}$

(e) $\sqrt{27}$

(f) $\sqrt{98}$

Answers

1. 4   **3.** $\dfrac{1}{4}$ **5.** 4   **7.** 19   **9.** 5   **11.** 6   **13.** No solution

15. 5   **17.** 7   **19.** $\dfrac{7}{2}$ **21.** $\dfrac{7}{3}$ **23.** 3   **25.** 4   **27.** 6   **29.** 7

31. $q = \dfrac{h^2}{p}$ **33.** $R = \dfrac{v^2}{2g}$ **35.** $S = 2\pi r^2$ **a.** 4   **b.** 7   **c.** 11

d. 3.464   **e.** 5.196   **f.** 9.899

9.6 Applications of the Pythagorean Theorem

9.6 OBJECTIVES

1. Apply the Pythagorean theorem in solving problems
2. Use the distance formula to find the distance between two points.

Perhaps the most famous theorem in all of mathematics is the **Pythagorean theorem.** The theorem was named for the Greek mathematician Pythagoras, born in 572 B.C.E. Pythagoras was the founder of the Greek society the Pythagoreans. Although the theorem bears Pythagoras's name, his own work on this theorem is uncertain because the Pythagoreans credited new discoveries to their founder.

Rules and Properties: The Pythagorean Theorem

For every right triangle, the square of the length of the hypotenuse is equal to the sum of the squares of the lengths of the legs.

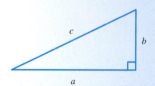

c is the hypotenuse

a and b are the legs.

NOTE Here we use c to represent the length of the hypotenuse.

$$c^2 = a^2 + b^2$$

OBJECTIVE 1   **Example 1   Verifying the Pythagorean Theorem**

Verify the Pythagorean theorem for the given triangles.

(a) For the following right triangle,

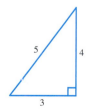

$$5^2 \overset{?}{=} 3^2 + 4^2$$
$$25 \overset{?}{=} 9 + 16$$
$$25 = 25$$

(b) For the following right triangle,

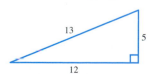

$$13^2 \overset{?}{=} 12^2 + 5^2$$
$$169 \overset{?}{=} 144 + 25$$
$$169 = 169$$

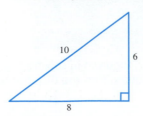

CHECK YOURSELF 1

Verify the Pythagorean theorem for the right triangle shown.

The Pythagorean theorem can be used to find the length of one side of a right triangle when the lengths of the two other sides are known.

Example 2   Solving for the Length of the Hypotenuse

Find length x.

NOTE x will be longer than the given sides because it is the hypotenuse.

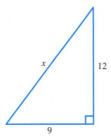

$$x^2 = 9^2 + 12^2$$
$$= 81 + 144$$
$$= 225$$

so

$$x = 15 \qquad \text{or} \qquad \underline{x = -15}$$

We reject this solution because a length must be positive.

CHECK YOURSELF 2

Find length x.

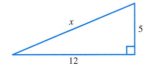

Sometimes, one or more of the lengths of the sides may be represented by an irrational number.

Example 3   Solving for the Length of the Leg

Find length x. Then use your calculator to give an approximation to the nearest tenth.

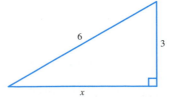

$$3^2 + x^2 = 6^2$$
$$9 + x^2 = 36$$
$$x^2 = 27$$
$$x = \pm\sqrt{27}$$

NOTE You can approximate $3\sqrt{3}$ (or $\sqrt{27}$) with the use of a calculator.

but distance cannot be negative, so

$$x = \sqrt{27}$$

So x is approximately 5.2.

CHECK YOURSELF 3

Find length x and approximate it to the nearest tenth.

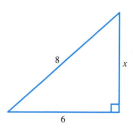

The Pythagorean theorem can be applied to solve a variety of geometric problems.

Example 4   Solving for the Length of the Diagonal

Find, to the nearest tenth, the length of the diagonal of a rectangle that is 8 cm long and 5 cm wide. Let x be the unknown length of the diagonal:

NOTE Always draw and label a sketch showing the information from a problem when geometric figures are involved.

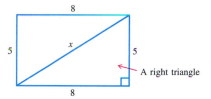

A right triangle

So

NOTE Again, distance cannot be negative, so we eliminate $x = -\sqrt{89}$.

$$x^2 = 5^2 + 8^2$$
$$= 25 + 64$$
$$= 89$$
$$x = \sqrt{89}$$

Thus

$$x \approx 9.4 \text{ cm}$$

CHECK YOURSELF 4

The diagonal of a rectangle is 12 in. and its width is 6 in. Find its length to the nearest tenth.

The application in Example 5 also makes use of the Pythagorean theorem.

Example 5   Solving an Application

NOTE Always check to see if your final answer is reasonable.

How long must a guywire be to reach from the top of a 30-ft pole to a point on the ground 20 ft from the base of the pole?

Again be sure to draw a sketch of the problem.

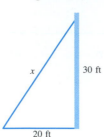

$$x^2 = 20^2 + 30^2$$
$$= 400 + 900$$
$$= 1300$$
$$x = \sqrt{1300}$$
$$\approx 36 \text{ ft}$$

 CHECK YOURSELF 5 _____

A 16-ft ladder leans against a wall with its base 4 ft from the wall. How far off the floor is the top of the ladder?

To find the distance between any two points in the plane, we use a formula derived from the Pythagorean theorem. First, we need an alternate form of the Pythagorean theorem.

Rules and Properties: Pythagorean Theorem

Given a right triangle in which *c* is the length of the hypotenuse, we have the equation

$$c^2 = a^2 + b^2$$

NOTE A distance is always positive, so we use only the principal square root.

We can rewrite the formula as

$$c = \sqrt{a^2 + b^2}$$

We use this form of the Pythagorean theorem in Example 6.

OBJECTIVE 2   ### Example 6   Finding the Distance Between Two Points

Find the distance from (2, 3) to (5, 7).

The distance can be seen as the hypotenuse of a right triangle.

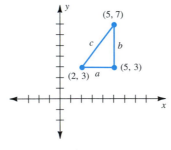

The lengths of the two legs can be found by finding the difference of the two *x*-coordinates and the difference of the two *y*-coordinates. So

$$a = 5 - 2 = 3 \qquad \text{and} \qquad b = 7 - 3 = 4$$

The distance c can then be found using the formula

$$c = \sqrt{a^2 + b^2}$$

or, in this case,

$$c = \sqrt{3^2 + 4^2}$$
$$c = \sqrt{9 + 16}$$
$$= \sqrt{25}$$
$$= 5$$

The distance is 5 units.

 CHECK YOURSELF 6

Find the distance between (0, 2) and (5, 14).

If we call our points (x_1, y_1) and (x_2, y_2), we can state the **distance formula.**

Definition: Distance Formula

The distance between points (x_1, y_1) and (x_2, y_2) can be found using the formula
$$d = \sqrt{(x_2 - x_1)^2 + (y_2 - y_1)^2}$$

Example 7   Finding the Distance Between Two Points

Find the distance between $(-2, 5)$ and $(2, -3)$. Simplify the radical answer.
    Using the formula,

$$d = \sqrt{[2 - (-2)]^2 + [(-3) - 5]^2}$$
$$= \sqrt{(4)^2 + (-8)^2}$$
$$= \sqrt{16 + 64}$$

NOTE $\sqrt{80} = \sqrt{16 \cdot 5} = 4\sqrt{5}$
$$= \sqrt{80}$$
$$= 4\sqrt{5}$$

 CHECK YOURSELF 7

Find the distance between (2, 5) and (−5, 2).

In Example 7, you were asked to find the distance between $(-2, 5)$ and $(2, -3)$.

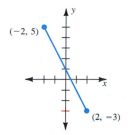

To form a right triangle, we include the point $(-2, -3)$

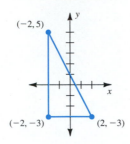

Note that the lengths of the two sides of the right triangle are 4 and 8. By the Pythagorean theorem, the hypotenuse must have length $\sqrt{4^2 + 8^2} = \sqrt{80} = 4\sqrt{5}$. The distance formula is an application of the Pythagorean theorem.

Using the square root key on a calculator, it is easy to approximate the length of a diagonal line. This is particularly useful in checking to see if an object is square or rectangular.

Example 8   Approximating Length With a Calculator

Approximate the length of the diagonal of the given rectangle. The diagonal forms the hypotenuse of a triangle with legs 12.2 in. and 15.7 in. The length of the diagonal would be $\sqrt{12.2^2 + 15.7^2} = \sqrt{395.33} \approx 19.88$ in. Use your calculator to confirm the approximation.

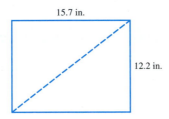

CHECK YOURSELF 8

Approximate the length of the diagonal of the rectangle to the nearest tenth.

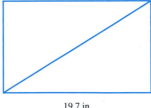

CHECK YOURSELF ANSWERS

1. $10^2 \stackrel{?}{=} 8^2 + 6^2$; $100 \stackrel{?}{=} 64 + 36$; $100 = 100$ **2.** 13   **3.** $2\sqrt{7}$; or approximately 5.3   **4.** The length is approximately 10.4 in.
5. The height is approximately 15.5 ft.   **6.** 13   **7.** $\sqrt{58}$ **8.** ≈ 24.0 in.

9.6 Exercises

Find the length *x* in each triangle. Express your answer in simplified radical form.

1.

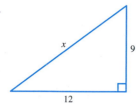

2.

3.

4.

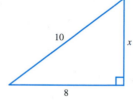

5.

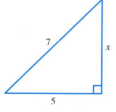

6.

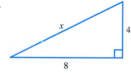

In exercises 7 to 12, express your answer to the nearest thousandth.

7. Geometry. Find the diagonal of a rectangle with a length of 10 cm and a width of 7 cm.

8. Geometry. Find the diagonal of a rectangle with 5 in. width and 7 in. length.

9. Geometry. Find the width of a rectangle whose diagonal is 12 ft and whose length is 10 ft.

10. Geometry. Find the length of a rectangle whose diagonal is 9 in. and whose width is 6 in.

11. Geometry. How long must a guywire be to run from the top of a 20-ft pole to a point on the ground 8 ft from the base of the pole?

ANSWERS

12. _____

13. _____

14. _____

15. _____

16. _____

17. _____

18. _____

12. Geometry. The base of a 15-ft ladder is 5 ft away from a wall. How high from the floor is the top of the ladder?

Find the altitude of each triangle.

13.

5   5

6

14.

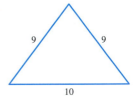

9   9

10

15. Geometry. A homeowner wishes to insulate her attic with fiberglass insulation to conserve energy. The insulation comes in 40-cm wide rolls that are cut to fit between the rafters in the attic. If the roof is 6 m from peak to eave and the attic space is 2 m high at the peak, how long does each of the pieces of insulation need to be? Round to the nearest tenth.

16. Geometry. For the home described in exercise 15, if the roof is 7 m from peak to eave and the attic space is 3 m high at the peak, how long does each of the pieces of insulation need to be? Round to the nearest tenth.

17. Geometry. A solar collector and its stand are in the shape of a right triangle. The collector is 5.00 m long, the upright leg is 3.00 m long, and the base leg is 4.00 m long. Because of inefficiencies in the collector's position, it needs to be raised by 0.50 m on the upright leg. How long will the new base leg be? Round to the nearest tenth.

18. Geometry. A solar collector and its stand are in the shape of a right triangle. The collector is 5.00 m long, the upright leg is 2.00 m long, and the base leg is 4.58 m long. Because of inefficiencies in the collector's position, it needs to be lowered by 0.50 m on the upright leg. How long will the new base leg be? Round to the nearest tenth.

Find the distance between each pair of points.

19. $(2, 0)$ and $(-4, 0)$

20. $(-3, 0)$ and $(4, 0)$

21. $(0, -2)$ and $(0, -9)$

22. $(0, 8)$ and $(0, -4)$

23. $(2, 5)$ and $(5, 2)$

24. $(3, 3)$ and $(5, 7)$

25. $(5, 1)$ and $(3, 8)$

26. $(2, 9)$ and $(7, 4)$

27. $(-2, 8)$ and $(1, 5)$

28. $(2, 6)$ and $(-3, 4)$

29. $(6, -1)$ and $(2, 2)$

30. $(2, -8)$ and $(1, 0)$

31. $(-1, -1)$ and $(2, 5)$

32. $(-2, -2)$ and $(3, 3)$

33. $(-2, 9)$ and $(-3, 3)$

34. $(4, -1)$ and $(0, -5)$

35. $(-1, -4)$ and $(-3, 5)$

36. $(-2, 3)$ and $(-7, -1)$

37. $(-2, -4)$ and $(-4, 1)$

38. $(-1, -1)$ and $(4, -2)$

39. $(-4, -2)$ and $(-1, -5)$

40. $(-2, -2)$ and $(-4, -4)$

41. $(-2, 0)$ and $(-4, -1)$

42. $(-5, -2)$ and $(-7, -1)$

Use the distance formula to show that each set of points describes an isosceles triangle (a triangle with two sides of equal length).

43. $(-3, 0)$, $(2, 3)$, and $(1, -1)$

44. $(-2, 4)$, $(2, 7)$, and $(5, 3)$

45. Geometry. The length of one leg of a right triangle is 3 in. more than the other. If the length of the hypotenuse is 15 in., what are the lengths of the two legs?

ANSWERS

19. _____

20. _____

21. _____

22. _____

23. _____

24. _____

25. _____

26. _____

27. _____

28. _____

29. _____

30. _____

31. _____

32. _____

33. _____

34. _____

35. _____

36. _____

37. _____

38. _____

39. _____

40. _____

41. _____

42. _____

43. _____

44. _____

45. _____

46. Geometry. The length of a rectangle is 1 cm longer than its width. If the diagonal of the rectangle is 5 cm, what are the dimensions (the length and width) of the rectangle?

Use the Pythagorean theorem to determine the length of each line segment. Where appropriate, round to the nearest hundredth.

47.

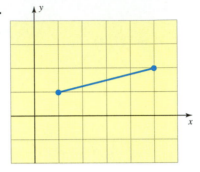

48.

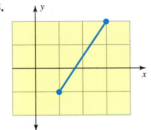

49.

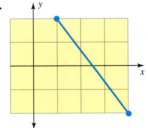

For each figure, use the slope concept and the Pythagorean theorem to show that the figure is a square. (Recall that a square must have four right angles and four equal sides.) Then give the area of the square to the nearest hundredth.

50.

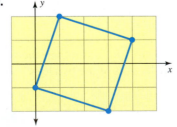

51.

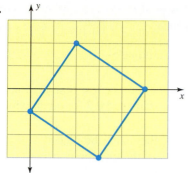

52.

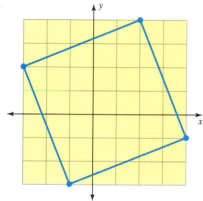

53. Business and finance. Your architectural firm just received this memo.

To:	Algebra Expert Architecture, Inc.
From:	Microbeans Coffee Company, Inc.
Re:	Design for On-Site Day Care Facility
Date:	Aug. 10, 2003

We are requesting that you submit a design for a nursery for preschool children. We are planning to provide free on-site day care for the workers at our corporate headquarters.

The nursery should be large enough to serve the needs of 20 preschoolers. There will be three child care workers in this facility. We want the nursery to be 3000 ft^2 in area. It needs a playroom, a small kitchen and eating space, and bathroom facilities. There should be some space to store toys and books, many of which should be accessible to children. The company plans to put this facility on the first floor on an outside wall so the children can go outside to play without disturbing workers. You are free to add to this design as you see fit.

Please send us your design drawn to a scale of 1 ft to 0.25 in., with precise measurements and descriptions. We would like to receive this design 1 week from today. Please give us some estimate of the cost of this renovation to our building.

Submit a design, keeping in mind that the design has to conform to strict design specifications for buildings designated as nurseries, including:

1. Number of exits: Two exits for the first 7 people and one exit for every additional 7 people.

2. Width of exits: The total width of exits in inches shall not be less than the total occupant load served by an exit multiplied by 0.3 for stairs and 0.2 for other exits. No exit shall be less than 3 ft wide and 6 ft 8 in. high.

3. Arrangements of exits: If two exits are required, they shall be placed a distance apart equal to but not less than one-half the length of the maximum overall diagonal dimension of the building or area to be served measured in a straight line between exits. Where three or more exits are required, two shall be placed as above and the additional exits arranged a reasonable distance apart.

4. Distance to exits: Maximum distance to travel from any point to an exterior door shall not exceed 100 ft.

Answers

1. 15   **3.** 15   **5.** $2\sqrt{6}$ **7.** ≈ 12.207 cm   **9.** ≈ 6.633 ft
11. ≈ 21.541 ft   **13.** 4   **15.** $8\sqrt{2} \approx 11.3$ m   **17.** ≈ 3.6 m   **19.** 6
21. 7   **23.** $3\sqrt{2}$ **25.** $\sqrt{53}$ **27.** $3\sqrt{2}$ **29.** 5   **31.** $3\sqrt{5}$
33. $\sqrt{37}$ **35.** $\sqrt{85}$ **37.** $\sqrt{29}$ **39.** $3\sqrt{2}$ **41.** $\sqrt{5}$
43. Sides have length $\sqrt{34}$, $\sqrt{17}$, and $\sqrt{17}$ **45.** 9 in., 12 in.   **47.** 4.12

49. 5   **51.** Length of sides: $\sqrt{13}$; slopes of sides: $-\dfrac{2}{3}$ and $\dfrac{3}{2}$; 13

53.

9 Summary

DEFINITION/PROCEDURE	EXAMPLE	REFERENCE
Roots and Radicals		**Section 9.1**
Square Roots $\sqrt{x}$ is the principal (or positive) square root of x. It is the non-negative number we must square to get x. $\quad -\sqrt{x}$ is the negative square root of x. The square root of a negative number is not a real number.	$\sqrt{49} = 7$ $-\sqrt{49} = -7$ $\sqrt{-49}$ is not a real number.	p. 749
Other Roots $\sqrt[3]{x}$ is the cube root of x. $\sqrt[4]{x}$ is the fourth root of x.	$\sqrt[3]{64} = 4$ because $4^3 = 64$. $\sqrt[4]{81} = 3$ because $3^4 = 81$.	p. 751
Simplifying Radical Expressions		**Section 9.2**
An expression involving square roots is in *simplest form* if **1.** There are no perfect-square factors in a radical. **2.** No fraction appears inside a radical. **3.** No radical appears in the denominator.		p. 760
To simplify a radical expression, use one of the following properties. The square root of a product is the product of the square roots. $$\sqrt{ab} = \sqrt{a} \cdot \sqrt{b}$$	$\sqrt{40} = \sqrt{4 \cdot 10}$ $\quad = \sqrt{4} \cdot \sqrt{10}$ $\quad = 2\sqrt{10}$ $\sqrt{12x^3} = \sqrt{4x^2 \cdot 3x}$ $\quad = \sqrt{4x^2} \cdot \sqrt{3x}$ $\quad = 2x \cdot \sqrt{3x}$	p. 760
The square root of a quotient is the quotient of the square roots. $$\sqrt{\dfrac{a}{b}} = \dfrac{\sqrt{a}}{\sqrt{b}}$$	$\sqrt{\dfrac{5}{16}} = \dfrac{\sqrt{5}}{\sqrt{16}} = \dfrac{\sqrt{5}}{4}$ $\sqrt{\dfrac{2y}{3}} = \dfrac{\sqrt{2y}}{\sqrt{3}} = \dfrac{\sqrt{2y} \cdot \sqrt{3}}{\sqrt{3} \cdot \sqrt{3}}$ $\quad = \dfrac{\sqrt{6y}}{\sqrt{9}} = \dfrac{\sqrt{6y}}{3}$	p. 762
Adding and Subtracting Radicals		**Section 9.3**
Like radicals have the same index and the same radicand (the expression inside the radical). $\quad$ Like radicals can be added (or subtracted) in the same way as like terms. Apply the distributive law and combine the coefficients.	$3\sqrt{5}$ and $2\sqrt{5}$ are like radicals. $2\sqrt{3} + 3\sqrt{3} = (2 + 3)\sqrt{3}$ $\quad = 5\sqrt{3}$ $5\sqrt{7} - 2\sqrt{7} = (5 - 2)\sqrt{7}$ $\quad = 3\sqrt{7}$	p. 769
Certain expressions can be combined after one or more of the terms involving radicals are simplified.	$\sqrt{12} + \sqrt{3} = 2\sqrt{3} + \sqrt{3}$ $\quad = (2 + 1)\sqrt{3}$ $\quad = 3\sqrt{3}$	p. 769
Multiplying and Dividing Radicals		**Section 9.4**
Multiplying To multiply radical expressions, use the first property of radicals in the following way: $$\sqrt{a} \cdot \sqrt{b} = \sqrt{ab}$$	$\sqrt{6} \cdot \sqrt{15} = \sqrt{6 \cdot 15} = \sqrt{90}$ $\quad = \sqrt{9 \cdot 10}$ $\quad = 3\sqrt{10}$	p. 776

Continued

Multiplying and Dividing Radicals		Section 9.4
The distributive property can also be applied in multiplying radical expressions.	$\sqrt{5}(\sqrt{3} + 2\sqrt{5})$ $= \sqrt{5} \cdot \sqrt{3} + \sqrt{5} \cdot 2\sqrt{5}$ $= \sqrt{15} + 10$	*p. 777*
The FOIL pattern allows us to find the product of binomial radical expressions.	$(\sqrt{5} + 2)(\sqrt{5} - 1)$ $= \sqrt{5} \cdot \sqrt{5} - \sqrt{5} + 2\sqrt{5} - 2$ $= 3 + \sqrt{5}$ $(\sqrt{10} + 3)(\sqrt{10} - 3)$ $= 10 - 9 = 1$	*p. 777*
Dividing To divide radical expressions, use the second property of radicals in the following way: $$\frac{\sqrt{a}}{\sqrt{b}} = \sqrt{\frac{a}{b}}$$	$\frac{\sqrt{50}}{\sqrt{2}} = \sqrt{\frac{50}{2}}$ $= \sqrt{25}$ $= 5$	*p. 778*
Solving Radical Equations		Section 9.5
Power Property of Equality To solve an equation involving radicals, apply the power property of equality: Given any two expressions a and b and any positive integer n, If $a = b$, then $a^n = b^n$. This property is used in the following algorithm.		*p. 783*
Solving Equations Involving Radicals 1. Isolate a radical on one side of the equation. 2. Raise each side of the equation to the smallest power that will eliminate the radical. 3. If any radicals remain in the equation derived in step 2, return to step 1 and continue the process. 4. Solve the resulting equation to determine any possible solutions. 5. Check all solutions to determine whether extraneous solutions may have resulted from step 2.	Solve $\sqrt{2x - 3} + x = 9$ $\sqrt{2x - 3} = -x + 9$ $2x - 3 = x^2 - 18x + 81$ $0 = x^2 - 20x + 84$ $0 = (x - 6)(x - 14)$ $x = 6$ or $x = 14$ By substitution, 6 is the only valid solution.	*p. 786*
Applications of the Pythagorean Theorem		Section 9.6
In words, for every right triangle, the square of the length of the hypotenuse is equal to the sum of the squares of the lengths of the legs. $c^2 = a^2 + b^2$	Find length x: $x^2 = 10^2 + 6^2$ $= 100 + 36$ $= 136$ $x = \sqrt{136}$ or $2\sqrt{34}$	*p. 789*

Summary Exercises

This summary exercise set is provided to give you practice with each of the objectives of this chapter. Each exercise is keyed to the appropriate chapter section. When you are finished, you can check your answers to the odd-numbered exercises against those presented in the back of the text. If you have difficulty with any of these questions, go back and reread the examples from that section. Your instructor will give you guidelines on how to best use these exercises in your instructional setting.

[9.1] Evaluate if possible.

1. $\sqrt{81}$

2. $-\sqrt{49}$

3. $\sqrt{-49}$

4. $\sqrt[3]{64}$

5. $\sqrt[3]{-64}$

6. $\sqrt[4]{81}$

7. $\sqrt[4]{-81}$

[9.2] Simplify each of the following radical expressions. Assume that all variables represent positive real numbers.

8. $\sqrt{50}$

9. $\sqrt{45}$

10. $\sqrt{7a^3}$

11. $\sqrt{20x^4}$

12. $\sqrt{49m^5}$

13. $\sqrt{200b^3}$

14. $\sqrt{147r^3s^2}$

15. $\sqrt{108a^2b^5}$

16. $\sqrt{\dfrac{10}{81}}$

17. $\sqrt{\dfrac{18x^2}{25}}$

18. $\sqrt{\dfrac{12m^5}{49}}$

19. $\sqrt{\dfrac{3}{7}}$

20. $\sqrt{\dfrac{3a}{2}}$

21. $\sqrt{\dfrac{8x^2}{7}}$

[9.3] Simplify by combining like terms.

22. $\sqrt{3} + 4\sqrt{3}$

23. $9\sqrt{5} - 3\sqrt{5}$

24. $3\sqrt{2} + 2\sqrt{3}$

25. $3\sqrt{3a} - \sqrt{3a}$

26. $7\sqrt{6} - 2\sqrt{6} + \sqrt{6}$

27. $5\sqrt{3} + \sqrt{12}$

28. $3\sqrt{18} - 5\sqrt{2}$

29. $\sqrt{32} - \sqrt{18}$

30. $\sqrt{27} - \sqrt{3} + 2\sqrt{12}$

31. $\sqrt{8} + 2\sqrt{27} - \sqrt{75}$

32. $x\sqrt{18} - 3\sqrt{8x^2}$

[9.4] Simplify each radical expression.

33. $\sqrt{6} \cdot \sqrt{5}$

34. $\sqrt{3} \cdot \sqrt{6}$

35. $\sqrt{3x} \cdot \sqrt{2}$

36. $\sqrt{2} \cdot \sqrt{8} \cdot \sqrt{3}$

37. $\sqrt{5a} \cdot \sqrt{10a}$

38. $\sqrt{2}(\sqrt{3} + \sqrt{5})$

39. $\sqrt{7}(2\sqrt{3} - 3\sqrt{7})$

40. $(\sqrt{3} + 5)(\sqrt{3} - 3)$

41. $(\sqrt{15} - 3)(\sqrt{15} + 3)$

42. $(\sqrt{2} + 3)^2$

43. $\dfrac{\sqrt{7x^3}}{\sqrt{3}}$

44. $\dfrac{18 - \sqrt{20}}{2}$

[9.5] Solve each of the following equations. Be sure to check your solutions.

45. $\sqrt{x-5} = 4$

46. $\sqrt{3x-2} + 2 = 5$

47. $\sqrt{y+7} = y - 5$

48. $\sqrt{2x-1} + x = 8$

[9.6] Find length x in each triangle. Express your answer in simplified radical form.

49.

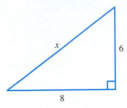

50.

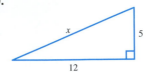

51.

52.

53.

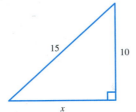

54.

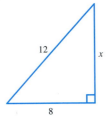

Solve each of the following word problems. Approximate your answer to one decimal place where necessary.

55. Find the diagonal of a rectangle whose length is 12 in. and whose width is 9 in.

56. Find the length of a rectangle whose diagonal has a length of 10 cm and whose width is 5 cm.

57. How long must a guywire be to run from the top of an 18-ft pole to a point on level ground 16 ft away from the base of the pole?

58. The length of one leg of a right triangle is 2 in. more than the length of the other. If the length of the hypotenuse of the triangle is 10 in., what are the lengths of the two legs?

Find the distance between each pair of points.

59. $(-3, 2)$ and $(-7, 2)$

60. $(2, 0)$ and $(5, 9)$

61. $(-2, 7)$ and $(-5, -1)$

62. $(5, -1)$ and $(-2, 3)$

63. $(-3, 4)$ and $(-2, -5)$

64. $(6, 4)$ and $(-3, 5)$

Self-Test for Chapter 9

The purpose of this self-test is to help you check your progress and to review for the next in-class exam. Allow yourself about an hour to take this test. At the end of that hour check your answers against those given in the back of the text. Section references accompany the answers. If you missed any questions, go back to those sections and reread the examples until you master the concepts.

ANSWERS

1. _____

2. _____

3. _____

4. _____

5. _____

Evaluate if possible.

1. $\sqrt{121}$

2. $\sqrt[3]{27}$

3. $\sqrt{-144}$

4. $-\sqrt[3]{-64}$

6. _____

7. _____

Simplify each of the following radical expressions.

5. $\sqrt{75}$

6. $\sqrt{24a^3}$

7. $\sqrt{\dfrac{16}{25}}$

8. $\sqrt{\dfrac{5}{9}}$

8. _____

9. _____

10. _____

11. _____

12. _____

Simplify by combining like terms.

9. $2\sqrt{10} - 3\sqrt{10} + 5\sqrt{10}$

10. $3\sqrt{8} - \sqrt{18}$

11. $2\sqrt{50} - \sqrt{8} - \sqrt{50}$

12. $\sqrt{20} + \sqrt{45} - \sqrt{5}$

13. _____

14. _____

15. _____

16. _____

17. _____

18. _____

Simplify each of the following radical expressions.

13. $\sqrt{3x} \cdot \sqrt{6x}$

14. $(\sqrt{5} + 3)(\sqrt{5} + 2)$

15. $\dfrac{\sqrt{7}}{\sqrt{2}}$

16. $\dfrac{14 + 3\sqrt{98}}{7}$

Solve the following equations for x.

17. $\sqrt{x - 2} = 9$

18. $\sqrt{3x + 4} + x = 8$

ANSWERS

19. ____________

20. ____________

21. ____________

22. ____________

23. ____________

24. ____________

25. ____________

Find length x in each triangle. Write the answer in simplified radical form.

19.

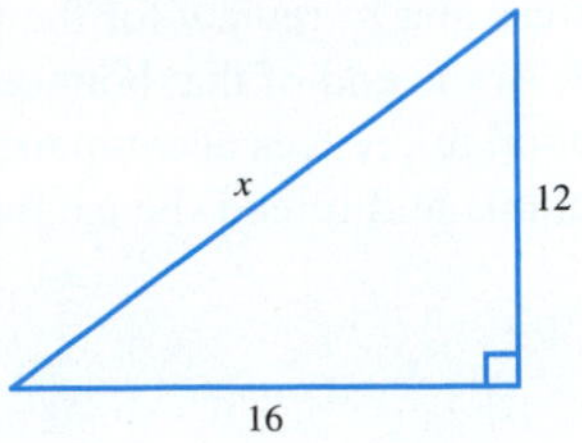

20.

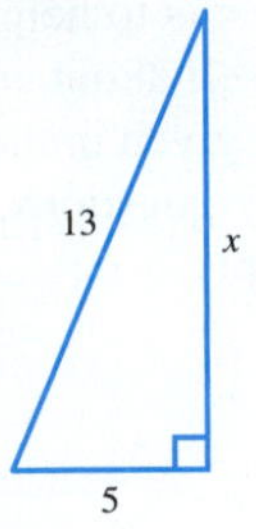

21.

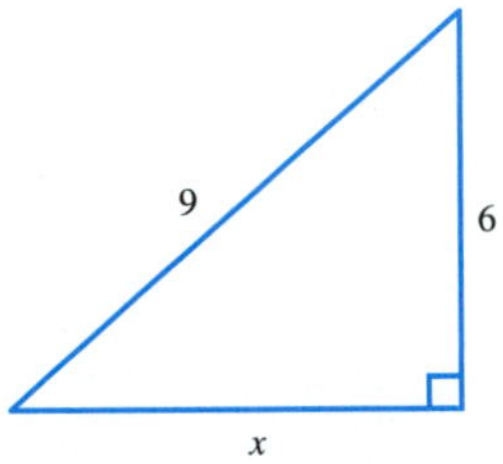

22.

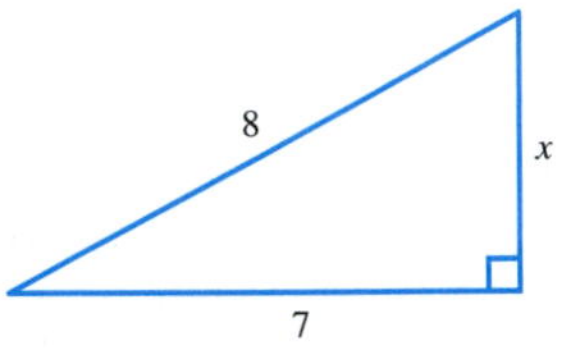

23. If the diagonal of a rectangle is 12 cm and the width of the rectangle is 7 cm, what is the length of the rectangle? Round to the nearest thousandth.

Find the distance between the two points.

24. $(-3, 7)$ and $(-12, 7)$

25. $(-2, 5)$ and $(-9, -1)$

ACTIVITY 7: THE SWING OF THE PENDULUM

Each activity in this text is designed to either enhance your understanding of the topics of the preceding chapter, provide you with a mathematical extension of those topics, or both. The activities can be undertaken by one student, but they are better suited for a small group project. Occasionally it is only through discussion that different facets of the activity become apparent. For material related to this activity, visit the text website at www.mhhe.com/streeter.

The action of a pendulum seems simple. Scientists have studied the characteristics of a swinging pendulum and found them to be quite useful. In 1851 in Paris, Jean Foucault (pronounced "Foo-koh") used a pendulum to clearly demonstrate the rotation of the Earth about its own axis.

A pendulum can be as simple as a string or cord with a weight fastened to one end. The other end is fixed, and the weight is allowed to swing. We define the **period** of a pendulum to be the amount of time required for the pendulum to make one complete swing (back and forth). The question we pose is: How does the *period* of a pendulum relate to the *length* of the pendulum?

For this activity, you need a piece of string that is approximately 1 m long. Fasten a weight (such as a small hexagonal nut) to one end, and then place clear marks on the string every 10 cm up to 70 cm, measured from the center of the weight.

1. Working with one or two partners, hold the string at the mark that is 10 cm from the weight. Pull the weight to the side with your other hand and let it swing freely. To estimate the period, let the weight swing through 30 periods, record the time in the given table, and then divide by 30. Round your result to the nearest hundredth of a second and record it. (*Note:* If you are unable to perform the experiment and collect your own data, you can use the sample data collected in this manner and presented at the end of this activity.)

    Repeat the described procedure for each length indicated in the given table.

Length of string, cm	10	20	30	40	50	60	70
Time for 30 periods, s							
Time for 1 period, s							

2. Let L represent the length of the pendulum and T represent the time period that results from swinging that pendulum. Fill out the table:

L	10	20	30	40	50	60	70
T							

3. Which variable, L or T, is viewed here as the independent variable?
4. On graph paper, draw horizontal and vertical axes, but plan to graph the data points in the first quadrant only. Explain why this is so.

5. With the independent variable marked on the horizontal axis, scale the axes appropriately, keeping an eye on your data.
6. Plot your data points. Should you connect them with a smooth curve?
7. What period T would correspond to a string length of 0? Include this point on your graph.
8. Use your graph to predict the period for a string length of 80 cm.
9. Verify your prediction by measuring the period when the string is held at 80 cm (as described in step 1). How close did your experimental estimate come to the prediction made in step 8?

You have created a graph showing T as a function of L. The shape of the graph may not be familiar to you yet. The shape of your pendulum graph fits that of a square root function.

Sample Data

Length of string, cm	10	20	30	40	50	60	70
Time for 30 periods, s	19	27	33	38	42	46	49

Name _____

Section _____ Date _____

The following exercises are presented to help you review concepts from earlier chapters. This is meant as a review and not as a comprehensive exam. The answers are presented in the back of the text. Section references accompany the answers. If you have difficulty with any of these exercises, be certain to at least read through the summary related to those sections.

Simplify the following expressions.

1. $8x^2y^3 - 5x^3y - 5x^2y^3 + 3x^3y$

2. $(4x^2 - 2x + 7) - (-3x^2 + 4x - 5)$

Evaluate each expression when $x = 2$, $y = -1$, and $z = -4$.

3. $2xyz^2 - 4x^2y^2z$

4. $-2xyz + 2x^2y^2$

Solve the following equations for x.

5. $-3x - 2(4 - 6x) = 10$

6. $5x - 3(4 - 2x) = 6(2x - 3)$

7. Solve the inequality $3x - 11 < 5x - 19$.

Perform the indicated operations.

8. $2x^2y(3x^2 - 5x + 19)$

9. $(5x + 3y)(4x - 7y)$

Factor each of the following completely.

10. $36xy - 27x^3y^2$

11. $8x^2 - 26x + 15$

Perform the indicated operations.

12. $\dfrac{2}{3x + 21} - \dfrac{3}{5x + 35}$

13. $\dfrac{x^2 - x - 6}{x^2 - x - 20} \div \dfrac{x^2 + x - 2}{x^2 + 3x - 4}$

Graph each of the following:

14. $4x + 5y = 20$

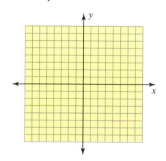

15. $5x - 4y \geq 20$

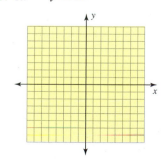

ANSWERS

1. _____
2. _____
3. _____
4. _____
5. _____
6. _____
7. _____
8. _____
9. _____
10. _____
11. _____
12. _____
13. _____
14. _____
15. _____

16. _____

17. _____

18. _____

19. _____

20. _____

21. _____

22. _____

23. _____

24. _____

25. _____

26. _____

27. _____

28. _____

29. _____

30. _____

16. Find the slope of the line through the points $(2, 9)$ and $(-1, -6)$.

17. Given that the slope of a line is $-\dfrac{3}{2}$ and the y-intercept is $(0, 5)$, write the equation of the line.

Solve each of the following systems. If a unique solution does not exist, state whether the system is inconsistent or dependent.

18. $4x - 5y = 20$
$\quad\;\; 2x + 3y = 10$

19. $4x + \;\;7y = 24$
$\quad\;\; 8x + 14y = 12$

Solve the following application. Be sure to show the system of equations used for your solution.

20. Amir was able to travel 80 mi downstream in 5 h. Returning upstream, he took 8 h to make the trip. How fast can he travel in still water, and what was the rate of the current?

Evaluate each root, if possible.

21. $\sqrt{144}$

22. $-\sqrt{144}$

23. $\sqrt{-144}$

24. $\sqrt[3]{-27}$

Simplify each of the following radical expressions.

25. $a\sqrt{20} - 2\sqrt{45a^2}$

26. $\dfrac{\sqrt{8x^3}}{\sqrt{3}}$

27. $\dfrac{12 - \sqrt{72}}{3}$

28. $\sqrt{98x^2}$

29. $\sqrt{150m^3 n^2}$

30. $\sqrt{\dfrac{12a^2}{25}}$

QUADRATIC EQUATIONS

INTRODUCTION

Large cities often commission fireworks artists to choreograph elaborate displays on holidays. Such displays look like beautiful paintings in the sky, in which the fireworks seem to dance to well-known popular and classical music. The displays are feats of engineering and very accurate timing. Suppose the designer wants a second set of rockets of a certain color and shape to be released after the first set reaches a specific height and explodes. He or she must know the strength of the initial liftoff and use a quadratic equation to determine the proper time for setting off the second round.

The equation $h = -16t^2 + 100t$ gives the height in feet t seconds after the rockets are shot into the air if the initial velocity is 100 feet per second. Using this equation, the designer knows how high the rocket will ascend and when it will begin to fall. The designer can time the next round to achieve the desired effect. Displays that involve large banks of fireworks in shows that last up to an hour are programmed using computers, but quadratic equations are at the heart of the mechanism that creates the beautiful effects.

Pre-Test Chapter 10

ANSWERS

1. _____

2. _____

3. _____

4. _____

5. _____

6. _____

7. _____

8. _____

9. _____

10. _____

11. _____

12. _____

13. _____

14. _____

15. _____

16. _____

17. _____

18. _____

This pre-test will provide a preview of the types of exercises you will encounter in each section of this chapter. The answers for these exercises can be found in the back of the text. If you are working on your own, or ahead of the class, this pre-test can help you find the sections in which you should focus more of your time.

Solve each of the equations for x.

[10.1] **1.** $x^2 = 17$ **2.** $x^2 - 12 = 0$

       3. $(x - 1)^2 = 5$ **4.** $9x^2 = 14$

Solve each of the equations by completing the square.

[10.2] **5.** $x^2 - 3x - 10 = 0$ **6.** $x^2 - 5x + 2 = 0$

       7. $x^2 - 4x - 4 = 0$ **8.** $2x^2 - 4x - 7 = 0$

Use the quadratic formula to solve each of the equations.

[10.3] **9.** $x^2 + 5x - 14 = 0$ **10.** $x^2 - 10x + 25 = 0$

      11. $x^2 + 3x = 5$ **12.** $2x^2 = 3x + 3$

      13. $3x = 4 + \dfrac{2}{x}$ **14.** $(x - 1)(x + 4) = 3$

Graph each quadratic equation after completing the given table of values.

[10.4] **15.** $y = x^2 + 2$ **16.** $y = x^2 + 4x$

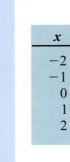

x	y
-2	
-1	
0	
1	
2	

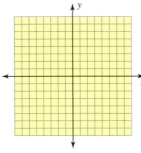

x	y
-4	
-3	
-2	
-1	
0	

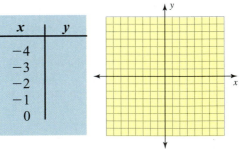

17. $y = x^2 + x - 6$ **18.** $y = -x^2 + 3$

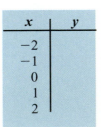

x	y
-2	
-1	
0	
1	
2	

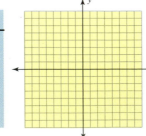

x	y
-2	
-1	
0	
1	
2	

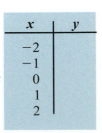

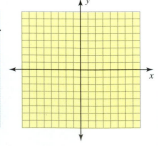

10.1 More on Quadratic Equations

OBJECTIVES

1. Solve equations of the form $ax^2 = k$
2. Solve equations of the form $(x - h)^2 = k$

We now have more tools for solving quadratic equations. In Sections 10.1 to 10.3 we will be using the ideas of Chapter 9 to extend our solution techniques.

In Section 4.8 we identified all equations of the form

$$ax^2 + bx + c = 0$$

as quadratic equations in standard form. In that section, we discussed solving these equations whenever the quadratic expression was factorable. In this chapter, we want to extend our equation-solving techniques so that we can find solutions for all such quadratic equations.

First, we review the factoring method of solution that we introduced in Chapter 4.

Example 1   Solving Quadratic Equations by Factoring

Solve each quadratic equation by factoring.

(a) $x^2 = -7x - 12$

First, we write the equation in standard form.

> **NOTE** Add $7x$ and 12 to both sides of the equation. The quadratic expression must be *set equal to 0*.

$$x^2 + 7x + 12 = 0$$

Once the equation is in standard form, we can factor the quadratic member.

$$(x + 3)(x + 4) = 0$$

Finally, using the zero-product rule, we solve the equations $x + 3 = 0$ and $x + 4 = 0$ as follows:

> **RECALL** These solutions can be checked by substitution into the original equation.

$$x = -3 \quad \text{or} \quad x = -4$$

(b) $x^2 = 16$

Again, we write the equation in standard form.

> **NOTE** Here we factor the quadratic member of the equation as a difference of squares.

$$x^2 - 16 = 0$$

Factoring, we have

$$(x + 4)(x - 4) = 0$$

Finally, the solutions are

$$x = -4 \quad \text{or} \quad x = 4$$

 CHECK YOURSELF 1

Solve each of the following quadratic equations.

(a) $x^2 - 4x = 45$ **(b)** $w^2 = 25$

Certain quadratic equations can be solved by other methods, such as the square root method. Return to the equation in part (b) of Example 1.

Beginning with

$$x^2 = 16$$

we can take the square root of each side, to write

$$\sqrt{x^2} = \sqrt{16}$$

From Section 9.1, we know that this is equivalent to

$$\sqrt{x^2} = 4$$

or

RECALL By definition $\sqrt{x^2} = |x|$

$$|x| = 4$$

Values for x of 4 or -4 will both satisfy this last equation, and so we have the two solutions

$$x = 4 \qquad \text{or} \qquad x = -4$$

We usually write the solutions as

NOTE $x = \pm 4$ is simply a convenient "shorthand" for indicating the two solutions, and we generally will go directly to this form.

$$x = \pm 4$$

Two more equations solved by this method are shown in Example 2.

OBJECTIVE 1

Example 2 Solving Equations by the Square Root Method

Solve each of the following equations by the square root method.

(a) $x^2 = 9$

By taking the square root of each side, we have

$$\sqrt{x^2} = \sqrt{9}$$
$$|x| = 3$$
$$x = \pm 3$$

(b) $x^2 = 5$

Again, we take the square root of each side to write our two solutions as

$$\sqrt{x^2} = \sqrt{5}$$
$$|x| = \sqrt{5}$$
$$x = \pm\sqrt{5}$$

CHECK YOURSELF 2

Solve.

(a) $x^2 = 100$ **(b)** $t^2 = 15$

You may have to add or subtract on both sides of the equation to write an equation in the form of those in Example 2, as Example 3 illustrates.

Example 3   Solving Equations by the Square Root Method

Solve $x^2 - 8 = 0$.

    First, add 8 to both sides of the equation. We have

$$x^2 = 8$$

Now take the square root of both sides.

$$x = \pm\sqrt{8}$$

RECALL

$\sqrt{8} = \sqrt{4 \cdot 2}$

$\phantom{\sqrt{8}} = \sqrt{4} \cdot \sqrt{2}$

$\phantom{\sqrt{8}} = 2\sqrt{2}$

Normally, the solutions are written in the simplest form. In this case we have

$$x = \pm 2\sqrt{2}$$

CHECK YOURSELF 3 _____

Solve.

(a) $x^2 - 18 = 0$ **(b)** $x^2 + 1 = 7$

NOTE In the form

$ax^2 = k$

a is the coefficient of x^2 and *k* is some number.

    To solve a quadratic equation of the form $ax^2 = k$, divide both sides of the equation by *a* as the first step. This is shown in Example 4.

Example 4   Solving Equations by the Square Root Method

Solve $4x^2 = 3$.

    Divide both sides of the equation by 4.

$$x^2 = \frac{3}{4}$$

Now take the square root of both sides.

$$x = \pm\sqrt{\frac{3}{4}}$$

RECALL

$\sqrt{\dfrac{3}{4}} = \dfrac{\sqrt{3}}{\sqrt{4}}$

$\phantom{\sqrt{\dfrac{3}{4}}} = \dfrac{\sqrt{3}}{2}$

Again write your result in the simplest form, so

$$x = \pm\frac{\sqrt{3}}{2}$$

CHECK YOURSELF 4 _____

Solve $9x^2 = 5$.

    Equations of the form $(x - h)^2 = k$ can also be solved by taking the square root of both sides. Consider Example 5.

OBJECTIVE 2   **Example 5   Solving Equations by the Square Root Method**

Solve $(x - 1)^2 = 6$.

    Again, take the square root of both sides of the equation.

$$x - 1 = \pm\sqrt{6}$$

Now add 1 to both sides of the equation to isolate x.

$$x = 1 \pm \sqrt{6}$$

 CHECK YOURSELF 5 _____

Solve $(x + 2)^2 = 12$.

Equations of the form $a(x - h)^2 = k$ can also be solved if each side of the equation is divided by a first, as shown in Example 6.

Example 6   Solving Equations by the Square Root Method

Solve $3(x - 2)^2 = 5$.

$$(x - 2)^2 = \frac{5}{3}$$

NOTE

$$\sqrt{\frac{5}{3}} = \frac{\sqrt{5}}{\sqrt{3}} \cdot \frac{\sqrt{3}}{\sqrt{3}} = \frac{\sqrt{15}}{3}$$

$$x - 2 = \pm\sqrt{\frac{5}{3}} = \frac{\pm\sqrt{15}}{3}$$

$$x = 2 \pm \frac{\sqrt{15}}{3}$$

$$x = \frac{6}{3} \pm \frac{\sqrt{15}}{3}$$

$$x = \frac{6 \pm \sqrt{15}}{3}$$

CHECK YOURSELF 6 _____

Solve $5(x + 3)^2 = 2$.

What about an equation such as the following?

$$x^2 + 5 = 0$$

If we apply the methods of Examples 4–6, we first subtract 5 from both sides, to write

$$x^2 = -5$$

Taking the square root of both sides gives

$$x = \pm\sqrt{-5}$$

But we know there are no square roots of -5 in the real numbers, so this equation has *no real number solutions.* You might work with this type of equation in your next algebra course.

CHECK YOURSELF ANSWERS _____

1. (a) $-5, 9$; **(b)** $-5, 5$ **2. (a)** ± 10; **(b)** $\pm\sqrt{15}$ **3. (a)** $\pm 3\sqrt{2}$; **(b)** $\pm\sqrt{6}$

4. $\pm\dfrac{\sqrt{5}}{3}$ **5.** $-2 \pm 2\sqrt{3}$ **6.** $\dfrac{-15 \pm \sqrt{10}}{5}$

10.1 Exercises

Solve each of the equations for x.

1. $x^2 = 5$

2. $x^2 = 15$

3. $x^2 = 33$

4. $x^2 = 43$

5. $x^2 - 7 = 0$

6. $x^2 - 13 = 0$

7. $x^2 - 20 = 0$

8. $x^2 = 28$

9. $x^2 = 40$

10. $x^2 - 54 = 0$

11. $x^2 + 3 = 12$

12. $x^2 - 7 = 18$

13. $x^2 + 5 = 8$

14. $x^2 - 4 = 17$

15. $x^2 - 2 = 16$

16. $x^2 + 6 = 30$

17. $9x^2 = 25$

18. $16x^2 = 9$

19. $49x^2 = 11$

20. $16x^2 = 3$

21. $4x^2 = 7$

22. $25x^2 = 13$

Solve each equation for x.

23. $(x - 1)^2 = 5$

24. $(x - 3)^2 = 10$

25. $(x + 1)^2 = 12$

26. $(x + 2)^2 = 32$

27. $(x - 3)^2 = 24$

28. $(x - 5)^2 = 27$

29. $(x + 5)^2 = 25$

30. $(x + 2)^2 = 16$

31. $3(x - 5)^2 = 7$

32. $2(x - 5)^2 = 3$

ANSWERS

1. _____   2. _____
3. _____   4. _____
5. _____   6. _____
7. _____   8. _____
9. _____   10. _____
11. _____ 12. _____
13. _____ 14. _____
15. _____ 16. _____
17. _____ 18. _____
19. _____
20. _____
21. _____
22. _____
23. _____
24. _____
25. _____
26. _____
27. _____
28. _____
29. _____
30. _____
31. _____
32. _____

ANSWERS

33.

_____

34.

_____

35.

_____

36.

_____

37.

_____

38.

_____

39.

_____

40.

_____

41.

_____

42.

_____

43.

_____

44.

_____

45.

_____

46.

_____

47.

_____

48.

_____

33. $4(x + 5)^2 = 9$

34. $16(x + 2)^2 = 25$

35. $-2(x + 2)^2 = -6$

36. $-5(x + 4)^2 = -10$

37. $-4(x - 2)^2 = -5$

38. $-9(x - 2)^2 = -11$

For exercises 39 and 40, fill in the blanks with *always, sometimes,* or *never.*

39. An equation of the form $(x - h)^2 = k$, where k is positive, _____ has two distinct solutions.

40. An equation of the form $x^2 = k$ _____ has real number solutions.

Solve each equation for x.

41. $x^2 - 2x + 1 = 7$
(*Hint:* Factor the left-hand side.)

42. $x^2 + 4x + 4 = 7$
(*Hint:* Factor the left-hand side.)

43. $(2x + 11)^2 + 9 = 0$

44. $(3x + 14)^2 + 25 = 0$

45. Number problems. The square of a number decreased by 2 is equal to the negative of the number. Find the numbers.

46. Number problems. The square of 2 more than a number is 64. Find the numbers.

47. Business and finance. The revenue (in dollars) for selling x units of a product is given by

$$R = x\left(5 - \frac{1}{10}x\right) \qquad 0 < x < 50$$

Determine the number of units that must be sold if the revenue is to be $60.

48. Number problems. The square of the sum of a number and 5 is 36. Find the numbers.

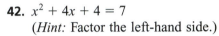

818 SECTION 10.1

49. In this section, you solved quadratic equations by "extracting roots," taking the square root of both sides after writing one side as the square of a binomial. But what if the algebraic expression cannot be written this way? Work with another student to decide what needs to be added to each of the following expressions to make it a "perfect square trinomial." Label the dimensions of the squares and the area of each section.

(a)

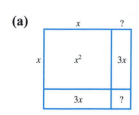

$$x^2 + 6x + \underline{\quad} = (x + ?)^2$$

(b)

$$n^2 + 10n + \underline{\quad} = (n + ?)^2$$

(c)

$$a^2 + a + \underline{\quad} = (a + ?)^2$$

(d)

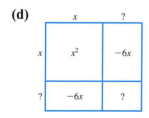

$$x^2 - 12x + \underline{\quad} = (x - ?)^2$$

(e) $x^2 + 20x + \underline{\quad} = (x + ?)^2$ **(f)** $n^2 - 16n + \underline{\quad} = (n - ?)^2$

Getting Ready for Section 10.2 [Section 3.5]

Multiply each of the following expressions.

(a) $(x + 1)^2$ (b) $(x + 5)^2$

(c) $(x - 2)^2$ (d) $(x - 7)^2$

(e) $(x + 4)^2$ (f) $(x - 3)^2$

(g) $(2x + 5)^2$ (h) $(2x - 1)^2$

Answers

1. $\pm\sqrt{5}$　　**3.** $\pm\sqrt{33}$　　**5.** $\pm\sqrt{7}$　　**7.** $\pm 2\sqrt{5}$　　**9.** $\pm 2\sqrt{10}$

11. ± 3　　**13.** $\pm\sqrt{3}$　　**15.** $\pm 3\sqrt{2}$　　**17.** $\pm\dfrac{5}{3}$　　**19.** $\dfrac{\pm\sqrt{11}}{7}$

21. $\dfrac{\pm\sqrt{7}}{2}$　　**23.** $1\pm\sqrt{5}$　　**25.** $-1\pm 2\sqrt{3}$　　**27.** $3\pm 2\sqrt{6}$

29. $-10, 0$　　**31.** $\dfrac{15\pm\sqrt{21}}{3}$　　**33.** $-\dfrac{13}{2}, -\dfrac{7}{2}$　　**35.** $-2\pm\sqrt{3}$

37. $\dfrac{4\pm\sqrt{5}}{2}$　　**39.** Always　　**41.** $1\pm\sqrt{7}$　　**43.** No real number solutions

45. $1, -2$　　**47.** $30, 20$　　**49.** **(a)** $9, 3$; **(b)** $25, 5$; **(c)** $\dfrac{1}{4}, \dfrac{1}{2}$; **(d)** $36, 6$;

(e) $100, 10$; **(f)** $64, 8$　　**a.** $x^2 + 2x + 1$　　**b.** $x^2 + 10x + 25$
c. $x^2 - 4x + 4$　　**d.** $x^2 - 14x + 49$　　**e.** $x^2 + 8x + 16$　　**f.** $x^2 - 6x + 9$
g. $4x^2 + 20x + 25$　　**h.** $4x^2 - 4x + 1$

10.2 Completing the Square

We can solve a quadratic equation such as

$$x^2 - 2x + 1 = 5$$

very easily if we notice that the expression on the left is a perfect-square trinomial. Factoring, we have

$$(x - 1)^2 = 5$$

so

$$x - 1 = \pm\sqrt{5} \qquad \text{or} \qquad x = 1 \pm \sqrt{5}$$

The solutions for the original equation are then $1 + \sqrt{5}$ and $1 - \sqrt{5}$.

It is true that every quadratic equation can be written in the form $x = h \pm \sqrt{k}$ (with a perfect-square trinomial on the left). That is the basis for the **completing-the-square method** for solving quadratic equations.

First, let's look at two perfect-square trinomials.

$$x^2 + 6x + \quad 9 = (x + 3)^2$$
$$x^2 - 8x + 16 = (x - 4)^2$$

There is an important relationship between the coefficient of the middle term (the x-term) and the constant.

In the first equation,

$$\left(\frac{1}{2} \cdot 6\right)^2 = 3^2 = 9$$

The x-coefficient   The constant

In the second equation,

$$\left[\frac{1}{2}(-8)\right]^2 = (-4)^2 = 16$$

The x-coefficient   The constant

It is always true that, in a perfect-square trinomial with a coefficient of 1 for x^2, the square of one-half of the x-coefficient is equal to the constant term.

OBJECTIVE 1

> ### Example 1 Completing the Square

(a) Find the term that should be added to $x^2 + 4x$ so that the expression is a perfect-square trinomial.

To complete the square of $x^2 + 4x$, add the square of one-half of 4 (the x-coefficient).

NOTE The coefficient of x^2 must be 1 before the added term is found.

$$x^2 + 4x + \left(\frac{1}{2} \cdot 4\right)^2 \qquad \text{or} \qquad x^2 + 4x + 2^2 \qquad \text{or} \qquad x^2 + 4x + 4$$

The trinomial $x^2 + 4x + 4$ is a perfect square because

$$x^2 + 4x + 4 = (x + 2)^2$$

(b) Find the term that should be added to $x^2 - 10x$ so that the expression is a perfect-square trinomial.

To complete the square of $x^2 - 10x$, add the square of one-half of -10 (the x-coefficient).

$$x^2 - 10x + \left[\frac{1}{2}(-10)\right]^2 \quad \text{or} \quad x^2 - 10x + (-5)^2 \quad \text{or} \quad x^2 - 10x + 25$$

Check for yourself, by factoring, that this is a perfect-square trinomial.

CHECK YOURSELF 1 _____

Complete the square and factor.

(a) $x^2 + 2x$ **(b)** $x^2 - 12x$

We can now use the process of Example 1 along with the solution methods of Section 10.1 to solve a quadratic equation.

OBJECTIVE 2   **Example 2   Solving a Quadratic Equation by Completing the Square**

Solve $x^2 + 4x - 2 = 0$ by completing the square.

NOTE Add 2 to both sides to remove -2 from the left side.

$$x^2 + 4x = 2$$

We find the term needed to complete the square by squaring one-half of the x-coefficient.

$$\left(\frac{1}{2} \cdot 4\right)^2 = 2^2 = 4$$

We now add 4 to both sides of the equation.

NOTE This *completes the square* on the left.

$$x^2 + 4x + 4 = 2 + 4$$

Now factor on the left and simplify on the right.

$$(x + 2)^2 = 6$$

Now solving as before, we have

$$x + 2 = \pm\sqrt{6}$$
$$x = -2 \pm \sqrt{6}$$

CHECK YOURSELF 2 _____

Solve by completing the square.

$$x^2 + 6x - 4 = 0$$

To complete-the-square in this manner, the coefficient of x^2 must be 1. Example 3 illustrates the solution process when the coefficient of x^2 is not equal to 1.

Example 3   Solving a Quadratic Equation by Completing the Square

Solve $2x^2 - 4x - 5 = 0$ by completing the square.

$$2x^2 - 4x - 5 = 0$$ Add 5 to both sides.

$$2x^2 - 4x = 5$$ Because the coefficient of x^2 is not 1 (here it is 2), divide every term by 2. This will make the new leading coefficient equal to 1.

$$x^2 - 2x = \frac{5}{2}$$

$$x^2 - 2x + 1 = \frac{5}{2} + 1$$ Complete the square and solve as before.

$$(x - 1)^2 = \frac{7}{2}$$

$$x - 1 = \pm\sqrt{\frac{7}{2}}$$

$$x - 1 = \pm\frac{\sqrt{14}}{2}$$ Simplify the radical on the right.

$$x = 1 \pm \frac{\sqrt{14}}{2}$$

NOTE

$$\sqrt{\frac{7}{2}} = \frac{\sqrt{7}}{\sqrt{2}} \cdot \frac{\sqrt{2}}{\sqrt{2}}$$
$$= \frac{\sqrt{14}}{2}$$

or

NOTE We have combined the terms on the right with the common denominator of 2.

$$x = \frac{2 \pm \sqrt{14}}{2}$$

CHECK YOURSELF 3

Solve by completing the square.

$$3x^2 - 6x + 2 = 0$$

The completing-the-square method is easiest to use when the coefficient of x is even. If it is odd, the method still works, but it will definitely involve fractions. Consider Example 4.

Example 4   Solving a Quadratic Equation by Completing the Square

Solve $2x^2 - 6x - 9 = 0$ by completing the square.

$$2x^2 - 6x - 9 = 0$$ Add 9 to both sides.
$$2x^2 - 6x = 9$$ Divide every term by 2.
$$x^2 - 3x = \frac{9}{2}$$

Be careful here. $\left[\dfrac{1}{2}(-3)\right]^2 = \left(-\dfrac{3}{2}\right)^2 = \dfrac{9}{4}$. So add $\dfrac{9}{4}$ to both sides.

$$x^2 - 3x + \frac{9}{4} = \frac{9}{2} + \frac{9}{4}$$

To factor the trinomial, remember how we got $\dfrac{9}{4}$: by squaring $-\dfrac{3}{2}$.

$$\left(x - \frac{3}{2}\right)^2 = \frac{27}{4}$$

$$x - \frac{3}{2} = \pm\sqrt{\frac{27}{4}}$$

$$x - \frac{3}{2} = \pm\frac{\sqrt{27}}{2}$$

$$x - \frac{3}{2} = \pm\frac{3\sqrt{3}}{2}$$

$$x = \frac{3}{2} + \frac{3\sqrt{3}}{2}$$

or

$$x = \frac{3 \pm 3\sqrt{3}}{2}$$

 CHECK YOURSELF 4 _____

Solve by completing the square.

$2x^2 + 10x - 3 = 0$

We summarize the steps used to solve a quadratic equation by completing the square.

Step by Step: Solving a Quadratic Equation by Completing the Square

Step 1   Write the equation in the form

$ax^2 + bx = k$

so that the variable terms are on the left side and the constant is on the right side.

Step 2   If the coefficient of x^2 is not 1, divide both sides of the equation by that coefficient.

Step 3   Add the square of one-half the coefficient of x to both sides of the equation.

Step 4   The left side of the equation is now a perfect-square trinomial. Factor and solve as before.

CHECK YOURSELF ANSWERS _____

1. **(a)** $x^2 + 2x + 1 = (x + 1)^2$; **(b)** $x^2 - 12x + 36 = (x - 6)^2$

2. $-3 \pm \sqrt{13}$ 3. $\dfrac{3 \pm \sqrt{3}}{3}$ 4. $\dfrac{-5 \pm \sqrt{31}}{2}$

10.2 Exercises

Determine whether each of the following trinomials is a perfect square.

1. $x^2 - 14x + 49$

2. $x^2 + 9x + 16$

3. $x^2 - 18x - 81$

4. $x^2 + 10x + 25$

5. $x^2 - 18x + 81$

6. $x^2 - 24x + 48$

Find the constant term that should be added to make each of the following expressions a perfect-square trinomial.

7. $x^2 + 6x$

8. $x^2 - 8x$

9. $x^2 - 10x$

10. $x^2 + 5x$

11. $x^2 + 9x$

12. $x^2 - 20x$

Solve each of the following quadratic equations by completing the square.

13. $x^2 + 4x - 12 = 0$

14. $x^2 - 6x + 8 = 0$

15. $x^2 - 2x - 5 = 0$

16. $x^2 + 4x - 7 = 0$

17. $x^2 + 3x - 27 = 0$

18. $x^2 + 5x - 3 = 0$

19. $x^2 + 6x - 1 = 0$

20. $x^2 + 4x - 4 = 0$

21. $x^2 - 5x + 6 = 0$

22. $x^2 - 6x - 3 = 0$

23. $x^2 + 6x - 5 = 0$

24. $x^2 - 2x = 1$

25. $x^2 = 9x + 5$

26. $x^2 = 4 - 7x$

27. $2x^2 - 6x + 1 = 0$

28. $2x^2 + 10x + 11 = 0$

29. $2x^2 - 4x + 1 = 0$

30. $2x^2 - 8x + 5 = 0$

31. $4x^2 - 2x - 1 = 0$

32. $3x^2 - x - 2 = 0$

ANSWERS

1. ____ 2. ____ 3. ____
4. ____ 5. ____ 6. ____
7. ____ 8. ____ 9. ____
10. ____ 11. ____
12. ____ 13. ____
14. ____ 15. ____
16. ____
17. ____
18. ____
19. ____
20. ____
21. ____ 22. ____
23. ____
24. ____
25. ____
26. ____
27. ____
28. ____
29. ____
30. ____
31. ____
32. ____

33. _____

34. _____

35. _____

36. _____

37. _____

38. _____

  a. _____

  b. _____

  c. _____

  d. _____

  e. _____

  f. _____

For exercises 33 and 34, fill in the blanks with *always, sometimes,* or *never.*

33. When completing the square for $x^2 + bx$, the number added is _____ negative.

34. A quadratic equation can _____ be solved by completing the square.

Solve the following problems.

35. Number problems. If the square of 3 more than a number is 9, find the number(s).

36. Number problems. If the square of 2 less than an integer is 16, find the number(s).

37. Business and finance. The revenue for selling x units of a product is given by $R = x\left(25 - \dfrac{1}{2}x\right)$. Find the number of units sold if the revenue is \$294.50.

38. Number problems. Find two consecutive positive integers such that the sum of their squares is 85.

Getting Ready for Section 10.3 [Section 1.5]

Evaluate the expression $b^2 - 4ac$ for each set of values.

(a) $a = 1, b = 1, c = -3$ (b) $a = 1, b = -1, c = -1$

(c) $a = 1, b = -8, c = -3$ (d) $a = 1, b = -2, c = -1$

(e) $a = -2, b = 4, c = -2$ (f) $a = 2, b = -3, c = 4$

Answers

1. Yes   **3.** No   **5.** Yes   **7.** 9   **9.** 25   **11.** $\dfrac{81}{4}$ **13.** $-6, 2$

15. $1 \pm \sqrt{6}$ **17.** $\dfrac{-3 \pm 3\sqrt{13}}{2}$ **19.** $-3 \pm \sqrt{10}$ **21.** 2, 3

23. $-3 \pm \sqrt{14}$ **25.** $\dfrac{9 \pm \sqrt{101}}{2}$ **27.** $\dfrac{3 \pm \sqrt{7}}{2}$ **29.** $\dfrac{2 \pm \sqrt{2}}{2}$

31. $\dfrac{1 \pm \sqrt{5}}{4}$ **33.** Never   **35.** $-6, 0$ **37.** 19, 31   **a.** 13

b. 5   **c.** 76   **d.** 8   **e.** 0   **f.** -23

10.3 The Quadratic Formula

We are now ready to derive and use the **quadratic formula,** which will allow us to solve all quadratic equations. We derive the formula by using the method of completing the square.

To use the quadratic formula, the quadratic equation you want to solve must be in *standard form.* That form is

$$ax^2 + bx + c = 0 \qquad \text{in which } a \neq 0$$

OBJECTIVE 1 | **Example 1** **Writing Quadratic Equations in Standard Form**

Write each equation in standard form.

(a) $2x^2 - 5x + 3 = 0$

The equation is already in standard form.

$$a = 2 \qquad b = -5 \qquad \text{and} \qquad c = 3$$

(b) $5x^2 + 3x = 5$

The equation is *not* in standard form. Rewrite it by adding -5 to both sides.

$$5x^2 + 3x - 5 = 0 \qquad \text{Standard form}$$

$$a = 5 \qquad b = 3 \qquad \text{and} \qquad c = -5$$

 CHECK YOURSELF 1 _____

Rewrite each quadratic equation in standard form.

(a) $x^2 - 3x = 5$ **(b)** $3x^2 = 7 - 2x$

Once a quadratic equation is written in standard form, we will be able to find any solutions to the equation. Remember that a solution is a value for x that will make the equation true.

What follows is the derivation of the quadratic formula, which can be used to solve quadratic equations.

Step by Step: Deriving the Quadratic Formula

Let $ax^2 + bx + c = 0$, in which $a \neq 0$.

$$ax^2 + bx = -c$$ Subtract c from both sides.

$$x^2 + \frac{b}{a}x = -\frac{c}{a}$$ Divide both sides by a.

$$x^2 + \frac{b}{a}x + \frac{b^2}{4a^2} = \frac{b^2}{4a^2} - \frac{c}{a}$$ Add $\frac{b^2}{4a^2}$ to both sides.

$$\left(x + \frac{b}{2a}\right)^2 = \frac{b^2 - 4ac}{4a^2}$$ Factor on the left and add the fractions on the right.

$$x + \frac{b}{2a} = \pm\sqrt{\frac{b^2 - 4ac}{4a^2}}$$ Take the square root of both sides.

$$x + \frac{b}{2a} = \pm\frac{\sqrt{b^2 - 4ac}}{2a}$$ Simplify the radical on the right.

$$x = -\frac{b}{2a} \pm \frac{\sqrt{b^2 - 4ac}}{2a}$$ Add $-\frac{b}{2a}$ to both sides.

$$x = \frac{-b \pm \sqrt{b^2 - 4ac}}{2a}$$ Use the common denominator, $2a$.

NOTE This is the completing-the-square step that makes the left-hand side a perfect square.

Definition: The Quadratic Formula

If $ax^2 + bx + c = 0$, then

$$x = \frac{-b \pm \sqrt{b^2 - 4ac}}{2a}$$

Let's use the quadratic formula to solve some equations.

OBJECTIVE 2   **Example 2   Using the Quadratic Formula to Solve an Equation**

Use the quadratic formula to solve $x^2 - 5x + 4 = 0$.
   The equation is in standard form, so first identify a, b, and c.

NOTE The leading coefficient is 1, so $a = 1$.

$$x^2 - 5x + 4 = 0$$

$a = 1 \quad b = -5 \quad c = 4$

We now substitute the values for a, b, and c into the formula.

$$x = \frac{-b \pm \sqrt{b^2 - 4ac}}{2a}$$

NOTE Simplify the expression.

$$= \frac{-(-5) \pm \sqrt{(-5)^2 - 4(1)(4)}}{2(1)}$$

$$= \frac{5 \pm \sqrt{25 - 16}}{2}$$

$$= \frac{5 \pm \sqrt{9}}{2}$$

$$= \frac{5 \pm 3}{2}$$

NOTE These results could also have been found by factoring the original equation. You should check that for yourself.

Now,

$$x = \frac{5 + 3}{2} \quad \text{or} \quad x = \frac{5 - 3}{2}$$

$$= 4 \qquad\qquad\qquad = 1$$

The solutions are 4 and 1.

 CHECK YOURSELF 2 _____

Use the quadratic formula to solve $x^2 - 2x - 8 = 0$. Check your result by factoring.

The main use of the quadratic formula is to solve equations that *cannot* be factored.

Example 3   Using the Quadratic Formula to Solve an Equation

Use the quadratic formula to solve $2x^2 = x + 4$.

First, the equation *must be written* in standard form to find a, b, and c.

$$2x^2 - x - 4 = 0$$

$$a = 2 \qquad b = -1 \qquad c = -4$$

$$x = \frac{-b \pm \sqrt{b^2 - 4ac}}{2a}$$

NOTE Substitute the values for a, b, and c into the formula.

$$= \frac{-(-1) \pm \sqrt{(-1)^2 - 4(2)(-4)}}{2(2)}$$

$$= \frac{1 \pm \sqrt{1 + 32}}{4}$$

$$= \frac{1 \pm \sqrt{33}}{4}$$

 CHECK YOURSELF 3 _____

Use the quadratic formula to solve $3x^2 = 3x + 4$.

Example 4   Using the Quadratic Formula to Solve an Equation

Use the quadratic formula to solve $x^2 - 2x = 4$.

In standard form, the equation is

$$x^2 - 2x - 4 = 0$$

$$a = 1 \qquad b = -2 \qquad c = -4$$

NOTE Again substitute the values into the quadratic formula.

$$x = \frac{-(-2) \pm \sqrt{(-2)^2 - 4(1)(-4)}}{2(1)}$$

$$= \frac{2 \pm \sqrt{20}}{2}$$

NOTE 20 has a perfect-square factor,
$$\sqrt{20} = \sqrt{4 \cdot 5}$$
$$= 2\sqrt{5}$$

Unless you are finding a decimal approximation, you should always write your solution in simplest form.

$$x = \frac{2 \pm 2\sqrt{5}}{2}$$

$$= \frac{2(1 \pm \sqrt{5})}{2} \qquad \text{Now factor the numerator and divide by the common factor 2.}$$

$$= 1 \pm \sqrt{5}$$

 CHECK YOURSELF 4

Use the quadratic formula to solve $3x^2 = 2x + 4$.

Sometimes equations have common factors. Factoring first simplifies these equations, making them easier to solve. This is illustrated in Example 5.

Example 5   Using the Quadratic Formula to Solve an Equation

Use the quadratic formula to solve $3x^2 - 6x - 3 = 0$.
Because the equation is in standard form, we could use

$$a = 3 \qquad b = -6 \qquad \text{and} \qquad c = -3$$

in the quadratic formula. There is, however, a better approach.
Note the common factor of 3 in the quadratic member of the original equation. Factoring, we have

$$3(x^2 - 2x - 1) = 0$$

and dividing both sides of the equation by 3 gives

$$x^2 - 2x - 1 = 0$$

NOTE The advantage to this approach is that these values will require much less simplification after we substitute into the quadratic formula.

Now let $a = 1$, $b = -2$, and $c = -1$. Then

$$x = \frac{-(-2) \pm \sqrt{(-2)^2 - 4(1)(-1)}}{2 \cdot 1}$$

$$= \frac{2 \pm \sqrt{8}}{2}$$

$$= \frac{2 \pm 2\sqrt{2}}{2}$$

$$= \frac{2(1 \pm \sqrt{2})}{2}$$

$$= 1 \pm \sqrt{2}$$

 CHECK YOURSELF 5

Use the quadratic formula to solve $4x^2 - 20x = 12$.

In applications that lead to quadratic equations, you may want to find approximate values for the solutions.

Example 6   Using the Quadratic Formula to Solve an Equation

Use the quadratic formula to solve $x^2 - 5x + 5 = 0$ and write your solutions in approximate decimal form.

Substituting $a = 1$, $b = -5$, and $c = 5$ gives

$$x = \frac{-(-5) \pm \sqrt{(-5)^2 - 4(1)(5)}}{2(1)}$$

$$= \frac{5 \pm \sqrt{5}}{2}$$

Use your calculator to find $\sqrt{5} \approx 2.236$, so

$$x \approx \frac{5 + 2.236}{2} \qquad \text{or} \qquad x \approx \frac{5 - 2.236}{2}$$

$$= \frac{7.236}{2} \qquad\qquad\qquad = \frac{2.764}{2}$$

$$= 3.618 \qquad\qquad\qquad\quad = 1.382$$

CHECK YOURSELF 6

Use the quadratic formula to solve $x^2 - 3x - 5 = 0$ and approximate the solutions in decimal form to the thousandth.

You may be wondering whether the quadratic formula can be used to solve all quadratic equations. It can, but not all quadratic equations will have real number solutions, as Example 7 shows.

Example 7   Using the Quadratic Formula to Solve an Equation

NOTE Make sure the quadratic equation is in standard form. $x^2 - 3x = -5$ is equivalent to $x^2 - 3x + 5 = 0$.

Use the quadratic formula to solve $x^2 - 3x = -5$.

Substituting $a = 1$, $b = -3$, and $c = 5$, we have

$$x = \frac{-(-3) \pm \sqrt{(-3)^2 - 4(1)(5)}}{2(1)}$$

$$= \frac{3 \pm \sqrt{-11}}{2}$$

In this case, there are no real number solutions because the radicand is negative.

CHECK YOURSELF 7

Use the quadratic formula to solve $x^2 - 3x = -3$.

We summarize the steps used for solving equations by the quadratic formula.

Step by Step: Solving Equations with the Quadratic Formula

Step 1   Rewrite the equation in standard form.

$$ax^2 + bx + c = 0$$

Step 2   If a common factor exists, divide both sides of the equation by that common factor.

Step 3   Identify the coefficients a, b, and c.

Step 4   Substitute values for a, b, and c into the formula

$$x = \frac{-b \pm \sqrt{b^2 - 4ac}}{2a}$$

Step 5   Simplify the right side of the expression formed in step 4 to write the solutions for the original equation.

Often, applied problems will lead to quadratic equations that must be solved by the methods of Section 10.2 or this section.

CHECK YOURSELF ANSWERS

1. **(a)** $x^2 - 3x - 5 = 0$; **(b)** $3x^2 + 2x - 7 = 0$

2. $x = 4, -2$ 3. $x = \dfrac{3 \pm \sqrt{57}}{6}$ 4. $x = \dfrac{1 \pm \sqrt{13}}{3}$

5. $x = \dfrac{5 \pm \sqrt{37}}{2}$ 6. $x \approx 4.193$ or -1.193 7. $\dfrac{3 \pm \sqrt{-3}}{2}$, no real solutions

10.3 Exercises

Use the quadratic formula to solve each of the following quadratic equations.

1. $x^2 + 9x + 20 = 0$

2. $x^2 - 9x + 14 = 0$

3. $x^2 - 4x + 3 = 0$

4. $x^2 - 13x + 22 = 0$

5. $3x^2 + 2x - 1 = 0$

6. $x^2 - 8x + 16 = 0$

7. $x^2 + 5x = -4$

8. $4x^2 + 5x = 6$

9. $x^2 = 6x - 9$

10. $2x^2 - 5x = 3$

11. $2x^2 - 3x - 7 = 0$

12. $x^2 - 5x + 2 = 0$

13. $x^2 + 2x - 4 = 0$

14. $x^2 - 4x + 2 = 0$

15. $2x^2 - 3x = 3$

16. $3x^2 - 2x + 1 = 0$

17. $3x^2 - 2x = 6$

18. $4x^2 = 4x + 5$

19. $3x^2 + 3x + 2 = 0$

20. $2x^2 - 3x = 6$

21. $5x^2 = 8x - 2$

22. $5x^2 - 2 = 2x$

23. $2x^2 - 9 = 4x$

24. $3x^2 - 6x = 2$

ANSWERS

1. _____ 2. _____

3. _____ 4. _____

5. _____ 6. _____

7. _____

8. _____

9. _____

10. _____

11. _____

12. _____

13. _____

14. _____

15. _____

16. _____

17. _____

18. _____

19. _____

20. _____

21. _____

22. _____

23. _____

24. _____

25. $3x - 5 = \dfrac{1}{x}$ **26.** $x + 3 = \dfrac{1}{x}$

27. $(x - 2)(x + 1) = 3$ **28.** $(x - 3)(x + 2) = 5$

Solve the following quadratic equations by factoring or by any of the techniques of this chapter.

29. $(x - 1)^2 = 7$ **30.** $(2x + 3)^2 = 5$

31. $x^2 - 5x - 14 = 0$ **32.** $9x^2 + 6x - 3 = 0$

33. $6x^2 - 23x + 10 = 0$ **34.** $x^2 + 7x - 18 = 0$

35. $2x^2 - 8x + 3 = 0$ **36.** $x^2 + 2x - 1 = 0$

37. $x^2 - 9x - 4 = 6$ **38.** $5x^2 + 10x + 2 = 2$

39. $4x^2 - 8x + 3 = 5$ **40.** $x^2 + 4x = 21$

Solve the following equations.

41. $\dfrac{3}{x} + \dfrac{5}{x^2} = 9$ **42.** $\dfrac{8}{x} - \dfrac{3}{x^2} = -6$

43. $\dfrac{x}{x + 1} + \dfrac{10x}{x^2 + 4x + 3} = \dfrac{15}{x + 3}$ **44.** $x - \dfrac{9x}{x - 2} = \dfrac{-10}{x - 2}$

For exercises 45 to 47, fill in the blanks with *always, sometimes,* or *never.*

45. The quadratic formula can _____ be used to solve a quadratic equation.

46. If the value of $b^2 - 4ac$ is negative, the equation $ax^2 + bx + c = 0$ _____ has real number solutions.

47. The solutions of a quadratic equation are _____ irrational.

ANSWERS

48. _____

49. _____

50. _____

51. _____

52. _____

48. **Construction.** A garden area is 30 ft long and 20 ft wide. A path of uniform width is set around the edge. If the remaining garden area is 400 ft^2, what is the width of the path?

49. **Crafts.** A solar collector is 2.5 m long by 2.0 m wide. It is held in place by a frame of uniform width around its outside edge. If the exposed collector area is 2.5 m^2, what is the width of the frame, to the nearest tenth of a centimeter?

50. **Crafts.** A solar collector is 2.5 m long by 2.0 m wide. It is held in place by a frame of uniform width around its outside edge. If the exposed collector is 4 m^2, what is the width of the frame to the nearest tenth of a centimeter?

51. The part of the quadratic formula, $b^2 - 4ac$, that is under the radical is called the **discriminant.** Complete the following sentences to show how this value indicates whether there are *no* solutions, *one* solution, or *two* solutions for the quadratic equation.

    (a) When $b^2 - 4ac$ is _____, there are no real number solutions because. . . .
    (b) When $b^2 - 4ac$ is _____, there is one solution because. . . .
    (c) When $b^2 - 4ac$ is _____, there are two solutions because. . . .
    (d) When $b^2 - 4ac$ is _____, there are two *rational* solutions because. . . .
    (e) When $b^2 - 4ac$ is _____, there are two *irrational* solutions because. . . .

52. Work with a partner to decide all values of b in the following equations that will give one or more real number solutions.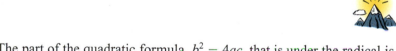

    (a) $3x^2 + bx - 3 = 0$
    (b) $5x^2 + bx + 1 = 0$
    (c) $-3x^2 + bx - 3 = 0$
    (d) Write a rule for judging if an equation has solutions by looking at it in standard form.

53. _____

54. _____

a. _____

b. _____

c. _____

d. _____

e. _____

f. _____

53. Which method of solving a quadratic equation seems simplest to you? Which method should you try first?

54. Complete this statement: "You can tell an equation is quadratic and not linear by. . . ."

Getting Ready for Section 10.4 [Section 1.5]

Evaluate each of the given expressions for the value of the variable given.

(a) $x^2 + 3x - 5$; $x = 3$

(b) $x^2 - 3x - 5$; $x = -2$

(c) $3x^2 + 4x - 6$; $x = -2$

(d) $-2x^2 - 5x + 3$; $x = 4$

(e) $-5x^2 - 5x + 6$; $x = -1$

(f) $\frac{2}{3}x^2 - \frac{1}{3}x + 5$; $x = 6$

Answers

1. $-4, -5$ **3.** $3, 1$ **5.** $-1, \frac{1}{3}$ **7.** $-4, -1$ **9.** 3

11. $\dfrac{3 \pm \sqrt{65}}{4}$ **13.** $-1 \pm \sqrt{5}$ **15.** $\dfrac{3 \pm \sqrt{33}}{4}$ **17.** $\dfrac{1 \pm \sqrt{19}}{3}$

19. No real number solutions   **21.** $\dfrac{4 \pm \sqrt{6}}{5}$ **23.** $\dfrac{2 \pm \sqrt{22}}{2}$

25. $\dfrac{5 \pm \sqrt{37}}{6}$ **27.** $\dfrac{1 \pm \sqrt{21}}{2}$ **29.** $1 \pm \sqrt{7}$ **31.** $-2, 7$

33. $\dfrac{1}{2}, \dfrac{10}{3}$ **35.** $\dfrac{4 \pm \sqrt{10}}{2}$ **37.** $-1, 10,$ **39.** $\dfrac{2 \pm \sqrt{6}}{2}$

41. $\dfrac{1 \pm \sqrt{21}}{6}$ **43.** 5 **45.** Always   **47.** Sometimes

49. ≈ 32.5 cm   **51.**   **53.**

a. 13 **b.** 5 **c.** -2 **d.** -49 **e.** 6 **f.** 27

10.4 Graphing Quadratic Equations

10.4 OBJECTIVES

1. Graph a quadratic equation by plotting points
2. Find the axis of symmetry of a parabola
3. Find the x-intercepts of a parabola

In Section 6.3 you learned to graph first-degree (or linear) equations. Similar methods allow you to graph quadratic equations of the form

$$y = ax^2 + bx + c \qquad a \neq 0$$

The first thing to notice is that the graph of an equation in this form is not a straight line. The graph is always a curve called a **parabola.**

Here are some examples:

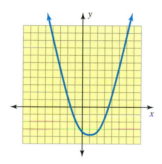

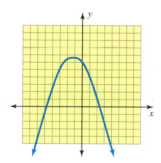

To graph quadratic equations, start by finding solutions for the equation. We begin by completing a table of values. Choose any convenient values for x. Then use the given equation to compute the corresponding values for y, as Example 1 illustrates this process.

OBJECTIVE 1   **Example 1   Completing a Table of Values**

Complete the ordered pairs to form solutions for the equation $y = x^2$. Then show these results in a table of values.

$(-2, \ \ ), (-1, \ \ ), (0, \ \ ), (1, \ \ ), (2, \ \ )$

For example, to complete the pair $(-2, \ )$, substitute -2 for x in the given equation.

$$y = (-2)^2 = 4$$

RECALL A solution is a pair of values that makes the equation a true statement.

So $(-2, 4)$ is a solution.

Substituting the other values for x in this manner gives the following table of values for $y = x^2$:

x	y
-2	4
-1	1
0	0
1	1
2	4

 CHECK YOURSELF 1

Complete the ordered pairs to form solutions for $y = x^2 + 2$ and form a table of values.

$(-2, \ ), (-1, \ ), (0, \ ), (1, \ ), (2, \ )$

We can now plot points in the cartesian coordinate system that correspond to solutions to the equation.

Example 2 Plotting Some Solution Points

Plot the points from the table of values corresponding to $y = x^2$ from Example 1.

x	y
-2	4
-1	1
0	0
1	1
2	4

Notice that the y-axis acts as a mirror. Do you see that any point graphed in quadrant I will be "reflected" in quadrant II?

CHECK YOURSELF 2 _____

Plot the points from the table of values formed in Check Yourself 1.

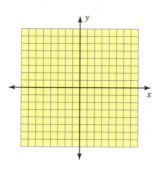

The graph of the equation can be drawn by joining the points with a smooth curve.

Example 3   Completing the Graph of the Solution Set

Draw the graph of $y = x^2$.

We can now draw a smooth curve between the points found in Example 2 to form the graph of $y = x^2$.

RECALL The graph must be a parabola.

NOTE Notice that a parabola **does not** come to a point.

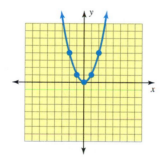

CHECK YOURSELF 3 _____

Draw a smooth curve between the points plotted in Check Yourself 2.

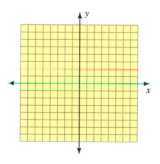

You can use any convenient values for x in forming your table of values. You should use as many pairs as are necessary to get the correct shape of the graph (a parabola).

Example 4 **Graphing the Solution Set**

Graph $y = x^2 - 2x$. Use integer values of x from -1 to 3.

First, determine solutions for the equation. For instance, if $x = -1$,

$$y = (-1)^2 - 2(-1)$$
$$= 1 + 2$$
$$= 3$$

and $(-1, 3)$ is a solution for the given equation.

Substituting the other values for x, we can form the following table of values. We then plot the corresponding points and draw a smooth curve to form our graph.

The graph of $y = x^2 - 2x$.

NOTE Any values can be substituted for x in the original equation.

x	y
-1	3
0	0
1	-1
2	0
3	3

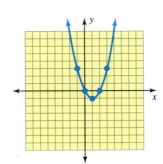

 CHECK YOURSELF 4

Graph $y = x^2 + 4x$. Use integer values of x from -4 to 0.

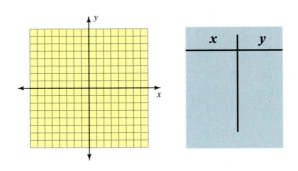

Choosing values for x is also a valid method of graphing a quadratic equation that contains three terms.

Example 5 **Graphing the Solution Set**

Graph $y = x^2 - x - 2$. Use integer values of x from -2 to 3. We show the computation for two of the solutions.

If $x = -2$:   If $x = 3$:

$$y = (-2)^2 - (-2) - 2 \qquad y = 3^2 - 3 - 2$$
$$= 4 + 2 - 2 \qquad\qquad = 9 - 3 - 2$$
$$= 4 \qquad\qquad\qquad = 4$$

You should substitute the remaining values for x into the given equation to verify the other solutions shown in the following table of values.

The graph of $y = x^2 - x - 2$:

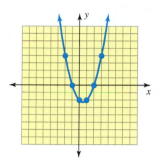

x	y
-2	4
-1	0
0	-2
1	-2
2	0
3	4

CHECK YOURSELF 5

Graph $y = x^2 - 4x + 3$. Use integer values of x from -1 to 4.

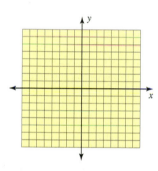

x	y

In Example 6, the graph looks significantly different from previous graphs.

Example 6 Graphing the Solution Set

Graph $y = -x^2 + 3$. Use integer values from -2 to 2.
Again we show two computations.

If $x = -2$:   If $x = 1$:

NOTE $-(-2)^2 = -4$

$$y = -(-2)^2 + 3 \qquad\qquad y = -(1)^2 + 3$$
$$= -4 + 3 \qquad\qquad\quad = -1 + 3$$
$$= -1 \qquad\qquad\qquad = 2$$

Verify the remainder of the solutions shown in the following table of values for yourself.

The graph of $y = -x^2 + 3$.

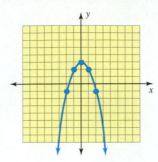

x	y
-2	-1
-1	2
0	3
1	2
2	-1

There is an important difference between this graph and the others we have seen. This time the parabola opens downward! Can you guess why? The answer is in the coefficient of the x^2-term.

If the coefficient of x^2 is *positive*, the parabola opens *upward*.

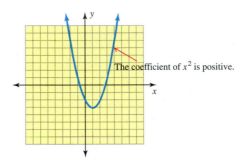

The coefficient of x^2 is positive.

If the coefficient of x^2 is *negative*, the parabola opens *downward*.

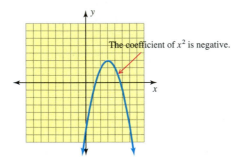

The coefficient of x^2 is negative.

 CHECK YOURSELF 6

Graph $y = -x^2 - 2x$. Use integer values from -3 to 1.

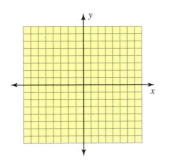

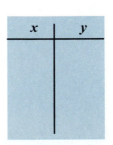

x	y

There are two other terms we would like to introduce before closing this section on graphing quadratic equations. As you may have noticed, all the parabolas that we graphed are symmetric about a vertical line. This is called the **axis of symmetry** for the parabola.

The point at which the parabola intersects that vertical line (this will be the lowest—or the highest—point on the parabola) is called the **vertex.**

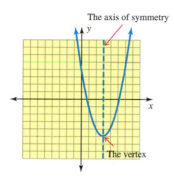

The axis of symmetry

The vertex

As we construct a table of values, we want to include x-values which will span the parabola's turning point: the vertex. For a quadratic equation $y = ax^2 + bx + c$, the following is an equation for the axis of symmetry:

RECALL $x = k$ (for any number k) is the equation of a vertical line.

$$x = \frac{-b}{2a}$$

Once we have this, we can choose x-values to either side of the axis of symmetry to help us build a table. Look at Example 7.

OBJECTIVE 2

Example 7   Using the Axis of Symmetry to Create a Table and Graph

Given $y = -x^2 - 8x - 12$, complete the following.

(a) Write the equation for the axis of symmetry.

Since $a = -1$ and $b = -8$, we have

$$x = \frac{-b}{2a} = \frac{-(-8)}{2(-1)} = \frac{8}{-2} = -4$$

Thus the vertical line $x = -4$ is the axis of symmetry.

(b) Use the equation to create a table of values for the quadratic equation.

Finding values on both sides of -4, we make the following table:

x	y
-6	0
-5	3
-4	4
-3	3
-2	0

(c) Sketch the graph. Show the axis of symmetry as a dotted vertical line.

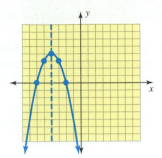

NOTE Here again the parabola opens downward.

 CHECK YOURSELF 7

For $y = x^2 - 7x + 4$, find the axis of symmetry, create a table of values, and sketch the graph.

The x-intercepts of a parabola are the points where the parabola touches the x-axis. We can locate these directly without actually graphing the parabola. The key is to set y equal to 0 and then solve for x.

OBJECTIVE 3   **Example 8   Finding the x-Intercepts of a Parabola**

Find the x-intercepts for each equation.

(a) $y = x^2 - 5x + 2$

We set $y = 0$, and solve.

$0 = x^2 - 5x + 2$ Use the quadratic formula.

$$x = \frac{5 \pm \sqrt{(-5)^2 - 4(2)}}{2} = \frac{5 \pm \sqrt{17}}{2}$$

NOTE If decimal approximations are desired, we have (0.44, 0) and (4.56, 0).

There are two x-intercepts: $\left(\dfrac{5 - \sqrt{17}}{2}, 0\right)$ and $\left(\dfrac{5 + \sqrt{17}}{2}, 0\right)$.

(b) $y = x^2 + 6x + 9$

$0 = x^2 + 6x + 9$

$0 = (x + 3)(x + 3)$

$x = -3$

There is only one x-intercept: $(-3, 0)$. The parabola touches (but does not cross) the x-axis at this point.

(c) $y = x^2 - 4x + 5$

$0 = x^2 - 4x + 5$

$$x = \frac{4 \pm \sqrt{(-4)^2 - 4(5)}}{2}$$

$$= \frac{4 \pm \sqrt{-4}}{2}$$

Because these values are not real numbers, there are no x-intercepts. The parabola does not touch the x-axis.

CHECK YOURSELF 8

Find the x-intercepts for each equation.

(a) $y = x^2 + 3x - 5$

(b) $y = x^2 - 10x + 25$

CHECK YOURSELF ANSWERS

1.

x	y
-2	6
-1	3
0	2
1	3
2	6

2.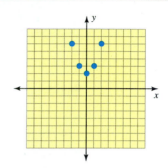

3. $y = x^2 + 2$

4. $y = x^2 + 4x$

x	y
-4	0
-3	-3
-2	-4
-1	-3
0	0

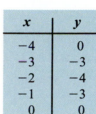

5. $y = x^2 - 4x + 3$

x	y
-1	8
0	3
1	0
2	-1
3	0
4	3

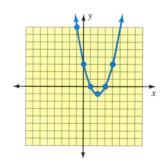

6. $y = -x^2 - 2x$

x	y
-3	-3
-2	0
-1	1
0	0
1	-3

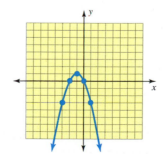

7. $y = x^2 - 7x + 4$; axis: $x = \dfrac{7}{2}$

x	y
1	-2
2	-6
3	-8
4	-8
5	-6
6	-2

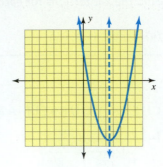

8. (a) $\left(\dfrac{-3 - \sqrt{29}}{2}, 0\right)$ and $\left(\dfrac{-3 + \sqrt{29}}{2}, 0\right)$; **(b)** $(5, 0)$

10.4 Exercises

Graph each of the following quadratic equations after completing the given table of values.

1. $y = x^2 + 1$

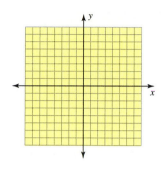

2. $y = x^2 - 2$

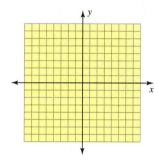

3. $y = x^2 - 4$

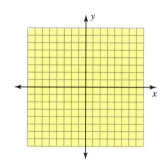

4. $y = x^2 + 3$

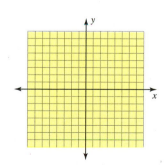

5. $y = x^2 - 4x$

x	y
0	
1	
2	
3	
4	

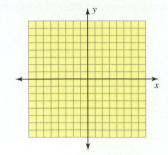

6. $y = x^2 + 2x$

x	y
−3	
−2	
−1	
0	
1	

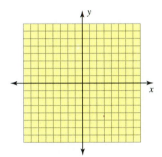

7. $y = x^2 + x$

x	y
−2	
−1	
0	
1	
2	

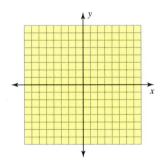

8. $y = x^2 - 3x$

x	y
−1	
0	
1	
2	
3	

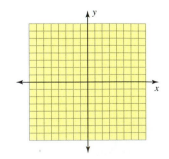

9. $y = x^2 - 2x - 3$

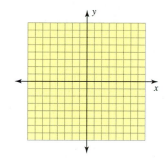

x	y
−1	
0	
1	
2	
3	

10. $y = x^2 - 5x + 6$

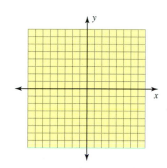

x	y
0	
1	
2	
3	
4	

11. $y = x^2 - x - 6$

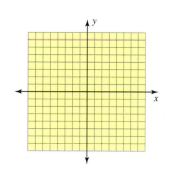

x	y
−1	
0	
1	
2	
3	

12. $y = x^2 + 3x - 4$

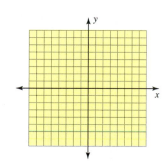

x	y
−4	
−3	
−2	
−1	
0	

ANSWERS

13. ______

14. ______

15. ______

16. ______

13. $y = -x^2 + 2$

x	y
−2	
−1	
0	
1	
2	

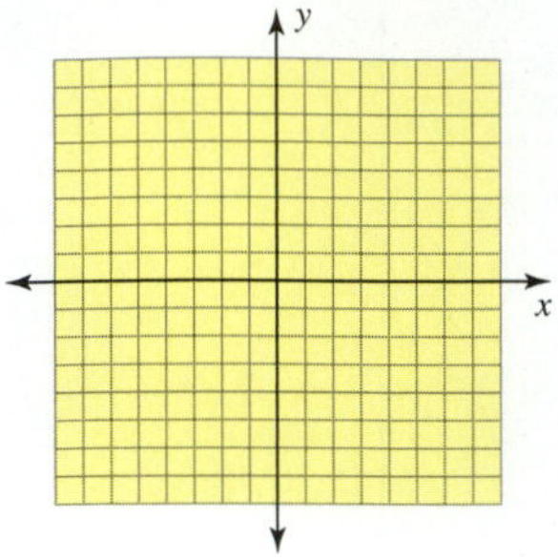

14. $y = -x^2 - 2$

x	y
−2	
−1	
0	
1	
2	

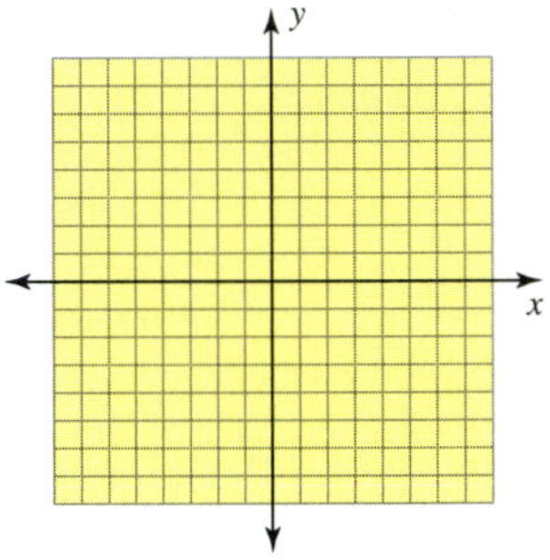

15. $y = -x^2 - 4x$

x	y
−4	
−3	
−2	
−1	
0	

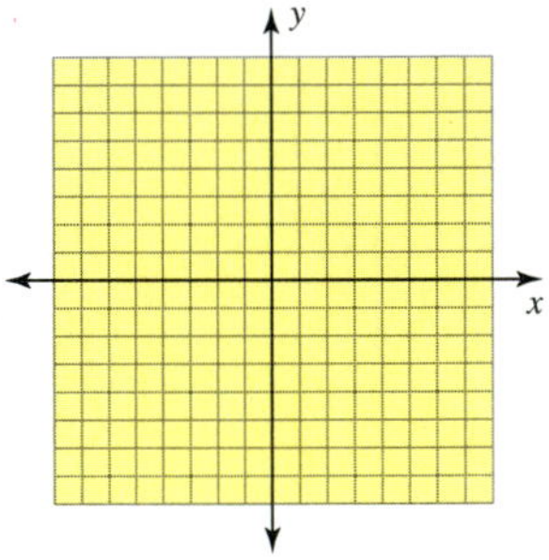

16. $y = -x^2 + 2x$

x	y
−1	
0	
1	
2	
3	

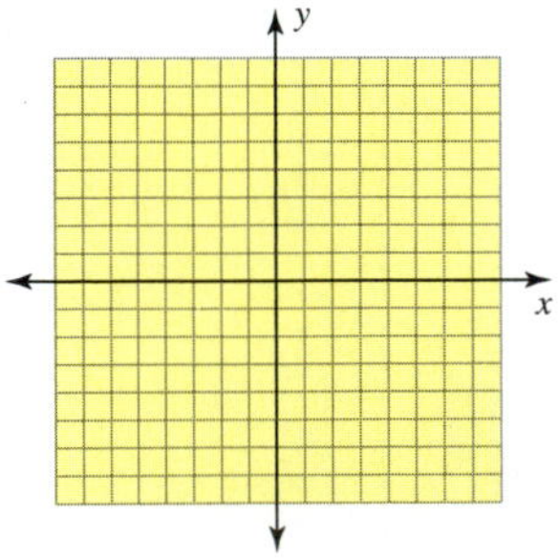

Match each graph with the correct equation on the right.

ANSWERS

17. _____

18. _____

19. _____

20. _____

21. _____

22. _____

23. _____

24. _____

17.

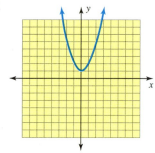

18.

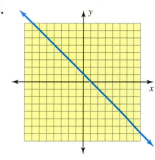

19.

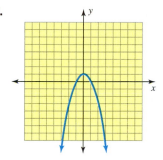

20.

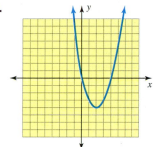

(a) $y = -x^2 + 1$

(b) $y = 2x$

(c) $y = x^2 - 4x$

(d) $y = -x + 1$

(e) $y = -x^2 + 3x$

21.

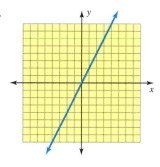

22.

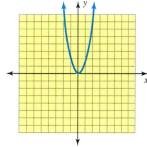

(f) $y = x^2 + 1$

(g) $y = x + 1$

(h) $y = 2x^2$

23.

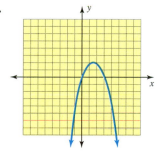

24.

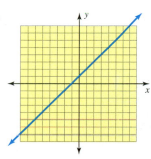

ANSWERS

25. _____

26. _____

27. _____

28. _____

For each equation in exercises 25 to 28, identify the axis of symmetry, create a suitable table of values, and sketch the graph (including the axis of symmetry).

25. $y = x^2 + 4x$

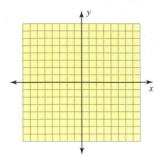

26. $y = x^2 - 5x + 3$

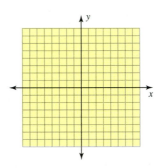

27. $y = -x^2 - 3x + 3$

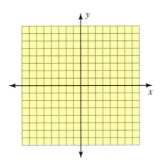

28. $y = -x^2 + 6x - 2$

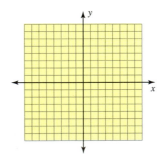

ANSWERS

29. _____

30. _____

31. _____

32. _____

33. _____

34. _____

35. _____

36. _____

37. _____

38. _____

39. _____

40. _____

For each equation in exercises 29 to 34, find the x-intercepts.

29. $y = x^2 - 2x - 8$

30. $y = x^2 + x - 6$

31. $y = x^2 + 3x - 6$

32. $y = x^2 - 5x - 4$

33. $y = x^2 - 2x + 1$

34. $y = x^2 - x + 2$

For exercises 35 to 40, fill in the blanks with *always, sometimes,* or *never.*

35. The vertex of a parabola is _____ located on the axis of symmetry.

36. The vertex of a parabola is _____ the highest point on the graph.

37. The graph of $y = ax^2 + bx + c$ _____ intersects the x-axis.

38. The graph of $y = ax^2 + bx + c$ _____ has more than two x-intercepts.

39. The graph of $y = ax^2 + bx + c$ _____ intersects the y-axis.

40. The graph of $y = ax^2 + bx + c$ _____ intersects the y-axis more than once.

Answers

1. $y = x^2 + 1$

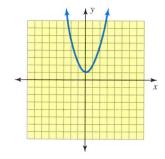

3. $y = x^2 - 4$

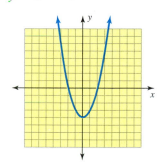

5. $y = x^2 - 4x$

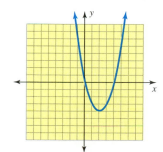

7. $y = x^2 + x$

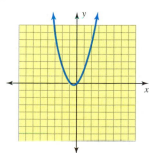

9. $y = x^2 - 2x - 3$

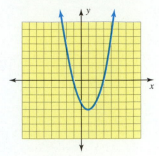

11. $y = x^2 - x - 6$

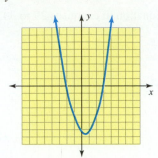

13. $y = -x^2 + 2$

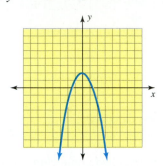

15. $y = -x^2 - 4x$

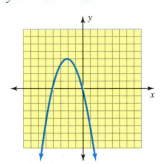

17. f   **19.** a   **21.** b   **23.** e

25. Axis: $x = -2$

x	y
-4	0
-3	-3
-2	-4
-1	-3
0	0

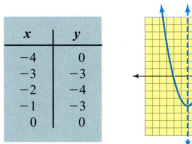

27. Axis: $x = -\dfrac{3}{2}$

x	y
-4	-1
-3	3
-2	5
-1	5
0	3
1	-1

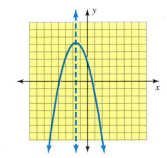

29. $(-2, 0), (4, 0)$ **31.** $\left(\dfrac{-3 - \sqrt{33}}{2}, 0\right), \left(\dfrac{-3 + \sqrt{33}}{2}, 0\right)$

33. $(1, 0)$ **35.** Always   **37.** Sometimes   **39.** Always

10.5 Applications of Quadratic Equations

10.5 OBJECTIVE

1. Solve a variety of applications involving quadratic equations

If you drop a pebble from the top of a building into a pond below, can you predict how long it will take to hit the water? If you fire a projectile directly upward with known velocity, can you predict when it will be at a certain height?

Many applied problem situations such as these can be modeled with quadratic equations, and, therefore, solved by techniques presented in this chapter. Such problems may arise in geometry, construction, physics, and economics, to name but a few.

RECALL The Pythagorean theorem was introduced in Section 9.6.

We begin with examples that rely on use of the Pythagorean theorem, studied previously.

OBJECTIVE 1

Example 1 Solving a Construction Application

How long must a guywire be to reach from the top of a 30-ft pole to a point on the ground 20 ft from the base of the pole? Round to the nearest tenth of a foot.

We start by drawing a sketch of the problem.

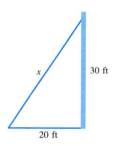

Using the Pythagorean theorem, we write

$$x^2 = 20^2 + 30^2$$
$$x^2 = 400 + 900$$
$$x^2 = 1300$$

This is a quadratic equation which can be solved using the square root method.

$$x = \pm\sqrt{1300}$$

NOTE Always check to see if your final answer is reasonable.

Since x must be positive, we reject $-\sqrt{1300}$, and keep $\sqrt{1300}$. To the nearest tenth of a foot, we have $x = 36.1$ ft.

CHECK YOURSELF 1

A 16-ft ladder leans against a wall with its base 4 ft from the wall. How far above the floor, to the nearest tenth of a foot, is the top of the ladder?

Example 2 illustrates a geometric problem using the Pythagorean theorem.

Example 2   Solving a Geometry Application

The length of one leg of a right triangle is 2 cm more than the other. If the length of the hypotenuse is 6 cm, what are the lengths of the two legs? Round to the nearest tenth of a foot.

Draw a sketch of the problem, labeling the known and unknown lengths. Here, if one leg is represented by x, the other must be represented by $x + 2$.

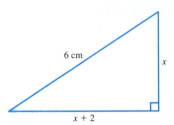

NOTE The sum of the squares of the lengths of the unknown legs is equal to the square of the length of the hypotenuse.

Use the Pythagorean theorem to form an equation.

$$x^2 + (x + 2)^2 = 6^2$$
$$x^2 + x^2 + 4x + 4 = 36$$
$$2x^2 + 4x - 32 = 0 \qquad \text{Divide both sides by 2.}$$
$$x^2 + 2x - 16 = 0$$

This equation can be solved by completing the square or with the quadratic formula. Completing the square is a good choice, since the coefficient of x^2 is 1 and the coefficient of x is even:

$$x^2 + 2x = 16$$
$$x^2 + 2x + 1 = 16 + 1$$
$$(x + 1)^2 = 17$$
$$x + 1 = \pm\sqrt{17}$$
$$x = -1 \pm \sqrt{17} \approx 3.1 \quad \text{or} \quad -5.1$$

We generally reject the negative solution in a geometric problem.

NOTE Be sure to include units with the final answer.

If $x \approx 3.1$, then $x + 2 \approx 5.1$. The lengths of the legs are approximately 3.1 and 5.1 cm.

 CHECK YOURSELF 2

The length of one leg of a right triangle is 1 in. more than the other. If the length of the hypotenuse is 3 in., what are the lengths of the legs? Round to the nearest thousandth of an inch.

In Example 3, we consider a problem arising in construction. Note the importance of drawing a sketch to set up the problem.

Example 3   Solving a Construction Application

A rectangular garden is to be surrounded by a walkway of constant width. The garden's dimensions are 5 m by 8 m. The total area, garden plus walkway, is to be 100 m². What must be the width of the walkway? Round to the nearest hundredth of a meter.

First we sketch the situation.

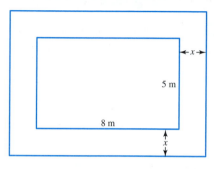

The area of the entire figure may be represented by

$$\underbrace{(8 + 2x)}_{\text{length}}\underbrace{(5 + 2x)}_{\text{width}} \qquad \text{(Do you see why?)}$$

Since the area is to be 100 m², we write

$$(8 + 2x)(5 + 2x) = 100$$
$$40 + 16x + 10x + 4x^2 = 100$$
$$4x^2 + 26x - 60 = 0$$
$$2x^2 + 13x - 30 = 0$$
$$x = \frac{-13 \pm \sqrt{13^2 - 4(2)(-30)}}{2(2)}$$
$$= \frac{-13 \pm \sqrt{409}}{4}$$
$$x \approx 1.8059\ldots \quad \text{or} \quad -8.3059\ldots$$

Since x must be positive, we have (rounded) $x = 1.81$ m.

 CHECK YOURSELF 3

A rectangular swimming pool is 16 ft by 40 ft. A tarp to go over the pool also covers a strip of equal width surrounding the pool. If the area of the tarp is 1100 ft², how wide is the covered strip around the pool? Round to the nearest tenth of a foot.

NOTE We are neglecting air resistance in this discussion.

When a ball is thrown directly upward, or if some sort of projectile is fired upward, the height of the object is approximated by a quadratic function of time. This concept is illustrated in Example 4.

Example 4   Solving a Science Application

Suppose that a ball is thrown upward with an initial velocity of 80 ft/s. (The initial velocity is the speed with which the ball leaves the thrower's hand. The speed will then decrease as the ball rises.) If the ball is released at a height of 5 ft, the height equation may be written as follows:

$$h = -16t^2 + 80t + 5$$

When, to the nearest hundredth of a second, will the ball be at a height of 93 ft?

Since the desired height is 93 ft, we write

$$93 = -16t^2 + 80t + 5$$

We must solve for t:

$93 = -16t^2 + 80t + 5$	We need a 0 on one side.
$16t^2 - 80t + 88 = 0$	Divide through by 8.
$2t^2 - 10t + 11 = 0$	Use the quadratic formula.

$$t = \frac{10 \pm \sqrt{(-10)^2 - 4(2)(11)}}{2(2)}$$

$$= \frac{10 \pm \sqrt{12}}{4} \approx 1.63 \quad \text{or} \quad 3.37$$

NOTE Use a calculator to check decimal results such as these.

The ball reaches a height of 93 ft twice: first, on the way up, at 1.63 s, and second, on the way down, at 3.37 s.

 CHECK YOURSELF 4 _____

Using the same height equation provided in Example 4, determine to the nearest hundredth of a second, when the ball will be at a height of 77 ft.

In Example 5, we look at a similar example. It highlights the importance of proper interpretation of results.

Example 5   Solving a Science Application

A projectile is fired upward from a platform in such a way that the object will miss the platform (fortunately!) on the way down. If the height of the platform is 20 ft, and the initial velocity is 200 ft/s, the height is given by

$$h = -16t^2 + 200t + 20$$

To the nearest tenth of a second, when will the projectile hit the ground?

The projectile will hit the ground when the height is 0, so we write

$0 = -16t^2 + 200t + 20$ Divide through by −4.

$0 = 4t^2 - 50t - 5$

$$t = \frac{50 \pm \sqrt{(-50)^2 - 4(4)(-5)}}{2(4)}$$

$$= \frac{50 \pm \sqrt{2580}}{8} \approx 12.599 \quad \text{or} \quad -0.099$$

The projectile will hit at about 12.6 s. We reject the negative solution, as it would indicate a time *before* the projectile was fired.

CHECK YOURSELF 5

A projectile fired upward from a platform follows the height equation

$$h = -16t^2 + 160t + 40$$

When will it hit the ground? Round to the nearest tenth of a second.

We close this section with an application that arises in the field of economics. Related to the production of a certain item, there are two important equations: a **supply** equation and a **demand** equation. Each is related to the price of the item. The **equilibrium** price is the price for which supply equals demand.

Example 6   Solving a Business Application

The demand equation for a certain computer chip is given by

$$D = -3p + 65$$

The supply equation is predicted to be

$$S = -p^2 + 36p - 120$$

Find the equilibrium price.

We wish to know what price p results in equal supply and demand, so we write

$$-p^2 + 36p - 120 = -3p + 65$$
$$0 = p^2 - 39p + 185$$

So

$$p = \frac{39 \pm \sqrt{(-39)^2 - 4(185)}}{2}$$

$$= \frac{39 \pm \sqrt{781}}{2}$$

$$= 5.53 \quad \text{or} \quad 33.47$$

NOTE You should confirm that when $p = 5.53$, supply and demand are positive.

Here we must be careful: supply and demand must be positive. When $p = 33.47$, they are negative, so the equilibrium price is $5.53.

CHECK YOURSELF 6

The demand equation for a certain item is given by

$$D = -2p + 32$$

The supply equation is predicted to be

$$S = -p^2 + 18p - 5$$

Find the equilibrium price.

CHECK YOURSELF ANSWERS

1. 15.5 ft   **2.** 1.562 in., 2.562 in.   **3.** 3.6 ft   **4.** 1.18 s, 3.82 s
5. 10.2 s   **6.** $2.06

10.5 Exercises

Solve each of the following applications. Give all answers to the nearest thousandth.

1. **Construction.** How long must a guywire be to run from the top of a 20-ft pole to a point on the ground 8 ft from the base of the pole?

2. **Construction.** How long must a guywire be to run from the top of a 16-ft pole to a point on the ground 6 ft from the base of the pole?

3. **Construction.** The base of a 15-ft ladder is 5 ft away from a wall. How far above the floor is the top of the ladder?

4. **Construction.** The base of an 18-ft ladder is 4 ft away from a wall. How far above the floor is the top of the ladder?

5. **Geometry.** The length of one leg of a right triangle is 3 in. more than the other. If the length of the hypotenuse is 8 in., what are the lengths of the two legs?

6. **Geometry.** The length of a rectangle is 1 cm longer than its width. If the diagonal of the rectangle is 4 cm, what are the dimensions (the length and the width) of the rectangle?

7. **Geometry.** The width of a rectangle is 3 ft less than its length. If the area of the rectangle is 75 ft^2, what are the dimensions of the rectangle?

8. **Geometry.** The length of a rectangle is 5 cm more than its width. If the area of the rectangle is 90 cm^2, find the dimensions.

9. **Geometry.** The length of a rectangle is 2 cm more than 3 times its width. If the area of the rectangle is 95 cm^2, find the dimensions of the rectangle.

10. **Geometry.** If the length of a rectangle is 3 ft less than twice its width and the area of the rectangle is 66 ft^2, what are the dimensions of the rectangle?

11. **Geometry.** One leg of a right triangle is twice the length of the other. The hypotenuse is 8 m long. Find the length of each leg.

12. **Geometry.** One leg of a right triangle is 2 ft longer than the shorter side. If the length of the hypotenuse is 18 ft, how long is each leg?

13. **Geometry.** One leg of a right triangle is 1 in. shorter than the other leg. The hypotenuse is 4 in. longer than the shorter side. Find the length of each side.

14. Geometry. The hypotenuse of a given right triangle is 6 cm longer than the shorter leg. The length of the shorter leg is 2 cm less than that of the longer leg. Find the lengths of the three sides.

15. Geometry. The sum of the lengths of the two legs of a right triangle is 30 m. The hypotenuse is 22 m long. Find the lengths of the two legs.

16. Geometry. The sum of the lengths of one side of a right triangle and the hypotenuse is 20 cm. The other leg is 5 cm shorter than the hypotenuse. Find the length of each side.

17. Construction. A rectangular field is 300 ft by 500 ft. A roadway of width x ft is to be built just inside the field. What is the widest the roadway can be and still leave 100,000 ft^2 in the region?

18. Construction. A rectangular garden is to be surrounded by a walkway of constant width. The garden's dimensions are 20 ft by 28 ft. The total area, garden plus walkway, is to be 1100 ft^2. What must be the width of the walkway?

19. Construction. A rectangular garden is to be surrounded by a walkway of constant width. The garden's dimensions are 15 ft by 25 ft. The total area, garden plus walkway, is to be 650 ft^2. What must be the width of the walkway?

20. Construction. A rectangular field is 200 m by 300 m. A roadway of width x m is to be built just inside the field. What is the widest the roadway can be and still leave 50,000 m^2 in the region?

21. Science and medicine. The equation $h = -16t^2 + 112t$ gives the height of an arrow, shot upward from the ground with an initial velocity of 112 ft/s, where t is the time after the arrow leaves the ground. Find the time it takes for the arrow to reach a height of 120 ft.

22. Science and medicine. The equation $h = -16t^2 + 112t$ gives the height of an arrow, shot upward from the ground with an initial velocity of 112 ft/s, where t is the time after the arrow leaves the ground. Find the time it takes for the arrow to reach a height of 180 ft.

23. Science and medicine. The equation $h = -16t^2 - 32t + 320$ gives the height of a ball, thrown downward from the top of a 320-ft building with an initial velocity of 32 ft/s. Find the time it takes for the ball to reach a height of 160 ft.

24. Science and medicine. The equation $h = -16t^2 - 32t + 320$ gives the height of a ball, thrown downward from the top of a 320-ft building with an initial velocity of 32 ft/s. Find the time it takes for the ball to reach a height of 64 ft.

ANSWERS

14.

15.

16.

17.

18.

19.

20.

21.

22.

23.

24.

25. **Science and medicine.** If a ball is thrown vertically upward from a height of 6 ft, with an initial velocity of 64 ft/s, its height h after t s is given by $h = -16t^2 + 64t + 6$. How long does it take the ball to return to the ground?

26. **Science and medicine.** If a ball is thrown vertically upward from a height of 6 ft, with an initial velocity of 64 ft/s, its height h after t s is given by $h = -16t^2 + 64t + 6$. How long does it take the ball to reach a height of 38 ft on the way up?

27. **Science and medicine.** If a ball is thrown vertically upward from a height of 5 ft, with an initial velocity of 96 ft/s, its height h after t s is given by $h = -16t^2 + 96t + 5$. How long does it take the ball to return to the ground?

28. **Science and medicine.** If a ball is thrown vertically upward from a height of 5 ft, with an initial velocity of 96 ft/s, its height h after t s is given by $h = -16t^2 + 96t + 5$. How long does it take the ball to pass through a height of 120 ft on the way back down to the ground?

29. **Science and medicine.** If a pebble is dropped toward a pond from the top of a 240-ft building, its height h after t s is given by $h = -16t^2 + 240$. How long does it take the pebble to pass through a height of 120 ft?

30. **Science and medicine.** If a pebble is dropped toward a pond from the top of a 240-ft building, its height h after t s is given by $h = -16t^2 + 240$. How long does it take the pebble to hit the water?

If a ball is thrown upward from the roof of a building 80 m tall with an initial velocity of 15 m/s, its approximate height h after t s is given by

$$h = -5t^2 + 15t + 80$$

The difference between this equation and the ones used in Examples 4 and 5 has to do with the units used. When we used feet, the t^2-coefficient was -16 (from the fact that the acceleration due to gravity is approximately 32 ft/s²). When we use meters as the height, the t^2-coefficient is -5 (that same acceleration becomes approximately 10 m/s²). Use this information to solve exercises 31 and 32.

31. **Science and medicine.** When will the ball reach a height of 85 m?

32. **Science and medicine.** How long does it take the ball to fall back to the ground?

Changing the initial velocity to 25 m/s will only change the t-coefficient. Our new equation becomes

$$h = -5t^2 + 25t + 80$$

Use this equation to solve exercises 33 and 34.

33. **Science and medicine.** How long does it take the ball to fall back to the ground?

34. **Science and medicine.** When will the ball reach a height of 95 m?

The only part of the height equation that we have not discussed is the constant. You have probably noticed that the constant is always equal to the initial height of the ball (80 m in our previous exercises). Now, let's have *you* develop a height equation.

A ball is thrown upward from the roof of a building 100 m tall with an initial velocity of 20 m/s. Use this information for exercises 35 to 38.

35. Science and medicine. Write the equation for the height of the ball.

36. Science and medicine. When will the ball reach a height of 80 m?

37. Science and medicine. How long does it take the ball to fall back to the ground?

38. Science and medicine. Will the ball ever reach a height of 125 m?

A ball is thrown upward from the roof of a 120-ft building with an initial velocity of 20 ft/s. Use this information for exercises 39 to 41.

39. Science and medicine. Write the equation for the height of the ball.

40. Science and medicine. How long does it take the ball to fall back to the ground?

41. Science and medicine. When will the ball reach a height of 80 ft?

42. Business and finance. Suppose that the cost C, in dollars, of producing x chairs is given by $C = 2x^2 - 40x + 2400$. How many chairs can be produced for $4650?

43. Business and finance. Suppose that the profit P, in dollars, of producing and selling x appliances is given by $P = -3x^2 + 240x - 1800$. How many appliances must be produced and sold to achieve a profit of $2325?

44. Business and finance. A small manufacturer's weekly profit in dollars is given by $P = -3x^2 + 270x$. Find the number of items x that must be produced to realize a profit of $5775

45. Business and finance. The demand equation for a certain computer chip is given by

$D = -4p + 50$

The supply equation is predicted to be

$S = -p^2 + 20p - 6$

Find the equilibrium price.

ANSWERS

35. _____

36. _____

37. _____

38. _____

39. _____

40. _____

41. _____

42. _____

43. _____

44. _____

45. _____

46. Business and finance. The demand equation for a certain type of printer is given by

$$D = -200p + 35,000$$

The supply equation is predicted to be

$$S = -p^2 + 400p - 20,000$$

Find the equilibrium price.

47. Business and finance. The demand equation for a certain type of printer is given by

$$D = -80p + 7000$$

The supply equation is predicted to be

$$S = -p^2 + 220p - 8000$$

Find the equilibrium price.

48. Business and finance. The demand equation for a certain computer chip is given by

$$D = -5p + 62$$

The supply equation is predicted to be

$$S = -p^2 + 23p - 11$$

Find the equilibrium price.

For exercises 49 and 50, fill in the blanks with *always, sometimes,* or *never.*

49. When solving an applied problem, an algebraic solution _____ represents a solution to the application.

50. A ball thrown vertically _____ reaches a specified height 2 times.

Answers

1. 21.541 ft **3.** 14.142 ft **5.** 3.954 in., 6.954 in.
7. 7.289 ft by 10.289 ft **9.** 5.304 cm by 17.912 cm **11.** 3.578 m, 7.155 m
13. 7.899 in., 8.899 in., 11.899 in. **15.** 19.123 m, 10.877 m **17.** 34.168 ft
19. 2.990 ft **21.** 1.321 or 5.679 s **23.** 2.317 s **25.** 4.092 s
27. 6.052 s **29.** 2.739 s **31.** 0.382 or 2.618 s **33.** 7.217 s
35. $h = -5t^2 + 20t + 100$ **37.** 6.899 s **39.** $h = -16t^2 + 20t + 120$
41. 2.325 s **43.** 25 or 55 appliances **45.** $2.62 **47.** $63.40
49. Sometimes

DEFINITION/PROCEDURE	EXAMPLE	REFERENCE
More on Quadratic Equations		**Section 10.1**
Solving Quadratic Equations by Factoring 1. Add or subtract the necessary terms on both sides of the equation so that the equation is in standard form (set equal to 0). 2. Factor the quadratic expression. 3. Set each factor equal to 0. 4. Solve the resulting equations to find the solutions. 5. Check each solution by substituting in the original equation.	To solve: $$x^2 + 7x = 30$$ $$x^2 + 7x - 30 = 0$$ $$(x + 10)(x - 3) = 0$$ $$x + 10 = 0 \quad \text{or} \quad x - 3 = 0$$ $x = -10$ and $x = 3$ are solutions.	*p. 813*
Square Root Property If $x^2 = k$, then $x = \sqrt{k}$ or $x = -\sqrt{k}$. If $(x - h)^2 = k$, then $x = h \pm \sqrt{k}$.	To solve: $$(x - 3)^2 = 5$$ $$x - 3 = \pm\sqrt{5}$$ $$x = 3 \pm \sqrt{5}$$	*p. 815*
Completing the Square		**Section 10.2**
Completing the Square To solve a quadratic equation by completing the square: 1. Write the equation in the form $$ax^2 + bx = k$$ so that the variable terms are on the left side and the constant is on the right side. 2. If the leading coefficient of x^2 is not 1, divide both sides by that coefficient. 3. Add the square of one-half coefficient of x to both sides of the equation. 4. The left side of the equation is now a perfect-square trinomial. Factor and solve as before.	To solve: $$2x^2 + 2x - 1 = 0$$ $$2x^2 + 2x = 1$$ $$x^2 + x = \frac{1}{2}$$ $$x^2 + x + \left(\frac{1}{2}\right)^2 = \frac{1}{2} + \left(\frac{1}{2}\right)^2$$ $$\left(x + \frac{1}{2}\right)^2 = \frac{3}{4}$$ $$x + \frac{1}{2} = \pm\sqrt{\frac{3}{4}} = \pm\frac{\sqrt{3}}{2}$$ $$x = \frac{-1 \pm \sqrt{3}}{2}$$	*p. 824*
The Quadratic Formula		**Section 10.3**
The Quadratic Formula To solve an equation by formula: 1. Rewrite the equation in standard form. $$ax^2 + bx + c = 0$$ 2. If a common factor exists, divide both sides of the equation by that common factor. 3. Identify the coefficients a, b, and c. 4. Substitute the values for a, b, and c into the quadratic formula. $$x = \frac{-b \pm \sqrt{b^2 - 4ac}}{2a}$$ 5. Simplify the right side of the expression formed in step 4 to write the solutions for the original equation.	To solve: $$x^2 - 2x = 4$$ Write the equation as $$x^2 - 2x - 4 = 0$$ $$a = 1 \qquad b = -2 \qquad c = -4$$ $$x = \frac{-(-2) \pm \sqrt{(-2)^2 - 4(1)(-4)}}{2(1)}$$ $$= \frac{2 \pm \sqrt{20}}{2}$$ $$= \frac{2 \pm 2\sqrt{5}}{2} = \frac{2(1 \pm \sqrt{5})}{2}$$ $$= 1 \pm \sqrt{5}$$	*p. 832*

Continued

Graphing Quadratic Equations		Section 10.4
To graph equations of the form $$y = ax^2 + bx + c$$ **1.** Find the axis of symmetry using $$x = \frac{-b}{2a}$$ **2.** Form a table of values by choosing x-values that span the axis of symmetry. **3.** Plot the points from the table of values. **4.** Draw a smooth curve between the points. The graph of a quadratic equation will always be a parabola. The parabola opens upward if a, the coefficient of the x^2-term, is positive.	$$y = x^2 - 4x$$ $$x = \frac{-(-4)}{2(1)} = \frac{4}{2} = 2$$ Choose x-values to the right and left of 2. <table><tr><th>x</th><th>y</th></tr><tr><td>-1</td><td>5</td></tr><tr><td>0</td><td>0</td></tr><tr><td>2</td><td>-4</td></tr><tr><td>4</td><td>0</td></tr><tr><td>5</td><td>5</td></tr></table> 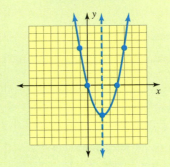	p. 842
The x-intercepts of a parabola are the points where the parabola intersects the x-axis. They may be found by setting y equal to 0, and then solving for x.	To find the x-intercepts of $$y = x^2 - 4x - 6$$ set $y = 0$ and solve. $$0 = x^2 - 4x - 6$$ $$x = \frac{4 \pm \sqrt{(-4)^2 - 4(-6)}}{2}$$ $$= \frac{4 \pm \sqrt{40}}{2} = \frac{4 \pm 2\sqrt{10}}{2}$$ $$= 2 \pm \sqrt{10}$$ The x-intercepts are $(2 - \sqrt{10}, 0)$ and $(2 + \sqrt{10}, 0)$.	p. 844

Summary Exercises

This summary exercise set is provided to give you practice with each of the objectives of this chapter. Each exercise is keyed to the appropriate chapter section. When you are finished, you can check your answers to the odd-numbered exercises against those presented in the back of the text. If you have difficulty with any of these questions, go back and reread the examples from that section. Your instructor will give you guidelines on how to best use these exercises in your instructional setting.

[10.1] Solve each of the following equations for x by the square root method.

1. $x^2 = 10$

2. $x^2 = 48$

3. $x^2 - 20 = 0$

4. $x^2 + 2 = 8$

5. $(x - 1)^2 = 5$

6. $(x + 2)^2 = 8$

7. $(x + 3)^2 = 5$

8. $64x^2 = 25$

9. $4x^2 = 27$

10. $9x^2 = 20$

11. $25x^2 = 7$

12. $7x^2 = 3$

[10.2] Solve each of the following equations by completing the square.

13. $x^2 - 3x - 10 = 0$

14. $x^2 - 8x + 15 = 0$

15. $x^2 - 5x + 2 = 0$

16. $x^2 - 2x - 2 = 0$

17. $x^2 - 4x - 4 = 0$

18. $x^2 + 3x = 7$

19. $x^2 - 4x = -2$

20. $x^2 + 3x = 5$

21. $x^2 - x = 7$

22. $2x^2 + 6x = 12$

23. $2x^2 - 4x - 7 = 0$

24. $3x^2 + 5x + 1 = 0$

[10.3] Solve each of the following equations using the quadratic formula.

25. $x^2 - 5x - 14 = 0$

26. $x^2 - 8x + 16 = 0$

27. $x^2 + 5x - 3 = 0$

28. $x^2 - 7x - 1 = 0$

29. $x^2 - 6x + 1 = 0$

30. $x^2 - 3x + 5 = 0$

31. $3x^2 - 4x = 2$

32. $2x - 3 = \dfrac{3}{x}$

33. $(x - 1)(x + 4) = 3$

34. $x^2 - 5x + 7 = 5$

35. $2x^2 - 8x = 12$

36. $5x^2 = 15 - 15x$

Solve by factoring or by any of the methods of this chapter.

37. $5x^2 = 3x$

38. $(2x - 3)(x + 5) = -11$

39. $(x - 1)^2 = 10$

40. $2x^2 = 7$

41. $2x^2 = 5x + 4$

42. $2x^2 - 4x = 30$

43. $2x^2 = 5x + 7$

44. $3x^2 - 4x = 2$

45. $3x^2 + 6x - 15 = 0$

46. $x^2 - 3x = 2(x + 5)$

47. $x - 2 = \dfrac{2}{x}$

48. $3x + 1 = \dfrac{5}{x}$

[10.4] Graph each quadratic equation after completing the table of values.

49. $y = x^2 + 3$

x	y
-2	
-1	
0	
1	
2	

50. $y = x^2 - 2$

x	y
-2	
-1	
0	
1	
2	

51. $y = x^2 - 3x$

x	y
-1	
0	
1	
2	
3	

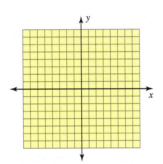

52. $y = x^2 + 4x$

x	y
-4	
-3	
-2	
-1	
0	

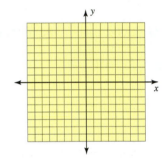

53. $y = x^2 - x - 2$

x	y
-1	
0	
1	
2	
3	

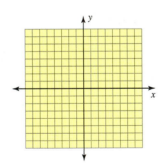

54. $y = x^2 - 4x + 3$

x	y
0	
1	
2	
3	
4	

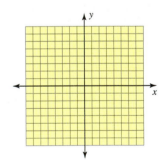

55. $y = x^2 + 2x - 3$

x	y
-3	
-2	
-1	
0	
1	

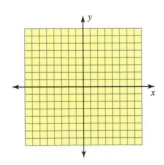 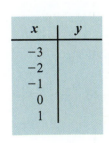

56. $y = 2x^2$

x	y
-2	
-1	
0	
1	
2	

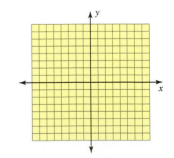 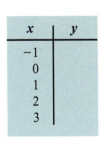

57. $y = 2x^2 - 3$

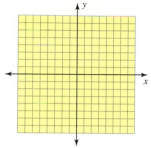

x	y
-2	
-1	
0	
1	
2	

58. $y = -x^2 + 3$

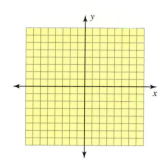

x	y
-2	
-1	
0	
1	
2	

59. $y = -x^2 - 2$

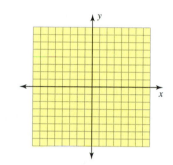

x	y
-2	
-1	
0	
1	
2	

60. $y = -x^2 + 4x$

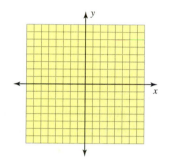

x	y
0	
1	
2	
3	
4	

For each equation, identify the axis of symmetry, create a suitable table of values, and sketch the graph.

61. $y = x^2 + 6x + 3$

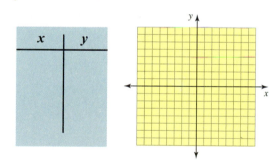

x	y

62. $y = -x^2 - 4x + 3$

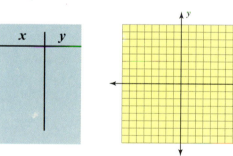

x	y

63. $y = -x^2 + 5x + 2$

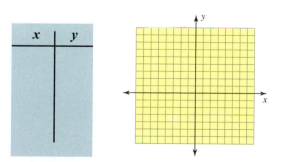

x	y

64. $y = x^2 - 3x + 4$

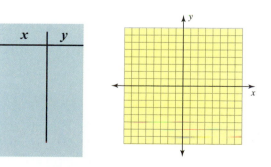

x	y

For each equation, find the x-intercepts of the graph.

65. $y = x^2 + 2x - 15$ **66.** $y = x^2 - 3x - 4$

67. $y = x^2 + 6x + 3$ **68.** $y = x^2 - 3x + 4$

69. $y = -x^2 + 2x - 3$ **70.** $y = -x^2 - 4x + 1$

[10.5]   Solve each of the following applications. Round results to the nearest hundredth.

71. Geometry. The longer leg of a right triangle is 3 cm less than twice the shorter leg. The hypotenuse is 18 cm long. Find the length of each leg.

72. Construction. A rectangular garden is to be surrounded by a walkway of constant width. The garden's dimensions are 15 ft by 24 ft. The total area, garden plus walkway, is to be 750 ft². What must be the width of the walkway?

73. Science and medicine. If a ball is thrown vertically upward from a height of 5 ft, with an initial velocity of 80 ft/s, its height h after t s is given by $h = -16t^2 + 80t + 5$. Find the time it takes for the ball to reach a height of 75 ft.

74. Science and medicine. If a ball is thrown vertically upward from a height of 5 ft, with an initial velocity of 80 ft/s, its height h after t s is given by $h = -16t^2 + 80t + 5$. How long does it take the ball to return to the ground?

75. Business and finance. Suppose that the cost C, in dollars, of producing x items is given by $C = 3x^2 - 50x + 1800$. How many items can be produced for $13,000?

76. Business and finance. The demand equation for a certain computer chip is given by

$D = -2p + 25$

The supply equation is predicted to be

$S = -p^2 + 16p + 5$

Find the equilibrium price.

Name _____

Section _____ Date _____

The purpose of this self-test is to help you check your progress and to review for the next in-class exam. Allow yourself about an hour to take this test. At the end of that hour check your answers against those given in the back of the text. Section references accompany the answers. If you missed any questions, go back to those sections and reread the examples until you master the concepts.

Solve each of the following equations for x.

1. $x^2 = 15$

2. $x^2 - 8 = 0$

3. $(x - 1)^2 = 7$

4. $9x^2 = 10$

Solve each of the following equations by completing the square.

5. $x^2 - 2x - 8 = 0$

6. $x^2 + 3x - 1 = 0$

7. $x^2 + 2x - 5 = 0$

8. $2x^2 - 5x + 1 = 0$

Solve each of the following equations by using the quadratic formula.

9. $x^2 - 2x - 3 = 0$

10. $x^2 - 6x + 9 = 0$

11. $x^2 - 5x = 2$

12. $2x^2 = 2x + 5$

13. $2x - 1 = \dfrac{4}{x}$

14. $(x - 1)(x + 3) = 2$

Graph each quadratic equation after completing the given table of values.

15. $y = x^2 + 4$

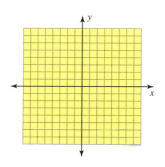

x	y
-2	
-1	
0	
1	
2	

1. _____

2. _____

3. _____

4. _____

5. _____

6. _____

7. _____

8. _____

9. _____

10. _____

11. _____

12. _____

13. _____

14. _____

15. _____

16. _____

17. _____

18. _____

19. _____

20. _____

16. $y = x^2 - 2x$

x	y
-1	
0	
1	
2	
3	

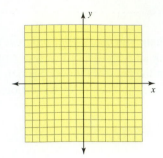

17. $y = x^2 + x - 2$

x	y
-2	
-1	
0	
1	
2	

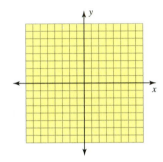

Give the axis of symmetry and the x-intercepts.

18. $y = x^2 - 4x - 21$

Solve the following applications. Round results to the nearest hundredth.

19. Geometry. If the length of a rectangle is 4 ft less than 3 times its width and the area of the rectangle is 96 ft^2, what are the dimensions of the rectangle?

20. Science and medicine. If a pebble is dropped toward a pond from the top of a 180-ft building, its height h after t s is given by $h = -16t^2 + 180$. How long does it take the pebble to pass through a height of 30 ft?

ACTIVITY 8: MONETARY EXCHANGE—PREDICTING THE FUTURE

Each activity in this text is designed to either enhance your understanding of the topics of the preceding chapter, provide you with a mathematical extension of those topics, or both. The activities can be undertaken by one student, but they are better suited for a small group project. Occasionally it is only through discussion that different facets of the activity become apparent. For material related to this activity, visit the text website at www.mhhe.com/streeter.

Although no one can predict the future completely, scientists, economists, and analysts often use data that has been gathered (usually over a time period) to determine whether a trend might exist. If such a trend is evident, one might use the graph of the data to help predict a future event. For example, if the value of a certain item has risen steadily for the past 10 years, one might predict that the value will continue to rise in the eleventh year.

In this activity, we will work with currency data (as we did in Activity 2). We use this data to explore linear as well as nonlinear graphs (as we demonstrated in Chapter 3). We are also going to try to make predictions based on our graphs.

I.

1. Connect to the Internet, open up a web browser, and go to the website containing exchange rate information that you *bookmarked* in Activity 2. [*Note:* If you did not do this in Activity 2, search the Web (see Activity 1) for a site containing currency exchange rate information.]
2. Select a country that you would like to visit (you may use your choice from Activity 2). Calculate how much you would receive if you exchanged US$1 (see Activity 2).
3. Calculate how much you would receive if you exchanged US$2, US$3, US$4, and US$5.
4. Make a scatterplot of the data found in steps 2 and 3. On the horizontal axis, set up a scale of {$1, $2, $3, $4, $5} and label the horizontal axis "US$." On the vertical axis, set up a nice scale based on the currency you chose, *in terms of U.S. dollars*. Label the vertical axis with the currency you chose, and plot the five points [as ordered pairs (x, y) = (US$, foreign currency equivalent)].
5. Write an equation using variables, but fixing the exchange at today's rate.
6. Describe the trend of your data points. Based on your graph and the equation given in step 5, predict how much of your currency is equivalent to US$6 and US$20.

II.

1. Search for information concerning past exchange rates between your chosen country (from part I) and U.S. dollars. Bookmark this site.
2. Find the average exchange rate for the years 1980, 1985, 1990, 1995, and 2000.
3. Make a scatterplot: Label the horizontal axis "Years" and the vertical axis "Exchange Rate." Scale the axes appropriately. Plot the points (Year, Exchange Rate).
4. Describe the trend of your data points. Based on your graph, predict how much of your currency is equivalent to US$1 in 1975, 1987, 1997, and 2005. Can you give an equation for the graph of this trend? Justify your answer.

Sample Data Set

I. The numbers below relate to Part I on page 873.

 2. We are using our choice of Canada from Activity 2 with the 12/14/01 exchange rate of 1.561.

 3. Remember to round to the nearest hundredth.

US$	1	2	3	4	5
CAN$	1.56	3.12	4.68	6.24	7.81

4.

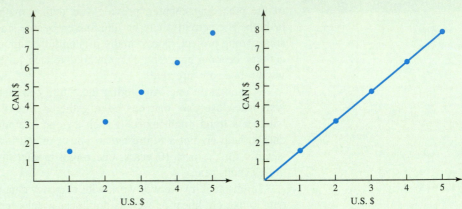

 5. Using the description in Activity 2, we have $y = 1.561x$.

 6. The data are increasing in a linear fashion. The line has a slope of 1.561 and a y-intercept at the origin. Exchanging US$6, we would receive CAN$9.37. We would receive CAN$31.22 for US$20.

II. The number below relates to Part II on page 873.

 2.

Years	1980	1985	1990	1995	2000
Rate	1.1692	1.3655	1.1668	1.3724	1.4851

Source: University of British Columbia.

Name _____

Section _____ Date _____

The following exercises are presented to help you review concepts from earlier chapters. This is meant as a review and not as a comprehensive exam. The answers are presented in the back of the text. Section references accompany the answers. If you have difficulty with any of these exercises, be certain to at least read through the summary related to those sections.

Simplify the following expressions.

1. $6x^2y - 4xy^2 + 5x^2y - 2xy^2$

2. $(3x^2 + 2x - 5) - (2x^2 - 3x + 2)$

Evaluate each expression when $x = 2$, $y = -3$, and $z = 4$.

3. $4x^2y - 3z^2y^2$

4. $-3x^2y^2z^2 - 2xyz$

5. Solve for x: $4x - 2(3x - 5) = 8$.

6. Solve the inequality $4x + 15 > 2x + 19$.

Perform the indicated operations.

7. $3xy(2x^2 - x + 5)$

8. $(2x + 5)(3x - 2)$

9. $(3x + 4y)(3x - 4y)$

Factor each of the following completely.

10. $16x^2y^2 - 8xy^3$

11. $8x^2 - 2x - 15$

12. $25x^2 - 16y^2$

Perform the indicated operations.

13. $\dfrac{7}{4x + 8} - \dfrac{5}{7x + 14}$

14. $\dfrac{5x + 5}{x - 2} \cdot \dfrac{x^2 - 4x + 4}{x^2 - 1}$

15. $\dfrac{3x^2 + 8x - 3}{15x^2} + \dfrac{3x - 1}{5x^2}$

16. _____

17. _____

18. _____

19. _____

20. _____

21. _____

22. _____

23. _____

24. _____

25. _____

Graph the following equations.

16. $3x - 2y = 6$

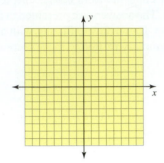

17. $y = 4x - 5$

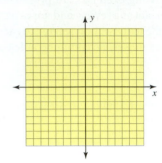

18. Find the slope of the line through the points $(2, 9)$ and $(-1, -6)$.

19. Given that the slope of a line is 2 and the y-intercept is $(0, -5)$, write the equation of the line.

20. Graph the inequality $x + 2y < 6$.

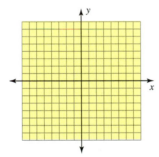

Solve each of the following systems. If a unique solution does not exist, state whether the system is inconsistent or dependent.

21. $2x - 3y = 6$
$x - 3y = 2$

22. $2x + y = 4$
$y = 2x - 8$

23. $5x + 2y = 8$
$x - 4y = 17$

24. $2x - 6y = 8$
$x = 3y + 4$

Solve each of the following applications. Be sure to show the system of equations used for your solution.

25. One number is 4 less than 5 times another. If the sum of the numbers is 26, what are the two numbers?

26. Receipts for a concert attended by 450 people were $2775. If reserved-seat tickets were $7 and general admission tickets were $4, how many of each type of ticket were sold?

27. A chemist has a 30% acid solution and a 60% solution already prepared. How much of each of the two solutions should be mixed to form 300 mL of a 50% solution?

Evaluate each root, if possible.

28. $\sqrt{169}$

29. $-\sqrt{169}$

30. $\sqrt{-169}$

31. $\sqrt[3]{-64}$

Simplify each of the following radical expressions.

32. $\sqrt{12} + 3\sqrt{27} - \sqrt{75}$

33. $3\sqrt{2a} \cdot 5\sqrt{6a}$

34. $(\sqrt{2} - 5)(\sqrt{2} + 3)$

35. $\dfrac{8 - \sqrt{32}}{4}$

Solve each of the following equations.

36. $x^2 - 72 = 0$

37. $x^2 + 6x - 3 = 0$

38. $2x^2 - 3x = 2(x + 1)$

Graph each of the following quadratic equations.

39. $y = x^2 - 2$

40. $y = x^2 - 4x$

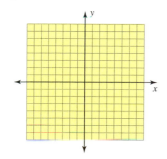

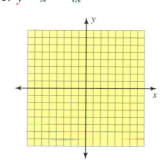

ANSWERS

26.

27.

28.

29.

30.

31.

32.

33.

34.

35.

36.

37.

38.

39.

40.

Solve the following applications.

41. The equation $h = -16t^2 - 64t + 250$ gives the height of a ball, thrown downward from the top of a 250-ft building with an initial velocity of 64 ft/s. Find the time it takes for the ball to reach a height of 100 ft.

42. The demand equation for a certain type of printer is given by

$$D = -120p + 16{,}000$$

The supply equation is predicted to be

$$S = -p^2 + 260p - 9000$$

Find the equilibrium price.

Final Examination

Name _____

Section _____ Date _____

Evaluate the given expressions.

1. $|-25| - |-11|$

2. $16 + (-22)$

3. $(-41) - (-15)$

4. $(-5)(-3)(-7)$

5. $\dfrac{3(-2) - 8}{-7 - (-4)(3)}$

6. $6 - 2^3 \cdot 5$

Evaluate the expressions for the given values of the variables.

7. $b^2 - 4ac$ for $a = -3$, $b = -4$, and $c = 2$

8. $-x^2 - 7x - 3$ for $x = -2$

9. Write the expression $9 \cdot p \cdot p \cdot p \cdot q \cdot q \cdot q \cdot q \cdot q$ in exponential form.

Simplify the expressions using the properties of exponents. Write all answers using positive exponents only.

10. $z^{-11}z^5$

11. $(5c^8d^7)^2$

12. $\dfrac{4x^8y^5z^3}{2x^6y^9z^7}$

Perform the indicated operations. Write each answer in simplified form.

13. $2x(x + 3) + 5$

14. $(6x^2 - 3x - 20) - 2(4x^2 - 16x + 11)$

15. $(7x - 9)(4x + 5)$

16. $(3x + 4y)(3x - 4y)$

17. $(5x - 2y)^2$

18. $x(x - 3y) - 2y(y + 6x)$

Solve the given equations.

19. $5x - 9 = -7x - 3$

20. $2x - 3(x - 2) = 8$

21. $\dfrac{5 - x}{-2} = 3x$

22. Solve the inequality $4x - 3 > 6x - 2$.

ANSWERS

1. _____ 2. _____

3. _____ 4. _____

5. _____ 6. _____

7. _____ 8. _____

9. _____

10. _____

11. _____

12. _____

13. _____

14. _____

15. _____

16. _____

17. _____

18. _____

19. _____

20. _____

21. _____

22. _____

ANSWERS

23. _____

24. _____

25. _____

26. _____

27. _____

28. _____

29. _____

30. _____

31. _____

32. _____

33. _____

34. _____

35. _____

36. _____

37. _____

38. _____

Factor each expression completely.

23. $10x^2 - 490$

24. $x^2 - 12x + 36$

25. $4p^2 - p - 18$

26. $3xy + 3xz - 5y - 5z$

Simplify the given expression.

27. $\dfrac{2y - 36}{3y^2 - 54y}$

28. $\dfrac{4z}{z + 8} + \dfrac{32}{z + 8}$

29. $\dfrac{6a}{a^2 - 9} - \dfrac{5a}{a^2 + a - 6}$

30. $\dfrac{y^2 + y - 2}{y + 5} \cdot \dfrac{3y + 3}{9y - 9}$

31. $\dfrac{x^2 - 3x - 10}{3x} \div \dfrac{5x - 25}{15x^2}$

Solve the given equation by the indicated method.

32. $3x^2 + 5x - 2 = 0$ by factoring

33. $x^2 + x - 3 = 0$ by using the quadratic formula

34. $x^2 + 6x = 5$ by completing the square

35. Solve the equation $\dfrac{1}{x - 1} + \dfrac{x + 1}{x^2 + 2x - 3} = \dfrac{1}{x + 3}$

36. Find the slope of the line through the points $(2, -3)$ and $(5, 9)$.

37. Find the slope of the line whose equation is $3x - 4y = 12$.

38. Find the equation of the line that passes through the point $(4, -2)$ and is parallel to the line $2x + y = 6$.

39. _____

40. _____

41. _____

42. _____

43. _____

44. _____

45. _____

46. _____

47. _____

39. Graph the line whose equation is $4x + 5y = 20$.

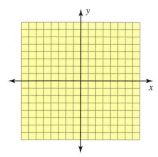

In exercises 40 to 43, perform the indicated operations and simplify the result.

40. $-\sqrt{\dfrac{64}{25}}$

41. $\sqrt{18x^5y^6}$

42. $3\sqrt{20} - 2\sqrt{125}$

43. $(\sqrt{5} + 2)(\sqrt{5} - 8)$

44. Solve the system of equations

$$2x + 3y = 4$$
$$4x - 2y = 8$$

45. Determine the equation of the line in the following graph.

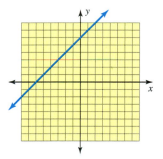

46. The length of a rectangle is 2 cm more than 3 times the width. The perimeter is 44 cm. Find the length.

47. Find the equation of the line that passes through the points $(-1, -3)$ and $(2, 6)$.

48. One number is 3 more than 6 times another. If the sum of the numbers is 38, find the two numbers.

49. A store marks up items to make a 30% profit. If an item sells for $3.25, what does it cost before the mark–up?

50. Graph the equation $y = x^2 - 3x + 2$.

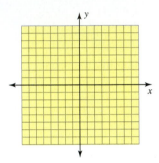

Answers to Pre-Tests, Summary Exercises, Self-Tests, and Cumulative Reviews

Pre-Test Chapter 0

[0.1] 1. 1, 2, 3, 6, 7, 14, 21, 42
2. Prime: 2, 3, 7, 17, 23; composite: 6, 9, 18, 21
3. (a) $2 \times 2 \times 3 \times 5$; (b) $2 \times 5 \times 5 \times 7$ **4.** (a) 4; (b) 6
5. (a) 20; (b) 180   **[0.2] 6.** $\frac{5}{4}$ **7.** $\frac{3}{2}$ **8.** $\frac{19}{12}$
9. $\frac{7}{18}$ **[0.3] 10.** 3.767   **11.** 22.8404   **[0.2] 12.** $12\frac{3}{5}$
13. $\frac{3}{4}$ **[0.3] 14.** 2.435   **[0.4] 15.** 6   **16.** 8
17. 13   **18.** 2   **[0.3] 19.** (a) $\frac{23}{100}$; (b) 0.23   **20.** $3\frac{1}{2}\%$

[0.5] 21.

22. $-4, -2, -1, 0, 1, 5$ **23.** Max: 7; min: -5
24. 5   **25.** 6   **26.** 6   **27.** 6   **28.** 6 **29.** 16
30. -23

Summary Exercises

1. 1, 2, 4, 13, 26, 52   **3.** 1, 2, 4, 19, 38, 76
5. Prime: 2, 5, 7, 11, 17, 23, 43; composite: 14, 21, 27, 39
7. $2^2 \cdot 3 \cdot 5 \cdot 7$ **9.** $2^2 \cdot 3^2 \cdot 5$ **11.** 1   **13.** 60
15. 16   **17.** 36   **19.** $\frac{6}{22}, \frac{9}{33}, \frac{12}{44}$ **21.** $\frac{3}{8}$ **23.** $\frac{1}{6}$
25. $\frac{5}{6}$ **27.** $\frac{31}{36}$ **29.** $\frac{7}{54}$ **31.** 4.637   **33.** $10\frac{2}{7}$
35. $4\frac{1}{3}$ **37.** $9\frac{3}{5}$ **39.** $1\frac{1}{2}$ **41.** 8.222   **43.** $\frac{1}{50}$
45. $\frac{3}{8}$ **47.** $2\frac{1}{3}$ **49.** 0.75   **51.** 0.0625   **53.** 0.006
55. 6%   **57.** 240%   **59.** 3.5%   **61.** 43%
63. 40%   **65.** $266\frac{2}{3}\%$ **67.** 75   **69.** 400   **71.** 25
73. 16   **75.** 27   **77.** 169   **79.** 15
81.

83. $-8, -7, -3, 0, 1, 2, 3, 7, 8$ **85.** Max: 8; min: -9 **87.** 63
89. 9   **91.** -9 **93.** 20   **95.** 6   **97.** 14

Self-Test for Chapter 0

[0.1] 1. Prime: 5, 13, 17, 31; Composite: 9, 22, 27, 45
2. $2 \cdot 2 \cdot 2 \cdot 3 \cdot 11$ **3.** 12   **4.** 8   **5.** 108   **6.** 36
[0.2–0.3] 7. $\frac{2}{7}$ **8.** $\frac{3}{4}$ **9.** $\frac{19}{12}$ **10.** $\frac{2}{21}$ **11.** 7.375
12. 3.884   **13.** 22.9635   **14.** 79.91   **15.** $3\frac{3}{7}$
16. $7\frac{11}{12}$ **17.** $1\frac{11}{18}$ **18.** 7.35   **[0.3] 19.** $\frac{7}{100}$
20. $\frac{72}{100}$ or $\frac{18}{25}$ **21.** 0.42   **22.** 0.06   **23.** 1.6

24. 3%   **25.** 4.2%   **26.** 40%   **27.** 62.5%
[0.4] 28. 4^4 **29.** 9^5 **30.** 3   **31.** 65   **32.** 144
33. 7   **34.** 8   **35.** 7
[0.5] 36.

37. $-6, -3, -2, 0, 2, 4, 5$ **38.** Max: 6; min: -5 **39.** 7
40. 7   **41.** 11   **42.** 11   **43.** -19 **44.** -40
45. 19

Pre-Test Chapter 1

[1.1] 1. $x - 8$ **2.** $\frac{w}{17x}$ **3.** Not an expression
4. An expression   **[1.2] 5.** Commutative prop. of mult.
6. Distributive property   **7.** Associative property of addition
[1.3–1.4] 8. -10 **9.** -1 **10.** -5 **11.** -1
12. -3 **13.** -19 **14.** 12   **15.** 0   **16.** 21
17. 14   **18.** $\frac{1}{2}$ **19.** 7   **[1.5] 20.** -3 **21.** 55
22. -7 **[1.6] 23.** $8w^2t$ **24.** $-a^2 + 4a + 3$
[1.7] 25. $12xy^2$

Summary Exercises

1. $y + 5$ **3.** $8a$ **5.** $5mn$ **7.** $17x + 3$ **9.** Yes
11. No   **13.** Associative property of addition
15. Associative property of multiplication   **17.** $72 = 72$
19. $20 = 20$ **21.** $80 = 80$ **23.** $3 \cdot 7 + 3 \cdot 4$
25. $4w + 4v$ **27.** $3 \cdot 5a + 3 \cdot 2$ **29.** -11 **31.** 0
33. -18 **35.** -4 **37.** -5 **39.** 17   **41.** 0
43. 3   **45.** 9   **47.** 1   **49.** 5   **51.** 8   **53.** 13
55. -70 **57.** 45   **59.** 0   **61.** $-\frac{3}{2}$ **63.** 5
65. 9   **67.** -4 **69.** -1 **71.** -7 **73.** 75
75. 400   **77.** 25   **79.** 16   **81.** 27   **83.** 169
85. 46   **87.** 80   **89.** -81 **91.** -6 **93.** -3
95. 6   **97.** $5x^2, -7x, 3$ **99.** $4ab^2, ab^2, -3ab^2$
101. $7x$ **103.** $3c$ **105.** $7ab^2$ **107.** $x + 5y$
109. $a^3 + 2a^2 + 3a$ **111.** $7x^2$ **113.** a **115.** m^4
117. $3x^4$ **119.** $12x^8$ **121.** $4ab^3c^4$ **123.** $36x^3y$
125. $10x^4y^4$ **127.** $23 - x$ **129.** $x + 5$ **131.** $x + 4$
133. $x, 25 - x$

Self-Test for Chapter 1

[1.1] 1. $a - 5$ **2.** $6m$ **3.** $4(m + n)$ **4.** $\frac{a + b}{3}$
[1.2] 5. Commutative property of multiplication
6. Distributive property   **7.** Associative property of addition
8. 21   **9.** $20x + 12$ **10.** Not an expression
11. Expression   **[1.3] 12.** -13 **13.** -3 **14.** -21
15. 1   **16.** -6 **17.** -21 **18.** 9   **19.** 0
20. 3   **21.** 1   **[1.4] 22.** -40 **23.** 63   **24.** -27
25. -24 **26.** 14   **[1.5] 27.** -25 **28.** 3
29. -5 **30.** Undefined   **31.** 17   **32.** 65
33. 144   **34.** -9 **35.** -4 **[1.6] 36.** $13a$

37. $19x + 5y$ **38.** $8a^2$ **[1.7] 39.** a^{14} **40.** $8x^7y^3$
41. $3x^6$ **42.** $4ab^3$ **43.** x^9 **44.** $2x - 8$
45. $2w + 4$

Pre-Test Chapter 2

[2.1] 1. No **2.** Yes **[2.2–2.3] 3.** 8 **4.** -12
5. 7 **6.** 35 **7.** -2 **8.** 2
[2.4] 9. $W = \dfrac{P - 2L}{2}$ or $\dfrac{P}{2} - L$

10. $y = \dfrac{5x - 14}{3}$ or $\dfrac{5}{3}x - \dfrac{14}{3}$ **[2.5] 11.** $4x + 5 = 17$

12. $4(y + 6) = 10y + 6$ **[2.7] 13.** $x \le 15$ **14.** $x \le -1$
[2.6–2.7] 15. 6 **16.** 15, 17 **17.** 4 cm $\times$ 13 cm
18. $540 **19.** $13,125 **20.** 15%

Summary Exercises

1. Yes **3.** Yes **5.** No **7.** 2 **9.** -7 **11.** 5
13. 1 **15.** -7 **17.** 7 **19.** -4 **21.** 32
23. 27 **25.** 3 **27.** -2 **29.** $\dfrac{7}{2}$ **31.** 18
33. 6 **35.** $\dfrac{2}{5}$ **37.** 6 **39.** 6 **41.** 4
43. 5 **45.** $\dfrac{1}{2}$ **47.** $\dfrac{P - 2W}{2}$ or $\dfrac{P}{2} - W$ **49.** $\dfrac{2A}{b}$
51. $mq + p$ **53.** 8 **55.** 17, 19, 21
57. Susan: 7 years, Larry: 9 years, Nathan: 14 years
59. 22% **61.** $18,200 **63.** $2800 **65.** 6.5%
67. $3150 before, $3276 after **69.** 500 s **71.** $114.50

73.
75.
77.
79.
81.
83.

Self-Test for Chapter 2

[2.1] 1. No **2.** Yes **[2.1–2.3] 3.** 11 **4.** 12
5. 7 **6.** 7 **7.** -12 **8.** 25 **9.** 3 **10.** 4
11. $-\dfrac{2}{3}$ **12.** $-\dfrac{9}{4}$ **[2.4] 13.** $\dfrac{C}{2\pi}$ **14.** $\dfrac{3V}{B}$
15. $\dfrac{6 - 3x}{2}$ **[2.7] 16.** $x \le 14$
17. $x < -4$
18. $x \ge \dfrac{4}{3}$

19. $x > -1$
[2.4–2.6] 20. 7 **21.** 21, 22, 23 **22.** Juwan, 6; Jan, 12;
Rick, 17 **23.** 10 in., 21 in. **24.** 5% **25.** $35,000

Cumulative Review Chapters 0–2

[1.3–1.5] 1. 4 **2.** -12 **3.** 8 **4.** 3 **5.** -18
6. 44 **7.** -5 **8.** 10 **9.** 0 **10.** Undefined
[1.5] 11. 20 **12.** -11 **13.** 27 **14.** -28
15. -4 **16.** 2 **[1.6–1.7] 17.** $3x^2y$ **18.** $6x^4 - 10x^3y$
19. $x - 2y + 3$ **20.** $12x^2 + 3x$ **[2.1–2.3] 21.** 5
22. -24 **23.** $\dfrac{5}{4}$ **24.** $-\dfrac{2}{5}$ **25.** 5 **[2.4] 26.** $\dfrac{I}{pt}$
27. $\dfrac{2A}{b}$ **28.** $\dfrac{c - ax}{b}$
[2.7] 29. $x < 3$
30. $x \le -\dfrac{3}{2}$
31. $x > 4$
32. $x \ge \dfrac{4}{3}$

[2.4–2.6] 33. 13 **34.** 42, 43 **35.** 7 **36.** $420
37. 5 cm, 17 cm **38.** 8 in., 13 in., 16 in. **39.** 2.5%
40. 7.5%

Pre-Test Chapter 3

[3.1] 1. x^{12} **2.** $8x^5y^7$ **3.** $3x^3y$ **[3.2] 4.** $4x^6y^8$
5. x^{16} **6.** $\dfrac{2y^3}{x^5}$ **[3.3] 7.** Binomial **8.** Trinomial
9. $2x^2 - 2x - 2$ **10.** $9x^2 - 11x + 6$
[3.4] 11. $12x^3y^3 - 6x^2y^2 + 21x^2y^4$ **12.** $6x^2 - 11x - 10$
[3.5] 13. $x^2 - 4y^2$ **14.** $16m^2 + 40m + 25$
15. $3x^3 - 14x^2y + 17xy^2 - 6y^3$ **16.** $9x^3 - 30x^2y + 25xy^2$
[3.6] 17. $4y - 5x^2y^3$ **18.** $x + 2$ **19.** $x - 3$
20. $3x - 5 - \dfrac{5}{x + 4}$

Summary Exercises

1. x^7 **3.** x **5.** $2p^2$ **7.** $5m^5n^2$ **9.** $8p^2q^2$
11. $4a^2b^2$ **13.** $72x^{12}y^8$ **15.** x **17.** $27y^{12}$
19. -4 **21.** -19 **23.** Binomial **25.** Trinomial
27. Binomial **29.** $9x, 1$ **31.** $x + 5, 1$
33. $7x^6 + 9x^4 - 3x, 6$ **35.** 1 **37.** 1 **39.** $\dfrac{1}{3^3}$
41. $\dfrac{4}{x^4}$ **43.** $\dfrac{1}{m^2}$ **45.** $\dfrac{x^5}{y^5}$ **47.** $\dfrac{1}{a^{18}}$
49. 5.1×10^4 cps **51.** 3.22×10^9 **53.** 2×10^{17}
55. $21a^2 - 2a$ **57.** $5y^3 + 4y$ **59.** $5x^2 + 3x + 10$
61. $x - 2$ **63.** $-9w^2 - 10w$ **65.** $9b^2 + 8b - 2$
67. $2x^2 - 2x - 9$ **69.** $5a^5$ **71.** $54p^5$ **73.** $15x - 40$
75. $-10r^3s^2 + 25r^2s^2$ **77.** $x^2 + 9x + 20$ **79.** $a^2 - 49b^2$
81. $a^2 + 7ab + 12b^2$ **83.** $6x^2 - 19xy + 15y^2$
85. $y^3 - y + 6$ **87.** $x^3 - 8$ **89.** $2x^3 - 2x^2 - 60x$

91. $x^2 + 14x + 49$ **93.** $4w^2 - 20w + 25$
95. $a^2 + 14ab + 49b^2$ **97.** $x^2 - 25$ **99.** $4m^2 - 9$
101. $25r^2 - 4s^2$ **103.** $2x^3 - 20x^2 + 50x$ **105.** $3a^3$
107. $3a - 2$ **109.** $-3rs + 6r^2$ **111.** $x - 5$
113. $x - 3 + \dfrac{2}{x - 5}$ **115.** $x^2 + 2x - 1 + \dfrac{-4}{6x + 2}$
117. $x^2 + x + 2 + \dfrac{1}{x + 2}$

Self-Test for Chapter 3

[3.1] 1. a^{14} **2.** $15x^3y^7$ **3.** $2x^3$ **4.** $4ab^3$
5. $27x^6y^3$ **6.** $\dfrac{4w^4}{9t^6}$ **7.** $16x^{18}y^{17}$ **8.** 6
9. Binomial   **10.** Trinomial
11. $8x^4 - 3x^2 - 7$; 8, -3, -7; 4   **[3.2] 12.** 1   **13.** 6
14. $\dfrac{1}{y^5}$ **15.** $\dfrac{3}{b^7}$ **16.** $\dfrac{1}{y^4}$ **17.** $\dfrac{1}{p^{10}}$
[3.3] 18. $10x^2 - 12x - 7$ **19.** $7a^3 + 11a^2 - 3a$
20. $3x^2 + 11x - 12$ **21.** $b^2 - 7b - 5$ **22.** $7a^2 - 10a$
23. $4x^2 + 5x - 6$ **24.** $2x^2 - 7x + 5$
[3.4] 25. $15a^3b^2 - 10a^2b^2 + 20a^2b^3$ **26.** $3x^2 + x - 14$
27. $2x^3 + 7x^2y - xy^2 - 2y^3$ **28.** $8x^2 - 14xy - 15y^2$
29. $12x^3 + 11x^2y - 5xy^2$ **[3.5] 30.** $9m^2 + 12mn + 4n^2$
31. $a^2 - 49b^2$ **[3.6] 32.** $2x^2 - 3y$ **33.** $4c^2 - 6 + 9cd$
34. $x - 6$ **35.** $x + 2 + \dfrac{10}{2x - 3}$
36. $2x^2 - 3x + 2 + \dfrac{7}{3x + 1}$ **37.** $x^2 - 4x + 5 + \dfrac{-4}{x - 1}$
[3.2] 38. 1.68×10^{20} **39.** 3.12×10^{-10} **40.** 5.2×10^{19}

Cumulative Review Chapters 0–3

[1.3–1.4] 1. 17   **2.** 6   **3.** 150   **4.** 4   **[1.5] 5.** 55
6. $-\dfrac{26}{21}$ **[3.1] 7.** $9x^{16}$ **8.** $\dfrac{x^{10}}{y^6}$ **9.** $8x^9y^3$
[3.2] 10. 7   **11.** 1   **12.** $\dfrac{1}{x^4}$ **13.** $\dfrac{3}{x^2}$ **14.** $\dfrac{1}{x^4}$
15. $\dfrac{1}{x^3y^3}$ **[1.6] 16.** $4x^5y$ **[3.3] 17.** $x^2 + 7x$
18. $2x - 2y$ **[3.4] 19.** $x^2 - 2x - 15$
[3.5] 20. $x^2 + 2xy + y^2$ **21.** $9x^2 - 24xy + 16y^2$
[3.6] 22. $x + 4$ **[3.5] 23.** $x^3 - xy^2$ **[2.3] 24.** -2
25. -2 **26.** 84   **27.** 1   **[2.4] 28.** $B = 2A - b$
[2.7] 29. $x \geq -2$ **30.** $x < -22$
[2.5] 31. Sam: $510; Larry: $250   **32.** 37, 39
33. $2120   **34.** $645

Pre-Test Chapter 4

[4.1] 1. $5(3c + 7)$ **2.** $4q^3(2q - 5)$ **3.** $6(x^2 - 2x + 4)$
4. $7cd\,(c^2d - 3 + 2d^2)$ **[4.2] 5.** $(b - 3)(b + 5)$
6. $(x + 4)(x + 6)$ **7.** $(x - 9)(x - 5)$
8. $(a + 3b)(a + 4b)$ **[4.3] 9.** $(3y - 4)(y + 3)$
10. $(5w + 3)(w + 4)$ **11.** $(3x + 7y)(2x - 3y)$
12. $x(2x + 3)(x - 5)$ **[4.4] 13.** $(b + 7)(b - 7)$
14. $(6p + q)(6p - q)$ **15.** $(3x - 2y)^2$
16. $3x(3y - 4x)(3y + 4x)$ **[4.8] 17.** 4, 7   **18.** $-2, 7$
19. $\dfrac{3}{5}, -2$ **20.** 0, 2

Summary Exercises

1. $6(3a + 4)$ **3.** $8s^2(3t - 2)$ **5.** $7s^2(5s - 4)$
7. $9m^2n(2n - 3 + 2n^2)$ **9.** $8ab(a + 3 - 2b)$
11. $(x + y)(2x - y)$ **13.** $(x + 4)(x + 5)$
15. $(a - 4)(a + 3)$ **17.** $(x + 6)(x + 6)$
19. $(b - 7c)(b + 3c)$ **21.** $m(m + 7)(m - 5)$
23. $3y(y - 7)(y - 9)$ **25.** $(3x + 5)(x + 1)$
27. $(2b - 3)(b - 3)$ **29.** $(5x - 3)(2x - 1)$
31. $(3y - 5z)(3y + 4z)$ **33.** $4x(2x + 1)(x - 5)$
35. $3x(2x - 3)(x + 1)$ **37.** $(p + 7)(p - 7)$
39. $(m + 3n)(m - 3n)$ **41.** $(5 + z)(5 - z)$
43. $(5a + 6b)(5a - 6b)$ **45.** $3w(w + 2z)(w - 2z)$
47. $2(m + 6n^2)(m - 6n^2)$ **49.** $(x + 4)^2$
51. $(2x + 3)^2$ **53.** $x(4x + 5)^2$ **55.** $(x - 4)(x + 5)$
57. $(3x + 2)(2x - 5)$ **59.** $x(2x + 3)(3x - 2)$
61. $1, -\dfrac{3}{2}$ **63.** 0, 10   **65.** $-3, 5$ **67.** $2, \dfrac{5}{4}$
69. 0, 3   **71.** $-4, 4$

Self-Test for Chapter 4

[4.1] 1. $6(2b + 3)$ **2.** $3p^2(3p - 4)$ **3.** $5(x^2 - 2x + 4)$
4. $6ab(a - 3 + 2b)$ **[4.4] 5.** $(a - 5)(a - 5)$
6. $(8m + n)(8m - n)$ **7.** $(7x + 4y)(7x - 4y)$
8. $2b(4a + 5b)(4a - 5b)$ **[4.2] 9.** $(a - 7)(a + 2)$
10. $(b + 3)(b + 5)$ **11.** $(x - 4)(x - 7)$
12. $(y + 10z)(y + 2z)$ **[4.5] 13.** $(x + 2)(x - 5)$
14. $(2x - 3)(3x + 1)$ **[4.6–4.7] 15.** $(2x - 1)(x + 8)$
16. $(3w + 7)(w + 1)$ **17.** $(4x - 3y)(2x + y)$
18. $3x(2x + 5)(x - 2)$ **[4.8] 19.** 3, 5   **20.** $-1, 4$
21. $-1, \dfrac{2}{3}$ **22.** 0, 3   **23.** 0, 4   **24.** $-3, 8$ **25.** 7

Cumulative Review Chapters 0–4

[1.3–1.4] 1. 17   **2.** -2 **[3.3] 3.** $9x^2 - x - 5$
4. $-4a^2 - 2a - 5$ **5.** $6b^2 + 8b - 3$
[3.4] 6. $15r^3s^2 - 12r^2s^2 + 18r^2s^3$ **7.** $6a^3 - 5a^2b + 3ab^2 - b^3$
[3.6] 8. $-y^2 + 3xy - 2x^2$ **9.** $3a + 2$
10. $x^2 - 2x + \dfrac{5}{2x + 4}$ **[2.3] 11.** $x = -2$
[2.7] 12. $x \leq \dfrac{33}{5}$ **[2.4] 13.** $t = \dfrac{2S - na}{n}$
[3.1] 14. x^{17} **15.** $6x^5y^7$ **16.** $9x^4y^6$
17. $4xy^2$ **18.** $108x^8$ **[4.7] 19.** $12w^4(3w - 4)$
20. $5xy(x - 3 + 2y)$ **21.** $(5x + 3)^2$
22. $4p(p + 6q)(p - 6q)$ **23.** $(a + 3)(a + 1)$
24. $2w(w^2 - 2w - 12)$ **25.** $(3x + 2y)(x + 3y)$
[4.8] 26. 3, 4   **27.** $-4, 4$ **28.** $\dfrac{2}{3}, -1$
29. 6   **30.** 5 in. by 21 in.

Pre-Test Chapter 5

[5.1] 1. $\dfrac{-3b^6}{5a^2}$ **2.** $\dfrac{x + 4}{2}$ **3.** $\dfrac{x - 1}{2x}$ **[5.3] 4.** $\dfrac{13a}{6}$
5. 5   **6.** $x + 6$ **[5.4] 7.** $\dfrac{5w - 6}{2w^2}$ **8.** $\dfrac{3b + 3}{b(b - 3)}$
9. $\dfrac{-11}{6(x - 1)}$ **10.** $\dfrac{10}{x - 5}$ **[5.2] 11.** $\dfrac{a}{2b^2}$ **12.** $\dfrac{x + 3}{2x}$

13. $2b^2$ **14.** $\dfrac{x + y}{4x}$ **[5.7] 15.** $\dfrac{3x}{2}$ **16.** $\dfrac{y}{2y + x}$

[5.5] 17. 3 **18.** $-2, 5$ **19.** 40 **20.** 6 **21.** 5

22. 7 **[5.6] 23.** 5, 20

24. 48 mi/h going; 40 mi/h returning **25.** 15 ft, 40 ft

Summary Exercises

1. $\dfrac{2}{3a}$ **3.** $\dfrac{w^2 - 25}{2w - 8}$ **5.** $\dfrac{-m - 1}{m + 3}$ **7.** $\dfrac{2}{3x}$

9. $\dfrac{x}{2}$ **11.** $\dfrac{2}{3p}$ **13.** $\dfrac{2x - 3}{x}$ **15.** $\dfrac{a + b}{4a}$ **17.** $\dfrac{x}{3}$

19. $\dfrac{11}{x + 2}$ **21.** $\dfrac{2r - s}{r}$ **23.** 2 **25.** $\dfrac{7x}{6}$

27. $\dfrac{5m - 6}{2m^2}$ **29.** $\dfrac{3x + 3}{x(x - 3)}$ **31.** $\dfrac{3w - 5}{(w - 5)(w - 3)}$

33. $\dfrac{-11}{6(x - 1)}$ **35.** $\dfrac{5a}{(a + 4)(a - 1)}$ **37.** $\dfrac{2}{3x}$

39. $\dfrac{y + x}{y - x}$ **41.** $\dfrac{n - m}{n + m}$ **43.** $\dfrac{a + 3}{a - 3}$ **45.** None

47. $-1, 2$ **49.** $-1, -2$ **51.** 7 **53.** 40 **55.** 6

57. 7 **59.** 8 **61.** 4, 8 **63.** 4, 12

65. 48 mi/h, 40 mi/h **67.** 120 mi/h **69.** 150 mL

Self-Test for Chapter 5

[5.1] 1. $\dfrac{-3x^4}{4y^2}$ **2.** $\dfrac{4}{a}$ **3.** $\dfrac{x + 1}{x - 2}$ **[5.3] 4.** a **5.** 2

6. 5 **[5.4] 7.** $\dfrac{17x}{15}$ **8.** $\dfrac{3s - 2}{s^2}$ **9.** $\dfrac{4x + 17}{(x - 2)(x + 3)}$

10. $\dfrac{15}{w - 5}$ **[5.2] 11.** $\dfrac{4p^2}{7q}$ **12.** $\dfrac{2}{x - 1}$ **13.** $\dfrac{3}{4y}$

14. $\dfrac{3}{m}$ **[5.7] 15.** $\dfrac{2}{3x}$ **16.** $\dfrac{n}{2n + m}$ **[5.5] 17.** 4

18. $-3, 3$ **[5.6] 19.** 36 **20.** 2, 6 **21.** 6

22. 4 **23.** 4, 12 **24.** 50 mi/h, 45 mi/h

25. 20 ft, 35 ft

Cumulative Review Chapters 0–5

[1.6–1.7] 1. $-2xy$ **2.** $\dfrac{4a^2}{3}$ **3.** $2x^2 - 5x + 6$

4. $3a^2 + 6a + 1$ **[0.4] 5.** 31 **[0.5] 6.** 1

[3.4] 7. $2x^2 - xy - 6y^2$ **8.** $x^2 + 11x + 28$

[3.6] 9. $2x - 1 + \dfrac{1}{x + 2}$ **10.** $x + 1 - \dfrac{4}{x - 1}$

[2.3] 11. 4 **12.** -2 **[4.7] 13.** $(x - 7)(x + 2)$

14. $3mn(m - 2n + 3)$ **15.** $(a + 3b)(a - 3b)$

16. $2x(x - 6)(x - 8)$ **[2.4] 17.** 7 **[5.6] 18.** 4

19. 264 ft/s **[4.8] 20.** 5 in. by 7 in. **[5.1] 21.** $\dfrac{m}{3}$

22. $\dfrac{a - 7}{3a + 1}$ **[5.4] 23.** $\dfrac{8r + 3}{6r^2}$ **24.** $\dfrac{x + 33}{3(x - 3)(x + 3)}$

[5.2] 25. $\dfrac{3}{x}$ **26.** $\dfrac{1}{3w}$ **[5.7] 27.** $\dfrac{x - 1}{2x + 1}$

28. $\dfrac{n}{3n + m}$ **[4.8] 29.** $\dfrac{6}{5}$ **30.** $\dfrac{-9}{2}, 7$

Pre-Test Chapter 6

[6.1] 1. $(15, 3); (18, 6)$ **2.** $(0, 3); (2, 0)$

3. $(1, 3); (0, 5); (-3, 11)$ **4.** Answers vary

[6.2] 5. $A(2, 3), B(0, -5), C(-3, 5)$

6. 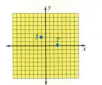 **[6.3] 7.**

8. **9.**

[6.4] 10. 1 **11.** -1 **12.** 8

[6.5] 13. 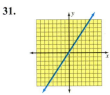 **14.** Predicted: 233 million pounds

Actual: 350 million pounds

[6.4] 15. \$1000

Summary Exercises

1. Yes **3.** Yes **5.** No **7.** $(6, 0), (3, -3), (0, -6)$

9. $(3, 0), (-3, 4), (0, 2)$ **11.** $(4, 4), (0, 8), (8, 0), (6, 2)$

13. $(3, 0), (6, -2), (9, -4), (-3, 4)$

15. $(0, 10), (2, 8), (4, 6), (6, 4)$

17. $(0, -2), (3, 0), (6, 2), (9, 4)$ **19.** $(4, 6)$ **21.** $(-1, -5)$

23–26. **27.**

29. 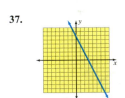 **31.**

33. **35.**

37. **39.**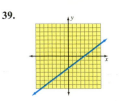

41.

43.

45.

47. -3 **49.** $\dfrac{2}{3}$

51. $-\dfrac{5}{2}$

53. Undefined

55. -2

57. $-\dfrac{2}{3}$

59.

61.

63. $\dfrac{5}{3}$

65.

67. 41,378   **69.** 52%
71. 24%   **73.** 7,029,990
75. 40%   **77.** 500%

79.

Income and Education

In thousands / Years of education

Self-Test for Chapter 6

[6.1] 1. $(3, 6), (9, 0)$ **2.** $(4, 0), (5, 4)$

3. $(3, 3), (6, 2), (9, 1)$ **4.** $(3, 0), (0, 4), \left(\dfrac{3}{4}, 3\right)$

5. Answers will vary   **6.** Answers will vary
[6.2] 7. $(4, 2)$ **8.** $(-4, 6)$ **9.** $(0, -7)$
10 – 12

[6.3] 13.

14.

15.

16.

17.

18.

[6.4] 19. 1   **20.** $\dfrac{3}{4}$

21. Undefined   **22.** 0
23. -7

24.

25. 5   **[6.5] 26.** 15%
27. $17,200,000   **28.** 2500
29. Decreased by 2000
30. 1980 to 1985; 2000

Cumulative Review Chapters 0 – 6

[1.3–1.4] 1. 3   **2.** 5   **3.** 37   **4.** -53 **5.** 69
6. -120 **7.** -5 **8.** 3   **[1.5] 9.** 108   **10.** 3

11. 69   **12.** 9   **[2.3] 13.** $-\dfrac{4}{3}$ **14.** $-\dfrac{8}{3}$ **15.** 10

16. 1   **17.** $C = \dfrac{5}{9}(F - 32)$ **[2.7] 18.** $x < 4$

19. $x \le 3$ **[3.1] 20.** $\dfrac{1}{x^4 y^6}$ **21.** $\dfrac{y^5}{x}$ **22.** 1

[3.3] 23. $5x^2 - 10$ **24.** $-7a^2 + 7a + 2$ **25.** 19
26. 26   **[3.4] 27.** $6x^2 + 2xy - 20y^2$
28. $6x^3 - 3x^2 - 45x$ **29.** $4a^2 - 49b^2$
[4.7] 30. $4pn^2(3p + 5 - 4n)$ **31.** $(y - 3)(y^2 - 5)$
32. $b(3a + 7)(3a - 7)$ **33.** $2(3x + 2)(x - 1)$

34. $(2a + 3b)(3a - b)$ **[4.8] 35.** $-3, 11$ **36.** $\dfrac{4}{5}, \dfrac{2}{7}$

[5.1] 37. $\dfrac{-5a^3}{3b^2}$ **38.** $\dfrac{w - 2}{w + 3}$ **[5.3] 39.** $\dfrac{a + 5}{a(a - 5)}$

40. $\dfrac{10}{w - 5}$ **[5.2] 41.** $\dfrac{3x^3}{4}$ **42.** $\dfrac{m - 4}{4m}$ **[5.5] 43.** 6

44. 5
[6.3] 45.

46.

47.

[6.4] 48. 2
49.

50. 30   **[2.5] 51.** Width: 5 in. Length: 7 in.
52. 41, 43, 45   **[2.6] 53.** $200

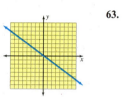

Pre-Test Chapter 7

[7.1] 1. Slope: -2; y-intercept: $(0, 3)$

2. Slope: $\dfrac{5}{4}$; y-intercept: $(0, -5)$

3. $y = 5x - 2$; See answers at the end of the text for graphs

4. $y = -4x + 6$; See answers at the end of the text for graphs

[7.2] 5. Parallel   **6.** Perpendicular

[7.3] 7. $y = -4x - 12$ **8.** $y = -\dfrac{5}{3}x + \dfrac{7}{3}$

9. $y = \dfrac{3}{5}x - 3$

[7.4] 10.

11.

[7.5] 12. 13; -3

Summary Exercises

1. 2   **3.** $-\dfrac{1}{2}$ **5.** 0   **7.** $\dfrac{1}{2}$

9. Slope: 2; y-intercept: $(0, 5)$

11. Slope: $-\dfrac{3}{4}$; y-intercept: $(0, 0)$

13. Slope: $-\dfrac{2}{3}$; y-intercept: $(0, 2)$

15. Slope: 0; y-intercept: $(0, -3)$

17. $y = 2x + 3$ **19.** $y = -\dfrac{2}{3}x + 2$

21. Perpendicular   **23.** Parallel

25. $y = -3$ **27.** $x = 4$ **29.** $y = -3$

31. $y = -\dfrac{4}{3}x - 2$ **33.** $x = -\dfrac{5}{2}$ **35.** $y = -\dfrac{1}{5}x + 4$

37. $y = -\dfrac{5}{4}x + 2$ **39.** $y = -\dfrac{5}{3}x + 3$

41. $y = \dfrac{4}{3}x + \dfrac{14}{3}$

43.

45.

47.

49. (a) -7 (b) -13 (c) -17

51. (a) -39 (b) -9 (c) -3

53. (a) 9   (b) -5 (c) -1

55. $f(x) = -7x - 3$

57. $f(x) = -\dfrac{3}{2}x - 6$

59.

61.

63.

65. 5, 2   **67.** 5, 9

69. $7a - 1, 21b - 1, 7x - 8$

Self-Test for Chapter 7

[7.1] 1. 1   **2.** $\dfrac{3}{4}$ **3.** Slope: $\dfrac{4}{5}$; y-intercept: $(0, -2)$

4. Slope: $-\dfrac{2}{3}$; y-intercept: $(0, -9)$

5. $y = -3x + 6$ **6.** $y = 5x - 3$

[7.2] 7. Parallel   **8.** Perpendicular

9. $y = \dfrac{1}{2}x + \dfrac{21}{2}$ **[7.3] 10.** $y = \dfrac{4}{3}x + \dfrac{25}{3}$

11. $y = -\dfrac{3}{2}x + \dfrac{5}{2}$ **12.** $y = -\dfrac{2}{3}x + 5$ **13.** $y = -8$

[7.4] 14.

15.

16.

[7.5] 17. $-5; 7$ **18.** $-15; -7$

19. $-15; 6$

20. $3a - 25; 3x - 28$

Cumulative Review Chapters 0–7

[3.3] 1. $x^2y^2 - 3xy$ **2.** $\dfrac{4m^3n}{3}$ **3.** $-x + 9$

4. $3z^2 - 3z + 5$ **[3.4] 5.** $2x^2 + 11x - 21$

6. $2a^2 + 6ab - 8b^2$ **[3.6] 7.** $x + 6 + \dfrac{20}{x - 3}$

8. $x^3 - 2x^2 + 4x - 10 + \dfrac{20}{x - 2}$ **[2.3] 9.** $-\dfrac{4}{3}$ **10.** -2

[4.7] 11. $(x - 8)(x + 7)$ **12.** $2x^2y(2x - y + 4x^2)$

13. $2a(2a + 3b)(2a - 3b)$ **14.** $3(5x - 2y)(x - y)$

[6.4] 15. 1   **16.** $-\dfrac{9}{4}$ **[5.2–5.4] 17.** $\dfrac{2x - 3}{x}$

18. $\dfrac{1}{a-7}$ **19.** $\dfrac{5m+6}{2m^2}$ **20.** $\dfrac{2x+6}{x(x-3)}$

21. $\dfrac{5y}{(y+4)(y-1)}$ **[5.5] 22.** -4 **23.** $7,-10$

[5.6] 24. 9   **25.** 49 mi/h going; 42 mi/h returning

26. 126 min   **[7.2] 27.** $y=-\dfrac{1}{7}x+2$

[7.1] 28. $y=-5x+3$

[7.4] 29.

[7.5] 30. -2

Pre-Test Chapter 8

[8.1] 1. Inconsistent   **2.** $(0,-2)$ **[8.2–8.3] 3.** $(6,-3)$
4. Dependent system   **5.** $(4,1)$ **6.** $(3,-1)$
7. 16, 24   **8.** 13 m, 17 m   **9.** 19 dimes, 26 quarters
10. Canoe: 7.5 mi/h; current: 1.5 mi/h
[8.4] 11.   **12.**

Summary Exercises

1. $(4,2)$ **3.** No solution   **5.** Infinite number of solutions
7. $(5,3)$ **9.** $(5,-2)$ **11.** $(3,-1)$ **13.** $(-4,2)$
15. $\left(2,-\dfrac{2}{3}\right)$ **17.** Inconsistent system   **19.** $\left(\dfrac{3}{5},2\right)$
21. $(6,-2)$ **23.** $(4,2)$ **25.** Inconsistent system
27. $\left(-\dfrac{4}{3},2\right)$ **29.** $(8,-2)$ **31.** $(4,-2)$ **33.** $(5,3)$
35. Inconsistent system   **37.** $(5,-4)$ **39.** $(-3,-2)$
41. $(-3,0)$ **43.** $(8,2)$ **45.** $(4,-6)$ **47.** $\left(0,\dfrac{4}{3}\right)$
49. $\left(\dfrac{1}{2},\dfrac{9}{2}\right)$ **51.** $(2,-1)$ **53.** 4, 13
55. Tablet \$1.50, pencil \$0.25
57. Speakers \$425, amplifier \$500
59. Width 14 cm, length 18 cm   **61.** 10 nickels, 20 quarters
63. 200 mL of 20%, 400 mL of 50%
65. \$10,000 at 11%, \$8000 at 7%
67. Plane 500 mi/h, wind 50 mi/h
69.   **71.**

73.   **75.**

Self-Test for Chapter 8

[8.1] 1. $(4,1)$ **2.** $(4,2)$ **3.** Inconsistent system
4. $(2,4)$ **[8.2] 5.** $(4,1)$ **6.** $(4,2)$ **7.** $(1,3)$
8. $\left(2,\dfrac{5}{2}\right)$ **9.** Dependent system   **10.** $\left(\dfrac{3}{4},-1\right)$
11. $(6,2)$ **12.** Inconsistent system   **[8.3] 13.** $(2,6)$
14. $(6,-3)$ **15.** $(6,2)$ **16.** $(5,4)$ **17.** $(-3,3)$
18. Inconsistent system   **19.** $(3,-5)$ **20.** $(3,2)$
21. 12, 18   **22.** 21 m, 29 m   **23.** Width 12 in., length 20 in.
24. 12 dimes, 18 quarters   **25.** Boat 15 mi/h, current 3 mi/h

[8.4] 26.   **27.**

28.

Cumulative Review Chapters 0–8

[3.3] 1. $8x^2-7x-4$ **2.** w^2-8w-4
[3.4] 3. $28x^3y^2-14x^2y^2+21x^2y^3$ **4.** $15s^2-23s-28$
[3.6] 5. $-x^2+2xy-3y$ **6.** $2x+4$ **[2.3] 7.** 9
[4.1–4.4] 8. $8a^2(3a-2)$ **9.** $7mn(m-3-7n)$
10. $(a+8b)(a-8b)$ **11.** $5p(p+4q)(p-4q)$
12. $(a-6)(a-8)$ **13.** $2w(w-7)(w+3)$
[4.8] 14. 4, 5   **15.** $-4,4$ **16.** 7   **17.** 5 in. by 17 in.
[5.1] 18. $\dfrac{m}{3}$ **19.** $\dfrac{a-7}{3a+1}$ **[5.2] 20.** $\dfrac{3}{x}$ **21.** $\dfrac{1}{3w}$

[7.1] 22.   **23.**

24.   **25.**

[6.4] 26. $\dfrac{10}{7}$ **[7.1] 27.** Slope: $\dfrac{5}{3}$; y-intercept: $(0,-5)$
28. $y=2x-5$
[7.4] 29.   **30.**

[8.1] 31. $(4, -3)$ **[8.2–8.3] 32.** $\left(7, -\dfrac{5}{2}\right)$

33. Dependent system   **34.** $(5, 0)$

35. Inconsistent system   **36.** $\left(\dfrac{3}{2}, -\dfrac{1}{3}\right)$ **37.** $5, 21$

38. VHS $4.50, cassette $1.50
39. 325 at $7, 125 at $4   **40.** $5000 at 6%, $7000 at 9%

Pre-Test Chapter 9

[9.1] 1. 12   **2.** 4   **3.** Not a real number   **4.** 5

[9.2] 5. $6\sqrt{2}$ **6.** $3x\sqrt{3x}$ **7.** $\dfrac{1}{2}$ **8.** $\dfrac{\sqrt{7}}{5}$

[9.3] 9. $7\sqrt{7}$ **10.** $3\sqrt{5}$ **11.** $3\sqrt{3}$ **12.** $\sqrt{2}$
[9.4] 13. $x\sqrt{10}$ **14.** $\sqrt{10} - 6$ **15.** $9 + 5\sqrt{3}$

16. $\dfrac{\sqrt{15}}{3}$ **17.** $4 - \sqrt{5}$ **[9.6] 18.** 15   **19.** 24

20. 8.49   **21.** 4.12   **22.** 19.21 cm   **23.** 14
24. 6.71   **[9.5] 25.** 84

Summary Exercises

1. 9   **3.** Not a real number   **5.** -4 **7.** Not a real number
9. $3\sqrt{5}$ **11.** $2x^2\sqrt{5}$ **13.** $10b\sqrt{2b}$ **15.** $6ab^2\sqrt{3b}$

17. $\dfrac{3x\sqrt{2}}{5}$ **19.** $\dfrac{\sqrt{21}}{7}$ **21.** $\dfrac{2x\sqrt{14}}{7}$ **23.** $6\sqrt{5}$

25. $2\sqrt{3a}$ **27.** $7\sqrt{3}$ **29.** $\sqrt{2}$ **31.** $2\sqrt{2} + \sqrt{3}$
33. $\sqrt{30}$ **35.** $\sqrt{6x}$ **37.** $5a\sqrt{2}$ **39.** $2\sqrt{21} - 21$

41. 6   **43.** $\dfrac{x\sqrt{21x}}{3}$ **45.** 21   **47.** 9   **49.** 10

51. 15   **53.** $5\sqrt{5}$ **55.** 15 in.   **57.** 24.1 ft   **59.** 4
61. $\sqrt{73}$ **63.** $\sqrt{82}$

Self-Test for Chapter 9

[9.1] 1. 11   **2.** 3   **3.** Not a real number   **4.** 4

[9.2] 5. $5\sqrt{3}$ **6.** $2a\sqrt{6a}$ **7.** $\dfrac{4}{5}$ **8.** $\dfrac{\sqrt{5}}{3}$

[9.3] 9. $4\sqrt{10}$ **10.** $3\sqrt{2}$ **11.** $3\sqrt{2}$ **12.** $4\sqrt{5}$

[9.4] 13. $3x\sqrt{2}$ **14.** $11 + 5\sqrt{5}$ **15.** $\dfrac{\sqrt{14}}{2}$

16. $2 + 3\sqrt{2}$ **[9.5] 17.** 83   **18.** 4   **[9.6] 19.** 20
20. 12   **21.** $3\sqrt{5}$ **22.** $\sqrt{15}$
23. Approximately 9.747 cm   **24.** 9   **25.** $\sqrt{85}$

Cumulative Review Chapters 0–9

[1.6] 1. $3x^2y^3 - 2x^3y$ **[3.3] 2.** $7x^2 - 6x + 12$
[3.1] 3. 0   **4.** -8 **[2.3] 5.** 2   **6.** 6
[2.7] 7. $x > 4$ **[3.4] 8.** $6x^4y - 10x^3y + 38x^2y$
9. $20x^2 - 23xy - 21y^2$ **[4.7] 10.** $9xy(4 - 3x^2y)$

11. $(4x - 3)(2x - 5)$ **[5.4] 12.** $\dfrac{1}{15(x + 7)}$

[5.2] 13. $\dfrac{x - 3}{x - 5}$

[6.3] 14.   **[7.4] 15.**

[6.4] 16. 5   **[7.3] 17.** $y = -\dfrac{3}{2}x + 5$ **[8.2] 18.** $(5, 0)$

19. Inconsistent system   **20.** Boat: 13 mi/h, current: 3 mi/h
[9.1] 21. 12   **22.** -12 **23.** Not a real number   **24.** -3

[9.2–9.4] 25. $-4a\sqrt{5}$ **26.** $\dfrac{2x\sqrt{6x}}{3}$ **27.** $4 - 2\sqrt{2}$

28. $7x\sqrt{2}$ **29.** $5mn\sqrt{6m}$ **30.** $\dfrac{2a\sqrt{3}}{5}$

Pre-Test Chapter 10

[10.1] 1. $\pm\sqrt{17}$ **2.** $\pm2\sqrt{3}$ **3.** $1 \pm \sqrt{5}$

4. $\dfrac{\pm\sqrt{14}}{3}$ **[10.2] 5.** $-2, 5$ **6.** $\dfrac{5 \pm \sqrt{17}}{2}$

7. $2 \pm 2\sqrt{2}$ **8.** $\dfrac{2 \pm 3\sqrt{2}}{2}$ **[10.3] 9.** $-7, 2$

10. 5   **11.** $\dfrac{-3 \pm \sqrt{29}}{2}$ **12.** $\dfrac{3 \pm \sqrt{33}}{4}$

13. $\dfrac{2 \pm \sqrt{10}}{3}$ **14.** $\dfrac{-3 \pm \sqrt{37}}{2}$

[10.4] 15.

x	y
-2	6
-1	3
0	2
1	3
2	6

16.

x	y
-4	0
-3	-3
-2	-4
-1	-3
0	0

17.

x	y
-2	-4
-1	-6
0	-6
1	-4
2	0

18.

x	y
-2	-1
-1	2
0	3
1	2
2	-1

Summary Exercises

1. $\pm\sqrt{10}$ **3.** $\pm2\sqrt{5}$ **5.** $1 \pm \sqrt{5}$ **7.** $-3 \pm \sqrt{5}$
9. $\dfrac{\pm3\sqrt{3}}{2}$ **11.** $\dfrac{\pm\sqrt{7}}{5}$ **13.** $-2, 5$ **15.** $\dfrac{5 \pm \sqrt{17}}{2}$

17. $2 \pm 2\sqrt{2}$ **19.** $2 \pm \sqrt{2}$ **21.** $\dfrac{1 \pm \sqrt{29}}{2}$

23. $\dfrac{2 \pm 3\sqrt{2}}{2}$ **25.** $-2, 7$ **27.** $\dfrac{-5 \pm \sqrt{37}}{2}$

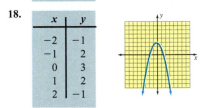

29. $3 \pm 2\sqrt{2}$ **31.** $\dfrac{2 \pm \sqrt{10}}{3}$ **33.** $\dfrac{-3 \pm \sqrt{37}}{2}$

35. $2 \pm \sqrt{10}$ **37.** $0, \dfrac{3}{5}$ **39.** $1 \pm \sqrt{10}$

41. $\dfrac{5 \pm \sqrt{57}}{4}$ **43.** $-1, \dfrac{7}{2}$ **45.** $-1 \pm \sqrt{6}$

47. $1 \pm \sqrt{3}$

49.

x	y
−2	7
−1	4
0	3
1	4
2	7

51.

x	y
−1	4
0	0
1	−2
2	−2
3	0

53.

x	y
−1	0
0	−2
1	−2
2	0
3	4

55.

x	y
−3	0
−2	−3
−1	−4
0	−3
1	0

57.

x	y
−2	5
−1	−1
0	−3
1	−1
2	5

59.

x	y
−2	−6
−1	−3
0	−2
1	−3
2	−6

61. Axis: $x = -3$

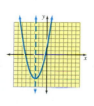

x	y
−5	−2
−4	−5
−3	−6
−2	−5
−1	−2

63. Axis: $x = \dfrac{5}{2}$

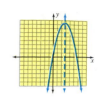

x	y
0	2
1	6
2	8
3	8
4	6
5	2

65. $(-5, 0), (3, 0)$ **67.** $(-3 - \sqrt{6}, 0), (-3 + \sqrt{6}, 0)$
69. None   **71.** 9.23 cm, 15.46 cm   **73.** 1.13 s and 3.87 s
75. 70 items

Self-Test for Chapter 10

[10.1] 1. $\pm\sqrt{15}$ **2.** $\pm 2\sqrt{2}$ **3.** $1 \pm \sqrt{7}$ **4.** $\dfrac{\pm\sqrt{10}}{3}$

[10.2] 5. $4, -2$ **6.** $\dfrac{-3 \pm \sqrt{13}}{2}$ **7.** $-1 \pm \sqrt{6}$

8. $\dfrac{5 \pm \sqrt{17}}{4}$ **[10.3] 9.** $-1, 3$ **10.** 3

11. $\dfrac{5 \pm \sqrt{33}}{2}$ **12.** $\dfrac{1 \pm \sqrt{11}}{2}$ **13.** $\dfrac{1 \pm \sqrt{33}}{4}$

14. $-1 \pm \sqrt{6}$
[10.4] 15.

x	y
−2	8
−1	5
0	4
1	5
2	8

16.

x	y
−1	3
0	0
1	−1
2	0
3	3

17.

x	y
−2	0
−1	−2
0	−2
1	0
2	4

18. Axis: $x = 2$; $(-3, 0)$ and $(7, 0)$
[10.5] 19. 6.36 ft by 15.08 ft   **20.** 3.06 s

Cumulative Review Chapters 0–10

[1.6] 1. $11x^2y - 6xy^2$ **2.** $x^2 + 5x - 7$ **[1.5] 3.** -480
4. -1680 **[2.3] 5.** 1   **[2.7] 6.** $x > 2$
[3.4–3.5] 7. $6x^3y - 3x^2y + 15xy$
8. $6x^2 + 11x - 10$ **9.** $9x^2 - 16y^2$
[4.1–4.4] 10. $8xy^2(2x - y)$ **11.** $(2x - 3)(4x + 5)$

12. $(5x - 4y)(5x + 4y)$ **[5.2, 5.4] 13.** $\dfrac{29}{28(x + 2)}$

14. $\dfrac{5(x - 2)}{x - 1}$ **15.** $\dfrac{(3x - 1)(x + 6)}{15x^2}$

[6.4] 16. 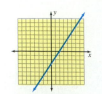   **17.**

[6.5] 18. 5   **[7.1] 19.** $y = 2x - 5$

[7.4] 20.

[8.2 – 8.3] 21. $\left(4, \dfrac{2}{3}\right)$

22. $(3, -2)$

23. $\left(3, \dfrac{-7}{2}\right)$

24. Dependent system

25. 5, 21

26. 325 reserved-seat; 125 gen adm

27. 100 mL of 30%; 200 mL of 60%   **[9.1] 28.** 13

29. -13 **30.** Not a real number   **31.** -4

[9.2 – 9.4] 32. $6\sqrt{3}$ **33.** $30a\sqrt{3}$ **34.** $-13 - 2\sqrt{2}$

35. $2 - \sqrt{2}$ **[10.1 – 10.3] 36.** $\pm 6\sqrt{2}$ **37.** $-3 \pm 2\sqrt{3}$

38. $\dfrac{5 \pm \sqrt{41}}{4}$

[10.4] 39.   **40.**

[10.5] 41. 1.657 s   **42.** $84.64

Index

A

Absolute value, 58, 68
Absolute value expressions, simplification of, 58
ac method of factoring, 390–393, 412
ac test, 387–389
  definition of, 387
  sign patterns for, 392
Addition
  associative property of, 87–88, 151
  commutative property of, 87, 151
  of decimals, 37, 67
  distributive property of multiplication over, 88–89, 151
    in multiplication of polynomials, 307–308
    in multiplying radical expressions, 777, 802
    solving linear equations with, 170–172
    solving linear inequalities with, 250–251
  of fractions, 22–23, 66
  of mixed numbers, 26, 66
  notation for, 78, 151
  of polynomials
    combining like terms in, 297–298
    overview of, 297, 336
    removing grouping signs in, 297–298
    using horizontal method, 298
    using vertical method, 301
  of radical expressions, 769–772, 801
  of rational expressions
    like, 437–440, 492
    unlike, 444–451, 492
    steps in, 445
  of real numbers
    additive identity property, 98
    additive inverse property, 98–99
    application of, 103
    different sign, 96–97, 151–152
    same sign, 95–96, 151
Addition method, for solving systems of linear equations in two variables, 690–696
  steps in, 694, 733–734
Addition property
  of equality
    definition of, 166, 259
    solving equations with, 167–172, 187–194
  of inequality
    definition of, 245
    solving with, 245–246
Additive identity, 98
Additive identity property, 98
Additive inverse (opposite)
  additive inverse property, 98–99
  definition of, 57
  of negative number, 57–58
Additive inverse property, 98–99
Algebra, history of, 1, 75
Algebraic equations. See Equation(s)
Algebraic expressions
  actions to perform on
    vs. equations, 458
    possible, 653

  evaluation of, 126–130, 152
    on calculator, 128, 130
    with fractions, 127, 129
    in function notation, 654–655
    in standard form, 653
    steps in, 126, 152
  grouping symbols in, 80
  radical. See Radical expressions
  rational. See Rational expressions
  terms of
    definition of, 136–137
    like
      combining, 137–139, 169–170, 181, 297–300
      definition of, 137
    numerical coefficient of, 137
Algebraic fractions. See Rational expressions
Algebraic notation. See Notation
Algorithm(s), definition of, 389
Amount, in percent expressions, 230, 260
Anxiety, overcoming
  preparing for class, 163
  procrastination and, 163
  syllabus familiarity, 77
  test preparation, 273
  textbook familiarity, 3
  working together, 349
Applications. See also Word problems
  of addition of real numbers, 103
  of decimals, 42
  of direct variation, 555–557
  of division of real numbers, 117
  of equations in two variables, 508
  of functions, 658
  of graphs, 501, 567–576
  of inequalities, 251–252
  of linear equations in two variables, 508
  of linear inequalities in one variable, 251–252
  of linear inequalities in two variables, 641–642
  of mean, 183
  of median, 183
  of multiplicative inverse, 471
  of percent, 231–234
  of perpendicular lines, 623
  of point-slope form of linear equation, 632–633
  of Pythagorean Theorem, 747, 790–794, 855–856
  of quadratic equations, 407, 855–859
  of rational expressions, 470–477
  of reciprocals, 471
  of scientific notation, 290–291, 292
  of slope-intercept slope form of linear equation, 608–609
  of subtraction of real numbers, 103
  of systems of linear equations in two variables, 696–703, 715–717
  of systems of linear inequalities in two variables, 725–726
  of zero-product principle, 404–407
Archimedes, 290–291
Ascending order, 56
Associative property
  of addition, 87–88, 151
  of multiplication, 87–88, 151

Average, 182
Axis (axes)
  scaling of, 536–537
  x-axis, 514, 581
  y-axis, 514, 581
Axis of symmetry, of parabola, 843–844, 866

B

Bar graphs, 571–572, 582
Bar notation, 36
Base
  in exponential notation, 47, 68
  in percent expressions, 230, 260
Binomial(s)
  definition of, 278, 335
  division of polynomials by, 327–330
  factoring, difference of two squares, 375–376, 411
  multiplication of
    by binomial, 309
      FOIL method for, 309–311, 336
      with vertical method, 311
    by binomial differing only in sign, 319–320
    by monomial, 308
    square of binomial, 318–319, 336, 369
    by trinomial, using vertical method, 312
  square of, 318–319, 336, 369
Boundary line, for half plane, 638–641
Brackets, in algebraic notation, 80

C

Calculator
  approximation of length with, 793
  change-of-sign key, 102
  division on, 116–117
  evaluation of algebraic expressions on, 128, 130
  evaluation of fractions on, 116–117
  exponents on, 49
  as function machine, 654
  graphing with, 595–598
    changing window, 596
    creating graph, 595
    home screen, 598
    zoom key, 596–597
  home screen of, 598
  order of operations on, 49, 130
  scientific notation on, 290
  square roots on, 750–751
  subtraction on, 102–103
  table feature on, 598
Cartesian coordinate system. See Rectangular coordinate system
Cells, of table, 567
Check digits, of International Standard Book Numbers, 343–344
Circle graphs. See Pie charts
Closed circle, 244, 260

Coefficient(s)
  of algebraic expression, 137
  of polynomial term, 277, 335
  of trinomial, identifying, 387
Combining of like terms, 137–139, 169–170, 181
  in addition of polynomials, 297–298
  in subtraction of polynomials, 299–300
Common divisors, 9
Common factors, 9
Common multiples, 12
Commutative property
  of addition, 87, 151
  of multiplication, 87, 151
Completing the square, 821–824
  steps in, 824, 865
Complex fractions
  definition of, 483
  simplifying, 483–486
Complex rational expressions, simplification of, 483–486, 492
Composite numbers
  definition of, 5, 65
  factoring of, 6–7
    into prime factors, 7–8
  identification of, 6
Compound interest, 240–241
Conjugates, 778
  product of, 778
Consecutive integers, 203
Consistent systems of linear equations in two variables, 681, 682
Constant of variation
  definition of, 555
  finding, 556
Continued division, factoring of number by, 8
Coordinate system. See Rectangular coordinate system
Counting numbers, 3
Cube root
  common, list of, 752
  definition of, 751, 801

D

Decimal equivalents, 35–36
Decimals
  addition of, 37, 67
  applications of, 42
  changing percent to, 40, 67
  changing to percent, 41, 68
  converting fractions to, 35–36, 66
  converting to fractions, 36–37, 67
  converting to/from scientific notation, 291
  division of, 38–39, 67
  irrational numbers as, 753
  multiplication of, 37–38, 67
  repeating, 36
  subtraction of, 37, 67
Degree
  of equation, 165
  of polynomial, 278–279, 335
Denominator, least common. See Least common denominator

Dependent systems, of linear equations in two variables, 681–682, 683, 694–695, 713
Descartes, René, 514
Descending-power form of polynomial, 279, 335
Difference of two squares
    definition of, 375
    factoring, 375–376, 411
Direct variation
    applications of, 555–557
    definition of, 555, 582
    graphing of, 556–557
    writing equation for, 556
Discriminant, of quadratic formula, 835
Distance, formula for vs. rate and time, 219, 472, 702
Distance formula, 793
Distributive property of multiplication over addition, 88–89, 151
    in multiplication of polynomials, 307–308
    in multiplying radical expressions, 777, 802
    solving linear equations with, 170–172
    solving linear inequalities with, 250–251
Division
    on calculator, 116–117
    of decimals, 38–39, 67
    of exponents, 145–147
    factoring of number by, 8
    of fractions, 21–22, 66, 432, 483
    of mixed numbers, 26
    of monomial, by monomial, 325
    notation for, 80–81, 151
    order of operations in, 116–117
    of polynomials
        by monomial, 325–326, 336
        by polynomials, 327–330
    of radical expressions, 778–779, 802
    of rational expressions, 432–433, 491
    of real numbers, 114–115, 152
        applications, 117
        division of/by zero, 115–116
Divisors, 4
    common, 9
Double (repeated) solution, to quadratic equations, 406

E
Elimination method. See Addition method
Ellipses (notation), 55
Encryption systems, 347
Equality
    addition property of
        definition of, 166, 259
        solving equations with, 167–172, 187–194
    multiplication property of
        definition of, 177, 259
        solving linear equations with, 177–184, 187–194
    power property of, 783–786, 802
Equation(s)
    addition property of equality for
        definition of, 166, 259

solving equations with, 167–172, 187–194
    definition of, 163, 259
    degree of, 165
    equivalent, 166, 259
    of horizontal line, 630, 632
    identities, 193
    of line. See Linear equations
    literal. See Literal equations
    multiplication property of equality for
        definition of, 177, 259
        solution of equations with, 177–184, 187–194
    in one variable, solutions of, 506
    proportion form of
        definition of, 464
        solving, 464–465, 475–477
    quadratic. See Quadratic equations
    radical. See Radical equations
    rational. See Rational equations
    solution of
        definition of, 164, 259
        multiple, 165
        verification of, 164
    in two variables
        applications, 508
        definition of, 504
        solution of
            definition of, 504, 581
            notation for, 505
            number of, 504
            solving, 504–507
    of vertical line, 630, 632
    without solution, 193
Equilibrium price, 859
Equivalent equations, 166, 259
Equivalent fractions, 19–20, 421
Equivalent inequalities, 245
Eratosthenes, sieve of, 5
Exponent, in exponential notation, 47, 68
Exponent(s)
    on calculator, 49
    division of, 145–147
    factors of, 47
    multiplication of, 144–145
    negative
        definition of, 288, 335
        simplification of, 288–290
    notation for, 47, 68
    order of operations and, 48, 68
    properties of, 144–147, 273–277, 335
        property 1, 144–145, 152, 273, 336
        property 2, 146–147, 152, 274, 336
        property 3, 274, 336
        property 4, 275, 336
        property 5, 276, 336
    simplification of
        negative, 288–290
        positive, 275–276
    zero, 287, 335
Exponential form, 47
Exponential notation, 47, 68
    with variables, 144
Expression(s)
    algebraic. See Algebraic expressions
    definition of, 79, 151
    grouping symbols in, 80
    radical. See Radical expressions
    rational. See Rational expressions

Extraneous solutions, of radical equations, 783
Extreme(s), of proportion, 464
Extreme values, of set, 57

F
Factor(s)
    common, 9
    definition of, 4, 65
    of exponent, 47
    finding, 4
    greatest common
        definition of, 9, 65
        finding, 9–11, 65
        in polynomial
            definition of, 350, 411
            factoring out, 350–352, 361, 370–371, 377, 392, 399
        in rational expressions, factoring out, 421–426
Factor trees, 7
Factorability, test for, 387–389
Factoring
    ac method of, 390–393, 412
    checking of, 386
    difference of two squares, 375–376, 411
    by grouping, 382–383, 411
    of number, 6–7
        by division, 8
        into prime factors, 7–8
    of perfect square trinomials, 369, 377–378, 411
    as reverse of multiplication, 349–350
    solving quadratic equations by, 404–407, 412, 813, 865
    steps in, 399–400, 412
    strategies in, 399–401
    trial and error method, 357–362, 366, 411
    of trinomials
        of form $ax^2 + bx + c$, 366–370
        of form $x^2 + bx + c$, 357–362
        identifying coefficients, 387
        perfect square, 369, 377–378, 411
        sign patterns for
            examples of, 366–369
            rules for, 366
        trial and error method, 357–362, 366, 411
        unfactorable trinomials, 362
Feasible regions, 641
Fifth property of exponents, 276, 336
First-degree equations, definition of, 165
First property of exponents, 144–145, 152, 273, 336
Fixed cost, 537, 608
FOIL method
    for binomials, 309–311, 336
    for radical expressions, 777–778, 802
Formulas. See Literal equations; Quadratic formula
Foucault, Jean, 807
Fourth property of exponents, 275, 336
Fourth root
    common, list of, 752
    definition of, 751, 801
FPF. See Fundamental principle of fractions

Fraction(s)
    addition of, 22–23, 66
    algebraic. See Rational expressions
    on calculator, 116–117
    changing percent to, 40, 67
    changing to percent, 41, 68
    complex
        definition of, 483
        simplifying, 483–486
    converting decimals to, 36–37, 67
    converting to decimals, 35–36, 66
    converting to/from mixed number, 24–25, 66
    division of, 21–22, 66, 432, 483
    equivalent, 19–20, 421
    fundamental principle of, 19, 65, 421, 483
    improper, 19
    like, definition of, 437
    multiplication of, 21, 65, 431
    proper, 19
    as ratio, 464
    simplest form of, 421, 491
    simplification of
        complex, 483–486
        standard, 20
    as subdivision of rational numbers, 753
    subtraction of, 23, 66
Fraction bar, as grouping symbol, 116
Function(s)
    applications of, 658
    converting linear equations in two variables to, 656
    definition of, 654
    evaluating, 654–655
    linear, graphing of, 657
    notation for, 654, 666
    substituting non-numerical values in, 657–658
Function machines, 654
Fundamental principle of fractions (FPF), 19, 65, 421, 483
Fundamental principle of rational expressions, 421
Fundamental Theorem of Arithmetic, 8

G
GCF. See Greatest common factor
Graph(s) and graphing
    applications of, 501, 567–576
    bar graphs, 571–572, 582
    with calculator, 595–598
        changing window, 596
        creating graph, 595
        home screen, 598
        zoom key, 596–597
    definition of graph, 569
    of direct variation, 556–557
    inequalities, 244–251, 260
    Internet tutorial on, 673
    line graphs
        creating, 574–575
        predictions using, 575–576
        reading of, 573–574, 582
    of linear equations in one variable
        horizontal line, 532–533
        vertical line, 531–532, 533
    of linear equations in two variables
        intercept method of, 533–535
        drawbacks of, 535
        steps in, 535

by plotting three points, 526–538
  steps in, 527, 581
selecting method for, 608
slope-intercept method of, 607
  steps in, 607
by solving for y, 535–536
parabolas, 837–846
pie charts, 569–570, 582
of points, 516–518
  steps in, 516, 581
predictions from line graphs, 575–576
of quadratic equations (parabolas), 837–846
reading values from, 567–576, 582
uses of, 501
Graphical method, for systems of linear inequalities in two variables, 679–683
  steps in, 682–683, 733
Graphing calculator. *See* Calculator
Greater than (>), 243, 260
Greater than or equal to (≥), 244, 260
Greatest common factor (GCF)
  definition of, 9, 65
  finding, 9–11, 65
  in polynomial
    definition of, 350, 411
    factoring out, 350–352, 361, 370–371, 377, 392, 399
  in rational expressions, factoring out, 421–426
Grouping, factoring by, 382–383, 411
Grouping symbols
  brackets, 80
  fraction bar, 116
  order of operations and, 48, 68
  parentheses, 80
  removal of
    in adding polynomials, 297–298
    in solving equation, 191, 215–219
    in subtracting polynomials, 299–300
  solving linear equations with, 215–219

H
Half plane
  boundary line for, 638–641
  definition of, 638
  equation for, 638–641
Horizontal change, of line, 549, 582
Horizontal line
  equation of, 630, 632
  graphing of, 532–533
  slope of, 552, 630, 632
Horizontal method, addition of polynomials using, 298

I
Identities, 193
Improper fractions
  converting to/from mixed number, 24–25, 66
  definition of, 19
Inconsistent systems, of linear equations in two variables
  definition of, 681, 683
  solving, 681, 695–696, 713–714
Indeterminate form, 116

Index of radical, 751
Inequalities
  addition property of
    definition of, 245
    solving with, 245–246
  applications of, 251–252
  definition of, 243, 260
  equivalent, 245
  graphing of, 244–251, 260
  linear. *See* Linear inequalities
  multiplication property of
    definition of, 247
    solving linear inequalities with, 247–250
  notation for, 243, 244, 260
  solution set of, 244
  solving, 244–251, 260
    with addition property, 245–246
    with distributive property, 250–251
    with multiplication property, 247–250
Insights, mathematical, value of, 389
Integers, 55–57, 68
  consecutive, 203
  as subdivision of rational numbers, 753
Intercept method of graphing linear equations, 533–535
  drawbacks of, 535
  steps in, 535
Interest
  compound, 240–241
  formula for, 701
International Standard Book Numbers (ISBNs), 343–344
Internet
  bookmarking of websites, 159
  graphing tutorial on, 673
  search engines, 159
  searches on, 159–160
Intersection, of lines, 620
Irrational numbers
  decimal representations of, 753
  definition of, 752
  identification of, 753
  as subdivision of real numbers, 753
ISBNs. *See* International Standard Book Numbers

L
LCD. *See* Least common denominator
LCM. *See* Least common multiple
Least common denominator (LCD)
  in addition of fractions, 22, 66
  definition of, 22
  for unlike rational expressions, 444–451, 492
Least common multiple (LCM)
  definition of, 12, 65
  finding, 12–13, 65
  in solving equations, 192–193
Legend, of bar graph, 571–572
Less than (<), 243, 260
Less than or equal to (≤), 244, 260
Light years, 292
Like radicals, 769
Like terms
  combining, 137–139, 169–170, 181
  in addition of polynomials, 297–298
  in subtraction of polynomials, 299–300
  definition of, 137

Line(s)
  equation for. *See* Linear equations
  parallel
    definition of, 618, 665
    determining, 618–619, 622
    finding, 622
    slopes of, 618–619
  perpendicular
    applications of, 623
    definition of, 620, 665
    equation of, 620–621
    finding, 622
    slopes of, 620–621, 665
  slope of. *See* Slope
Line graphs
  creating, 574–575
  predictions using, 575–576
  reading of, 573–574, 582
Linear equations
  addition property of equality for
    definition of, 166, 259
    solving equations with, 167–172, 187–194
  definition of, 165, 581
  degree of, 165
  forms of, 632
  multiplication property of equality for
    definition of, 177, 259
    solution of equations with, 177–184, 187–194
  solving
    with addition property of equality, 167–172, 187–194
    definition of, 166
    with distributive property, 170–172
    with multiplication property of equality, 177–184, 187–194
    with fractions, 179–181, 192–193
    removing parentheses in, 191, 215–219
    steps in, 194, 259
  standard form of, 632
  without solution, 193
Linear equations in one variable
  definition of, 165
  graphing of, 531–533
  solutions of, 506
Linear equations in two variables
  applications of, 508
  consistent systems of, 681, 682
  converting to function form, 656
  definition of, 504, 527
  forms of, 632
  graphing
    intercept method of, 533–535
    drawbacks of, 535
    steps in, 535
    by plotting three points, 526–538
    steps in, 527, 581
    selecting method for, 608
    slope-intercept form, 607
    steps in, 607
    by solving for y, 535–536
  inconsistent systems of
    definition of, 681, 683
    solving, 681, 695–696, 713–714
  point-slope form
    applications, 632–633
    definition, 629, 632, 666
    finding, 629–630, 631–632
  slope-intercept form of, 605–609, 665

applications of, 608–609
  definition of, 605, 632
  finding, 605–607
  graphing, 607
solution of
  definition of, 504, 581
  notation for, 505
  number of, 504
solving, 504–507
standard form of, 632
systems of. *See* Systems of linear equations in two variables
Linear functions, graphing of, 657
Linear inequalities in one variable
  applications, 251–252
  definition of, 245
  graphing of, 244–251, 260
  solving, 244–251, 260
    with addition property, 245–246
    with distributive property, 250–251
    with multiplication property, 247–250
Linear inequalities in two variables. *See also* Systems of linear inequalities
  applications, 641–642
  definition of, 638
  graphing, 638–641
    steps in, 641
  test point method for, 638–641
Linear systems of equations in two variables. *See* Systems of linear equations in two variables
Literal equations (formulas)
  definition of, 198, 259
  distance formula, 793
  for distance, rate and speed, 219, 472, 702
  for interest, 701
  solving, 198–206
    steps in, 199, 259

M
Magic squares, 134–135
Marginal cost, 608
Maximum, of set, 57
Mean(s)
  of proportion, 464
  of set
    applications of, 183
    definition of, 182
    finding, 182–183
    notation for, 182
Median, of set
  applications of, 183
  definition of, 99
  finding, 99–100
Minimum, of set, 57
Mixed numbers
  addition of, 26, 66
  converting to/from improper fraction, 24–25, 66
  definition of, 24, 66
  division of, 26
  multiplication of, 25
  subtraction of, 27, 66
Modulars, 344
Monomial(s)
  definition of, 278, 335
  division of, by monomial, 325
  division of polynomials by, 325–326, 336
  factoring out of polynomial, 350–352, 411

Monomial(s)—Cont.
  multiplication of
    by binomial, 308
    by monomial, 307
    by polynomial, 307–308, 336
Motion, formula for distance, rate
    and speed, 219, 472, 702
Multiples
  common, 12
  definition of, 11
  least common (LCM)
    definition of, 12, 65
    finding, 12–13, 65
    in solving equations, 192–193
Multiplication
  associative property of, 87–88, 151
  of binomials
    by binomial, 309
      FOIL method for,
        309–311, 336
      with vertical method, 311
    by binomial differing only in
      sign, 319–320
    by monomial, 308
    square of binomial, 318–319,
      336, 369
    by trinomial, using vertical
      method, 312
  commutative property of, 87, 151
  of decimals, 37–38, 67
  distributive property over addition,
    88–89, 151
    in multiplication of polynomials,
      307–308
    in multiplying radical
      expressions, 777, 802
  solving linear equations with,
    170–172
  solving linear inequalities with,
    250–251
  of exponents, 144–145
  of fractions, 21, 65, 431
  of mixed numbers, 25
  of monomials
    by binomial, 308
    by monomial, 307
    by polynomial, 307–308, 336
  notation for, 79, 151
  of polynomials
    by monomial, 307–308, 336
    special products, 318–320, 336
      binomials differing only in
        sign, 319–320
      square of binomial, 318–319,
        336, 369
    using vertical method, 312
  of radical expressions, 776–778,
    801–802
  of rational expressions,
    431–432, 491
  of real numbers
    with different signs, 111, 152
    multiplicative identity
      property, 112
    multiplicative inverse property,
      113–114
    multiplicative property of
      zero, 113
    with same sign, 112, 152
Multiplication property
  of equality
    definition of, 177, 259
    solving linear equations with,
      177–184, 187–194
  of inequality, 247
    solving linear inequalities with,
      247–250

Multiplicative identity property, 112
Multiplicative inverse (reciprocal),
    113–114
  applications, 471
  solving linear equations with,
    180–181
Multiplicative inverse property,
    113–114
Multiplicative property of zero, 113

N

Natural numbers, 3
Negative exponents
  definition of, 288, 335
  simplification of, 288–290
Negative numbers, 54–55, 68
  multiplication of two, 112, 118
  negative of, 57–58
  roots of, 751–752
  square root of, 750
Negative, of negative number, 57–58
Negative powers
  definition of, 288, 335
  simplification of, 288–290
Notation
  absolute value, 58, 68
  addition, 78, 151
  closed circle, 244, 260
  division, 80–81, 151
  ellipses (...), 55
  exponents, 47, 68
  for functions, 654, 666
  greater than (>), 243, 260
  greater than or equal to (≥),
    244, 260
  inequalities, 243, 244, 260
  less than (<), 243, 260
  less than or equal to (≤), 244, 260
  mean, 182
  multiplication, 79, 151
  open circle, 244, 260
  ordered pairs, 505, 581
  radical sign (√), 749
  repeating decimals, 36
  roots, 749
  sets, 3
  subtraction, 78, 151
  summation (Σ), 130
  translating words into, 202
Number(s)
  composite
    definition of, 5, 65
    factoring of, 6–7
      into prime factors, 7–8
    identification of, 6
  counting, 3
  integers, 55–57, 68
    consecutive, 203
    as subdivision of rational
      numbers, 753
  irrational
    decimal representations of, 753
    definition of, 752
    identification of, 753
    as subdivision of real
      numbers, 753
  mixed
    addition of, 26, 66
    converting to/from improper
      fraction, 24–25, 66
    definition of, 24, 66
    division of, 26
    multiplication of, 25
    subtraction of, 27, 66

natural, 3
  negative, 54–55, 68
    multiplication of two, 112, 118
    roots of, 751–752
    square root of, 750
  of ordinary arithmetic, 19
  positive, 54–55, 68
  prime
    definition of, 4, 65
    identification of, 5
  rational
    changing to percent form, 229
    definition of, 752
    identification of, 753
    as subdivision of real
      numbers, 753
  real. See Real number(s)
  signed, 55
  whole, 3–4
Number line
  closed circle on, 244, 260
  graphing inequalities on,
    244–251, 260
  integers on, 55–56
  open circle on, 244, 260
  origin of, 4
  real, 753
  whole numbers on, 4
Numerical coefficient
  of algebraic expression, 137
  of polynomial term, 277, 335
  of trinomial, identifying, 387

O

One (1)
  as multiplicative identity, 112
  as neither prime nor composite,
    6, 65
Open circle, 244, 260
Opposite (additive inverse)
  additive inverse property, 98–99
  definition of, 57
  of negative number, 57–58
Order of operations, 48–50, 68
  calculators and, 49, 130
  in division, 116–117
Ordered pair(s)
  completing, 506–507
  graphing of, 516–518, 581
  notation for, 505, 581
  quadrants of rectangular coordinate
    system and, 514–515
  as solution to function, 655–656
Ordered sets, 56
Origin
  of number line, 4
  in rectangular coordinate system,
    514, 581

P

Parabola
  axis of symmetry, 843–844, 866
  equation for, 837
    upward vs. downward
      orientation, 842
  graphing of, 837–846
  vertex of, 843
  x-intercepts of, 844–845, 866
Parallel lines
  definition of, 618, 665
  determining, 618–619, 622

finding, 622
  slopes of, 618–619
Parentheses
  in algebraic notation, 80
  removal of, in solving equation,
    191, 215–219
Percent
  applications of, 231–234
  changing decimals to, 41, 68
  changing fractions to, 41, 68
  changing rational numbers to, 229
  changing to decimal, 40, 67
  changing to fraction, 39–40, 67
  definition of, 39, 67, 229
Percent expressions, parts of,
    229–230, 260
Percent proportion, definition of,
    231, 260
Perfect square terms, identification
    of, 375
Perfect square trinomials, factoring,
    369, 377–378, 411
Period, of pendulum, 807
Perpendicular lines
  applications of, 623
  definition of, 620, 665
  equation of, 620–621
  finding, 622
  slopes of, 620–621, 665
Pie charts, 569–570, 582
Point(s), graphing of, 516–518
  steps in, 516, 581
Point-slope form of linear equations
  applications of, 632–633
  definition of, 629, 632, 666
  finding, 629–630, 631–632
Polynomial(s). See also Binomial(s);
    Monomial(s); Trinomial(s)
  addition of
    combining like terms in,
      297–298
    overview of, 297, 336
    removing grouping signs in,
      297–298
    using horizontal method, 298
    using vertical method, 301
  definition of, 277, 335
  degree of, 278–279, 335
  descending-power form of, 279, 335
  division of
    by monomial, 325–326, 336
    by polynomials, 327–330
  evaluation of, 279–280
  factoring. See Factoring
  greatest common factor of
    definition of, 350, 411
    factoring out, 350–352, 361,
      370–371, 377, 392, 399
  multiplication of
    by monomial, 307–308, 336
    special products, 318–320, 336
      binomials differing only in
        sign, 319–320
      square of binomial, 318–319,
        336, 369
    using vertical method, 312
  subtraction of
    combining like terms in,
      299–300
    overview of, 336
    removing grouping signs in,
      299–300
    using horizontal method,
      299–300
    using vertical method, 301
  terms of
    coefficient of, 277, 335

combining, in addition and
 subtraction, 297–300
definition of, 277, 335
missing, 329–330
Positive numbers, 54–55, 68
Power, of exponent, 47
Power property, of equality,
 783–786, 802
Prediction, using line graphs,
 575–576
Prime factorization, 7–8
Prime number
 definition of, 4, 65
 identification of, 5
Principal (positive) square root,
 749–750
Product property, of radicals,
 760–762, 801
Proper fractions, 19
Property 1
 of exponents, 144–145, 152,
  273, 336
 of radical expressions
  (product property),
  760–762, 801
Property 2
 of exponents, 146–147, 152,
  274, 336
 of radical expressions (quotient
  property), 762–764, 801
Property 3 of exponents, 274, 336
Property 4 of exponents, 275, 336
Property 5 of exponents, 276, 336
Proportion
 definition of, 464
 extremes of, 464
 means of, 464
 properties of, 464
 solving for unknown, 464–465,
  475–477
Proportion form of equation
 definition of, 464
 solving, 464–465, 475–477
Pythagoras, 789
Pythagorean Theorem
 alternative form, 792
 applications of, 747, 790–794,
  855–856
 standard form, 789, 802
 verification of, 789–790

Q
Quadrants of rectangular coordinate
 system
 definition of, 514
 signs of ordered pairs in, 514–515
Quadratic equations
 applications of, 407, 855–859
 graphs of. See Parabola
 solution to
  number of, 406
  repeated (double), 406
 solving
  by completing the square,
   821–824
   steps in, 824, 865
  by factoring, 404–407, 412,
   813, 865
  with quadratic formula,
   828–832
   steps in, 832, 865
  square root method, 814–816
 standard form of, 404, 813, 827
 zero-product principle for, 404–407

Quadratic formula
 definition of, 828
 derivation of, 828
 discriminant of, 835
 main use for, 829
 solving quadratic equations with,
  828–832
  steps in, 832, 865
Quotient property, of radicals,
 762–764, 801

R
Radical(s), like, 769
Radical equations, solving,
 783–786
 extraneous solutions, 783
 steps in, 786, 802
Radical expressions
 adding and subtracting,
  769–772, 801
 dividing, 778–779, 802
 evaluating, 754
 multiplying, 776–778, 801–802
 parts of, 751
 property 1 (product property) of,
  760–762, 801
 property 2 (quotient property) of,
  762–764, 801
 simplest form of
  definition of, 760, 801
  writing expressions in, 760–764
 simplifying, 760–764, 779, 801
Radical sign (√), 749, 751
Radicand, 751
Range, of set, 102
Rate
 formula for, vs. distance and time,
  219, 472, 702
 in percent expressions, 230, 260
Ratio, definition of, 464
Rational equations
 definition of, 347
 proportion form of, 464
  solving, 464–465, 475–477
 solving, 457–465
  finding excluded values of x,
   458–459
  steps in, 460, 465, 492
 without solution, 462
Rational expressions
 addition of
  like, 437–440, 492
  unlike, 444–451, 492
   steps in, 445
 applications of, 470–477
 complex, simplification of,
  483–486, 492
 division of, 432–433, 491
 fundamental principle of, 421
 like
  addition and subtraction of,
   437–440, 492
  definition of, 437
 multiplication of, 431–432, 491
 simplest form of
  in addition and subtraction,
   438–440
  writing expressions in,
   421–426, 491
 simplification of
  complex expressions,
   483–486, 492
  standard expressions,
   421–426, 491

subtraction of
 like, 437–440, 492
 unlike, 444–451, 492
  steps in, 445
unlike
 addition and subtraction of,
  444–451, 492
  steps in, 445
 definition of, 444
Rational numbers
 changing to percent form, 229
 definition of, 752
 identification of, 753
 as subdivision of real numbers, 753
Real number(s)
 addition of
  additive identity property, 98
  additive inverse property,
   98–99
  application of, 103
  different sign, 96–97, 151–152
  same sign, 95–96, 151
 division of, 114–115, 152
  applications, 117
  division of/by zero, 115–116
 multiplication of
  with different signs, 111, 152
  multiplicative identity
   property, 112
  multiplicative inverse property,
   113–114
  multiplicative property of
   zero, 113
  with same sign, 112, 152
 properties of, 87–90, 151
 subsets of numbers in, 753
 subtraction of, 100–101, 152
  application of, 103
Real number line, 753
Reciprocal (multiplicative inverse),
 113–114
 applications of, 471
 solving linear equations with,
  180–181
Rectangular coordinate system,
 514–518, 581
 axes in
  scaling of, 536–537
  x-axis, 514, 581
  y-axis, 514, 581
 on calculator, adjustment of,
  596–597
 origin, 514, 581
 quadrants of
  definition of, 514
  signs of ordered pairs in,
   514–515
Relatively prime numbers, 11
Repeated (double) solution, to
 quadratic equations, 406
Repeating decimals, 36
Root(s)
 cube
  common, list of, 752
  definition of, 751, 801
  evaluating, 752
 fourth
  common, list of, 752
  definition of, 751, 801
 of negative numbers, 751–752
 notation for, 749
 square. See Square root
Root (of equation). See Solution(s)
Rounding
 of decimal equivalent, 35–36
 in division of decimals, 39
RSA encryption systems, 347

S
Scientific notation
 applications of, 290–291, 292
 on calculator, 290
 conversion to/from decimal
  form, 291
 definition of, 291, 335
Search engines, Internet, 159
Second property of exponents,
 146–147, 152, 274, 336
Set(s)
 average of, 182
 definition of, 3
 extreme values of, 57
 maximum of, 57
 mean of
  applications of, 183
  definition of, 182
  finding, 182–183
  notation for, 182
 median of
  applications of, 183
  definition of, 99
  finding, 99–100
 minimum of, 57
 notation, 3
 ordered, 56
 range of, 102
 summation of, 130
Sieve of Eratosthenes, 5
Signed numbers, 55
Silent negative one, 171–172
Simplest form
 of fraction, 421, 491
 of radical expression
  definition of, 760, 801
  writing expressions in, 760–764
 of rational expression
  in addition and subtraction,
   438–440
  writing expressions in,
   421–426, 491
Simplification
 of complex fractions, 483–486
 of complex rational expressions,
  483–486, 492
 of exponents
  negative, 288–290
  positive, 275–276
 of fractions
  complex, 483–486
  standard, 20
 of radical expressions, 760–764,
  779, 801
 of rational expressions
  complex, 483–486, 492
  standard, 421–426, 491
Slope, 549–557. See also Point-slope
  form of linear equations;
  Slope-intercept form of
  linear equations
 definition of, 549, 582
 of equation y = mx, 555
 finding
  given two point, 549–553
  from graph, 553–555
 of horizontal line, 552, 630, 632
 negative, 551
 of parallel lines, 618–619
 as steepness, 549, 551
 undefined, 552, 630, 632
 of vertical line, 552, 630, 632
 zero slope, 551–552, 630, 632
Slope-intercept form of linear
  equations, 605–609, 665
 applications, 608–609
 definition of, 605, 632

Slope-intercept form of linear
  equations—*Cont.*
  finding, 605–607
  graphing, 607
Solution(s)
  double (repeated), of quadratic
    equations, 406
  of equation
    definition of, 164, 259
    multiple, 165
    verification of, 164
  equations without, 193, 462
  extraneous, for radical
    equations, 783
  of function, 655–656
  of linear equation, 166
  of linear equation in one
    variable, 506
  of linear equation in two variables
    definition of, 504, 581
    notation for, 504
    number of, 504
  to quadratic equations
    number of, 406
    repeated (double), 406
  of systems of linear equations,
    definition of, 679, 733
Solution set, of inequality, 244
Solving of equations for a variable
  definition of, 199
  examples of, 198–206
  steps in, 199, 259
Speed, formula for, 219, 472, 702
Square, of binomial, 318–319,
    336, 369
Square root
  on calculator, 750–751
  common, list of, 752
  definition of, 749, 801
  finding, 749–750
  negative, 750
  of negative number, 750
  notation for, 749
  principal (positive), 749–750
Square root method of solving
  quadratic equations,
    814–816
Standard form
  of linear equations, 632
  of quadratic equations, 404,
    813, 827
Substitution method, for nonlinear
  systems of equations in two
  variables, 711–714
  steps in, 714, 734
Substitution, solving word problems
  by, 201
Subtraction
  on calculators, 102–103
  of decimals, 37, 67
  of fractions, 23, 66
  of mixed numbers, 27, 66
  notation for, 78, 151
  of polynomials
    combining like terms in,
      299–300
    overview of, 336
    removing grouping signs in,
      299–300

using horizontal method,
    299–300
  using vertical method, 301
  of radical expressions,
    769–772, 801
  of rational expressions
    like, 437–440, 492
    unlike, 444–451, 492
    steps in, 445
  of real numbers, 100–101, 152
    application of, 103
Summation notation, 130
Supply and demand, 859
Syllabus familiarity, importance
  of, 77
Symbols. *See* Notation
Systems of linear equations
  definition of, 679, 733
  solution for the system, defined,
    679, 733
Systems of linear equations in two
  variables
  applications, 696–703, 715–717
  consistent, definition of, 681, 682
  dependent, 681–682, 683,
    694–695, 713
  inconsistent
    definition of, 681, 683
    solving, 681, 695–696, 713–714
  solving
    addition method, 690–696
      steps in, 694, 733–734
    graphical method, 670–683
      steps in, 682–683, 733
    method selection, 714–715
    by substitution method, 711–714
      steps in, 714, 734
Systems of linear inequalities in two
  variables
  applications of, 725–726
  solving of, by graphing,
    724–725, 734

T
Table(s)
  on calculator, 598
  cells of, 567
  definition of, 567, 582
  reading of, 567–569
Table of values, in graphing,
    837–838
Term(s)
  of algebraic expressions, 136–137
  like
    combining, 137–139,
      169–170, 181, 297–300
    definition of, 137
    numerical coefficient of, 137
  perfect square, identification
    of, 375
  of polynomial
    coefficient of, 277, 335
    combining, in addition and
      subtraction, 297–300
    definition of, 277, 335
    missing, 329–330

Test preparation methods, 273
Textbook familiarity, approach to, 3
Third property of exponents,
    274, 336
TI-83 Plus calculator, graphing with,
    595–598
Time, *vs.* rate and speed, formula for,
    219, 472, 702
Trial-and-error method of factoring,
    357–362, 366, 411
Trinomial(s)
  ac test for, 387–389
    definition of, 387
    sign patterns for, 392
  definition of, 278, 335
  factorability of, test for, 387–389
  factoring
    of form $ax^2 + bx + c$, 366–370
    of form $x^2 + bx + c$, 357–362
    identifying coefficients, 387
    perfect square, 369,
      377–378, 411
    sign patterns for
      examples of, 366–369
      rules for, 366
    trial and error method, 357–362,
      366, 411
    unfactorable trinomials, 362
  multiplication of, by binomials,
    using vertical method, 312
  perfect square, factoring, 369,
    377–378, 411

U
Unlike rational expressions,
  definition of, 444

V
Variable(s)
  definition of, 77
  letters used to represent, 81
Variable cost, 537, 608
Variation, direct
  applications of, 555–557
  definition of, 555, 582
  graphing of, 556–557
  writing equation for, 556
Vertex (vertices), of parabola, 843
Vertical change, of line, 549, 582
Vertical line
  equation of, 630, 632
  graphing of, 531–532, 533
  slope of, 552, 630, 632
Vertical method
  addition of polynomials with, 301
  multiplication of binomial by
    binomial with, 311
  multiplication of polynomials
    with, 312
  multiplication of trinomials by
    binomials with, 312
  subtraction of polynomials
    with, 301

W
Wallis, John, 288
Websites, evaluation of, 673
Whole numbers, 3–4
Word problems. *See also*
    Applications
  history of, 201
  solving of
    algorithm for, 201, 696, 734
    examples, 202–206
    by substitution, 201
    translating into algebraic
      notation, 202
Working together, benefits of, 349
World Wide Web, searches on,
    159–160

X
x-axis
  in rectangular coordinate system,
    514, 581
  scaling of, 536–537
x-coordinate
  definition of, 505, 581
  identification of, 515–516
x-intercept(s)
  of line, 534
  of parabola, 844–845, 866

Y
y-axis
  in rectangular coordinate system,
    514, 581
  scaling of, 536–537
y-coordinate
  definition of, 505, 581
  identification of, 515–516
y-intercept
  definition of, 534
  finding, 605–606

Z
Zbox feature, on graphing
    calculator, 597
Zero
  as additive identity, 98
  division of/by, 115–116
  multiplicative property of, 113
  as neither prime nor composite,
    6, 65
Zero exponent, 287, 335
Zero power, definition of, 287, 335
Zero-product principle
  application of, 404–407
  definition of, 404
Zero slope, 551–552, 630, 632
Zoom key, on graphing calculator,
    596–597